OPTO-MECHATRONIC SYSTEMS HANDBOOK

Techniques and Applications

The Mechanical Engineering Handbook Series

Series Editor
Frank Kreith
Consulting Engineer

Published Titles

Air Pollution Control Technology Handbook
Karl B. Schnelle, Jr. and Charles A. Brown
Computational Intelligence in Manufacturing Handbook
Jun Wang and Andrew Kusiak
Fuel Cell Technology Handbook
Gregor Hoogers
Handbook of Heating, Ventilation, and Air Conditioning
Jan F. Kreider
Hazardous and Radioactive Waste Treatment Technologies Handbook
Chang Ho Oh
Inverse Engineering Handbook
Keith A. Woodbury
Opto-Mechatronic Systems Handbook: Techniques and Applications
Hyungsuck Cho
The CRC Handbook of Mechanical Engineering
Frank Kreith
The CRC Handbook of Thermal Engineering
Frank Kreith
The Handbook of Fluid Dynamics
Richard W. Johnson
The MEMS Handbook
Mohamed Gad-el-Hak

Forthcoming Titles

Biomedical Technology and Devices Handbook
James Moore and George Zouridakis
Handbook of Mechanical Engineering, Second Edition
Frank Kreith and Massimo Capobianchi
Multi-Phase Flow Handbook
Clayton T. Crowe
Shock and Vibration Handbook
Clarence W. de Silva

OPTO-MECHATRONIC SYSTEMS HANDBOOK

Techniques and Applications

Edited by
Hyungsuck Cho

CRC PRESS

Boca Raton London New York Washington, D.C.

Library of Congress Cataloging-in-Publication Data

Opto-mechatronic systems handbook : techniques and applications / edited by
Hyungsuck Cho.
p. cm.
Includes bibliographical references and index.
ISBN 0-8493-1162-4 (alk. paper)
1.Mechatronics—Handbooks, manuals, etc. 2. Electrooptics—Industrial
applications—Handbooks, manuals, etc. I. Cho, Hyung Suck.

TJ163.12 .O66 2002
621—dc21 2002073354

Direct all inquiries to CRC Press LLC, 2000 N.W. Corporate Blvd., Boca Raton, Florida 33431.

Visit the CRC Press Web site at www.crcpress.com

International Standard Book Number 0-8493-1162-4
Library of Congress Card Number 2002073354
Printed in the United States of America 1 2 3 4 5 6 7 8 9 0
Printed on acid-free paper

Preface

The past two decades have witnessed an unrelenting technological evolution that has produced advancements in devices, machines, processes, and systems by enhancing their performance and creating new value and function. Mechatronic technology, integrated by mechanical, electronic, and computer technologies, has certainly played an important role in this evolution.

In recent years optical technology has been increasingly incorporated, and at an accelerated rate, into mechatronic systems, and vice versa. This may be attributable to the fact that optically integrated technology provides solutions to complex problems by achieving a desirable function or performance that mechatronic technology alone cannot solve. As a result, mechatronic or optical products, machines, and systems have further evolved toward a state of precision, reduced size, and greater intelligence and autonomy. In the future this trend will continue to map out the direction of next-generation technologies associated with most mechatronic- and optical-engineering-related fields.

We now refer to the technology fusion in this new paradigm as "opto-mechatronic technology," which is an integration of optical and mechatronic technologies. This handbook terms the devices, products, machines, processes, and systems developed by this technology "opto-mechatronic systems." This definition has never before appeared in the literature, although such systems have long been commercially available. As the above discussion makes clear, this optically and mechatronically integrated technology is certainly multifaceted in nature. However, to date, despite the nature of opto-mechatronic systems technology, little effort has been made to bring together researchers from optical engineering and mechatronics to collaborate on entire projects, from concept generation to design fabrication. As a result, few attempts have been made at a concurrent design approach that would produce desired functionalities and performance.

It is sincerely hoped that this handbook will serve as a building block that might eliminate the formidable barriers that exist between the two technological communities. The handbook, therefore, has three major objectives. The first is to present the definition, fundamentals, and application aspects of the technology. The second is to provide readers with an integrated view of opto-mechatronic systems, thereby enabling them to begin to understand how optical systems and devices can be fused or integrated with mechatronic systems starting at the design and manufacturing stages. The third goal is to help readers learn about the roles of optical systems in overall system performance and to understand their synergistic effect.

The *Opto-Mechatronic Systems Handbook: Techniques and Applications* is a collection of 23 chapters organized in five parts covering the fundamental elements and applications of opto-mechatronic systems technology. The chapters are written by an international group of leading experts from academia and industry and representing North America, Asia, and Europe.

Part I: Understanding Opto-Mechatronic Technology and Its Applications

Part I is composed of two chapters that provide the reader with the definition, fundamental functions, and classification of opto-mechatronic systems technology by illustrating and classifying practical

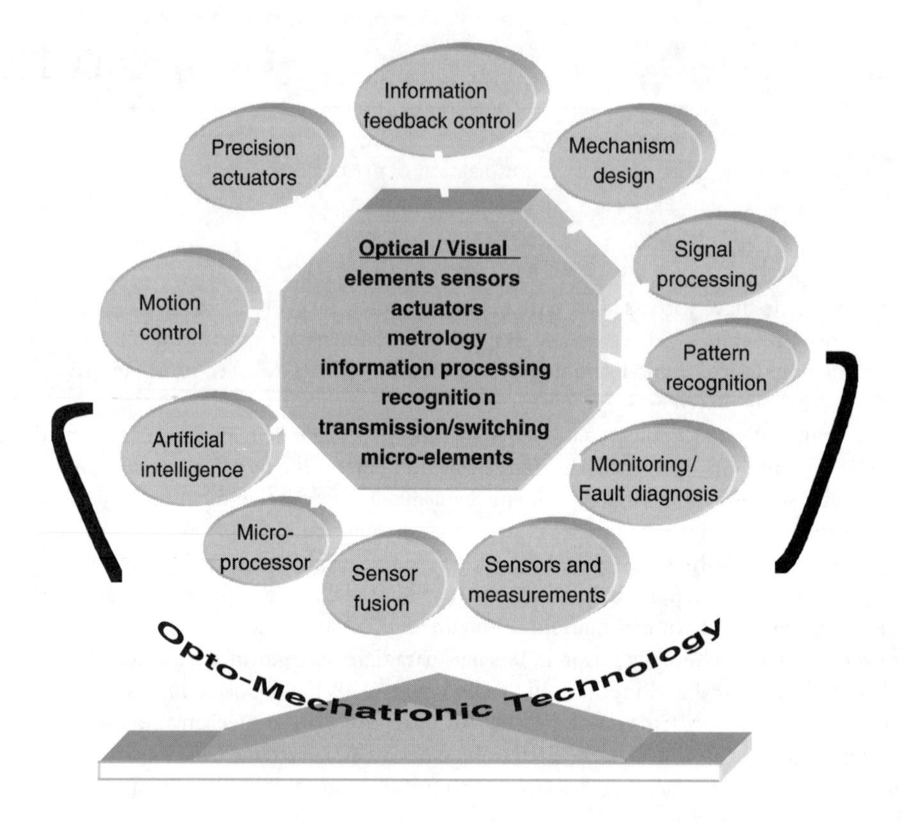

examples of such systems, and providing fundamental considerations needed for the design of opto-mechatronic products and systems.

Part II: Optical Elements, Sensors, and Measurements

Part II comprises Chapters 3 through 7, which focus on basic principles, theory, and applications of lasers, optical sensors, and distributed optical fiber sensing. Chapter 6 describes how biological-based optical sensors and transducers can be used to perform simple signal-detection and signal-conversion operations. Optical metrology essential for opto-mechatronic applications is presented in brief; the focus here will be on displacement sensing, which serves as the basis for the construction of three-dimensional geometry and shapes. Chapter 7 discusses the fundamentals of visual sensing and image-processing techniques and their importance for real mechatronic applications.

Part III: Optical Information Processing and Recognition

The technology associated with optical information processing and recognition is an essential part of opto-mechatronic systems technology. In fact, the systems that perform optical information and pattern recognition belong, by definition, to opto-mechatronic systems since most such systems are composed not only of optical but also mechatronic elements. Part III, which consists of Chapters 8 through 12, treats processing and recognition of the raw optical signals/data/information obtained by sensors. It presents basic concepts and principles, theories, and applications in volume holography in optical imaging, real-time image feature extraction, real-time image recognition, and optical pattern recognition. To aid understanding of the recognition aspect of opto-mechatronic systems technology generic pattern-recognition methods are surveyed, with an emphasis on utilization of artificial intelligence.

Part IV: Opto-Mechatronic Systems Control

Control of either optical systems or optical-based mechatronic systems permeates every area of opto-mechatronic systems, which makes them adaptive or reconfigurable in accordance with changes in operating conditions. Part IV, which consists of Chapters 13 through 16, addresses control issues associated with opto-mechatronic systems and presents basic control methodologies applicable to various systems. In addition, visual feedback control, which is widely used in a number of tasks, is extensively treated with respect to feature extraction and selection and servoing theory. Principles and concepts of in-process monitoring and control associated with optical-based processes are treated in some detail.

Part V: Opto-Mechatronic Processes and Systems

After a presentation of the basic technologies in Parts I through IV, Part V illustrates underlying practical applications to various processes and systems. Part V contains Chapters 17 through 23 and includes discussions of semiconductor fabrication processes, inspection and control of surface-mount processes for electronic part assembly, optical-based manufacturing processes, and systems such as optical data-storage systems and opto-skill-capturing systems for service robots. The final chapters, Chapters 22 and 23, discuss issues related to optical MEMs, particularly optical array sources, and the design and control of vision-based microassembly systems.

Acknowledgments

I would like to express my heartfelt thanks to all the contributors of this handbook for their time and effort in preparing their chapters. Their excellent work is very much appreciated. I also wish to express a special thanks to Cindy Renee Carelli, CRC Press Acquisitions Editor, for her assistance, advice, and patience during the editing phase of this handbook. Her enthusiasm and encouragement provided me with the stimulus and motivation to launch and continue moving this project ahead. In addition, many thanks to Helena Redshaw, CRC Supervisor, Editorial Project Development, for ensuring that all manuscripts were ready for production, and Won Sik Park, my Ph.D. student, for managing the incoming draft manuscripts.

Finally, many thanks to Young Jun Roh, my Ph.D. student, and Hwa Yong Lim, my secretary, for their tremendous help in preparing the manuscripts and art work.

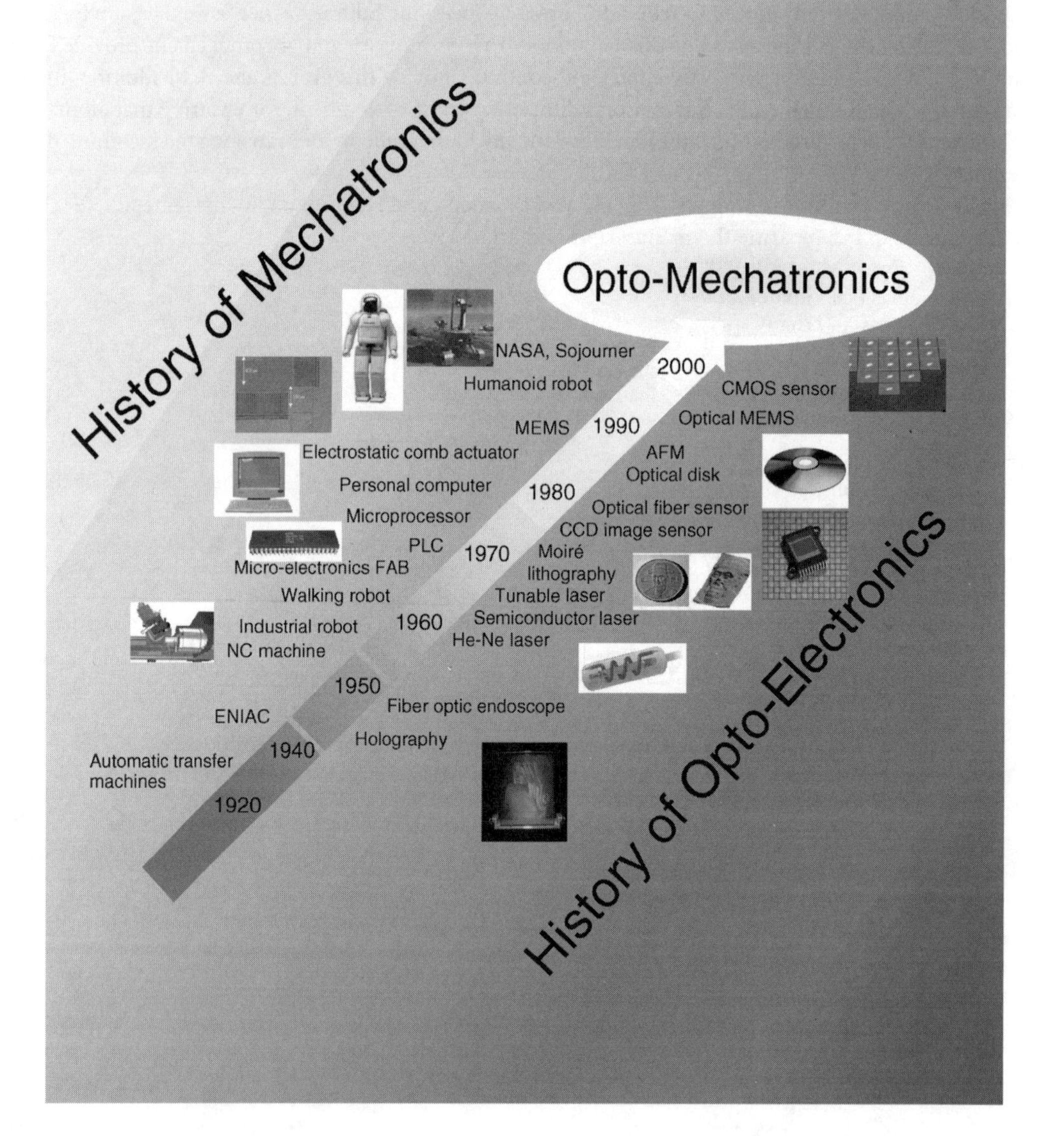
Opto-Mechatronic Systems
History of Mechatronics
Opto-Mechatronics
NASA, Sojourner
2000
Humanoid robot
CMOS sensor
MEMS
1990
Optical MEMS
Electrostatic comb actuator
AFM
Optical disk
Personal computer
1980
Optical fiber sensor
Microprocessor
CCD image sensor
PLC
1970
Moiré
Micro-electronics FAB
lithography
Walking robot
Tunable laser
Industrial robot
1960
Semiconductor laser
NC machine
He-Ne laser
1950
ENIAC
Fiber optic endoscope
1940
Holography
Automatic transfer machines
1920
History of Opto-Electronics

About the Editor

Hyungsuck Cho, Ph.D., received his B.S. in 1971 from Seoul National University, Korea, his M.S. in 1973 from Northwestern University, Evanston, Illinois, U.S.A., and his Ph.D. in 1977 from the University of California at Berkeley, U.S.A. From 1977 to 1978 he was a postdoctoral fellow in the Department of Mechanical Engineeering, University of California at Berkeley.

Since 1978 Dr. Cho has been a professor in the Department of Production Engineering, Department of Automation and Design, Seoul Campus. He is also currently a member of the Department of Mechanical Engineering, Korea Advanced Institute of Science and Technology (KAIST), Taejeon, Korea. From 1984 to 1985 he was a visiting scholar at the Institut für Producktionstechnik and Automatisierung (IPA), Germany, where he conducted research on robot-based assembly. He has been invited as a short-term visiting scholar to several universities, including Ritsumeikan University, Japan (1987); the University of Paderborn, Germany (1992); and the New Jersey Institute of Technology, U.S.A. (1998). From 1995 to 1996 he was a visiting professor in graduate school education for the Advanced Manufacturing Program (AMP) of the University of California at San Diego, U.S.A.

Dr. Cho's research interests focus on the areas of environmental perception and recognition for mobile robots, machine vision and pattern classification, and application of artificial intelligence/machine intelligence. He has published seven book chapers and more than 377 research papers (307 papers published in international journals or presented at conferences and 70 papers published in Korean-language journals). He serves on the editorial boards of five international journals: *Journal of Robotic Systems, Robotica, Control Engineering Practice (IFAC), Journal of Advanced Robotics*, and *Journal of Engineering Manufacture (PIME)*. In 1998 he served as a guest editor for a special issue on "Intelligent Robotic Assembly" for the journal *Robotica.*

In addition to his academic activities, Dr. Cho has served on technical committees on robotics, advanced manufacturing, and measurements on robotics for IFAC and IMEKO. He has organized and participated in international symposia and conferences, and has served on the program committees of several international conferences, including IEEE R&A, and IEEE/RSJ IROS, IFAC, ASME, and SPIE. He founded the conference Opto-Mechatronic Systems of SPIE (ISAM). He has chaired or co-chaired several conferences and symposia, including two symposia of the ASME Winter Annual Meeting (1991, 1993); IFAC Workshop on Intelligent Manufacturing (1997); IEEE/RSJ IROS (1999); International Workshop on Mechatronics Technology (1999); and SPIE-Opto-Mechatronic Systems (2000, 2001).

Dr. Cho received the fellowship of the Alexander von Humboldt Award of Germany in 1984. He received the Best Paper Award at an ISAM conference in 1994. In 1998 he was awarded the Thatcher Bros Prize from the Institution of Mechanical Engineers, U.K. for his achievements in research in the fields of robotics and automation. He served as President of the Institute of Control, Automation, and Systems Engineers in Korea for 2001.

Contributors

George Barbastathis
Department of Mechanical Engineering
Massachusetts Institute of Technology
Cambridge, Massachusetts

Amarjeet S. Bassi
Department of Chemical and Biochemical Engineering
University of Western Ontario
London, Ontario, Canada

Hyungsuck Cho
Department of Mechanical Engineering
Korea Advanced Institute of Science and Technology
Taejeon, South Korea

Roy Davies
Machine Vision Group
Department of Physics
Royal Holloway
University of London
Egham, Surrey, U.K.

Kazuhiro Hane
Department of Mechatronics and Precision Engineering
Tohoku University
Sendai, Japan

Farrokh Janabi-Sharifi
Department of Mechanical, Aerospace, and Industrial Engineering
Ryerson University
Toronto, Ontario, Canada

Yoshitada Katagiri
NTT Telecommunications Energy Labs
Kanagawa, Japan

Jideog Kim
Samsung Advanced Institute of Technology
Yongin, Kyungghi-Do, South Korea

Yongkwon Kim
School of Electrical Engineering and Computer Science
Seoul National University
Seoul, South Korea

George K. Knopf
Department of Mechanical and Materials Engineering
University of Western Ontario
London, Ontario, Canada

Sukhan Lee
System and Control Sector
Samsung Advanced Institute of Technology
Yongin, Kyungghi-Do, South Korea

King Pui Liu
City University of Hong Kong
Kowloon, Hong Kong, China

Carl G. Looney
Department of Computer Science
University of Nevada
Reno, Nevada

Osamu Matsuda
Sony Corporation
Tokyo, Japan

Katsunori Matsuoka
National Institute of Advanced Industrial Science and Technology
Human Stress Signal Research Center
Osaka, Japan

Gary S. May
School of Electrical and Computer Engineering
Georgia Institute of Technology
Atlanta, Georgia

Takeshi Mizuno
Sony Corporation
Tokyo, Japan

Bradley J. Nelson
Department of Mechanical Engineering
University of Minnesota
Minneapolis, Minnesota

Noriaki Nishi
Sony Corporation
Kanagawa, Japan

Francesca Odone
Department of Informatics and Information Science
University of Genova
Genova, Italy

Alan Rogers
Department of Electronic Engineering
University of Surrey
Guildford, Surrey, U.K.

Minoru Sasaki
Department of Mechatronics and Precision Engineering
Tohoku University
Sendai, Japan

Masatake Shiraishi
Department of Systems Engineering
Ibaraki University
Ibaraki, Japan

Shiu Kit Tso
City University of Hong Kong
Kowloon, Hong Kong, China

Alessandro Verri
Department of Informatics and Information Science
University of Genova
Genova, Italy

Barmeshwar Vikramaditya
Seagate Technology
Bloomington, Minnesota

Contents

Part III: Optical Information Processing and Recognition

Part IV: Opto-Mechatronic Systems Control

Illustrative Applications

Part V: Opto-Mechatronic Processes and Systems

I

Understanding Opto-Mechatronic Technology and Its Applications

1

Characteristics of Opto-Mechatronic Systems

Hyungsuck Cho
Korea Advanced Institute of Science and Technology
Taejeon, South Korea

1.1 Introduction

Most engineered devices, products, machines, processes, or systems have moving parts and require manipulation and control of their mechanical or dynamic constructions to achieve a desired performance. This involves the use of modern technologies such as mechanism, sensor, actuators, control, microprocessors, optics, software, communication, and so on. In the early days, however, these were operated mostly via mechanical elements or by devices that caused inaccuracy and inefficiency, making the achievement of a desired performance difficult.

Figure 1.1 shows how the key technologies contributed to the evolution of machines/systems in terms of value or performance with the passage of time. As seen from the figure, tremendous efforts have been made to enhance system performance by combining electrical and electronic hardware with mechanical systems. A typical example is a gear-trained mechanical system controlled by a hard-wired controller. This mechanical and electronic configuration, called a "mechatronic" configuration, consists of two kinds of components: mechanism and electronics and electric hardware. Due to the hard-wired structural limitations of this early mechatronic configuration, flexibility was not embedded in most systems of that time [Ishi, 1990]. This kind of tendency lasted until the mid-1970s, when microprocessors came into use for industrial applications.

0-8493-1162-4/03/$0.00+$1.50

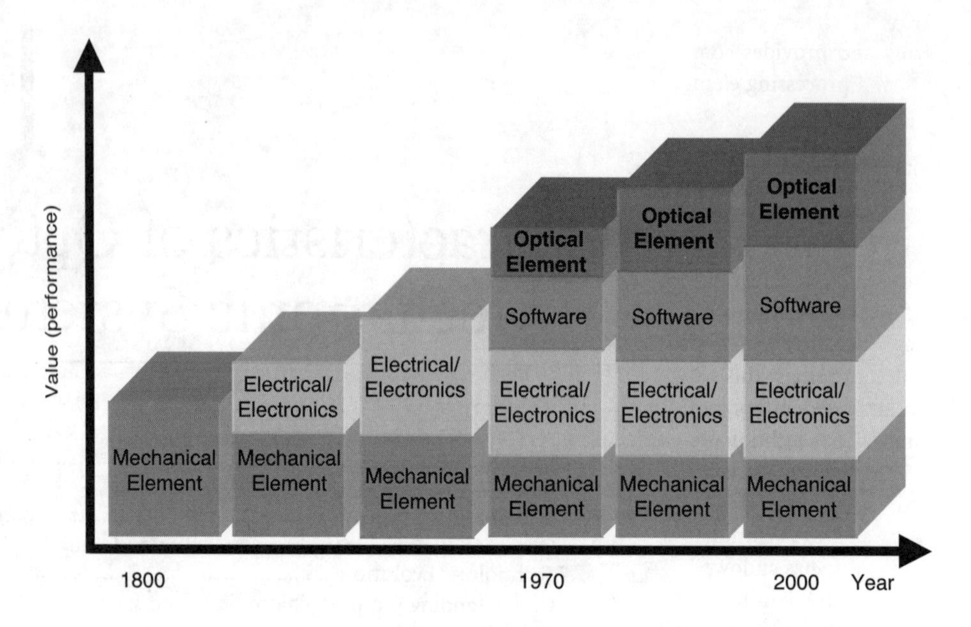

FIGURE 1.1 Key component technologies contributed to system evolution (not to scale).

The development of microprocessors has provided a new catalyst for industrial evolution. This brought about a major development—the replacement of many mechanical functions with electronic ones. This evolutionary change has opened up the era of mechatronics and has increased the autonomy of machines/systems while at the same time increasing versatility and flexibility. The flexibility and autonomy achieved thus far, however, have a limited growth, as both the hardware/software of the developed mechatronic systems have not been developed to the point where they can perform many complicated functions autonomously while adapting to changing environments. In addition, information structures have not been developed to have real-time access to appropriate system data/information.

There may be several reasons for such delayed growth. The first one may be that, in many cases, mechatronic components alone cannot fulfill a desired function or achieve a desired performance as specified for a system design. The second one is that, although mechatronic components alone can fulfill a desired function or achieve a desired performance, the achieved results are not satisfactory because of their low perception and execution capabilities and also their inadequate incorporation between hardware and software. In fact, in many cases the measurements are difficult or not even feasible due to the inherent characteristics of the systems. In some other cases, the measurement data obtained by conventional sensors are not accurate or reliable enough to be used for further processing. They can be noisy, necessitating some means of filtering or signal conditioning. The difficulties listed here may limit the enhancement of the functionality and performance of the mechatronic systems. This necessitates the integration of the mechatronic technology with others.

In recent years, optical technology has been incorporated into mechatronic systems at an accelerated rate, and, as a result, a great number of mechatronic products—machines/systems with smart optical components—have been introduced into markets. As shown in Figure 1.1, the presence of optical technology is increasing; this enhances system value and performance because optical elements incorporating mechatronic elements embedded in the system provide some solutions to difficult technical problems.

As discussed at the Forum of Opto-Mechatronic Systems Conference, SPIE, ISAM in 2000 and 2001, this emerging trend shows that optically integrated technology provides enhanced characteristics such that it creates new functionalities that are not achievable with conventional technology alone, exhibits higher functionalities because it makes product/systems function in an entirely different way or in a more efficient manner, and produces high precision and reliability because it can facilitate or enable in-process monitoring and control of the system state. Besides, the technology makes it feasible to achieve dimensional downsizing and provides compactness for the system, as it has the capability of integrating sensors, actuators, and processing elements into one tiny unit.

1.2 Historical Background of Opto-Mechatronic Technology

Optically integrated mechatronic technology has its roots in the technological developments of mechatronics and opto-electronics. Figure 1.2 shows the chronology of those developments.

The real electronic revolution occurred in the 1960s with the integration of transistor and other semiconductor devices into monolithic circuits, which had been made possible by the invention of the transistor in 1948. Then, the microprocessor was invented in 1971 with the aid of semiconductor fabrication technology; this had a tremendous impact on a broad spectrum of technology-related fields. In particular, the development created a synergistic fusion of hardware and software technologies by combining various technologies with computer technology. The fusion made it possible for machines to transform analog signals into digital signals, to perform calculations, to draw conclusions based on the computed results and on software algorithms, and, finally, to take proper action according to those conclusions and to accumulate knowledge/data/information within their own memory domain. This new functionality has endowed machines/systems with characteristics such as flexibility and adaptability [Kayanak, 1996]. Accordingly, the importance of this concept has been recognized among industrial sectors, which accelerated the pace of development of applications that had an ever-wider number of uses. In the 1980s, semiconductor technology also created microelectromechanical systems (MEMS), and this brought a new dimension to machines/systems, microsizing their dimension.

Another technological revolution, known as opto-electronic integration, has continued over the last 40 years, since the invention of the laser in 1960. This was made possible with the aid of advanced fabrication methods such as chemical vapor deposition, molecular-beam epitaxy, and focused-ion-beam micromachining. These methods enabled the integration of optical, electro-optic, and electronic components into a single compact device. The CCD image sensor developed in 1974 not only engendered computer vision technology, but also, starting in 1976, opened up a new era of optical technology and optical fiber sensors. The developed optical components and devices possessed a number of favorable characteristics. These components did not involve contact and were noninvasive, were easy to transduce,

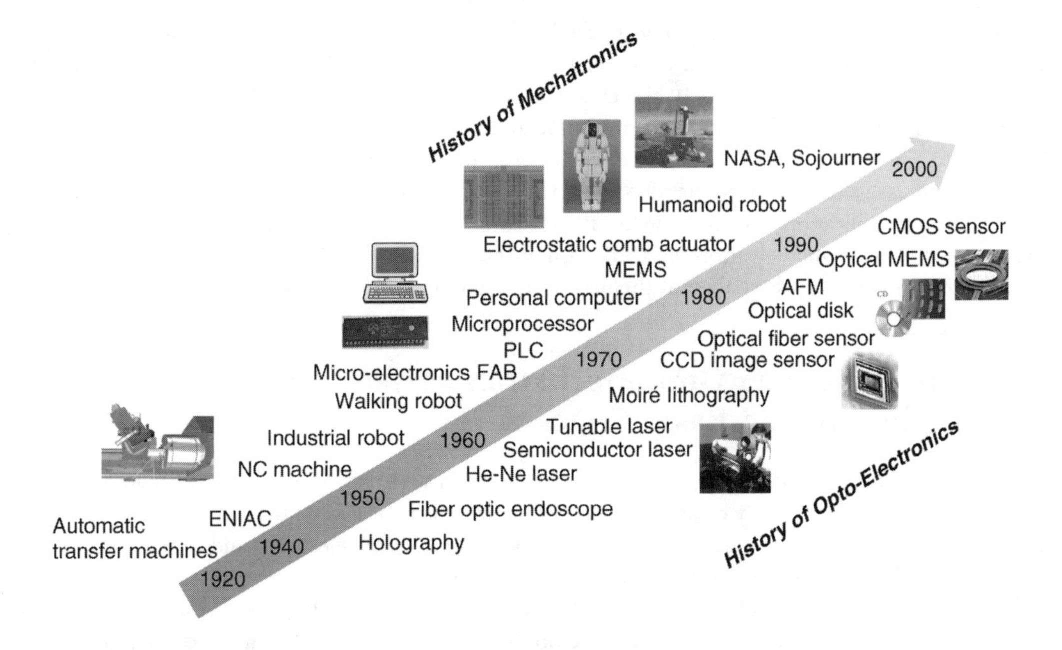

FIGURE 1.2 Historical background of development of opto-mechatronics.

had a wide sensing range, were insensitive to electrical noises, had distributed-sensing and communication, and had high bandwidth.

Naturally, these favorable optical characteristics began to be integrated with those of mechatronic elements, and the integration thus helped achieve systems of higher performance. When a system or machine is integrated in this way, i.e., optically, mechanically, and electronically, it is called an opto-mechatronic system. Figure 1.2 shows several practical applications of this system. The lithography tool that fabricates integrated circuits (ICs) and other semiconductor devices belongs to this system category [Geppert, 1996]: It functions due to a series of elaborate mirrors in addition to a light beam, optical units, and a stepper-servo mechanism that shifts the wafer from site to site with great precision. Another representative system is the optical pick-up device, which went into mass production in 1982. The pick-up system reads information off a spinning disk by controlling both the up-and-down and side-to-side tracking of a read head that carries a low-power diode laser beam focused onto the pits of the disk. Since then, a great number of opto-mechatronic products, machines, or systems have been released at an accelerated rate, because it is possible to achieve significant results with optical components. As shown in the figure, atomic force microscopy (AFM), optical MEMS, humanoid robots, and Sojourner were born in recent years.

1.3 Opto-Mechatronic Integration As a Prime Moving Technology Evolution

As seen from a historical perspective, the electronic revolution accelerated the integration of mechanical and electronic components, and later the optical revolution effected the integration of optical and electronic components. This trend enabled a number of conventional systems that had low-level autonomy and very low-level performance to evolve into ones having improved autonomy and performance.

Figure 1.3 illustrates the practical systems currently in use that evolved from their original old version. Until recently, the PCB inspection was carried out by human workers with the naked eye aided by a microscope, but it was enabled by a visual inspection technique. The functions of the chip mounter, which originally were mostly mechanical, are now being carried out by integrated devices and include part position estimators, visual sensors, and servo control units. The coordinate measuring machine (CMM) appeared as a contact for the first time, followed by noncontact digital electromagnetic, and then the optical type. In recent years, the CMM has been intensely researched in the hope that it may be introduced as an accurate, reliable, and versatile product that uses sensor integration technique. This is shown in the figure. The washing machine shown in Figure 1.10 also evolved from a mechanically operated machine into one with optical sensory feedback and intelligent control function. We shall elaborate more on this issue and make reference to a number of practical systems to characterize the opto-mechatronic technology.

The introductory chapter of this handbook has two main objectives: (1) to present the definition, basic properties, and application aspects of the technology; and (2) to give readers an integrated view of opto-mechatronic systems, thereby affording them insight into how optical systems/devices can be fused or integrated with mechatronic systems at the design and manufacturing stages.

1.4 Understanding Opto-Mechatronic Systems: Definition and Basic Concept

As mentioned in the introduction, the technology associated with the developments of machines/processes/systems has continuously evolved to enhance their performance and to create new value and new function. Mechatronic technology integrated by mechanical, electronic/electrical, and computer technologies has certainly played an important role in this evolution.

However, to make them further evolve toward systems of precision, intelligence, and autonomy, optics and optical engineering technology had to be embedded into mechatronic systems, compensating for the existing functionalities and creating new ones. The opto-mechatronic system is, therefore, a system

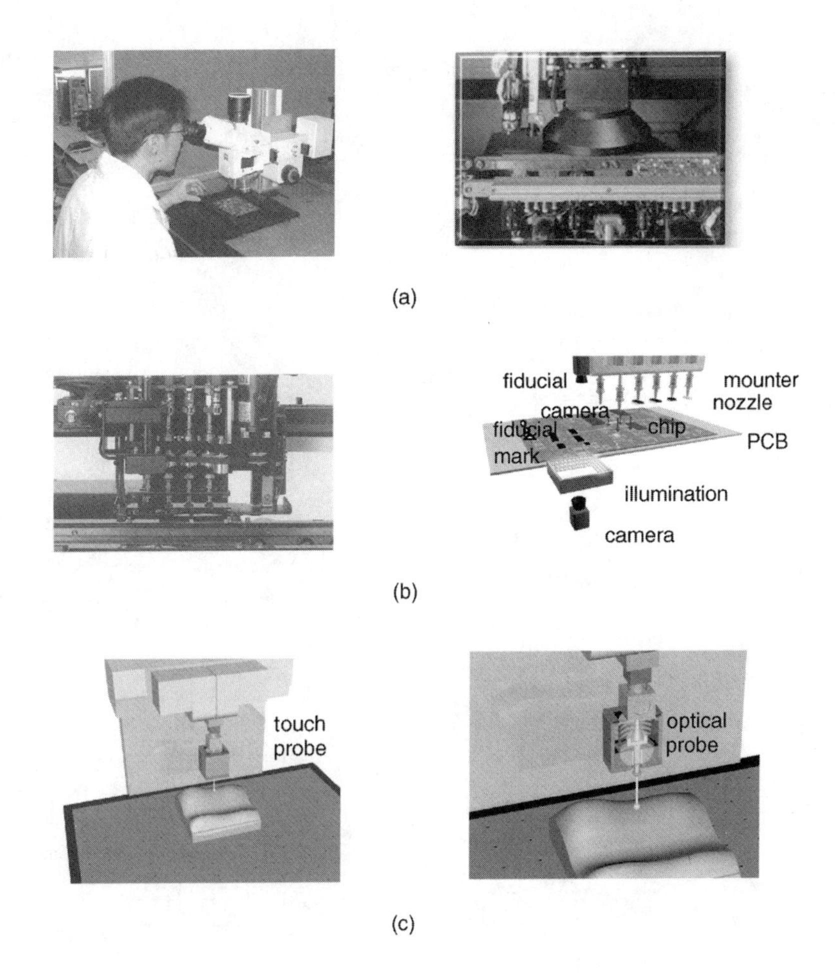

FIGURE 1.3 Illustrative evolutions: (a) PCB inspection; (b) chip/SMD mounting; (c) coordinate measuring machine.

integrated with optical elements, mechanical elements, electrical/electronic elements, and a computer system. In this section, to provide a better understanding of and insight into the system, we will briefly review the basic roles of optical technology and illustrate a variety of opto-mechatronic systems being used today.

1.4.1 Basic Roles of Optical Elements

The major functions and roles of optical components or elements in opto-mechatronic systems can be categorized into several technological domains, as shown in Figure 1.4:

1. *Illumination*: Illumination shown in Figure 1.4(a) provides the source of photometric radiant energy incident to object surfaces. In general, it produces a variety of different reflective, absorptive, and transmissive characteristics depending on the material properties and surface characteristics of objects to be illuminated.
2. *Sensation*: Optical sensors provide fundamental information on physical quantities such as force, temperature, pressure, and strain as well as geometric quantities such as angle, velocity, etc. This information is obtained by optical sensors using various optical phenomena such as reflection, scattering, refraction, interference, diffraction, etc. Conventionally, optical sensing devices are

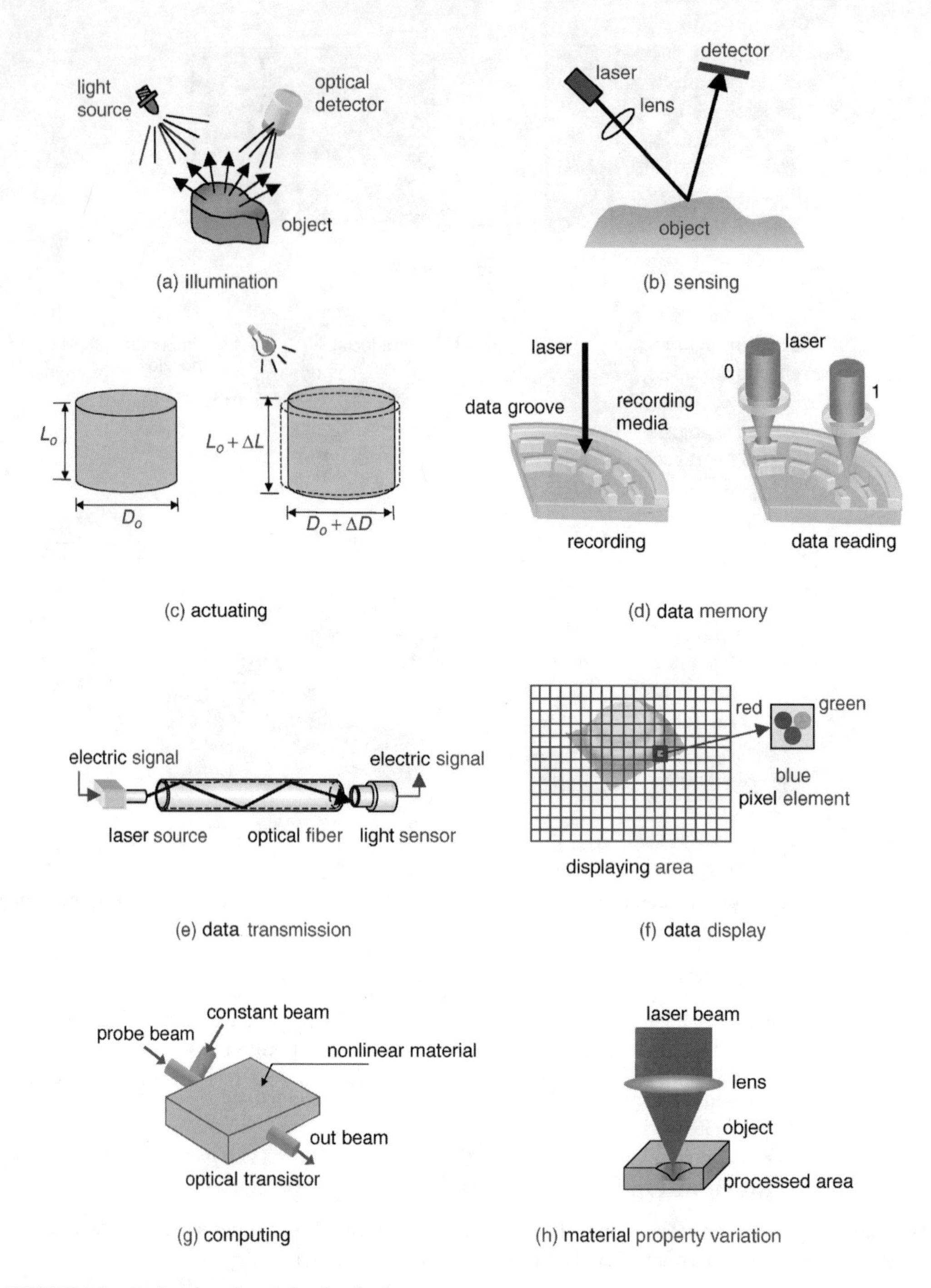

FIGURE 1.4 Basic roles of optical technology.

composed of a light source and photonic sensors and optical components such as lenses, beam splitters, and optical fibers, as shown in Figure 1.4(b). Recently, numerous sensors have been developed that make use of the advantages of optical fiber in various applications. Optical technology can also contribute to material science. The composition of chemicals can be analyzed by spectrophotometry, which recognizes the characteristic spectrum of light that can be reflected, transmitted, and radiated from the material of interest.

3. *Actuation*: Light can change the physical properties of materials by increasing the temperature of the material or by affecting the electrical environment. The materials that can be changed by light are PZT and shape memory alloy (SMA). As shown in Figure 1.4(c), the PZT is composed of ferroelectric material, in which the polar axis of the crystal can be changed by applying an electric field. In optical PZT, the electric field is induced in proportion to the intensity of light. The SMA is also used as an actuator. When the SMA is illuminated by light, the shape of the SMA is changed as a memorized shape due to the increase in temperature. On the other hand, when the temperature of the SMA is decreased, the shape of the SMA is recovered. The SMA is used in a variety of actuators, transducers, and memory applications.
4. *Data (signal) storage*: Digitized data composed of 0 and 1 can be stored in media and read optically, as illustrated in Figure 1.4(d). The principle of optical recording is to use light-induced changes in the reflection properties of a recording medium. That is to say, the data are carved in media by changing the optical properties in the media with laser illumination. Then, data reading is achieved by checking the reflection properties in the media using optical pickup sensors.
5. *Data transmission*: Because of its inherent characteristics such as high bandwidth coupled with imperviousness to external electromagnetic noise, light is a good medium for delivering data. Laser, a light source used in optical communication, has high bandwidth and can hold a lot of data at one time. In optical communication, the digitized raw data, such as text or pictures, are transformed into light signals, delivered to the other side of the optical fiber, and decoded as the raw data. As indicated in Figure 1.4(e), the light signal is transferred within the optical fiber with minimal, if any, loss by total internal reflection.
6. *Data display*: Data are effectively understood by end users through visual information. In order to transfer to users in the form of an image or graph, various displaying devices are used such as CRT (cathode ray tube), LCD (liquid crystal display), LED (light-emitting diode), PDP (plasma display panel), etc. As illustrated in Figure 1.4(f), they all are made of pixel elements composed of three basic cells that emit red, green, and blue light. Arbitrary colors can be made by combining these three colors.
7. *Computation*: Optical computing is performed by using switches, gates, and flip-flops in their logic operation just as in digital electronic computing. Optical switches can be built from modulators using opto-mechanical, opto-electronic, acousto-optic, and magneto-optic technologies. Optical devices can switch states in about a picosecond, or one thousandth of one billionth of a second. An optical logic gate can be constructed from the optical transistor. For an optical computer, a variety of circuit elements besides the optical switch are assembled and interconnected, as shown in Figure 1.4(g). Light alignment and waveguide are two major problems in the actual implementation of an optical computer [Higgins, 1995a].
8. *Material property variation*: When a laser is focused on a spot using optical components, the laser power is increased on a small focusing area. This makes the highlighted spot of material change in the material state as shown in Figure 1.4(h). Laser material processing methods utilize a laser beam as the energy input and can be categorized into two groups: (1) a method of changing the physical shape of the materials, and (2) a method of changing the physical state of the materials.

1.4.2 Practical Opto-Mechatronic Systems

Examples of opto-mechatronic systems are found in many control and instrumentation, inspection and testing, optical, manufacturing, consumer, and industrial electronics products as well as in MEMS, automotive, bioapplications, and many other fields of engineering. Below are some examples of such applications.

Cameras and motors are typical devices that are operated by opto-mechatronic components. For example, a smart camera is equipped with an aperture control and a focusing adjustment with an illuminometer designed to perform well independent of ambient brightness. With this system configuration, new functionalities are created for the performance enhancement of modern cameras. As shown

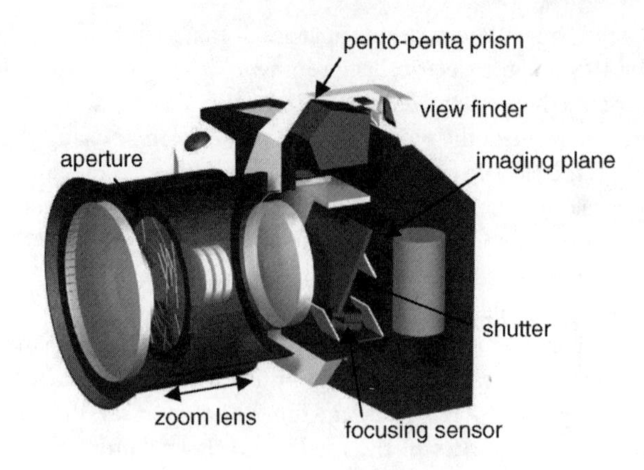

FIGURE 1.5 Camera.

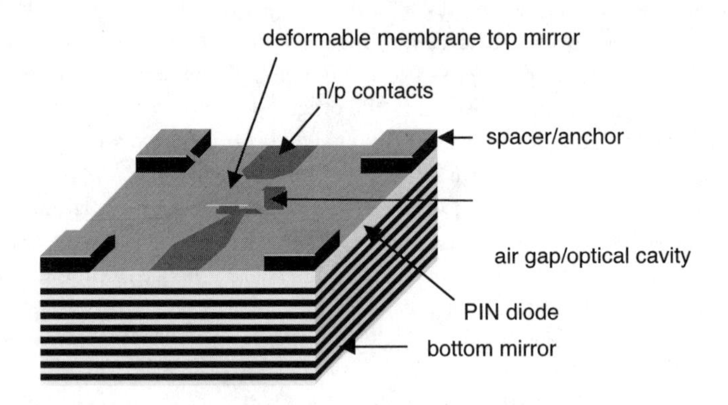

FIGURE 1.6 Tunable laser. (From Larson, M. C., Tunable optoelectronic devices, http://www-snow.stanford.edu/~larson/research.html, 2000. With permission.)

in Figure 1.5, the main components of a camera are lenses, one aperture, one shutter, and film or an electrical image cell such as CCD or CMOS [Canon Co., 2002]. Images are brought into focus and exposed on the film or electrical image cell via a series of lenses that zoom and focus on the object. Changing the position of the lenses with respect to the imaging plane results in a change in magnification and focal points. The amount of light that enters through the lenses is detected by a photo sensor and adequately controlled by changing either the aperture or the shutter speed. Recently, photo sensors or even CMOS area sensors have been used for auto-focusing with a controllable focusing lens.

The tunable laser shown in Figure 1.6 is a typical opto-electronic device [Larson and Harris, 1996, 2000]. This device has a deformable membrane (top mirror) and a GaAs/AlAS distributed Bragg reflector (bottom mirror), which form an optical cavity. The movement of the top mirror directly tunes the wavelength ranges as large as 40 nm in the near-infrared (IR) at an operating voltage of 0–20 V. The gap should be appropriately controlled so as to obtain the desired wavelength of laser beam.

A number of optical fiber sensors employ opto-mechatronic technology whose sensing principle is based on the detection of modulated light in response to changes in the measurand of interest. The optical pressure sensor uses this principle, i.e., the principle of reflective diaphragm, in which deflection of the diaphragm under the influence of pressure changes is used to couple light from an input fiber to an output fiber.

An atomic force microscope (AFM) composed of several opto-mechatronic components, a cantilever probe, a laser source, a position-sensitive detector (PSD), a piezoelectric actuator and a servo controller, and an x-y servoing stage are shown in Figure 1.7. The principle of this AFM employs a constant-force mode. In this case, the deflection of the cantilever is used as input to a feedback controller, which, in turn,

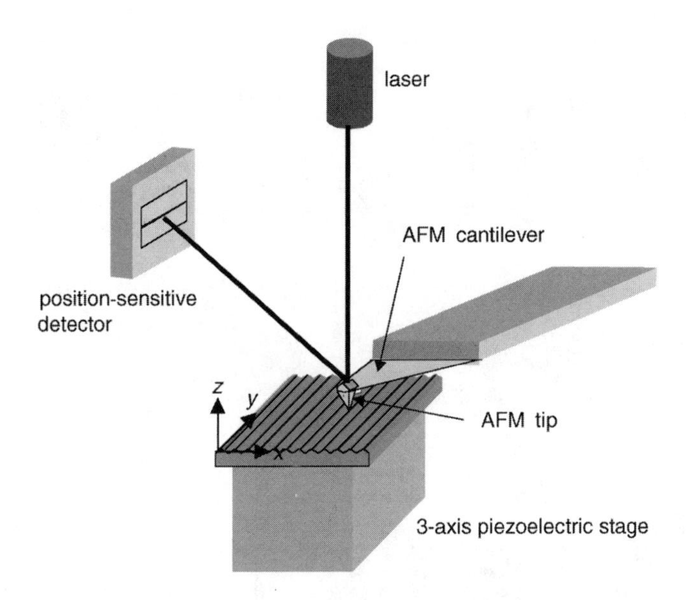

FIGURE 1.7 Atomic force microscope.

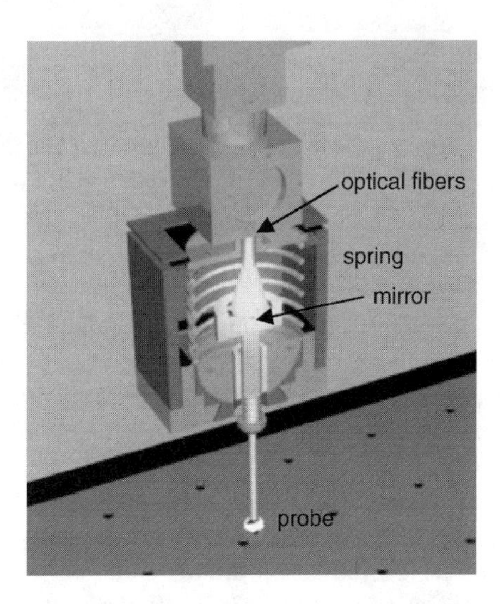

FIGURE 1.8 Optical-based coordinate measuring machine.

moves the piezoelectric element up and down along *z*, responding to the surface topography by holding the cantilever deflection constant. This motion yields positional variation of the light spot at the PSD, which detects the *z*-motion of the cantilever. The position-sensitive photodetector provides a feedback signal to the piezo motion controller. Depending upon the contact state of the cantilever, the microscope is classified as one of three types: (1) contact AFM, (2) intermittent AFM, and (3) noncontact AFM [Veeco Co., 2000].

A three-dimensional contacting probe or sensor, commonly known as a coordinate measuring machine (CMM), measures the three-dimensional geometric surface dimensions of objects. The CMM shown in Figure 1.8 uses the principle of typical opto-mechatronic technology: it adopts an opto-mechatronic sensing principle with the aid of a device that scans the motion of the measuring machine [Butler and Yang, 1993]. The probe is composed of a seven-fiber array and a concave mirror mounted on a stylus that contacts the object to be measured. The infrared light emitted by one of the fibers is reflected by the mirror. The reflected light is then detected by the optical fiber array and modulated by the position and orientation of the mirror, which changes with the position of the stylus. Nano CMM [Takamasu, 2001] is another typical example of

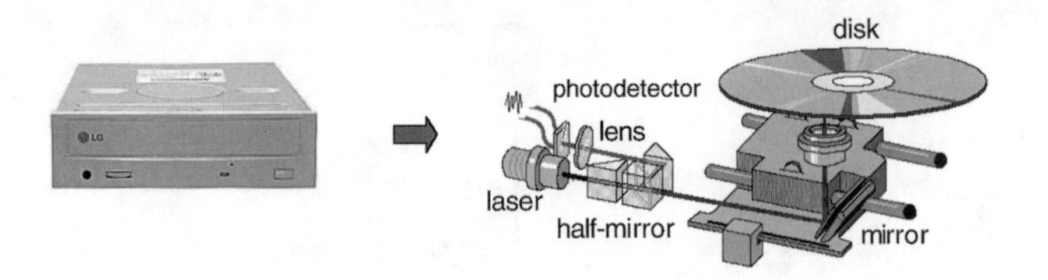

FIGURE 1.9 Optical storage disk.

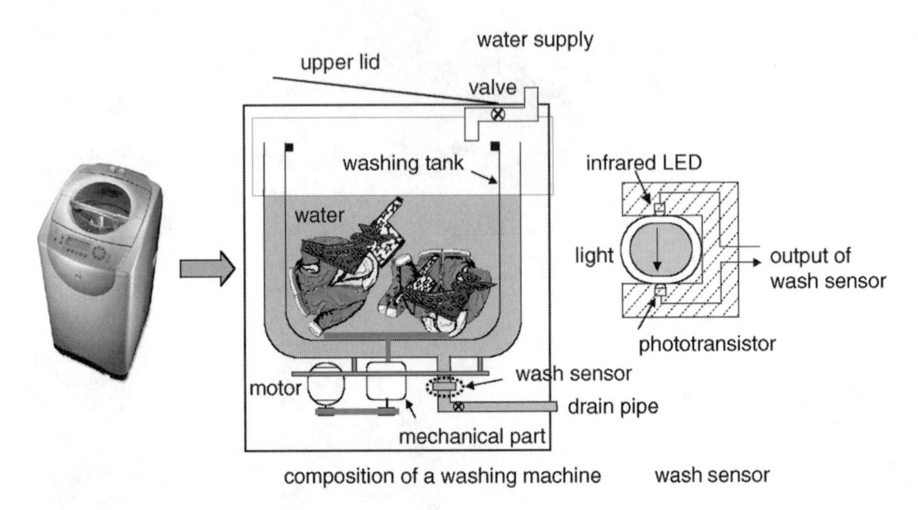

FIGURE 1.10 Modern washing machine with optical sensory feedback.

a system that accommodates optical elements into the conventional system to enhance system measurement performance.

The optical disk drive (ODD) is an opto-mechatronic system; it is shown in Figure 1.9. The ODD is composed of an optical head that carries a laser diode, a beam focus servo that dynamically maintains the laser beam in focus, and a fine-track VCM servo that accurately positions the head at a desired track [Andenovic and Uttamchandani, 1989; Higgins, 1995b]. The disk substrate has an optically sensitive medium protected by a dielectric overcoat and rotates under a modulated laser beam focused through the substrate to a diffraction-limited spot on the medium.

Present-day washing machines effectively utilize opto-electronic components to improve washing performance. They have the ability to feedback-control the water temperature in the washing drum and adjust wash cycle time, depending on how dirty the water inside the wash drum is. As shown in Figure 1.10, the machine is equipped with an opto-mechatronic component to achieve this function. To detect water contamination a light source and a photodetector are installed at the drain port of the water flowing out from the wash drum, and this information is fed back to the fuzzy controller to adjust washing time or water temperature [Wakami et al., 1996].

Figure 1.11 shows an optically ignited mechatronic weapons system [Krupa, 2000]. A laser ignition system mounted in the breech of a gun was the first major change in the past 100 years in the way guns were fired. The system replaces the conventional firing method that utilized an igniter material to ignite the propellant charge. The laser system, however, does not require any primer or igniter material and eliminates the need for a soldier to stand behind the weapon. It only needs an inexpensive but reliable sapphire window through which the laser could be fired.

The precision mini-robot equipped with a vision system is carried by an ordinary industrial robot (coarse robot) as shown in Figure 1.12 [Chen and Hollis, 1999]. Its main function is the fine positioning

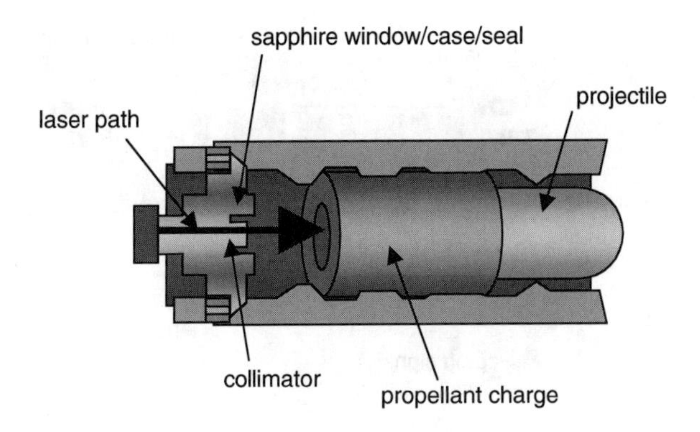

FIGURE 1.11 Optically ignited mechatronic weapon system. (From Krupa, T. J., *Optics Photonics News*, 16–39, June 2000. With permission.)

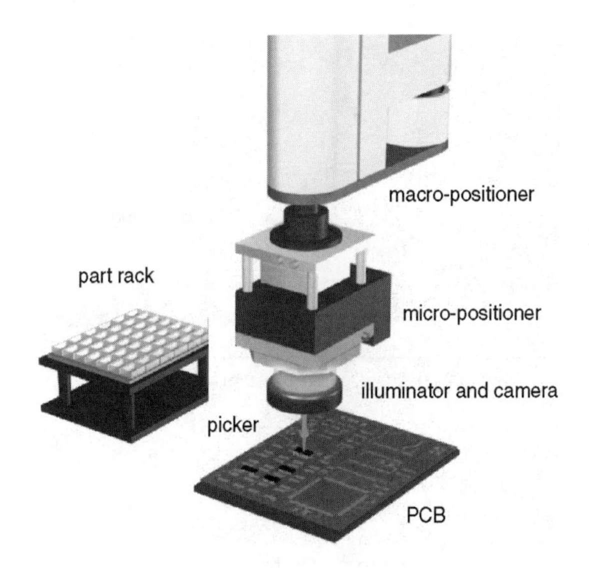

FIGURE 1.12 Vision-guided micro-positioning system.

of an object or part to be placed in a designated location. This vision-guided precision robot is directly controlled by visual information feedback, independently of the course of robot motion. The robot is flexible and inexpensive, being easily adaptable to changes in batch run size, unlike expensive, complex, and mass-production equipment. This system can be effectively used to assemble wearable computers that require the integration of greater numbers of heterogeneous components in an even more compact and lightweight arrangement.

Optical microelectromechanical (MEM) components, shown in Figure 1.13 [Robinson, 2001], are miniature mechanical devices capable of moving and directing light beams. The tiny structures (optical devices such as mirrors) are actuated by electrostatics, electromagnetics, and thermal actuating devices. If the structure is an optical mirror, the device can move and manipulate light. In optical networks, optical MEMS can dynamically attenuate, switch, compensate, combine, and separate signals, all optically. Applications for optical MEMS are increasing and are classified into five main areas: optical switches, optical attenuators, wavelength tunable devices, dynamic gain equalizers, and optical add/drop multiplexes.

Figure 1.14 illustrates a fine image fiberscope device, which can perform active curvature operations for inspecting tiny, confined areas such as a microfactory [Tsuruta et al., 1999]. A shape memory alloy

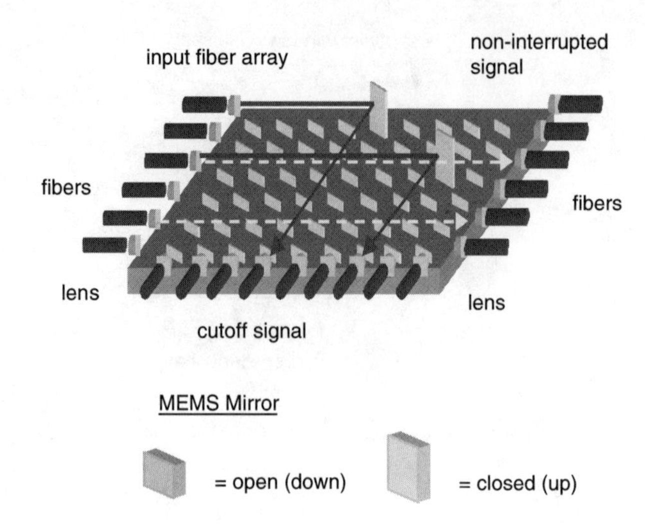

FIGURE 1.13 An n × n optical switching system.

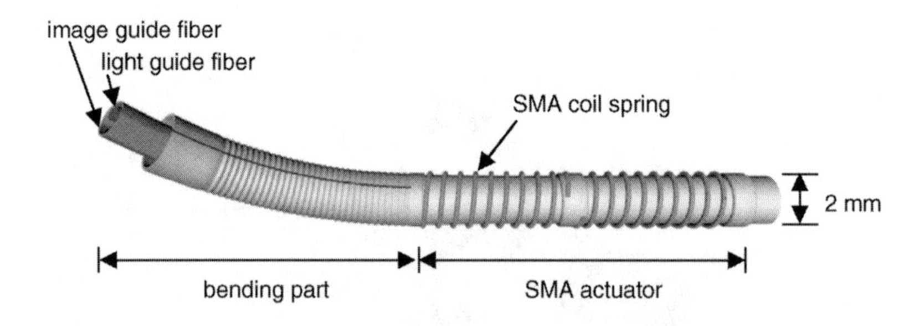

FIGURE 1.14 Fiber scope device for inspection for microfactory. (From Tsuruta, K. et al., *Sensor Rev.*, 19, 37–42, 1999. With permission.)

(SMA) coil actuator enables the fiberscope to move through a tightly curvatured area. The device has a fine-image fiberscope 0.2 mm in outer diameter with light guides and 2000 pixels.

Laser-based rapid prototyping (RP) is a technology that produces prototype parts in a much shorter time than traditional machining processes. An example of this technology includes the stereo-lithography apparatus (SLA). Figure 1.15 illustrates an SLA that utilizes a visible or ultraviolet laser and a position servo mechanism to solidify selectively liquid photo-curable resin [Cho et al., 2000]. The process machine forms a layer with a cross-sectional shape that has been previously prepared from computer-aided design (CAD) data of the product to be produced. Through repeating the forming layers in a specified direction, the desired three-dimensional shape is constructed layer by layer. This process solidifies the resin to 96% of full solidification. After building, the part is placed in an ultraviolet oven to be cured up to 100%, i.e., a post-curing process.

A number of manufacturing processes require feedback control of in-process state information that must be detected by opto-electronic measurement systems. One such process is illustrated here to aid readers in understanding the concept of the opto-mechatronic systems. Figure 1.16 shows a pipe-welding process that requires stringent weld quality control [Ko et al., 1998]. A structured laser triangulation system achieves this by detecting the shape of a weld bead in an online manner and feeding back this information to a weld controller. The weld controller adjusts weld current, depending on the shape of the element being made. In this situation, no other effective method of instantaneous weld quality measurement can replace the visual in-process measurement and feedback control described here.

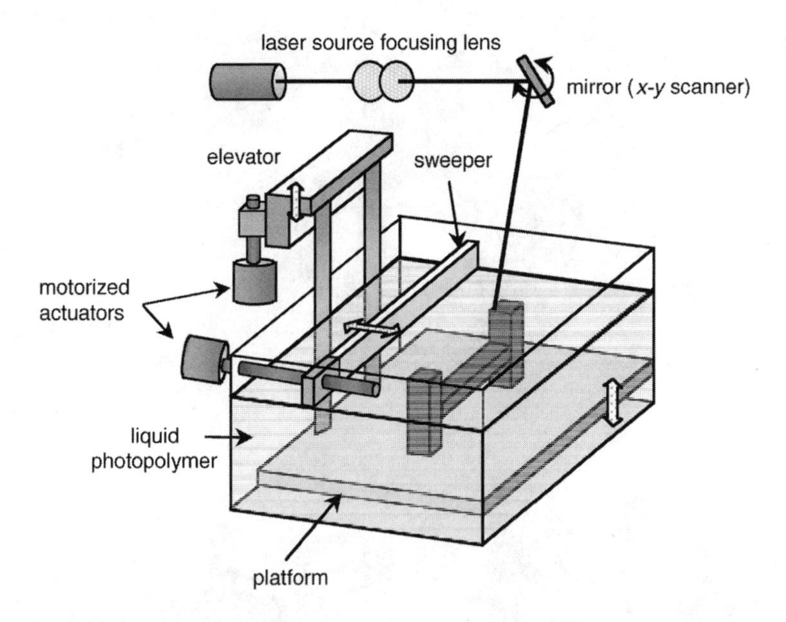

FIGURE 1.15 Rapid prototyping process. (From Tsuruta, K. et al., *Sensor Rev.*, 19, 37–42, 1999. With permission.)

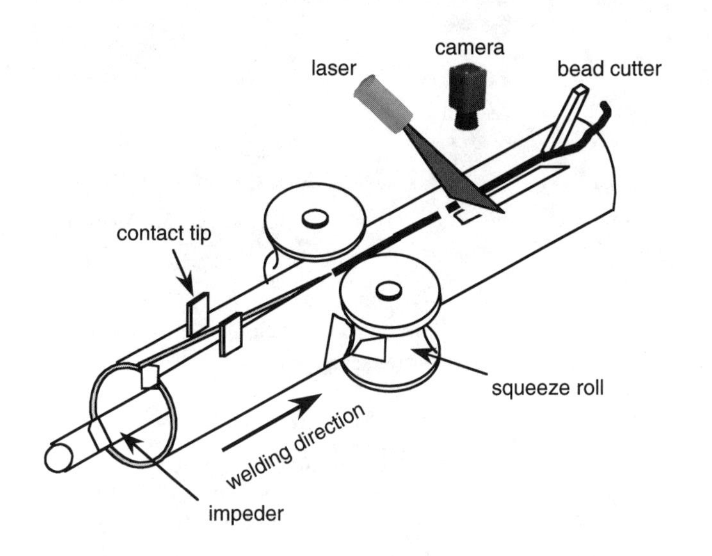

FIGURE 1.16 Pipe welding process.

1.4.3 Types of Opto-Mechatronic Systems

As we have seen from the various illustrative examples of opto-mechatronic systems discussed, the systems can be categorized into three classes, depending on how optical elements and mechatronic components are integrated. As indicated in Figure 1.17, the classes may be described as follows.

1.4.3.1 Opto-Mechatronically Fused System

In this system, optical and mechatronic elements are not separable in the sense that, if optical or mechatronic elements were removed from the system they constitute, the system could not function properly. This implies that those two separate elements are functionally and structurally fused together

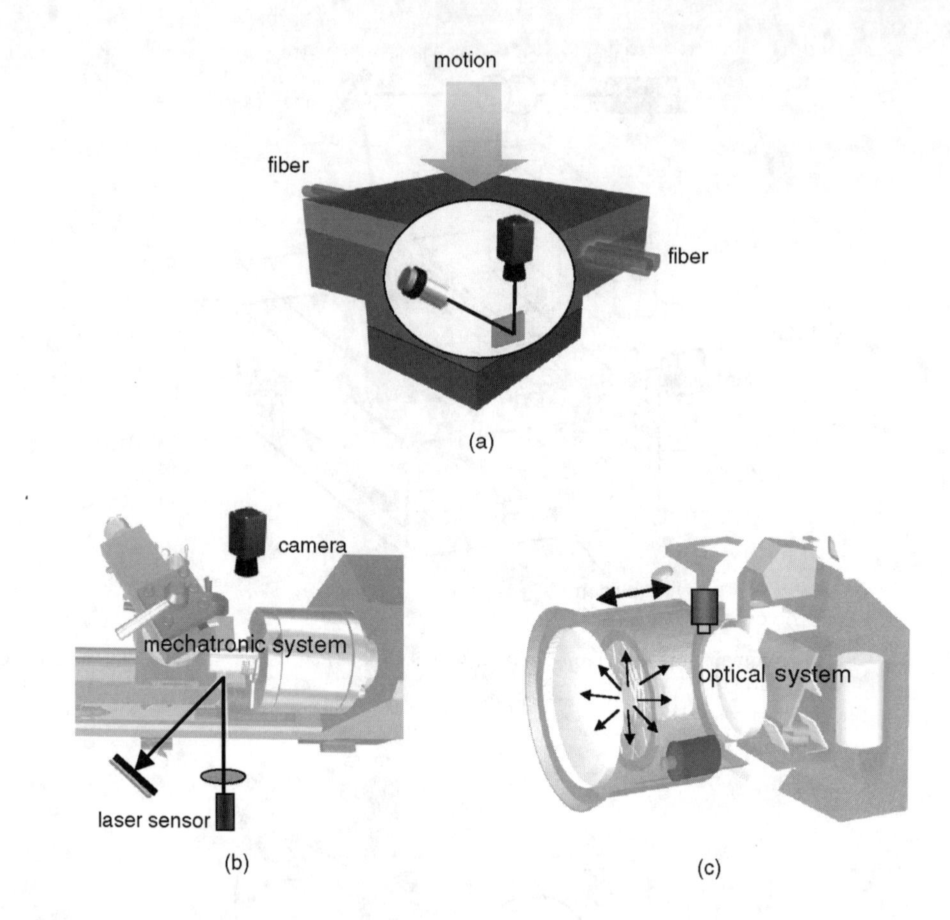

FIGURE 1.17 Types of opto-mechatronic system: (a) opto-mechatronically fused system; (b) optically embedded mechatronic system; (c) mechatronically embedded optical system.

to achieve a desired system performance. The systems that belong to this class are auto camera, adaptive mirror, tunable laser, CD-pickup, optical pressure sensor, etc.

1.4.3.2 Optically Embedded Mechatronic System

This system is basically a mechatronic system that is mainly composed of optical, mechanical, and electrical/electronic elements. In this system, the optical element is embedded onto the mechatronic system. The optical element is separable from the system, and yet the system can function but with a decreased level of performance. The majority of engineered opto-mechatronic systems belong to this category: washers, vacuum cleaners, monitoring and control systems for machines/manufacturing processes, robots, cars, servomotors, etc.

1.4.3.3 Mechatronically Embedded Optical System

This system is basically an optical system whose construction is integrated with mechanical and electrical components. Many optical systems require positioning or servoing optical elements/devices to manipulate and align the beam and control polarization of the beam. Typical systems that belong to positioning or servoing include cameras, optical projectors, galvanometers, series parallel scanners, line scan polygons, optical switches, fiber squeezer polarization controllers, etc. In some other systems acoustic wave generators driven by a piezo element are used to create a frequency shift of the beam. Several applications include fiber optic communication and sensors. Another typical system is the passive (offline) alignment of optical fiber–fiber or fiber–wave guide attachments. In this system, any misalignment is either passively or actively corrected by using a micropositioning device to maximize their coupling between fibers or between the fiber and wave guide.

It is noted here that in the present discussion the word "system" includes elements/devices/ machines/ processes/actual systems/products. Each of these systems may possess only one distinct "system" type but in some other cases a system may have a combination of the three types.

1.5 Fundamental Functions of Opto-Mechatronic Systems

Having illustrated various types of opto-mechatronic systems, we now need to provide some details about its enabling technologies and fundamental functions.

1.5.1 Elements of Opto-Mechatronic Technology

Using the discussions in the previous sections as our point of departure, we can now elaborate on what type of fundamental functions opto-mechatronic systems can perform. There are a number of distinct functions that grew out of the basic roles of optical elements and mechatronic elements. When these are combined together, the combined results generate the fundamental functions of the opto-mechatronic system.

Figure 1.18 illustrates the component technologies needed to produce opto-mechatronic systems. In the center of the figure, those related to optical elements are shown, while mechatronic elements and artificial intelligence are listed in its periphery. These enabling technologies are integrated together to form five engines that drive the system technology. These include: (1) sensor module, (2) actuator module, (3) control module, (4) signal/information processing module, and (5) decision-making module. The integration of these gives a specified functionality with certain characteristics. A typical example is the sensing of an object's surface with an atomic force microscope (AFM). The AFM requires a complicated interaction between the sensing element (laser sensor), actuator (piezo element), controller, and other relevant software. This kind of interaction between modules is very common in opto-mechatronic systems and produces the characteristic property of the systems.

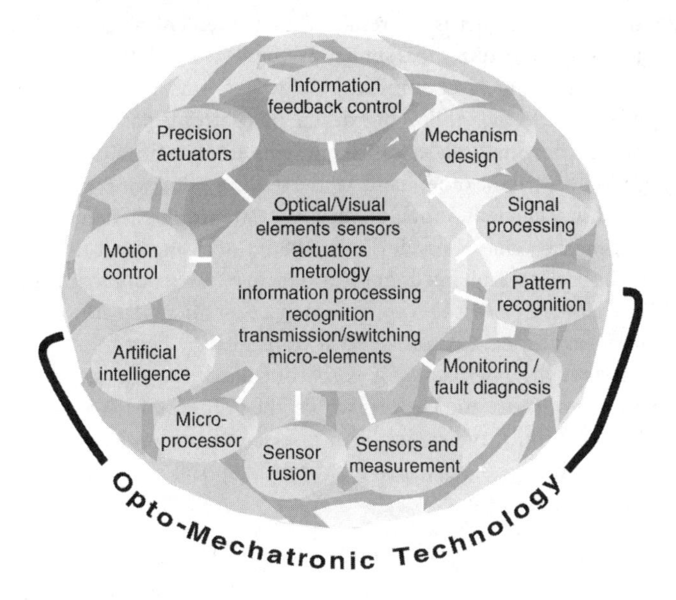

FIGURE 1.18 Enabling technologies for opto-mechatronic systems.

1.5.2 Fundamental Functions

A number of fundamental functions can be generated by fusing optical elements with mechatronics elements. These include (1) illumination control, (2) sensing, (3) actuating, (4) optical scanning, (5) motion control, (6) visual/optical information feedback control, (7) data storage, (8) data transmission/switching, (9) data display, (10) inspection, (11) monitoring/control/diagnosis, (12) three-dimensional shape reconstruction, (13) optical property variation, (14) sensory-feedback-based optical system control, (15) optical pattern recognition, (16) remote monitoring/control, and (17) material processing.

Various practical systems employing those functionalities are listed and classified in Table 1.1. The table illustrates the technology transition flow for achieving each functionality.

1.5.2.1 Illumination Control

Illumination needs to be adjusted depending on optical surface characteristics and the surface geometry of objects in order to obtain good quality images. The parameters to be adjusted include incident angle and the distribution and intensity of light sources. Figure 1.19 illustrates the typical configuration of such an illumination control system [Mitutoyo Co., 2001]. The system consists of a guardant ring fiber light source, a paraboloid mirror, a toroid mirror, and a positioning mechanism to adjust the distance between the object's surface and the parabola mirror. The incident angle as shown in the figure is controlled by adjusting the distance, while the intensity of the light source at each region is controlled independently.

1.5.2.2 Sensing

Various types of opto-mechatronic systems are used to take measurements of various types of physical quantities such as displacement, geometry, force, pressure, target motion, etc. The commonly adopted feature of these systems is that they are composed of optical, mechanical moving, servoing, and electronic elements. The optical elements are usually the sensing part divided into optical fiber and nonfiber optical transducers. Recently these elements are accelerating the sensor fusion, in which other types of sensors are fused together to obtain necessary information.

When the three-dimensional objects to be measured are stationary, as shown in Figure 2.20(a), the three-dimensional scanning motion needs to be intentionally generated. Two methods are frequently used: (1) a sensor head (light source plus detector) is made to travel, or in some cases light sources or detectors can be moved while the object is standing stationary during measurement; or (2) the objects move sequentially relative to a stationary sensor head. If a moving object is being measured, the motion is not provided intentionally, but rather is the result of the movement of the mechanical part.

Optical-fiber-based sensing also utilizes this similar principle in which the motion of a sensing element (part) is provided either intentionally or unintentionally. The sensor shown in Figure 2.20(b) uses a modulated light amplitude due to the motion of external mechatronic parts. Microbending fiber sensors and the pressure sensors employing a deflective diaphragm are examples of an unintentionally driven sensing element being adopted. On the other hand, tactile sensing using servoing is an example of utilizing an intentionally driven sensing element, which can be found when grasping an object with a complicated surface where visual sensors cannot work well [Pugh, 1986].

Figure 1.20(c) shows a six-degree-of-freedom (6-D) sensory system employing the opto-mechatronic principle [Park and Cho, 1999]. The sensor is composed of a 6-D microactuator, a laser source, three PSD sensors, and a three-facet tiny crystal (or mirror) that transmits the light beam into three PSD sensors. The sensing principle is that the microactuator is so controlled that the top of tiny mirror is always positioned within the center of the beam. In other words, the output of each PSD sensor is kept identical regardless of the object's motion. A typical problem associated with optical sensing is focusing, which needs to be adjusted depending on the distance between the optical lens and the surface of the objects to be measured [Zhang and Cai, 1997].

1.5.2.3 Actuating

Two types of optical actuating principles are shown in Figure 1.21(a) and (b). Figure 1.21(a) adopts the photostrictive phenomenon, in which optical energy is converted to mechanical displacement of

TABLE 1.1 Classification and Technology Transition State of Various Functionalities

Functionality	Illustrative Techniques/System	System Type	Technology Transition State
Sensing	Optical-fiber-based sensor: pressure, temperature, displacement, acceleration Optical/sensor: three-dimensional imaging, optical motion capture, confocal sensor, AFM, camera, adaptive mirror CMM: nano-probe system, fiber optic CMM	MO/OPME	ME→O→O
Optical scanning	Optical scanning device: galvanometer, resonant scanner, acoustics-optic scanner, polygon mirror, pan-tilt mechanism Optical/visual scanning system: navigation/ surveillance robot, image recognition system, PCB inspection system, wafer inspection system	MO/OPME	OE→ME→O
Optical actuator	Hyper-sensitive light-driven device	OPME	OE→ME
Visual inspection	Devices: endoscope, ersascope Inspection: PCB pattern/PCB solder joint inspection, weld seam pattern	OPME	ME→OE→OE
Motion control	Vision-guided machine/robot: weld seam tracker, mobile robot, navigation, visual serving end-effectors Optical-based motion control: inspection head, auto focusing, optical-based dead reckoning, lithography state	OM	OE→ME
Data storage	Optical disk drive, DVD	MO	OE→O→ME→O
Data transmission	Optical switch, optical filter, optical modulator, optical attenuator	MO	O→ME→O
Data display	Digital micromirror array (DMD)	MO	O→ME→O
Monitoring/control/ diagnosis	Machining process, smart structure, assembly process, semiconductor process, monitoring/control, laser material process, welding process, textile fabric fabrication process, metal forming process	OM/MO	O→ME→M
Three-dimensional shape reconstruction	X-ray tomography, x-ray radiography	OPME	E→O→ME→O
Optical property variation	Tunable laser, fiber loop polarization, frequency modulator, tunable wave length (Raman oscillator)	OPME	E→M→O
Laser material processing	Laser cutting, laser drilling, laser welding, laser grooving, laser hardening, laser lithography, stereo lithography	OPME	OE→O→M
Optical pattern recognition	Target recognition, target tracking, vision-based navigation	MO	OE→ME→O
Optical/visual information feedback control	Pipe welding process, washing machine, arc welding process, laser surface hardening, SMD mounting, camera, smart car	OM/MO	OE→ME→M
Illumination control	Intelligent illumination system	MO	OE→ME→O
Remote monitoring/ control	Internet-based sensing/control, remote monitoring/diagnosis, remote visual feedback control	OM/MO	(O) →ME→OE→O
Information-based visual/optical system control	Supervisory visual control, camera zoom control	MO	ME→ME→O

Note: OM, optical embedded mechatronics system; OPME, opto-mechatronically fused system; MO, mechatronically embedded optical system; O, optical; OE, opto-electronic; ME, mechatronic; M, mechanical.

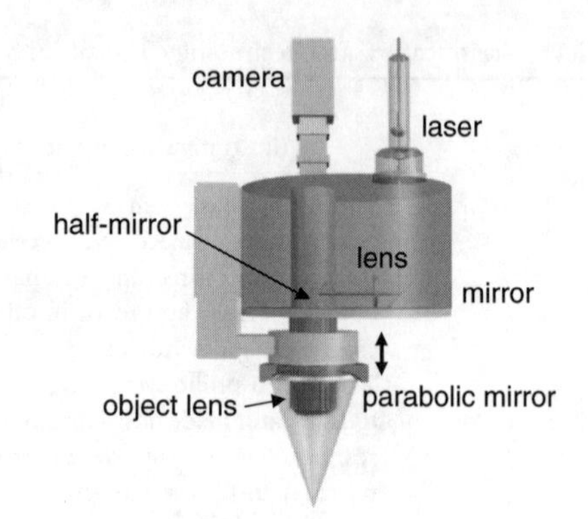

FIGURE 1.19 Illumination control.

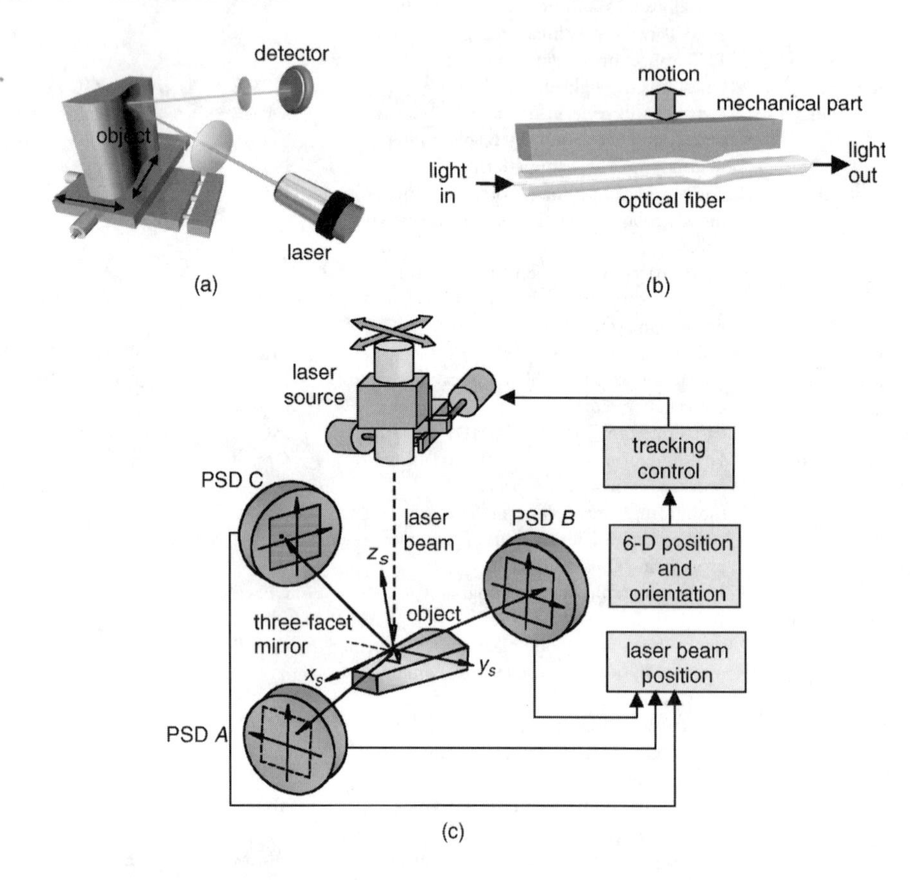

FIGURE 1.20 Typical sensing configurations: (a) scanning-based sensing; (b) optical-fiber-based sensing; (c) measurement of six degrees of freedom of arbitrary objects by mounting a three-facet mirror.

piezoelectric material. In Figure 1.21(b), optical energy is used as a heat generator, which causes a temperature-sensitive material (e.g., shape memory alloy) to move. These actuators can be used to accurately control the micromovement of mechanical elements.

Figure 1.21(c) depicts the configuration of an optical gripper using the bimorph optical piezoelectrical actuators [Fukuda et al., 1993]. The element of the optical piezoelectrical actuator (PLZT) is a ceramic polycrystalline material made of (Pb, La)(Zr, Ti)O_3. When the ultraviolet (UV) beam irradiates from the surface of the PLZT, the PLZT is elongated due to the photostrictive phenomenon, which is the conversion of an electromotive force to strain by the piezoelectric effect. As shown in the figure, each actuator is composed of an optical-fiber carrying UV beam, a beam-directing mirror, and a piezoelectric

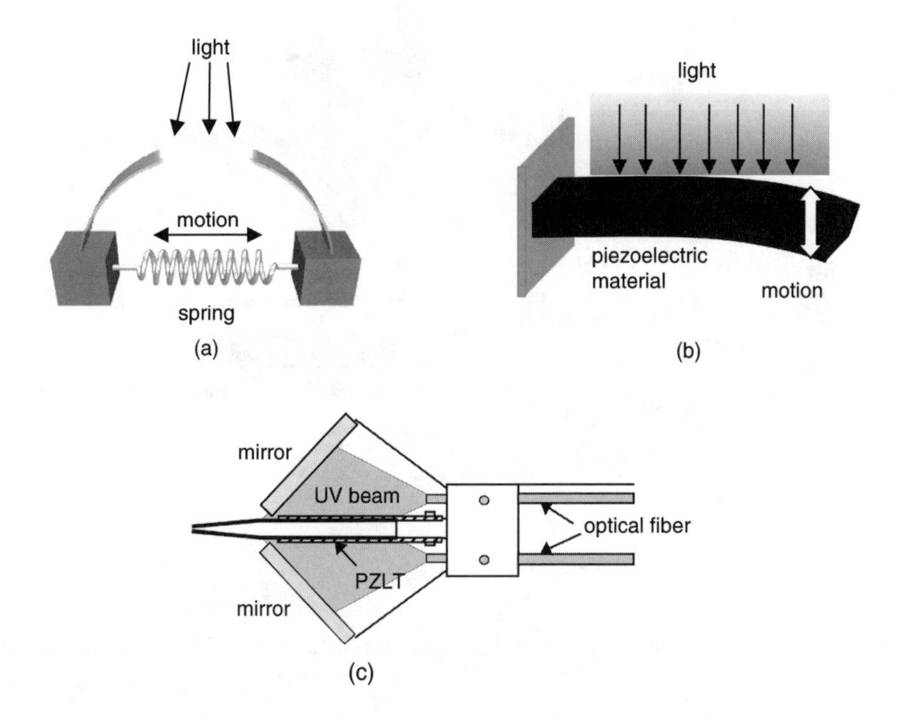

FIGURE 1.21 Two types of optical actuating, (a) shape memory-based actuating and (b) piezo-based actuating, and (c) an optically actuated gripper.

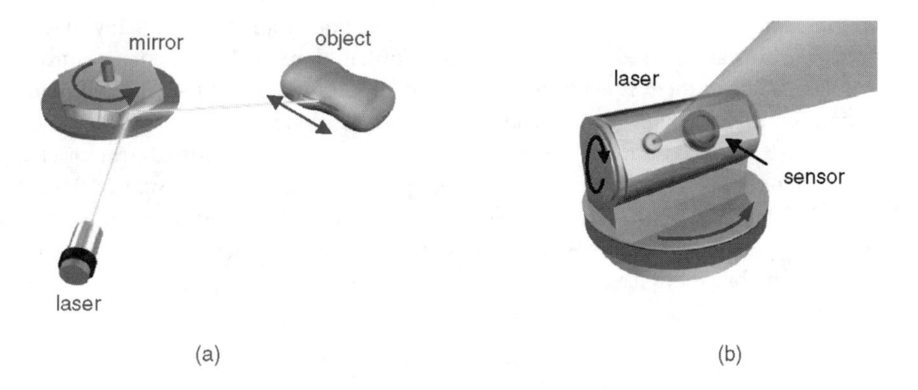

FIGURE 1.22 Optical-based scanning: (a) mirror-based scanning; (b) pan-tilt mechanism.

element. When the UV beam is irradiated by the PLZT surface, the gripper can be open or closed due to the actuator of the PLZT. The displacement of the gripper at the tip is 100 μm.

1.5.2.4 Optical Scanning

Optical scanning generates a sequential motion of optical elements such as optical-based sensors and light sources. For high-speed applications, the polygon mirror or galvanometer [Zankowsky, 1996] can be used effectively, as shown in Figure 1.22(a). Another device can be found from active optical-fiber scanners [Backmutsky and Vaisman, 1996]. The pan-tilt device shown in Figure 1.22(b) scans a certain range of sensing area with a servoing mechanism embedded with optical sensors such as a CCD camera, fiber-displacement sensors, etc. In the case of low- and medium-speed applications with the sensor servoing system, the pan-tilt device is conventionally used. To operate the scanning system, the scanning step and the scanning angle must be considered carefully. These parameters depend on the field of view

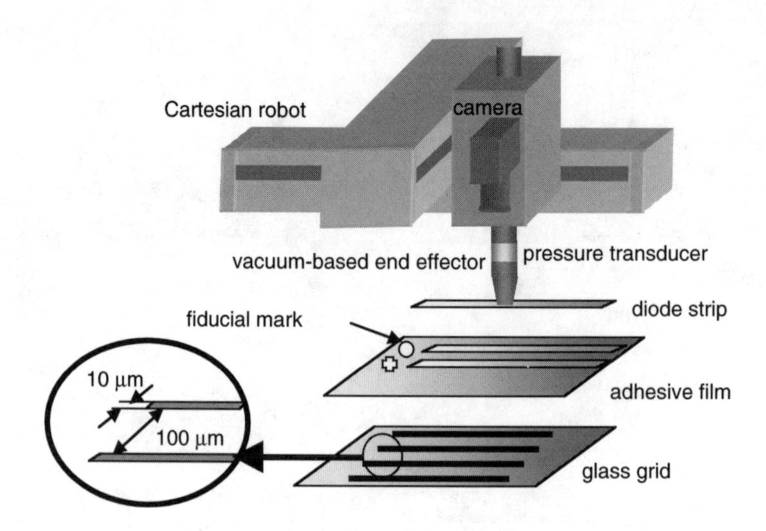

FIGURE 1.23 Motion control: placing diode strip into a glass.

(FOV) of the sensing device, the sensing resolution specification, and the field of interest (FOI) defined by the operator.

1.5.2.5 Motion Control

A variety of opto-mechatronic systems need motion control. To have this function the systems need to be equipped with sensors, actuators, and controllers. The systems may be optical systems or mechatronic systems. Sensors and actuators also may be of the optical or nonoptical type. In many practical uses, however, optical sensors are utilized for acquiring the information needed for motion control, and the information obtained from them is fed back to a servo controller to achieve the required motion. Two types of motion control are (1) positioning and (2) tracking. Positioning is to control the position of an object, moving it to a specified location. When a visual sensor is used, the motion control is called "visual servoing." An example is the accurate placement of diode strips into a glass grid using a robot, as shown in Figure 1.23 [Wilson, 2001]. Two ways of tracking motion are by following the motion of a moving object and following a desired path based on the optical-sensor information. A number of typical examples related to mobile robot navigation can be found in the literature. Optical-based dead reckoning [Borenstein, 1998], map building, and vision-based obstacle avoidances are a few examples.

1.5.2.6 Visual/Optical Information Feedback Control

Visual/optical information is very useful in the control of machines, processes, and systems. The distinction from the fundamental function of motion control is that the information obtained by optical sensors is utilized for changing other variables operating the systems besides motion. A number of opto-mechatronic systems require this type of information feedback control. As shown in Figure 1.16, a pipe-welding process necessitates optical information feedback control. The objective of this process control is to regulate the electric power needed for welding in order to attain a specified weld quality. Therefore, the process state relevant to the quality is detected and its instantaneous information is fed back in real time. This principle is also illustrated in the washing machine shown in Figure 1.10. In this system, the optical sensor detects the dirtiness of water and this information is fed back to a controller to adjust washing time or to adjust the water temperature inside the drum.

1.5.2.7 Data Storage

Data storage and retrieval are performed by a spinning optical disk and controlled optical units whose main functions are beam focusing and track following, as illustrated in Figure 1.24. The storage and retrieval principles of the optical disk were explained in Section 1.4.2. Conventionally, the recording

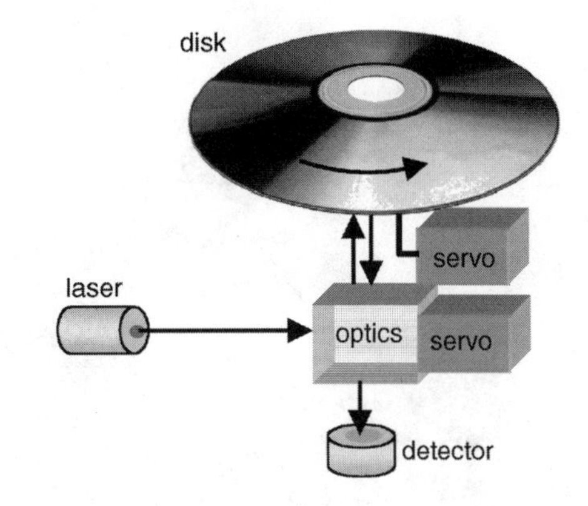

FIGURE 1.24 Data storage/retrieval.

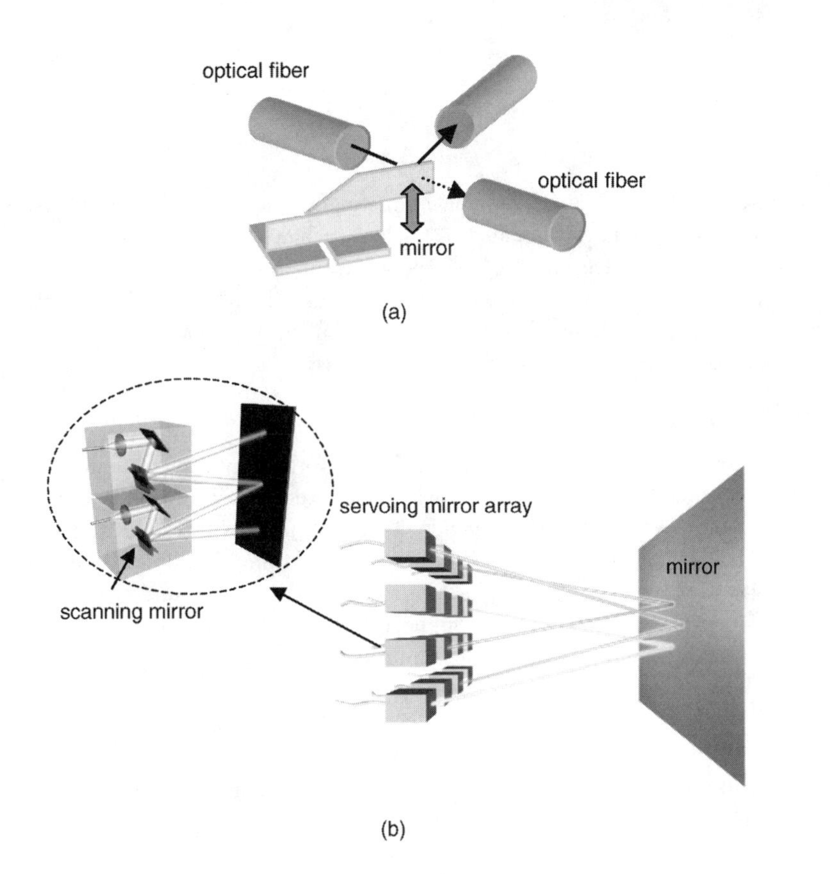

FIGURE 1.25 Data transmission by optical switching: (a) principle of optical switching; (b) optical cross connect operation by mirror servo control.

density is limited by the spot size and wavelength of the laser source. Recently, new approaches to increasing the recording density have been researched. Examples include near-field optical memory and holographic three-dimensional storage.

1.5.2.8 Data Transmission/Switching

Optical data switching is achieved by an all-optical network to eliminate the multiple optical-to-electrical-to-optical (O–E–O) conversions in conventional optical networks [Robinson, 2001]. Figure 1.25 illustrates a simple two-dimensional optical MEMS switch to illustrate an all-optical network. In this configuration,

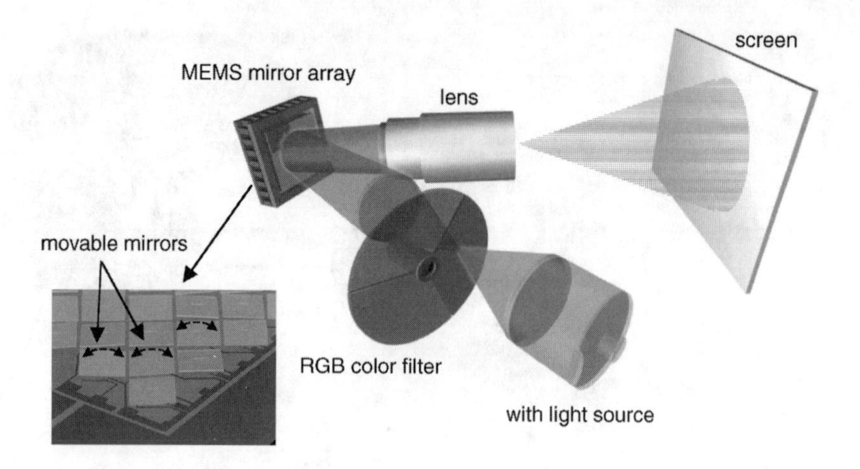

FIGURE 1.26 Optical projection via MEMS.

the system consists of a mirror, an actuator, a collimating lens, and input/output fibers through which light comes in from a fiber and is collimated. Then, by switching the action, light is directed or switched from one fiber to another. Precision pointing accuracy of the mirrors is required by analog control, as shown in Figure 1.25(b). Here, mirrors are controlled to connect any light path to any port through the use of servo controllers. This kind of system configuration is also applied to switch microreflective or deflective lenses actuated in front of optical filters [Toshyoshi et al., 1990].

1.5.2.9 Data Display

Digital micromirror devices (DMDs) make projection displays by converting white-light illumination into full-color images via spatial light modulators with independently addressable pixels. As schematically illustrated in Figure 1.26, the DMD developed by Texas Instruments [McDonald and Yoder, 1998] is a lithographically fabricated MEMS composed of thousands of titling aluminum alloy mirrors (16 μm × 16 μm) that work as pixels in the display. Each mirror is attached to an underlying subpixel called the yoke, which, in turn, is attached to a hinge support post by way of an aluminum torsion hinge. This allows the mirror to rotate about the hinge axis until the landing tips touch the landing electrical. This switching action occurs from +10° to −10°C and takes place in several microseconds.

1.5.2.10 Inspection

Inspection is an integrated action composed of measurement, motion control, and synthesis of the measured data. Unlike monitoring, this task is normally done offline except for an online inspection. In general, the task objectives are given prior to the beginning of the inspection. As indicated in Figure 1.27, the optical sensor provides a measuring capability, the motion control provides the ability to search the area to be inspected in a designated sequence, and the synthesizing unit processes the measured data and analyzes the processed information. The inspection work is usually followed by such tasks as pattern classification and pattern recognition using conventional algorithms or artificial intelligence techniques.

1.5.2.11 Monitoring/Control/Diagnosis

Monitoring requires real-time identification or estimation of the characteristic changes of devices, machines, process, or systems based on the evaluation of their signature without interrupting normal operations. When this is done, a series of tasks, such as sensing, signal processing, feature extraction and selection, pattern classification, and pattern recognition, need to be performed. A number of opto-mechatronic systems perform monitoring of the in-process condition in situations where conventional sensors cannot be used or are otherwise unavailable. Three typical types of monitoring configurations

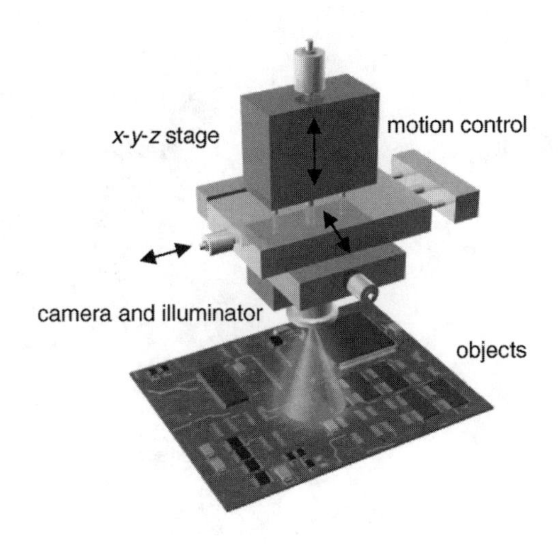

FIGURE 1.27 Inspection.

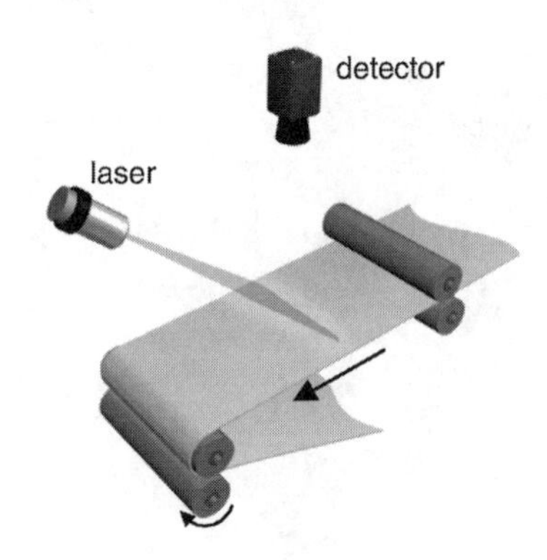

FIGURE 1.28 Monitoring/control (stationary head) based on optical scanning.

are illustrated here:

1. Monitoring head, built-in moving system
2. Monitoring head, stationary relative to moving system
3. Monitoring system, embedded into system to be monitored and compensated

A variety of monitoring systems belong to category (1) and can be found in manufacturing processes, robot surveillance, and task condition monitoring. Category (2) is also practiced in a variety of actual manufacturing and robotic systems, as shown in Figure 1.28. In some cases in category (3) monitoring and control units are embedded into the structure of a system to be monitored. They are embedded in a kind of layer and are distributed in nature; an example of this is the smart structure.

Figure 1.29 shows a typical arrangement of a machining system in which an opto-electronic device is embedded in a mechatronic machine (lathe) [Shiraishi, 1979]. Control of the machined workpiece's qualities, such as dimension and surface roughness, is ultimately important in the cutting operation. This can be effectively achieved only by in-process monitoring and control. Monitoring methods other than optical do not seem to be effective in terms of measurement accuracy and cost because the workpiece is rotating and the conditions for sensing are not favorable due to chip formation and the presence of cutting oil. The measurement principle employed here is to use a laser (He–Ne gas) sensor and a photodetector. Due to the cutting operation the workpiece dimension (diameter) is continuously varied and the optical sensor system detects this varying dimension: As soon as a laser spot (0.5 mm) hits the workpiece surface, the reflected laser light is magnified by the receiving optical system and an image is made on the photodetector on the focal plane.

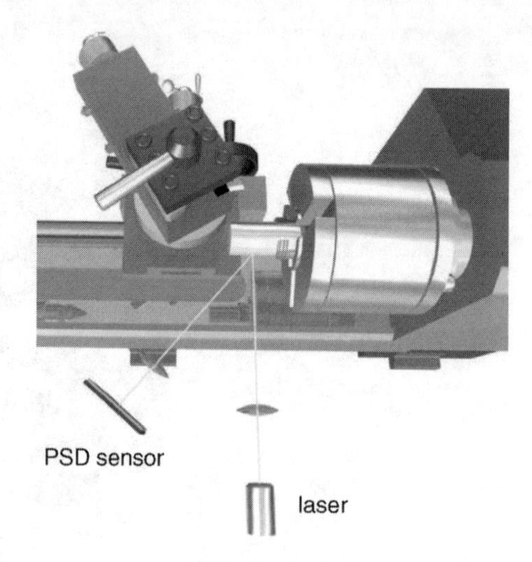

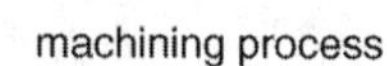

FIGURE 1.29 Optical-based monitoring and control of manufacturing process.

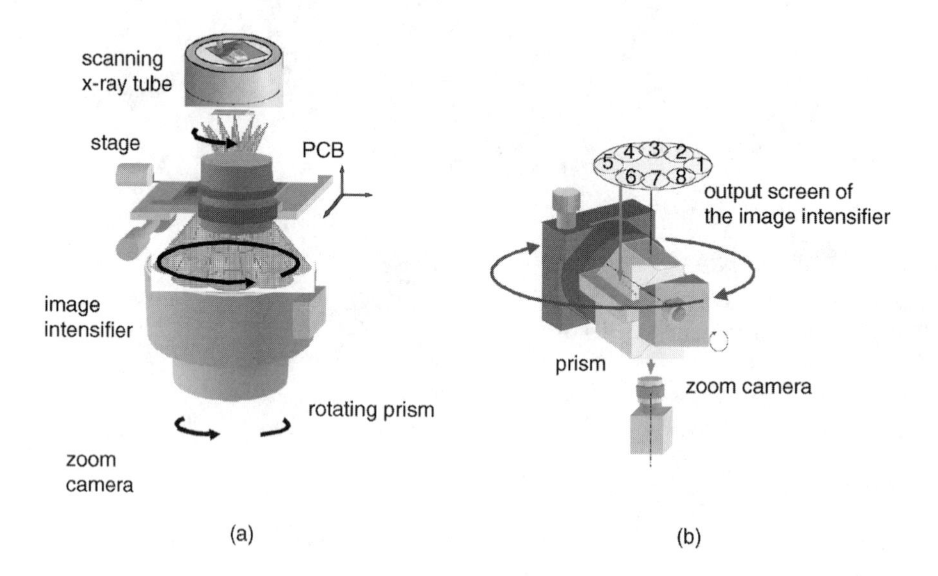

FIGURE 1.30 X-ray three-dimensional imaging system: (a) system configuration; (b) rotating prism.

1.5.2.12 Three-Dimensional Shape Reconstruction

Three-dimensional reconstruction of the geometric shape and dimension of objects and their surrounding environments is an integrated task that is combined with the scanned measurement over the region of interest and display of the measured data in three-dimensional space. This can be achieved in various ways by using an opto-mechatronic system. Measurement of the three-dimensional geometry of objects can be made either by scanning three-dimensional optical-based sensors or by stationary sensors. Unlike other situations, such as inspection, monitoring, and motion control, reconstruction requires the entire range of data, which are sometimes too large to store, process, and reconstruct. Therefore, the reconstruction algorithm is crucial for accurately obtaining the three-dimensional shape in a short time.

Three-dimensional volume reconstruction is very important in x-ray imaging technology. Figure 1.30 shows a specialized x-ray three-dimensional imaging system, which consists of a scanning x-ray tube, an image intensifier, a rotating prism, and a camera equipped with a zoom lens [Roh et el., 2000]. The scanning x-ray tube is designed to electrically control the position of an x-ray spot instead of the mechanical

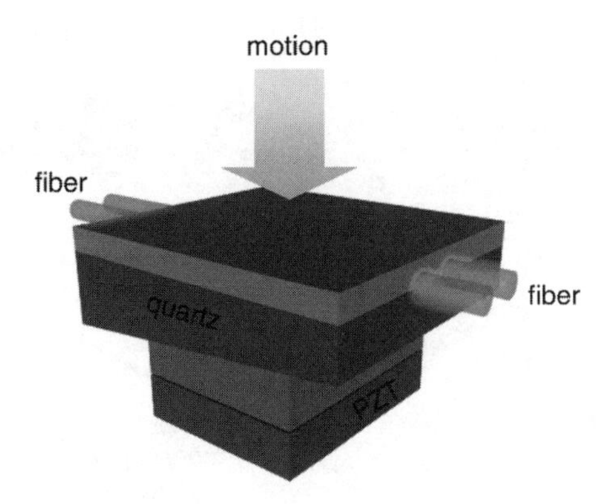

FIGURE 1.31 Optical wave length frequency shifter.

movement of the x-ray's source to get x-ray images from different angles. Attenuated x-rays passing through the object are collected by an image intensifier and converted into a visible image on the instrument's output screen. The object is projected on a circular trajectory on the image intensifier as the x-ray is steered. Then, eight or more images are sequentially acquired by the zoom camera through the rotating prism, which rotates in synchrony with the x-ray-steered position. The sequential images thus captured are stored in the digital memory of a computer and then processed to reconstruct a three-dimensional image of the object. Here, volume reconstruction of three-dimensional objects can be performed with the aid of an x-ray radiography imaging system and the reconstruction algorithm called algebraic reconstruction technique (ART). Here, the three-dimensional object is represented by a finite number of volume elements, or voxels, with their own density values. Then, the density values of the voxels are determined based on the information of their projections by the reconstruction algorithm.

1.5.2.13 Optical Property Variation

As shown and discussed in Section 1.2, laser wavelength is tuned according to the movement of a diaphragm, which is a mechatronic element driven by a servo principle.

Figure 1.31 illustrates a frequency shifter, which consists of an acoustic wave generator, a piezoelectric element, and a high birefringence fiber [Andonovic and Uttamchandari, 1989]. Inline fiber optic frequency shifters utilize a traveling acoustic wave to couple light between the two polarization modes of a high-birefringence fiber, with the coupling accompanied by a shift in the optical frequency.

1.5.2.14 Sensory-Feedback-Based Optical System Control

In many cases, optical or visual systems are operated based on the information provided by external sensory feedback, as shown in Figure 1.32. The systems that require this configuration include (1) a zoom and focus control system for video-based eye-gaze detection using an ultrasonic distance measurement [Ebisawa et al., 1996]; (2) laser material processing systems for which the laser power or laser focus is real-time controlled depending on the process monitoring [Haran et al., 1996]; (3) visually assisted assembly systems that cooperate with other sensory information such as force, displacement, and tactility [Zhou et al., 2000]; (4) sensor fusion systems in which optical/visual sensors are integrated with other sensory systems; and (5) visual/optical systems that need to react to acoustic sound and other sensory information such as tactility, force, displacement, velocity, etc.

1.5.2.15 Optical Pattern Recognition

As shown in Figure 1.33, a three-dimensional pattern recognition system uses a laser and a photorefractive crystal as the recording media to record and read holograms [Shin, 2001]. The photorefractive crystal stores the intensity of the interference fringe constructed by the beam reflected from a three-dimensional object and a plane-wave reference beam. The information on the shape of the object is stored in the photorefractive crystal at this point. And this crystal can be used as a template to be compared with another three-dimensional object. To recognize an arbitrary object the technique of optical correlation

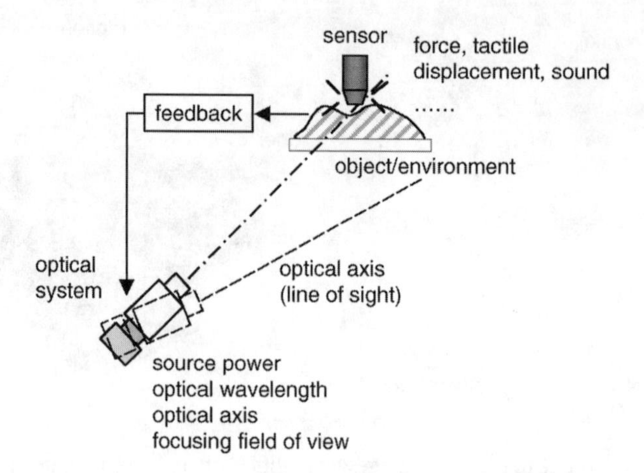

FIGURE 1.32 Reconfigurability of optical/visual systems based on sensory information.

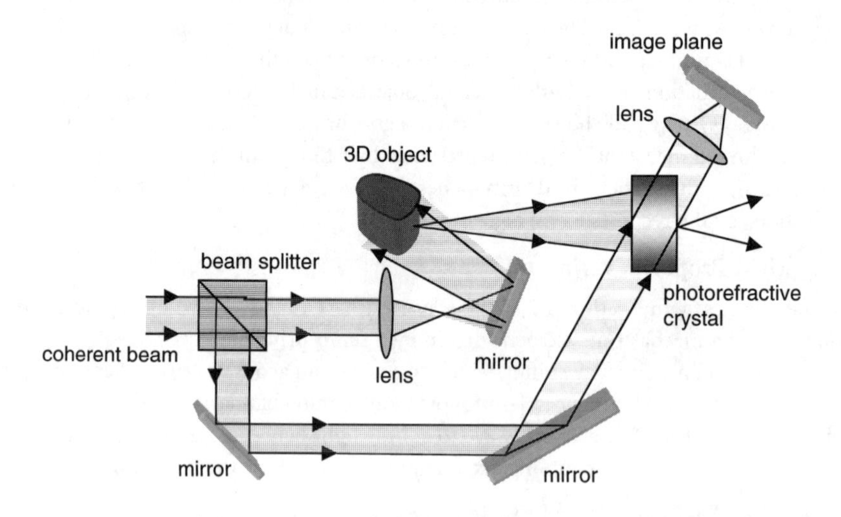

FIGURE 1.33 Optical pattern recognition system.

processing is employed. When an object to be recognized is replaced at the right position of the original object having the same orientation, the recorded hologram diffracts the beam reflected from the object to be compared, and the diffracted wave propagating to the image plane forms an image that represents the correlation between the Fourier transforms of the template object beam and the object to be compared. If a robot or a machine tries to find targets in unknown environments, this system can be effectively used to recognize and locate them in real time. In this case a scanned motion of the optical recognition system is necessary over the entire region of interest.

1.5.2.16 Remote Operation Via Optical Data Transmission

Optical data transmission is widely used when data/signals obtained from sensors are subject to external electrical noise or when the amount of data needed to be sent is vast, or when an operation is being performed at a remote site. Operation of systems at a remote site is ubiquitous now. In particular, Internet-based monitoring, inspection, and control are becoming pervasive in many practical systems. Visual servoing of a robot operated in a remote site is a typical example of such a system, as shown in Figure 1.34 [Kamiya, 1998]. In the operation room, the robot controls the position of a vibration sensor to monitor

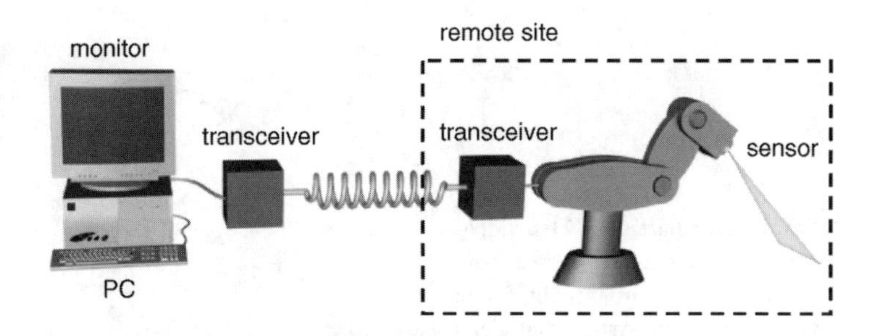

FIGURE 1.34 Data transmission for remote operations.

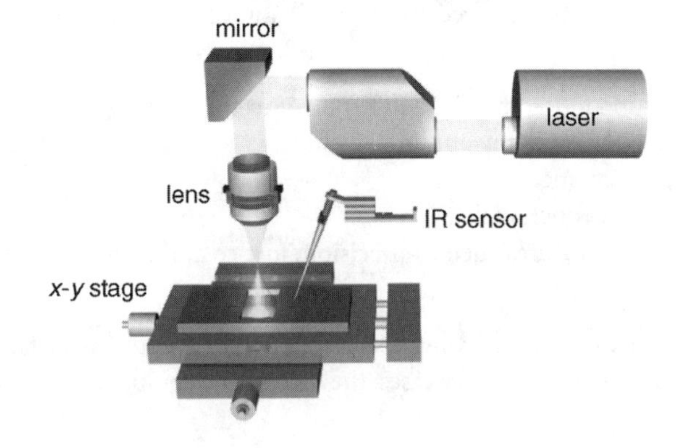

FIGURE 1.35 Material processing via high-power laser.

the vibration signal with the aid of visual information and transmits this signal to the transceiver. Another typical example is the operation of visual servoing of mobile robots over the Internet [Han et al., 2001].

1.5.2.17 Material Processing

Material processing can be achieved by the integration of the laser optical source and the mechatronic servo mechanism. The system produces material property change or a cut surface and the heat-treated surface of work pieces. In addition to conventional laser machining, laser micromachining is becoming popular due to reduced costs and increased accuracy. MEMS fabrication, drilling, and slotting in medicine and wafer dry cleaning and ceramic machining employ such micromachining technology.

Figure 1.35 shows a laser surface-hardening process, which is a typical example of an optical-based monitoring and control system [Woo and Cho, 1998]. The high-power, 4-kW laser is focused through a series of optical units, hits the surface of a workspace, and changes the material state of the workpiece. Maintaining a uniform thickness of the surface coating (less than 1 mm) is not an easy task because it depends heavily on the workpiece travel speed and laser power and the surface properties of the material. Here, the indirect measurement of the coating thickness is made by an infrared temperature sensor and feedback to the laser power controller.

1.6 Synergistic Effects of Opto-Mechatronic Systems

In the previous sections, the fundamental functions of opto-mechatronic systems were discussed and some illustrative systems were used as example. As we have already seen from those systems, the fusing or integration of optical technology into mechatronic systems brings about significant synergy in several ways.

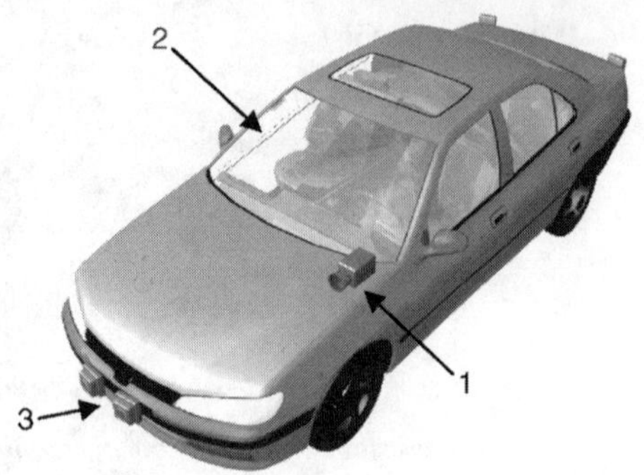

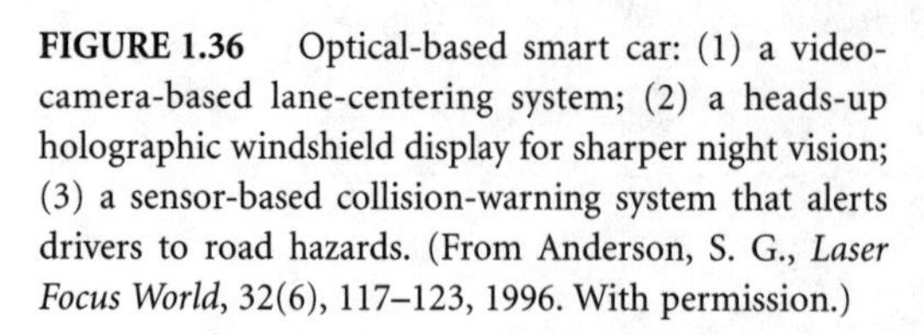
FIGURE 1.36 Optical-based smart car: (1) a video-camera-based lane-centering system; (2) a heads-up holographic windshield display for sharper night vision; (3) a sensor-based collision-warning system that alerts drivers to road hazards. (From Anderson, S. G., *Laser Focus World*, 32(6), 117–123, 1996. With permission.)

The synergistic effects that can be achieved from this fusion or integration may be illustrated by highlighting the following aspects:

1. Creating new functionalities, thereby adding them to the existing one
2. Increasing the level of autonomy and intelligence
3. Achieving high performance
4. Enhancing the level of functionality
5. Achieving a variety of features such as precision, low cost, robustness, distributed aspect, miniaturization, etc.

We shall see that in some cases mechatronic systems alone cannot possess such functionality or achieve such a level of system performance. In other cases, they cannot be equipped with a high level of autonomy and intelligence.

1.6.1 Creating New Functionalities

One of the significant contributions of combining optical elements with those of mechatronics is that it creates new functionalities for a system. This type of contribution can be found in a number of practical systems. One typical system is introduced below with a brief explanation.

A car equipped with opto-electronic units, called a "smart" car, has or will have a variety of functions governing safety and navigation. As shown in Figure 1.36 [McCarty, 2001; Anderson, 1996] such cars have a heads-up holographic windshield display for sharper vision, a machine vision system for monitoring the location of lane boundaries, an infrared scanning laser radar system for monitoring the distance to vehicles in front, a vehicle vision navigation system for self-localization or traffic management, IR sensors for detecting pedestrians, interior CCD and an IR emitter for monitoring drowsiness of drivers, CCD cameras mounted on the ends of bumpers that offer drivers wide-angle views at blind intersections, and a headlight system for adjusting the illumination pattern based on traffic and road conditions detected by the CCD camera.

Part feeders in current use feed parts only in a specified orientation. If the parts are not fed in the correct orientation, they are rejected and dropped from the feed track. This system is very inefficient because at the final exit stage they are dropped onto the bottom of the bowl. The feeder shown in Figure 1.37 can detect the orientation of all parts transported onto the track by a set of optical fiber sensors. The information on the identified part orientation is then used for sorting, depending on the orientation for subsequent tasks [Park and Cho, 1989].

1.6.2 Increasing the Level of Autonomy

The level of autonomy is not easy to measure exactly but can be qualitatively measured by such factors as flexibility, adaptability, intelligence, and agility. Thus, autonomy of a variety of practical opto-mechatronic systems can be qualitatively evaluated in this vague sense.

The assembly system shown in Figure 1.38 consists of two machine (robot) assembling parts; a sensory system, such as visual/optical sensors, force/torque sensors, and other sensors; a software system, including

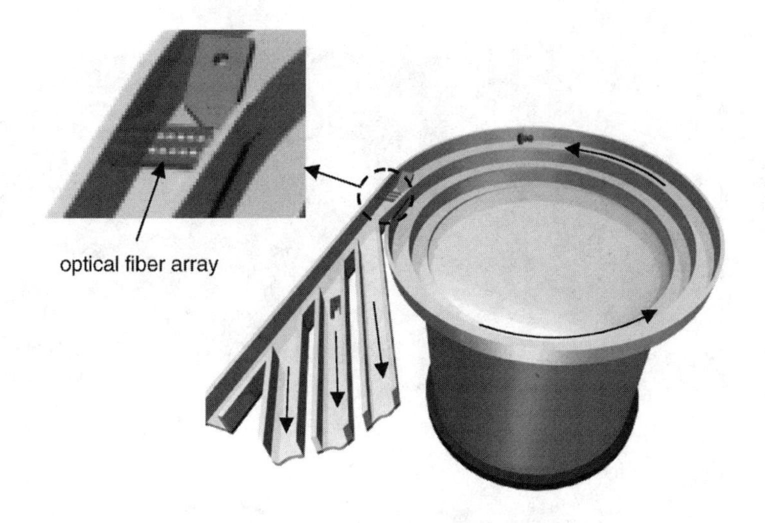

FIGURE 1.37 Intelligent bowl feeder.

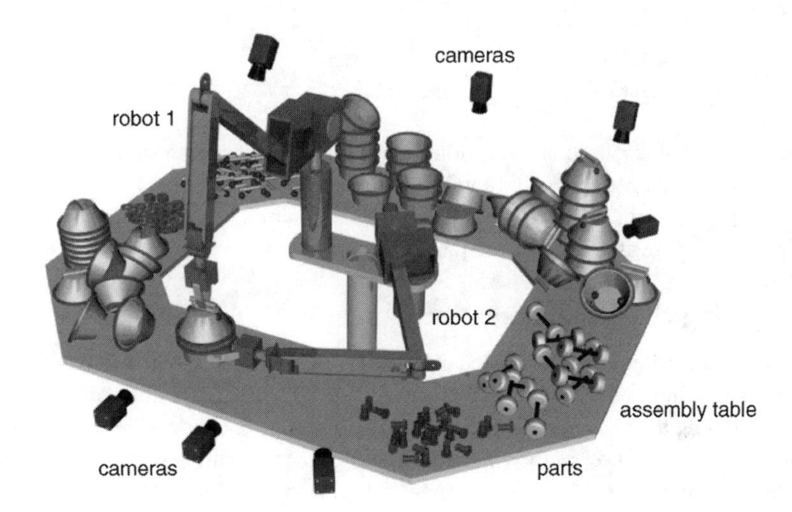

FIGURE 1.38 Intelligent flexible assembly station.

part recognition software, assembly planning, and path planner; and a control and assembly table. The drastic difference in its system configuration from that of the current automatic assembly system that simply employs a robot is that this system's configuration is completely unstructured and does not incorporate conveyors, feeders, and magazines and thus does not require an orderly arrangement of assembly equipment. The core technology involved with this system is the visual recognition and servo control that enable it to identify parts, move them while avoiding collision, and mate them into an already aggregated subassembly according to the assembly sequence. In this sense, the system is intelligent because it dose not require any structured assembly environment.

Another example of a system that incorporates various mechatronic components is a mobile robot (robotic machine) navigating in a task environment. In any mobile robot navigation, the optical/visual perception unit plays a crucial role because it provides the environment information for navigation. As shown in Figure 1.39, the robot can autonomously perceive, model, and map the environment through which it navigates to carry out certain tasks. In other words, it can sense three-dimensional shapes and the dimensions of objects, construct a map of the environment, and recognize the objects within the

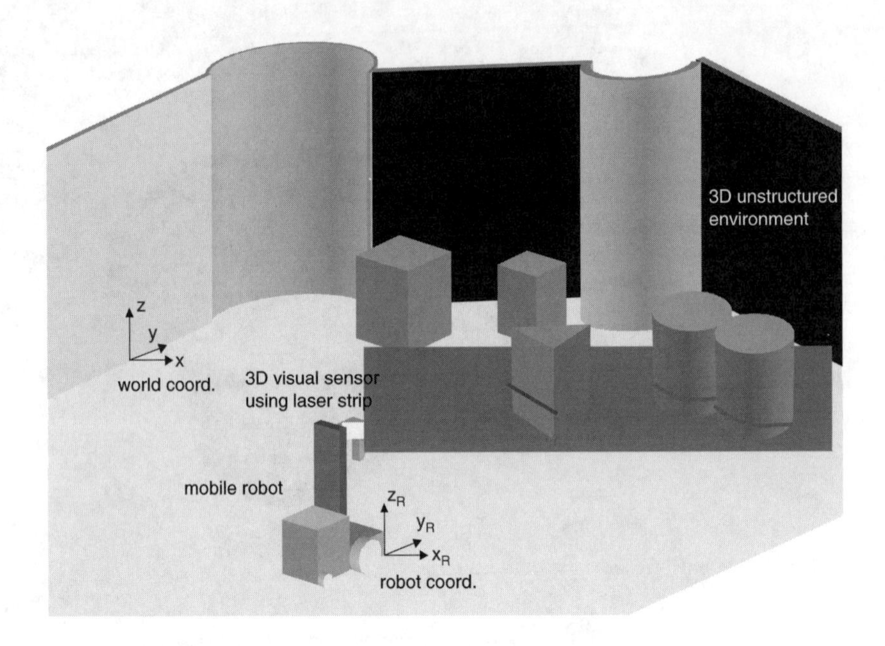

FIGURE 1.39 View planning robot.

environment. This autonomous task execution is performed by an optical/visual sensor, an optical scanning system, a robot controller, and a multifunctional software system.

The third example is a robot that automatically adapts its sensor morphology by controlling the rotation of 16 long tubes equipped with light sensors [McKee, 2000].

1.6.3 Enhancing the Level of Performance

The performance of a system may be evaluated in a meaningful way by measuring a variety of factors such as quality of system output, productivity, effectiveness, and cost. A high level of system performance can be achieved by adopting the principle of opto-mechatronic fusion. A variety of practical systems come to achieve an increased performance level as a result of this fusion. Monitoring and control are one area; optical sensing is another. Optical sensing is otherwise impossible or impractical by conventional sensors and feeds the acquired information back to the controller part, thereby increasing performance level.

Let us illustrate a practical system that can be placed in this category. The intelligent welding system shown in Figure 1.40 is equipped with an optical sensory system composed of a laser triangulation sensor and an noncontacting IR temperature sensor, a welding robot, hardware control modules, and a software system that contains various algorithms, including modules of weld seam identification, a path-control algorithm, and weld quality assessment based on the neural network. The system can identify the weld seam pattern and weld path; this enables it to control weld torch speed and track the desired identified path and evaluate the weld quality. Seam tracking is achieved by the seam identifier and seam tracker [Kim et al., 1996]. At the same time, weld quality is measured instantaneously by multipoint surface temperature measurements of the weld pool [Lim and Cho, 1993]. Based on this temperature information, weld pool control can be regulated online to the desired size [Boo and Cho, 1994]. Thus, the welding system possesses intelligence because it has the ability to autonomously identify the shape of the weld seam, follow its path, carry out welding, and control the quality of the welded material in real time.

High-performance measuring can be obtained when a coordinate-measuring machine (CMM) is equipped with multiple sensory heads [Mahr Co., 2001]. The CMM system shown in Figure 1.41 possesses a dynamic touch probe with integrated probe changer, a CCD camera with automatic optical filters, and a laser with autofocus and scanning capability. The touch probe offers application flexibility because of the probe-changing system; the vision system provides multiwindow techniques and high-speed image

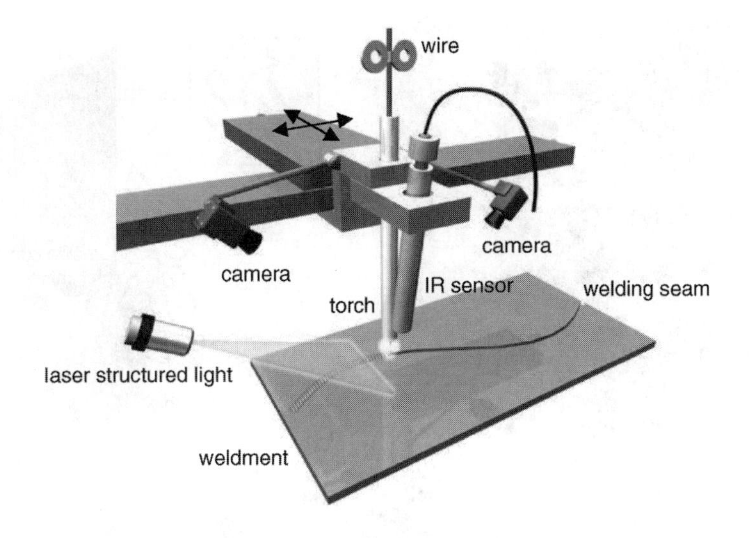

FIGURE 1.40 Intelligent welding system.

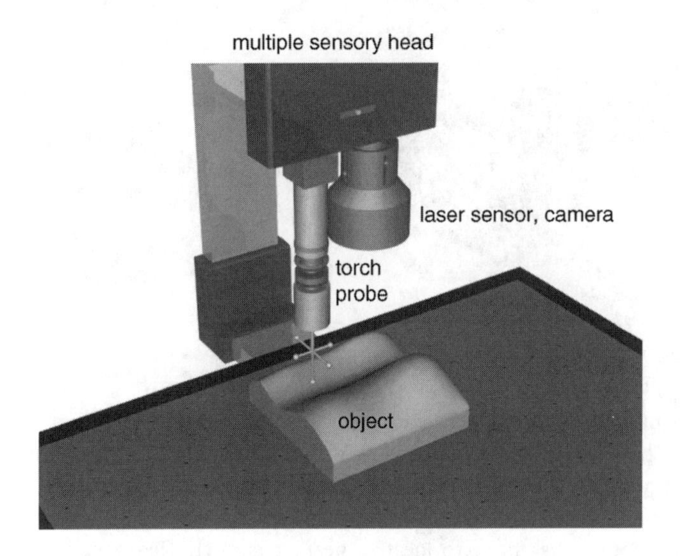

FIGURE 1.41 Sensor-integrated CMM.

processing, while the laser sensor enables measurements even on polished glass, ceramic, or metal surfaces where conventional triangular laser systems fail. This configuration can provide a sensory fusion function, which is generated by integrating the information obtained by each sensor.

1.6.4 Achieving High Functionality

System performance dependent on how high the functionalities are that can be generated in the system. When a system possesses high functionality, the system is regarded as having high performance. Measurement or detection is one important functionality for visual servoing. In some cases, however, measurement by conventional sensors often yields inaccurate, unreliable information, which retards processing or causes the failure of the system operation to be performed.

In an assembly operation or target tracking, visual information is vital for accurately locating the position of parts to be mated, although some other means of measurement such as force and noncontact displacement sensing are sometimes used. Figure 1.42 indicates a unidirectional visual sensing unit

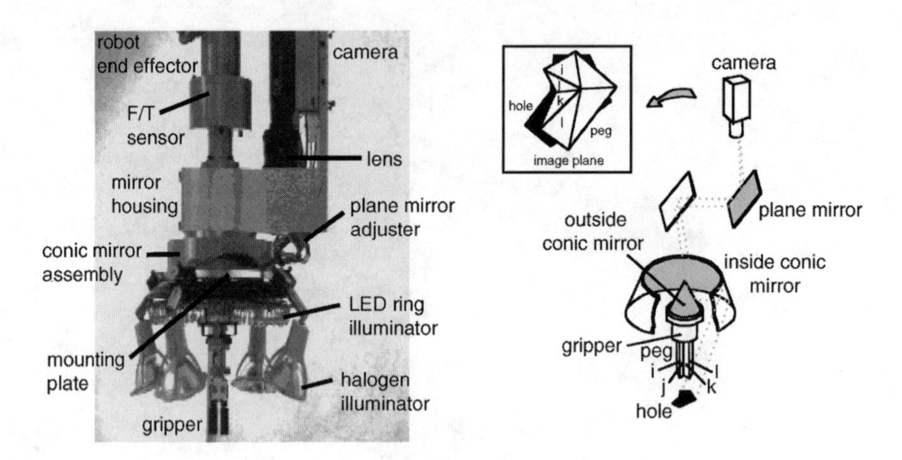

FIGURE 1.42 Omni-directional vision for assembly.

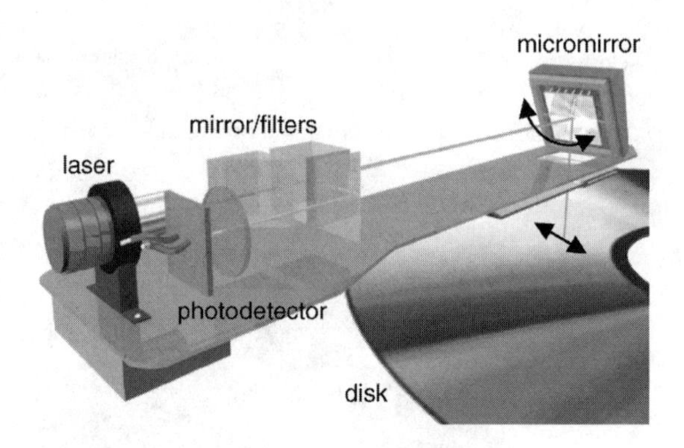

FIGURE 1.43 Fine track searching for optical disk.

embedded in a robot end-effector, which sends the measured visual information to the robot [Kim and Cho, 1998, 2000]. The powerfulness of this sensor is that it has no occlusion or unseen area. Therefore, the robot can gain omni-directional information on the objects or parts of interest so that it can react accurately and quickly to carry out a given task. Figure 1.43 shows a fine-positioning mechanism with a micromirror for an optical disk drive to locate a beam accurately [Jeong et al., 1999]. This is needed because the aerial density in ODDs has a limit due to the optical diffraction limit. Here, a micro-scale mirror actuator with vertical comb is capable of fine-positioning the beam with high servo bandwidth.

1.6.5 Distributed Functionalities

Most machines/systems have a lumped characteristic in that they possess sensor actuators and controllers at discrete locations. It is desirable for machines/systems to have distributed characteristics that would allow them to obtain the necessary information distributed spatially and generate necessary actions according to this distributed configuration.

Embedding sensors and actuators into some delicate parts or structures of a machine or system is one way of monitoring and controlling its dynamic state.

In Figure 1.44, a smart structure or skin is shown, which can easily be embedded into systems. Here, the optical sensor layers provide some information needed to generate an action by actuators that are also embedded into the structure with sensors, controllers, and electronic components. Optical tactile sensors embedded in the fingers of a robot and so-called "smart structures" such as aircraft and concrete infrastructure are typical examples, as indicated in the figure [Rogers, 1995].

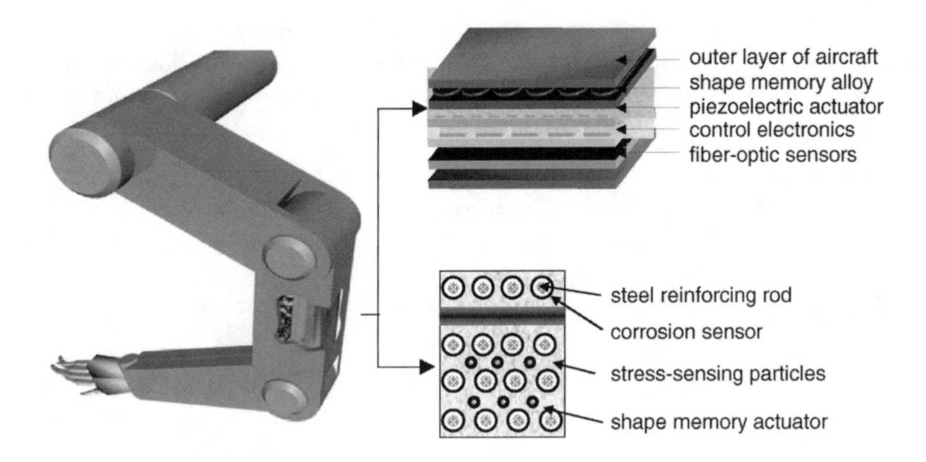

FIGURE 1.44 Smart skin or structures.

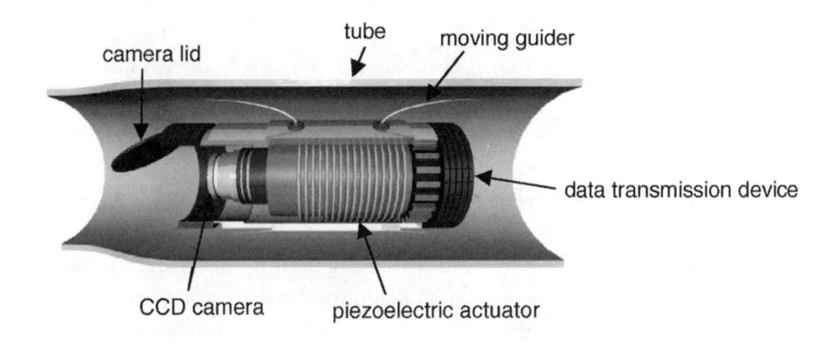

FIGURE 1.45 Micromachine for pipe inspection. (From Tsuruta, K. et al., *Sensor Rev.*, 19, 37–42, 1999. With permission.)

1.6.6 Miniaturization

Inspection or measurement of very small closed areas is often very difficult due to the dimension problem of the sensing and actuating units. Micro sizing of such machines or systems can be achieved with the aid of inherent optical sensing. A micromachine that inspects the inner surface of tubes in complex piping systems such as power plants is shown in Figure 1.45 [Tsuruta, 1999]. The machine consists mainly of a micro-CCD camera (10 mm in diameter) for visual inspection, a piezoelectric driving actuator, and microwave-based energy supply and data transmission devices. It can travel in a straight and curved metal tube with a minimum diameter of 10 mm.

1.7 Summary

In recent years, integration of optical elements into mechatronic systems has been accelerated because it produces a synergistic effect, creating new functionalities for the systems or enhancing the performance of the systems. This trend will certainly be the future direction of mechatronic technology and will contribute to the advent of a new technological paradigm.

This chapter has focused on helping readers understand opto-mechatronic technology and opto-mechatronic systems, in which optical, mechanical electrical/electronics, and information engineering technologies are integrated. In particular, the definition and fundamental concepts of the technology were established. Using these concepts, it was possible to identify different types of opto-mechatronic systems, depending on how their component technologies were integrated and organized together to produce their basic functions. The fundamental functions that can be created by the integration were described by defining and illustrating them from a number of practical systems currently being used. A total of 17 functions were identified. Finally, the synergistic effects that can be achieved because of the integration were identified to be:

1. Creating new functionalities, thus adding them to the existing one
2. Increasing the level of autonomy and intelligence
3. Achieving high performance
4. Enhancing the level of functionality
5. Achieving a variety of features such as precision, low cost, robustness, distributed aspect, miniaturization, etc.

Future developments of technologies such as mechatronics, instrumentation and control, MEMS, and biosystems as well as information processing and storage and communication aim at achieving these characteristic effects. Therefore, opto-mechatronics is becoming a prime moving technological element that will dictate the future direction of various technologies.

Defining Terms

distributed function: System function that is generated in a spatially distributed fashion (e.g., smart structure, image acquisition of a CCD camera, etc.).
MOEMS: Micro-optical electron mechanical system.
ODD: Optical device drive.
optical pattern recognition: Optical means of processing optical images and then recognizing patterns or objects based on the processed results.
optical polarization: An optical process that makes the orientation of the electric field of light dependent only on spatial coordinates, although its magnitude and sign vary in time.
optical switch: Optical device that can direct light from one fiber to another.
opto-mechatronic technology: Technology that creates a synergistic integration of optical elements with mechatronic ones.
opto-mechatronically-fused system: System in which optical or mechatronic elements are not separable from the system to achieve a certain function.
sensor fusion: A fusion technique that can fuse the information obtained by multiple sensors.
tunable laser: Laser whose wavelength can be tuned.

References

Anderson, S. G., Smart cars take the high-tech road, *Laser Focus World*, 32(6), 117–123, 1996.

Andonovic, I. and Uttamchandani, D., *Principles of Modern Optical Systems*, Artech House, Norwood, MA, 1989.

Backmutsky, V. and Vaisman, G., Theoretical and experimental investigation of new mechatronic scanners, in *Proc. IEEE 19th Convention Electrical Electron. Engineers Israel*, Israel, 1996, pp. 363–366.

Borenstein, J., Experimental evaluation of a fiber optics gyroscope for improving dead-reckoning accuracy in mobile robots, in *Proc. IEEE Int. Conf. Robotics Automation*, Vol. 4, Leuven, Belgium, 1998, pp. 3456–3461.

Butler, C. and Yang, Q., A fibre-optic three-dimensional analog probe for component scanning on coordinate-measuring machines, *Sensor Actuator A*, 37–38, 473–479, 1993.

Cannon Co., http://www.canon.com, 2002.

Chen, M. and Hollis, R., Vision-guided precision assembly, http://www-2.cs.cmu.edu/afs/cs/project/msl/www/tia/tia_desc.html, 1999.

Cho, H. S. and Park, W. S., Determining optimal parameters for stereo-lithography process via genetic algorithm, *Manuf. Sys. J.*, 19(1), 18–27, 2000.

Ebisawa, Y., Ohtani, M., and Sugioka, A., Proposal of a zoom and focus control method using an ultrasonic distance-meter for video-based eye-gaze detection under free-head conditions, *18th Annu. Conf. IEEE Eng. Med. Biol. Soc.*, Amsterdam, the Netherlands, 2, 1996.

Fukuda, T., Hattori, S., Arai, F., Matsuura, H., Hiramatsu, T., Ikeda, Y., and Maekawa, A., Characteristics of optical actuator-servomechanisms using bimorph optical piezo-electric actuator, *Proc. IEEE Int. Conf. Robotics Automation*, 2, 618–623, 1993.

Geppert, L., Semiconductor lithography for the next millennium, *IEEE Spectrum*, 33–38, April 1996.

Han, K., Kim, S., Kim Y., and Kim, J., Internet control architecture for internet-based personal robot, *Autonomous Robots*, 10, 135–147, 2001.

Haran, F. M., Hand, D. P., Peters, C., and Jones, J. D. C., Real-time focus control in laser welding, *Measurement Sci. Technol.*, 7, 1095–1098, 1996.

Higgins, T. V., Optical storage lights the multimedia future, *Laser Focus World*, 31(9), 103–111, 1995a.

Higgins, T. V., Computing with photons at the speed of light, *Laser Focus World*, 31(12), 72–77, 1995b.

Ishi, T., Future trends in mechatronics, *JSME Int. J.*, 33(1), 1–6, 1990.

Jeong, H. M., Choi, J. J., Kim, K. Y., Lee, K. B., Jeon, J. U., and Pak, Y. E., Milli-scale mirror actuator with bulk micromachined vertical combs, *Proc. Transducer*, Sendai, Japan, 1999, pp. 1006–1009.

Kamiya, M., Ikeda, H., and Shinohara, S., Data collection and transmission system for vibration test, *Proc. Ind. Appl. Conf.*, 3, 1679–1685, 1998.

Kayanak, M. O., The age of mechatronics, *IEEE Trans. Ind. Electron.*, 43(1), 2–3, 1996.

Kim, W. S. and Cho, H. S., A novel sensing device for obtaining an omnidirectional image of three-dimensional objects, *Mechatronics*, 10, 717–740, 2000.

Kim, J. S. and Cho, H. S., A robust visual seam tracking system for robotic arc welding, *Mechatronics*, 6(2), 141–163, 1996

Kim, W. S. and Cho, H. S., A novel omnidirectional image sensing sytem for assembling parts with arbitrary cross-sectional shapes, *IEEE/ASME Tran. Mechatronics*, 3(4), 275–292, 1998.

Ko, K. W., Cho, H. S., Kim, J. H., and Kong, W. I., A bead shape classification method using neural network in high frequency electric resistance weld, *Proc. World Automation Congr.*, Anchorage, Alaska, 1998.

Krupa, T. J., Optical R&D in the army research laboratory, *Optics Photonics News*, 16–39, June 2000.

Larson, M. C., Tunable optoelectronic devices, http://www-snow.stanford.edu/~larson/research.html, 2000.

Larson, M. C. and Harris, J. S., Wide and continuous wavelength tuning in a vertical cavity surface emitting laser using a micromachined deformable membrane mirror, *Appl. Phys. Lett.*, 68, 893, 1996.

Lim, T. G. and Cho, H. S., Estimation of weld pool sizes in GMA welding process using neural networks, *Proc. Inst. Mech. Engineers*, 207, 15–26, 1993.

Mahr Co., Multiscope 250/400, http://www.mahr.com/en/content/products/mess/mms/ms250.html, 2001.

McCarthy, D. C., Hands on the wheel, cameras on the road, *Photonics Spectra*, 78–82, April 2001.

McDonald, T. G. and Yoder, L. A., Digital micromirror devices make projection displays. Imaging Handbook, supplement to *Laser Focus World*, 33(8), 55–58, 1997.

McKee, G., Robotics and machine perception, *SPIE Int. Tech. Group Newslett.*, 9(2), 1–11, 2000.

Mitutoyo Co., Quick Vision, http://www.mitcat.com/e-02.htm, 2001.

Park, I. O., Cho, H. S., and Gweon, D. G., Development of programmable bowl feeder using a fiber optic sensor, *10th Int. Conf. Assembly Automation*, Tokyo, Japan, 1989.

Park, W. S., Cho, H. S., Byun, Y. K., Park, N. Y., and Jung, D. K., Measurement of three-dimensional position and orientation of rigid bodies using a 3-facet mirror, *SPIE Int. Symp. Intelligent Sys. Advanced Manuf.*, Boston, MA, 1999, Vol. 3835, pp. 2–13.

Pugh, A., *Robot Sensors*, Vol. 2, *Tactile and Non-Vision*, Springer-Verlag, Berlin, 1986.

Robinson, S. D., MEMS technology—micromachines enabling the "all optical network," *Electron. Components Technol. Conf.*, Orlando, FL, June 2001, pp. 423–428.

Rogers, C. A., Intelligent materials, *Sci. Am.*, 122–127, September 1995.

Roh, Y. J., Cho, H. S., and Kim, J. H., Three-dimensional volume reconstruction of an object from x-ray images, *Conf. SPIE Opto-Mechatronic Syst.*, part of *Intelligent Systems Advanced Manuf.*, Boston, MA, 2000, Vol. 4190, pp. 181–191.

Shin, S. H. and Javidi, B., Three-dimensional object recognition by use of a photorefractive volume holographic processor, *Optics Lett.*, 26(15), 1161–1163, 2001.

Shiraishi, M., In-process control of workpiece dimension in turning, *Ann. CIRP*, 28(1), 333–337, 1979.

Takamasu, K., Development of a nano-probe system, *Q. Mag. Micromachine*, 35, 6, May 2001.

Toshiyoshi, H., Su, J. G., LaCosse, J., and Wu, M. C., Micromechanical lens scanners for fiber optic switches, *Proc. 3rd Int. Conf. Micro Opto Electro Mechanical Sys.*, MOEMS '99, August, Mainz, Germany, 1999, pp. 165–170.

Tsuruta, K., Mikuriya, Y., and Ishikawa, Y., Micro sensor developments in Japan, *Sensor Rev.*, 19, 37–42, 1999.

Veeco Co., http://www.tmmicro.com/tech/modes/contact.htm, 2000.

Wakami, N., Nomura, H., and Araki, S., Fuzzy logic for home appliances, in *Fuzzy Logic and Neural Networks*, McGraw-Hill, New York, 1996, pp. 21.1–21.23.

Wilson, A., Machine vision speeds robot productivity, *Vision Systems Design*, October 2001.

Woo, H. G. and Cho, H. S., Estimation of hardened layer dimensions in laser surface hardening processes with variations of coating thickness, *Surface Coatings Technol.*, 102(3), 205–217, 1998.

Zankowsky, D., Applications dictate choice of scanner, *Laser Focus World*, 32(12), 99–105, 1996.

Zhang, J. H. and Cai, L., An autofocusing measurement system with a piezoelectric translator, *IEEE/ASME Trans. Mechatronics*, 2(3), 213–216, 1997.

Zhou, Y., Nelson, B. J., and Vikramaditya, B., Integrating optical force sensing with visual servoing for microassembly, *J. Intelligent Robotic Sys.*, 28, 259–276, 2000.

2

Opto-Mechatronic Products and Processes: Design Considerations

George K. Knopf
University of Western Ontario
London, Ontario, Canada

2.1 Introduction

As an engineering discipline, mechatronics strives to optimally integrate mechanical, electronic, and computer technologies in order to create innovative products and processes. Optical sensors and actuators are being incorporated at an accelerated rate into mechatronic systems because these lightwave technologies provide components for high precision, rapid data processing, flexible circuits, and circuit miniaturization. Opto-mechatronics, a subset of mechatronic system design, focuses on the tools and technologies needed to create intelligent systems from optical **transducers** and **embedded control** systems.

At first glance the concept of opto-mechatronics may appear to duplicate the goals of the more established area of opto-mechanics, but this is a false impression. The role of opto-mechanical design is to maintain the proper shapes and positions of the various optical components that comprise precision instruments such as telescopes, microscopes, metrology instruments, and eyeglasses [Ahmed, 1997].

By contrast, opto-mechatronics stresses technology integration for enhanced system performance. Starting with a clear problem definition and continuing through to product manufacture, the opto-mechatronic design process works toward the efficient utilization of available technologies to produce quality systems in a timely manner with features the customer wants. In essence, it is not a specific technology but rather a **design** philosophy that promotes the creation of high-quality "smart" products and process.

The difference between an opto-mechatronic approach and a more traditional systems engineering approach to design is not the constituent parts that embody the solution; rather, it is the method in which the various components are developed. Most complex **systems** are created using sequential design practice, where the roles of the various engineering disciplines are well defined and, often, narrow in

0-8493-1162-4/03/$0.00+$1.50

focus. Each constituent discipline in the design exercise contributes to the solution by employing a collection of analytical procedures and heuristic guidelines unique to the practitioners of that discipline. In contrast, opto-mechatronic design follows the principles of multidisciplinary **concurrent engineering,** resulting in systems with performance characteristics better than the sum of the individual technologies that comprise the solution. The opto-mechatronic approach to design can achieve impressive results, greater productivity, higher quality, and increased product reliability by innovatively incorporating leading-edge lightwave and embedded control technologies.

In this chapter, the unique and beneficial characteristics of opto-mechatronic system design are introduced. First, the difference between traditional and opto-mechatronic design is illustrated by looking at a simple consumer product. Then, the design process is investigated and the key technologies for designing opto-mechatronic solutions such as optical transducers, embedded controllers, and **integrated optics** are described. Finally, the effect of opto-mechatronic design philosophy on the development of products and processes is illustrated by several practical system designs.

2.2 Traditional vs. Opto-Mechatronic Designs

Consider several alternative designs of a user-friendly bathroom scale. A mechanical solution to the problem (Figure 2.1) would involve the compression of leaf springs and a mechanism to convert the downward motion of the person standing on the scale into rotation of a shaft and, hence, movement of a pointer across a visual display [Bolton, 1999]. One problem with a purely mechanical design is that the weight displayed on the dial should not depend on the person's position on the scale. The consumer may perceive the product as being of poor quality if the scale measure is inconsistent or fluctuates while the person attempts to read the dial.

A mechatronic solution to the bathroom scale problem is to replace the compression springs by load cells containing strain gauges, as shown in Figure 2.2. The electrical signal produced by the strain gauge is received by a microcontroller that sends a control signal to a stepper motor in order to provide a read-out of the individual's weight. The stepper motor rotates a mechanical pointer on a marked display. In addition to controlling the visual display, the microcontroller performs simple mathematical calculations to average the measured weight values taken over a fixed time period in an effort to minimize the effects of sudden changes in the strain gauge output due to slight shifting of the body weight. The proposed mechatronic

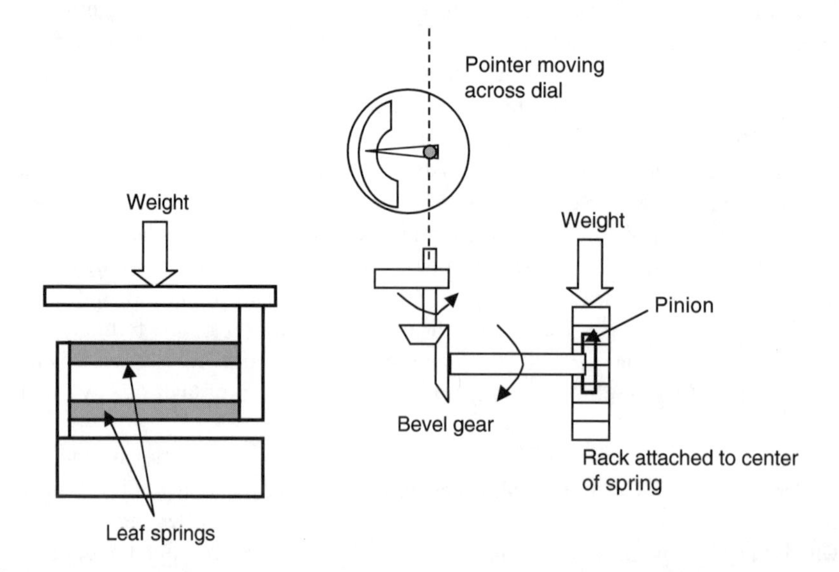

FIGURE 2.1 Illustration of a mechanical solution to the bathroom scale problem [Bolton, 1999].

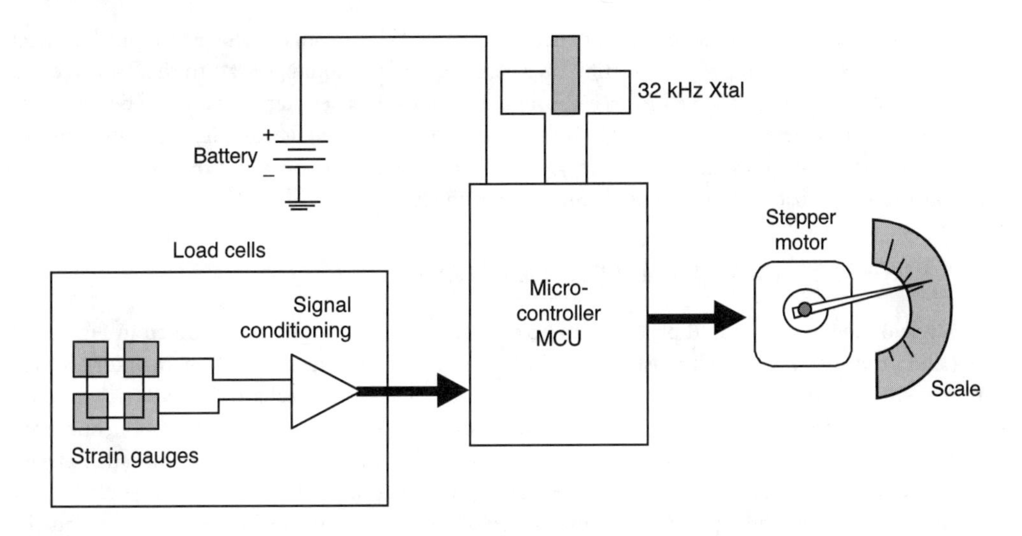

FIGURE 2.2 Mechatronic solution to the bathroom scale problem.

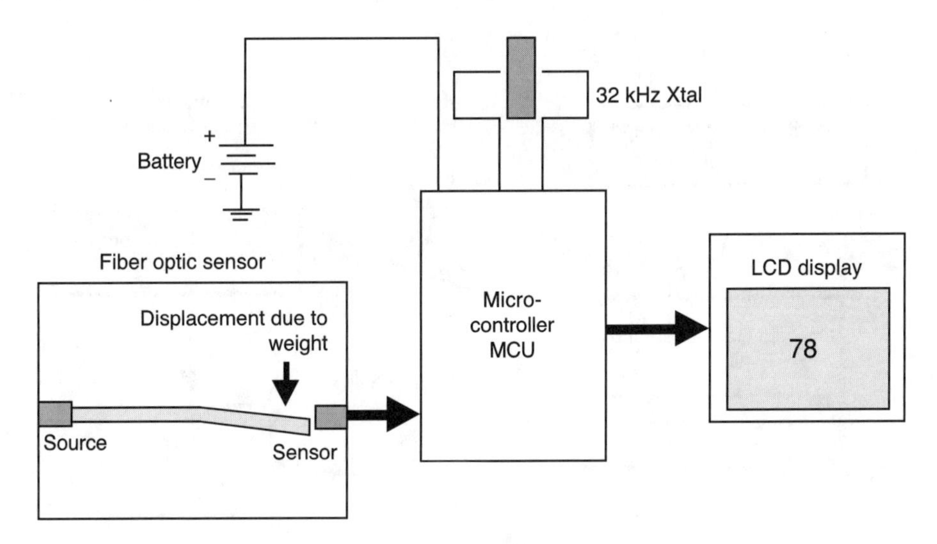

FIGURE 2.3 Opto-mechatronic solution to the bathroom scale problem.

solution has transferred a great deal of the system complexity to software, resulting in a household bathroom scale that is mechanically simpler and involves fewer costly components and moving parts.

To reduce cost and improve performance, an opto-mechatronic solution to the bathroom scale problem is to incorporate the latest lightwave technologies into the product, shown in Figure 2.3. An optical fiber displacement sensor based on the principle of light intensity modulation replaces the highly sensitive strain gauge. A load placed on the scale causes the gap along the optic fiber to widen and proportionately reduce the amount of light transmitted between the light source and sensor. Although the fiber optic sensor in this application does not improve measurement accuracy, it does reduce the sensitivity of the sensing element to natural variations in the environment. The fiber optic displacement sensor can be easily redesigned to be more or less sensitive to the natural fluctuations in a person's weight, thereby simplifying the signal-processing algorithms and microcontroller computations.

Furthermore, the proposed opto-mechatronic solution reduces product power consumption by using a liquid crystal display (LCD) panel to indicate the measured weight. The LCD panel does not emit light; rather, it controls the medium through which the light travels. This is achieved through the use of a layer of liquid crystals sandwiched between two polarization plates. When a small voltage is applied between the polarization plates, it changes the direction of polarization of the liquid crystals so that light is either reflected from the crystals or allowed to pass through. By controlling selected areas of the liquid crystal panel, characters can be formed to display information. A key advantage of this innovative display technology is that it consumes very little battery power. Electrical power consumption is often a limiting design constraint in most battery-operated consumer products.

2.3 Opto-Mechatronic Design Process

The life cycle of any manufactured product can be divided into eight stages, as illustrated in Figure 2.4. Product and production designers must consider all aspects of the product's life cycle, including service and retirement, in order to make the appropriate decisions. The opto-mechatronic approach to engineering design involves structured design methodologies and software tools that support concurrent engineering design practise. These methodologies promote a knowledge-based approach, called **Design for Excellence** (DFX) [Bralla, 1996], that invokes a series of guidelines, principles, recommendations, or rules of thumb for designing a product that maximizes all desirable characteristics such as high quality,

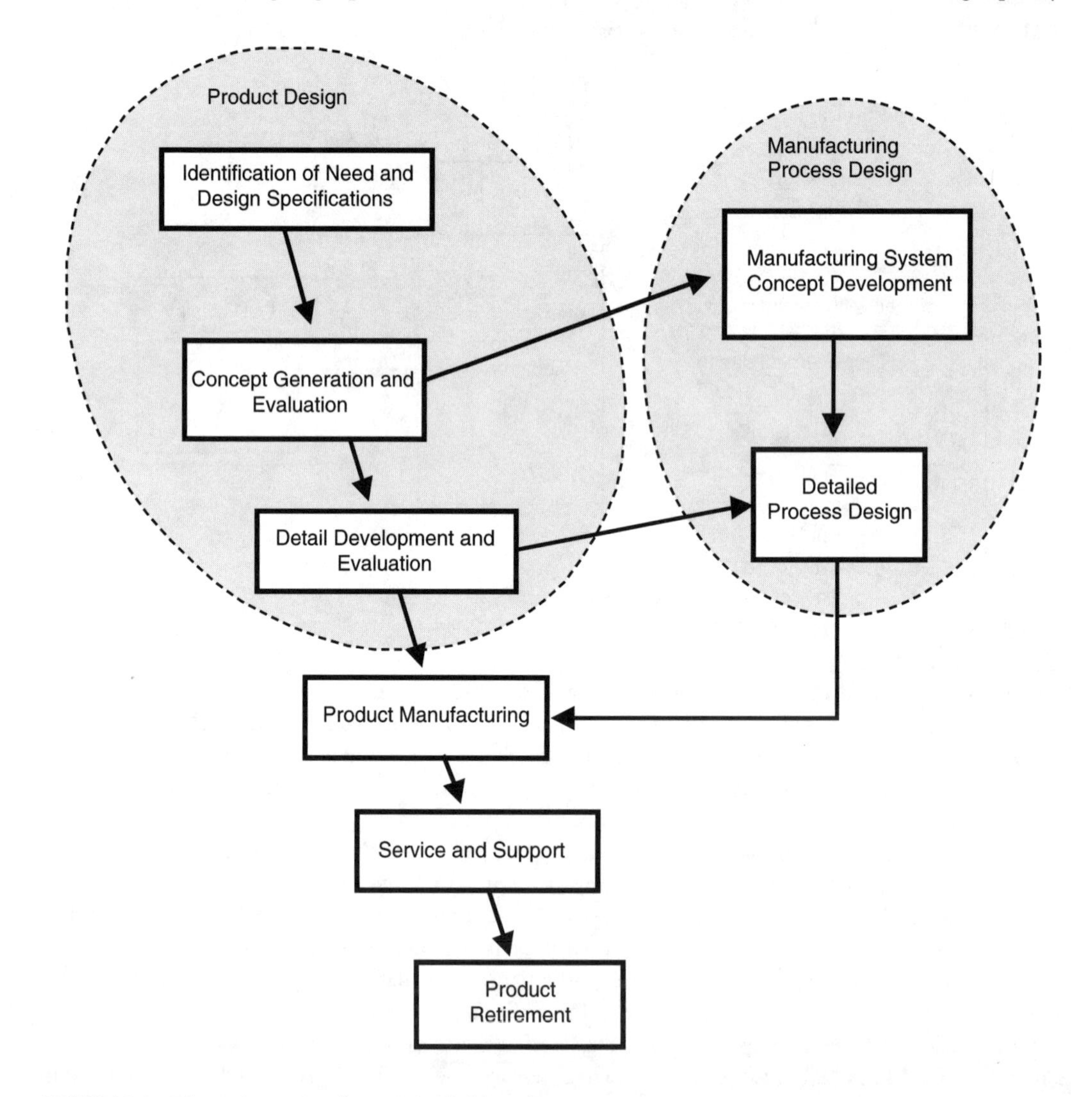

FIGURE 2.4 The various stages in a product's life cycle.

reliability, serviceability, safety, user friendliness, and short time-to-market in the product while at the same time minimizing lifetime costs (including manufacturing costs). The key to effective concurrent engineering is the flow of design information. Project plans, concept sketches, meeting notes, and detailed drawings must be shared with the right people at the right time. Innovative software tools and recent advances in product data exchange are enabling design teams to function both locally and globally [McMahon and Browne, 1993].

The exercise of doing design is the process of solving open-ended, ill-defined problems that have multiple satisfactory solutions. The importance of design activity can be illustrated by the changes in the photocopy industry over the last 30 years. In the 1960s and 1970s, Xerox was the dominant manufacturer of photocopiers in the world [Ullman, 1997]. However, by the early 1980s there were over 40 different manufacturers of photocopiers in the world, mainly from Japan, competing for this lucrative market. Consequently, Xerox's share of the market dropped significantly. Xerox studied the changes in the market and came to the realization that one of its problems was *how* the product had been designed. The Japanese competitors could produce photocopiers of similar functionality for 50% less than Xerox because their material costs were 10% less, fabrication costs were 15% less, and—most important—nearly 25% of the overall savings was the result of *how* the competition originally designed the product. In essence, the competition had focused on improving the way the design was created by emphasizing the simultaneous development of the product and the manufacturing process.

Another study based on data collected from the Ford Motor Company [Ullman, 1997] demonstrated that only 5% of the product development cost was associated with performing design activities, but the decisions made during the design exercise influenced 50 to 75% of the final manufactured cost. Critical decisions about the material to be used, components to be purchased, method of assembly and, hence, level of automation, and manufacturing layout were made during the early stages of the design process. Clearly, the early design decisions had the greatest influence on the final product cost for the least investment. The key stages of early product design are: identification of need and design specifications, concept generation and evaluation, and detail (prototype) development and evaluation.

2.3.1 Identification of Need and Design Specifications

A structured approach to tackling design problems ensures that all critical aspects have been thoroughly analyzed and design decisions are justified prior to committing expensive resources to product manufacture. The pivotal point for all system designs is a clear understanding of the customer's need for the product. This is perhaps the most important stage of the design process because a poorly defined problem will waste expensive development time on concepts that will not fulfill the design objectives.

A useful technique for establishing design specifications is the Quality Function Deployment (QFD) method [Ullman, 1997]. QFD is a systematic approach for improving product quality by concentrating on what the customer wants and will continue to buy in the product. The design specifications will state the problem, any constraints placed on the solution, and the criteria that will be used to judge the appropriateness of the proposed solution. In stating the problem, all the functions required of the design, together with any desirable features, must be explicitly specified. Important specifications that must be determined when designing an opto-coupled system are summarized in Figure 2.5 and include optic fiber or free-space communication link; simplex (one-way) or full-duplex (two-way) communication; required form of signal modulation (analog or digital); choice of multiplexing strategy; desired wavelength of operation; specification of fiber optic link (size, material, ease of coupling, attenuation); choice of connectors and splicing techniques; and choice of photodetectors (response characteristics, wavelength). Other design specifications for opto-mechatronic systems may include physical mass, geometric dimensions and tolerances, types and range of motion required, accuracy, input and output requirements of constituent elements, interfaces, power requirements, constraints due to operating environment, relevant standards, and codes of practice.

Setting down the design specifications is often a difficult task because not all of the performance characteristics are known in advance. This leads to a design paradox where the flexibility to make design

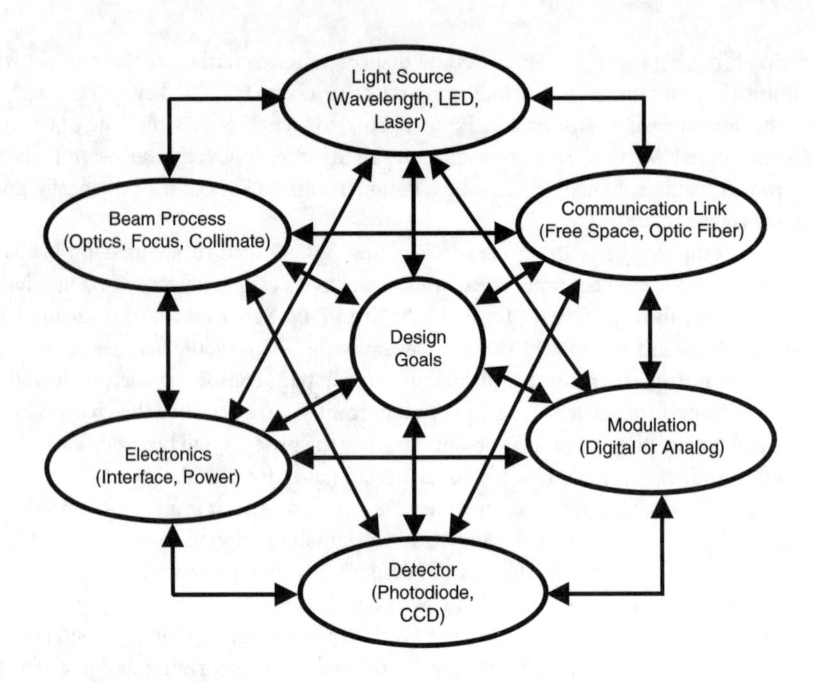

FIGURE 2.5 Typical design specifications for an opto-coupled system.

changes diminishes as knowledge about the problem increases [Bralla, 1996]. The process of detailing specifications is, however, a valuable exercise in all design problems because it will help pinpoint some of the unknowns that must be investigated throughout the course of the design.

2.3.2 Concept Generation and Evaluation

During the concept-generation stage a large variety of solutions are proposed, and sufficient detail is provided as to how each of the required functions can be achieved. Techniques that help generate novel concepts for poorly understood problems include functional decomposition, morphological analysis, and brainstorming [Dieter, 2000; Ullman, 1997]. The various proposed solutions must be analyzed using structured methods such as feasibility judgement, technology readiness assessment, go/no-go screening, and decision matrices [Dieter, 2000; Otto and Wood, 2001; Ullman, 1997]. Decision matrices are very useful because they compare all proposed concepts with the customer's requirements. The best solution, or a hybrid of several solutions, is moved forward into detail development.

2.3.3 Detail Development and Evaluation

Engineering details of the most viable concept must be created, mathematical analysis of subsystems performed, prototypes created for experimental testing, production of mechanical drawings and circuit diagrams, and writing of software code. If the opto-mechatronic system utilizes extensive optical networks, then **ray sketching** of an unfolded system will provide useful information about the placement of components on the optic axis and paraxial **ray traces** will help identify all required stops, pupils, and windows [O'Shea, 1985]. The optoelectronic hardware that is needed to interface the optical subsystem with the data analyzer or microprocessor must also be thoroughly evaluated. For example, when selecting a photodetector for an opto-mechatronic application it is necessary to ensure that the wavelength characteristics of the detector match the light source and that the photodetector is sensitive enough for the proposed application.

2.4 Opto-Mechatronic Technologies

Opto-mechatronic systems will often exhibit a number of important characteristics such as the functional interaction between optical, electronic, and mechanical components; spatial integration of subsystems into a single physical unit; utilization of multifunctional devices; and exploitation of embedded control. The hardware subsystems that comprise a typical opto-mechatronic product or process are shown in Figure 2.6. Optical sensors and actuators are robust, low-cost solutions that enable high-precision and rapid signal-processing operations to be performed. The inclusion of a programmable controller in an opto-mechatronic system provides a mechanism for embedded artificial intelligence. In addition, this computing device will greatly enhance the designer's ability to expand the functionality of the product without raising the manufacturing costs. This section summarizes some of the unique and important opto-electronic technologies that contribute to the performance of an opto-mechatronic system. Table 2.1 lists many of the optical transducers and controllers discussed in this chapter.

2.4.1 Optical Transducers

Optical transducers can be classified as either sensors or actuators. The use of a lightwave transducer requires a source of light radiation, a medium through which the light travels, and a detector to convert the light energy into another measurable form. The processing of the light energy is accomplished by controlling any combination of the parts in the system. Because both the source and the transmission medium determine the amount of light received by the detector, it is possible to convey information

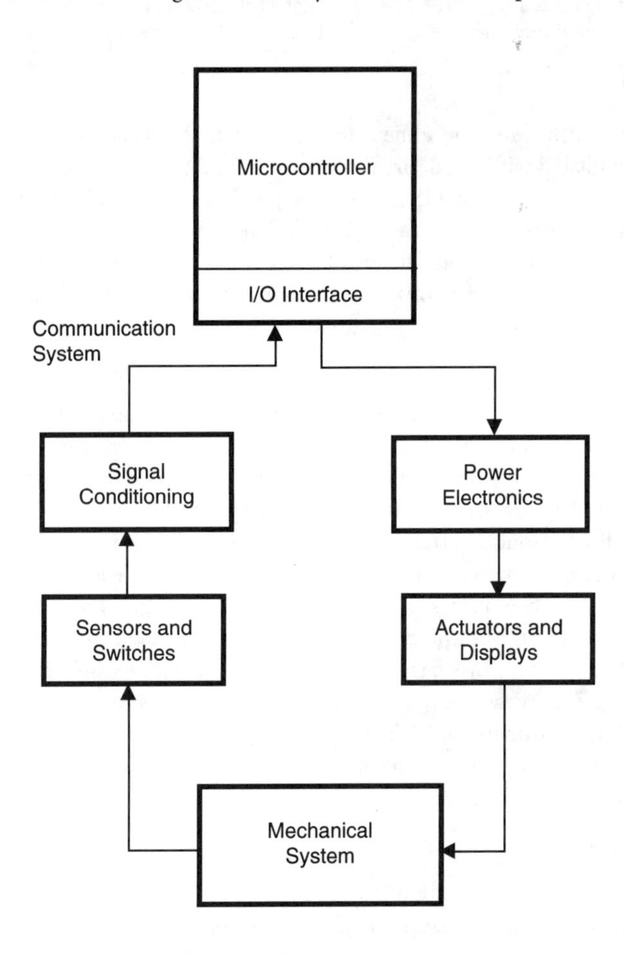

FIGURE 2.6 Subsystems of a typical opto-mechatronic product or process.

TABLE 2.1 Summary of Optical Transducers and Controllers Used in Opto-Mechatronic Systems

Optical Device	References
Optical Encoders	
Incremental and absolute encoders	[Bolton, 1999; Necsulescu, 2002; Shetty and Kolk, 1997]
Grating transducers	[Saleh and Teich, 1991; Shetty and Kolk, 1997]
Fiber Optic Sensors	
Intensity modulation	[Allard, 1990; Palais, 1998]
Reflective	[Allard, 1990]
Microbending	[Allard, 1990]
Optical Range Sensors	
Triangulation	[Shetty and Kolk, 1997]
Light-spot scanning	[Shetty and Kolk, 1997]
Light-stripe sensing	[Knopf and Kofman, 1998]
Direct Optical Actuators	
Silicon microactuators	[Goldfarb, 1999; Tabib-Azar, 1998]
Optically generated electron hole pair	[Tabib-Azar, 1998]
Indirect Optical Actuators	
Thermo expansion of fluid	[Tabib-Azar, 1998]
Thermo expansion of solid	[Yoshizawa et al., 2001]
Integrated Optics	
Optical transmitter and receiver	[Saleh and Teich, 1991]
Microbeam actuator	[Tabib-Azar, 1998]
Optical Controllers	
Opto-electronic fuzzy controller	[Itoh et al., 1997; Zhang and Karim, 1999]
Multidimensional optical controller	[Gur et al., 1998; Zalevsky et al., 2000]

about either the source or the medium to the detector. If the medium characteristics do not change, then the source can be modulated with the desired information and the sensor designed to detect the information. Figure 2.7 is an illustration of a through-beam photoelectric sensor system. The most important functional specifications for selecting a transducer are linearity, sensitivity, resolution, range, accuracy, repeatability, response time, bandwidth, bandwidth–length product, optic power, and signal-to-noise ratio. The following is a brief overview of key technologies such as optical encoders, fiberoptic sensors, optical range sensing systems, and optical actuators.

2.4.1.1 Optical Encoders

Optical encoders are commonly used for measuring angular or linear position, velocity, and direction of movement. A typical optical encoder consists of a light source, a disk on which a pattern is etched, and a sensing head as illustrated in Figure 2.8. The coded disk is positioned between the focused light source and photodetector and is mounted on a shaft or translating member. Inscribed onto the surface of the coded disk are alternating opaque and transparent sections. When the opaque section of the wheel passes in front of the light beam, the detector is turned off and no signal is generated. Alternatively, when the transparent section of the wheel passes in front of the projected light, the detector is turned on. The result is a series of pulse signals that correspond to the rotation of the disk. A simple counter is used to keep track of how much the coded disk has rotated. Velocity information can be determined by calculating the time difference between measured pulses. The main advantages of optical encoders are low cost, simplicity, accuracy, high sensitivity, and reliability. These encoders are often used as precision measurement devices in a variety of products and machine tools. However, optical encoders, like any transducer, must be calibrated prior to use for a specific application. This is important because of the differing sizes and resolution requirements.

There are two classes of optical encoders: incremental and absolute [Bolton, 1999; Necsulescu, 2002]. The incremental encoder generates a simple pulse each time the object being measured has moved a fixed distance. To perform angular measurements, the disk of the incremental encoder attached to the shaft is divided into an equal number of sectors on the circumference. In a similar fashion, the coded

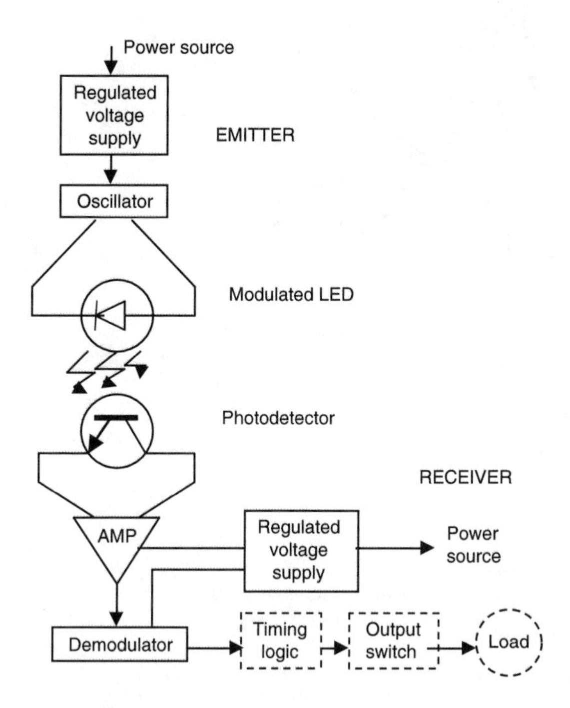

FIGURE 2.7 Basic components of an opto-coupled sensor system consisting of an emitter and receiver pair. When the beam of the emitter (modulated LED) is broken, the output oscillations generated by the receiver unit will change proportionately.

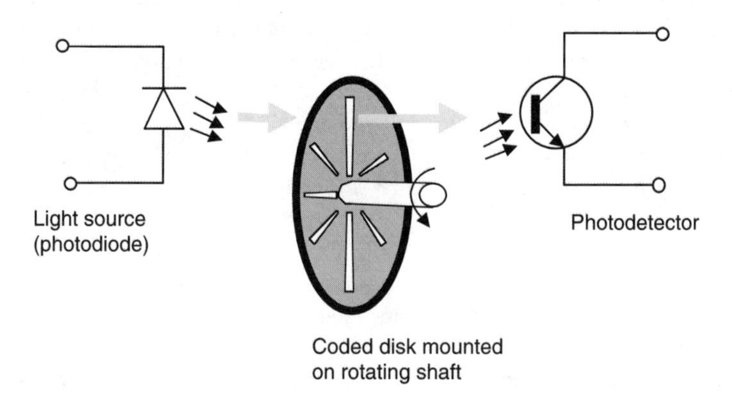

FIGURE 2.8 The operating principle of an incremental optical encoder.

disk of the linear encoder is divided into an equal number of segments over the length of movement. Because the optical encoder counts lines, a higher measurement resolution can be obtained by etching more lines on the disc. This specification is an important factor in encoder selection and is often expressed as pulses per revolution.

Accurate incremental encoders may utilize simple grating patterns to determine the distance moved. Figure 2.9 illustrates the transducer principle based on translating grating patterns. As the moving grating pattern is shifted relative to a fixed grating, the pulses are counted and the position calculated. These optical gratings are used in both linear and radial forms. Recently, reflection gratings with steel-backings have been used in the machine tool industry. During operation the light passes from the source to the

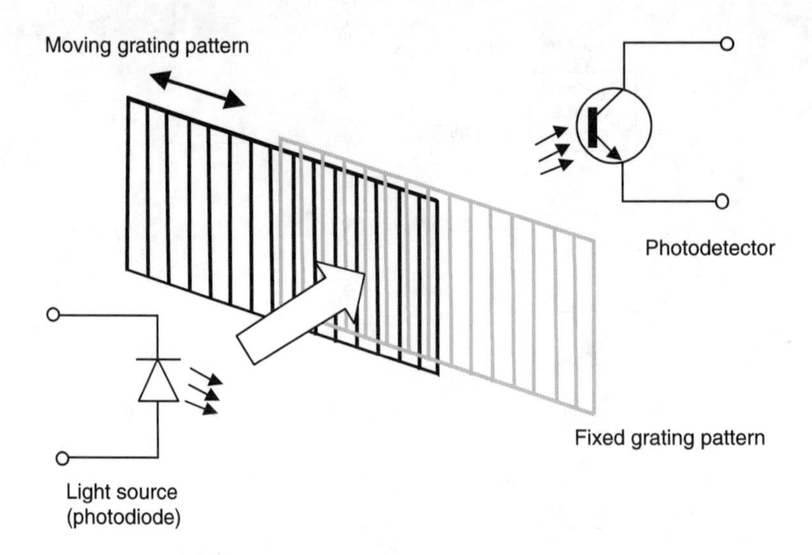

FIGURE 2.9 An illustration of how the grating principle is used in an optical encoder.

scale grating through the index grating and is reflected back through the index grating to the photodetector. An important advantage is that the reflection gratings can be precisely positioned on the moving member of the instrument.

An absolute optical encoder has a similar construction to incremental encoders except that a unique binary word, or code, corresponds to each possible position. Instead of counting pulses from a datum as is done in incremental encoders, the absolute optical encoder reads a system of coded tracks to establish the position. This enables the encoder to retain the current shaft or translation position even if the power to the system is turned off. In addition, absolute encoders are less susceptible to electrical noise or stray signals. These features make absolute optical encoders useful for a variety of time-delayed applications such as satellite-tracking antennas, where the slow-moving antenna may be inactive for long periods of time and only occasional position verification is required [Shetty and Kolk, 1997].

The Moiré fringe principle has also been used for optical transducers and encoders. An essential element of the transducer design is that the optical gratings consist of a regular succession of opaque lines separated by transparent spaces of equal width. The opaque lines are at right angles to the length of the grating. If the two grating sections are superimposed with the lines at a slight angle to each other, a Moiré fringe pattern can be created from the integrated interference effects at the line intersections on each grating, as shown in Figure 2.10. When one grating pattern is moved with respect to the other at right angles to its lines, the Moiré fringe pattern travels at right angles to the direction of movement.

The measure of movement depends on the relative distance traveled by the gratings. Analysis of the geometric relationship between the Moiré fringes and the grating pair enables displacement to be computed using [Shetty and Kolk, 1997]:

$$\gamma = \frac{\rho_A \rho_B}{\left[\rho_A^2 \sin^2\alpha + \left(\rho_A \cos\alpha - \rho_B\right)^2\right]^{\frac{1}{2}}} \tag{2.1}$$

where γ is the fringe separation or period; ρ_A and ρ_B are the pitches of the A and B gratings, respectively; α is the acute angle formed by the intersecting gratings; and β is the acute or obtuse angle between the lines of the first gratings and the fringe (as shown in Figure 2.10). This principle has been applied to measuring length, angle, straightness, and circularity of motion. In addition, it can supply information about the variable required, and it is relatively unaffected by external effects.

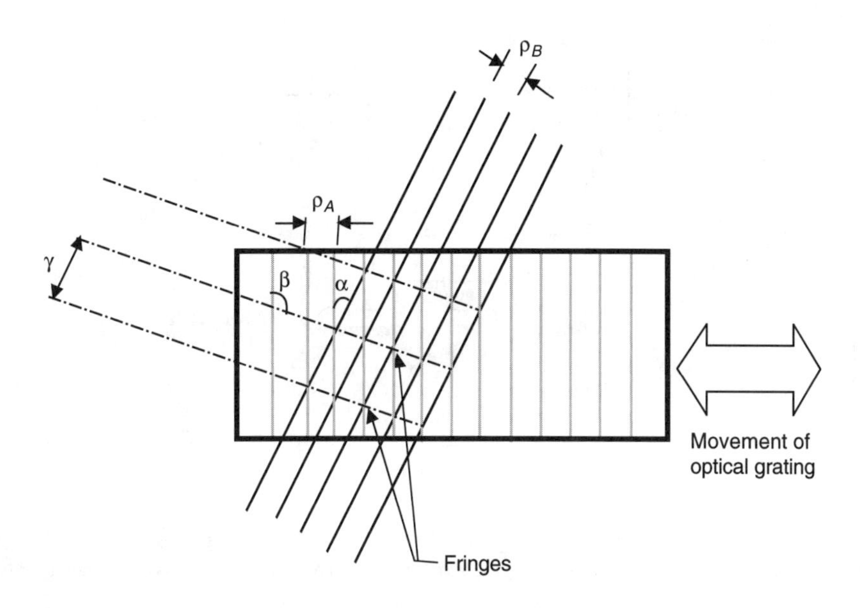

FIGURE 2.10 Moiré fringe principle used for an optical encoding.

2.4.1.2 Fiber Optic Sensors

There has been a significant increase in the use of optical sensors in the past 20 years because light can be modulated and transmitted over large distances using fiber optic bundles. Optical fibers are basically a cylindrical guidance system with high internal reflective characteristics where light transmission is achieved through multiple reflections at the cylinder walls. These internal reflections of light rays obey Snell's law in geometric optics [O'Shea, 1985; Palais, 1998]. If a light beam in a transparent medium strikes the surface of another transparent medium, a portion of the light will be reflected and the remainder may be transmitted (refracted) into the second medium. The information can be transmitted as either intensity or phase-modulated signals.

Based on the technique used for obtaining measurements, fiber optic sensors are classified as either *intrinsic sensors* or *extrinsic sensors*. Intrinsic fiber optic sensors carry the light signal and form part of the optical sensing element. In contrast, extrinsic fiber optic sensors use the fiber only as a transmission medium for an optical signal. This is the approach used by lightwave communication systems. The most widely used fiber optic sensors are based on modulating the intensity light by a variety of means, including cutting the light between two fibers (*shutters*), reflecting the light emitted by one fiber and collected by another one (*Y-guide probe*), varying the coupling between two fibers, and changing the attenuation losses in fiber with small movements, called *microbending*. Several examples of intensity-modulated fiber optic sensors are summarized below. These are all very simple, highly reliable devices that can be produced economically.

2.4.1.2.1 Intensity Modulation: Fiber Hydrophone

The first example of an intensity-modulated fiber optic sensor is the hydrophone [Allard, 1990] illustrated in Figure 2.11. The intrinsic sensor is used to measure acoustic disturbances in the water. The fiber is not a continuous waveguide but has a small break in it. At the gap, one end of the optic fiber is fixed and the other end is attached to a speaker diaphragm. A sound wave vibrates the diaphragm, thereby displacing the movable optic fiber. The coupling efficiency changes according to the amplitude and frequency of the displacement. The power delivered to the receiver is then a measure of the frequency and amplitude of the acoustic wave. In this particular system design, the optic fiber acts as both the sensing element and the transmission channel for the information.

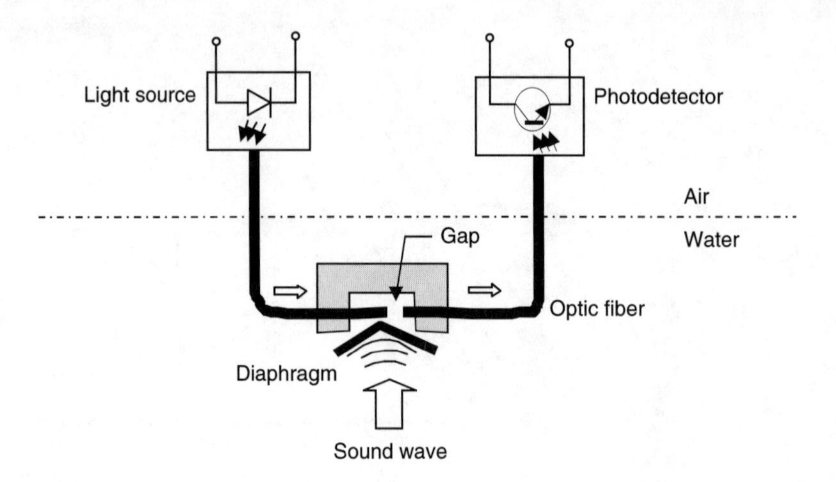

FIGURE 2.11 A schematic diagram of the optic fiber hydrophone. The fiber on the left is displaced when an acoustic wave is present, changing the amount of light coupled across the gap. The size of the gap is exaggerated for the reader's interpretation.

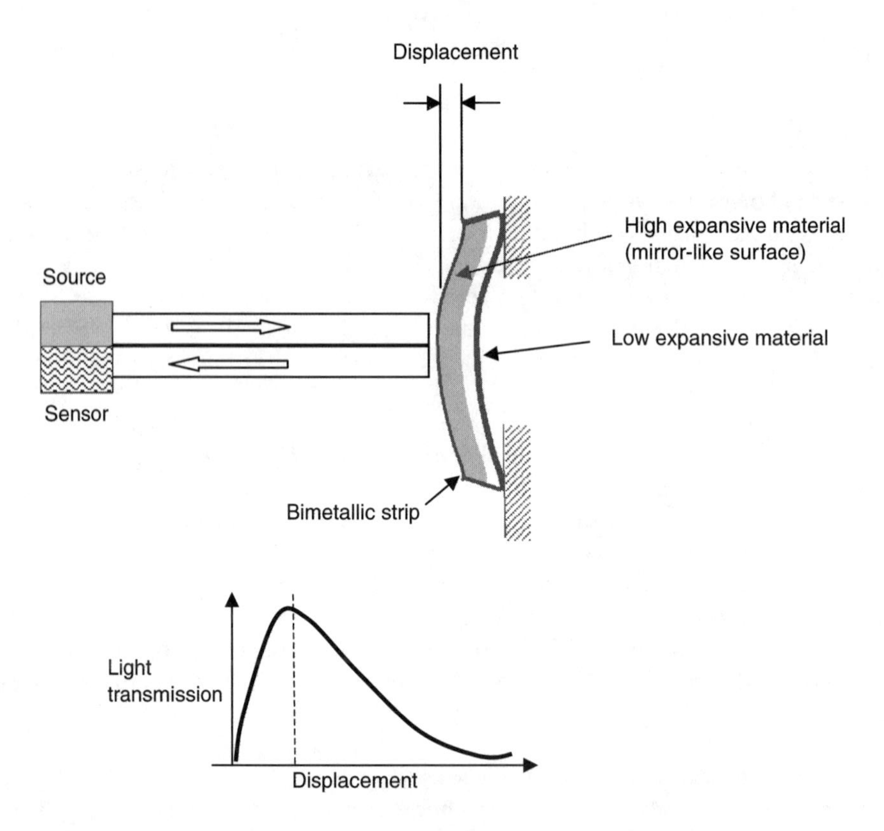

FIGURE 2.12 Fiber optic temperature sensor based on reflective properties of a bimetallic strip.

2.4.1.2.2 Reflective: Fiber Optic Temperature Sensor

One of the simplest designs for measuring temperature is an extrinsic reflective fiber optic sensor (see Figure 2.12), which uses the displacement of a bimetallic element as the transducer for indicating temperature variations. The optic waveguide is used to transmit the light from the source to the bimetallic temperature element and receives the reflective light for transmission to the photodetector. In addition,

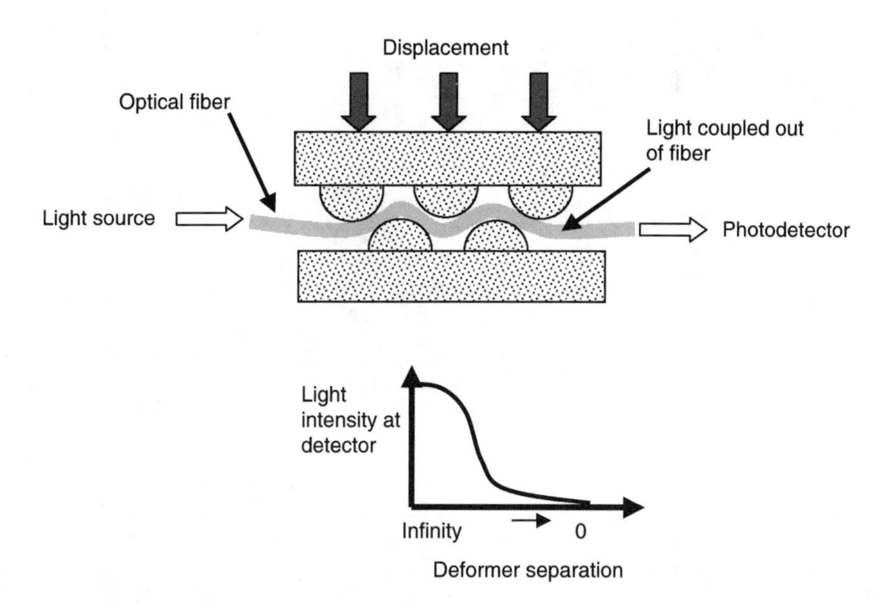

FIGURE 2.13 An illustration of a fiber optic load sensor based on the principle of microbending.

the incident light is transmitted back from the bimetallic object. The analysis and comparison of transmitted and reflected intensities are done separately to give a measure of the displacement. The intensity of the reflected light captured depends on the distance of the reflecting target from the inspection probe. One disadvantage of this type of sensor is that it is sensitive to the orientation of the reflective surface and to contamination.

2.4.1.2.3 Microbending: Load Sensor

Another example of a fiber optic displacement sensor is based on the concept of microbending, shown in Figure 2.13. In this example, fiber optic bundles are squeezed between two mechanical deformers. The external force influences the total internal reflection of the fibers. Instead of reflecting, the light beam moves orthogonally and refracts through the fiber wall. The modulation of the light intensity by the applied pressure gives a measure of the applied force (load). The amount of light received at the detector compared to the light source is a measure of the physical property influencing the bend. Microbend fiber optic strain gauges have application in the areas of tactile sensing and vibration monitoring.

2.4.1.3 Optical Range Sensors

Optical range sensing devices have been used in a variety of opto-mechatronic products and processes including automatic 35-mm cameras, vehicle guidance systems, visual servoing, online product inspection, precision measurement, and machine condition monitoring. Most range measurement devices consist of a small light source and a position-sensitive detector (PSD) such as a one-dimensional photodiode array or two-dimensional CCD matrix (see Figure 2.14). The light-emitting diode and collimating lens transmit a pulse in the form of a narrow light beam that strikes the surface of the object and is reflected back through a focusing lens to the PSD array.

The PSD array is a silicon device that generates current values proportional to the distance, x, of the incident light spot from the array center. The photoelectric current produced at each terminal, i_1 and i_2, is a function of the resistance between the electrode and the point of incidence. If i is the total current produced by the detected light, and i_1 is the current at one output electrode, then the current produced at each terminal will be a function of the corresponding distance between the point of incidence light and electrode. Thus, the current at each terminal can be computed with respect to distances [Shetty and

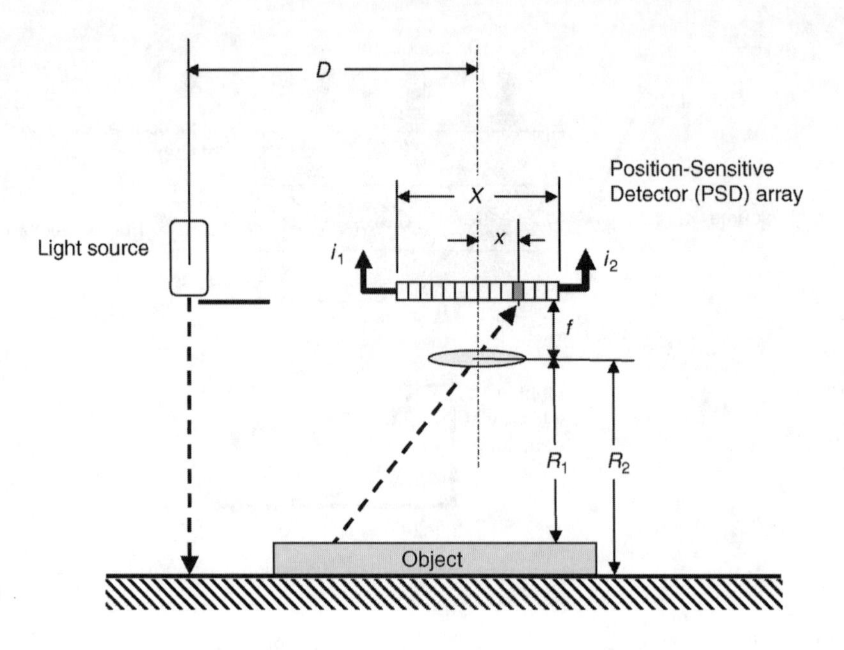

FIGURE 2.14 The triangulation principle applied to the position-sensitive detector (PSD).

Kolk, 1997] using,

$$i_1 = i\frac{(X-x)}{X}, \quad i_2 = i\frac{x}{X} \tag{2.2}$$

where X is the distance between terminals i_1 and i_2. Because the ratio of the resulting current values at the two terminals is

$$Q = \frac{I_1}{I_2} = \frac{X}{x} - 1 \tag{2.3}$$

the distance x can be determined using

$$x = \frac{X}{Q+1} \tag{2.4}$$

Finally, the range distance R_1 from the object surface to the activated element in the PSD can be calculated using simple trigonometric principles as

$$R_1 = f\frac{D}{x} = f\frac{D}{X}(Q+1) \tag{2.5}$$

where f is the focal length of the lens, and D is the fixed baseline distance between the optical axis of the light projector and the center of the PDS array.

2.4.1.3.1 Measuring Material Thickness

To further illustrate the concept of triangulation, consider the problem of measuring the thickness of a material sheet. The light source projects a beam onto the planar material surface and the reflected spot of light is detected by the PSD array. Because the photodetector is positioned at a fixed distance from the base or table surface, R_2, the thickness of the material, d_{th}, can be determined using:

$$d_{th} = R_2 - R_1 = R_2 - D\tan\phi \tag{2.6}$$

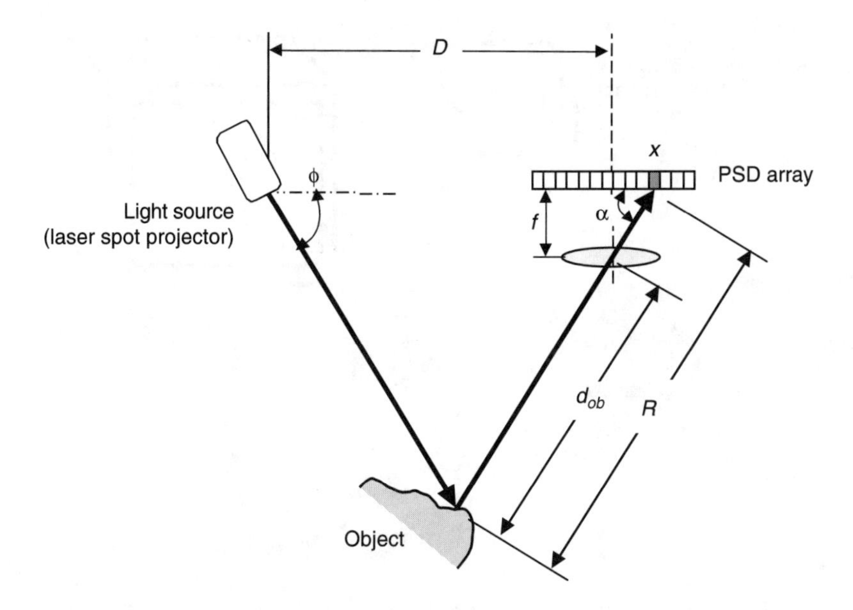

FIGURE 2.15 Range sensing using a moving laser spot projector.

where f is the focal length of the lens, and D is the fixed distance between the optical axis of the light source and center of the PSD array.

2.4.1.3.2 Three-Dimensional Range Sensing

The digitization of three-dimensional objects can be achieved by moving a projected light spot over the entire surface and then performing a range calculation at each point in the scanned area. Many commercially available three-dimensional digitizers utilize a rotating mirror to traverse the beam from right to left and top to bottom over the scanned region. The projected beam and reflected light spot at any instant in time creates a triangle that can be used to calculate the range R. The distance of the brightest image point, x, from the center of the sensor array can be determined as described above. Figure 2.15 is a simple range sensing system that utilizes a moving laser spot projector and a one-dimensional PSD. Because the focal length of the lens, f, and the baseline distance between the projector and optic axis of the PSD are fixed, it is possible to calculate R for different ϕ as the beam traverses the object surface. The angle α can be calculated as

$$\alpha = \tan^{-1}\frac{f}{x} \tag{2.7}$$

Depending on whether the detect point is to the right (+) or the left (−) of the center of the PSD array, x can be positive or negative.

The angle the light projector makes, ϕ, is known, and from the above information it is possible to calculate the range value, R, using the law of sines [Shetty and Kolk, 1997]:

$$\frac{R}{\sin\phi} = \frac{(D+x)}{\sin[180-(\phi+\alpha)]} \tag{2.8}$$

$$R = \frac{(D+x)\,\sin\phi}{\sin[180-(\phi+\alpha)]} \tag{2.9}$$

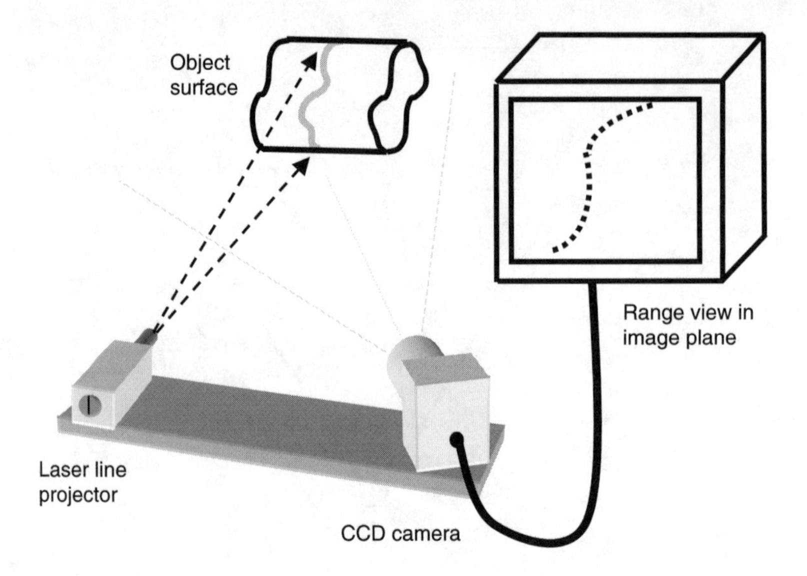

FIGURE 2.16 Measurement of range or distance using line stripe method.

Thus, the range or distance of a surface point with respect to the optic axis of the lens is given by two parameters (α, d_{ob}), where d_{ob} is

$$d_{ob} = R - f\sin\alpha \tag{2.10}$$

The basic principle of triangulation can be extended to light-stripe or projected-pattern range sensing. In general, a structured light pattern is projected onto an object, and a charge-coupled device (CCD) camera captures a two-dimensional image of the deformed pattern formed on the object surface (see Figure 2.16). The two-dimensional pixel coordinates of the captured light pattern in the image plane are used to determine the three-dimensional (x, y, z) object coordinates in the world space. When only a single light line is projected, it generates a plane of light through space that forms a profile on the object surface. Because the profile lies in the plane of the projected light, the measured object points are planar and correspond directly to the two-dimensional image coordinates. The three-dimensional reconstruction of object points is therefore simplified and can be performed after the relationship between image-plane and object-plane coordinates is determined through range-sensor calibration methods [Knopf and Kofman, 1998]. The complete surface geometry of the object surface is achieved by translating the light plane perpendicularly to new known positions along one of the world coordinates and performing the measurement and reconstruction using the same image-plane to object-plane relationship.

Calibration is essential for this range-sensing method to work. It is often carried out using the known optical and geometric parameters of the laser-camera system, or by extracting the parameters by sampling object and image points. However, these calibration methods rely on valid models of the system geometry and optics, which must account for alignment of system components and lens distortion. Alternatively, methods to map image-plane coordinates to object-plane coordinates without the use of sensor optical and geometric models have been proposed [Knopf and Kofman, 1998]. One limitation of the light stripe or projected pattern approach is the poor depth resolution for surfaces parallel to the light plane. This limitation can be overcome by scanning the image in two directions, one perpendicular to the other. The benefit of light-striping methods is that they are relatively simple and fast compared to light-spot scanning. Furthermore, the light stripe can assist the process of image segmentation.

2.4.1.4 Optical Actuators

Optical actuators are the least developed of all optical components. However, light can be either directly or indirectly transformed into mechanical deformations that generate small displacements in the micro- or nanometer range [Goldfarb, 1999]. Direct optical microactuators use the variations in light intensity

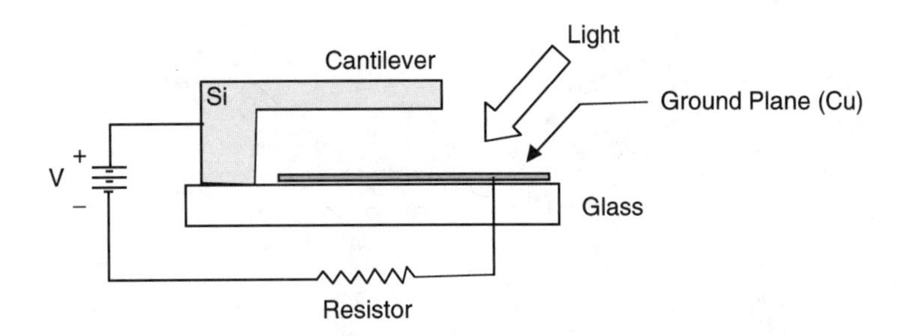

FIGURE 2.17 Schematic diagram of a direct optical microactuator described by Tabib-Azar [1998].

to generate photoelectrons that actuate a silicon microcantilever beam. Indirect optical actuation methods use the properties of light to heat gases or solids that upon expansion cause the desired movement. Indirect optical methods are usually less complicated and often generate more actuation power than direct optical actuation methods. An important feature of the optical actuators, both direct and indirect, is that they can be easily interfaced with fiber optics and other waveguides found in integrated optic circuits. This feature makes optical actuators an ideal component for smart structures and nanofabrication.

2.4.1.4.1 Direct Optical Actuators

Silicon microactuators can be excited directly by an optical light signal using a number of different techniques. One method uses photo-generated electrons to change the electrostatic pressure on a cantilever beam that forms a parallel plate capacitor with a ground plate. The structure of a typical optical microactuator is shown in Figure 2.17. A silicon (Si) cantilever beam forms the top plate of a capacitor that deflects when a potential is applied. Tabib-Azar [1998] describes a microactuator that uses a $600 \times 50 \times 1\ \mu m^3$ cantilever beam with a gap of 12 μm. A bias voltage of 6 V and optical power less than 0.1 mW/cm^2 is used to move the cantilever 4 μm in approximately 0.1 ms. Because this actuation scheme involves a conduction current, a battery or other current source is needed if more than one cycle is to be performed. Continuously charging the capacitor with a current $i \leq \frac{i_{max}}{2}$, where i_{max} is determined by the battery circuit, allows light-controlled actuation in either direction. A continuous photon flux, $\Phi < \frac{i}{\eta}$, short-circuits the capacitor more slowly than the battery charges it, causing a charge build-up, which closes the plates. A photon flux $\Phi > \frac{i}{\eta}$ causes an opposing photocurrent greater than the charging current. The net charge then decreases, and the capacitor plates relax open [Tabib-Azar, 1998].

The other direct methods of microactuation use optically generated electron hole pair screening in a semi-insulating layer, permitivity modulation of a gas by light, and a solar cell that provides the power for actuation [Tabib-Azar, 1998]. Several practical benefits distinguish direct optical methods from indirect radiative-thermal processes. These microactuators can be designed for a specific application using the proven fabrication techniques of semiconductor doping and etching. Furthermore, direct optical actuation can be much faster than indirect processes and often require significantly less power to function. This enables the opto-mechatronic system to be designed smaller and more versatile.

2.4.1.4.2 Indirect Optical Actuators

The optical heating and subsequent expansion of gases and solids can be used in microactuator designs. The optically actuated silicon diaphragm valve shown in Figure 2.18 is used to control the flow of a gas. The cavity is filled with a gas that expands when heated from the light source. As the diaphragm expands it produces the desired deflection. These microfabricated devices are capable of controlling both gas and liquid. These microactuated flow controllers [Tabib-Azar, 1998] have speeds of 21 ms in air flow and 67 ms in oil flow and showed sensitivities of 304 Pa/mW and 75 Pa/mW, respectively.

Other approaches to indirect microactuation involve the expansion of solids that experience a discontinuous change in their volume near the phase-transformation temperature. Shape memory alloys (SMAs) such as NiTi exhibit changes that are significantly larger than the linear volume changes caused by the

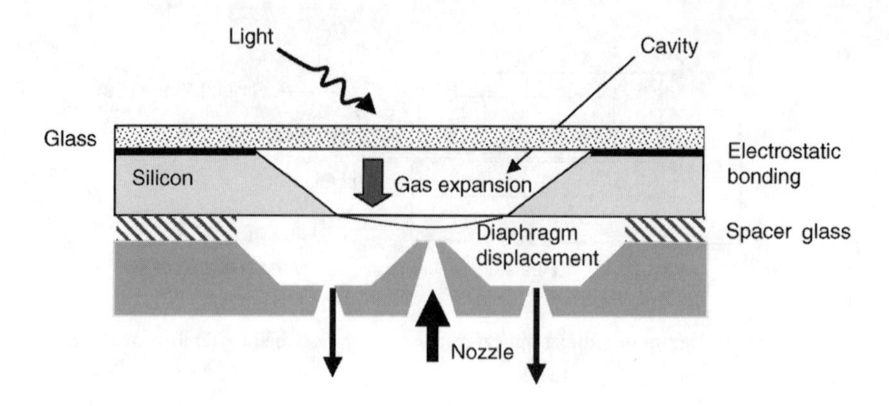

FIGURE 2.18 Schematic of an optically actuated silicon diaphragm used to control the flow of gases.

simple thermal expansion of other solid materials. An example of a light-driven walking machine that employed SMA actuators is described by Yoshizawa et al. [2001]. The miniaturized machine consists of two parts: a body made of SMA and springs, and feet made of magnets and temperature-sensitive ferrites. By repeatedly switching the projected light beam ON and OFF, the shape memory alloy produced the stretching and shrinking action required to move the machine. Furthermore, the switching of the light source activated the temperature-sensitive ferrites to create a magnetic force sufficient to enable the machine to adhere to the carbon steel floor.

2.4.1.5 Integrated Optics

The negative effects of current leakage and power loss are crucial design constraints in developing viable nanotechnology for product miniaturization. Optical circuits and devices have numerous advantages over conventional electronics because they can be activated by photons instead of currents and voltages. Optical systems are free from current losses, resistive heat dissipation, and friction forces that greatly diminish the performance and efficiency of conventional electronic systems. The goal of *integrated optics* is the miniaturization of optic circuits in much the same way that large-scale integrated circuits have miniaturized modern electronics.

Integrated optics is the technology of creating various optical devices and components for the generation, focusing, splitting, combining, isolation, polarization, coupling, switching, modulation, and detection of light — all on a single substrate. Implementation of such circuits is based on the deposition of thin film waveguides on the surface of a substrate or buried inside the substrate material [Chaimowicz, 1989; Saleh and Teich, 1991; Tabib-Azar, 1998]. Optical waveguides provide the connections between the components of the optic system. The waveguides could have the planer channel in coupled or branching form made of glass on amorphous, crystalline dielectric, or semiconducting material. The nature of the material used is very important because it determines the functionality of the photonic part created on the circuit. In addition to the semiconducting properties of the substrate, the electro-optic, acousto-optic, and nonlinear optic properties of the material are also used for developing integrated optic circuits.

A simplified illustration of an optical receiver and transmitter circuit described by Saleh and Teich [1991] is given in Figure 2.19. The received light that carries the information is coupled into a waveguide and directed toward the photodetector where the signal is detected. The light from the laser source is transmitted on the waveguide, modulated, and then coupled into the optic fiber for transmission.

2.4.1.6 Embedded Control

In addition to optical sensors and actuators, an opto-mechatronic design will often include an embedded controller for enhanced functionality. Embedded controllers are self-contained information-processing devices that are programmed for specific tasks and have no operator input/output (I/O) interface [Necsulescu, 2002].

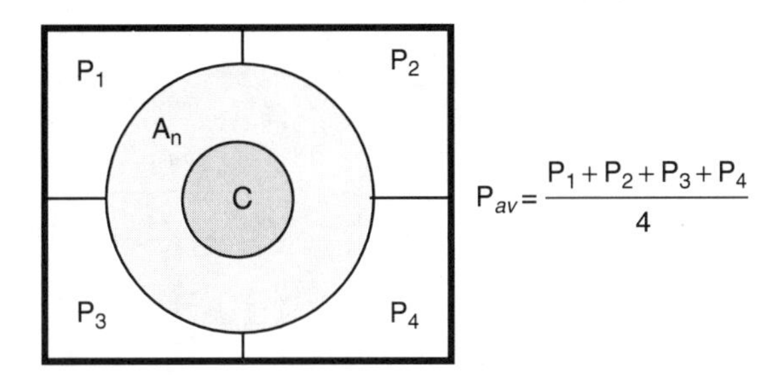

FIGURE 2.21 The basic arrangement of photodetectors used by the meter sensor to determine exposure time [Bolton, 1999]. The average value of the peripheral detectors is given by P_{av}.

a software algorithm are used to analyze the central detector (C), annular detector (A_n), and the average value of the peripheral detectors (P_{av}). This comparison operation is necessary to identify whether the scene has uniform illumination or a bright central zone surrounded by a dark background, as one would expect to find during close-up photography. The "best" exposure time is then determined using simple **IF-THEN** rules [Bolton, 1999] such as:

$$R_i: \ \textbf{IF}\ A_n = C\ \textbf{AND}\ (P_{av} - A_n) < 0,\ \textbf{THEN}\ t_{\text{Exposure}} = f(C)$$
$$R_{i+1}: \ \textbf{IF}\ A_n = C\ \textbf{AND}\ (P_{av} - A_n) = 0,\ \textbf{THEN}\ t_{\text{Exposure}} = P_{av}$$

where t_{Exposure} is the exposure time and $f(.)$ is the function used to compute this time value. The aperture control is achieved by a diaphragm drive system that uses a stepper motor to open and close a set of diaphragm blades.

While adjusting the shutter speed and aperture based on current lighting conditions, the microcontroller in the camera body determines whether the projected image is focused on the photographic film. After passing through the camera lens, the light reflected from the object in the field-of-view strikes the two 48-bit linear arrays of photodetectors. The spacing of the signals measured by the detector arrays will be at a known value when the image is correctly focused on the photographic film. However, if the image is out of focus, then the measured spacing will deviate from the expected value and the magnitude of the deviation is used to generate an error signal. The error signal is fed to the lens microcontroller in order to produce a control signal to drive the DC motors that adjust the focus of the lens assembly.

The precise movement of the focusing lens along the optical axis is accomplished by an arc-form drive and ultrasonic drive systems [Bolton, 1999]. The arc-form drive system employs a brushless permanent magnet DC motor, mechanical gears, and Hall sensors to detect the position of the rotor on the motor. The ultrasonic drive system uses a motor containing a series of piezoelectric elements in the form of a ring. When a current is supplied to a piezoelectric element it expands or contracts according to the polarity of the current. By switching the current to the piezoelectric drive elements in the appropriate sequence, a displacement wave can be created that travels around the ring of elements in either a clockwise or counterclockwise direction. Consequently, the rotor that is in contact with the surface of the piezoelectric ring will rotate and drive the focusing system. A separate encoder provides feedback from the lens assembly to the main microcontroller in the camera body so that the device knows when the focus adjustment is complete.

When the photographer fully depresses the shutter button (final position), the microcontroller in the camera body sends a signal to drive the mirror up and open the shutter for the necessary exposure time, and when the shutter has closed completely the microcontroller will advance the film for the next photograph.

2.5.2 Intelligent Washing Machine

One of the key objectives in opto-mechatronic system design is to optimally integrate embedded controllers with peripheral optical transducers in an effort to improve the performance of mechanical systems. To help further illustrate this point, consider the job of efficiently laundering dirty and oily clothes.

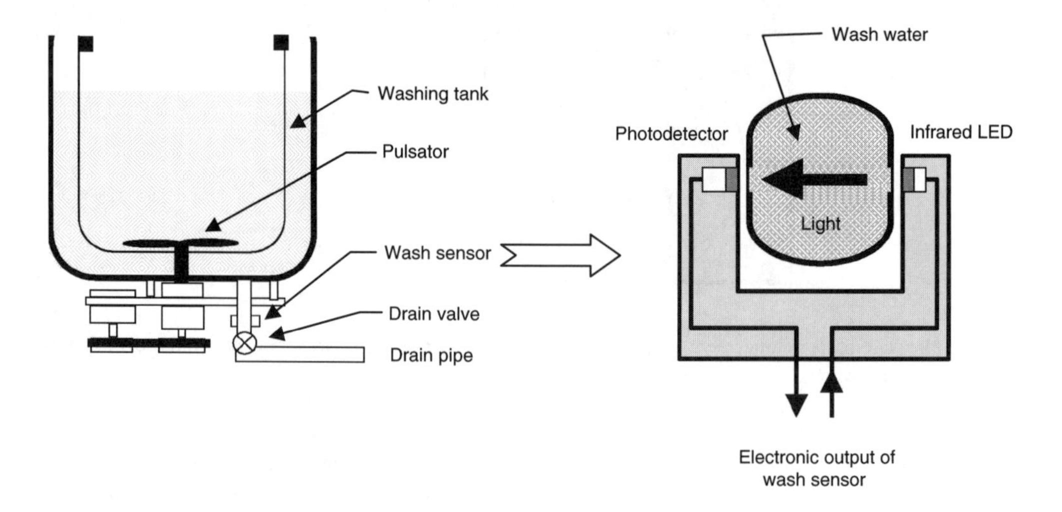

FIGURE 2.22 Schematic diagram of opto-coupled sensor system and location of the sensor system in the washing machine [Wakami et al., 1996].

Wakami et al. [1996] describe an intelligent household washing machine that uses a unique optoelectronic sensor and a fuzzy logic controller to automatically adjust the washing time based on the type and amount of dirt on the clothes.

The sensor system is installed near the drain valve (Figure 2.22) and detects the level of dirtiness by measuring the light transmission of the wash water. The opto-coupled system consists of an infrared light-emitting (LED) source and a phototransistor detector. The light beam generated by the infrared LED passes through the water in the pipe and strikes the phototransistor, which generates a voltage value proportional to the intensity of the light received. As the dirt is removed from the clothes during the cleaning process the wash water becomes cloudy, causing the light transmission properties of the water to decrease. The rate of decrease in light transmission depends on the type of material being removed, as illustrated in Figure 2.23. A fast change in the measured light reduction occurs for muddy clothes because the mud particles and debris are removed by the mechanical agitation of the rotating pulsator. In contrast, oily substances are removed from the clothes only after the detergent takes effect. Therefore, the rate of change in the measured light will decrease more slowly.

It is difficult to experimentally obtain the relationship between level of dirtiness and optimum washing time because of the large variety of material that can be deposited on clothes and the sheer difficulty in collecting detailed experimental data for all possible situations. In addition, it is difficult to obtain a unique mathematical formula relating wash time to type and amount of dirt because the process is nonlinear. In order to determine the wash time from the output of the sensor system, Wakami et al. [1996] employed a microcontroller programmed to function as a knowledge-based controller with fuzzy inference rules. The fuzzy inference engine enables the complex relationship between the level of dirtiness and required wash time to be expressed as a set of linguistic rules. As a result of using fuzzy rules and an inference mechanism, a savings in both energy and time can be achieved because inadequate or excessive washing times are significantly reduced.

A series of linguistically defined **IF-THEN** rules are used to describe the I/O mapping of the system. The inputs to the system are given as the amount of light transmission at saturation, τ_s, and the time taken to reach saturation, t_s. The level of light transmission at saturation is related to the amount of dirt present in the clothes, and the time for saturation to occur is a function of the type of dirt. The membership functions of the fuzzy input variables are summarized in Figure 2.24. These membership functions enable a simplified fuzzy inference procedure, in which the consequent part of the fuzzy rule (required wash time, t_w) is expressed by a real number, t_i, where the subscript i indicates the fuzzy rule in the rule base. The fuzzy approach to designing the control strategy makes it possible to simplify the inference operations and reduce

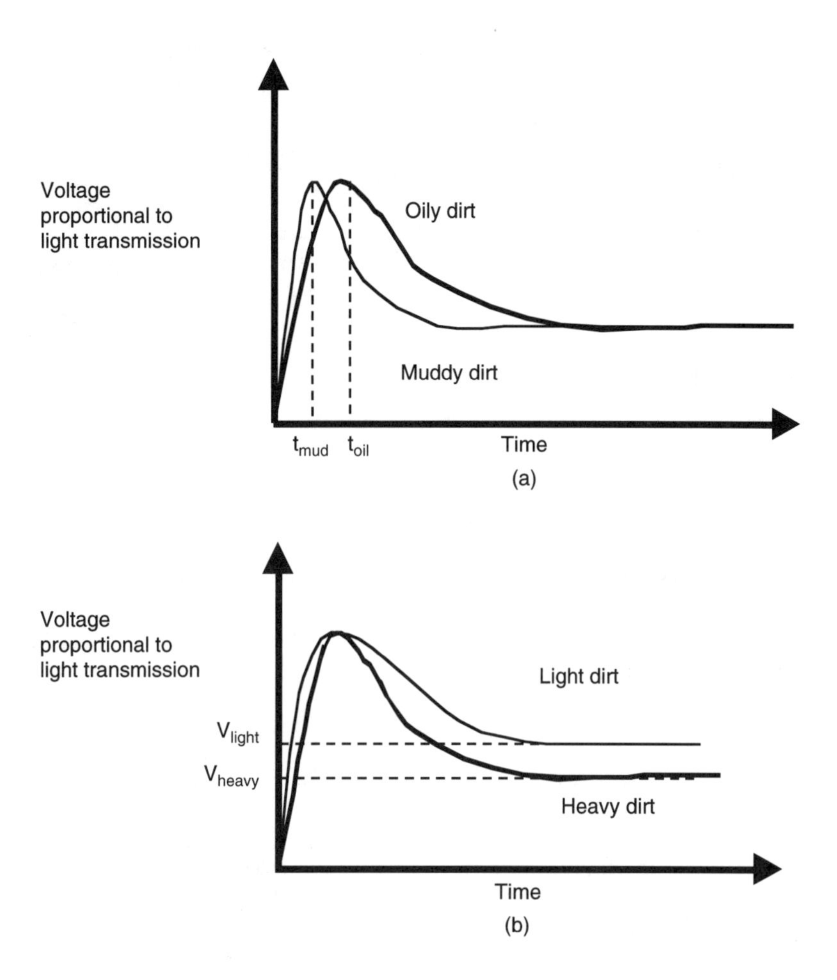

FIGURE 2.23 An illustration of the response curves generated by the opto-coupled sensor system proposed by Wakami et al. [1996].

the amount of required memory [Kosko, 1997]. In this situation, the fuzzy rules were created by observing the performance of a skilled expert and then fine-tuning the membership functions. The fuzzy rules used to determine the required wash time, t_w, are [Wakami et al., 1996]:

R_1: **IF** τ_s is low **AND** t_s is short, **THEN** $t_w = t_1$
R_2: **IF** τ_s is middle **AND** t_s is short, **THEN** $t_w = t_2$
R_3: **IF** τ_s is high **AND** t_s is short, **THEN** $t_w = t_3$
R_4: **IF** τ_s is low **AND** t_s is long, **THEN** $t_w = t_4$
R_5: **IF** τ_s is middle **AND** t_s is long, **THEN** $t_w = t_5$
R_6: **IF** τ_s is high **AND** t_s is long, **THEN** $t_w = t_6$

2.5.3 Optical Implementation of a SISO Rule-Based Controller

Fuzzy logic techniques are used to control electro-mechanical components in a variety of consumer products and industrial plants. However, these controllers are often slow because of the extensive signal-processing and algorithmic computations required to arrive at a satisfactory conclusion from numerous rules. To increase the processing speed of rule-based controllers, several researchers have proposed an optical solution [Gur et al., 1998; Itoh et al., 1997; Zalevsky et al., 2000; Zhang and Karim, 1999]. Itoh et al. [1997] describe an opto-electronic controller that performs fuzzy logic operations in real time by

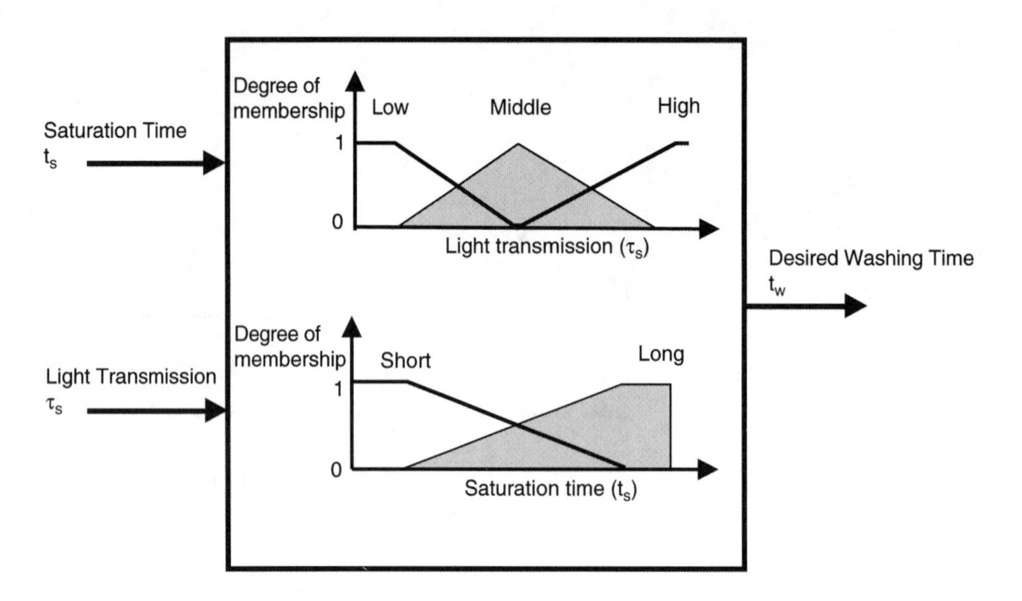

FIGURE 2.24 The knowledge-based controller used to determine the desired wash time (t_w) from the sensor system output as described by Wakami et al. [1996]. The controller inputs, (τ_s, t_w), are fuzzy variables and the output (t_w) is a real number.

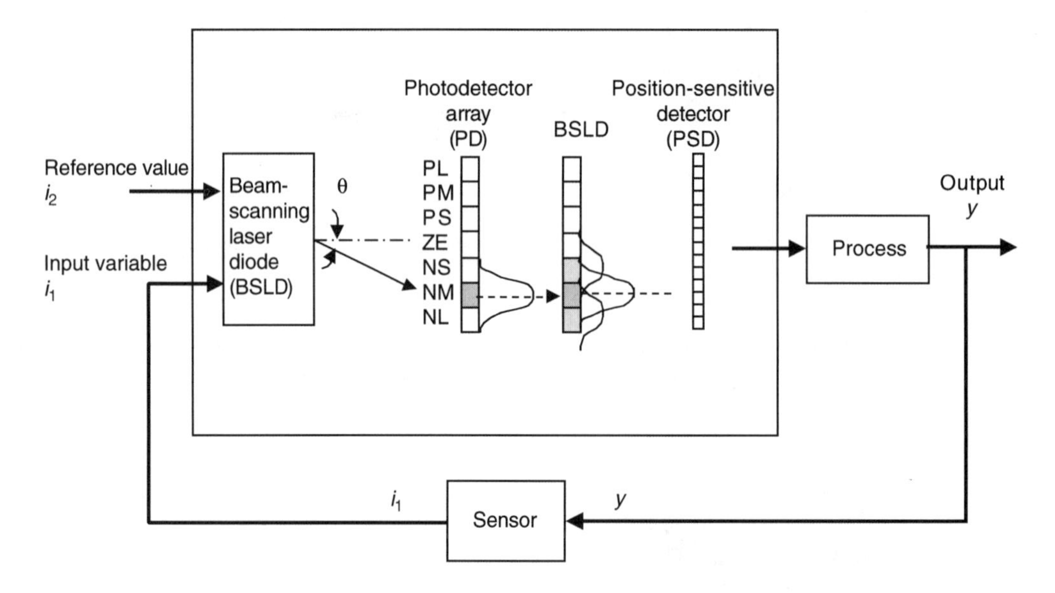

FIGURE 2.25 The opto-electronic implementation of a SISO fuzzy logic controller as proposed by Itoh et al. [1997]. In the above illustration the fuzzy membership functions are PL, positive large; PM, positive medium; PS, positive small; ZE, zero; NS, negative small; NM, negative medium; and NL, negative large.

utilizing a beam-scanning fuzzy inference architecture. The proposed architecture, Figure 2.25, uses a product-sum-gravity method with Gaussian membership functions instead of the conventional min-max gravity method with triangular membership functions [Kosko, 1997].

The beam-scanning laser diode (BSLD) of the single-input, single-output (SISO) system receives a current signal as an input, i_1, from a sensor monitoring the response of a physical process and a reference

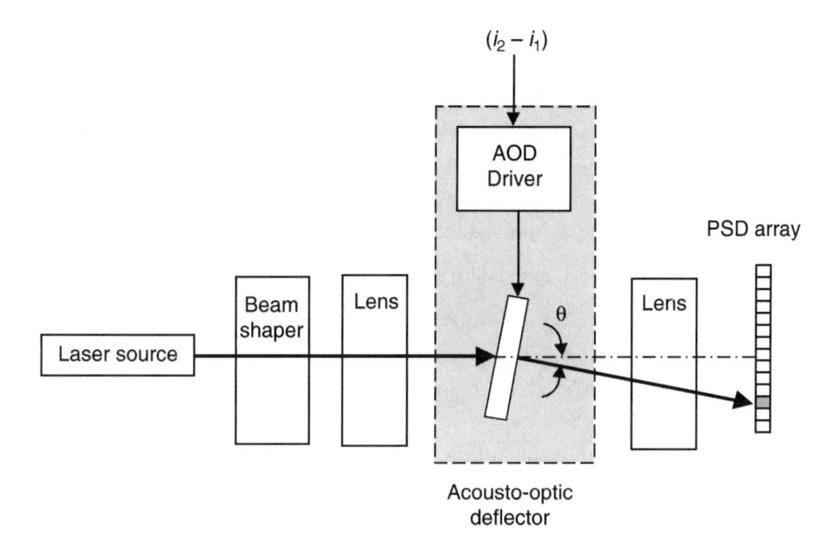

FIGURE 2.26 An acousto-optic deflector used for the light beam displacement [Gur et al., 1998] in an opto-electronic implementation of a fuzzy logic controller.

input given by the current signal, i_2. The light beam emitted by the BSLD diverges by the angle θ [Itoh et al., 1997]:

$$\theta = \beta(i_2 - i_1) \tag{2.11}$$

where β is the fixed system gain. The angle θ represents the basic *premise* for each rule in the fuzzy inference engine.

The angular shift in the beam direction can also be achieved using an acousto-optic deflector (AOD) or a spatial light modulator (SLM) [Gur et al., 1998]. For the first alternative, the laser light beam must be reshaped and Fourier-transformed prior to entering the acousto-optic cell located in the Fourier plane (Figure 2.26). The transformed light beam is then multiplied by an acoustic wave that is orthogonal to the beam-propagation direction. The frequency of the acoustic wave is proportional to the input ($i_2 - i_1$), resulting in a deflected beam angle of θ. In contrast, the spatial light modulator can be placed at either the laser output or at the Fourier plane of a 4*f* setup and used to directly shift the angle of the beam. In many control applications a low-resolution SLM might be sufficient; however, the SLM's slow response time is often the limiting factor in control speed.

Once the beam is deflected based on the input difference it strikes an array of photodetectors (PDs). Each PD represents a membership function for the angle θ. For example, the central PD represents angles close to zero, and the edge PDs represent large angles, either positive large (PL) or negative large (NL). Each PD produces a current whose value is proportional to the match between θ and the PD location.

Each constituent PD in the array drives a separate beam-scanning laser diode. When activated each BLSD emits a Gaussian beam with an intensity profile proportional to the driving current generated by the activated PD. The resulting beams from neighboring laser diodes project onto a position-sensitive detector (PSD). The output current of the PSD is proportional to the center of gravity (CoG) of the incident beams. Several Gaussian beams that originated from different inputs can reach the single-output PSD, and the detector will determine the total CoG for all inputs [Gur et al., 1998].

The main advantage of the optical inference engine proposed by Itoh et al. [1997] is its simplicity and modularity for extending the basic structure for multiple inputs. For example, several SISO controllers can be placed on top of each other to simultaneously control several systems. Furthermore, the simple optical and electronic components used by this computing architecture should enable several controllers to be developed on an integrated optic circuit for product miniaturization.

The basic controller design has several limitations. First, the beam-scanning laser diode and other signal deflection methods generate beams with Gaussian profiles. Reshaping the beam [Leger, 1997] to represent other types of membership functions is not a trivial task. Solutions to this problem include the use of an amplitude mask, reshaping the beam in the Fourier domain with a phase-only filter, or placing an amplitude-coded mask in the Fourier plane. A second lim-tation to this optical implementation of a fuzzy logic controller is that this optic circuit requires optical–electronic–optical conversions. As a result of these signal conversions, the original information can easily become distorted and lead to control error.

2.6 Conclusions

This chapter described the process of opto-mechatronic design. Opto-mechatronics stresses technology integration for enhanced system performance. In essence, it is not a specific technology but rather a design philosophy that promotes the creation of high-quality "smart" products and processes. Opto-mechatronic systems will often exhibit a number of important characteristics such as the functional interaction between optical, electronic, and mechanical components; spatial integration of subsystems into a single physical unit; utilization of multifunctional devices; and exploitation of embedded control. This chapter summarized some of the unique and important features of optical and electronic technologies that contribute to the performance of an opto-mechatronic system. Specifically, optical sensors and actuators were examined because these devices provide a mechanism for robust, low-cost solutions that enable high-precision and rapid signal-processing operations. Several examples were provided to illustrate how these simple devices could perform complex functions.

Defining Terms

concurrent engineering: The simultaneous evolution of the product and the manufacturing process required to produce it.

design: The creative process used to solve open-ended or ill-defined problems with numerous satisfactory solutions.

Design for Excellence (DFX): A knowledge-based approach that attempts to design products or processes by maximizing all desirable characteristics such as high quality, reliability, serviceability, safety, user friendliness, and short time-to-market while, at the same time, minimizing lifetime costs.

embedded controller: A microprocessor with combined memory and various input/output (I/O) features on a single integrated chip. These devices do not have an operator I/O interface.

integrated optics: Analogous to integrated electronic circuits but where the movement of photons replaces electrons.

ray sketching: A method for rapidly drawing light rays to determine approximate distances and dimensions of a proposed optical system.

ray tracing: A method used to evaluate the performance of an optical system by calculating the paths of one or more light rays through the constituent components.

system: A term used to describe a product or process that is viewed as a box with inputs and outputs; also, a mathematical function describing the relationship between inputs and outputs.

transducer: A device that transforms energy from one form into another; it does not matter whether the energy belongs to different domains or the same domains. Transducers may be sensors or actuators.

References

Ahmed, A., *Handbook of Optomechanical Engineering*, CRC Press, Boca Raton, FL, 1997.

Allard, F. C., *Fiber Optics Handbook for Engineers and Scientists*, McGraw-Hill, New York, 1990.

Bolton, W., *Electronic Control Systems in Mechanical Engineering*, Addison-Wesley-Longman, New York, 1999.

Bralla, J. G., *Design for Excellence*, McGraw-Hill, New York, 1996.

Chaimowicz, J. C. A., *Lightwave Technology: An Introduction*, Butterworths, London, 1989.

Dieter, G. E., *Engineering Design: A Materials and Processing Approach*, 3rd ed., McGraw-Hill, New York, 2000.

Goldfarb, M., Microsensors and microactuators, in *Mechatronics in Engineering Design and Product Development*, Popovic, D. and Vlacic, L., Eds., Marcel Dekker, New York, 1999, pp. 31–61.

Gur, E., Mendlovic, D., and Zalevsky, Z., Optical implementation of fuzzy logic controllers: Part I, *Applied Optics*, 37(29), 6937–6945, 1998.

Itoh, H., Yamada, T., Mukai, S., Watanabe, M., and Brandl, D., Optoelectronic implementation of real-time control of an inverted pendulum by fuzzy-logic-control units based on a light-emitting-diode array and a position-sensing device, *Appl. Opt.*, 36(4), 808–812, 1997.

Knopf, G. K. and Kofman, J., Range sensor calibration using a neural network, in *Intelligent Engineering Systems through Artificial Neural Networks*, Vol. 8, Dagli, C. H. et al., Eds., ASME Press, New York, 1998, pp. 491–496.

Kosko, B., *Fuzzy Engineering*, Prentice-Hall, Upper Saddle River, NJ, 1997.

Leger, J. R., Laser beam shaping, in *Micro-Optics: Elements, Systems and Applications*, Herzig, H. P., Ed., Taylor & Francis, London, 1997, pp. 223–257.

McMahon, C. and Browne, J., *CADCAM: From Principles to Practice*, Addison-Wesley, Wokingham, U.K., 1993.

Necsulescu, D., *Mechatronics*, Prentice-Hall, Upper Saddle River, NJ, 2002.

O'Shea, D. C., *Elements of Modern Optical Design*, Wiley, New York, 1985.

Otto, K. and Wood, K., *Product Design: Techniques in Reverse Engineering and New Product Development*, Prentice-Hall, Upper Saddle River, NJ, 2001.

Palais, J. C., *Fiber Optic Communications*, Prentice-Hall, Upper Saddle River, NJ, 1998.

Saleh, B. E. A. and Teich, M. C., *Fundamentals of Photonics*, John Wiley & Sons, New York, 1991.

Shetty, D. and Kolk, R. A., *Mechatronics System Design*, PWS Pub., Boston, 1997.

Tabib-Azar, M., *Microactuators: Electrical, Magnetic, Thermal, Optical, Mechanical, Chemical, and Smart Structures*, Kluwer Academic, Norwell, MA, 1998.

Ullman, D. G., *The Mechanical Design Process*, McGraw-Hill, New York, 1997.

Wakami, N., Nomura, H., and Araki, S., Fuzzy logic for home appliances, in *Fuzzy Logic and Neural Network Handbook*, Chen, C. H., Ed., McGraw-Hill, New York, 1996, pp. 21.1–21.23.

Yoshizawa, T., Hayashi, D., Yamamoto, M., and Otani, Y., A walking machine driven by a light beam, in *Opto-Mechatronic Systems II*, Cho, H. Y., Ed., Proceedings of SPIE, Vol. 4564, 2001, pp. 229–236.

Zalevsky, Z., Mendlovic, D., and Gur, E., Discussion on multidimensional fuzzy control, *Appl. Opt.*, 39(2), 333–336, 2001.

Zhang, S. and Karim, M. A., Optical triangular-partition fuzzy systems with on-memory-matrix fuzzy associative memory, *Appl. Opt.*, 24(7), 484–486, 1999.

For Further Information

Information on mechatronic system design and opto-mechatronics is included in several professional society journals and conference proceedings. A variety of the articles describing interesting applications are found in *Mechatronics, Journal of Robotics and Mechatronics, IEEE Transactions on Mechatronics, Journal of Micromechatronics, Journal of the Optical Society of America*, and *Optical Engineering*. The proceedings of the **Opto-Mechatronic Systems Conference** are published annually by SPIE, the International Society for Optical Engineering. These proceedings document the latest developments in the field of optical-based products and processes each year.

A number of introductory texts and reference books on mechatronics systems have been published in recent years. Two reference books that provide a unique perspective are:

HMT Limited, *Mechatronics and Machine Tools*, McGraw-Hill, New York, 1999

Popovic, D. and Vlacic, L., *Mechatronics in Engineering Design and Product Development*, Marcel Dekker, New York, 1999

II

Optical Elements, Sensors, and Measurements

3

Principles of Semiconductor Lasers and Their Applications

Yoshitada Katagiri
NTT Telecommunications Energy Labs
Kanagawa, Japan

3.1 Introduction

Coherent light has existed virtually since 1960, when laser oscillation was demonstrated for the first time using a ruby crystal in a Fabry–Perot resonator. This innovation spurred many people to create various innovations in lasers using various media. These innovations include semiconductor lasers, which have undergone much development since the laser oscillation was achieved at room temperature using double-heterojunction structures. These semiconductor lasers are very small and operate under low-driving power conditions and, hence, show great potential for a wide variety of applications in various fields including information, measurement, communications, etc. In this chapter, the fundamentals of such semiconductor lasers are briefly explained, and then several practical applications are described.

3.2 Fundamentals of Semiconductor Lasers

3.2.1 Light-Emission Processes in General Materials

We begin our discussion with the emission mechanism of light in general materials, which include semiconductors. The emission mechanism is explained using a level diagram with two-level systems (see Figures 3.1 and 3.3) [Berestetskii et al., 1990]. The three major optical processes are absorption, spontaneous emission, and stimulated emission. Absorption means that the atom in the ground state (state 1) makes a transition to the excited state (state 2) by absorbing the photon energy. The photon is required to have a larger energy than the gap between these states for the absorption process. Spontaneous emission means

0-8493-1162-4/03/$0.00+$1.50

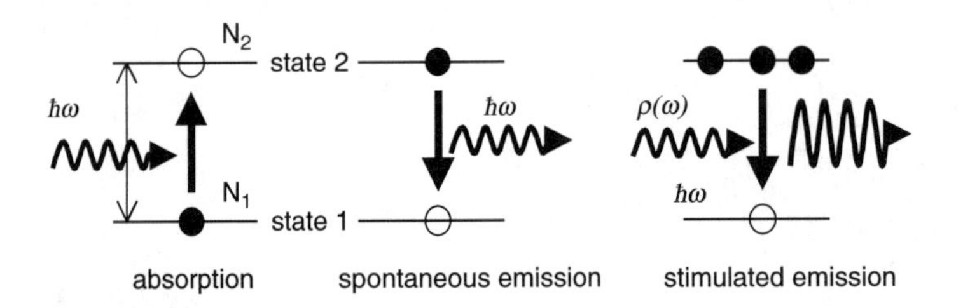

FIGURE 3.1 Level diagrams of atomic systems for exhibiting the three basic optical processes of absorption, spontaneous emission, and stimulated emission.

that the excited state makes a transition to the ground state according to a natural probability, independent of external triggers, with the simultaneous emission of a photon whose energy corresponds to the energy gap. Stimulated emission means that the transition with this kind of photon emission is triggered by the emission field according to the probability proportional to the emission field density. The emitted photon is cooperative with the existing field. Hence, this stimulated emission enhances the emission field.

The natural probabilities for these three emission processes are numerically evaluated using a rate equation given by

$$\frac{d}{dt}N_1 = -\frac{d}{dt}N_2 = N_2A_{21} + \rho(\omega)(-N_1B_{12} + N_2B_{12}) \tag{3.1}$$

with the following values:

$\rho(\omega)$: the energy density of the radiation field
N_1: the population of state 1
N_2: the population of state 2

Here A_{21}, Einstein's A-coefficient, is a transition probability from state 2 to state 1. The parameter B_{12}, Einstein's B-coefficient, means a transition probability from state 2 to state 1 under the emission field.

Consider an electromagnetic field in thermal states. Because the time-dependent terms are negligible and the population ratio of the two states follows the Boltzmann distribution given by

$$\frac{N_2}{N_1} = \exp(-\beta\eta\omega), \qquad \beta = \frac{1}{K_BT}, \tag{3.2}$$

the energy density of the emission field becomes

$$\rho(\omega) = \frac{A_{21}\exp(-\beta\eta\omega)}{B_{12} - B_{21}\exp(-\beta\eta\omega)}, \tag{3.3}$$

with the value K_B as Boltzmann's constant. This equation gives the relationship between Einstein's two coefficients

$$\frac{A_{21}}{B_{12}} = \frac{\hbar\omega^3}{c^3\pi^2}, \tag{3.4}$$

The light amplification is theoretically possible if the stimulated emission is used; however, it is inhibited by the absorption dominance in the medium because of $N_2 < N_1$ for every temperature. In order to obtain a net gain for the light amplification, $N_2 > N_1$ is required. This situation means that the reverse population nominally exhibits a negative temperature. It is of great importance that all lasers create the mechanisms for yielding such a reverse population.

3.2.2 The Laser Oscillation Mechanism in Semiconductors [Verdeyen, 1994; Casey and Panish, 1978]

Semiconductors have a characteristic electronic structure (see Figure 3.2), i.e., a band structure consisting of valence and conduction bands. The emission processes of light in semiconductors are similar to the three-level model as described above. Electrons in a valence band are thermally excited to a conduction band while they leave positive holes in the valence band. The excited electrons can be recombined with the holes. Relaxation processes of this recombination include photon emissions whose energy $\hbar\omega$ corresponds to the band-gap energy ΔE. This band-gap energy is strongly dependent on the materials (see Figure 3.3). Semiconductor lasers can be obtained in various emission wavelength bands—from visible to infrared ranges—if appropriate materials are selected for constructing devices of interest.

In order to achieve continuous photon emission by current injection, we have used a P–N-junction structure, as shown in Figure 3.4. This structure consists of two types of semiconductors involving impurities.

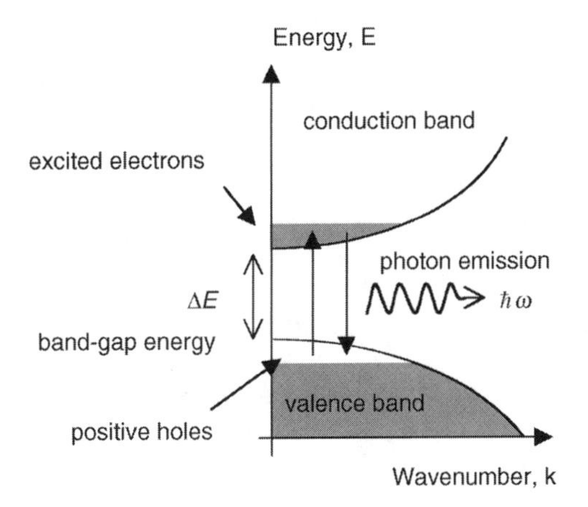

FIGURE 3.2 Emission of photons in semiconductors.

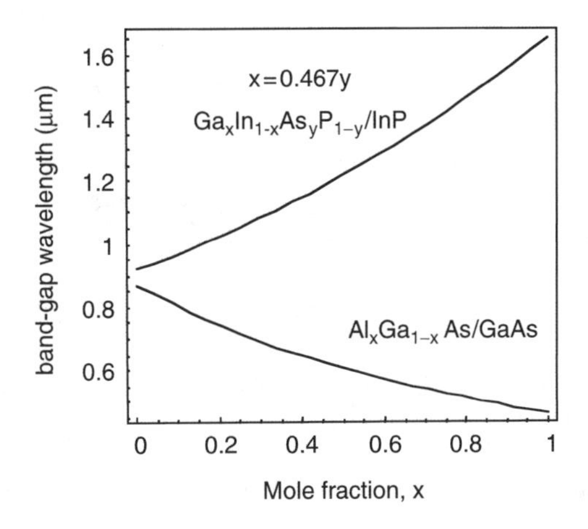

FIGURE 3.3 Band-gap wavelengths dependent on chemical composition of semiconductors. The wavelengths cover a wide range from the visible to the near-infrared region.

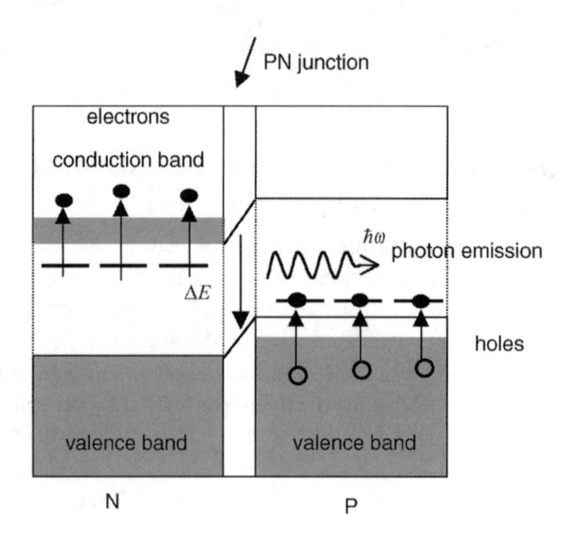

FIGURE 3.4 Emissions of photons at P–N junctions.

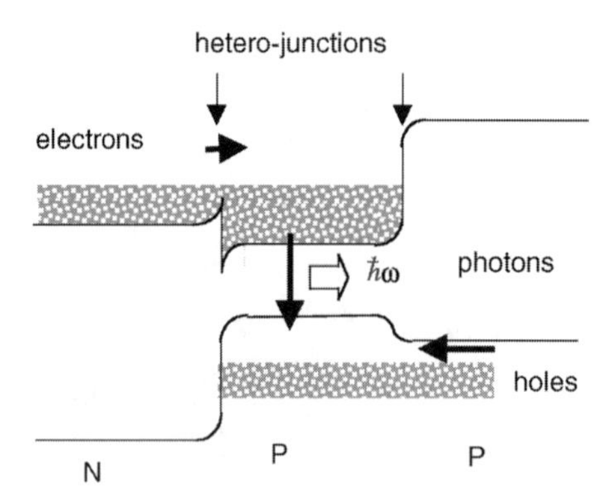

FIGURE 3.5 Double-hetero-junction structure for lasers.

One is an N-type semiconductor, where impurities provide electrons for the conduction band. The other is a P-type semiconductor, where impurities accept excited electrons into the valence band to create positive holes. When these two types of semiconductors are joined together, the interface region can include both electrons and holes. Hence, when we make a closed-loop circuit by applying a positive voltage to the P-type semiconductor while connecting the N-type semiconductor to the ground level, we obtain a stationary charge flow according to the amount of recombination in the junction. Therefore, we obtain stationary photon emissions.

We present a double-hetero-junction structure to achieve inverse populations in semiconductors, as shown in Figure 3.5. The hetero-junction involves connecting two kinds of semiconductors with different chemical elements. Such a junction offers high potential barriers for carriers. We have achieved strong carrier confinement by a sandwich structure using two heterojunctions. This structure, called the double-hetero-junction (DH) structure, is widely used for achieving inverse populations in semiconductor devices. In order to realize laser oscillation, however, we further need an optical confinement structure to increase the laser field intensity. Thus, optical waveguides, where the core region shows a higher refractive index while the outside cladding layers show a lower index, are used to confine the laser light in the core.

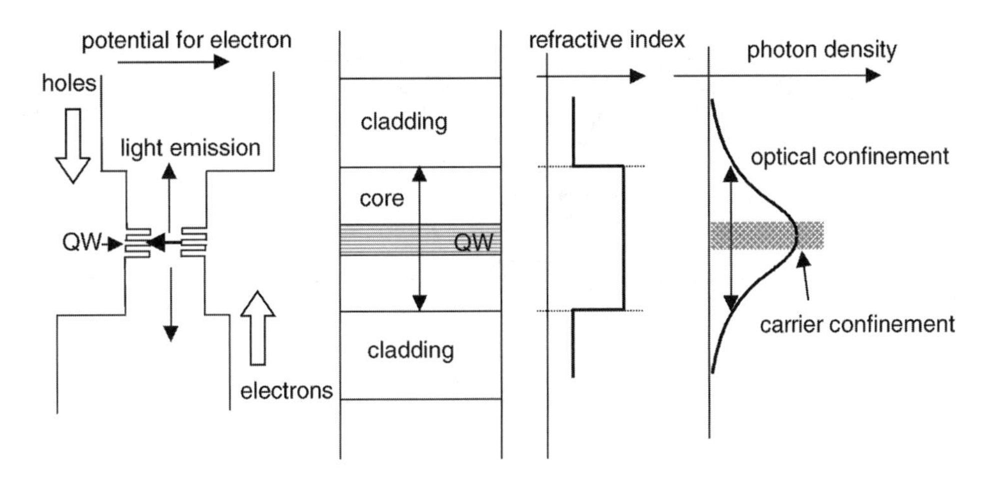

FIGURE 3.6 Schematic of separated confinement heterostructure (SCH) lasers with quantum wells (QWs) for electron confinement sandwiched by semiconductor cladding regions having a wide bandgap to offer transparency for emitted photons at the wells.

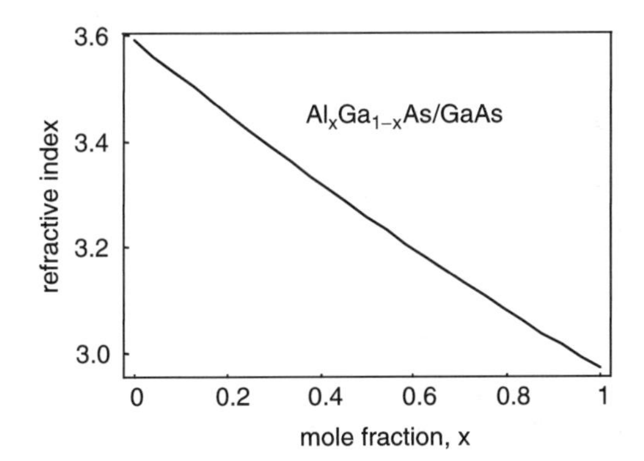

FIGURE 3.7 Refractive index of semiconductors.

The lasers with DH structures have a shortcoming — the core region produces a rather large propagation loss because the gain medium simultaneously acts as an absorption medium. Separated confinement heterostructure (SCH) lasers are used to eliminate this problem (see Figure 3.6). The SCH lasers have an electron confinement region that usually consists of very thin layers of several nanometers to construct quantum wells (QWs) and barriers, which are included in the core region. Because the core region is designed to have a wider bandgap compared with the energy of the emitted photons in the QWs, the lightwaves guided in the core show a low-loss propagation performance.

Practical lasers need three-dimensional confinement structures for both carriers and lightwaves. These structures are designed and realized based on the relationship between the chemical composition of the materials and the corresponding refractive index (see Figure 3.7).

We have developed a stripe-geometry laser structure, as shown in Figure 3.8, that basically consists of a waveguide with a layered vertical confinement structure and a buried lateral guide structure. This laser structure also has an optical resonance function that uses two facet mirrors to enhance the field intensity.

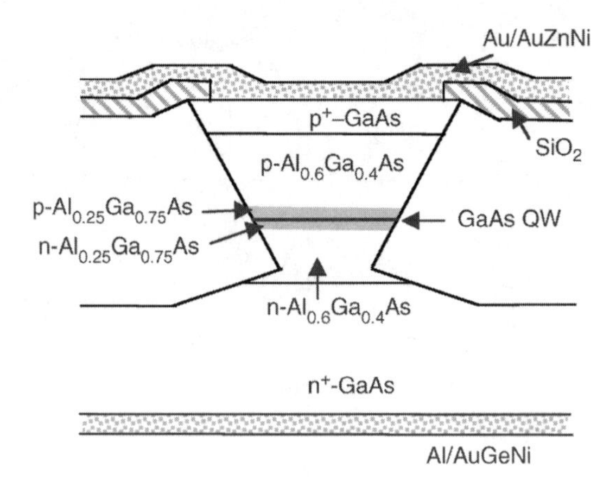

FIGURE 3.8 Cross-sectional view of separated confinement heterostructure laser with single-quantum well active region. The lateral confinement is achieved by a buried heterostructure (BH) configuration.

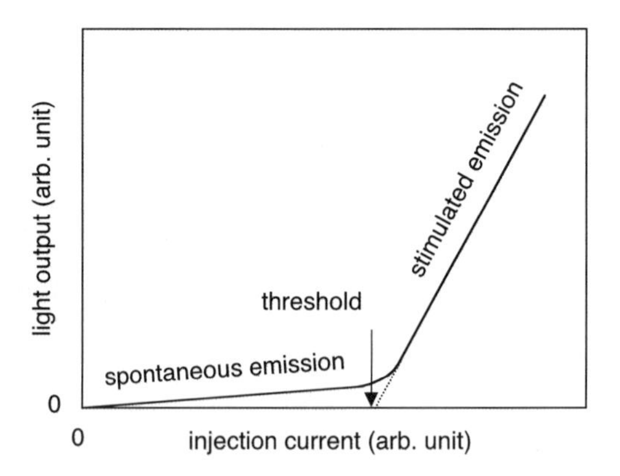

FIGURE 3.9 Injection current vs. light output curve for typical Fabry–Perot laser.

3.2.3 Threshold Condition and Oscillation Mode

As the injection current to semiconductor lasers increases, so too does the population of electrons in the conduction band. In the lower-current injection region, the light amplification in the semiconductor media is so insufficient that the light attenuates while circulating in the laser cavity. However, the light is dramatically amplified above a level of injection current that would give a round-trip with no attenuation. This condition is called the threshold condition for laser oscillation. Consequently, a characteristic injection-current vs. light-output curve is given, as shown in Figure 3.9. This feature is explained by the fact that electrons injected into semiconductor lasers are used for increasing the optical gain below the threshold, while the injected electrons are used to produce light by stimulated emission under high efficiency.

We evaluate the threshold using a simple stripe-geometry Fabry–Perot laser model (see Figure 3.10). The laser consists of two-facet mirrors and a waveguide with structural parameters, including a cavity length of L and facet reflectivity of R_1 and R_2. Using gain and loss coefficients g and γ, an amplitude round-trip gain a is defined as

$$a = \sqrt{R_1 R_2}\, \exp[(g-\gamma)L + 2ikL] \tag{3.5}$$

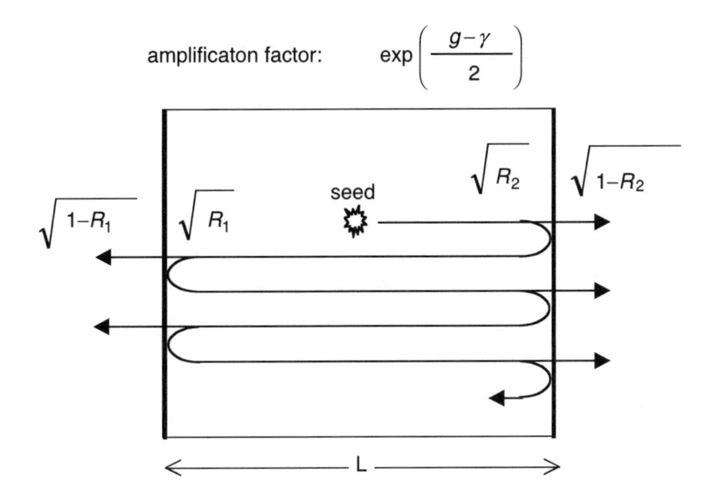

FIGURE 3.10 Model of Fabry–Perot laser.

If $a = 1$ under oscillation, we ca1n determine the wavenumber k and gain coefficient g_{th}, which give the oscillation conditions as

$$\left.\begin{aligned} k &= \frac{\pi}{nL} m \quad m = 1, 2, 3, \cdots \\ g_{th} &= \gamma + \frac{1}{2L} \log \frac{1}{R_1 R_2} \end{aligned}\right\} \tag{3.6}$$

The first condition corresponds to the resonant modes of the Fabry–Perot cavity, while the second condition gives the threshold condition necessary for laser oscillation.

Consider the second condition for evaluating the threshold current condition. The gain coefficient g is given as a function of injection-current density J

$$g = \beta\left(\frac{J}{d} - J_0\right), \tag{3.7}$$

where J_0 is the nominal current, d the thickness of the active region, and β a constant parameter. Substituting this equation into Equation (3.6), we find the threshold current density as

$$J_{th} = \frac{d}{\beta}\left(\gamma + \frac{1}{2L} \log \frac{1}{R_1 R_2}\right) + J_0 d \tag{3.8}$$

We also discuss the laser oscillation mode using the first oscillation condition. The laser eigenmodes determined by Equation 3.6 include several candidates for laser oscillation. However, a unique mode with a maximum gain coefficient remains while the others fade out during circulation in the cavity, as shown in Figure 3.11(a).

For more high-performance applications, the semiconductor lasers are desired for showing single-mode oscillation performance. This oscillation condition is readily obtained by using a grating as a mode selector to construct a laser resonator (see Figure 3.11(b)).

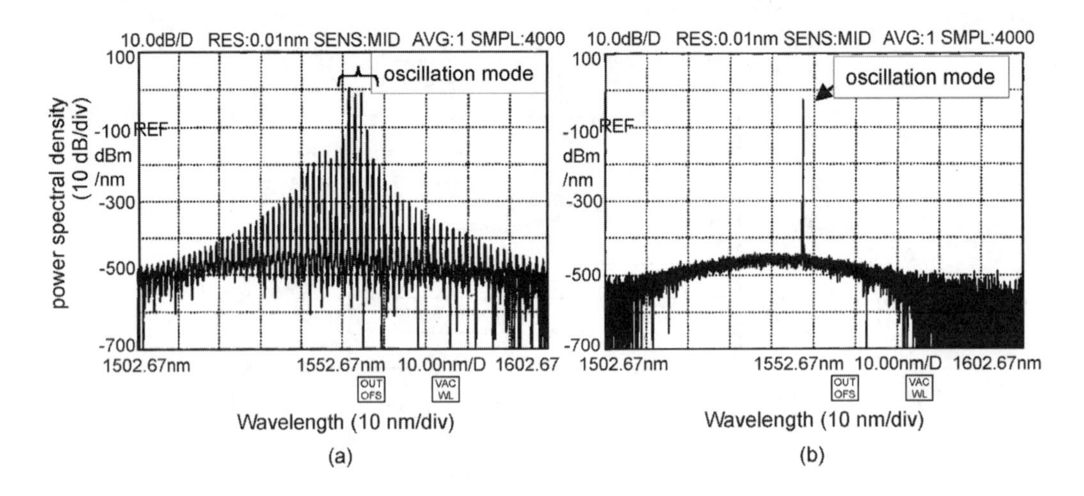

FIGURE 3.11 Oscillation mode spectra of semiconductor lasers: (a) conventional Fabry–Perot laser; (b) single-mode laser using grating mirror for mode selection.

3.3 Applications

3.3.1 Optically Switched Lasers and Their Applications to Data Storage

3.3.1.1 The Concept of Optically Switched Lasers

Performance of optically switched lasers is explained as a function of light-feedback effects of semiconductor lasers. Consider a coupled-cavity laser consisting of a Fabry–Perot laser diode with an external mirror in the proximity of a laser facet (see Figure 3.12). Generally the light feedback in such coupled-cavity lasers induces cooperative changes in photons and carriers in the laser medium. However, their oscillation performance can be discussed in the stationary condition as being available in many micro-opto-mechatronics applications. The discussions are performed with an effective reflectivity replacing the external cavity formed between the laser facet and the external mirror. Assuming a negligible facet reflectivity on the external mirror side, achieved through the use of a high-quality antireflection coating, the effective reflectivity is almost equivalent to that of the external mirror. Such substitution makes it possible to consider the coupled-cavity laser as a conventional Fabry–Perot laser. We only take into account the oscillation condition for amplitude as

$$I_{th} = I_0 + \frac{1}{\chi}\left(\alpha + \frac{1}{2L_c}\ln\frac{1}{R_1 R_3}\right) \tag{3.9}$$

with the following values:

I_0: the current at which the gain of the laser diode becomes zero
χ: coefficient of the gain as a linear function of injection current I
L_c: external cavity length between laser facet and external mirror
R_3: reflectivity of external mirror

The above equation numerically simulates an interesting effect, i.e., laser switching controlled by external-mirror reflectivity [Ukita et al., 1989]. For simplification, we assume that the mirror consists of two kinds of regions with a higher and a lower reflectivity. The coupled-cavity laser shows two laser-oscillation threshold currents corresponding to the reflectivity shown in Figure 3.13. Hence, it is switched between stimulated and spontaneous emission states responding to the regions of the external mirror under the bias condition between the two thresholds.

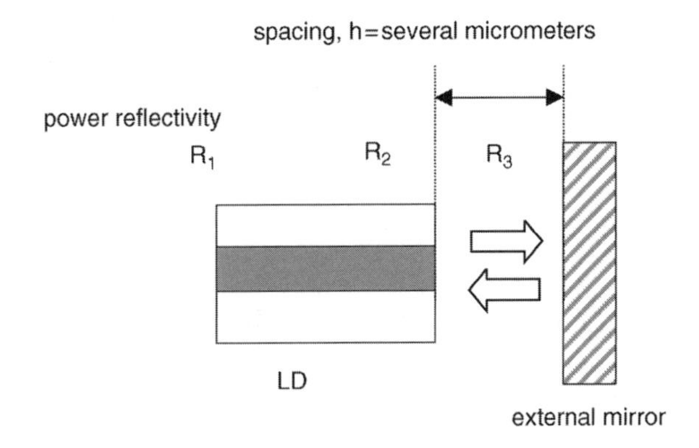

FIGURE 3.12 Schematics of laser with light feedback from external mirror.

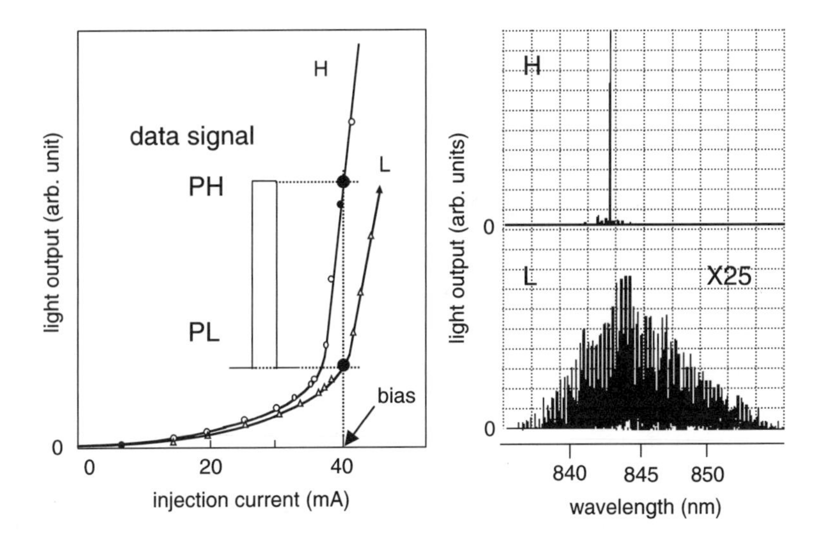

FIGURE 3.13 Optical switching performance. The laser is switched between lasing and nonlasing states under conditions of a constant bias current in accordance with reflectivity of an external mirror.

3.3.1.2 Optical Heads Based on Optically Switched Lasers

Great interest has been focused on high-performance optical disk memories for multimedia applications that handle large amounts of audio and video data. Conventional optical disk systems, however, suffer from slow access and low data transmission rates due to relatively large optical heads consisting of individual optical components. The concept of the optically switched lasers is well applied to a simple approach for eliminating these issues [Katagiri and Ukita, 1995]. This approach is based on a direct coupling scheme that offers extremely small optical heads simply equipped with a laser diode optically coupled to a recording medium acting as an external mirror without any lens [Ukita et al., 1994].

The heads detect data signals by monitoring the light output changed by the switching performance. Such data detection performance is, however, disturbed by fluctuations in the external-cavity length because of an interference undulation coming from a residual reflectivity of the antireflection-coated laser facet (see Figure 3.14). It will also be difficult to maintain such an extremely short external-cavity constant in the system with a disk rotating at a high speed of above 6000 rpm, if conventional servo focus-control methods are used. To eliminate this problem, the heads employ a method based on an air-bearing technique. A typical head configuration consists of a flying slider and a laser diode monolithically

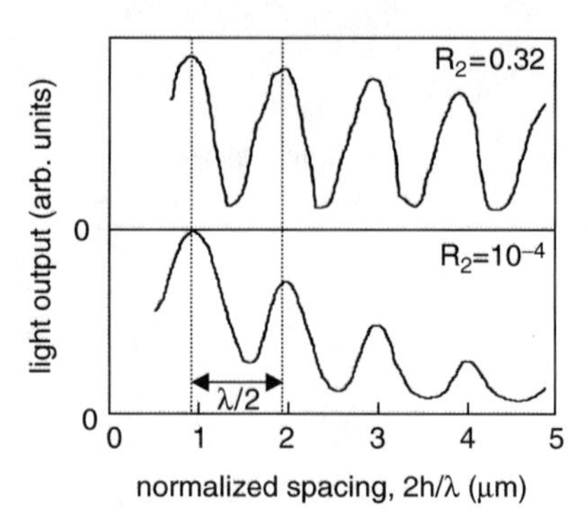

FIGURE 3.14 Interference undulation by light feedback from external mirror in proximity to laser facet. The oscillation wavelength is around 0.83 μm.

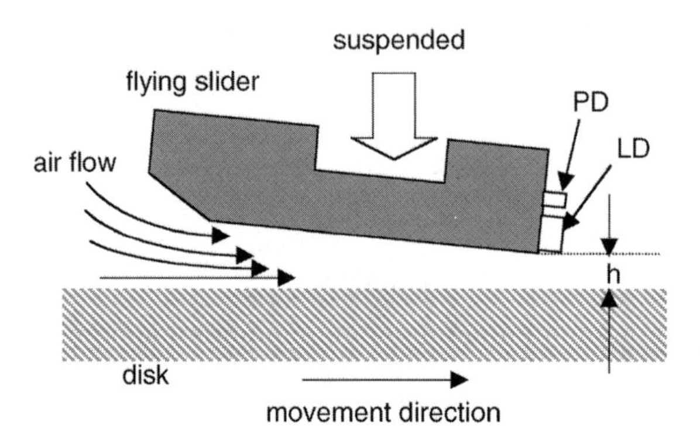

FIGURE 3.15 Configuration of an optically switched laser head mounted to a flying slider.

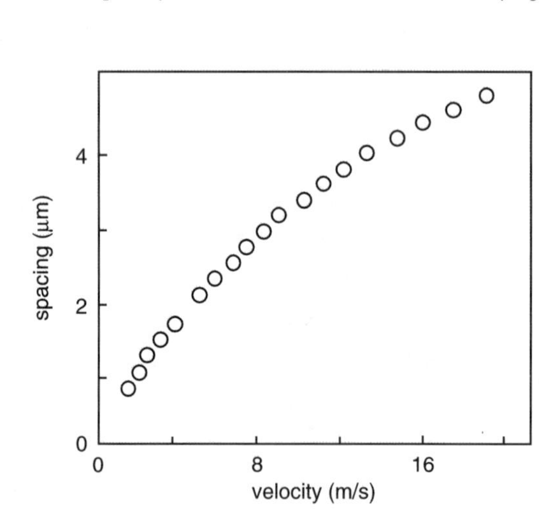

FIGURE 3.16 Spacing as a function of relative disk speed for a flying slider.

integrated with a photodiode (see Figure 3.15) [Ukita et al., 1991a,b]. The slider is designed to automatically maintain a constant spacing of a few micrometers based on air dynamics (see Figure 3.16). The optical heads equipped with such a mechanism follow a scheme—widely used in magnetic-disk systems—that requires a much smaller clearance of submicrometers (see Figure 3.17) [Hara et al., 1993].

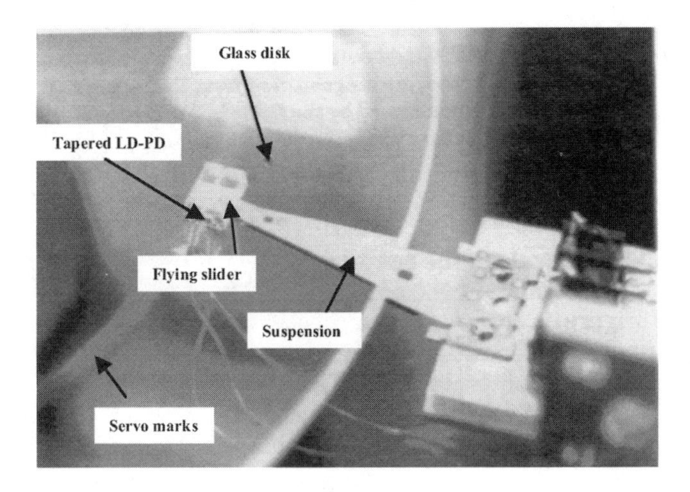

FIGURE 3.17 Setup of optically switched laser head.

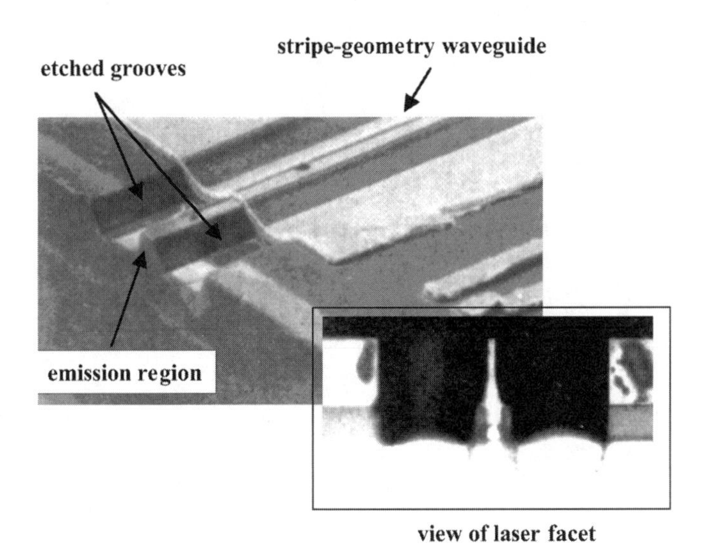

FIUGRE 3.18 Tapered laser diode fabricated by reactive ion-beam etching.

Achievement of a small beam spot is another way of detecting small bit marks on the disk. Conventional edge-emitting laser diodes provide a narrow beam profile equivalent to the diffraction limit; however, they have no way of preventing beam divergence without lenses. A tapered laser has been proposed to satisfy the requirements imposed on the beam size. This laser has a taper-ridge structure on the top of the laser facet that acts as an aperture (see Figure 3.18) [Uenishi et al., 1988]. The beam-spot size is determined in the near field by this aperture. Using such tapered lasers we can achieve wavelength-independent, high spatial resolution beyond the diffraction limit (see Figure 3.19).

How such flying heads are track controlled must also be clarified for practical implementation of disk systems. A sampled servo tracking technique is available based on the optically switched laser performance. Experiments using a glass disk with a clock and wobbling marks with reflectivity higher than that of the glass surface have successfully exhibited on- and off-tracking signals from which we readily derive tracking error signals and, hence, confirm the availability of the tracking method (see Figure 3.20).

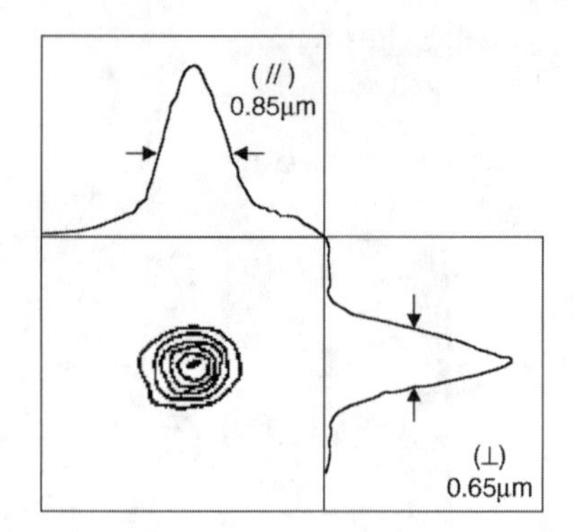

FIGURE 3.19 Near-field pattern of tapered laser diode. The wavelength is in the 1.3-μm range.

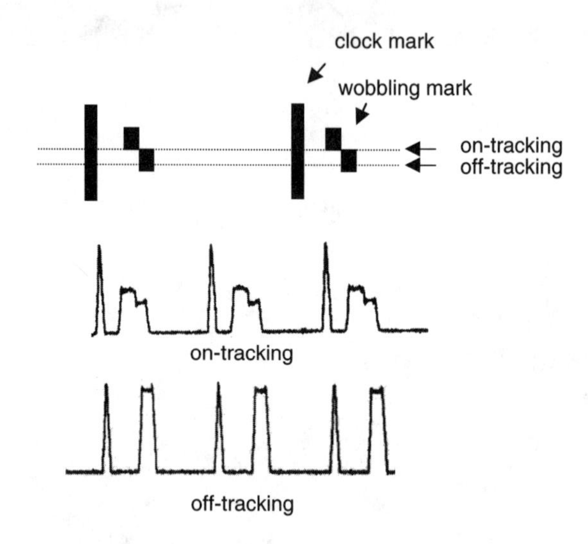

FIGURE 3.20 Scheme of tracking based on sampled servo control.

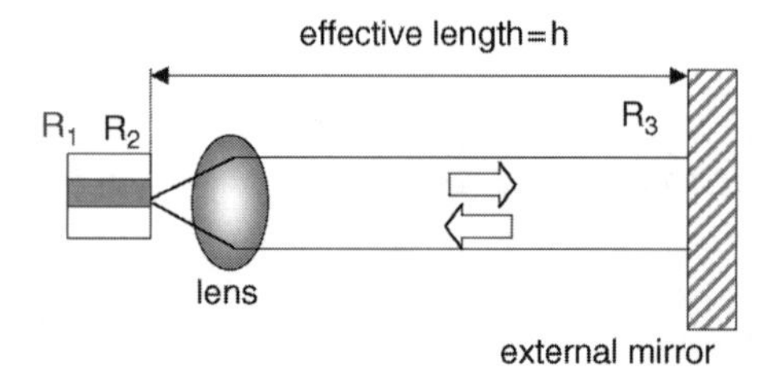

FIGURE 3.21 Schematics of laser with light feedback from external mirror.

3.3.2 Interference Undulations in Coupled-Cavity Lasers and Their Applications for Optical Measurements

3.3.2.1 Characterization of Coupled-Cavity Lasers

Coupled-cavity lasers with a simple configuration consisting of a Fabry–Perot laser diode and an external mirror (see Figure 3.21) have been the subject of intense investigation because of their attractive oscillation performance, which is readily controlled by external mirrors [Morikawa et al., 1971; Voumard et al., 1977; Lang and Kobayashi, 1980; Fleming and Mooradian, 1981; Acket et al., 1984]. Substitution of the external cavity with an effective mirror having a complex reflectivity is also available to characterize the performance, although the external cavity is relatively long compared to optically switched lasers. The reflectivity is represented considering multiple reflections in the external cavity as

$$r_{eff} = \sqrt{R_2} - \eta(1-R_2)\frac{\sqrt{R_2R_3}\exp(i\Omega\tau)}{1-\sqrt{R_2R_3}\exp(i\Omega\tau)}, \qquad \tau = \frac{2h}{c} \tag{3.10}$$

with the following values:

Ω: oscillation angular frequency
c: light speed in a vacuum
h: external-cavity length
R_2: reflectivity of laser facet facing external cavity
R_3: reflectivity of external mirror

The oscillation frequency is pulled into one of the eigenmodes of the laser diode owing to the strong optical gain; hence, it is represented as

$$\Omega = \omega_0 + m\frac{\pi c}{nL_0} \tag{3.11}$$

with the following values represented:

m: longitudinal-mode number
L_0: length of laser diode
n: refractive index of laser diode
ω_0: optical angular frequency of an eigenmode

We can readily derive the interference undulation with a period of half the wavelength from the representation of the effective reflectivity, assuming a single-mode oscillation with no mode-hopping. However, the undulation actually exhibits various increased spatial frequencies such as $\lambda/4, \lambda/6, \lambda/8$, etc., according to the external-cavity length (see Figure 3.22) [Katagiri and Hara, 1994]. Such behavior of the coupled-cavity lasers is explained by a mode selection rule according to which one of the eigenmodes with maximum reflectivity, corresponding to the minimum threshold, is selected as the oscillation mode.

The substitution of the external cavity with the effective reflectivity is insufficient for discussing asymmetric sawtooth undulations. We must take into account the time-dependent field component of

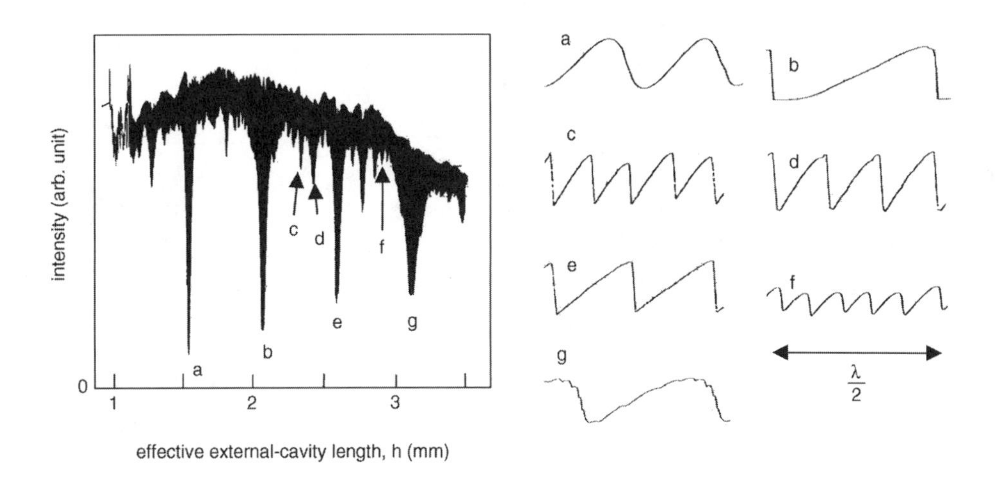

FIGURE 3.22 Interference undulations by light feedback from external mirror located several millimeters away from the laser facet.

the light in the external-cavity laser. Assuming the stationary condition, this consideration results in the oscillation frequency change being dependent on the external-cavity length as given by

$$\Omega = \omega_m + \eta(1 - R_2)\sqrt{\frac{R_3}{R_2}}\frac{c}{2nL_0}\sin(\tau\Omega) \tag{3.12}$$

where η is the coupling efficiency of the optical feedback from the external mirror to the laser diode and L_0 is the cavity length of the laser diode. Assuming that the equation gives a unique solution for Ω, the frequency change $\Omega - \omega_m$ corresponds to the phase change in the interference undulations at a period of half the wavelength. This phase change generates the sawtooth undulation curves [Spano et al., 1984; Olesen et al., 1986].

3.3.2.2 Displacement Detection Principle [Katagiri and Itaoh, 1998]

The measurement of a small displacement based on the coherence of a single-mode laser light has features of both high resolution and sensitivity; hence, it has been widely performed in various fields. However, conventional measurement schemes need stable, narrow-linewidth light sources consisting of relatively large and expensive optical components; thus, they are inadequate for general use. The coupled-cavity lasers in a simple configuration exhibiting the interference performance are generally expected to be used for this sort of displacement measurement. The problem is how to remove the oscillation instability at the same time as the displacement of the external cavity is induced for the measurement.

The paradoxical problem is eliminated by using a coupled-cavity laser stabilized by a mechanical negative-feedback loop circuit. In the loop the position of the laser diode is controlled along the axial direction by using a high-resolution actuator to cancel the transient displacement of the external mirror while the light output is monitored (see Figure 3.23). Consider the state of the laser on the sawtooth undulation curve with linear portions in every interval of half the wavelength. The initial state is defined at the halfway point on the curve, where the light output corresponds to P_0. A small temporal displacement Δh of the external mirror is detected by a photodiode as a differential signal $P - P_0$. This signal is filtered, amplified, and added to the control signal of the actuator to cancel the differential signal through an integrator. Consequently, a negative-feedback loop is formed to maintain a constant external-cavity length. The external-cavity laser controlled in this loop is thereby stabilized. We can know both the transient and the total displacement from the initial position by evaluating, respectively, the differential signal and the integral of the signals. The absolute displacement performance of the employed actuator is readily calibrated and maintained over a long period of time. This promises an accurate displacement measurement, independent of the oscillation performance of the employed laser diode and its driving condition.

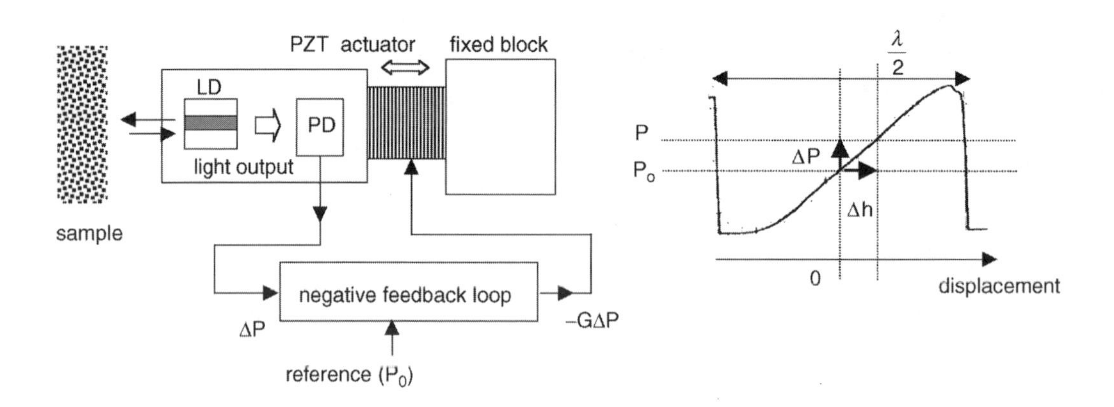

FIGURE 3.23 Schematic diagram for explaining a coupled-cavity laser displacement sensor (CCL sensor).

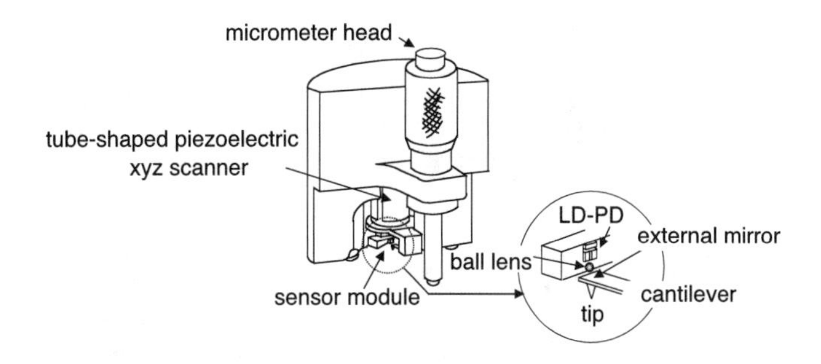

FIGURE 3.24 Schematic illustration of a CCL sensor. A monolithic LD-PD with a ball lens is employed.

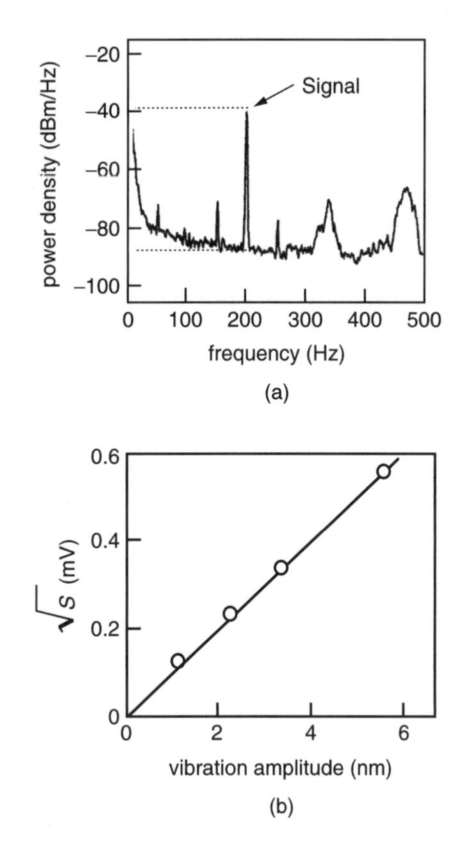

FIGURE 3.25 Spatial resolution measurement for a CCL sensor. The vibration amplitude is calibrated at 200 Hz.

3.3.2.3 Displacement Sensor Performance

A small displacement sensor fabricated for use in practical implementations consists of a monolithic laser diode-photodiode device, a TaF_3 ball lens 600 μm in diameter, and a piezoelectric-transducer actuator, all of which are assembled on a substrate (see Figure 3.24). The InP-based device emits a laser beam at 1.3 μm. The sensitivity of such a displacement sensor is readily estimated numerically by a slight vibration of the external mirror. The vibration generates a modulated signal whose amplitude is linearly related to the vibration amplitude. The corresponding frequency spectrum exhibits a sharp peak at the frequency of the vibration readily discriminated from broadband noises (see Figure 3.25). Maximum sensitivity for detecting displacement is estimated when the signal is equal to the noise floor. This minimal vibration is usually too small to detect and so is estimated at larger vibration regions based on the linear relationship. A typical maximum sensitivity is 0.02 nm/$\sqrt{Hz}$ at 200 Hz for an existing sensor.

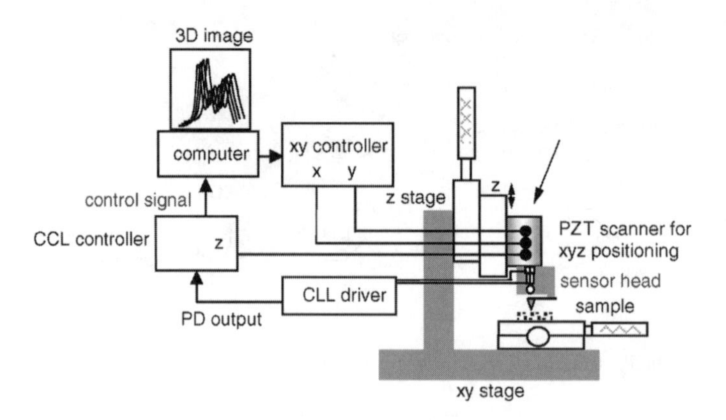

FIGURE 3.26 Configuration of scanning probe microscope with CCL displacement sensor.

An effective sensitivity level is also defined at the point where the signal equals the total noise power given by an appropriate integral range. A typical value is given as 0.8 nm in the range 0 to 500 Hz, which covers most micro-opto-mechatronic applications. This level of sensitivity substantially shows the spatial resolution of the displacement sensors.

3.3.2.4 Examples of Practical Implementation

3.3.2.4.1 Application to Scanning Probe Microscope [Katagiri and Hara, 1998]

Scanning probe microscopes (SPMs), including atomic-force microscopes using a cantilever with a sharp tip, are a powerful tool for imaging surface topography. Because the surfaces of interest have nanometer-scale structures, measurement of the resulting extremely small distortions of the cantilever is required. The displacement sensor, which uses the coupled-cavity lasers, has shown potential for taking such measurements. Figure 3.26 shows a schematic diagram of an SPM system equipped with a detecting head with a displacement sensor. As the head approaches a sample on a mechanical stage, the tip makes contacts with the sample surface. Further displacement of the head in the same direction generates a distortion of the cantilever. Because the cantilever works together with an external mirror of the displacement sensor, the distortion is translated to the photodiode output variance. The absolute distortion of the cantilever, which is derived from the detected signal, corresponds to the load of the tip to the surface. The numerical evaluation of the load is based on a calculation using Hooke's law with values of stiffness of the cantilever. Hence, an optimum load is adjustable according to the hardness of the sample. Once an optimum load is determined, the controller of the displacement sensor works to cancel a temporal distortion of the cantilever, while the external-cavity length remains constant by working a tube actuator along the z-axis. While the head is raster-scanned in the x–y plane by the tube actuator, we obtain images of the sample surface using the control signal in the z-axis direction.

A typical example is shown in Figure 3.27 for an optical-disk surface. The measured image faithfully reflects the surface with shallow tracking grooves 0.1 μm in depth and 1.6 μm in spacing.

3.3.2.4.2 Application to Dynamic Detection of Small Forces [Katagiri and Itaoh, 1998]

Dynamic measurement of small forces is of great importance for mechanical systems. In the field of high-precision information instruments, including hard-disk and optical-disk systems that must be miniaturized, the forces of interest become smaller as the system's dimensions are reduced. For optimum design of the systems such forces must be temporally measured under operating conditions. Such small forces can be measured from the distortion of a cantilever as described above. Stiff cantilevers with a high resonance frequency are needed for measuring such forces dynamically over a wide frequency range. The distortion is obviously small, so a highly sensitive distortion measurement is essential. The displacement sensor in a coupled-cavity laser configuration is the most suitable because it exerts negligible mechanical influence on the cantilever and enables highly sensitive detection.

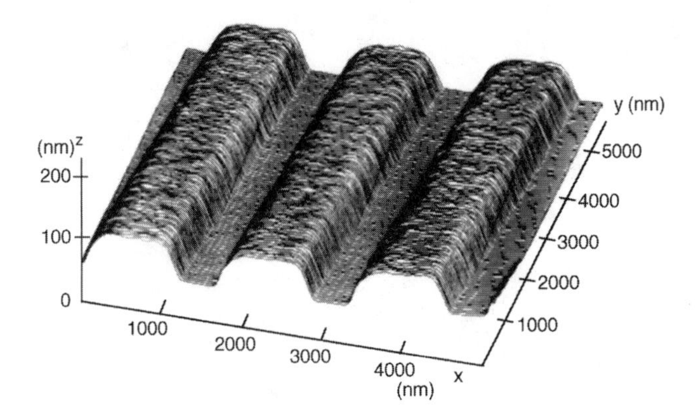

FIGURE 3.27 Application of a CCL sensor to scanning probe microscope imaging. The image corresponds to a surface of optical disk with 1.2-μm-pitch grooves.

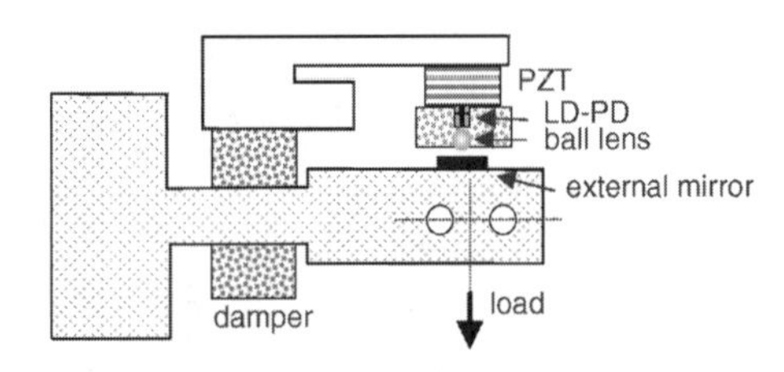

FIGURE 3.28 Schematic diagram of small-force detection sensor based on coupled-cavity laser performance.

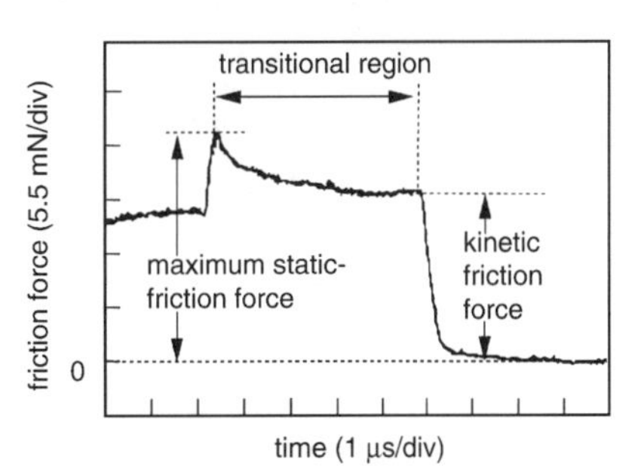

FIGURE 3.29 Application of a CCL sensor to detection of small forces.

A typical example is presented using a simple system of measuring the dynamic friction force to verify the force-measurement scheme. The system has a cantilever with a stiffness of 500 mN μm^{-1} (see Figure 3.28). This cantilever provides a resonant frequency of around 3 kHz and a minimum detectable force of 0.4 mN. The force-detection sensor is readily calibrated under the constant-load conditions. Figure 3.29 shows a temporal trace of the friction force that occurs when a small object is dragged over a rough surface. The trace conclusively reveals the transition process through the maximum friction-force state to the kinetic-force state. The measured friction values agree with those measured by conventional techniques with an inclined plane.

3.3.3 Mode-Locked Lasers and Their Tuning Schemes by Micro-Mechanism

3.3.3.1 Principle of Mode-Locking in Semiconductor Lasers

(See Derickson et al., 1986; Takada et al., 1994; Tucker et al., 1985; Morimoto et al., 1988; Bowers et al., 1989.) Mode-locking is the coupling of the eigenmodes of semiconductor lasers with the same mode spacing and the same phase. Such a mode-locked state is represented using a field component for the *m*th eigenmode:

$$E_m = \rho_m \cos[(\omega_0 + 2\pi m f)t + \phi]. \tag{3.13}$$

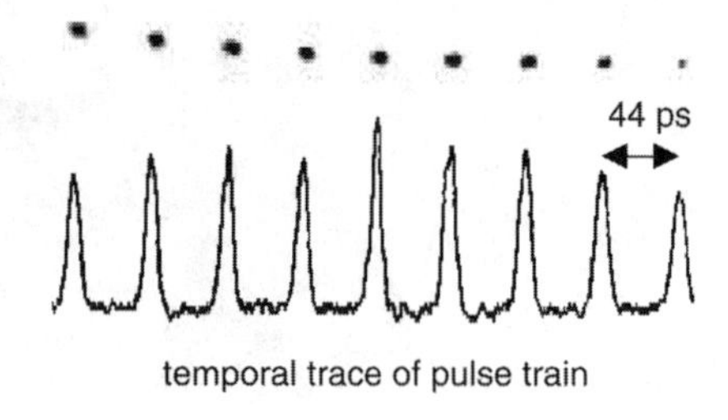

FIGURE 3.30 Example of direct temporal measurement of mode-locked pulses at around 22 GHz by using a synchro-scanned-streak camera.

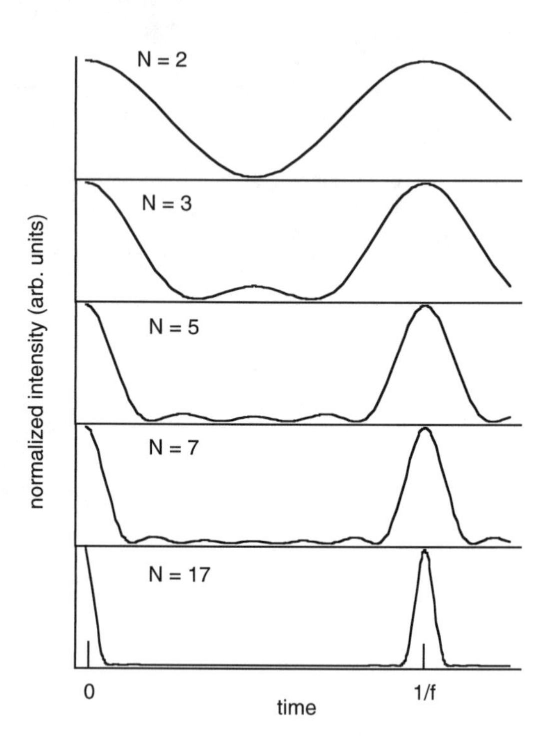

FIGURE 3.31 Various temporal waveforms of mode-locked pulses as a function of the number (N) of contributing modes.

Here, ρ_m represents the modal distribution with a central angular frequency ω_0 in the k space. The frequency f is close to the mode-spacing as

$$f \cong \frac{c}{2nL}. \tag{3.14}$$

The coupled modes perform as a single mode, which is called the *supermode.*

An interesting effect is that the light power pulsates in this mode-locked state (see Figure 3.30). The repetition rate of the pulses corresponds to the frequency f. As the number to be coupled increases, the pulse becomes increasingly sharper (see Figures 3.31). The temporal pulse profile is related to the envelope produced by connecting the points $\{\omega_0 + 2\pi f, \rho_m\}$ in the k space through a Fourier transform (see Figure 3.32). However, the minimum width of the profile derived from the Fourier transform of the optical spectrum is limited (transform limit; TL). This means that a product of the spectral width and pulse width has a minimal value. The product is, for example, around 0.43 for conventional mode-locked pulses with Lorentzian profiles. Here, the temporal traces can be measured by using an autocorrelator based on a second-harmonic generation (see Figure 3.33).

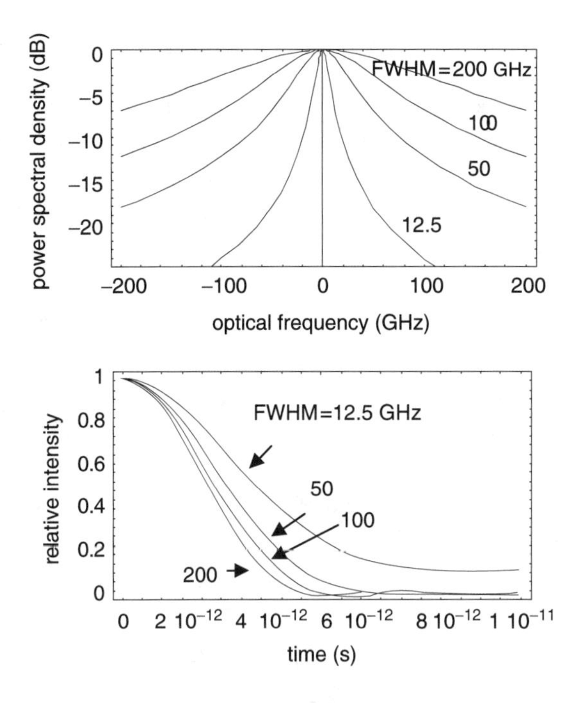

FIGURE 3.32 Temporal pulse profile and corresponding spectrum.The envelope of the spectral mode distribution is related to the temporal profile by Fourier transform.

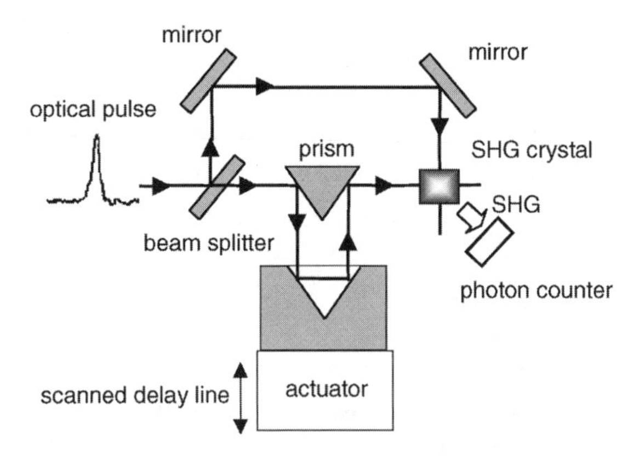

FIGURE 3.33 Autocorrelator for time-domain pulse measurement.

There are three major methods of achieving mode-locking in semiconductor lasers: active mode-locking, passive mode-locking, and hybrid mode-locking. Multisegment laser diodes, including a DBR segment for limiting undesirable spectral broadening, are generally used for the three mode-locking schemes mentioned above (see Figure 3.34).

Active mode-locking is achieved by modulating an injection current to the end segment of the multisegment lasers. The other segments are DC-biased to obtain an optical gain for laser oscillation and adjust the natural round-trip frequency with the modulation frequency. Passive mode-locking is achieved by reverse-biasing the end segment without any electrical signal. The end segment acts as a saturable absorber (SA); hence, ultra-short pulses in the femto-second range can be expected. The repetition rate is dependent only on the effective cavity length of the laser; hence, ultra-fast optical pulses with a repetition rate of above hundreds of gigahertz can be obtained. Hybrid mode-locking is achieved by

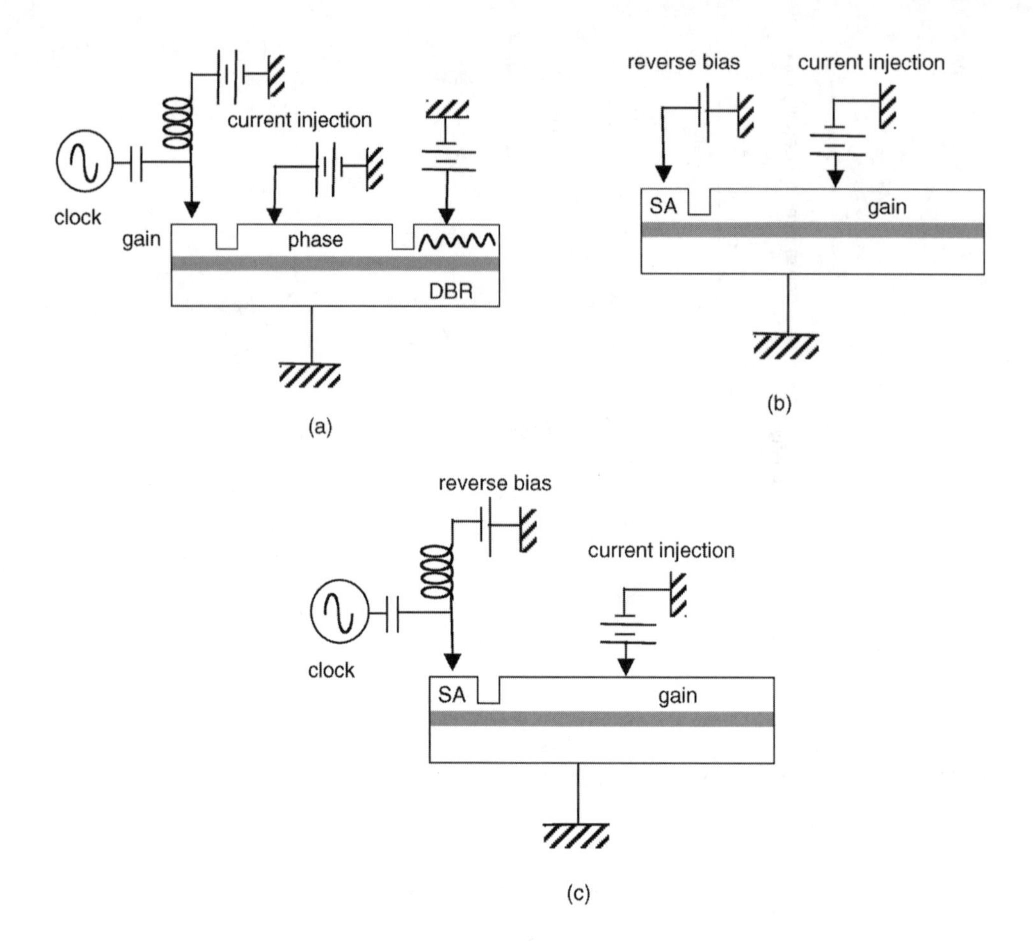

FIGURE 3.34 Mode-locking schemes: (a) active mode-locking; (b) passive mode-locking; (c) hybrid mode-locking.

adding an RF signal to the reverse voltage for the end segment as an SA. We can expect stable ultra-short optical pulses synchronized to an electrical signal.

In order to achieve mode locking with higher repetition rates, shortening the cavity length of the lasers is essential. However, this shortening has a drawback — the optical gain decreases with a decreasing cavity length. Harmonic mode locking is a state in which multiple pulses circulate in the laser cavity, so that the repetition rate is multiplied by the number of circulating pulses. Hence, high-repetition-rate pulse generation is expected even for lasers with a longer cavity length when the harmonic mode-locking scheme is used.

Such harmonic mode locking includes colliding-pulse mode-locking (CPM)[Chen and Wu, 1992]. The CPM laser has an SA segment in the center of the cavity and allows two pulses to circulate. Because these pulses collide at the SA segment, the effect of the SA is emphasized (see Figure 3.35). The concept of the CPM can also be extended to harmonic CPM operation based on simultaneous pulse collisions in all SA segments [Katagiri and Takada, 1997]. Figure 3.36 shows a typical temporal SH trace and the corresponding optical spectrum of multiple CPM lasers with a repetition rate of about 192 GHz.

3.3.3.2 Synchronization of Passively Mode-Locked Pulses by Phase-Locked Loop

Passively mode-locked lasers can be synchronized to an electrical signal by using a phase-locked loop (PLL) technique [Helkey et al., 1992; Buckman et al., 1993]. Although this technique is essentially available over a wide repetition-rate range, it suffers from the electrical limitation imposed on the available range

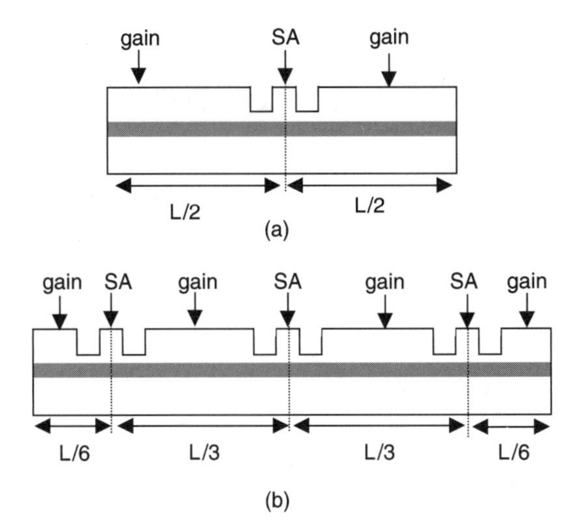

FIGURE 3.35 Schematic of mode-locked semiconductor lasers with multisegment configurations. Saturable absorbers (SAs) for mode-locking are formed by reverse-biasing corresponding segments while other segments are biased for current injection. These multisegment mode-locked lasers are categorized by three major types that provide (a) colliding-pulse mode-locking, and (b) multiple-colliding-pulse mode-locking.

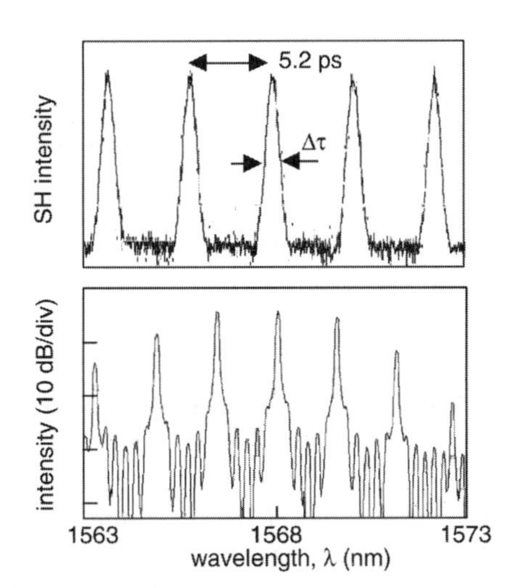

FIGURE 3.36 Example of SH trace and corresponding optical spectrum of mode-locked pulses. The measured pulses were generated by a multiple-CPM laser with a 192-GHz repetition rate.

by the maximal response frequency of photodiodes to monitor the pulse timing. However, a photonic down-conversion technique based on the generation of modulation sidebands eliminates this limitation. Consequently, the available range of the PLL is extended to sub-THz regions [Katagiri and Takada, 1996; Hashimoto et al., 1998].

Consider mode-locked pulses with repetition rate F. These optical pulses exhibit a comb-shaped mode distribution with a spacing equal to the rate in the optical-frequency domain. Intensity modulation produces sidebands around the original light as

$$|E(t)|^2 = |E_0|^2 \left| \exp(-i\omega t) + \sum_n \delta_n \exp[-i(\omega - 2\pi n f_m)t] \right|^2 \tag{3.15}$$

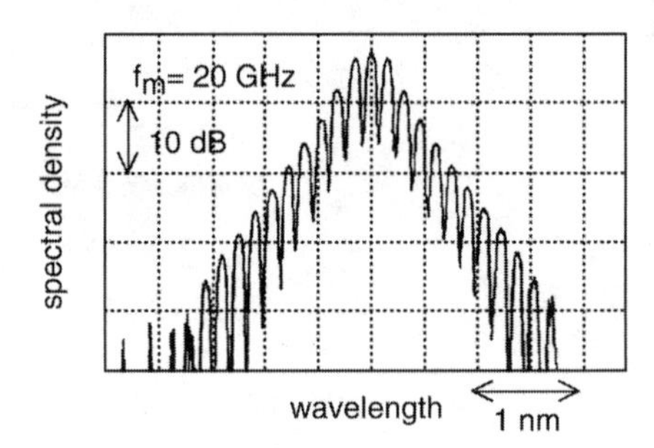

FIGURE 3.37 Optical spectrum of single-mode laser modulated at 20 GHz exhibiting corresponding sidebands.

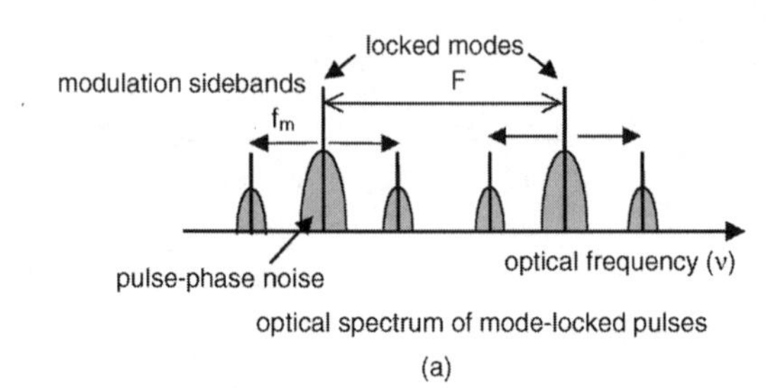

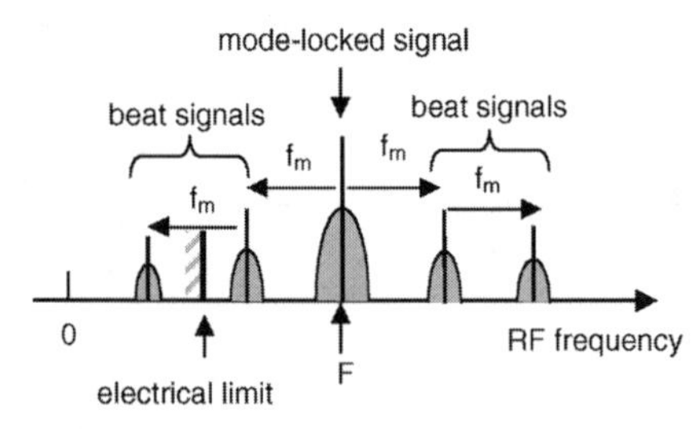

FIGURE 3.38 Mechanism of phase detection for ultra-fast optical pulses using modulation sidebands: (a) optical spectrum of intensity modulation of optical pulses; (b) RF spectrum of modulated optical pulses measured using a photodiode.

where $E(t)$ is the temporal electric field component with a constant amplitude and E_0 and $\{\delta_n\}$ are the parameters related to the distortion of the modulated curve.

Figure 3.37 shows a spectrum of typical sidebands generated by modulating a single-mode laser light at 20 GHz. Hence, when these pulses are directly modulated by an intensity modulator, modulation sidebands are generated around these locked modes at every modulation frequency f_m. Beat signals between the locked modes and their sidebands are thus generated at every modulation frequency. Although the repetition frequency F is above the maximal response frequency of the photodiode, the beat signals at lower frequencies can be readily detected (see Figure 3.38). These beat signals maintain the phase information of the original pulse train and are estimated as replicas of the mode-locked signals. A phase error signal is obtained by mixing one of the beat signals with an electrical signal from a reference oscillator. This error signal is filtered, amplified, and passed to the SA so as to minimize the error. Under the appropriate loop gain, synchronization is achieved (see Figure 3.39).

A mode-locked laser with a higher repetition rate is used to show that the PLL using the photonic down-conversion is feasible. Figure 3.40 shows the mode-locked signal and beat signals. The signal M at around 45 GHz corresponds to the mode-locked signal, and both the N_1 and N_2 signals are noises of the laser. The signal R at around 19 GHz down-converted from the mode-locked signal by intensity

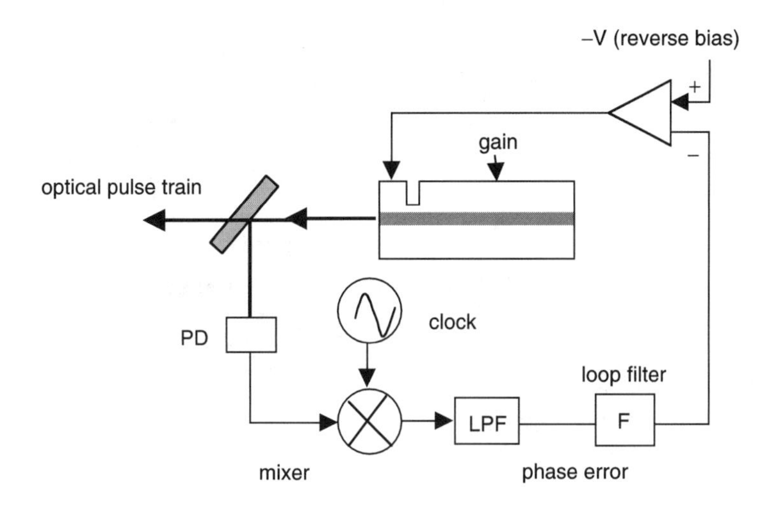

FIGURE 3.39 A synchronization scheme for mode-locked lasers based on optical phase-locked loop circuits. The scheme is available basically independent of repetion rates of optical pulses above the electrical measurement limit of around 50 GHz.

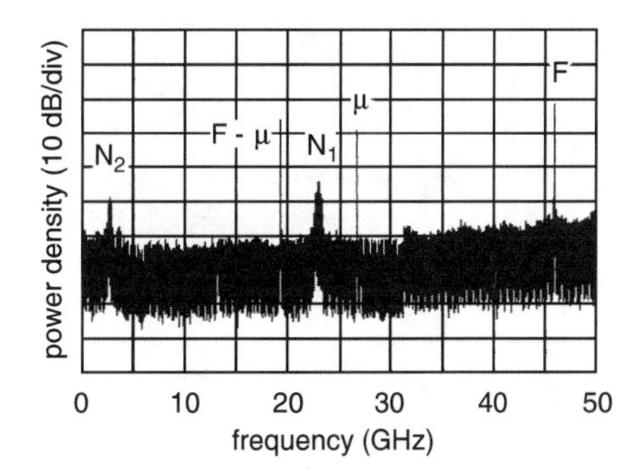

FIGURE 3.40 Example of modulation sidebands of passively mode-locked lasers at a repetiton rate of around F = 45 GHz. Modulation frequency μ is 26.5 GHz, and its sideband appears at $F - \mu$ = 19.5 GHz.

modulation at f_m = 26.5 GHz is used for the PLL synchronization. The power spectrum of the pulses from the mode-locked laser in the loop shows a mode-locked signal with a dramatically reduced linewidth as narrow as the electrical signal (see Figure 3.41). This confirms the synchronization.

3.3.3.3 Micromechanically Tunable Mode-Locked Lasers [Katagiri et al., 1996]

Small, low-driving power optical-pulse sources are desired in various practical fields including optical communication systems and measurement. The requirements of practical pulse sources include sufficient tunability at a repetition rate suitable for the application systems. However, while conventional monolithic mode-locked lasers with a fixed cavity length offer excellent stability, they suffer from instability in mode-locked oscillation as the discrepancy between the natural repetition rate and the electrical signal frequency increases. This problem will remain unresolved until the natural repetition rate is changed to reduce the discrepancy. External-cavity lasers with a mechanically changeable cavity length are suitable for repetition-rate-tunable mode-locked operations. The problem is that the current configuration, which uses separate components such as an actuator for displacing the external mirror and lenses for generating a collimated beam in the external cavity, has poor mechanical stability, and its large size limits the maximum possible repetition rate. Hence, a simple external-cavity laser configuration consisting of a laser diode and a moving external mirror located very close to the laser facet has been proposed and demonstrated for repetition-rate tunable mode-locked lasers.

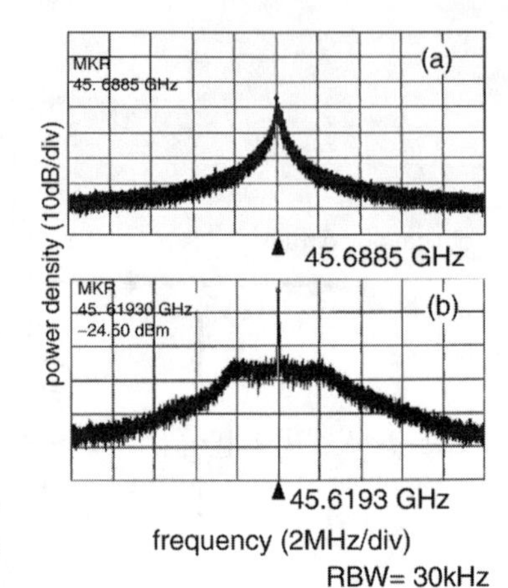

FIGURE 3.41 Example of synchronization of passively mode-locked-lasers at a repetiton rate of around 45 GHz: (a) free-running state; (b) synchronized state.

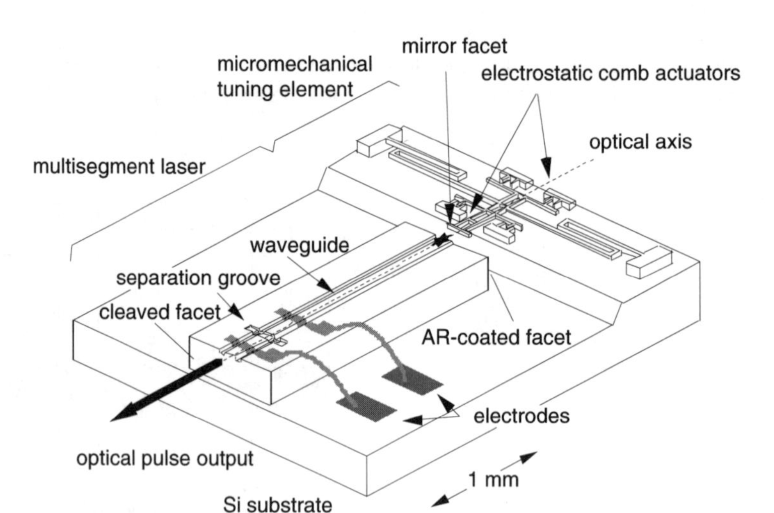

FIGURE 3.42 Micromechanically tunable mode-locked laser diode for repetition rate control. The laser has an external-cavity laser configuration with an antireflection-coated facet on the mirror side. The mirror is electrostatically displaced by comb actuators using beam tortion.

The mode-locked pulses circulate in the external-cavity laser passing through the laser facet on the mirror side whose reflectivity is reduced by an antireflection coating. Thus, the repetition rate is given assuming that the external-cavity length h is much smaller than L

$$F \cong F_0\left(1 - \frac{h}{nL}\right) \tag{3.16}$$

where F_0 is the repetition rate when $h = 0$. This equation means that the repetition rate is linearly changed by ΔF according to the the linear displacement of the mirror Δh as

$$\Delta F = -F_0 \frac{\Delta h}{nL} \tag{3.17}$$

A micromechanical tuning mechanism is used together with a multisegment laser diode for realizing such a tunable external-cavity laser configuration (see Figure 3.42). An essential element of the tuning

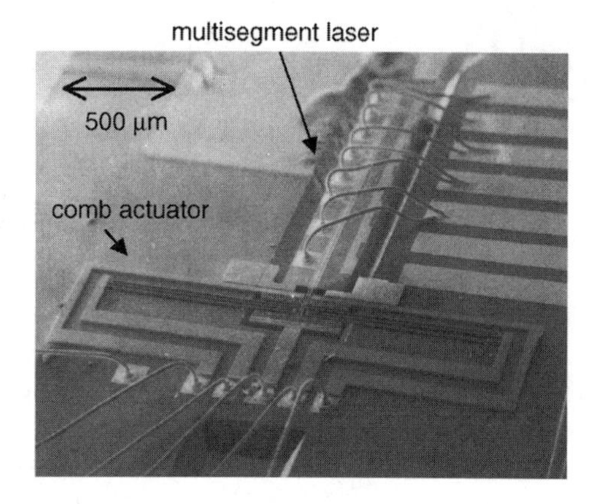

FIGURE 3.43 SEM image of micromechanically tunable mode-locked laser diode. The moving monolithic micro-mirror fabricated on an Si substrate is combined with a multisegment laser diode.

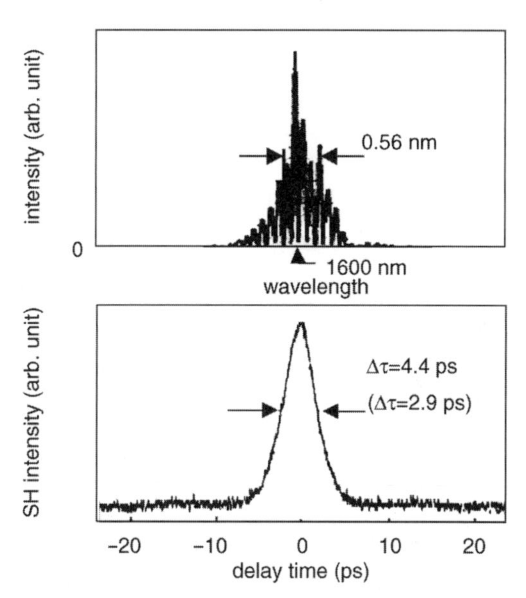

FIGURE 3.44 Optical spectrum and corresponding SH trace of pulses at a rate of 22 GHz from a micromechanically tunable mode-locked laser.

mechanism is a monolithic micro-moving mirror controlled by electrostatic comb drives. The moving mirror uses a side wall of a triple-fold-beam thin-film spring suspended over a substrate and separated by a narrow air gap. The differential (push–pull) manner of mirror displacement is achieved using two comb drives. These comb drives consist of interdigitated fingers and stationary electrodes attached to the substrate. A typical example of such a mechanism is fabricated with Ni films 20 μm thick on a silicon substrate [Uenishi et al., 1996]. The static displacement of the mirror is determined from the electrostatic force of the comb drive and the stiffness of the fold beam according to Hooke's law; thus, it is a linear function of the squared driving voltage. Such a tuning mechanism can be combined with a multisegment laser diode to construct a mode-locked laser. Figure 3.43 shows a typical example of such micromechanically tunable mode-locked lasers. The constructed laser has a 1950 μm-long effective cavity with an entire tuning span of around 10 μm; thus, it produces mode-locked pulses with a tunable repetition rate of around 22 GHz (see Figure 3.44). Figure 3.45 shows a repetition rate vs. squared-comb voltage corresponding to mirror displacement. Effects of the repetition-rate tuning performance on the mode-locked operation are typically shown for stabilizing the passively mode-locked pulses. Although the pulses are synchronized to an electrical signal optimized for the natural repetition frequency of the laser by

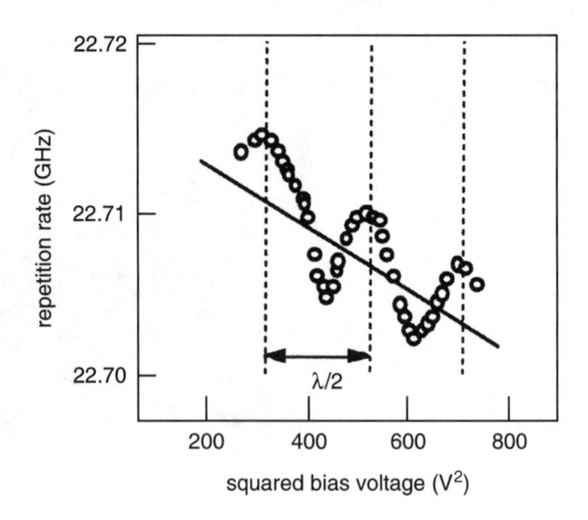

FIGURE 3.45 Repetition-rate controlled by comb actuator.

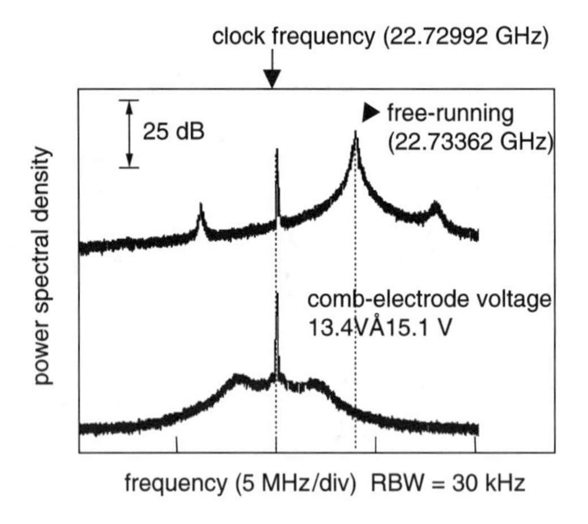

FIGURE 3.46 Effect of micro-mechanical cavity-length adjustment on synchronization of mode-locked lasers.

using a phase-locked loop, the synchronization is readily broken even by a slight change in signal frequency. However, synchronization is reestablished at the new electrical signal frequency by adjusting the mirror position so as to reduce the frequency discrepancy (see Figure 3.46) [Katagiri et al., 1998a].

3.3.4 Wavelength-Tunable Ring Lasers with Semiconductor Optical Amplifiers

3.3.4.1 Characterization of Ring Lasers

Just as light makes round trips in Fabry–Perot resonators, optical resonance is similarly achieved by circulating light in ring cavities (see Figure 3.47). Ring lasers use such cavities with optical gain media. The oscillation condition is derived from an amplitude increase factor per round:

$$\sqrt{\Gamma} E \exp\left(\frac{g_{th} - \alpha}{2} L\right) \exp(ikL) = 1 , \tag{3.18}$$

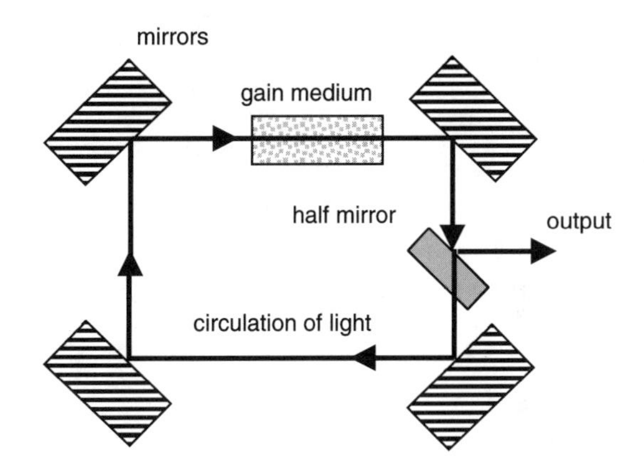

FIGURE 3.47 Fundamental configuration of ring-cavity lasers.

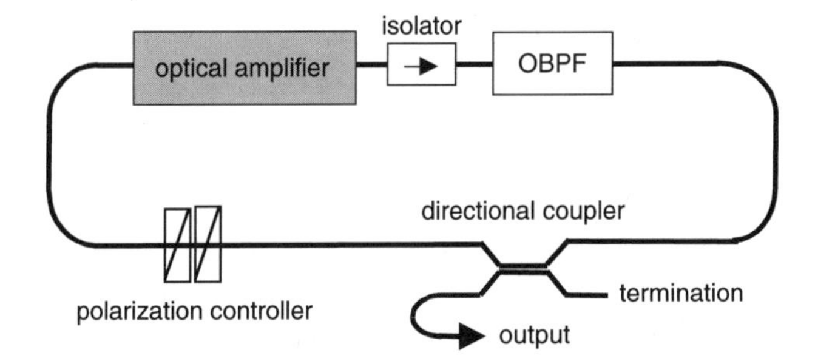

FIGURE 3.48 Schematic of wavelength-tunable fiber-ring laser with semiconductor optical amplifier. The wavelength is tuned by an optical bandpass filter (OBPF).

where g is the total loss of the ring except for that in the gain medium. This equation gives individual oscillation conditions imposed on gain and wavelength as

$$\left.\begin{aligned} g_{th} &= \gamma + \frac{1}{L}\log\frac{1}{\Gamma} \\ \frac{1}{\lambda_m} &= \frac{m}{nL} \end{aligned}\right\} \tag{3.19}$$

The value m is an arbitrary integer, but it must be determined so that the corresponding wavelength λ_m is in the gain band. The most remarkable difference between Fabry–Perot lasers and ring lasers is that the ring lasers can operate in traveling-wave modes producing spatially uniform electromagnetic-field intensity, while the Fabry–Perot lasers operate in standing-wave modes producing intensity inequality at every half wavelength. However, the possibility of laser oscillation in a standing-wave mode still remains for the ring lasers when bidirectional circulation is allowed. Hence, unidirectional circulation of light using optical isolators is essential for ensuring the traveling-mode operation of ring lasers.

There are many ways to construct the ring lasers, but it is convenient for the construction based on fiber optics because of simple assembly of a suit of components with fiber interfaces. These components typically include optical amplifiers as a gain medium, isolators for unidirectional lightwave circulation, directional couplers for obtaining laser output, polarization controllers, and an optical bandpass filter for selecting a unique oscillation mode (see Figure 3.48).

Both erbium-doped fiber amplifiers (EDFAs) and semiconductor optical amplifiers (SOAs) can be used as the gain medium for ring lasers (see Figure 3.49); however, they have their own merits and

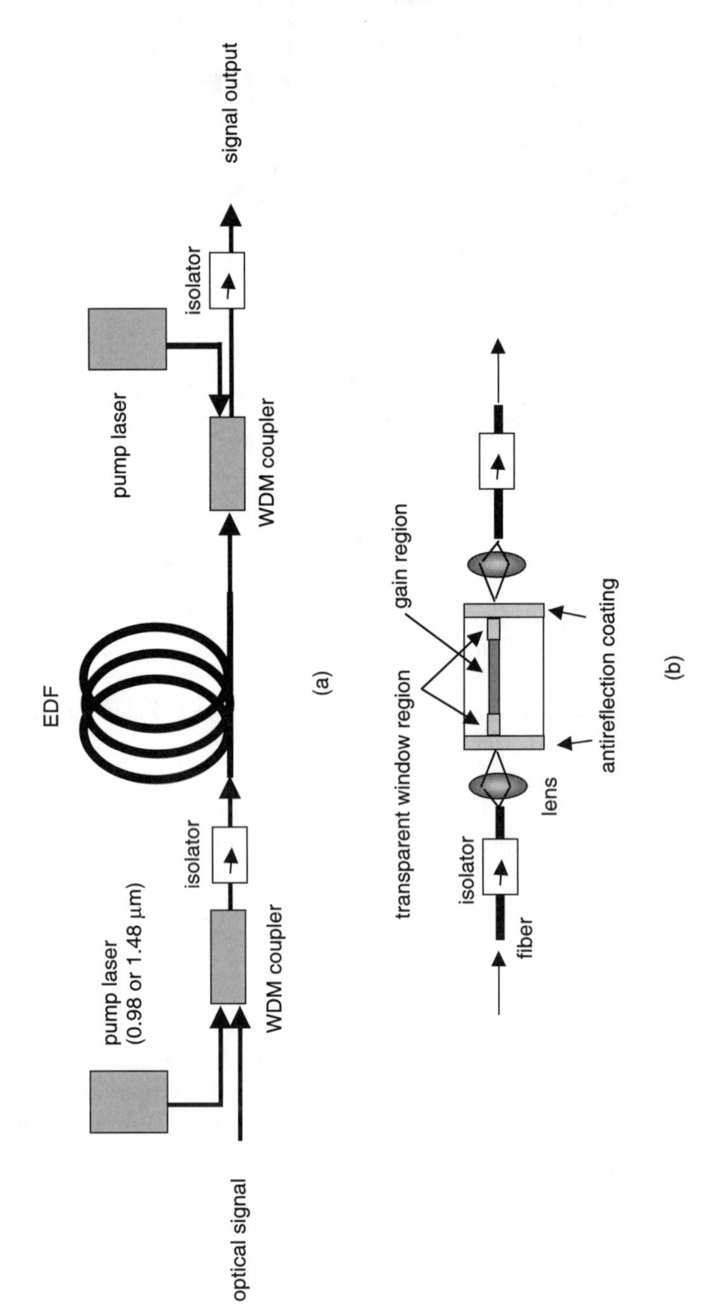

FIGURE 3.49 Schematic diagrams of optical amplifiers: (a) erbium-doped fiber amplifier; (b) semiconductor optical amplifier.

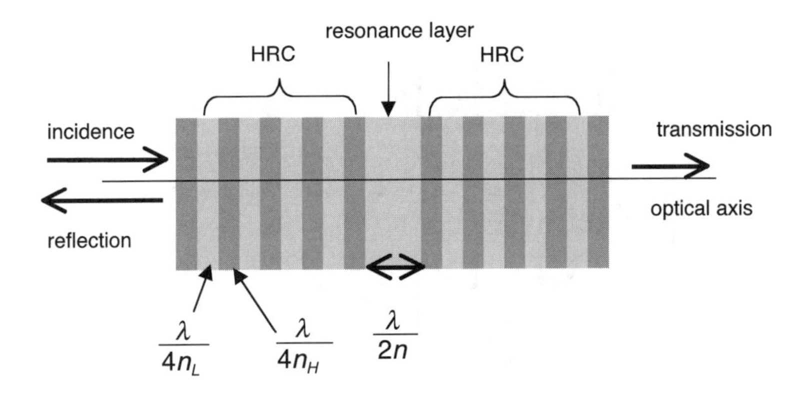

FIGURE 3.50 Structure of optical bandpass filter using multiple dielectric layers.

demerits. The EDFAs provide high-output performance but suffer from complicated structures consisting of many components. They also need a long path for obtaining sufficient optical gain and thus unnecessarily shorten the mode spacing of the ring lasers. On the other hand, the SOAs are as small as laser diodes, while they have polarization-dependent and saturable performances.

3.3.4.2 Wavelength-Tuning Mechanism [Katagiri et al., 1998b]

Wavelength-tunable lasers are achieved by simply inserting wavelength-tunable optical bandpass filters in the ring lasers. For use in lasers they are expected to have low-loss, high-resolution tenability in wide ranges and polarization independence. Although there are many candidates for such filters, few of them can satisfy all of the above requirements. An exception is the dielectric filter, to be introduced here.

The dielectric interference filter generally consists of a resonance layer half the wavelength in thickness sandwiched by multiple pairs of quarter-wavelength-thick layers with different indices of refraction (Figure 3.50), and transmit only the light that matches with the resonance condition that gives the transmission spectrum with a Lorentzian profile as

$$T(\lambda) = \frac{(\Delta\lambda)^2}{(\Delta\lambda)^2 + 4(\lambda - \lambda_c)^2} \tag{3.20}$$

Here, λ_c is the transmission center given by

$$\lambda_c = \frac{2nh}{m}, \qquad m = 1,2,3,\ldots \tag{3.21}$$

where h is the thickness of the resonance layer, n is its index of refraction, and $\Delta\lambda$ is the transmission bandwidth at 3 dB.

Directly changing the thickness of the resonance cavity may be reasonable for achieving wide tunability; however, conventional methods of using thermal or mechanical expansion of dielectrics are completely inadequate for the above purpose. One of the most effective ways uses a wedged structure for the resonance layer (see Figure 3.51). The thickness of the resonance layer is effectively changed according to the beam position. Such a tuning mechanism is realized using a circularly wedged, disk-shaped optical bandpass filter and a rotary positioning system (see Figure 3.52) [Thelen, 1965; Apfel, 1965]. Digital marks are drawn on the fringe of the filter disk and are read by a sensor to produce encoded signals including conventional Z, A, and B signals (see Figure 3.53). These signals are detected and immediately analyzed by a programmable logic gate circuit to determine the absolute position and relative displacement of the disk. This information is processed by a CPU to control the ultrasonic motor as a rotary actuator.

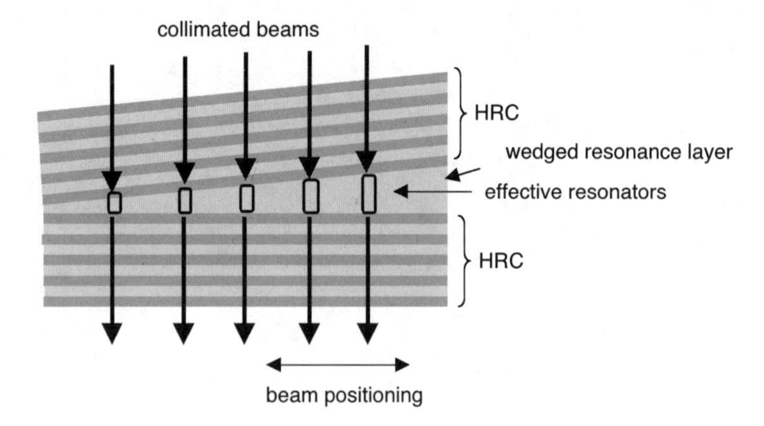

FIGURE 3.51 Control of length of effective resonators formed in a collimated beam by light-beam positioning based on wedged-layer structures.

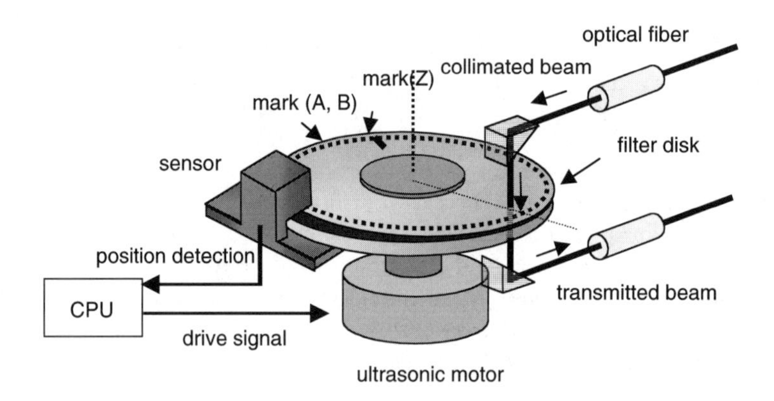

FIGURE 3.52 Disk-shaped wavelength-tunable optical bandpass filter. The filter disk has a circularly wedged cavity region and the center wavelength is determined by positioning the rotation angle of the disk. The position control uses an ultrasonic motor under the control with a high-resolution rotary encoder system.

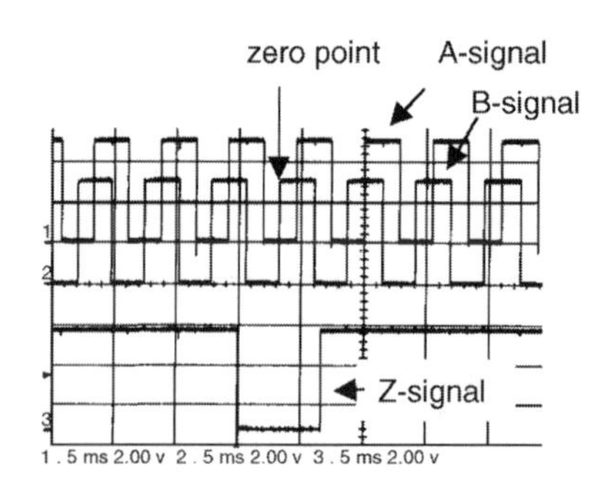

FIGURE 3.53 Encoded signals for determining the absolute position on a disk.

Both precise scanning and positioning are possible for the disk according to the preinstalled programs. The optical system maintains vertical incidence throughout such tuning operations. This enables precise wavelength tuning performance over a wide range while maintaining constant total and low polarization-dependent losses [Hashimoto and Katagiri, 2001]. The precise wavelength tuning is based on the wavelength calibration using a relationship between the center wavelength and the digitized position (see Figure 3.54). Figure 3.55 shows an example of such a tuning mechanism for the disk-shaped filter.

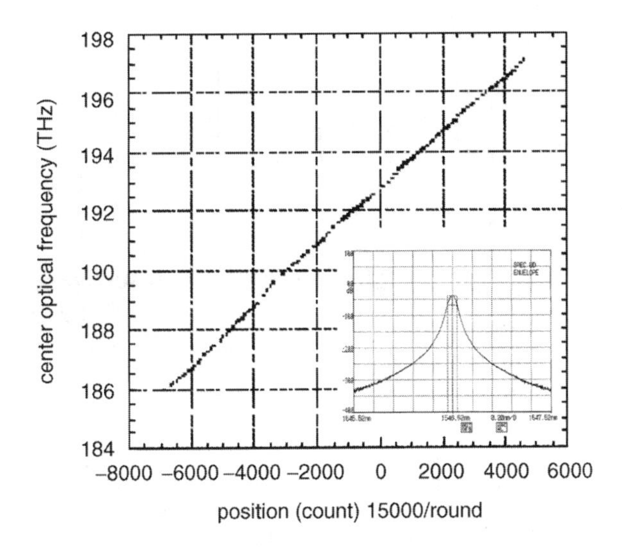

FIGURE 3.54 Tunable performance of a disk-shaped optical bandpass filter.

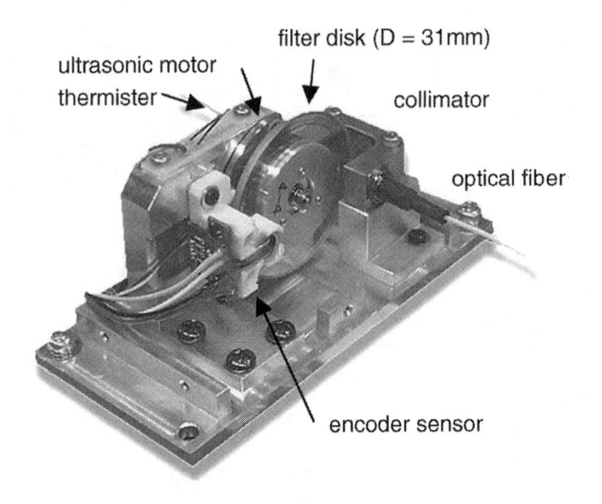

FIGURE 3.55 Mechanism of a disk-shaped wavelength-tunable filter.

A laser with a wide wavelength-tuning range is constructed using a disk-shaped, wavelength-tunable bandpass filter. The intensity of the laser light generated is stabilized by an SOA under gain-saturated conditions. Applying this stabilization method to the ring laser with an optimal filter disk and SOA as a gain medium, wide tenability in the 1530- to 1590-nm range is achieved in the communication bands under sufficient intensity-stabilized operating conditions (see Figure 3.56). A wavelength-scanning laser, which is of great use for optical measurement, can also be realized by rotating the disk synchronously with an electrical signal. When the scanning speed is optimized so that laser oscillation instability is suppressed, this kind of synchronous laser-wavelength scanner is achieved, and it maintains a line width of below 1 GHz. Figure 3.57 shows a typical example of how to measure the spectral response of an acetylene gas cell that has been used for absolute-wavelength calibration.

3.4 Conclusions

This chapter describes the principles of semiconductor lasers and their applications. Because semiconductor lasers are extremely small light sources that can produce an almost coherent light, they have proven useful for a wide variety of practical applications.

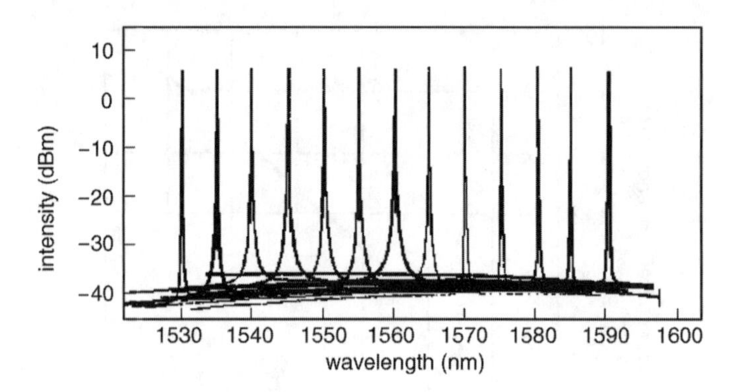

FIGURE 3.56 Tuning range of a wavelength-tunable ring laser.

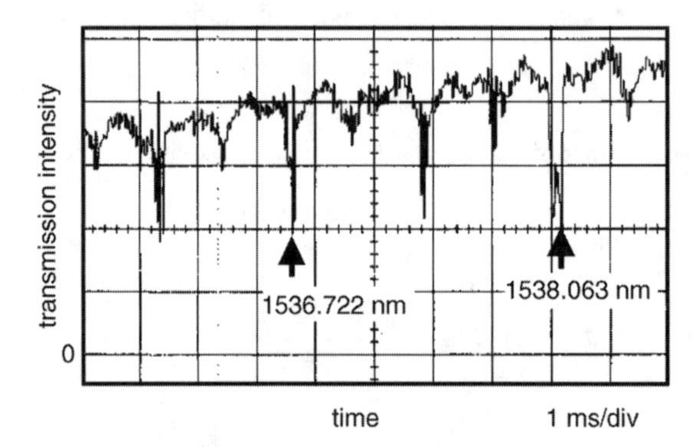

FIGURE 3.57 Application of wavelength-tunable laser to optical measurement. The laser is used in the scanning mode. The spectrum shows the absorption lines of C_2H_6 gas.

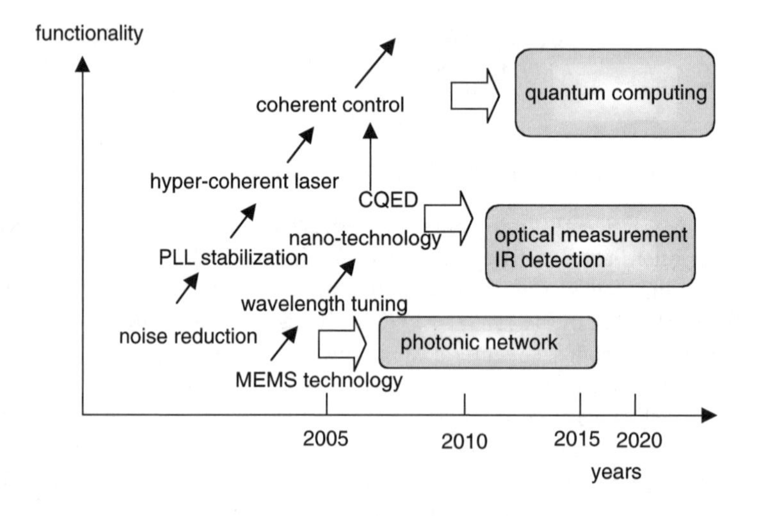

FIGURE 3.58 Future prospects of high-performance semiconductor lasers.

Semiconductor lasers are now evolving toward future high-performance applications, as shown in Figure 3.58. For example, ultra-narrow-linewidth semiconductor lasers realized by reducing the phase noise are now used for detecting gravitons in geophysics. By reducing the noise even further, we can realize a pure quantum-mechanical state of light. Such light will be useful for quantum-mechanical computing and communications. We can also produce ultra-high-frequency electromagnetic waves in the millimeter-frequency range as another attractive implementation of semiconductor lasers that is based on strict control over wavelength and phase of the lasers. Although conventional IR-sensing applications have been satisfactory sources for this, novel applications are being developed in communication-related fields.

Consequently, we all accept the fact that semiconductor lasers are very attractive because their laser oscillation performances are readily controlled by a wide variety of schemes using physical and mechanical techniques. The future no doubt promises further development in the area of semiconductor lasers supported by novel nano-technology as applied to the field of microscopic material processing and positioning technologies [Chang and Campillo, 1996].

Defining Terms

antireflection coating: Conventional semiconductor lasers have a facet reflectivity of 32%, which comes from the refractive index of semiconductor media of around 3.5. Antireflection (AR) coatings are used to reduce laser-facet reflectitivity with a residual reflectivity of less than 0.1%. The AR coatings consist of single or multiple dielectiric layers. An AR coating with a single layer has a thickness of a quarter wavelength and an optimum refractive index equal to the square root of the media. The optimum value for the laser media is therefore around 1.7, taking into account the above index. Materials suitable for AR coatings for laser media include glass films such as SiO_2 (refractive index $r_f = 1.54$), $Si_3N_4(r_f = 1.98)$ or their compound. An AR coating with multiple layers consisting of two kinds of films with different indices can reduce the laser-facet reflectivity in a wide wavelength range. AR coatings are used to construct external-cavity lasers in which the light circulates, passing through this kind of AR-coated facet.

multisegment lasers: Semiconductor lasers can be divided into independent segments, each of which is electrically isolated by the formation of etched grooves. Multisegment lasers consist of these isolated segments. When the depth of the grooves is optimized, electrical isolation is achieved while lightwaves traveling along laser waveguides feel negligible effects of the grooves. The grooves are usually several micrometers deep and around 20 mm wide.

quantum well: Electrons are electrically confined by potential barriers. Using two planar barriers, electrons are two-dimensionally confined. This kind of confinement structure exhibits a well shape. When the spacing of the two barriers becomes extremely small, the electrons confined in the well exhibit wave-like properties in accordance with quantum mechanics. The distinguishing feature of electrons in these quantum wells is their discrete energy levels. This discreteness facilitates light emission at a particular wavelength corresponding to an appropriate energy-level difference; hence it improves the laser oscillation performance.

recombination: When a light is absorbed in laser media, pairs of an electron and hall are produced. Although they may remain in the media, they are usually coupled together quickly and emit light. Such coupling is called recombination.

relaxation process: In semiconductor media, electrons are excited from a ground level to upper-energy levels by absorbing a light power. These excited electrons may return to the ground level by emitting excess energy corresponding to the light power. This kind of energy-level transition is called relaxation. Generally, relaxations have different processes. One is a photon-emission process, which means light emission, while the other is an electron-emission process in which an electron carries away the energy as kinetic energy. The former process is useful for laser oscillation, while the latter process is undesirable.

saturable absorber: Semiconductor laser media act as lightwave absorbers by applying a reverse-bias voltage to the media. While the absorption coefficient remains constant for the lower input light power, it is dramatically reduced when the input light power increases above an appropriate power level, depending on media materials. Due to such saturation the semicondcutor laser media are called saturable absorbers. A typical feature of the saturable absorbers is pulse-width narrowing.

When optical short pulses having a temporal profile with pedestals pass through such saturable absorbers, the pedestal portion with its lower power is whittled away, while their steeple-top portion with its higher power remains. Consequently, the pulses are narrowed.

ultrasonic motor: Ultrasonic motors are formed with arrayed vibrators that cooperatively move to generate a traveling wave. Objects attached to these vibrators are driven along the direction of the traveling wave by friction forces produced between the object and vibrator surfaces. Linear or circular drives are possible depending on configurations of the vibrators. The vibrators are typically made from piezoelectric thin films and the vibration frequency is in the several-megahertz range, equivalent to the ultrasonic frequency range. These ultrasonic motors are remarkable for their small size and high-torque performances and hence are suitable for opto-mechatronic applications.

References

Acket, G., Lenstra, D., Boef, A., and Verbeek, B., The influence of feedback intensity on longitudinal mode properties and optical noise in index-guided semiconductor lasers, *IEEE J. Quantum Electron.*, QE-20, 1163–1169, 1984.

Agrawal, G., Line narrowing in a single-mode injection lasers due to external optical feedback, *IEEE J. Quantum Electron.,* QE-20, 468–471, 1984.

Apfel, J. H., Circular wedged optical coatings. II. Experimental, *Appl. Opt.*, 4, 983–985, 1965.

Berestetskii, V. B., Lifshitz, E. M., and Pitaevskii, L. P., *Quantum Electrodynamics*, 2nd ed, Springer-Verlag, Berlin, 1990.

Bowers, J. E., Morton, P. A., Mar, A., and Corzine, S. W., Actively mode-locked semiconductor lasers, *IEEE J. Quantum Electron.*, 25, 1426–1439, 1989.

Buckman, L. A., Georges, J. B., Park, J., Vassilovski, D., Kahn, J. M., and Lau, K. Y., Stabilization of millimater-wave frequencies from passively mode-locked semiconductor lasers using an optoelectronic phase-locked loop, *IEEE Photon. Technol. Lett.*, 5, 1137–1140, 1993.

Casey, H. C., Jr. and Panish, M. B., *Heterostructure Lasers*, Academic Press, New York, 1978.

Chang, R. K. and Campillo, A. J., *Optical Processes in Microcavities*, World Scientific, Singapore, 1996.

Chen, Y. K. and Wu, M. C., Monolithic colliding-pulse mode-locked quantum-well laser, *IEEE J. Quantum Electron.*, 28, 2176–2185, 1992.

Derickson, D. J., Helkey, R. J., Mar, A., Karin, J. R., Wasserbauer, J. G., Bowers, J. E., Short pulse generation using multisegment mode-locked semiconductor lasers, *IEEE J. Quantum Electron.*, 28, 2186–2201, 1986.

Fleming, M. and Mooradian, A., Spectral characteristics of external-cavity controlled semiconductor lasers, *IEEE J. Quantum Electron.*, QE-17, 44–59, 1981.

Hara, S., Nakada, H., Sawada, R., and Isomura, Y., High precision bonding of semiconductor laser diodes, *Int. J. Jpn. Soc. Prec. Eng.*, 27, 49–53, 1993.

Hashimoto, E. and Katagiri, Y., 10-GHz-spacing DWDM channel selector using disk filter cascade with distributed amplification, *ECOC 2001*, Th. F3, 80–1, 2001.

Hashimoto, E., Takada, A., and Katagiri, Y., Synchronisation of subterahertz optical pulse train from PLL-controlled colliding pulse mode-locked semiconductor laser, *Electron. Lett.*, 34, 580–582, 1998.

Helkey, R. J., Derikson, D. J., Mar, A., Wasserbauer, J. G., Bowers, J. E., and Thornton, R. L., Repetition frequency stabilization of passively mode-locked semiconductor lasers, *Electron. Lett.*, 28, 1920–1922, 1992.

Katagiri, Y. and Hara, S., Increased spatial frequency in interferential undulations of coupled-cavity lasers, *Appl. Opt.*, 33, 5564–5570, 1994.

Katagiri, Y. and Hara, S., Scanning-probe microscope using an ultra-small coupled-cavity laser distortion sensor based on mechanical negative-feedback stabilization, *Meas. Sci. Technol.*, 9, 1441–1445, 1998.

Katagiri, Y. and Itaoh, K., Dynamic microforce measurement by distortion detection with a coupled-cavity laser displacement sensor stabilized in a mechanical negative-feedback loop, *Appl. Opt.*, 37, 7193–7199, 1998.

Katagiri, Y. and Takada, A., Synchronised pulse-train generation from passively mode-locked semiconductor lasers by a phase-locked loop using optical sidebands, *Electron. Lett.*, 32, 1892–1894, 1996.

Katagiri, Y. and Takada, A., A harmonic colliding-pulse mode-locked semiconductor laser for stable sub-THz pulse generation, *IEEE Photon. Technol. Lett.*, 9, 1442–1444, 1997.

Katagiri, Y. and Ukita, H., Optical heads based on coupled-cavity laser diode, *SPIE*, 2514, 100–111, 1995.

Katagiri, Y., Takada, A., Nishi, S., Abe, H., Uenishi, Y., and Nagaoka, S., Repetition-rate tunable micromechanical passively mode-locked semiconductor laser, *Electron. Lett.*, 32, 2354–2355, 1996.

Katagiri, Y., Takada, A., Nishi, S., Abe, H., Uenishi, Y., and Nagaoka, S., Passively mode-locked micromechanically-tunable semiconductor lasers, *IEICE Trans. Electron.*, E81-C, 151–159, 1998a.

Katagiri, Y., Tachikawa, Y., Aida, K., Nagaoka, S., and Ohira, F., Synchro-scanned rotating tunable optical disk filter for wavelength discrimination, *IEEE Photon. Technol. Lett.*, 10, 400–402, 1998b.

Lang, R. and Kobayashi, K., External optical feedback effects on semiconductor injection laser properties, *IEEE J. Quantum Electron.*, QE-16, 347–355, 1980.

Meystre, P. and Sargent, M., III, *Elements of Quantum Optics*, Springer-Verlag, Berlin.

Morikawa, T., Mitsuhashi, Y., and Shimada, J., Return-beam-induced oscillations in self-coupled semiconductor lasers, *Electron. Lett.*, 12, 435–436, 1971.

Morimoto, A., Kobayashi, T., and Sueta, T., Active mode locking of lasers using an electrooptic deflector, *IEEE J. Quantum Electron.*, QE-24, 94–98, 1988.

Olesen, H., Henrik, J., and Tromborg, B., Nonlinear dynamics and spectral behavior for an external cavity laser, *IEEE J. Quantum Electron.*, QE-22, 762–773, 1986.

Spano, P., Piazzolla, S., and Tamburrini, M., Theory of noise in semiconductor lasers in the presence of optical feedback, *IEEE J. Quantum Electron.*, QE-20, 350–357, 1984.

Takada, A., Sato, K., Saruwatari, M., and Yamamoto, M., Pulse width tunable subpicosecond pulse generation from an actively mode-locked monolithic MQW laser/MQW electroabsoption modulator, *Electron. Lett.*, 30, 898–900, 1994.

Thelen, A., Circularly wedged optical coatings. I. Theory, *Appl. Opt.*, 4, 977–981, 1965.

Tucker, R. S., Korotky, S. K., Eisenstein, G., Koren, U., Stulz, L. W., and Aveselka, J. J., 20-GHz active mode-locking of a 1.55 μm InGaAsP laser, *Electron. Lett.*, 21, 239–240, 1985.

Uenishi, Y., Isomura, Y., Sawada, R., Ukita, H., and Toshima, T., Beam converging laser diode by taper ridged waveguide, *Electron. Lett.*, 24, 623–624, 1988.

Uenishi, Y., Homma, K., and Nagaoka, S., Tunable laser diode using a nickel micromachined external mirror, *Electron. Lett.*, 32, 1207–1208, 1996.

Ukita, H., Katagiri, Y., and Uenishi, Y., Readout characteristics of micro-optical heads operated in bistable mode, *Jpn. J. Appl. Phys.*, 26(4, Suppl. 26), 111–116, 1989.

Ukita, H., Katagiri, Y., and Nakada, H., Flying head read/write characteristics using monolithically integrated laser diode/photodiode at a wavelength of 1.3 mm, *SPIE*, 1499, 248–262, 1991a.

Ukita, H., Sugiyama, Y., Nakada, H., and Katagiri, Y., Read/write performance and reliability of a flying optical head using a monolithically integrated LD-PD, *Appl. Opt.*, 30, 3770–3776, 1991b.

Ukita, H., Uenishi, Y., and Katagiri, Y., Applications of an extremely short strong-feedback configuration of an external-cavity laser diode system fabricated with GaAs-based integration technology, *Appl. Opt.*, 33, 5557, 1994.

Verdeyen, J. T., *Laser Electronics*, 3rd ed., Prentice-Hall, Englewood Cliffs, NJ, 1994.

Voumard, C., Salathe, R., and Weber, H., Resonance amplifier model describing diode lasers coupled to short external resonators, *Appl. Phys.*, 12, 369–378, 1977.

4

Optical Sensors and Their Applications

Kazuhiro Hane
Tohoku University
Sendai, Japan

Minoru Sasaki
Tohoku University
Sendai, Japan

4.1 Introduction

Optical sensors are indispensable for the precise control of mechatronic systems as well as other electronic and mechanical sensors. Optical sensors are especially preferable in noncontact measurements because light transmission and reflection can be used without contacting the object surface. Moreover, measurements using light as a means of detection are essentially the fastest in response because of their inherent propagation velocity, although the response time is limited by the detection electronics. In the case of displacement sensing using light waves, optical sensors are essentially high precision as the wavelength can be utilized as a unit length for the measurement, which is in the submicron region of visible light.

Several optical sensing techniques have been proposed for mechanical applications in industry. Table 4.1 shows the optical sensing technique used for mechatronics. They are categorized into several groups based on the sensing principle. The basic techniques consist of using the intensity of reflected light, straightness of light propagation, and interference of superimposed light beams. The position of the object can be detected based on the intensity of light reflected from the object. A high sensitivity can be obtained within a short range by using light focused with a microscopic objective. Based on the straightness of a light beam, the deviation from the optical axis can be sensed with simple semiconductor position sensors.

The highest precision is generally obtained using the optical interference technique because the wavelength of the light is used as the standard of length. With linear displacement measurement along a laser beam axis, a laser with high temporal coherence is used. The signal processing technique—sophisticated for fringe interpretation—has already been developed using microcomputers. The Fourier transform method and phase-shift technique are powerful tools for processing interference images.

With the time-of-flight technique, the distance between the light source and the target is directly measured from the round-trip time of the pulsed light. Thanks to the development of high-frequency electronics, measurement accuracy was improved recently to less than 1 cm in commercial distance meters.

0-8493-1162-4/03/$0.00+$1.50

TABLE 4.1 Optical Sensing Techniques Used for Mechatronics

Category	Principle	Measurement Techniques	Light Sources/Detectors
Basic techniques	Intensity	• Rough position measurements • Surface roughness measurements	LED, LD/PD
	Directionality	• Alignment • Object-shape measurements • Small displacement measurement vertical to optical axis	He–Ne Laser, LD/PD, position sensor
	Interference		
	Spatial coherence	• Holographic measurements (vibration, deflection) • Interferometric encoder	Ar laser, YAG laser LD/PD, CCD
	Temporal coherence	• Interferometer for displacement measurements • Precise displacement measurements	He–Ne (Zeeman laser), LD/PD, CCD
	Light velocity	• Long-distance measurements	LD/PD
Combined techniques and systems	Triangulation	• Short-range distance sensors • 3D-shape measurements by scanning object with laser beam • Surface roughness measurements	LED, LD/array PD
	Moiré	• Optical encoder • 3D-shape measurements using projected fringes	LED, LD white light/CCD
	Image processing	• Distance images • Alignment	White light/CCD

Triangulation is also a simple conventional technique used in a wide range of distance measurements, from 1 cm to 100 m. There are several versions of triangulation for distance measurements. The simple sensing system for triangulation can be constructed by combining only a light-emitting diode as a light source and a linear photodetector (PD) array. Image-detector and light-beam scanning can also be used for obtaining distant images.

The Moiré effect is convenient for magnifying small displacements and deformations using superimposed gratings. The phase shift between the gratings is visualized with the magnification depending on the angles of the grating lines. The Moiré effect is often used in optical encoders, which are conventional displacement sensors used in mechatronics. Moreover, the Moiré technique is useful for capturing the three-dimensional shape of an object by projecting the grating pattern.

Image processing using microcomputers and two image sensors of a charge-coupled device (CCD) is also an advanced method for sensing environmental conditions in robotics and mechatronics.

4.2 Optical Sensors for Displacement Sensing

Optical sensors are responsible for precise measurements of displacement, distance, position, and angles in several mechanical systems. In industrial mechatoronics applications, three kinds of optical sensors that employ the sensing principle are often used. The first is triangulation based on ray optics, the second is interferometry, and the third is the Moiré effect used for optical encoders.

The laser is a novel light source used in precise optical sensors for mechatronics. The laser had long been used only in laboratory measurements because of its sensitivity to environmental conditions such as temperature and vibration. Lately, the semiconductor laser diode (LD) has been improved and is now stable enough to be used in industrial environments. In addition, it is small enough to be installed in mechanical systems and has a lifetime long enough for industrial use. The price of the LD is low because huge numbers are used in audio systems that use optical disks. Although light-emitting diodes (LEDs) are widely used in optical sensors in industry, the number of LDs used in optical sensors for other fields will increase in the future. The coherent length of an LD (<1 m) is inherently shorter than that of He–Ne lasers (>1 m), which has been a light source for long-distance measurements based on interferometry. Therefore, LD is still suitable for short-distance sensors.

On the other hand, the sensing system must be made small enough for the sensors to be installed in mechanical systems. Integration of the optical sensors is also an important consideration in developing the built-in sensing systems. In sensors integrated for pressure and acceleration, the mass-production and assembly of the sensors have been achieved by silicon micromachining. With micromachining technology, the mechanical parts of the sensors are fabricated by semiconductor lithographic processes in a similar way to integrated circuits. The miniaturization of sensors and their integration with electronics are also important for the development of sensors for mechatronics. The microsystems fabricated by silicon micromachining are called microelectromechanical systems (MEMS), whose use has been widely expanded to a broad range of fields, from mechanical engineering to electronics, chemistry, bioengineering, etc. The technology is also merging with optics, which is categorized as optical MEMS. Several optical sensors and optical components have been proposed in which optical detectors, optical components, and mechanical structures are fabricated simultaneously by lithographic processes without the individual components being assembled. Those technologies will become important in the future evolution of integrated sensors.

4.3 Basic Principles and Methodologies

Straightness of light-beam propagation is a simple and useful property of light for alignment and precise displacement measurement. Figure 4.1 shows the schematic diagram of the position measurement using a laser beam and quadrant-cell position sensor. Four photodiodes are installed adjacently and the deviations of the laser spot from the center of the quadrant-cell along the x and y axes (U_X,U_Y) are obtained from the photocurrents (I_1, I_2, I_3, and I_4) of the cells according to the equations:

$$\begin{aligned} U_X &= (I_1 + I_3) - (I_2 + I_4) \\ U_Y &= (I_1 + I_2) - (I_3 + I_4) \end{aligned} \tag{4.1}$$

The linearity of the displacement measurement is limited to a region smaller than the beam spot size. The alignment between the laser axis and the center of the position sensor is carried out. The displacement signals U_X, U_Y become zero when the two centers are aligned. The aiming stability of the laser source is as good as 1 μ radian. The measurement sensitivity is also limited by the fluctuation of the atmospheric index due to temperature and airflow. Under the environmental conditions of a quiet laboratory, the displacement sensitivity is around 1 μm, or less, for the alignment of a 1-m-long light beam.

In addition to the measurement of displacement vertical to the optical beam axis, the displacement along the optical axis can also be performed by using the quadrant-cell position sensor. In an optical

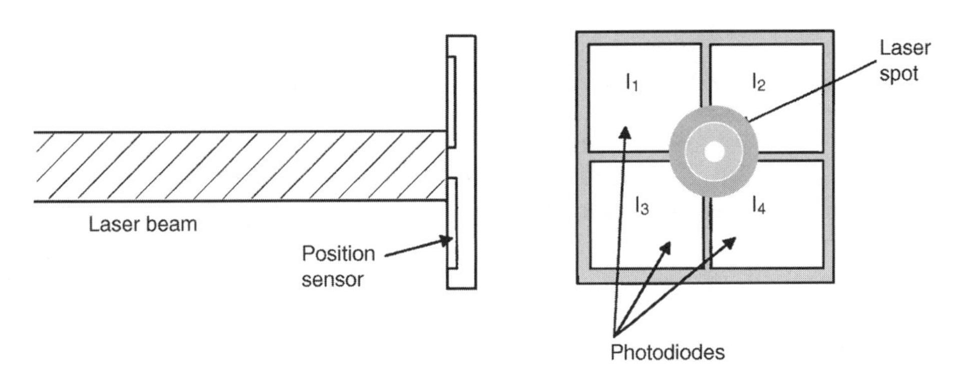

FIGURE 4.1 Schematic diagram of the position measurement using a laser beam and quadrant-cell position sensor.

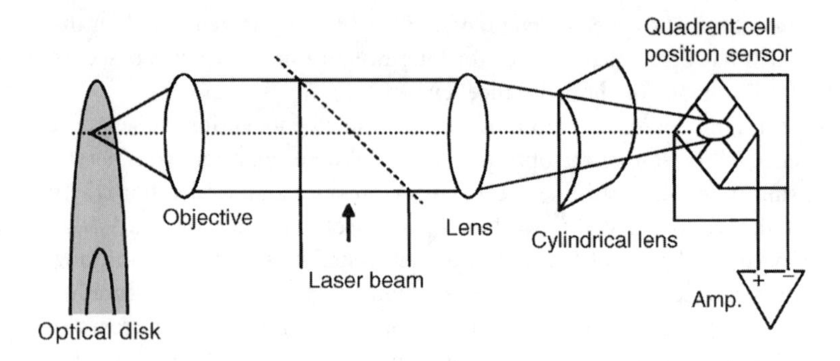

FIGURE 4.2 Focal position sensing in optical disk system.

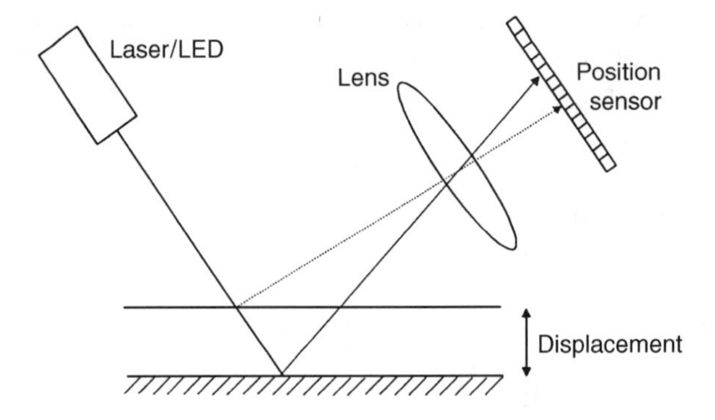

FIGURE 4.3 Schematic diagram of sensors using triangulation.

disk system, the laser light should be focused on the surface of the rotating disk during operation. The deviation of the focal position from the disk is detected by an optical system shown in Figure 4.2. The reflected laser beam is focused through a convex lens and a cylindrical lens. Because of the cylindrical lens, the focal length in the plane where the cylindrical lens works is shorter than that in the other plane. Therefore, the laser spot on the quadrant-cell position sensor is circular only at the position where the laser beam is just focused on the disk surface. Otherwise, the laser spot on the position sensor is an ellipse. Therefore, the deviation from the focal position can be detected based on the difference between the two sets of the signals obtained from the two photodiodes located diagonally to one another.

Triangulation is also a simple conventional technique for measuring distance. The absolute value of the distance can be obtained after calibration. The schematic diagram of the triangulation is shown in Figure 4.3. The light emitted from the light source impinges on the object surface. We assume here that the surface is optically rough. The light reflected from the surface is scattered widely around the reflection angle. (If the surface is optically flat, the reflection angle is equal to the incident angle, and the reflected light is not always received by the photodetector.) The reflected light converges on the position sensor through a convex lens. The light spot on the object surface is imaged on the sensor plane by the lens. The object position is determined from the spot on the sensor. Figure 4.4 shows the principle of triangulation. Knowing the length of the base line L and the values of the two angles α and β, the distance h from the base line is obtained from the equation

$$h = \frac{L \tan\alpha \tan\beta}{\tan\alpha + \tan\beta} \tag{4.2}$$

Based on the output of the position sensor, the direction in which the beam is reflected is measured. The triangulation is widely used in the autofocus of compact cameras. One example of the optical system

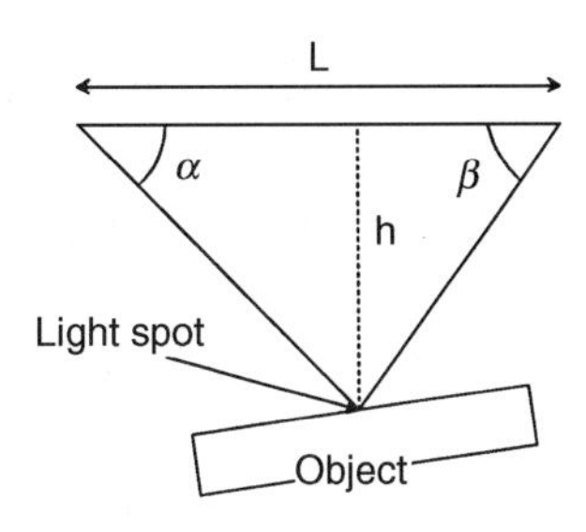

FIGURE 4.4 Principle of triangulation.

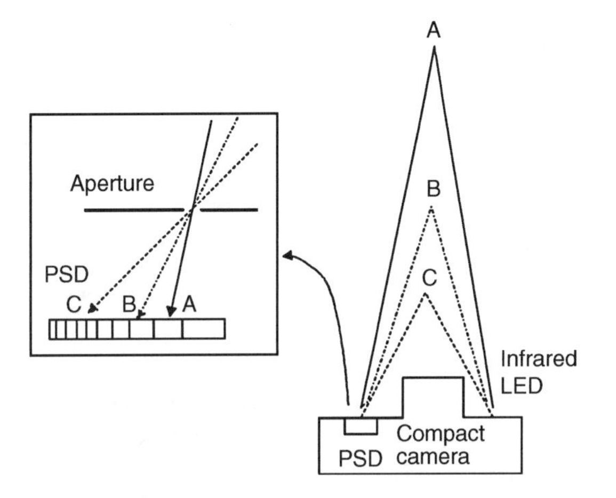

FIGURE 4.5 Triangulation used in compact camera.

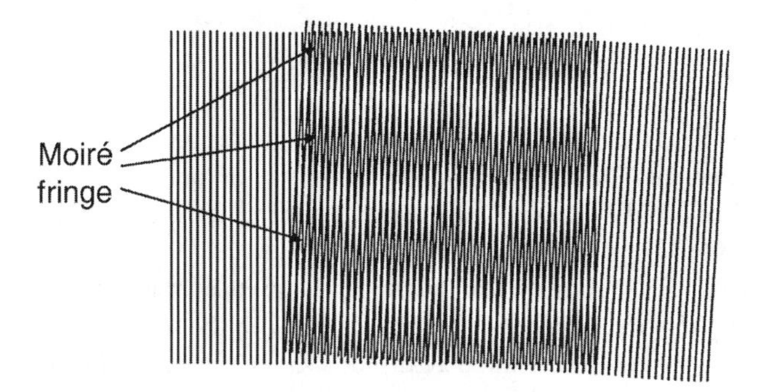

FIGURE 4.6 Moiré fringe generated by superimposed gratings.

is shown in Figure 4.5. An infrared light source is used and the reflected light is detected through an aperture or lens. The aperture works to determine the direction of the reflected light by combining the position sensor as shown in Figure 4.5.

In the case of digital cameras and videocassette recorders, a convenient technique for autofocus is image processing. The high-frequency component of the image is extracted from the image detected by the CCD. Under good focus conditions, the high-frequency component of the image is maximized, and the lens position is adjusted to obtain the maximum image size.

The Moiré technique is traditionally used for precise displacement measurements. Figure 4.6 shows the Moiré fringe generated by superimposing two gratings. The period of the Moiré fringe is much larger than that of the gratings. When the angle between the grating lines is increased, the period of the Moiré fringe decreases. Translating a grating in a direction perpendicular to the grating lines, the Moiré fringe

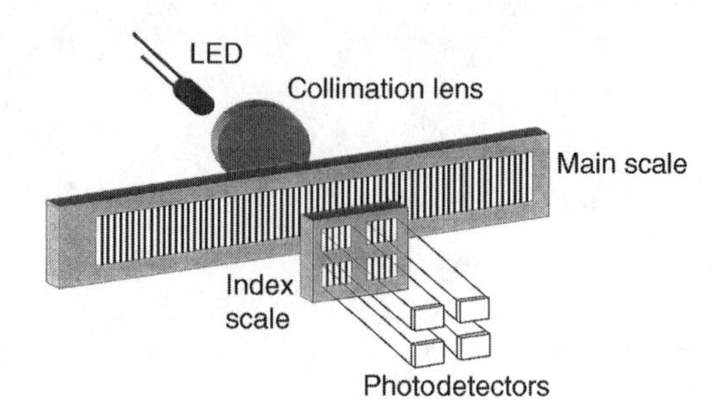

FIGURE 4.7 Optical configuration of a conventional optical encoder.

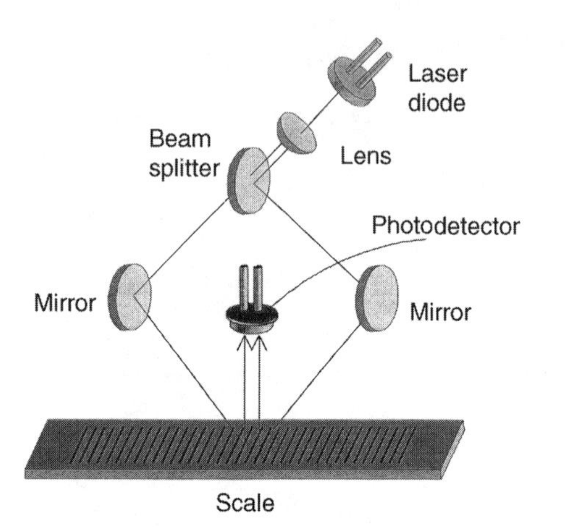

FIGURE 4.8 Interferometric encoder.

moves in the direction parallel to the grating lines with a magnified displacement. When the two gratings are parallel, the magnification of the displacement is maximized.

Optical encoders are common displacement sensors that utilize the Moiré effect. Figure 4.7 shows the optical configuration of a conventional optical encoder. The light from the LED is collimated through a convex lens and passes through the scale grating and index grating. The transmitted light is detected by a photodiode placed behind the index grating. The signals from the photodiodes vary nearly sinusoidally with increasing displacement. The index grating usually consists of four phase-shifted gratings for obtaining the phase-shifted sinusoidal signals. Those signals are used for the interpolation of the sinusoidal curve and the detection of the direction of motion.

When the period (p) of the grating used in the conventional encoder is decreased, the diffraction effect from the scale grating becomes significant. The gap (z) between the scale grating and the index grating has to be small ($z < p^2/\lambda$, where λ is the wavelength) for keeping the signal contrast high. The diffraction effect is advantageously used for a novel encoder. Figure 4.8 shows the optical encoder using the diffraction effect of the grating and the interference of the diffracted beams. Fine grating generates the diffraction beams when irradiated by a laser beam. The phases of the diffracted beams are shifted by the translation of the grating. The value of the phase shift θ is given by

$$\theta = 2\pi \frac{d}{p} N \tag{4.3}$$

where d and N represent displacement of the grating and the order of the diffraction beam. In Figure 4.8, the $\pm$first-order diffraction beams interfere to obtain the sinusoidal signal as a function of the displacement. The total phase difference between the interfering beams is twice as long as the value of Equation 4.3; thus, the period of the displacement signal is half the period of the grating.

There is another way to avoid the decrease of the signal contrast when the gap between the gratings is wide. Grating-image-type encoders are based on a different principle [Hane et al., 2002], although the

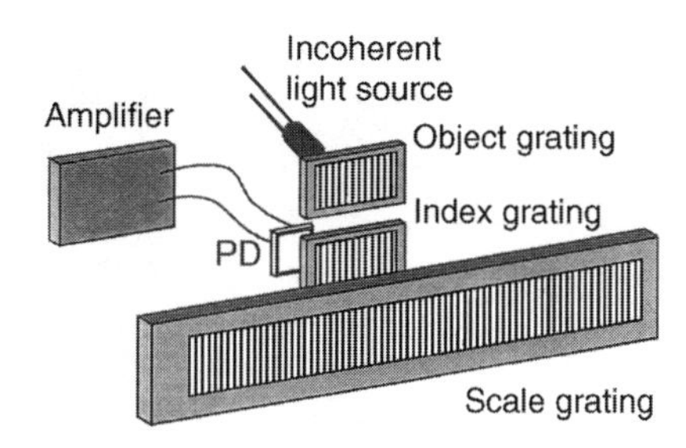

FIGURE 4.9 Grating-image-type encoder.

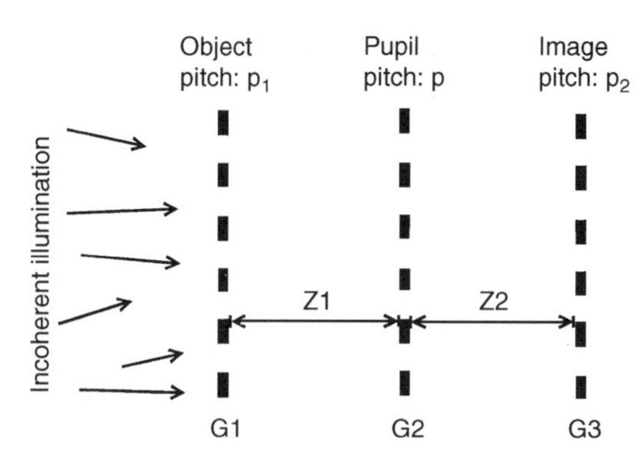

FIGURE 4.10 Optical configuration of grating image.

optical configuration is similar to the conventional one. Figure 4.9 shows the schematic diagram of the reflection-type encoder using grating imaging. Unlike the conventional Moiré encoder, object grating is placed in front of the light source. The object grating illuminated incoherently with the light source (here, LED) is imaged through the scale grating onto the index grating by the diffraction effect of the scale grating. Figure 4.10 shows the equivalent optical system of the grating-image-type encoder. The optical system consists of three gratings placed in tandem. Central grating acts as a pupil in the sense that the grating-like image is generated on the plane of the index grating. The superposition of the image and the index grating produces the Moiré fringe. The grating-like image is essentially generated by the diffraction of the central grating. The image is generated under spatially incoherent illumination. Due to the imaging effect, the gap between the gratings is set to be much wider than that used in the conventional Moiré encoder, where the simple shadow of the scale grating is used. To make a high signal contrast obtainable, the combinations of the periods of three gratings are p_1:p:p_2 = 2p:p:2p and p:p:p for the optical configuration shown in Figure 4.10. In the case of the combination p:p:p, the image contrast is not degraded by the polychromaticity of the light source. Therefore, white light can be used for the encoder.

The interferometer is also a valuable instrument for displacement measurement and has been studied for a long time. Figure 4.11 shows the basic optical configuration of the interferometer (Michelson interferometer). The laser beam is divided into two beams by a beam splitter. One beam is reflected by the reference mirror fixed on an optical bench. The other beam is reflected by the mirror fixed on the object for measuring the displacement. Two reflected beams interfere with each other after being combined on the photodetector. The interference intensity I is given by the equation,

$$I = I_1 + I_2 + 2\sqrt{I_1 I_2}\cos(k(L_1 - L_2) + \delta_1 - \delta_2) \tag{4.4}$$

where I_1 and I_2 are the intensity of the respective beams. The symbol k represents a wave number equal to $2\pi/\lambda$, where λ is the wavelength. The round-trip optical paths of the two arms in the interferometer are indicated by L_1 and L_2, respectively. The initial phases of the respective beams are represented by δ_1 and δ_2. The intensity I varies sinusoidally as a function of the optical path difference $L_1 - L_2$. Therefore,

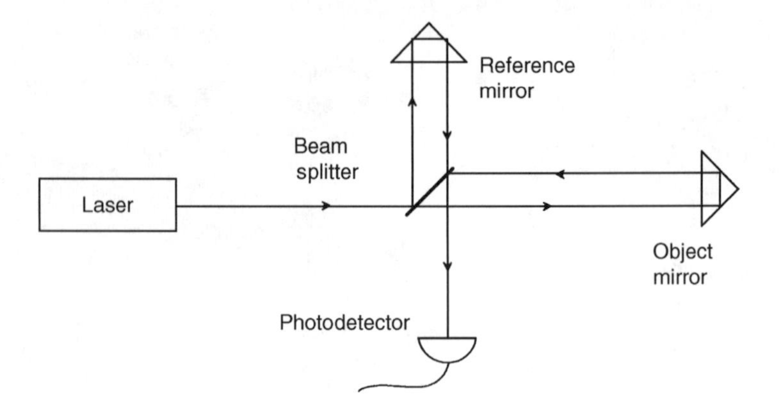

FIGURE 4.11 Optical interferometer used for displacement measurement.

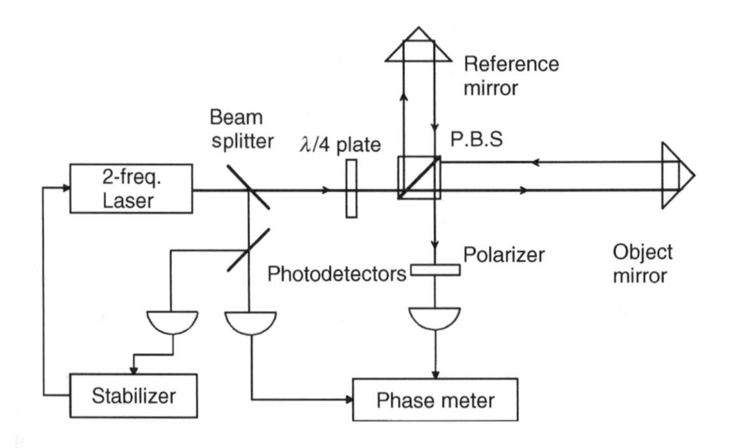

FIGURE 4.12 Heterodyne interferometer for displacement measurement.

the displacement $d(=\Delta L_1/2)$ is measured by $\lambda/2$ by feeding I into the up-down counter connected to the photodetector. The phase-shifted displacement signal, which is needed for the interpolation and determination of the direction of motion, may be obtained by tilting the reference mirror to generate a fringe where the phase-shifted signals are spatially generated.

The contrast V of the interference is defined by the equation:

$$V = \frac{I_{max} - I_{min}}{I_{max} + I_{min}} = \frac{2\sqrt{I_1 I_2}}{I_1 + I_2} \tag{4.5}$$

where I_{max} and I_{min} are the maximum and minimum values of the interference intensity I. The maximum contrast is obtained when the intensities of the two beams are equal to each other and the laser has high temporal coherence.

The heterodyne interferometer is convenient instrument for obtaining high resolution and sensitivity in displacement measurement. Figure 4.12 shows the schematic diagram of the heterodyne interferometer for displacement measurement. Unlike the interferometer shown in Figure 4.11, a two-frequency laser, generally the Zeeman He–Ne laser, is used. Applying a magnetic field causes a Zeeman splitting of the laser level. Two linearly polarized laser beams with a frequency difference of around 2 MHz or 100 kHz are generated depending on the direction of the applied magnetic field. The light beams reflected from the mirrors are interfered with after passing through the polarizer. Under static conditions, the beat of

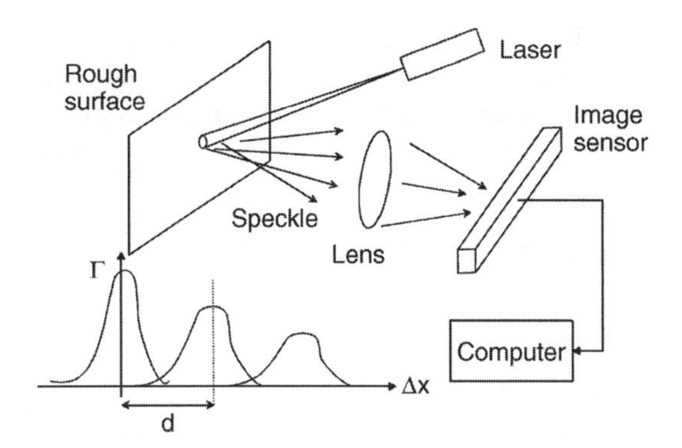

FIGURE 4.13 Displacement sensor using laser speckle.

the two-frequency lasers is obtained at a difference frequency. The interference intensity is given by the equation,

$$I = I_1 + I_2 + 2\sqrt{I_1 I_2}\cos(\Delta\omega t + k\Delta L + \delta) \tag{4.6}$$

where $\Delta\omega$ is equal to the difference of the two angular frequencies of the lasers. The wave number k is approximately given by the averaged value for the two laser lights. ΔL is equal to $L_1 - L_2$. The initial phase is represented by δ. The phase difference caused by the displacement ΔL or the object velocity $\nu(\Delta L = \nu t$, where ν is the object velocity) can be measured easily by an electronic phase meter. Because of progress in the design and manufacture of electronic instruments, the displacement and velocity can be monitored precisely in the heterodyne interferometer. A sensitivity of better than 10 nm is obtained in a commercialized heterodyne interferometer. The heterodyne interferometer can also be constructed by using the acoustic optical modulators for generating two frequencies from a single-frequency laser.

Techniques using laser speckles are also novel for sensing the displacement because the motion of an object with a rough surface is detected using simple optics. A speckle is the light intensity distribution generated by the interference of multiple beams reflected from the rough surface. Speckle characteristics such as average diameter and statistical distribution of intensity are not dependent on the surface properties when the surface is rough enough compared with the optical wavelength. Figure 4.13 shows the schematic diagram of the displacement measurement method using the laser speckle. The optical configuration is very simple. The surface is irradiated by a laser beam, and the scattered light that is the speckle field is received by an image sensor such as a CCD. Translating the object, the speckle field shifts at a magnification that depends on the optical configuration. The speckle field changes gradually as displacement increases. When the correlation between the speckle images before and after the movement is calculated, the correlation peak shifts according to the displacement as shown in the inset in Figure 4.13. The correlation can be calculated by the equation,

$$\Gamma(\Delta x, \Delta y) = \frac{1}{S}\iint_S I_1(x,y) I_2(x+\Delta x, y+\Delta y)\,dx\,dy \tag{4.7}$$

where Γ is the correlation function, $I_1(x, y)$ is the light intensity distribution before the movement, $I_2(x + \Delta x, y + \Delta y)$ is that after the movement. S is the area where the integration in the image plane takes places and it includes many speckles. When the object displacement increases, the part of the area in the laser spot leaves the irradiated area and part of the area outside the spot enters the irradiated area. Therefore, the correlation between the areas irradiated by laser light before and after the displacement decreases as displacement increases. Displacement is measured step by step in the region where the correlation is maintained. To calculate the image correlation, an image sensor and computer are needed, but the measurement is carried out with very simple optics.

For some industrial applications the object frequently has a rough surface, which can cause random scattering. A speckle interferometer has been already developed for this requirement. A kind of Moiré fringe is generated between a pair of the speckle patterns obtained before and after the object displacement. Statistical image processing, however, is usually necessary to extract the displacement.

4.4 Novel Applications to Metrological Sensing

4.4.1 New Position Sensors for Straightness Measurements [Sasaki et al., 1999a]

Position sensors consisting of photodiodes located adjacently to one another are conveniently applicable to mechatronics as described above. Here a new version of position sensor is described that is suitable for detecting the accurate straightness reference of a laser beam. This function is realized by fabricating the four-cell-type photodiode inside the Si micromesh structure by Si micromachining. The sensor does not disturb the direction or the wave front of the reference laser beam, absorbing only a part of the incident light and transmitting the rest down stream. Figure 4.14 shows the straightness measurement using the proposed sensors. The straightness of a structure 1 m in size can be measured with 1-μm accuracy at many points by placing sensors in series.

Like the conventional position sensor, the position sensor is based on the quadrant-cell photodiode. By comparing the magnitude of four signals obtained from each cell, the relative position between the sensor and the beam spot is measured.

Figure 4.15 shows the fabrication process. The Si substrate (n-type (100), 5 to 8 Ω-cm) used here is 200 μm thick. First, the substrate is oxidized, and a 40-μm-thick Si diaphragm is fabricated by etching the wafer backside anisotropically. The oxide film is then patterned to open the window through which the boron is implanted (100 keV as BF_2, 2.0 k $\times$ 10^{14} atoms/cm^2) to make the four-cell-type photodiode. The photodiode diaphragm is patterned as the mesh and the through-holes of the mesh are made by the reactive ion etching. One more oxidization process is carried out for reducing the leak current due to the surface state at the sidewall of the through-hole. After the contact holes are etched for electrical interconnections, an aluminum layer is deposited, patterned, and sintered.

Figure 4.16 shows the fabricated transmission-type position sensor, which has a honeycomb-like mesh structure. Letters printed on the paper behind the sensor are seen through the mesh. Within this Si mesh, four square photodiode cells are fabricated. The mesh dimensions are 150 μm in pitch, and 20 μm in beam width. The honeycomb mesh is 5 $\times$ 5 μm^2 in area.

To demonstrate the function of the proposed multipoint position detection, the transmission-type position sensors were placed on the translation stage of a machining machine, and the deviation from the straight movement was measured. Two transmission-type position sensors are placed as shown in Figure 4.17. When the stage is translated, the deviations are measured along the x and y axes. The deviations along the x and y axes are shown in Figures 4.18(a) and (b), respectively, as a function of the displacement in the range from 0 to 140 μm. The two sensors are denoted by No. 1 and 2, respectively. In order to show the repeated measurement, two curves corresponding to the forward and backward

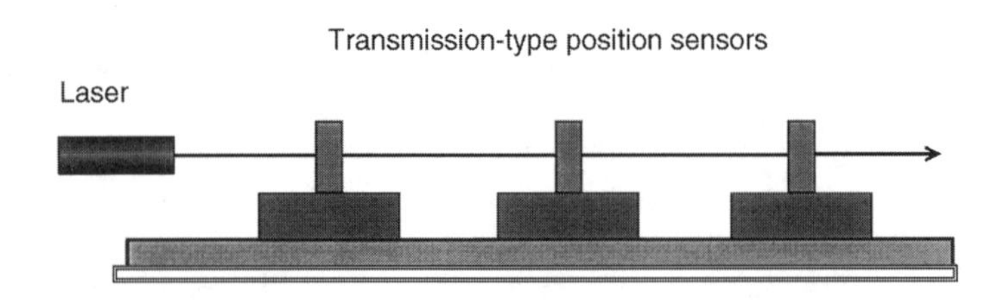

FIGURE 4.14 Alignment using transmission-type position sensors.

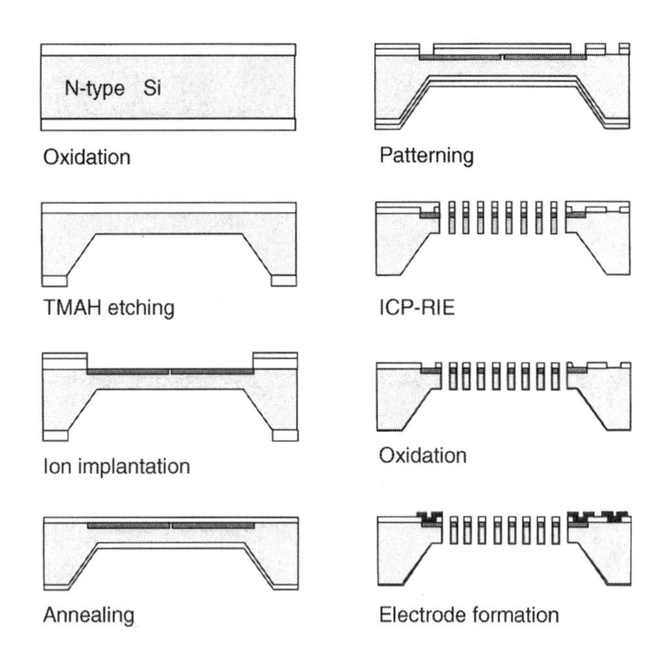

FIGURE 4.15 Fabrication process.

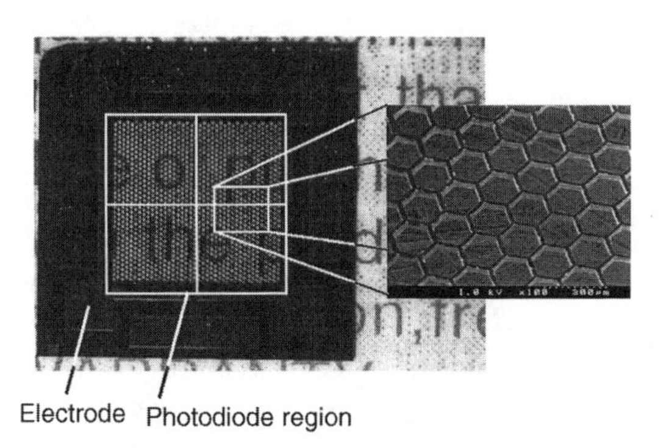

FIGURE 4.16 Fabricated transmission-type position sensor.

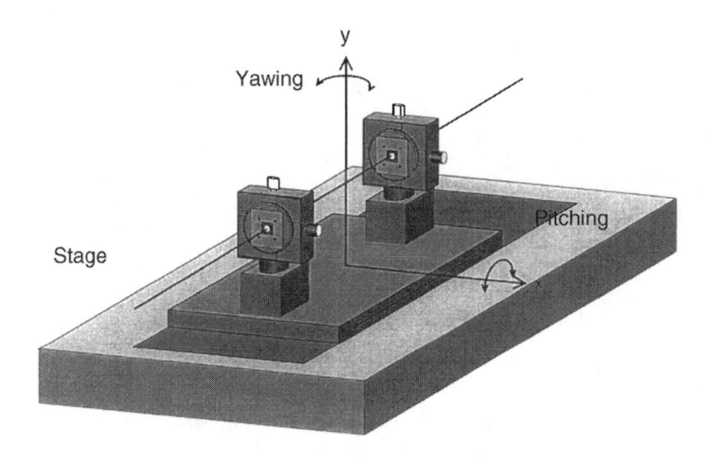

FIGURE 4.17 Translation measurement of stage.

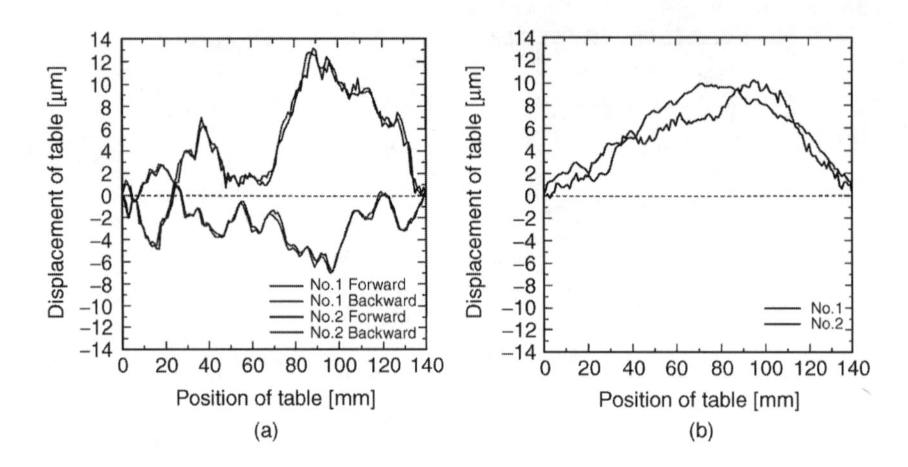

FIGURE 4.18 (a) Displacement x, and (b) y measured as a function of stage position.

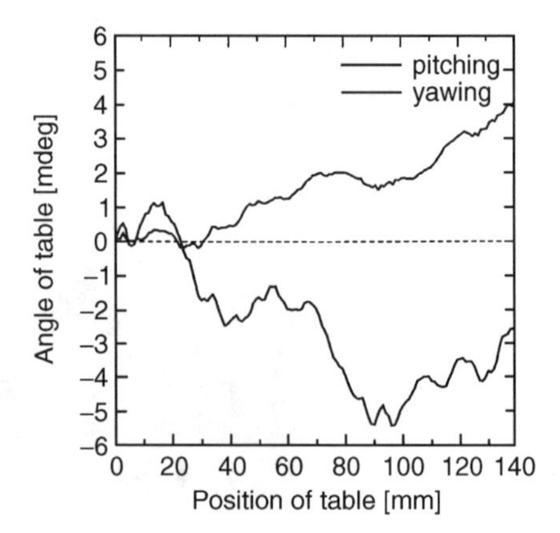

FIGURE 4.19 Angular deviation as a function of position.

movements are shown in Figure 4.18(a). As shown in Figures 4.18(a) and (b), the deviations of the stage from the laser beam are easily measured. Because two sensors were placed at the two ends of the stage, the values of yawing and pitching were also calculated from the displacements. Figure 4.19 shows those values measured in the experiment. The two values are measured simultaneously as a function of the stage displacement. Due to the simple optical configuration, deviations within 6 degrees of freedom were easily obtained.

4.4.2 Optical Encoder with Pitch-Modulated Photodiode Array

Because of its simple optical configuration, the Moiré encoder shown in Figure 4.7, in which two gratings are superimposed with an air gap, is still important in many practical applications. Recently, for this simple Moiré encoder it was proposed that the pitch-modulated gratings be used to suppress the harmonic noise [Ohashi et al., 1999]. The output of the encoder measured as a function of the lateral displacement includes some higher order distortions in addition to the sinusoidal signal. The harmonic noise is attributed to the harmonic components of the grating transmittance. Therefore, these harmonic noises cause a considerable measurement error when we obtain a displacement smaller than the pitch of the grating by the interpolation, assuming that the signal is sinusoidal.

It has been proposed that an index grating with a modulated pitch can be used for decreasing the higher order distortions in the output of the encoder. In the demonstration, the third- and fifth-order distortions of the displacement signal were decreased by a factor of more than 10.

A schematic diagram of the proposed index grating is shown in Figure 4.20. The encoder consists of a periodic grating with a constant pitch, p, as the main scale and a pitch-modulated grating with an

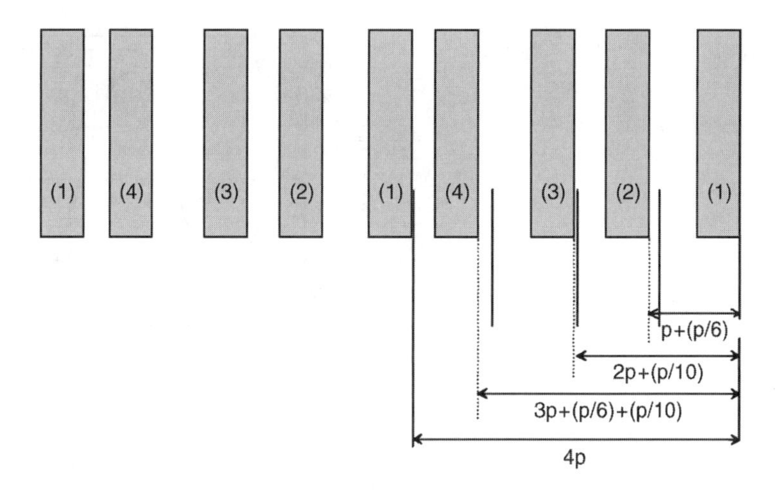

FIGURE 4.20 Schematic diagram of a pitch-modulated grating.

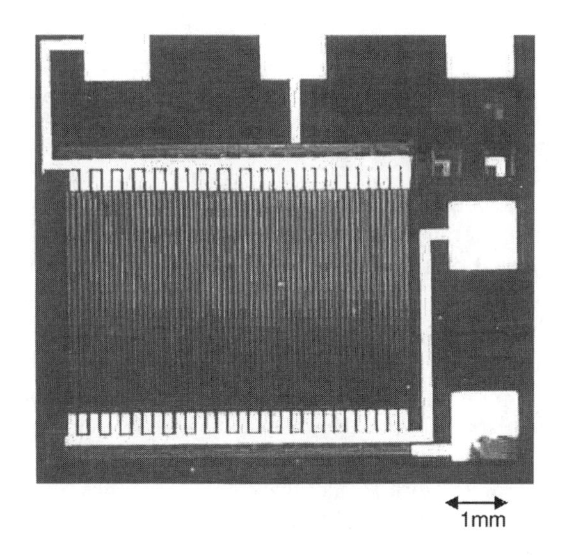

FIGURE 4.21 Photomicrograph of the pitch-modulated photodiode array.

averaged pitch p as the index scale; these are fixed parallel to the main scale. The schematic diagram of the pitch-modulated grating, which actually employs a unit structure of photodiodes with different phases, is shown in Figure 4.20. This unit structure is repeated by the phase differences of p/6, p/10, and p/15 with respect to the averaged pitch p. The phase difference between the set of the grating lines (1) and that of (2) is designed to be p + p/6 to compensate the third-order harmonics by inverting their phases. Similarly, the sets of grating lines of (3) and (4) are also arranged to have a phase difference of p + p/6. Furthermore, the phase difference between the combination of ([1] + [2]) and that of ([3] + [4]) is set to be 2p + p/10, thereby eliminating their fifth harmonics. Therefore the pitch-modulated grating shown in Figure 4.20, which is used as an index grating, suppresses both the third and fifth harmonics noises in the encoder signals as a function of displacement.

The pitch-periodic and pitch-modulated photodiode arrays with multichannels were fabricated for the encoder experiment by using integrated circuit process technology. The multichannels were used for obtaining four photocurrent signals of the phase-shifted photodiodes independently. Figure 4.21 shows a photomicrograph of the pitch-modulated photodiode array with the four electrodes for respective

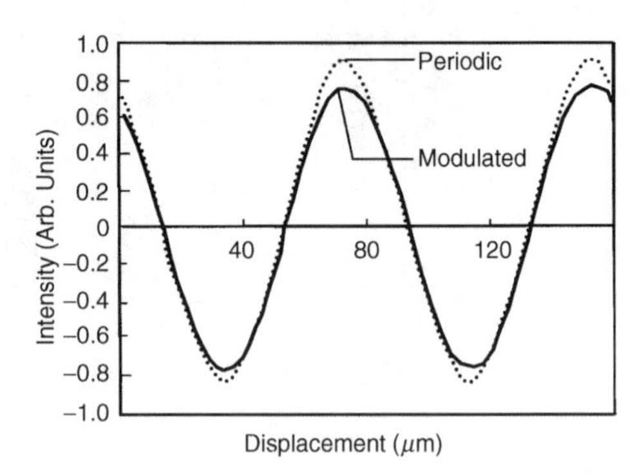

FIGURE 4.22 Light intensity with relative displacement.

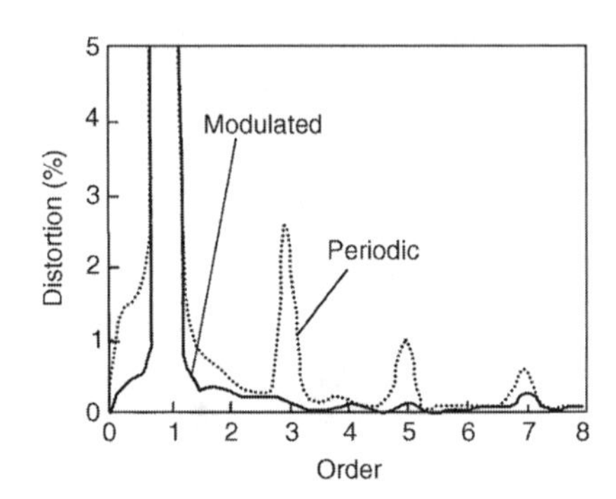

FIGURE 4.23 Fourier spectrums of light intensity.

channels. The substrate was a 300-μm-thick, n-type (100) silicon wafer. The p–n junction of the photodiodes was formed by doping boron ions. Then, aluminum metal film was patterned and etched for electrodes. The electrodes are fabricated as three-dimensional wiring. The width of the photodiodes in the index grating is 40 μm, with an averaged pitch of 80 μm.

The air gap z between the scale and index gratings is equal to 2.8 mm, and the typical light intensities of the two types obtained from the encoder experiment are shown in Figure 4.22. The intensity curve for the pitch-periodic type was the quasi-sinusoidal containing some higher harmonics, while the intensity curve for the pitch-modulated type was approximated by the sinusoidal. The Fourier spectra of the intensity from the encoder experiment are shown in Figure 4.23. The third- and fifth-order distortions for the pitch-modulated type were much smaller than those for the pitch-periodic type. Moreover, for the pitch-periodic type, the change of the air gap had a large effect on the third- and fifth-order distortions. Thus, high resolution can be obtained in the interpolation with the proposed encoder system.

The interpolation error was estimated as follows. The photocurrents can be transformed into voltages, which are approximated by sine and cosine curves. Then, the phase angle, i.e., the relative position between the scales, is directly calculated from these two voltages by applying the arc tangent function. The maximum third- and fifth-order distortions in the experiment using pitch-periodic type grating are 4.35 and 1.45%, respectively. On the other hand, when we use the pitch-modulated type, the maximum third- and fifth-order distortions are 0.44 and 0.17%, respectively, in the experiment. Then the interpolation errors are calculated at 0.11 and 0.043 μm. Moreover, the third- and fifth-order distortions are almost independent of the air gap. Thus the pitch-modulated photodiode array can reduce the interpolation errors by a factor of more than ten.

The optical encoder with the slit-like pitch-modulated photodiode array as the index scale has been developed to integrate the gratings and the detectors. The photodiode array was specially designed to decrease higher harmonics distortions of the displacement signal and to obtain four signals simultaneously. Subsequently, the photodiode array for the index grating was fabricated using integrated circuit process technology. The third- and fifth-order distortions have been reduced by a factor of more than ten in this new system. Moreover, these distortions were found to be independent of the air gap between the index scale and main scales.

4.4.3 Integrated Grating-Image-Type Encoder

In an optical system that is similar to the conventional Moiré encoder, the grating imaging effect was used previously for a precise displacement measurement [Hane and Grover, 1987a]. In the encoder, the signal was insensitive to the change of the air gap between the two gratings under incoherent illumination. The grating imaging effect was further investigated on the basis of the optical transfer function for the displacement measurement [Hane and Grover, 1987a,b].

More recently, an integrated grating-image-type encoder was proposed, in which the transmission grating was fabricated by silicon micromachining [Hane et al., 2001, 2002]. Two gratings, photodetectors (two line photodiodes on each grid), LEDs, and a preamplifier circuit chip were integrated by stacking them. The integrated optical encoder is described below.

Figure 4.24 shows the proposed integration of the encoder. Compared with the conventional grating image encoder shown in Figure 4.9, the optical system is integrated. The five components of the conventional encoder are stacked. The integrated encoder consists of the two gratings, an incoherent light source, and the photodetectors. The photodiodes are installed in the respective grids of the index grating. The scale grating is assumed to be an amplitude-reflection grating. The optical configuration in reflection makes the encoder system as compact as the conventional Moiré encoders. The light source used in the proposed encoder is assumed to be polychromatic and incoherent as an LED. The index grating is fabricated from a silicon wafer and consists of transmission grids. In each grid, which is a thin silicon beam, photodiodes are installed using semiconductor microfabrication technology.

Figure 4.25 shows the schematic front views of the designed index grating. The Si substrate is etched through to form the index gratings. Figure 4.25 shows two kinds of phase-shifted line photodiodes as installed in each grid, which is a thin Si beam. In all, four 90° phase-shifted photodiodes, in which the spatial phase differences are 90°, are needed to obtain four sinusoidal signals. They are used for eliminating

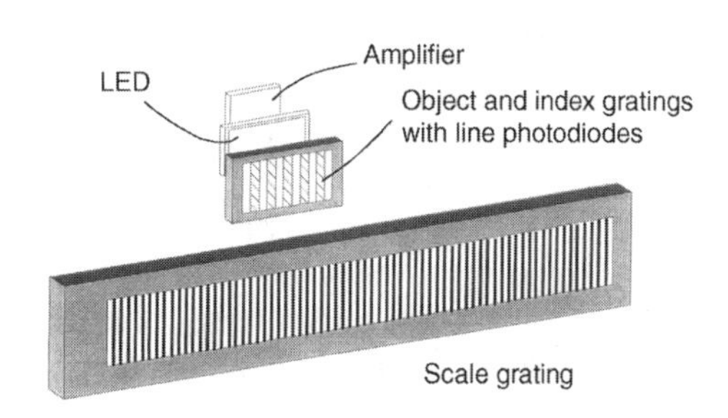

FIGURE 4.24 Schematic diagram of the grating-image-type optical encoder.

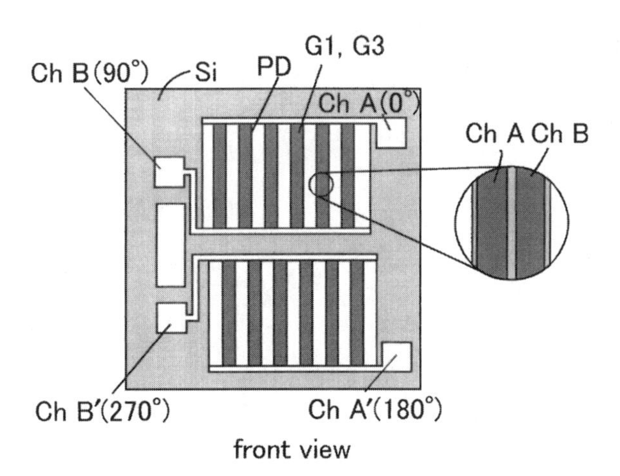

FIGURE 4.25 Designed index gratings consisting of Si grids with line photodiodes.

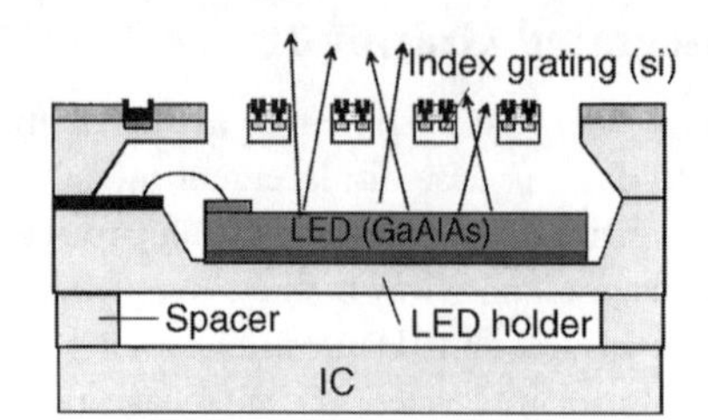

FIGURE 4.26 Cross-sectional view of the designed encoder.

DC offset of signals, for sensing direction, and for interpolating the signal. As shown in Figure 4.25, because the two kinds of 90° phase-shifted photodiodes are installed in each Si beam alternately and closely in space, the photodiode sensitivities are nearly equal and the average intensities of signals are not affected by the light intensity distribution in the large region. The index grating is illuminated from behind and the light passes through the slits of the grating. The grating periods used in the experiment were 80 and 40 μm.

Figure 4.26 shows the schematic diagram of the encoder cross-section (in which the scale grating is not shown). The index grating fabricated from the Si substrate is fixed to the LED holder to encapsulate the LED. The electrodes for the LED are patterned on the surface of the LED holder. Light is emitted from the LED through the Si index grating. The light reflected from the scale grating is detected by the photodiode fabricated on the Si grating. A chip of signal amplifier is fixed to the LED holder with a polymer spacer. The electronic circuits may be fabricated on the side area of the index grating if the fabrication facility can accept both processes.

As shown in Figures 4.9 and 4.10, the three gratings placed in tandem describe an equivalent optical system for this encoder. When the first grating (object grating) is irradiated with spatially incoherent light, the object-grating pattern is transferred by the center grating (reflection scale grating) onto the image plane (which is equal to the plane of the object and the index gratings). The center grating of the three gratings works as a pupil for imaging. The encoder optics has been analyzed by using optical transfer function. Based on the results of the analysis [Hane and Grover, 1987b], the grating period and distance between the gratings have been determined.

The grating imaging for this encoder receives a brief theoretical explanation below. The essential optical system of the encoder is described by the three gratings placed in tandem at the same distances z between the gratings, as shown in Figure 4.10. The grating-like image is formed by the slit array of the pupil grating under Fresnel diffraction. To understand the grating imaging, it may be easier to consider that each slit of the pupil grating images the object grating on the plane of the index grating, and the superposition of the images generated by the respective slits of the grating produces a grating-like image, if they are in phase. The optical transfer function of the pupil grating is obtained under our experimental conditions as follows [Hane and Grover, 1987b],

$$F(\sigma) = \Pi(2\sigma p)\exp(i\pi 2\sigma mp)\left[1-\left(\sigma-\frac{mp}{\lambda z}\right)\frac{\lambda z}{2\varepsilon}\right]\times \mathrm{sinc}\left\{4\varepsilon\sigma\left[1-\left(\sigma-\frac{mp}{\lambda z}\right)\frac{\lambda z}{2\varepsilon}\right]\right\} \tag{4.8}$$

$$\text{for } \frac{mp}{\lambda z} \le \sigma < \frac{mp}{\lambda z}+\frac{2\varepsilon}{\lambda z}$$

$$F(\sigma) = \Pi(2\sigma p)\exp(i\pi 2\sigma mp)\left[1+\left(\sigma-\frac{mp}{\lambda z}\right)\frac{\lambda z}{2\varepsilon}\right]\times \mathrm{sinc}\left\{4\varepsilon\sigma\left[1+\left(\sigma-\frac{mp}{\lambda z}\right)\frac{\lambda z}{2\varepsilon}\right]\right\} \tag{4.9}$$

$$\text{for } \frac{mp}{\lambda z}-\frac{2\varepsilon}{\lambda z} \le \sigma < \frac{mp}{\lambda z}.$$

Here, σ is the image frequency; p and 2ε are, respectively, the pitch and slit width of the pupil grating; z is the distance between the gratings; λ is the wavelength of light; and m is an integer. The function $\Pi(x)$ represents a comb function, which becomes a unit when x is equal to integers. When the image frequency is equal to the object frequency ($\sigma = 1/p$) and the slit width is assumed to be half of the pitch ($\varepsilon = p/4$), then Eqs. 4.8 and 4.9 are simplified to

$$F\left(\sigma=\frac{1}{p}\right)=(-1)^m\left(1-2\xi+2m\right)\times \text{sinc}\left(\frac{1-2\xi+2m}{2}\right) \tag{4.10}$$

$$\text{for}\quad m\leq\xi<m+\frac{1}{2}$$

$$F\left(\sigma=\frac{1}{p}\right)=(-1)^m\left(1+2\xi-2m\right)\times \text{sinc}\left(\frac{1+2\xi-2m}{2}\right) \tag{4.11}$$

$$\text{for}\quad m-\frac{1}{2}\leq\xi<m,$$

where ξ is the distance z normalized by $p^2/(2\lambda)(\xi = 2z\lambda/p^2)$. Therefore, the grating condition under which the periods of the three gratings are equal to each other corresponds to the second imaging condition ($x = 2$). The image contrast calculated as a function of the normalized distance ξ is shown in Figure 4.27. As shown in Figure 4.27, the image contrast is always positive under the optical conditions, although the contrast varies periodically with increasing ξ. The period of the image contrast as a function of z is equal to $p^2/(2\lambda)$ (i.e., $\xi = 1$), which is dependent on the wavelength λ of light. Therefore, in the case of polychromatic illumination (white light), the images generated by the respective wavelengths are superimposed constructively, and the image contrast is not degraded by the polychromaticity of the light source when $x = 2$. The contrast under white light illumination is schematically shown in Figure 4.27. Moreover, the maximum value of the image contrast does not decrease when the distance between the gratings increases, thus the displacement can be measured at a distance z larger than that used in the conventional Moiré encoder.

Because the imaging effect of the grating is used in the encoder, the relative displacement d of the object grating generates that of an image on the plane of the index grating in the opposite direction. Therefore, the encoder signal varies by two periods for the relative displacement of the grating equal to a single grating period. The sensitivity of the displacement detection in this encoder is improved by a factor of two by this phenomenon over the conventional Moiré encoder.

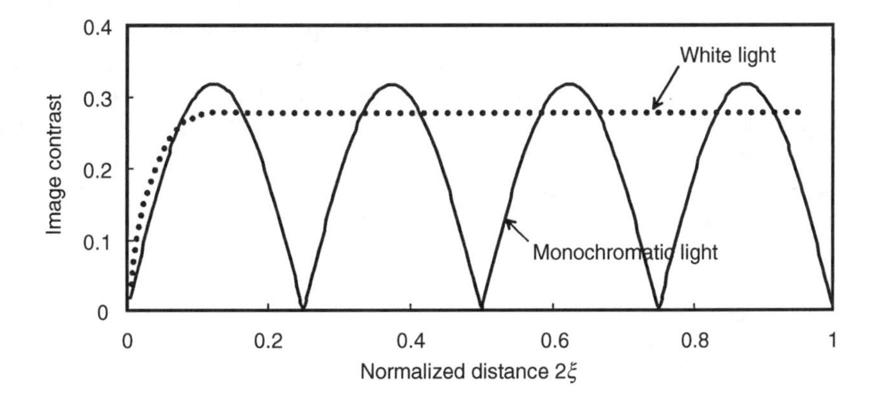

FIGURE 4.27 Image contrast calculated as a function of normalized distance for a single wavelength.

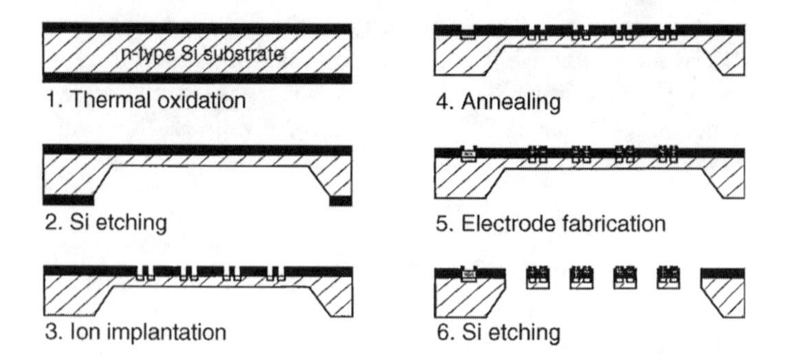

FIGURE 4.28 Lithographic process for fabricating the index grating.

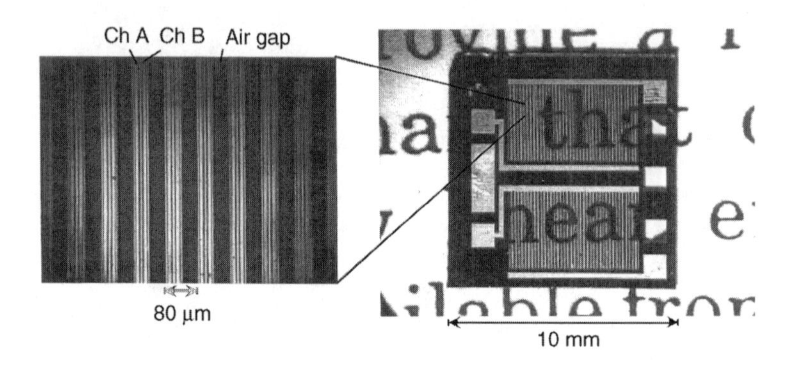

FIGURE 4.29 Fabricated index grating.

Figure 4.28 shows the lithographic processes for fabricating the index grating. Starting from an n-type Si substrate (200 μm thick and 1 to 10 Ω-cm), a 500-nm-thick SiO_2 film is formed by wet oxidation (1). From the rear surface, the wafer is etched with TMAH (2). The etched area is 3.5 mm × 5.7 mm and 40 μm thick. Next, the line photodiode is fabricated on the etched area by implanting B ions ($2 \times 10^{14}/cm^3$ at 120 keV) (3). After annealing (4) at a temperature of 1000°C, the Al electrode is patterned (5). The gratings with the line photodiodes are then fabricated by etching them through with inductively coupled reactive plasma (6).

A spatially incoherent light source is needed for this encoder and a 3-mm-long and 1-mm-wide GaAlAs LED was specially designed for this purpose. For further integrating the encoder, a preamplifier was designed for obtaining the two phase-shifted signals without DC offset from the four channel photodiodes, which consisted of eight operational amplifiers. Because all the operational amplifiers were fabricated on one chip and the photodiodes for sensing the light intensities were located closely in space, a signal intensity nearly equal to each other was obtained, which was effective for a high-precision interpolation.

The fabricated index grating with a period of 80 μm is shown in Figure 4.29. As shown in Figure 4.29, the duty ratio between the widths of Si grid and the slit is nearly unity. The two line photodiodes are installed as shown in the magnified image of the grating in Figure 4.29. The photocurrent of the diode was measured to be 500 nA with a 10-nA dark current using a standard light source of 12.5 lx. The cut-off frequency response of the fabricated photodiode was around 200 kHz.

Figure 4.30 shows the encoder signals measured as a function of displacement and the Lissajour figure of the two 90° phase-shifted signals. The gap between the index grating and the scale grating is 3 μm under experimental conditions. In this experiment, the index grating shown in Figure 4.29 was illuminated with the white light from a halogen lamp through a fiber bundle. The light reflected from the scale grating is detected with the photodiodes installed on the grids of the index grating. Therefore, the areas of light emission and detection are nearly the same. Two line photodiodes are installed on one grating line to

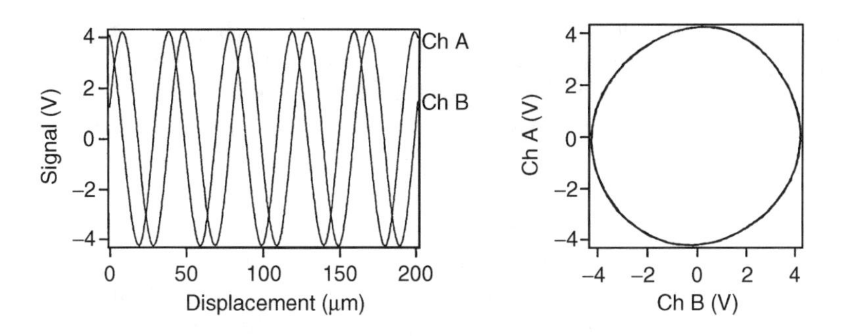

FIGURE 4.30 Encoder signal measured as a function of displacement and its Lissajour figure.

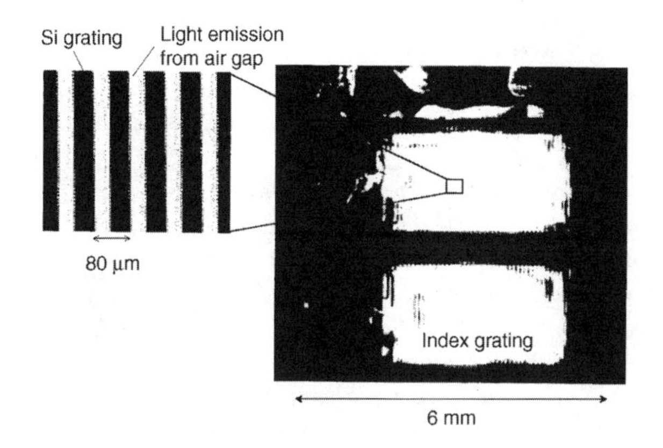

FIGURE 4.31 Light emission from the integrated sensor.

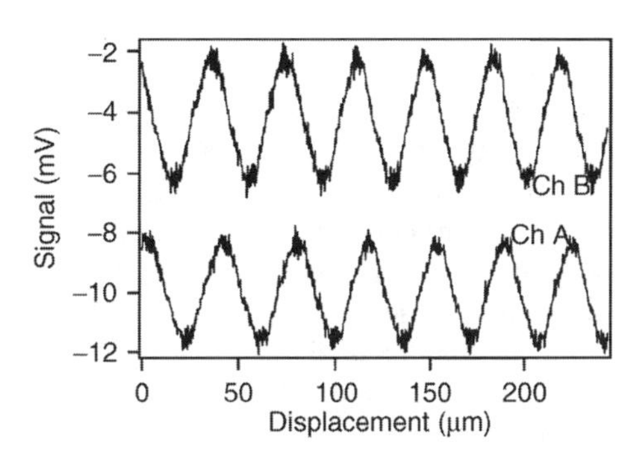

FIGURE 4.32 Encoder signals from an integrated sensor.

obtain the 90° phase-shifted signals. Another set of the phase-sifted signals is obtained from the index grating located below, as shown in Figure 4.29. As shown in Figure 4.30, two sinusoidal signals are obtained and the phase difference between the signals is nearly 90°. Because the Lissajour figure is almost a circle, the signal includes little harmonic noise. The signal contrast was measured as a function of the gap between the index and the scale gratings. The signal contrast was kept constant at large air gaps from 1 mm to 30 mm.

After testing the fabricated index grating as described above, the integrated encoder sensor was tested. The index grating, LED, LED holder, and the IC chip were integrated, as shown in Figure 4.26, by stacking them with epoxy resin. The integrated encoder sensor was 1.2 mm thick. Figure 4.31 shows the optical micrograph of the light emission from the fabricated encoder sensor. As shown in Figure 4.31, a grating-like emission through the Si grating is obtained. Therefore, the Si grating on which photodiodes are installed works simultaneously as a transmission object grating. Figure 4.32 shows the encoder signals from the two channels. Although some noise is superimposed on the signals, sinusoidal encoder signals are obtained, as shown in Figure 4.32. The low signal-to-noise ratio is mainly due to the low intensity of the fabricated LED. No significant influence of the heat generated by the LED and IC chip on the encoder signal was observed in the experiments.

4.4.4 Integrated Interferometric Sensors

Laser interferometry is a well-developed technique for displacement measurement as it has high resolution and utilizes a noncontact method as described in Section 4.3. In most interferometers, a half mirror is used for making sensing and reference beams from one laser beam and for superimposing these beams after their travels through interferometer arms. The bulky interferometer is used only in expensive and extremely large pieces of equipment (e.g., stepper, electron beam drawer).

Compact optical interferometric sensors are advantageous for installing in many pieces of industrial equipment because of its small mass and size, its stability, the fact that it requires no alignment adjustment, and its low cost. Although there is great demand for integrated displacement sensors, the practical displacement sensor is still poor in signal amplitude, contrast, and dynamic range of the measurable distance.

The approach to building integrated interferometers has been based on the miniaturization of two-beam interferometers using the waveguide on a plane surface [Suhara et al., 1995]. The optical feedback technique has been studied as a compact interferometric sensor [Merlo and Donati, 1997; Kato et al., 1995]. The principle involves mixing between the back-reflected sensing beam from a moving sample and the optical field inside the laser diode. This technique gives a very simple optical configuration as only one optical arm is needed. The dynamic range of the measurement, however, is limited by the mode-hopping of the laser diode and the lack of stable operation.

Recently, a compact interferometer based on standing-wave detection using a thin-film photodiode has been developed [Sasaki et al., 1999b; Mi et al., 2001]. The key device is an ultra-thin-film photodiode. The optical configuration is as simple as that of the optical feedback technique. When a coherent light beam is normally incident on a reflection mirror, the standing wave is generated in front of the reflection mirror. The use of an ultra-thin-film photodiode to detect the standing wave eliminates the need for a beam splitter and an optical arm for reference, whereas the usual two-beam interferometer requires two arms. The interferometer is essentially stable, with the potential for integration and suitability for use as a small sensor. In this section, interferometers based on standing-wave detection are described for developing a compact interferometric displacement sensor.

A schematic diagram of the proposed interferometer based on standing-wave detection using a thin-film photodiode is shown in Figure 4.33. The interferometer consists of a laser, an isolator, and a newly developed thin-film photodiode. The reflection mirror is on the moving object. The active layer of the ultra-thin-film photodiode, which absorbs photons, is designed to transmit most of the incident light beam. In our design, the absorption ratio to incident light power is estimated to be less than 1%. The almost incident light beam transmits through the photodiode before being absorbed and travels to the mirror on the moving object and returns to the thin-film photodiode. The incoming and reflection light beams superimpose each other and produce a standing wave that penetrates the active layer. The period of the standing wave is equal to $\lambda/(2n)$(λ is the wavelength of the incident light, n is the refractive index of the media). Because a node is generated on the reflection mirror, the intensity profile of the standing wave is spatially fixed to the reflection mirror. If the active layer has the appropriate thickness, which is thinner than the period of the standing wave of $\lambda/2n$, the intensity profile of the standing wave can be resolved. When the photodiode is located at the node of the standing wave, the signal decreases, and, when the photodiode is at the antinode, the signal increases. The relative position between the thin-film photodiode and the mirror determines the photodiode signal. This interference signal gives the displacement between the thin-film photodiode and the moving sample using the standing wave as the standard scale. In this interferometer, neither a reference mirror nor a beam splitter is necessary. The wavefront of the incoming laser beam is the reference. Because a laser beam with a large diameter can be used inside

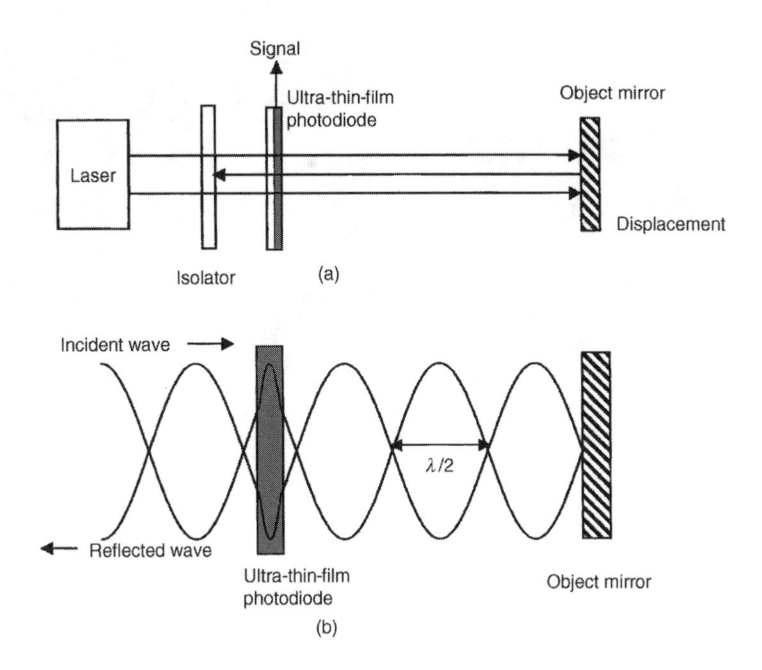

FIGURE 4.33 (a) Interferometer using an ultra-thin-film photodiode; (b) principle of standing-wave detection.

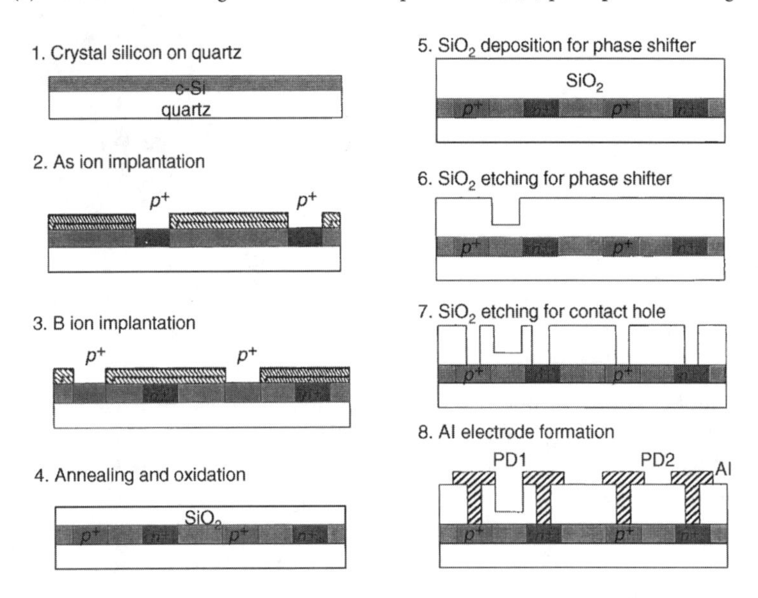

FIGURE 4.34 Fabrication sequence of thin film photodetector.

the sensor system, beam expansion due to diffraction is small. Measurement of the long distance up to the coherent length of the light source will be possible.

The active layer thickness is designed to be 40 nm in order to obtain enough spatial resolution to the standing wave [Sasaki et al., 1999b]. The absorbed power of the incident light is estimated to be about 0.4%. Inside such a thin Si film it is difficult to create a p–n junction in the thickness direction, which is the structure of the conventional photodiode. In the thin-film photodiode, the p–n junction is designed in the lateral direction in the Si film. The depletion region grows laterally in the Si film.

Figure 4.34 shows the fabrication sequence. The initial Si-on insulator wafer is prepared by direct wafer bonding between Si and quartz. The Si layer is thinned to 60 nm. The comb-shaped windows are

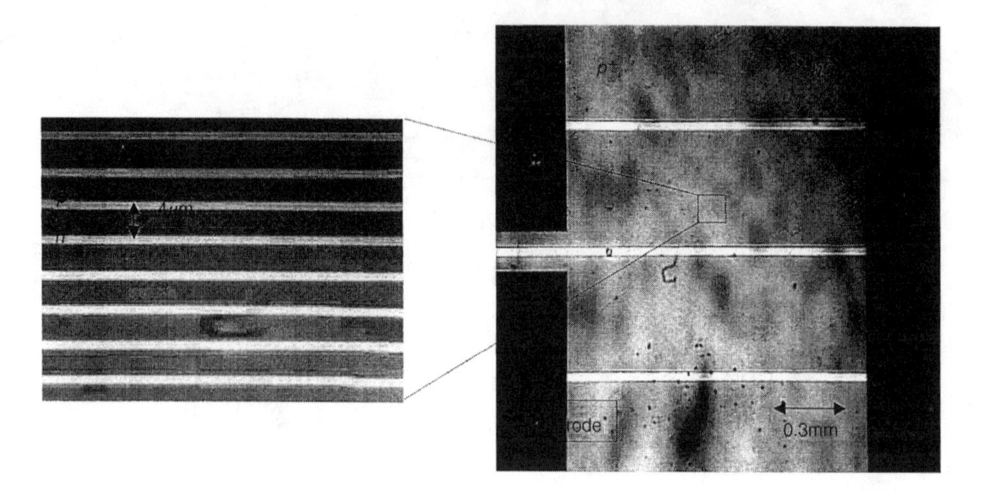

FIGURE 4.35 Fabricated thin-film photodiodes.

FIGURE 4.36 Si-on insulator substrate with four arrayed small photodiodes.

opened by lithography, through which As and B ions are implanted to make a p–n junction. After annealing and SiO_2 deposition, the contact hole and Al electrode are formed. Finally, the phase shifter may be fabricated by etching the deposited SiO_2 layer. The phase shifter can be used for obtaining the phase-shifted sinusoidal signals.

Figure 4.35 shows the fabricated thin-film photodiode with a 4-μm pitch comb shape. The p^+ and n^+ regions have a comb shape. The depletion region grows laterally between the p^+ and n^+ regions. This design is to lengthen the depleted region and to gather as many photocarriers as possible. The estimated thickness of the Si active layer is 35~40 nm. The overall transmission rate reaches 70% in power, including the reflections at the interfaces between Si and SiO_2, and between SiO_2 and air. Figure 4.36 shows the fabricated photodiode array on the substrate. (In this case, the photodiode area is smaller than that shown in Figure 4.35 and a whole substrate is shown.) As shown in this figure, the sensor is transparent.

The sensitivity of the thin-film photodiode is about 0.75 mA/W and about three orders smaller than that of the usual bulk Si photodiode. The measured series resistance and equivalent capacitance of the photodiode are 100 KΩ and 15 pF, respectively. Due to the small cross-sectional area of the thin-film photodiode, the series resistance is large, whereas the parallel capacitance is small. The photodiode itself is considered to have a response speed of up to 600 KHz.

Figure 4.37(a) shows a schematic diagram for the sensor package combined with a can-type laser diode. The displacement sensor using a thin-film photodiode is packaged by stacking the thin-film

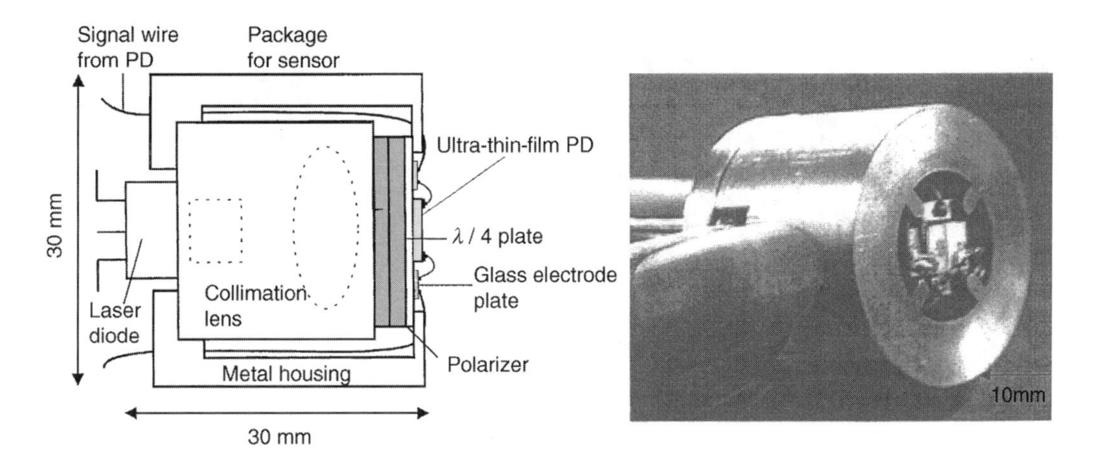

FIGURE 4.37 (a) Schematic diagram of a sensor package; (b) view of a packaged sensor.

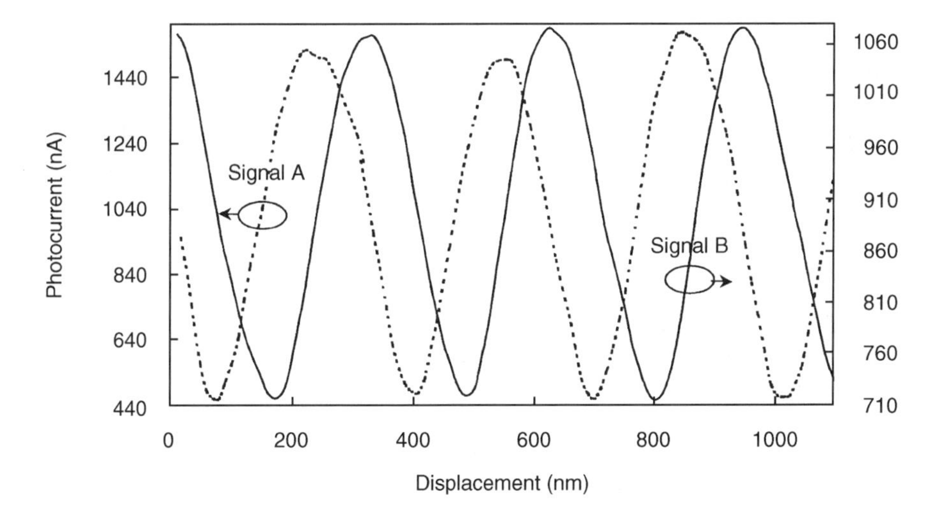

FIGURE 4.38 Two interference signals obtained from a dual thin-film photodiode combined with a phase shifter.

photodiode and isolator onto the collimation lens. All the optical elements are fixed within a metal housing. Figure 4.37(b) shows the fingertip-sized sensor package.

The experimental setup is the same as that shown in Figure 4.33(a). The red (632.8 nm, 7 mW) He–Ne laser is used as the light source and a fabricated photodiode having a 40-μm pitch of p^+ and n^+ regions is used as the detector. The reflection mirror is placed ~45 mm away from the thin-film photodiode and moved by a piezoactuator. The signal period agrees well with $\lambda/2$. A signal amplitude of over 1 μA is obtained. The contrast reaches a maximum level of 77%. At a normal incidence on the thin-film photodiode and the mirror, the Fabry–Perot effect occurs due to multiple reflections. When the thin-film photodiode is slightly slanted (~0.5 mrad) toward the wavefront of the incident laser beam, the interference signal becomes sinusoidal. Figure 4.38 shows two interference signals obtained from the dual thin-film photodiode combined with a phase shifter. The phase shift agrees well with the designed value of $\pi/2$. Using this phase relation, the moving direction of the mirror can be determined.

Figure 4.39 shows the interference signals obtained from the packaged sensor, in which a laser diode (655.2 nm, 10 mW) is used as the light source and a fabricated photodiode having a 4-μm pitch of p^+ and n^+ regions is used as the detector. The contrast is around 25%. It results mainly from the inaccurate alignment between the thin-film photodiode and the laser beam.

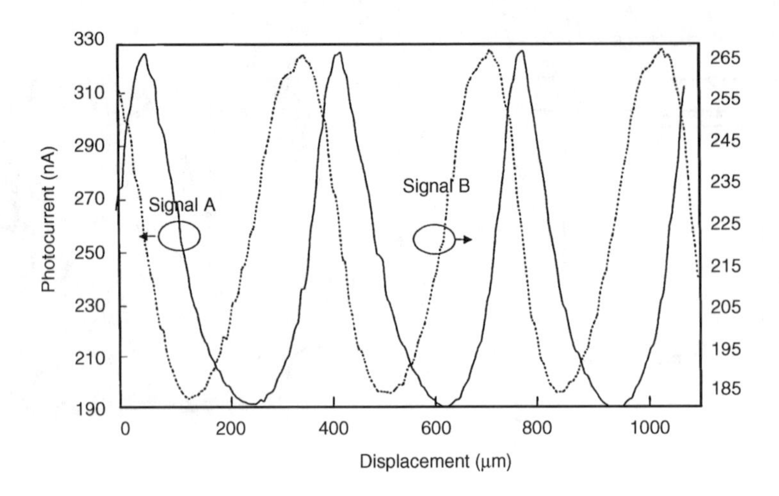

FIGURE 4.39 Interference signal obtained from a packaged sensor.

The sensors are suitable for integration due to their simple optical setup. The principle of the interferometer is based on standing-wave detection. This technique is compatible with other recently developed techniques. The ultra-thin-film photodiode makes it possible to use optical standing waves in metrological applications. The sensor package is constructed using a can-type laser diode, an isolator (combination of a polarizer and a wave plate), and an ultra-thin-film photodiode. The size is as small as a fingertip.

4.5 Conclusions

In the first part of this chapter, the general principles of the optical sensors used in mechatronics were given. For industrial applications, simple and reliable techniques are preferable. The principle of position sensors and its application were described. The tracking and focusing sensors of optical disks serve as examples. Triangulation is a simple method but is widely applicable to several sensors used in industry. Triangulation is commonly used for distance measurement in digital cameras. On the other hand, optical encoders are often used for measuring linear displacements and rotational angles in industrial mechanical systems. The principles of the three kinds of optical encoders were given. Furthermore, the principles of the optical interferometers were described from the point of view of displacement measurements. The highest precision is generally obtained with the optical interferometer, although it is very sensitive to environmental conditions.

In the latter part of the chapter advanced optical sensors were introduced. New versions of the position sensor, optical encoders, and optical interferometers were explained. In the new optical position sensors, the transmission structures have been fabricated by Si micromachining. The transmission-type position sensors are placed in tandem to measure displacements at multiple positions and are applied to the straightness measurement of the table translation in machine tools. The proposed optical encoder has also been fabricated by Si micromachining, in which some components for the sensor are integrated. The interpolation of the encoder signal has been improved by suppressing the harmonic noise of the encoder output. A large working distance between the scale and index gratings is a useful property of the grating-image-type encoder. Integration of the grating-image-type encoder has also been demonstrated. In addition, a new optical interferometer is also explained for a precise displacement measurement. The interferometer consists of an ultra-thin-film photodetector, by which the intensity distribution of the standing wave is monitored. Due to its simple optical configuration (i.e., single optical arm instead of the conventional two arms of the interferometer), the proposed interferometer is compact enough to be installed in mechanical systems. An integrated displacement sensor using the new interferometry has been reported.

References

Hane, K. and Grover, C. P., Magnified grating images used in displacement sensing, *Appl. Opt.*, 26, 2355–2359, 1987a.

Hane, K., and Grover, C. P., Imaging with rectangular transmission gratings, *J. Opt. Soc. Am.*, A4, 706–711, 1987b.

Hane, K., Endo, T., Ito,Y., and Sasaki, M., A compact optical encoder with micromachined photodetector, *J. Opt. A: Pure Appl. Opt.*, 3, 191–195, 2001.

Hane, K., Endo, T., Ishimori, M., Ito, Y., and Sasaki, M., Integration of grating-image-type encoder using Si micromachining, *Sensors Actuators A*, 97–98, 139–146, 2002.

Kato, J., Kikuchi, N., Yamaguchi, I., and Ozono, S., Optical feedback displacement sensor using a laser diode and its performance improvement, *Meas. Sci. Technol.*, 6, 45–52, 1995.

Merlo, S. and Donati, S., Reconstruction of displacement waveforms with a single-channel laser-diode feedback interferometer, *IEEE J. Quantum Electron.*, 33, 527–531, 1997.

Mi, X., Sasaki, M., Hirano, T., and Hane, K., Interferometers based on standing wave detection using thin film photodiodes, *Trans. IEE Jpn.*, 121E, 489–495, 2001.

Ohashi, T., Hirano, T., Ieki, A., Matsui, K., Nashiki, M., Sasaki, M., and Hane, K., Optical encoder having pitch-modulated photodiode array, *Trans. IEE Jpn.*, 119E, 86–93, 1999.

Sasaki, M., Takebe, H., and Hane, K., Transmission-type position sensors for the straightness measurement of a large structure, *J. Micromech. Microeng.*, 9, 429–433, 1999a.

Sasaki, M., Mi, X., and Hane, K., Standing wave detection and interferometer application using a photodiode thinner than optical wavelength, *Appl. Phys. Lett.*, 75, 2008–2010, 1999b.

Suhara, T., Taniguchi, T., Uemukai, M., Nishihara, H., Hirata, T., Iio, S., and Suehiro, M., Monolithic iterated-optic position/displacement sensor using waveguide gratings and QW-DFB laser, *IEEE Photo. Technol. Lett.*, 7, 1195–1197, 1995.

5

Distributed Optical-Fiber Sensing

Alan Rogers
University of Surrey
Guildford, Surrey, U.K.

5.1 Introduction

Distributed optical-fiber sensing (DOFS) offers an extra dimension for the monitoring and diagnosis of large structures: the use of a one-dimensional, passive, dielectric measurement medium, flexible enough to be installed, with minimum intrusion conveniently and (if necessary) retrospectively, on extended structures such as dams, bridges, oil wells, aircraft, spacecraft, industrial pressure vessels and boilers, power generation and chemical plants, and mining installations and equipment is attractive as a means for offering the measurement of, for example, temperature and strain distributions. It can offer both a continuous monitor for early detection of anomalous (perhaps potentially destructive) conditions so that corrective action may be taken and for improving the detailed understanding of behavior (especially under extreme conditions) for use in the next generation of design. The use of these techniques promises significant improvements in structural integrity over a broad range of industries.

DOFS comprises both quasi-distributed systems, which allow for the determination of a measurand field at specific, predetermined positions along the fiber, and fully distributed systems, with a capability for measurement at any point along the length of the fiber. Primary among measurands of industrial interest are temperature and strain/pressure, and several systems, both fully and quasi-distributed, have been reported for such measurements. These include commercially available systems for temperature measurement.

This review of DOFS involves an examination of the subject's origins and primary motivations, a description of the principles upon which it based and by means of which it has evolved, and a detailed look at the particular way in which it has developed. This will be followed by a review of current activity and application areas, closing with some projections as to possible future technical and commercial progressions.

DOFS is a particular development within the more general field of optoelectronics, a subject that effectively began with the invention of the laser in 1960. Moreover, the evolution of DOFS tracks quite

0-8493-1162-4/03/$0.00+$1.50

accurately the evolution of opto-electronics as a whole, almost every development in the latter finding valuable application in DOFS.

The laser provides a source of quasi-monochromatic, highly collimated, intense, coherent light. It thus provides an optical source onto which information can readily be impressed and from which it may be extracted. In 1960 for the first time a source of optical frequencies was available that compared favorably with those at radio and microwave frequencies for purposes of information transfer.

Awareness of these possibilities stimulated a flurry of activity in pursuit of optical communications, for the very high carrier frequency offered by light (~10^{14} Hz) implied an increase in communications bandwidth of many orders of magnitude. However, there was a rapid realization that light radiation propagating freely through the atmosphere could not practically sustain communications systems with paths greater than about 1 km, owing to atmosphere attenuation, especially in the presence of rain, fog, or snow [Hogg, 1964]. Hence attention was turned toward guiding protected paths for the light transmission.

The first ideas along these lines, in the early 1960s, involved the use of long pipes around 150 mm in diameter [Gouban and Christian, 1964]. These pipes were to be laid underground (to minimize transverse thermal gradients) and filled with a dust-free gas (e.g., N_2) that had no absorption bands in the optical range of interest. The optical radiation was then to be transmitted through this pipe, periodically refocused by lenses at intervals of ~500 m, and deflected around corners by mirrors. It soon became clear that such a system would be expensive to install and difficult to maintain in the face of drift in the optical alignment resulting from environmental changes.

Also around this time interest was being focused on optical fibers. These thin wires of glass had been used almost as interesting curiosities during the 1950s for very specialized short-range applications where light had to be delivered along circuitous paths, in microscopes or medical instruments, for example; the attenuation in the silica-based glass then available was much too great (~250 dB km^{-1}) for paths greater than a few meters to be contemplated.

However, a major breakthrough occurred in 1996 when Kao and Hockham [1996] in England and Werts [1966] in France demonstrated quite convincingly that the high losses in the fibers were largely due to removable impurities such as heavy-metal ions. This awareness initiated intense activity directed toward removing these impurities, leading to a major milestone in 1970 when fiber loss fell to 20 dB km^{-1} [Kapron et al., 1970], at which point fiber communications became competitive with coaxial cable systems. This result led to an explosion of optical-fiber communications activity. Optical fibers are now available with attenuations of only 0.15 dB km^{-1}, close to the theoretical limit imposed by Rayleigh scattering.

The convenient availability of low-loss optical fibers began to stimulate thoughts that they might be put to other uses. Primary among these was that of measurement sensing.

It was clear that the propagation of light in an optical fiber was influenced in a variety of ways by its external environment, thus allowing the possibility of measurement of external fields using a medium that was passive, insulating, flexible, and easily installed in existing structures.

Optical-fiber sensor technology began with very simple sensors such as the "breakpoint" temperature sensor (Figure 5.1), where the exiting light level depends upon the alignment of two fiber ends, which

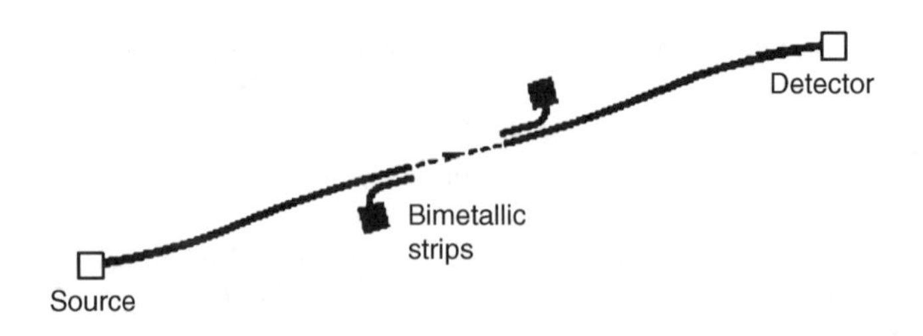

FIGURE 5.1 The "breakpoint" temperature sensor.

is itself controlled by the external temperature via the bending of bimetallic strips. Such a device can provide an inexpensive, simple fiber alarm.

The subject progressed through much more sophisticated "point" sensors such as the optical-fiber gyroscope [Vali and Shorthill, 1976; Lefèvre, 1993], which made use of optical interference, and the current-measurement device [Rogers, 1979; Smith, 1980], which used the light's polarization properties. Thus, the steady realization was growing that a variety of optical effects could be utilized in the measurement function via the facilities offered by laser light in association with optical fibers.

In parallel with these sensing developments optical-fiber communications had been proceeding apace, and by the mid-1970s operational systems were in place. These operational systems led to a requirement for a variety of diagnostic equipments. Among these was the requirement for equipment that could locate fiber breaks, regions of anomalously large loss, and bad joints. For this the optical time-domain reflectometer (OTDR) was born [Personick, 1977] (Figure 5.2).

The OTDR comprises what is, effectively, one-dimensional optical radar. An optical pulse from a laser is launched into the fiber, and light is continuously backscattered from it as it propagates, as a result of Rayleigh scatter from the small ($\ll \lambda$) inhomogeneities and impurities in the amorphous silica of which the fiber is composed. The backscattered light power emerges at the launch end and is time-resolved (Figure 5.2b) to provide a differential map of the spatial distribution of the optical attenuation along the fiber.

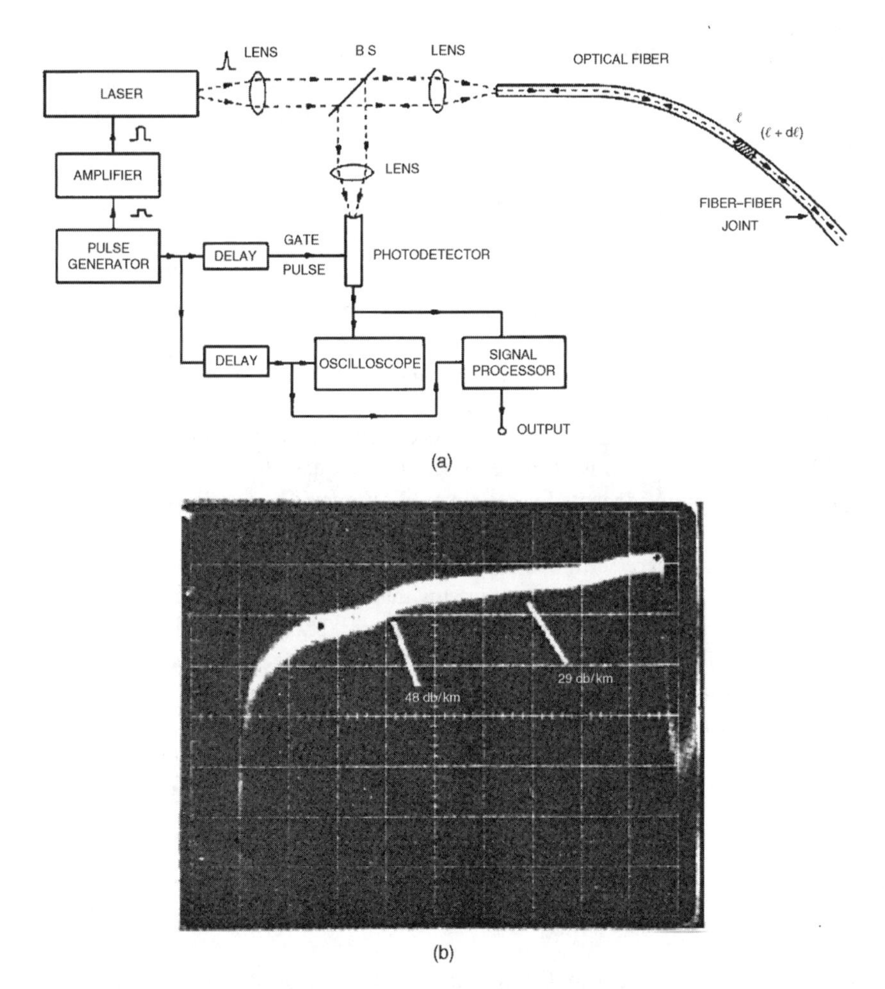

FIGURE 5.2 Optical time-domain reflectometry: (a) a schematic diagram of OTDR; (b) a typical OTDR trace. (Courtesy of Aurora Instruments, Inc., Ambler, PA.)

The OTDR thus effectively comprised the first DOFS system, as it allowed spatial measurement of any external parameter that was capable of influencing the fiber attenuation, such as those that might cause a bend or a break, for example.

Spatially distributed measurement was not the primary purpose of the OTDR, but it stimulated thought in this direction. Clearly, if distributed measurement were the primary purpose of a sensor design along the lines of OTDR, use should be made of a parameter that is sensitive to the desired measurand field, the latter being, for example, temperature, pressure, electric field, magnetic field, nuclear radiation level, etc. The fiber attenuation has only limited sensitivity to such fields, so other properties needed to be sought. Of the parameters that characterize a light wave, the most sensitive to external perturbation are those that depend on phase, for this is sensitive to variations in the optical path of the order of a wavelength. Thus, we look naturally toward properties such as polarization state (which depends upon differential phase), coherence, and phase itself (via optical interference phenomena).

The first arrangement to be examined seriously for DOFS was polarimetric in form. It was analogous to OTDR but it time-resolved the backscattered light's polarization state rather than its power level. It was designated "polarization-optical time-domain reflectometry" (POTDR) [Rogers, 1980], and it mapped the spatial distribution of the fiber's polarization properties and thus, correspondingly, the distribution of any external measurand field that modified them: the polarization properties of the fiber are sensitive to many external fields of interest (e.g., pressure, strain, temperature, electric field, magnetic field). POTDR is described in more detail in Section 5.4.1.

It soon became evident that DOFS offered an extra dimension to the monitoring and diagnosis of large structures. The use of a one-dimensional, flexible, passive, dielectric measurement medium, flexible enough to be installed, with minimum intrusion conveniently and retrospectively on structures such as bridges, dams, aircraft, spacecraft, industrial pressure vessels and boilers, and power generation and chemical plant and mining equipment was attractive as a means for offering, by measurements of strain and temperature distributions, for example, both a continuous monitor for detection of anomalous (perhaps potentially catastrophic) conditions and for improving the understanding of behavior (especially under extreme conditions) for use in a new design.

As awareness of the possibilities that were being offered began to gain strength, the subject developed structure, and separate strands became established. First, there was diversification into fully distributed DOFS systems, which have the ability to make a field measurement at any point along the length of the fiber, and quasi-distributed DOFS systems, which are able to make the measurement only at particular predetermined positions along the fiber, by dint of special treatment or arrangement at those positions. Clearly, the fully distributed system is more versatile; but the quasi-distributed system is often simpler and less expensive and is quite adequate for certain applications. (The quasi-distributed arrangement is sometimes referred to as a "multiplexed system," but this is not preferred as it should properly be applied to those systems that do not necessarily contain a linear array of sensing points.)

Second, there was diversification into those systems that use linear optical effects and those that use nonlinear effects, either of which may be fully or quasi-distributed. The linear effects are simpler to implement with respect to source and detector requirements, but the nonlinear effects offer extra degrees of freedom in their diversity and in their specific responses to measurement fields [Rogers, 1992a,b].

The subject of DOFS is expanding in all its strands with applications in a variety of industrial and commercial areas. Some of these are opening new windows for monitoring and control, such as the application to so-called "smart" structures, where the acquired measurement information is fed back to allow the structure to adapt continuously to a changing environment. This will lead eventually to a new generation of "intelligent" extended structures, but in the shorter term the primary advantages of the extra dimension in information gathering are large cost reductions consequent upon operation of extended, critical structures at optimum conditions and a deeper understanding of behavior, all of which will lead to improved design.

This chapter begins with a more detailed look at the principles on which DOFS is based; it will then review the development of quasi-distributed systems, followed by a review of fully distributed systems, both linear and nonlinear.

Current and possible future application areas will then be discussed, with an emphasis on the special problems that each one presents. Finally, there will be a discussion of the entire field of endeavor—and of what the future may hold.

5.2 Basic Principles

5.2.1 Optical-Fiber Basics

As might be expected, the basic principles underlying DOFS rely heavily on the basic principles of light propagating in optical fibers, so that these will now be summarized briefly, with an emphasis on those aspects that are especially important for DOFS.

An optical fiber is a thin wire of glass. Its overall thickness is about 100 μm—almost the same as that of a human hair. When light is launched in at one end of the fiber, it travels along to the other end by means of a series of reflections from the sides. This guiding action is maintained even if the fiber is bent quite sharply, so that the fiber can be handled in much the same way as copper wire.

To understand the various ways in which optical fibers may be used as sensors, it is necessary to look rather more closely at the way in which light propagates down a fiber. An optical fiber has the cross-sectional structure shown in Figure 5.3(a): a central core, normally of glass or silica, is surrounded by a cladding that is also glass or silica but that has a slightly lower refractive index. The consequence of this is that there will exist an angle θ_c within the central core, for which a light ray striking the core-cladding interface will be totally internally reflected. This ray and all rays at angles greater than θ_c will thus be guided down the fiber. The dimensions of the fiber are comparable to the wavelength of light, so that if a reasonably monochromatic light source is being used, we can expect recognizable interference patterns to be formed within the fiber core, as the many bouncing waves interact with each other. Such interference patterns are indeed set up and are referred to as *modes of propagation* (Figure 5.3(b)). The allowed modes will evidently depend on the fiber's geometry and on the wavelength of the light used; the distribution of light power among the allowed modes will depend on the original conditions for the launching of the light and on any geometrical or other perturbations that are capable of transferring power from mode to mode.

It is already clear, then, that the mode power distribution will be sensitive to any changes in the geometry of the fiber imposed by external influences. But the full sensitivity contains an added subtlety. For we know from elementary optics (Fresnel's laws) that, when a light ray suffers reflection at the boundary between two media, its polarization state is also altered. This is because the electric field component in the plane of incidence will experience a different phase shift from that of components at right angles to it. Consequently, each of the allowed modes will have associated with it definable polarization characteristics. And, consequently, the polarization behavior of the light in a fiber with a large number of allowable modes will be complex and difficult to use in a measurement system. It is, however, possible to arrange the geometry of the fiber in relation to the wavelength of light used, so that only one mode is allowable—just one angle of total internal reflection. This important case (Figure 5.3(c)), known as monomode (or single-mode) propagation, ensures that the light exists in a single, definable polarization state at each point in the fiber.

Now we see that the mode structure will be sensitive not only to imposed changes in the geometry of the fiber but also to any external agent that might be capable of altering the phase (and hence the polarization behavior) of the propagating light.

Consider first just how the geometry of the fiber might be changed. A variation in temperature will expand the fiber and alter the refractive indices of core and cladding; pressure may alter the shape of the cross-section, converting it from circular to one of elliptical section; and a physical displacement may bend or twist the fiber. By observing the changes in the mode structures thus produced we can see already how we might make measurements of temperature, pressure (including sound waves), displacement, and strain. But how may we alter the polarization of the light?

When light passes through any solid material such as glass, a semiconductor, or crystalline material, it does so by interacting with the atoms or molecules that constitute that material. But we may influence

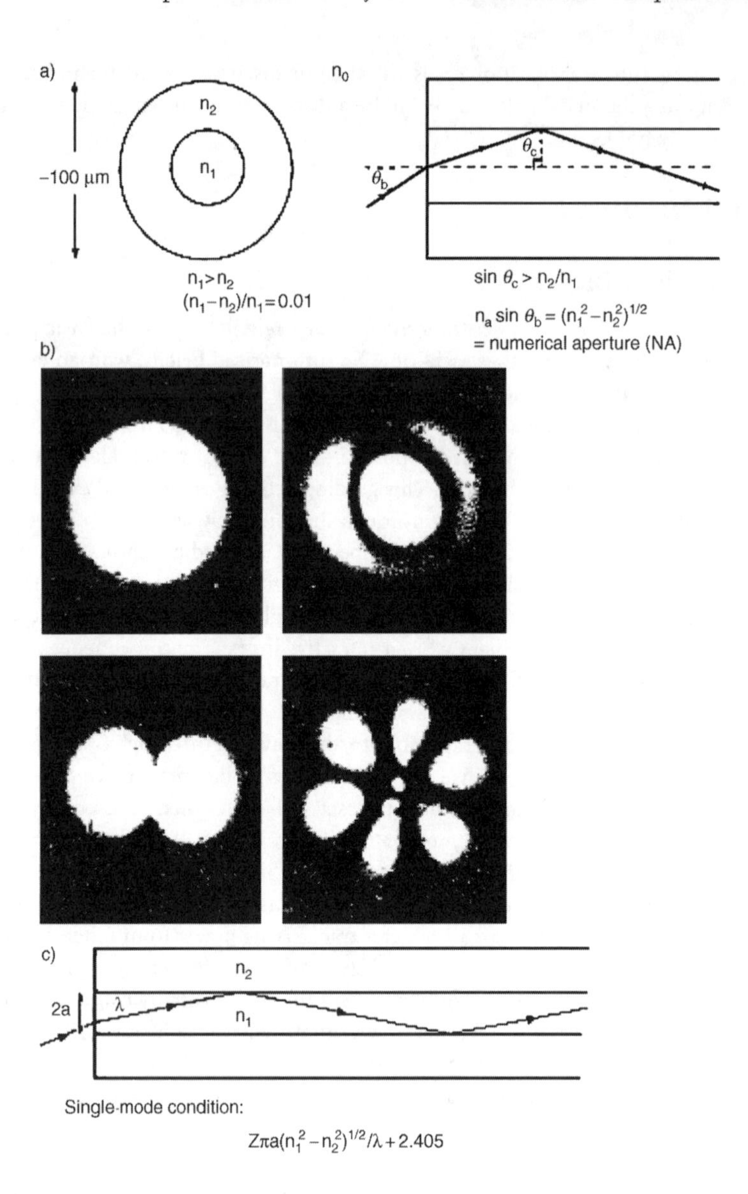

FIGURE 5.3 Optical-fiber basics: (a) the fiber cross-section; (b) modal patterns; (c) the monomode condition.

the properties of these atoms and molecules by applying external electric or magnetic fields; these fields restrict the directional movements of the atoms and molecular electrons. It follows that these external fields will also affect any light that is propagating through the material, and they will affect, in particular, its polarization state, as it is this that is determined by any structural directionality. Thus, we see how we may measure electric and magnetic fields and, hence, also measure the voltages and currents that give rise to them. (From what has previously been said it is clear, however, that we shall probably need monomode fibers for such measurements.)

There are other natural consequences of these interactions. The first is that light energy is absorbed by the molecules on which it is incident; in the process the molecules gain energy, later to be released either in the form of heat or reradiated at the same frequency as the incident light. In this latter effect, known as Rayleigh scattering, the light can be reemitted in any direction, some of it opposite to the original propagating direction; this is known as Rayleigh backscatter. Clearly, both of these interactive effects deplete the energy of the forward-propagating light, thus comprising a source of attenuation.

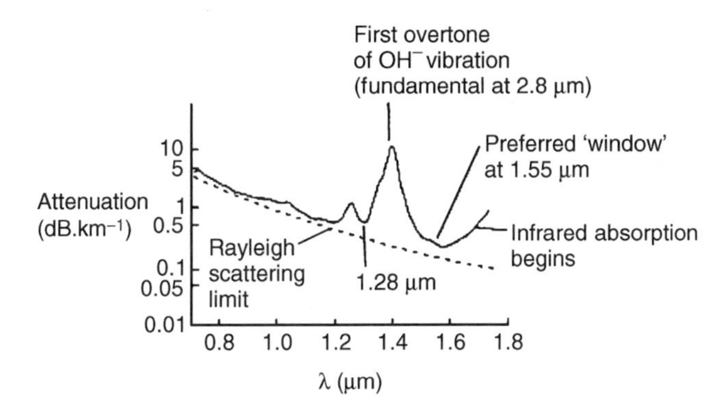

FIGURE 5.4 The absorption spectrum of a silica fiber.

This attenuation will be optical-frequency dependent, as the absorption process will depend on the molecules' natural resonances, and the scatter efficiency will depend on the relationship between the wavelength of the light and the scatter cross-section (in fact, scatter efficiency $\sim 1/\lambda^4$). Hence the fiber will be characterized by an attenuation spectrum that, for a typical telecommunications-grade fiber, will take the form shown in Figure 5.4.

The wavelength dependence of the light–matter interaction has another effect. The light is impeded in its forward passage to an extent that depends on wavelength. The result of this is that the light's phase velocity depends on wavelength, and thus the refractive index is a function of wavelength. This latter effect is known as chromatic (or material) dispersion, and it is extremely important in communications systems where the spread of wavelengths, which is a necessary property of all real optical sources, will lead to a blurring of the modulation information in the presence of such dispersion (e.g., a broadening of the pulses in a pulse-code system) and to a loss of information bandwidth. (Alleviative methods are introduced to combat this loss of bandwidth.)

In sensor systems these problems of attenuation and dispersion are less severe than in the case of telecommunications applications because the distances involved are two or three orders of magnitude smaller. Nevertheless, they must be considered when optimizing a sensing system, so that the operating wavelength and the spectral spread of the sources must be chosen carefully in relation to the fiber that is used.

With regard to communications systems, there are several special problems that must be considered when designing sensing systems. One obvious problem immediately arises. Because the fibers are sensitive to so many external influences, how do we ensure that a given fiber is sensitive to only one wanted parameter at a time? And a related question: Is this large range of sensitivities compatible with the much-vaunted interference-immunity of optical-fiber telecommunications? The first question is crucial for the design of optical-fiber sensing systems. To answer it for any given measurement function means that the designer must use considerable opto-electronic engineering skill and ingenuity. He must also rely heavily on a basic understanding of behavior and on the ability of the technologist to provide fiber with carefully controlled, and predefined, properties.

The answer to the second question is more straightforward. To avoid any sensitivity to external influences one must ensure that the optical detection system is sensitive not to mode structure but only to total received light power. With care this can be arranged.

The skill and ingenuity of the sensor system designer can be assisted considerably by specialized fibers. In addition to the basic fiber properties already described, special properties, of use in telecommunications and sensing technologies, can be added to the fibers. One of the processes by which this is done is the doping of the fiber with, for example, elements such as neodymium, erbium, praseodymium, yttrium, or germanium. These elements provide, variously, temperature-dependent absorption, fluorescence spectra, or enhanced Raman-scatter coefficients. (The fluorescence properties of the rare-earth dopant erbium are famously useful in fiber lasers and fiber amplifiers [Digonnet, 1993].)

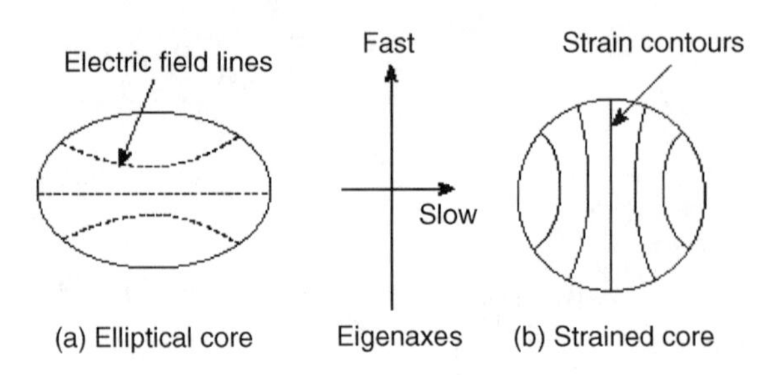

FIGURE 5.5 Types of high-birefringence fiber: (a) with an elliptical core and (b) with a strained core.

Another important special fiber is the so-called high-birefringence, or hi-bi, fiber. In the hi-bi fiber, cross-sectional asymmetry is introduced deliberately into the fiber core. This may be done either by making the core elliptical [Dyott, 1995] (Figure 5.5(a)) or by straining the fiber preferentially in one (transverse) direction (Figure 5.5(b)) [Birch et al., 1982]. The consequence of this is that light that is linearly polarized along one of the preferred axes (strain axis or ellipse axis) will travel at a different velocity than light linearly polarized in the orthogonal direction. Thus, the fiber exhibits birefringence, in this case linear birefringence. An important characterizing parameter for a hi-bi fiber is the length of fiber over which the two polarization components will slip in phase by 2π. This is known as the beat length and, for a typical proprietary hi-bi fiber, is of order a few millimetres. Clearly, the greater the birefringence, the smaller the beat length. The two linear polarization states along the preferred axes will propagate unchanged as linear states, while any other polarization state (including linear states not aligned with the axes) will change continuously as they propagate. The two states that propagate without change of form are called the polarization eigenstates or eigenmodes of the fiber. In the case described these states are linear, but they will generally be elliptically polarized states. This will be the case when a linearly birefringent fiber is twisted. For example, the twist adds circular birefringence to the linear birefringence already present to give a resultant elliptical birefringence.

Finally, there are fibers with special coatings. Telecommunications fibers need coatings to make them more rugged. A soft primary coating is applied to the fiber as it is drawn from the preform melt in order to protect it from atmospheric attack (especially from moisture) while its surface is bare and vulnerable. A much harder, secondary coating is added subsequently to provide strength for handling and duct installation. A sensor fiber also needs to be strengthened, but it also needs a coating that will enable it to survive in the measurement environment and to interact optimally and consistently with the measurand field. Some special sensor coatings are already available: metal coatings, polyimide coatings, and carbon surface-impregnation, for example. However, much more needs to be done in this area if fiber-sensing systems are to be matched properly to their measurand environments.

5.2.2 Basics of DOFS

The basic principles of distributed optical fiber sensing have been covered in earlier reviews [Rogers, 1986; Dakin, 1987; Kersey and Dandridge, 1988], but for convenience they will be summarized here. DOFS systems offer a unique measurement capability—that of making a spatially distributed measurement of a measurand field with a spatial resolution of 0.1 to 1 m, and a measurement accuracy of ~1%, over widely varying distances, according to the measurement method and application, from ~10 m to ~100 km.

The ability to determine the spatial and temporal features of a measurand field with a medium that is nonintrusive, dielectric, passive, flexible, and easy to install—even retrospectively—offers a new dimension in the monitoring, diagnosis, and control of large extended structures of all kinds. No conventional measurement techniques can compete effectively.

The singular problem, then, to be solved for DOFS systems, when compared with almost all other types of measurement systems, is to determine the value of a measurand continuously as a function of position along the length of an optical fiber with some definable spatial resolution and sensitivity. This implies that each spatially resolved measurement that is made must be identified in some way with a particular fiber section whose position is known. Clearly, in doing this it is not possible to identify the position with some kind of active, coded transmitter, if the important advantages of the fiber as a passive, dielectric medium are to be retained. Hence, the identification should be made from one or the other of the fiber ends; in some cases both ends are used. There are several ways in which this might be done.

As mentioned above, the subject of DOFS was stimulated by the OTDR technique, which uses the temporal resolution of light continuously Rayleigh-backscattered from an optical pulse propagating in the optical fiber [Barnoski and Jensen, 1976]. Clearly, if the delay between the launch of the pulse and the time at which the backscattered light is received is τ, then the fiber section from which the backscatter occurred is identified as that which lies at distance s from the launch end of the fiber, where:

$$s = \frac{c\tau}{2} \tag{5.1}$$

and c is the velocity of light in the fiber. Such temporal resolution can be used in both quasi-distributed (Figure 5.6(a)) and fully distributed (Figure 5.6(b)) arrangements.

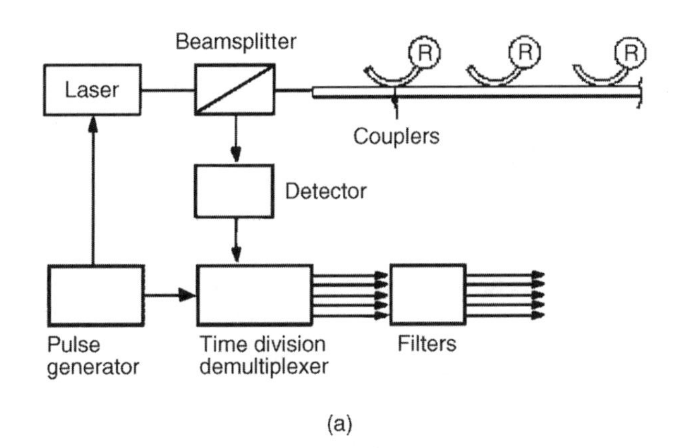

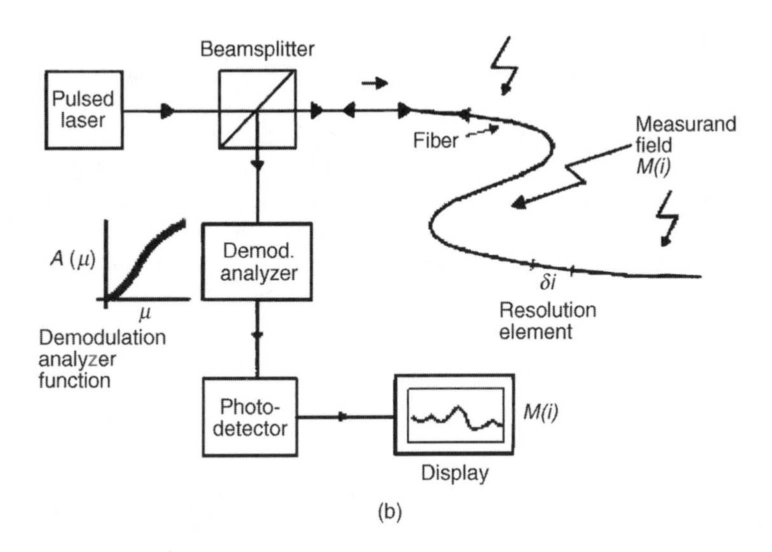

FIGURE 5.6 Schemes of distributed sensing: (a) quasi-distributed and (b) fully distributed.

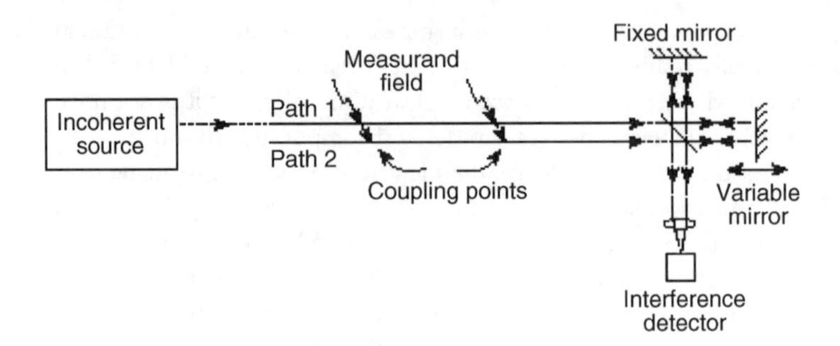

FIGURE 5.7 Distributed coherence sensing.

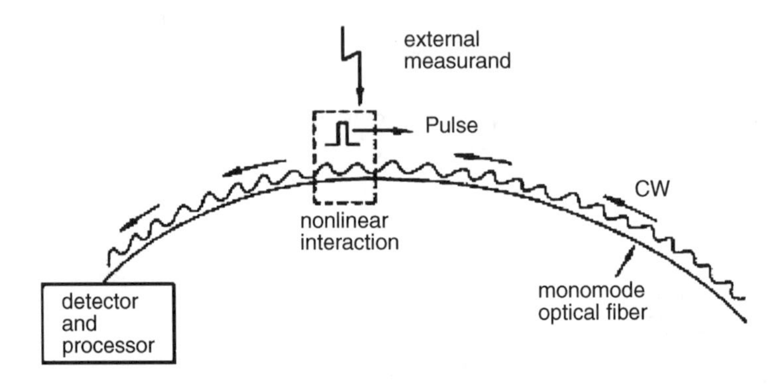

FIGURE 5.8 Basics of forward-scattering DOFS.

However, all measurement technology embraces the art of compromise and trade-off. In particular circumstances, this temporal resolution technique may not be optimal; for example, it may not provide the sensitivity required as backscatter power levels are very low (~10^{-6} of the pulse power per meter of fiber).

As a result, other methods also are used. For example, the individual sensors in a quasi-distributed system may be wavelength selective and can thus be interrogated with a broadband continuous wave (CW) source. The identification in this case is made in the frequency domain, via a detection grating, prism, or tunable filter.

A rather more subtle method for positional coding is illustrated in Figure 5.7. This arrangement involves two optical paths with differing effective light velocities. (These might, for example, comprise the two polarization modes of a high birefringence fiber.) The effect of the measurand field is to couple light from one path to the other. Light of low coherence is launched into one of the paths. When the measurand field causes coupling into the other path at a particular point, the two components then travel at different velocities to the exit end and experience a relative delay that renders them mutually incoherent. Optical interference between them occurs at the exit end only if a delay is inserted between them of just the right amount to correspond to their travel delay, which identifies the position at which the coupling occurs. Hence a variable delay at the exit will allow the fiber couplings—and the measurand field—to be scanned along the fiber length.

So far, we have considered only the possibilities offered by linear optical systems; but there is also a class of distributed sensors that use nonlinear effects, and these effects provide another option. This option is that of forward-scatter pulse–wave or pulse–pulse interaction. Consider the arrangement shown in Figure 5.8. In this case a pulse of light with high peak power is launched into a fiber, and it generates a local nonlinear effect as it propagates. A counter-propagating CW will experience the nonlinearity as the pulse passes through it, and it will be modulated in a way that depends upon the nature of the nonlinearity. Upon emergence, the CW's temporal variation will map the passage of the pulse through it,

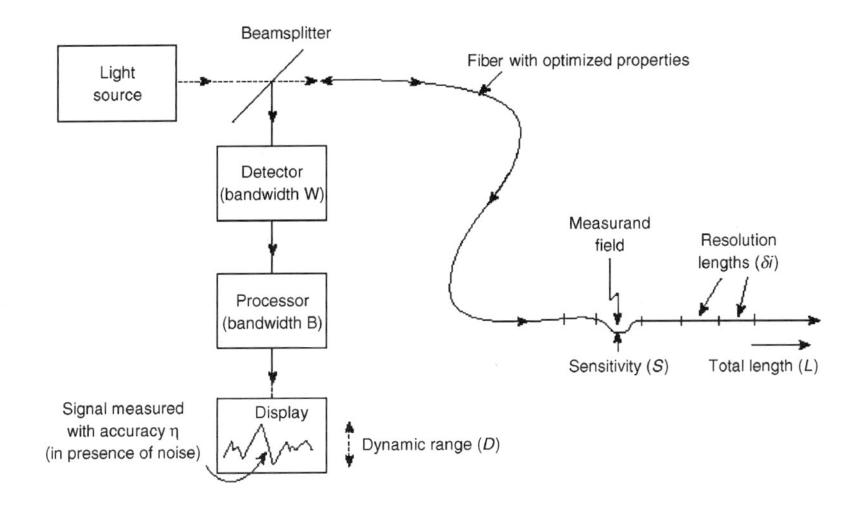

FIGURE 5.9 DOFS performance parameters.

so that, if the nonlinear interaction is influenced by an external field, this field will be correspondingly mapped along the fiber. Such systems possess the considerable advantage of temporal resolution without the low sensitivity inherent in backscatter methods. Their disadvantage is that they require a high-power, pulsed laser source in order to enter the nonlinear regime.

5.2.3 DOFS Performance Parameters

The performance parameters that characterize any given DOFS system, many of which are common to all measurement systems, may be stated as follows (see Figure 5.9):

1. The spatial resolution ($\delta\ell$) is the smallest fiber length over which any sensible change in the spatial variation of the measurand can be detected. Effectively, the measurement information is presented as a series of values, each value representing the magnitude of the measurand averaged over a section of fiber of length $\delta\ell$. In a sense, then, the system multiplexes $L/\delta\ell$ sensors in a linear array, where L is the total length of the measurement fiber (units: meters).
2. Sensitivity (S) is defined as the change in optical power, at the detector, produced by unit change of measurand field per unit length of fiber. (units: watts. field^{-1}. meters^{-1}).
3. Measurement bandwidth (B) is the bandwidth over which the changes in measurand field can be measured for the full fiber length, L (Hz).
4. System bandwidth (W) is the bandwidth that the detector must possess in order to operate for the system. It must be such as to allow the detector to respond to the passage of the optical pulse along one resolution length, $\delta\ell$ (Hz).
5. Dynamic range (D) is the ratio of maximum to minimum values of the measurand field that allow each to be measured to the required accuracy (dB).
6. Accuracy of measurement is the accuracy with which the output power at the detector can be measured in the face of system noise levels (%).
7. The system specifications must include a specification of the fiber in use, so as to allow any necessary modification and to facilitate an understanding of anomalous indications. These should include:

- The attenuation spectrum: the attenuation as a function of wavelength around the operating point.
- The dispersion spectrum: the refractive index as a function of wavelength around the operating point.

- The modal type: monomode or multimode (if the latter, number of modes).
- Geometrical properties: core diameter, cladding diameter, refractive index profile, cut-off wavelength for monomode operation, normal or hi-bi.
- Coating properties: primary coating type, secondary coating type, tensile strength, maximum operating temperature, etc.

As in all measurement systems there is always a strong trade-off among the above parameters in order to optimize for any given application. Examples of these trades-offs will abound when particular systems are later described. A simple, pervading example is that of sensitivity vs. spatial resolution: as defined, the sensitivity (S) will be greater, the greater the resolution length of fiber over which the measurement is made; however, if the spatial resolution is to be good, this length ($\delta\ell$) must be small. Thus, we encounter a sensitivity/resolution trade-off, in common with all measurement systems.

The ways in which the positional identification methods have been used, and the trades-off effected, will be amply illustrated as the development of the subject is considered.

5.3 Quasi-Distributed Systems

A quasi-distributed, optical-fiber sensing (QDOFS) system is one in which only prescribed sections of the fiber are sensitive to the measurand field. These have two main advantages. First, only those prescribed sections need have any sensitivity to the measurand field; and, second, their positions are known, so that they merely have to be identified somehow via an individual signature; the disadvantage is that the required measurement points in the field must be known in advance.

A schematic of a QDOFS system is shown in Figure 5.10(a). In these cases the identification of the individual sensors is performed in the time domain via backscatter [Dakin, 1992]. In the schematic in Figure 5.10(b), the identification is made in the optical frequency domain [Mallalieu et al., 1986].

Practical systems that have been studied use a variety of identification techniques [Brooks et al., 1983; Chen et al., 1990]. A much-studied example of one such system will now be described.

5.3.1 Bragg-Grating QDOFS

In 1978 photosensitivity was discovered in optical fibers [Hill et al., 1978]. This is a phenomenon whereby the refractive index of a fiber material (i.e., doped silica) can be modified (permanently or semipermanently) by exposure to ultraviolet (UV) light. The mechanisms of the effect are various, complex, and as yet incompletely understood, although it is known that they involve either electron traps created by the impurities in the material structure (type-I gratings, semipermanent [Mizrahi et al., 1993]) or actual physical damage to the core–cladding interface (type II, permanent [Archambault et al., 1993]). However, the methods by which photosensitivity can be induced in fibers are now well tried and tested.

Optical-fiber gratings are finding a variety of applications as components in optical communications systems (e.g., wavelength filters, selective reflectors in fiber lasers, and dispersion compensators), but they are also extremely useful in QDOFS systems. An illustration of their use is shown in Figure 5.11. Here, a number of sinusoidal gratings are arranged along a monomode fiber, each grating selectively reflecting a different wavelength according to its spatial period. The center wavelength of the reflection is given by:

$$\lambda_B = 2n\Lambda \tag{5.2}$$

where n is the refractive index of the fiber material and Λ is the spatial period (the "grating spacing"). If the system is now interrogated with a broadband source, the reflection spectrum consists of a series of peaks, each one corresponding to a particular grating, which is thereby identified. Now λ_B is dependent upon external fields that may vary n or Λ. The most important of these are temperature and strain, each of which modifies both n and Λ. Clearly, a series of Bragg gratings such as this can act as a quasi-distributed measurement system for temperature alone in a strain-free arrangement or simultaneously for temperature

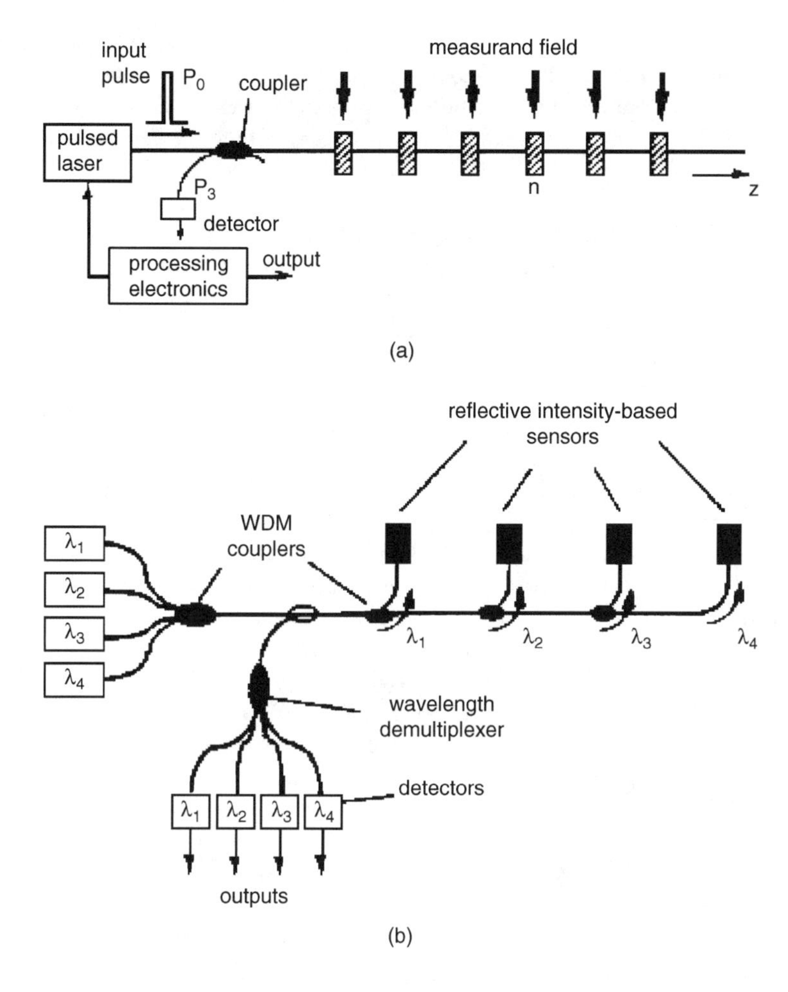

FIGURE 5.10 Quasi-distributed sensing: (a) backscatter time-domain arrangement; (b) frequency-domain arrangement.

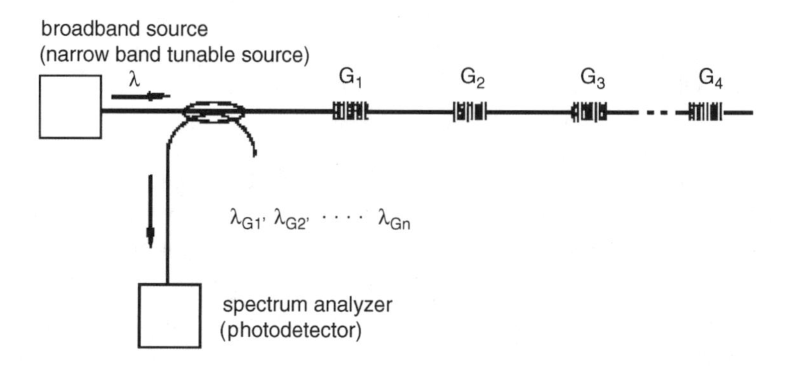

FIGURE 5.11 Interrogation of a Bragg-grating array.

and strain, provided that the two can be discriminated; much attention has been given to effective discriminatory procedures. These include the use of gratings with different spacings overlaid at the same position [Kanellopoulos et al., 1995], the use of two spatial propagation modes [Kanellopoulos et al., 1994], and the use of two polarization modes [Handerek and Rogers, 1993]. The writing process for the gratings gives easy control over their length, which can be as small as 1 to 2 cm. Hence, the spatial resolution is now of this order, which is very high by DOFS standards. As the number of gratings along the fiber increases, the design of an effective interrogation system becomes more difficult. The source cannot be too broad in bandwidth, for this creates difficulties with regard both to the launching of power into the fiber and to the fiber's attenuation spectrum and dispersion characteristics. If a single source is scanned in wavelength, the scanning range will be limited. There will also be difficulties with regard to the wavelength analysis of the returning light. Considerable attention has been paid to this problem [Kersey and Dandridge, 1993; Jackson et al., 1993], and the current limitation is at about 30 gratings. The durability, flexibility, and versatility of fiber gratings will ensure that development in this area will continue and that this technology will have a significant role to play in the monitoring and diagnostics of strain and temperature in extended structures well into the future. Already there have been many successful field trials.

5.3.2 Summary

The QDOFS systems are valuable because by presensitizing specific regions of the fiber to the wanted measurand field several advantages are conferred. First, the sensitization can be highly specific to the wanted measurand, leading to a good signal-to-noise ratio at the detector and thus a good measurement sensitivity. Second, the sensitive region can be made arbitrarily small, leading to good spatial resolution. Third, the sensitized regions (transducers) can be encoded to allow easy identification at the detector (the particular grating period of a fiber grating leading to a known frequency of optical reflection is a good case in point) and thus to simple detection electronics. However, there are disadvantages. Because the transducer points are fixed, the crucial regions in the measurand field must be known in advance, and they must remain fixed. This is by no means always convenient. Further, the sensitization of the fiber at fixed points usually leads to large attenuation of the interrogating light and thus to limited numbers of transducers and limited dynamic range. Finally, the necessity for decoding, in addition to other, more standard, requirements, tends to lead to a burdensome complexity in the electronics.

Fully distributed optical-fiber sensing possesses none of the above disadvantages and is generally much more versatile. Of course, it has its own special problems; these will become clear as we describe these systems.

5.4 Fully Distributed Systems

The fully distributed optical-fiber sensing (FDOFS) system possesses the singular advantage of allowing a measurement of the external field to be made at any point along the length of the fiber, within the limitation of the spatial resolution interval. Thus, it may be possible to make a distributed measurement of a field over 2 km with a resolution interval of 1 m, giving 1000 measurement points and, as a result, effectively linearly multiplexing 1000 transducers.

Most FDOFS systems use time-domain techniques, rather than (the equivalent) frequency domain techniques, owing to their considerably reduced complexity and increased system bandwidth. Hence, this review will concentrate on these. FDOFS systems fall into three primary subclasses:

1. *Linear backscatter*: In this class the propagating optical pulse lies within the linear regime, and light backscattered from the pulse is time-resolved and analyzed to provide the spatial distribution of the measurand field (Figure 5.12(a)).
2. *Nonlinear backscatter*: The difference here is that the optical pulse has sufficient peak power to enter the nonlinear regime, and the (linear) backscattered power has to be analyzed differently (Figure 5.12(b)). The advantages of entering the nonlinear regime are that there is a diverse range of nonlinear optical effects offering specific responses to external measurands and ready discrimination at the detector. The main disadvantage is that the magnitude of the effect is strongly dependent upon optical power and, therefore, can vary significantly along the fiber as a result of attenuation.

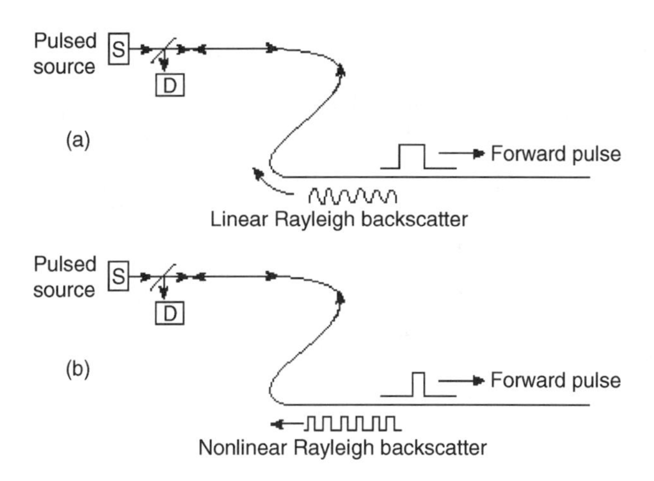

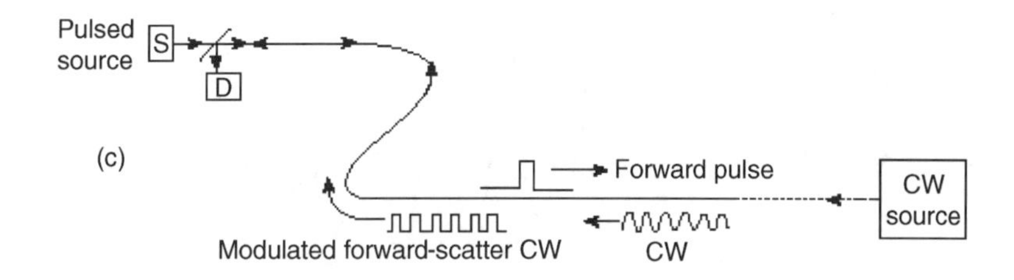

FIGURE 5.12 Schemes for fully distributed sensing: (a) linear backscattering; (b) nonlinear backscattering; (c) nonlinear forward-scattering.

3. *Nonlinear forward-scatter*: Another advantage of the nonlinear regime is that it allows independent optical signals to interact. Thus, it is possible for counter-propagating (CP) radiations (e.g., a pulse and a CP continuous wave or two CP pulses) to interact (see Figure 5.12(c)). When the interaction is influenced by the external field, the field can be mapped by a forward-scattered (as opposed to a backscattered) light propagation. However, the same disadvantage of strong power dependence also applies, of course, to this mode of operation.

Linear systems are less complex; in particular, they are less demanding with respect to source requirements and fiber properties. Nonlinear backscatter systems require high-power pulse sources and fibers appropriate for the nonlinear effect in question, but they do provide a broader range of measurand interactions and a ready discrimination at the detector. Nonlinear forward-scatter systems possess the same advantages and disadvantages as nonlinear backscatter systems but have the added advantage of a much higher signal level, which means a larger signal-to-noise ratio, and the added disadvantage of requiring two high-performance optical sources and, in most cases, access to both ends of the fiber.

An example of a linear, fully distributed optical-fiber measurement system will now be described.

5.4.1 Polarization-Optical Time-Domain Reflectometry (POTDR)

Polarization-optical time-domain reflectometry (POTDR) was, in fact, the first fully distributed optical-fiber measurement method to be studied in the laboratory [Rogers, 1980, 1981]. It is a polarimetric extension of OTDR. Whereas in OTDR the power level of the Rayleigh-backscattered radiation, from a propagating optical pulse, is time-resolved to provide the distribution of attenuation along the length of the fiber, in POTDR it is the polarization state of the backscattered light that is time-resolved; this provides the spatial distribution of the fiber's polarization properties. Only monomode fibers can be

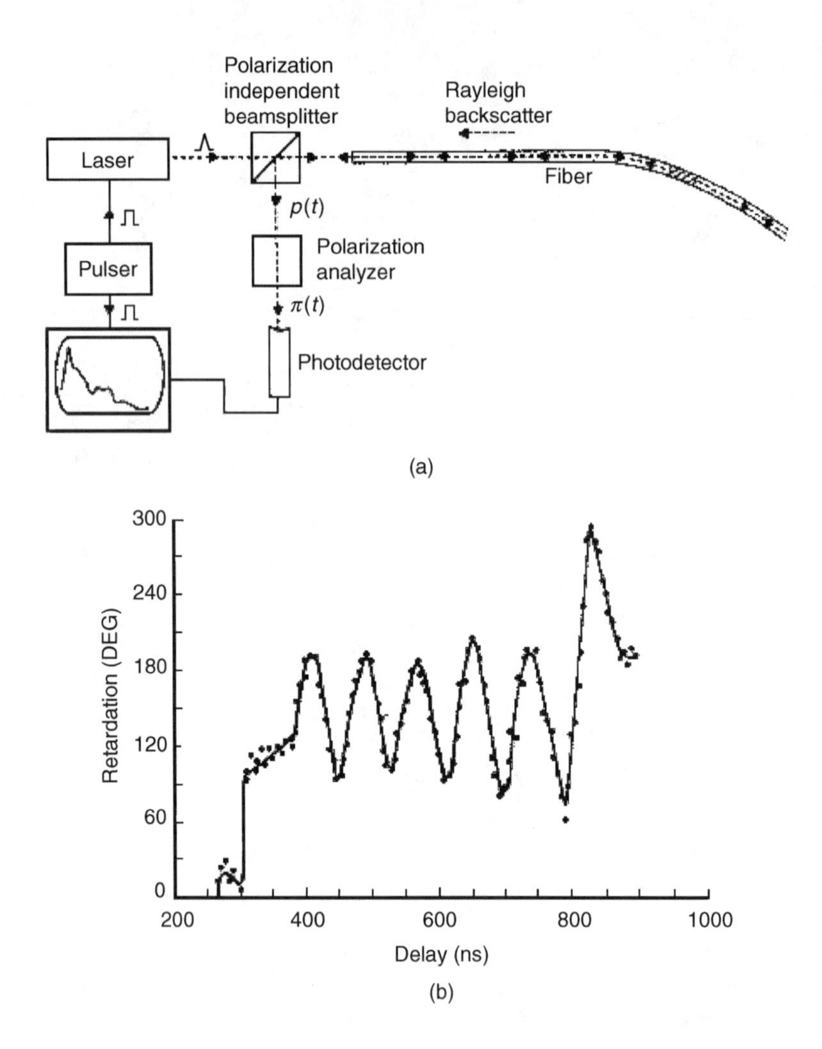

FIGURE 5.13 Polarization-optical time-domain reflectometry (POTDR): (a) basic arrangement; (b) a trace for bend birefringence on a drum-wound fiber.

involved in POTDR, because only with monomode propagation can there be a single, determinable polarization state of the light at any point in space or time within the fiber; each mode in multimode propagation will possess its own independent polarization state.

With the determination of the spatial distribution of the polarization properties of the fiber comes a capability for measurement of the distribution of any external field that modifies those properties. These include strain, pressure, temperature, electric field, magnetic field, etc.

Figure 5.13(a) shows the basic arrangement for POTDR, and Figure 5.13(b) shows the result obtained for the distribution of strain induced in a fiber when wound on a drum [Ross, 1981]. The measurement gave an accuracy of 1% for measurement of 3-m strain over 0.1 m of spatial resolution. The drum diameter was 185 mm. But the technique does possess several disadvantages. First, it cannot discriminate among the various effects (e.g., simultaneous temperature and strain), all of which are capable of modifying the polarization properties. Second, polarization information is lost in backscatter. This is most clearly appreciated by considering the propagation of light in an optically active (i.e., circularly birefringent) crystal. Any rotation of the polarization state that occurs on the forward passage of light through the crystal is canceled on back reflection through the crystal. As a result, all knowledge of a pure rotation is lost in backscatter. Consequently, the loss of information prevents full knowledge of the distribution of fiber polarization properties. There are two possible approaches to the solution of this problem: either some prior knowledge of the fiber's polarization properties must be available (as is the case in a hi-bi fiber, for example) or more sophisticated interrogation and processing techniques must be used. This problem now needs a convenient and practical solution in two arenas as interest in POTDR as a diagnostic tool has developed recently as a result of the polarization mode dispersion (PMD) problem in optical communications [Ono et al., 1994; Poole et al., 1986]. This problem derives from the fact that asymmetries in communications-grade fiber lead to small, generally elliptical, local birefringences in the fiber, which vary randomly or quasi-randomly with axial position. The result is that an optical pulse (in a digital communications system) will suffer from accumulated differential group delay (DGD) and thus become broadened by the experience, leading to a reduction in bandwidth. POTDR provides the means by which the local value of DGD can be mapped along the length of a fiber so that sections of fiber with large values can be identified and replaced [Gisin et al., 1991]. It also provides the means for comparing—and so improving—fiber fabrication processes with respect to their effectiveness in minimizing PMD in manufactured fibers [Ellison and Siddiqui, 1998].

As a result of these requirements various improvements in the POTDR technique have been proposed [Gisin et al., 1994; Zhou et al., 1997], and these have clear implications for improved DOFS systems.

In general, however, the lack of discrimination among the many effects that can modify the polarization properties of a monomode fiber has led to investigations of techniques that are more measurand specific. This leads naturally to chemical and to nonlinear methods.

5.5 Summary and Conclusions

Distributed optical-fiber sensor systems undoubtedly will have a large part to play in the monitoring and diagnostics of critical extended structures. This is especially true for the new generation of self-adjusting, self-monitoring "intelligent," or "smart," structures.

Many methods and systems have been studied, and representative examples of these have been reviewed in this chapter. The quasi-distributed systems are best suited to specific, custom-built applications over relatively short distances because attenuation of the interrogating light can be large and the multiplexing/decoding problems severe. Each of the three main classes of fully distributed systems (linear backscatter, nonlinear backscatter, nonlinear forward-scatter) has its own special advantages and disadvantages that offer matches with certain niche application areas, but they are generally more versatile than the quasi-distributed systems and operate over much larger distances (up to 100 km). On the other hand, they are more complex in design and implementation, especially in relation to sources and signal-processing requirements.

Commercialization of the ideas emerging (largely from the universities) has been slow. There are several reasons for this. First, optical fiber technology is, understandably, driven by optical telecommunications requirements, and the particular needs of the sensing technology have not been catered to specifically, although undoubtedly the rapid development of telecommunications has provided a range of fibers and components that have been very helpful to sensor researchers. For rapid, effective development and application of the sensor systems, however, there is a real need to meet their specific requirements, especially in relation to special fibers and fiber coatings.

This last point highlights the second obstacle to more effective deployment: the interface problem, i.e., that of ensuring an optimized interaction between the fiber system and the measurand field that is uniform and constant in both space and time. This problem is especially severe for fully distributed systems. The least-demanding measurand in this respect is temperature because all materials come naturally to thermal equilibrium with their environment. It is for this reason that the most successful commercialization of a fully distributed system to date has been the Raman differential temperature system [Dakin et al., 1985]. If, however, commercialization is to proceed apace, the interface problem must be tackled vigorously, and this means that a good deal of research attention must be paid to the fiber coatings that can optimize this interface.

Finally, there is the problem of the natural (and wholly understandable) conservatism of those industries that are responsible for measuring and monitoring large critical structures. This conservatism can only be broken down by effective, long-term demonstration of the advantages that DOFS has to offer. But it is sometimes difficult even to promote the advantages through demonstration: the classic, and pervasive, chicken-and-egg syndrome.

On the more positive side, the rapid advance of optical-fiber telecommunications has given rise to a large range of high-performance components and fiber types that have assisted considerably in the advance of the fiber sensor technology. For further progress, and for successful application and commercialization, it has now become very necessary to give some developmental emphasis to the specific needs of this sensor technology. But there can be little doubt that, as the requirement for ever greater understanding, monitoring, and control of large structures increases its demands on sensor technology, more and more technical and commercial attention will be paid to the powerful advantages offered by distributed optical-fiber sensing methods.

Defining Terms

Bragg grating: Comprises a (usually) sinusoidal variation in reflectivity offered to an electromagnetic wave by medium. The result is a highly wavelength-selective reflection, with a wavelength equal to twice the separation of the peak reflectivity. The idea was first used in 1912 by W. L. Bragg for selective reflection of x-rays from the periodic arrangements in crystals.

nonlinear effects: Result from the alteration of refractive index of the medium in which light is traveling by the passage of the light itself. This optical effect occurs when the intensity of the light is sufficiently large for its electric field to become comparable with the atomic electric field in the medium. Examples include Raman effect, Brillouin effect, Kerr effect, and four-wave mixing (FWM).

opto-electronics: The subject that has arisen as a result of the conjunction of the advantages of optics and electronics.

polarization: The ordered, transverse directionality of the electric and magnetic components of an electronmagnetic wave. For example, in linear polarization, the electric vector moves only in one fixed direction in space.

quasi-monchromatic: The attribute of a very narrow spectral width.

Rayleigh scattering: The scattering of light by particles very much smaller than its wavelength. In an optical fiber, these are composed of small inhomogeneities and defects in the silica structure.

References

Archambault, J. –L., Reekie L., and Russell, P. St. J., 100 percent reflectivity Bragg reflectors produced in optical fibres by single excimer laser pulses, *Electron. Lett.*, 29, 453–455, 1993.

Barnoski, M. K. and Jensen, S. M., Fibre waveguides: a novel technique for investigating attenuation characteristics, *Appl. Opt.*, 15, 2112–2115, 1976.

Birch, R. D., Payne, D. N., and Varnham, M. P., Fabrication of polarization-maintaining fibres using gas-phase etching, *Electron. Lett.*, 18, 1036–1038, 1982.

Brooks, J. L. et al., Coherence multiplexing of fibre-optic interferometric sensors, *IEEE JLT*, 1062–1072, 1983.

Chen, S. et al., A novel long-range opto-electronic scanner for coherence-multiplexed optical-fibre quasi-distributed sensors, *Proc. OFS7*, Sydney, 365–368, 1990.

Dakin, J. P., Multiplexed and distributed optical-fibre sensor systems, *J. Phys. E.*, 20(8), 954–967, 1987.

Dakin, J. P., Distributed optical-fibre sensors, *Proc. SPIE Conference on Distributed and Multiplexed Fibre-Optic Sensors II*, Vol. 1797, 1992.

Dakin, J. P., Pratt, D. J., Ross, J. N., and Bibby, G. W., Distributed anti-Stokes Raman thermometry, *Proc. OFS 33*, San Diego, post-deadline paper, 1985.

Digonnet, M. F. J., Ed., *Rare-Earth Doped Fibre Lasers and Amplifiers*, Marcel Dekker, New York, 1993.

Dyott, R. B., *Elliptical Fibre Waveguides*, Artech House, Norwood, MA, 1995.

Ellison, J. G. and Siddiqui, A. S., Estimation of linear birefringence suppression in spun fibre using POTDR, *Conference on Lasers and Electro-Optics (CLEO 1998)*, paper Cth 057, 1998.

Gisin, N. et al., Polarization mode dispersion of short and long single-mode fibres, *IEEE JLT*, Vol. 9, 821–827, 1991.

Gisin, N., Passy, R., and von der Weld, J. P., Definition and measurements of PM D: interferometric versus fixed analyzer methods, *IEEE PTL*, 6(6), 730–732, 1994.

Gouban, G. and Christian J. R., Some aspects of beam waveguide for long-distance transmission of optical frequencies, *IEEE Trans. on Microwave Theory and Devices MTT*, 12, 212, 1964.

Handerek, V. A., and Rogers, A. J., Static and dynamic fibre polarization grating couplers for sensing applications, *Proc. SPIE Symposium on Optical Tools for Manufacturing and Advanced Instrumentation*, Vol. 2071, paper 2071–18, 221–228, 1993.

Hill, K. O., Fujii Y., Johnson, D. C., and Kawasaki, B. S., Photosensitivity in optical-fibre waveguides, *Appl. Phys. Lett.*, 32, 647, 1978.

Hogg, D. C., Scattering and attenuation due to snow at optical wavelengths, *Nature*, 203, 396, 1964.

Jackson, D. A. et al., Simple multiplexing scheme for a fibre-optic grating sensor network, *Opt. Lett.*, 18, 1193, 1993.

Kanellopoulos, S. E., Handerek, V. A., and Rogers, A. J., Simultaneous strain and temperature sensing using a photogenerated polarization coupler and low order modes in an elliptically cored optical fibre, *Electron. Lett.*, 30(21), 1786–1787, 1994.

Kanellopoulos, S. E., Handerek, V. A., and Rogers, A. J., Simultaneous strain and temperature measurement sensing with photo-generated in-fibre gratings, *Opt. Lett.*, 20(3), 333–335, 1995.

Kao, K. C. and Hockham G. A., Dielectric fibre surface waveguides for optical frequencies, *Proc. IEEE*, 133, 1151–1158, 1996.

Kapron, F. P., Keck, D. B., and Maurer, R. D., Radiation losses in glass optical waveguides, *Appl. Phys. Lett.*, 17, 423–425, 1970.

Kersey, A. D. and Dandridge, A., Distributed and multiplexed fibre-optic sensor systems, *JIERE*, 58, S99, 1988.

Kersey, A. D. et al., Multiplexed fibre grating Bragg sensor system with a fibre Fabry–Perot wavelength filter, *Opt. Lett.*, 18, 1370, 1993.

Lefèvre, H., *The Fibre-Optic Gyroscope*, Artech House Books, Norwood, MA, 1993.

Mallalieu, K. I. et al., FMCW of optical source envelope modulation for passive multiplexing of frequency-based fibre-optic sensors, *Electron. Lett.*, 22, 809, 1986.

Mizrahi, V. et al., Ultraviolet-laser fabrication of ultrastrong optical-fibre gratings and of germania-doped channel waveguides, *Appl. Phys. Lett.*, 63, 1727–1729, 1993.

Ono, T., Yamazaki, S., Shimuzu, H., and Emura, K, Polarization control method for suppressing polarization mode dispersion influence in optical transmission systems, *IEEE JLT*, 12(5), 891–898, 1994.

Personick, S. D., Photon probe: an optical time-domain reflectometer, *BSTJ*, 56, 355–366, 1977.

Poole, C. and Wagner, R., Phenomenological approach to polarization dispersion in long single-mode fibres, *Electron. Lett.*, 22(19), 1029–1030, 1986.

Rogers, A. J., Optical measurement of current and voltage on power systems, *IEEE J. Elect. Power Appl.*, 2(4), 120–124, 1979.

Rogers, A. J., Polarization-optical time-domain reflectometry, *Electron. Lett.*, 16(13), 489–490, 1980.

Rogers, A. J., Polarization-optical time domain reflectometry: a new technique for the measurement of field distributions, *Appl. Opt.*, 20(6), 1060–1074, 1981.

Rogers, A. J., Distributed optical fibre sensors, *J. Phys D.*, 19, 2237–2255, 1986.

Rogers, A. J., Prospects for non-linear forward scatter distributed optical-fibre sensing, *Opt. Lasers Eng.*, 16(2), 179–192, 1992a.

Rogers, A. J., Non-linear distributed optical-fibre sensing, *Proc. SPIE Conference on Distributed and Multiplexed Sensors II*, invited paper, Vol. 1797, 50–62, 1992b.

Ross, J. N., Birefringence measurements in optical fibres by POTDR, *Appl. Opt.*, 21(19), 3489–3495, 1981.

Smith, A. M., Optical fibre for current measurement applications, *Opt. Laser Techn.*, 12(1), 25–29, 1980.

Vali, V. and Shorthill, R.W., Fibre ring interferometer, *Appl. Opt.*, 15(5), 1099–1100, 1976.

Werts, A., Propagation de la lumière cohérente dans les fibres optiques, *L'Onde Electrique*, 46, 967–980, 1966.

Zhou, Y. R., Handerek, V. A., and Rogers, A. J., Computational POTDR for measurement of the spatial distribution of PMD in optical fibre, *Proc. OFMC 97*, 126–129, 1997.

6

Biological-Based Optical Sensors and Transducers

George K. Knopf
University of Western Ontario
London, Ontario, Canada

Amarjeet S. Bassi
University of Western Ontario
London, Ontario, Canada

6.1 Introduction

Mechatronic products exploit the performance capabilities of optical sensors and actuators because these simple light-controlled devices enable high precision, rapid information processing, flexible circuit designs, and discrete component miniaturization. Optical devices are often lightweight and compact, have high noise immunity, and allow for a high information-transfer capacity. All-optical circuits and devices have numerous advantages over conventional electronics because they can be activated by photons instead of currents and voltages. Consequently, optical systems are free from electrical current losses, resistive heat dissipation, and friction forces that greatly diminish the performance and efficiency of conventional electronic systems. Unfortunately, most commercially available opto-electronic and photonic devices require significant power sources to operate properly. Supplying the required power to circuitry is often a limiting factor in system design. Furthermore, the negative effects of current leakage and power loss are crucial design constraints in developing viable nanotechnology for product miniaturization.

To create miniature devices that reliably process low-amplitude signals, researchers have begun to explore the possibility of using biological molecules that can act as signal wave guides, switches, transistors, and digital logic gates. In a broad context, biomolecular electronics is defined as technology that uses **chromophore** and **protein** molecules to encode, manipulate, and retrieve information at the molecular or macromolecular level. The approach is in sharp contrast to current microchip technology that exploits the lithographic manipulation of bulk silicon materials to generate integrated electronic and opto-electronic circuits. However, according to the highly quoted **Moore's law** [Birge et al., 1999] these conventional silicon chip designs are reaching their capacity to process information and make computations and are expected to reach their physical limits in the next decade. Exploiting biomolecular electronics can significantly reduce feature size by several orders of magnitude and decrease gate propagation delays because devices can be fabricated atom by atom. Biomolecular electronics also provides

0-8493-1162-4/03/$0.00+$1.50

an opportunity for designers to create new hybrid technologies and computing architectures that can perform tasks more energy efficiently.

From this perspective, a number of integrated optical devices have been proposed in the literature that exploit the light generation or optical transmission properties of biomolecular substances. **Biosensors** are a large class of sensors that utilize a biochemical reaction to determine the presence of a specific compound. The biosensor is typically an immobilized enzyme or cell that is combined with an electrochemical, electronic, or opto-electronic **transducer** to monitor a specific change in the measurand. One light-emitting organism that has been extensively studied for optical biosensor applications is the *Vibrio fisheri* bacterium, a substance found naturally in deep-water squid. The level of luminescence produced by the bacteria is a function of a number of factors including the level of oxygen and the degree of toxicity in the environment.

Another biomolecular material that has been investigated for signal processing and computing applications is thin bacteriorhodopsin (bR) films and coatings. When absorbing light energy, the photochromic protein molecules of the bR film undergo a complex photocycle, characterized by several spectroscopically distinct intermediate states. The bR protein's ability to generate a measurable electric signal upon photoconversion provides a mechanism for interfacing the biological transducer to conventional electronic circuits. Furthermore, the **light transmittance** properties of the bR film can be controlled by modifying either the wavelength or intensity of the incident light sources. In essence, this complementary suppression-modulated transmission of the incident light enables the bR film to act as a **spatial light modulator** (SLM). SLMs are the basic information-processing elements of nearly all optical systems and provide real-time input–output interfaces with peripheral electronic circuitry. Based on this simple biological mechanism, a large number of optical devices have been proposed in the literature. Table 6.1 is a summary of several applications of biomaterials to common optical sensing and signal processing tasks.

This chapter describes how biological-based optical sensors and transducers can be used to perform simple signal detection and signal conversion operations in a variety of applications. First, the light-producing bacterium *Vibrio fisheri* is introduced and described as the sensing element for detecting airborne toxins. Next, the biological material bacteriorhodopsin is discussed, and its application to optical signal processing is examined. The simple light-suppression property of the biomaterial enables bR films

TABLE 6.1 Sample Applications of Biologically Based Optical Sensors and Transducers

Biological Device	References
Optical Biosensors	
Monitoring airborne toxins	[Kelly et al., 1999; Knopf et al., 2000; Sandstrom and Turner, 1999]
Biological warfare agents	[Aston, 2001; Golden et al., 1997]
Optical Biotransducers	
Electronic ink	[Kolodner et al., 1997]
Electronic displays	[Sanio et al., 1999]
Spatial light modulators	[Birge, 1995; Oesterhelt and Stoeckenius, 1971; Song et al., 1993]
Signal conditioning	[Okamoto et al., 1997; Werner et al., 1992]
Image extraction	[Min et al., 2001]
Photonic transistors	[Zhang et al., 2000]
Photodiodes	[Rayfield, 1994]
Color image detection	[Choi et al., 2001]
Optical binary logic gates	[Collier et al., 1999; Gu et al., 1996; Rao et al., 1996; Zhang et al., 2000]
Nonlinear coherent image processing	[Downie, 1995]
Programmable spatial filter	[Storrs et al., 1996]
Gray-level image subtraction	[Gu et al., 1996]
Optical wavelet-matched filters	[Chen et al., 1997]
Optical computing	[Birge, 1992; 1995]
Holographic associative memories	[Birge, 1992; Millerd et al., 1995]
3D optical memories	[Birge, 1992; Birge et al., 1999]
Optical data storage	[Timucin and Downie, 1997]
Opto-electronic neural network synapses	[Shelton, 1997; Taki et al., 1991]

to function as all-optical transducers for spatial-light modulation and optical switching. Although the examples are often described as bench-top systems, the key sensing and light transducing properties of the biomaterials are retained when reduced to the micro-scale.

6.2 Biological-Based Optical Sensors

Biosensors are devices that utilize a biochemical reaction in order to detect a specific chemical compound. These sensor systems involve a biological recognition component such as receptors, nucleic acids, antibodies, or **enzymes** that are in direct contact with an electrochemical, electronic, or opto-electronic transducer. A variety of signal parameters such as changes in pH, oxygen consumption, ion concentrations, potential difference, current, resistance, or optical properties can be measured by an appropriate transducer.

Biosensors are divided into categories based on the method of signal transduction such as mass, electrochemical, thermal, or optical [Ivnitski et al., 1999]. Furthermore, biosensors can be classified as either direct-detection or indirect-detection systems. Direct-detection biosensors are designed such that the specific biochemical reaction, or target analyte, is measured directly by the transducer. In contrast, indirect-detection biosensors are those in which a preliminary biochemical reaction takes place and the products of this reaction are detected by the transducer.

Optical biosensors are an attractive solution for directly detecting infectious diseases, pathogens, and toxins. Some of these optical sensors are able to detect minute changes in the refractive index or material thickness that occur when cells bind to the immobilized receptors on the transducer surface. Several optical techniques have been reported for the detection of bacterial pathogens including monomode dielectric waveguides, surface plasmon resonance, ellipsometry, the resonant mirror, and the interferometer [Ivnitski et al., 1999]. Kelly et al. [1999] describe a simple, optical waveguide sensor for the detection of biological toxins. The biosensor works by optically tagging toxin receptors within a fluid phospholipid bilayer membrane that is formed on the surface of a planar optical waveguide. The process of toxin detection involves measuring the ratio of emission intensity from the donor–acceptor pair of fluorophores that are tagged onto the receptors. The ratio of fluorescent emission intensity depends on the concentration of toxin. The biosensor appears to be very sensitive with a high degree of specificity.

Recent advances in bio-analytical sensors have exploited the ability of certain enzymes to emit photons as a by-product of their reactions. This phenomenon, known as bioluminescence, can be used to detect the presence and physiological condition of cells. The concept of bioluminescence and the utilization of the light-emitting bacterium *Vibrio fisheri* for monitoring airborne toxins are presented below.

6.2.1 Bioluminescent Light Sources

The use of light for optical sensing, actuating, or communication requires a source of light radiation, a medium through which the light travels, and a detector to convert the light energy to another measurable form such as current or voltage. The transmission of information embedded in the light signal is accomplished by controlling any combination of the parts that comprise the system. Because both the source and the transmission medium determine the amount of light received by the detector, it is possible to convey information about both the source and the medium to the detector. If the source provides illumination while the medium is modulated or interrupted, then the detector can be designed to capture the modulated light while suppressing the effects of the ambient light conditions.

Any mechanism that causes an electron to vibrate will emit a stream of electromagnetic waves. If the electron is vibrated fast enough so that the wavelengths are in the 330- to 770-nm range of the electromagnetic spectrum (see Figure 6.1), then visible light is emitted. These electromagnetic waves radiate in every direction away from the point of origin. The observed light has both a wavelength and intensity. The frequency, f, of the electromagnetic wave is related to the vacuum wavelength, λ_0, and is given by

$$f\lambda_0 = c \tag{6.1}$$

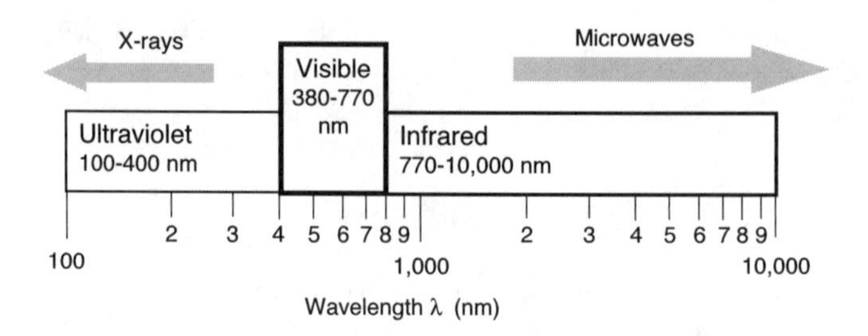

FIGURE 6.1 Optical portion of the electromagnetic spectrum.

where c is the speed of light in a vacuum ($\sim 3 \times 10^8$ m/s). Common artificial light sources include incandescent lamps, flourescent lamps, tungsten lamps, light-emitting diodes (LEDs), and lasers.

Some mechanisms can produce visible light through chemiluminescence, bioluminescence, and cathodoluminescence. Chemiluminescence occurs when the electron excitation energy necessary for photon emission is supplied by a chemical reaction. Phosphorous glow through oxidation in the air is one example. Bioluminescence is a subdivision of chemiluminescence and occurs in living organisms such as fireflies and glow-worms. Cathodoluminescence occurs when the excitation energy is supplied by an accelerated electron colliding with atoms. This causes an electron in the atomic structure to move from one orbit to another, which produces light. An example is the cathode ray tube used in television receivers, video terminals, and oscilloscopes.

6.2.2 *Vibrio fisheri* Bacteria

A variety of natural organisms emit visible light. These include squids, fish, insects, algae, and bacteria. These organisms can be found in a wide range of environments, from marine and freshwater areas to terrestrial habitats. Of the various light-emitting organisms, bacteria are the most abundant and widespread throughout the world. All luminescent bacteria are known to be Gram-negative mobile rods. These bacteria are mainly found in oceans living freely, symbiotically, saprophytically, or in a parasitic relationship with other higher-order organisms.

The *Vibrio fisheri* bacterium is one marine species found in the Pacific Ocean around Hawaii and the coastal areas of California. The most common function of *Vibrio fisheri* is to be a light source for other organisms. The squid *Euprymma scolopes* exploits *Vibrio fisheri* in a symbiotic relationship by allowing the bacteria to grow uninhibited in its light organ. Scientists have discovered that without the presence of high concentrations of *Vibrio fisheri* bacteria in immature squids, the light organ does not fully develop. The squid is very specific in the type of *Vibrio fisheri* it allows to inoculate its light organ by using sophisticated epithelial structure to lure the bacteria. In return for infecting the light organ, the squid provides nutrients and protection to the *Vibrio fisheri.*

The luminescence of the bacteria appears as a faint glow and can only be observed in a dark environment. The level of bioluminescence exhibited by the bacteria is highly dependent on cell density. This dependency is linked to the production of a chemical compound named the *lux autoinducer,* which provides communication between cells and allows individual bacteria to sense the response of the entire population. Both cellular and environmental factors control the bioluminescence reaction. These factors include the nutrient or growth medium, environmental toxicity, exposure to oxygen, and cell concentrations.

In *Vibrio fisheri*, the autoinducer is termed *N*-(3-oxohexanoyl homoserine lactone). When *Vibrio fisheri* bacteria live freely in the ocean, there are approximately 10^2 cells/ml. The autoinducer is diffused out of the cell because of the naturally small concentration of bacteria. When the bacteria are present in high concentrations, such as on the light organ of the squid (10^{10} to 10^{11} cells/ml), the autoinducer accumulates until it reaches a critical concentration of about 5 to 10 n*M*—the amount required to activate the luminescence gene transcription, which triggers the specific luminescence enzymes. There are numerous enzymes involved in the light-emitting process of *Vibrio fisheri*. The luminescent reaction is catalyzed by the luciferase, which involves the oxidation of long-chain aldehyde and reduced flavin mononucleotide ($FMNH_2$) and releases light:

$$FMNH_2 + RCHO + O_2 \rightarrow FMN + RCOOH + H_2O + \textit{light}\ (490\ \text{nm}) \tag{6.2}$$

6.2.3 Toxin Biosensor—An Illustrative Example

Many environmental monitoring systems employ a biological sensing element to detect the presence of toxins [Sandstrom and Turner, 1999]. Biosensor mechanisms that have been used for environmental applications include enzymes, antibodies, and microorganisms. These biological recognition elements can be interfaced to electrochemical, optical, or acoustic signal transducers. Currently, a laboratory-based MICROTOX^{TM} assay that utilizes bioluminescent bacteria is being used for a large number of applications. However, a self-contained biosensor that can be easily transported to the investigated site is preferred for field applications that are inaccessible for the technician to gather samples or that have a high degree of risk due to the level of toxicity present in the environment.

The biological sensing elements are immobilized luminescent bacteria whose response to the toxins in the environment can be quantitatively measured. The level of bioluminescence exhibited by the bacteria is highly dependent on cell density. The bacterial luminescence reaction involves the oxidation of the long-chain aliphatic aldehyde and reduced flavin mononucleotide ($FMNH_2$) with the liberation of excess free energy in the form of a blue-green light at 490 nm. Both cellular and environmental factors control the bioluminescence reaction. These factors include the nutrient or growth medium, environmental toxicity, exposure to oxygen, and cell concentrations. Figure 6.2 is a plot of the normalized luminescence of the bacteria before and after the introduction of the toxin acetone. As soon as acetone was added, the luminescence of the bacteria decreased sharply but started to increase after absorbing the shock of being exposed to the toxin. It should also be noted that at the end of 10 min the bacteria were never able to fully recover from the toxin.

A simple opto-mechatronic device has been proposed by Knopf et al. [2000] that exploits the light-emission characteristics of *Vibrio fisheri* bacteria to measure the degree of toxicity in the surrounding air. The bacteria are immobilized on polyvinyl alcohol gel capsules and placed inside a specially constructed, miniature light-sealed chamber. Enclosed along with the inoculated gel is an opto-electronic transducer that produces a train of pulses with a frequency proportional to the amount of light being emitted by

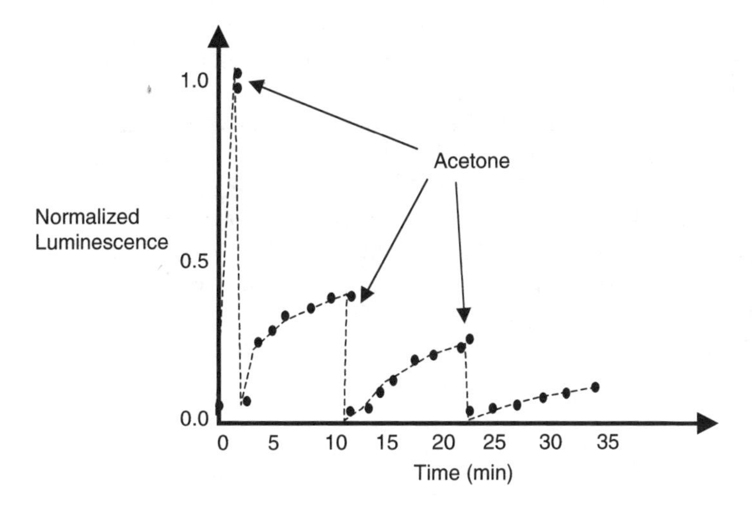

FIGURE 6.2 Normalized luminescence of immobilized *Vibrio fisheri* bacteria subjected to acetone at various times. (Adapted from Knopf, G. K. et al., in *OptoMechatronic Systems*, Cho, H. S. and Knopf, G. K., Eds., Proc. Soc. Photo-Opt. Instrum. Eng., Vol. 4190, pp. 9–19, 2000.)

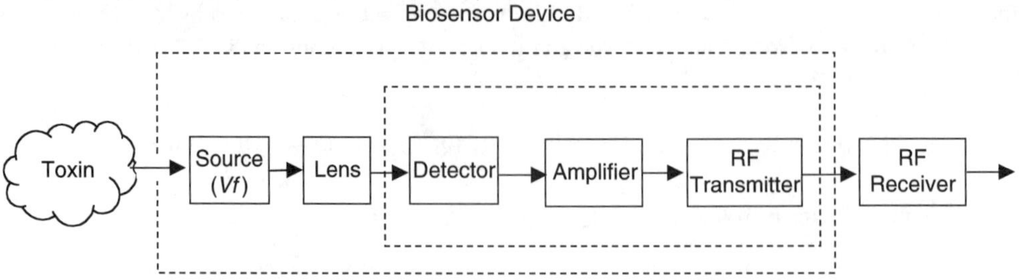

Biosensor Device

Toxin → Source (Vf) → Lens → Detector → Amplifier → RF Transmitter → RF Receiver →

FIGURE 6.3 Block diagram of the biosensor telemetry system [Knopf et al., 2000].

the bacteria. The transducer response is amplified and broadcast to a distant site by a wireless radio-frequency (RF) transmitter. The RF receiver can be placed at a remote location for data reception, storage, and display. The simple design will enable the biosensor to be implemented in the field with very little preparation time and minimal operator training. Figure 6.3 is a block diagram showing the information flow through the basic components of the biosensor telemetry device. The level of bioluminescence exhibited by the bacteria population is proportional to the degree of toxicity, cell density, and airflow in the chamber. The intensity of bioluminescent light is focused onto a photodetector by means of a convex lens. The opto-electronic transducer converts the light intensity into a voltage signal that undergoes amplification and signal conditioning. The conditioned signal is converted to a pulse frequency signal and broadcast as a wireless signal via an RF transmitter. At a remote site the RF signal is picked up by a receiver and reconstructed for further processing, analysis, or data storage.

6.3 Protein-Based Optical Transducers

6.3.1 Bacteriorhodopsin Films

Bacteriorhodopsin (bR) is a photochromic retinal protein found in the purple membrane of a salt marsh bacterium called *Halobacterium halobium.* Under restricted oxygen conditions, the bR protein molecules function as a light-driven proton pump that transports protons across the cell membrane. The proton gradient across the cell membrane generates an electrochemical potential that is used by the bacterium to generate adenosine triphosphate (ATP). As a consequence, the bR protein molecule converts sunlight directly into chemical energy.

In addition to transporting protons across the membrane, the protein molecule undergoes a complex photocycle when it absorbs light. The photocycle (Figure 6.4) involves several intermediate states with specific absorption maxima. In the absence of light the bR protein molecule is at the B state with an absorption spectrum peaked at around 570 nm. When the protein molecule absorbs a photon of light around this particular wavelength, it quickly transforms through several intermediate states by means of thermal relaxation until it reaches the M state. The transformation from B to M occurs because of light activation and is complete in about 50 μs.

The photochemistry of this transformation from the initial dark-adapted B state to the longest-lived M intermediate state, and the reverse reaction, is summarized as:

$$B(\lambda_{max} = 570 \text{ nm}) \underset{\Phi_2 = 0.65}{\overset{\Phi_1 = 0.65}{\rightleftarrows}} M(\lambda_{max} = 410 \text{ nm}) \tag{6.3}$$

where Φ_1 represents the quantum efficiency of the forward reaction and Φ_2 represents the quantum efficiency of the reverse reaction. Quantum efficiency is a measure of the probability that the specified

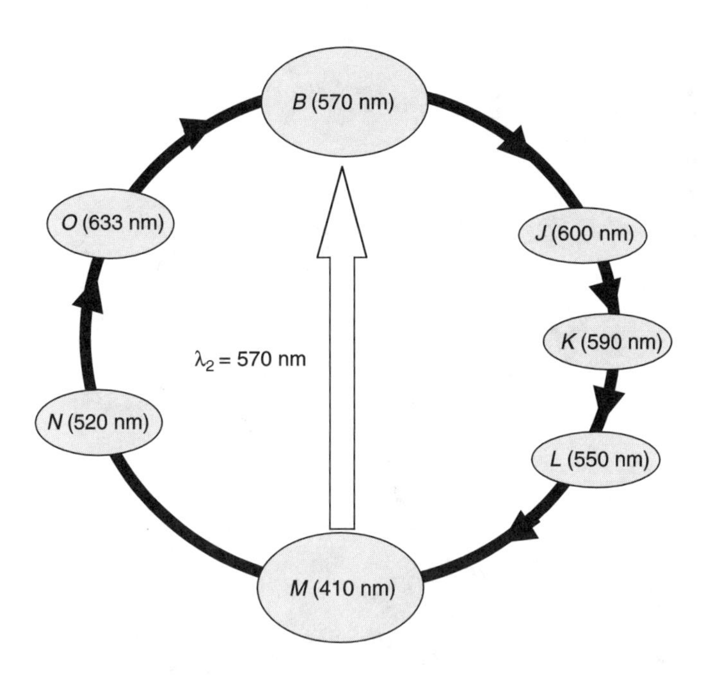

FIGURE 6.4 The basic photocycle of the thin bacteriorhodopsin film. The sequence of structural changes induced by light in the bR film allows for controlled light-transmission properties.

reaction will take place after the absorption of a photon of light. Thus, a quantum efficiency of unity would imply perfect efficiency. An important attribute of bR film is the high efficiency with which it converts light into the state change.

In the *M* state the absorption spectrum is shifted approximately 160 nm toward the blue range with a peak absorption wavelength of approximately 410 nm. The length of time that the molecule remains in the *M* state is typically around 10 ms. Although the life span of the *M* state is relatively long, it is possible to create bR films that hold the *M* state for as short as a millisecond or as long as several hours because the *M*-state life span is dependent upon the relative humidity and pH level of the environment during the film fabrication. The protein molecules in the *M* state transform back to the dark-adapted *B* state through thermal relaxation or by absorbing light near the blue wavelength at 410 nm. The bR molecule has a wavelength-dependent photoabsorption profile for each of the intermediate states in the photocycle. Although the spectra of both the *B* and *M* states are relatively broad (see Figure 6.5), there is only a small amount of overlap within the absorption spectrum. It is important to realize that absorption maxima for the various intermediate states are only approximate because these values can be altered by environmental, chemical, and genetic modification.

The absorption characteristic, and hence the local transmittance of the thin bR film, takes on a value that is dependent on the intensity of the incident light source. The intensity-dependent transmittance of a thin bR film [Storrs et al., 1996] can be described by

$$\tau(I) = \exp[-\alpha(I)\delta] \tag{6.4}$$

where τ is the local intensity transmittance of the bR film, $\alpha(I)$ is the intensity-dependent local absorption coefficient, and δ is the thickness of the film. The absorption of light and modification of light-transmittance properties make bR films an ideal material for a light-activated optical transducer. This is possible because the bR photocycle continues to operate after the protein molecules have been extracted from the bacteria in the form of two-dimensional crystalline sheets and suspended in various solutions during the preparation of films.

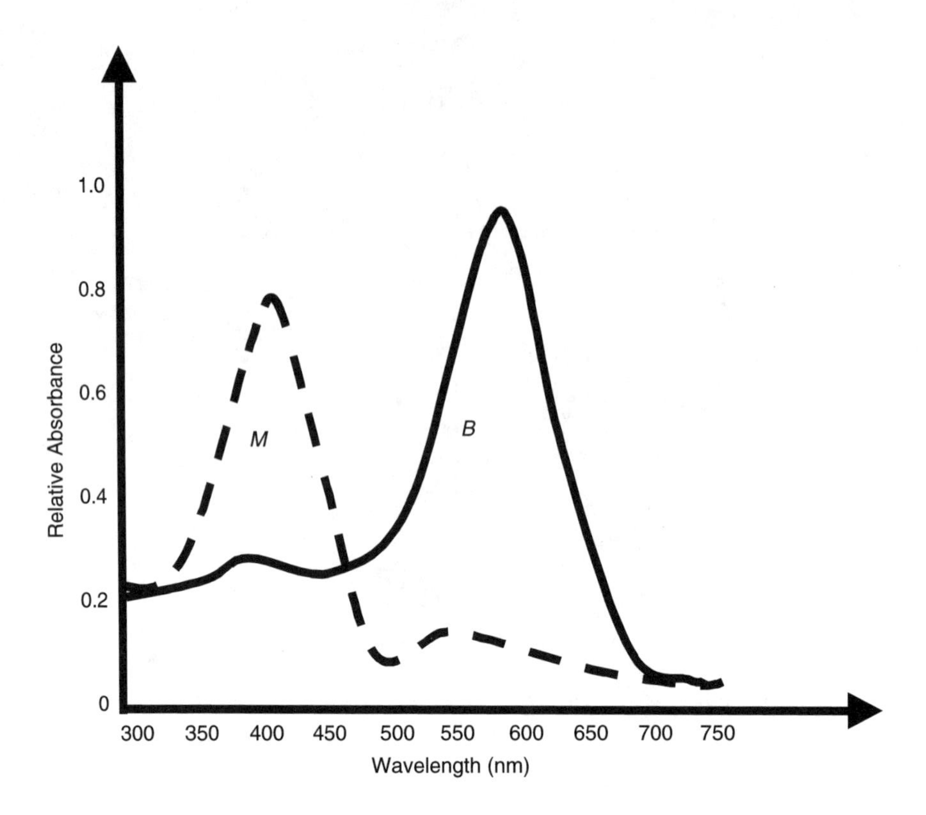

FIGURE 6.5 Simplified absorption spectra for *M* and *B* states [Gu et al., 1996].

One method of controlling the life span of the *M* state is to apply a voltage across the thin bR film. A voltage drop will affect the state transition because the primary biological function of the native bacteriorhodopsin is to pump protons across the 40-nm cell membrane. The transfer of protons across the membrane creates the intermediate *M* state. The subsequent photochemical reactions that transform the protein molecule back to the *B* state are also very sensitive to the applied voltage. Consequently, the speed of the *M*-to-*B* thermal reaction can be increased or decreased by adjusting the electric field across the bR film. However, several problems remain in the development of reliable voltage-controlled bR-based devices. First, the applied electric field necessary to generate a two-order-of-magnitude increase in the life span of the *M* state approaches 10^5 V/cm. This is a very high voltage value that approaches the failure limit of the polymer matrix that holds the bacteriorhodopsin protein. Second, while it is possible to generate thin films with excellent orientation properties, the applied electric fields tend to reorient the protein molecules over extended periods of time. These problems require new polymer matrices that resist voltage breakdown over time.

The bacteriorhodopsin protein molecules have several advantages over other biological substances for creating optical devices. The thin bR film exhibits high sensitivity in the range of 30 to 80 mJ/cm^2, a fast response time of around 50 μs, and resistance to thermal and photochemical degradation, and it is reusable with more than 1×10^6 cycles without observable degradation. The high forward and reverse quantum yields of the protein substance permit it to be used as a switch at low light levels. Experiments have also shown that the thin bR films undergo local changes in refractive index upon illumination (see Figure 6.6). The graph is based on work by Birge [1992] and shows the change in refractive index as a function of wavelength for the $B \rightarrow M$ photoisomerization of a 30-μm bR film with an optical density (OD) of approximately 3. This large change in refractive index allows the thin bR film to exhibit good holographic efficiencies. Furthermore, the bR protein's ability to generate photoelectric signals upon photo-conversion produces measurable electronic signals. These properties can be selectively enhanced

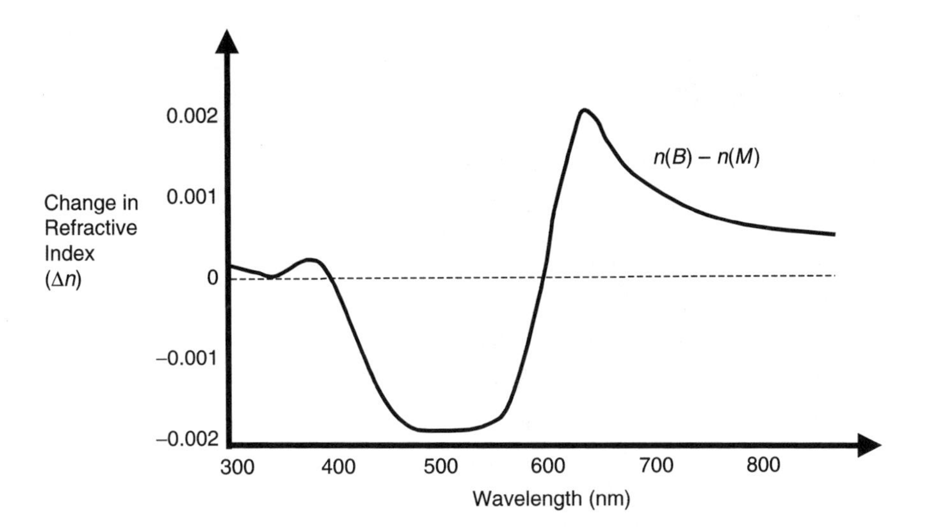

FIGURE 6.6 The graph shows the change in refractive index associated with the $B \rightarrow M$ photoisomerization. (Modified from Birge, R. R., *IEEE Computer*, November, 56–67, 1992.)

for specific applications by using chemical additives, substituting different chromophores, or using genetic engineering to change the protein's amino acid structure.

6.3.2 Optical Transducers

An important class of opto-electronic components that are used to construct mechatronic products is *transducers*. A transducer is a device that transforms energy from one form to another, where the form of energy may belong to different domains or the same domain. The protein-based optical transducers created from bacteriorhodopsin can be activated by either applying a voltage drop across the bR film or by varying the intensity of the illuminating light sources. The focus of this section is to summarize the light-suppression mechanism exhibited by the bR film and to describe how this biomaterial can be used to create all-optical transducers. All-optical transducers are free from current losses, resistive heat dissipation, and friction forces that greatly diminish the performance and efficiency of conventional electronic systems. Furthermore, transducers are free from electrical noise that can greatly distort small-amplitude signals. The negative effects of current leakage, power loss, and electrical noise are crucial design constraints in product miniaturization. Common all-optical transducers are spatial light modulators, optical switches, and optical logic gates.

6.3.2.1 Spatial Light Modulator

A spatial light modulator (SLM) is an optical device that controls the spatial distribution of the intensity, phase, and polarization of transmitted light as a function of electrical signals or a secondary light source. Spatial light modulators are the basic information-processing elements of nearly all optical systems and provide real-time input–output interfaces with peripheral electronic circuitry. Most commercially available SLMs have problems with high resolution, large bandwidth, long-term stability, high speed, and cost. Many of the shortcomings are directly related to the physical limitations of the materials used in the device. The observation that a thin bR film acts as either a voltage-controlled bistable optical device or a photochromatic bistable optical device suggests that the material can be used as the active medium in an SLM. These protein-based SLMs can take advantage of the large change in the absorption wavelength and refractive index, differential absorptivity, or potential gradient that accompanies the $B \Leftrightarrow M$ photoreaction. The information-processing and computational operations performed by the proposed mechanism are based only on light-waves that are different from the commercially available electro-optical or piezoelectric SLMs. As a result, there should be a higher level of parallel processing and system interconnection.

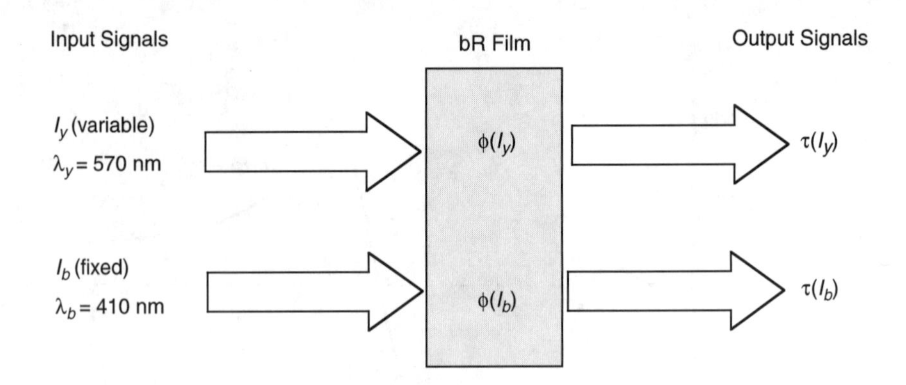

FIGURE 6.7 Diagram of "colored" light transmission through thin bR film.

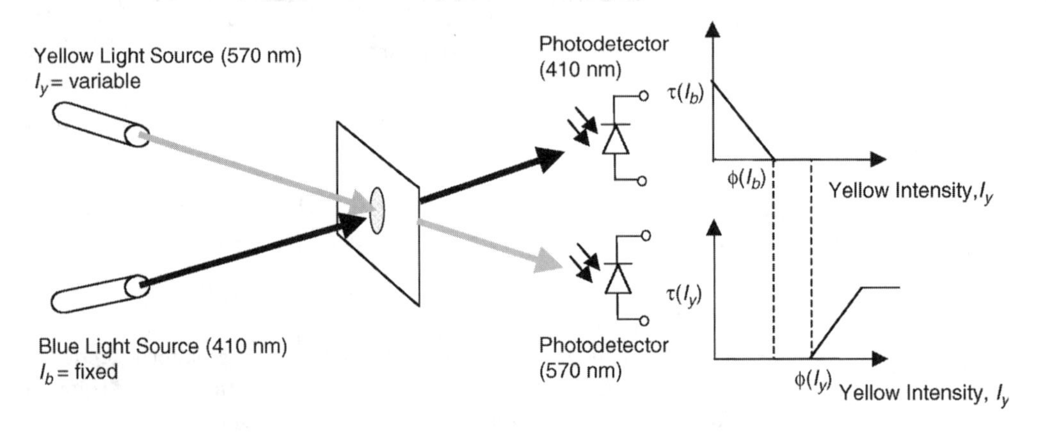

FIGURE 6.8 The transmission property of a thin bR film with fixed-intensity blue light and a variable-intensity yellow light source, I_y [Zhang et al., 2000]. The transmission intensity is given by $\tau(I_b)$ and $\tau(I_y)$, for the blue and yellow light beams, respectively.

A viable SLM can be created by taking advantage of the light-modulated transmission mechanism exhibited by bR film. The basic principle of the proposed optical SLM is shown in Figure 6.7. When a yellow beam of light with a single wavelength of approximately 570 nm, I_y, and a second deep-blue beam at 410 nm, I_b, illuminate the same region of the bR film, the two beams will mutually suppress the optical transmission properties of the bacteriorhodopsin film and reduce the intensity of the light output. Under the simultaneous illumination by yellow and blue light, the exposed region of the bR film will exhibit two threshold intensities. The threshold intensity of the bR film to the blue beam is $\phi(I_b)$, while that to the yellow beam is $\phi(I_y)$. The SLM outputs are the transmitted intensities of the blue and yellow light given by $\tau(I_b)$ and $\tau(I_y)$, respectively. The light transmission property of the bR film is optically controlled by fixing the intensity of one input. The second input is a variable that carries the information to be modulated.

The transmission property of the film exposed to fixed blue and variable yellow light is illustrated in Figure 6.8. As the intensity of the yellow light I_y varies in the region below threshold $\phi(I_b)$, the transmitted yellow beam $\tau(I_y)$ remains very low. However, the blue light is transmitted through the film but decreases linearly as the intensity of the yellow light increases and reaches the threshold $\phi(I_b)$. Both transmitted beams $\tau(I_y)$ and $\tau(I_b)$ are suppressed when $\phi(I_b) < I_y < \phi(I_y)$. As the intensity of the yellow light increases above $\phi(I_y)$, the transmitted blue light is almost suppressed and the transmitted yellow light increases linearly with the intensity of the yellow light source I_y. The transmitted intensity for the blue light is

$$\tau(I_b) \approx \max(K_b(I_y - \phi(I_b)), 0) \tag{6.5}$$

where K_b is the system gain for light transmission prior to the blue threshold $\phi(I_b)$. Similarly, the transmitted intensity for the yellow light is given by

$$\tau(I_y) \approx \max(0, K_y(I_y - \phi(I_y))) \tag{6.6}$$

where K_y is the gain for the bR film transmission property after the yellow threshold $\phi(I_y)$. Finally, the threshold $\phi(I_b)$ is determined by the wavelength and relative intensities of the incident light beams. Increasing the life span of the M intermediate state will narrow the region between $\phi(I_b)$ and $\phi(I_y)$.

6.3.2.2 Optical Switch

The two-state, or binary, optical switch is a simple transducer that is used in a variety of optical systems to carry out logical AND or NAND functions, to perform Fourier pattern recognition and optical associative memory. The optical switch is essentially a threshold SLM that responds to a light flux with a nonlinear response. An ideal threshold SLM will have a transmittance near 0% at intensities below the threshold intensity level and near 100% light transmission above the threshold intensity level.

One proposed bR-based threshold SLM operates by using the visible light in the green-red region of the spectrum to drive the $B \rightarrow M$ photochemistry. For example, if a two-dimensional image is imposed on the bR film by monochromatic light from a helium-neon laser (632.8 nm), the light will be absorbed strongly by the molecules in the B state but not by those in the M state. In regions of the image exposed to higher light intensity, a greater fraction of the molecules in the B state are driven to the M state, thereby increasing the overall percent of light transmitted. The spatial intensity distribution of light that exits from the thin bR film will be modified by the intensity pattern so that high-intensity segments are enhanced relative to lower-intensity segments due to the photochromic processes. In this way, the thin bR film acts as a threshold device with response properties determined by the lifetime of the M intermediate state. Figure 6.9 shows a typical response of a bR-based threshold SLM [Birge, 1992].

The two-state switch is a simple transducer used in a variety of mechatronic designs. Binary switches can be either *normally open* (NO) or *normally closed* (NC). If the input is less than a threshold for a NO switch, then the output is suppressed. As the input surpasses the threshold the output increases to a nonzero value. In terms of a NO optical switch based on the bR film (see Figure 6.10(a)), the input I_y is

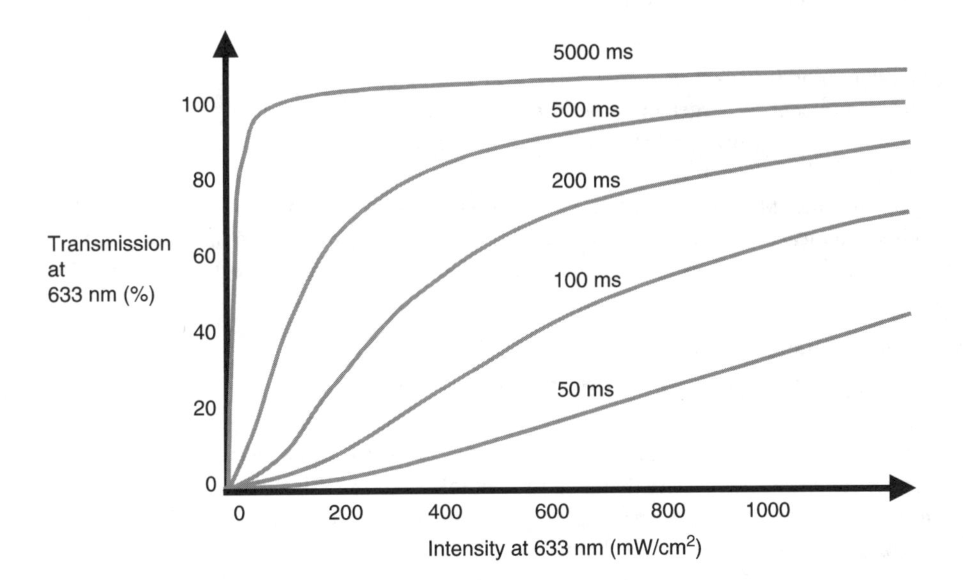

FIGURE 6.9 Simplified diagram showing the light-transmission response of a thin bR film as a function of the M-state lifetime. The bacteriorhodopsin film is 30 μm with an optical density of 2.5. (Modified from Birge, R. R., *IEEE Computer*, November, 56–67, 1992.)

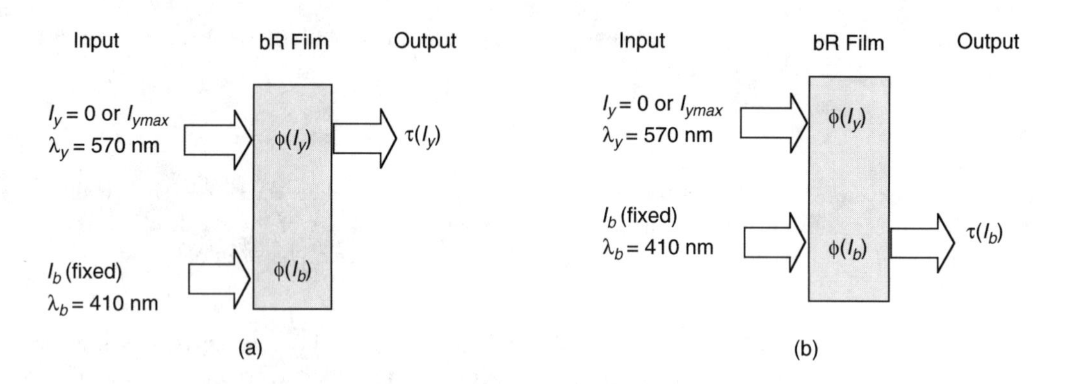

FIGURE 6.10 (a) Normally open (NO) optical switch. (b) Normally closed (NC) optical switch. Illustrations of how the threshold functions of the bR film can be used to create binary optical switches.

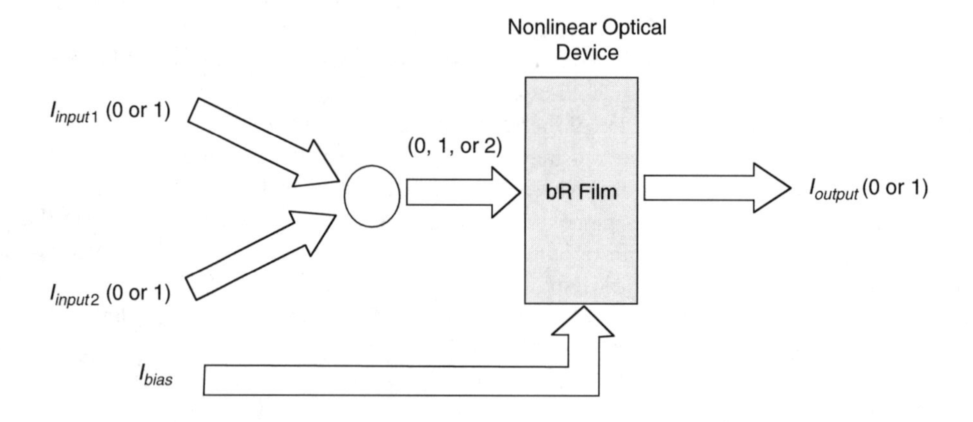

FIGURE 6.11 Diagram of a binary optical logic gate.

a binary value (0 or I_{ymax}) and the output is also binary (0 or $\tau(I_{ymax})$). The intensity I_{ymax} is a constant value that is sufficient to trigger the desired response from the film. Similarly, it is possible to create a NC optical switch, as shown in Figure 6.10(b). In this example, the desired output is the amount of transmitted blue light $\tau(I_b)$ at $I_y = (0 \text{ or } I_{ymax})$.

The proposed bR optical switch can be easily interfaced with conventional electronic circuitry by using a photodiode to measure the desired light output. However, it is important to make sure that the selected photonic device has spectral response characteristics and speed of response that match the yellow or blue transmitted light.

6.3.2.3 Optical Logic Gates

Optical logic gates (see Figure 6.11) are the elementary components of digital optical computing. Logic gates are often implemented using a nonlinear device. The logic states of 1 and 0 can be represented by "low" and "high" light intensity. Because bR film acts as two-state photochromic system, it is possible to construct an all-optical logic device that is controlled solely by light. Zhang et al. [2000] describe a simple but effective way of generating a variety of double variable logic functions using the light transmission property of bR film and different bias lights.

In the above system, the intensities of the two input light sources represent the two binary variables. The input light may be either yellow or blue. A third light source, called the bias light, is used to control the transmission property of the bR film. The transmitted intensity of yellow, blue, or a combination of yellow and blue is used as the logic output. In other words, two blue beams may be selected as the inputs

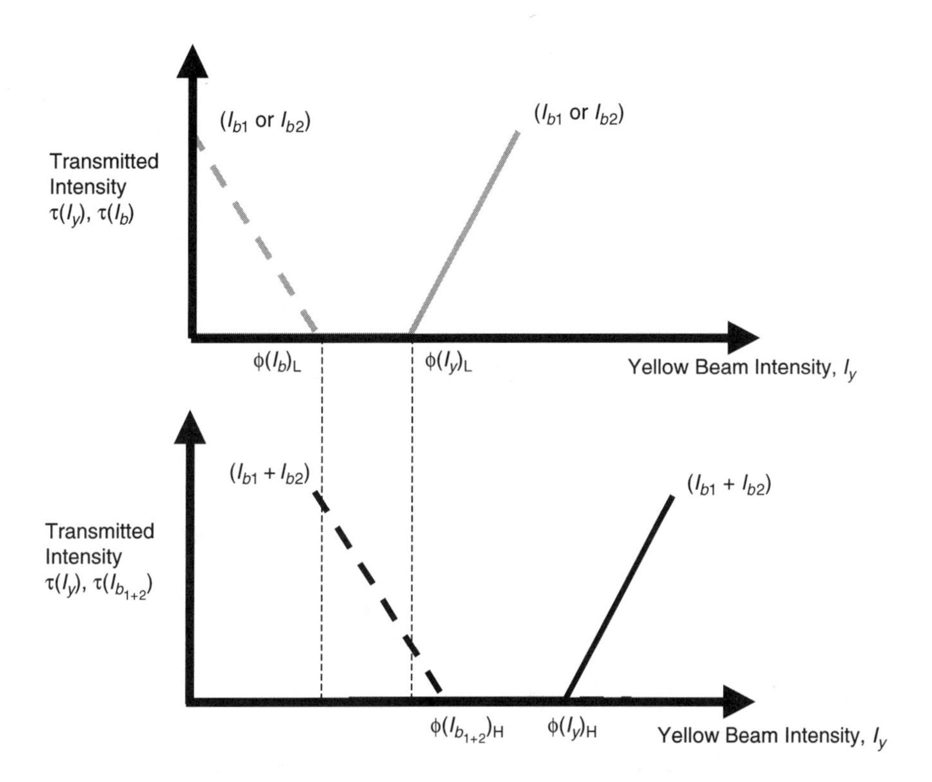

FIGURE 6.12 Mutually suppressed transmission property of the bR film as an all-optical transducer with two intensity-fixed blue beams and a variable-intensity yellow beam [Zhang et al., 2000]. The two sets of curves are for the one blue beam (I_{b_1} or I_{b_2}) and two blue beams (I_{b_1} and I_{b_2}).

and a yellow beam as the bias light. Depending upon the desired logic function, the output may be selected as $\tau(I_b)$, $\tau(I_y)$, or $\tau(I_b) + \tau(I_y)$.

For illustration purposes, two blue beams with fixed intensities and one yellow beam with variable intensity I_y are selected. The mutually suppressed transmission property for the bR film, with the preceding three beams, is shown in Figure 6.12. Let $\phi(I_b)_L$ be the threshold intensity of I_{b_1} or I_{b_2} and $\phi(I_b)_H$ be the threshold intensity of $(I_{b_1} + I_{b_2})$. The threshold intensity of the yellow beam is $\phi(I_y)_L$ and $\phi(I_y)_H$, respectively.

Consider the following two cases for illustrating this mechanism.

Case 1: The transmission property of the bR film with one yellow beam and one blue beam (I_{b_1} or I_{b_2}) is summarized as follows:

IF $I_y < \phi(I_b)_L$ **THEN** $\tau(I_y) = 0$ **AND** $\tau(I_b) = -K_b\, I_y$
IF $\phi(I_b)_L < I_y < \phi(I_y)_L$ **THEN** $\tau(I_y) = 0$ **AND** $\tau(I_b) = 0$
IF $I_y > \phi(I_y)_L$ **THEN** $\tau(I_y) = K_y I_y$ **AND** $\tau(I_b) = 0$

where K_b and K_y are gains.

Case 2: The transmission property of the bR film with one yellow beam and two blue beams (I_{b_1} and I_{b_2}) is summarized as follows:

IF $I_y < \phi(I_{b_{1+2}})_H$ **THEN** $\tau(I_y) = 0$ **AND** $\tau(I_{b_{1+2}}) = -K_b\, I_y$
IF $\phi(I_{b_{1+2}})_H < I_y < \phi(I_y)_H$ **THEN** $\tau(I_y) = 0$ **AND** $\tau(I_{b_{1+2}}) = 0$
IF $I_y > \phi(I_y)_H$ **THEN** $\tau(I_y) = K_y\, I_y$ **AND** $\tau(I_{b_{1+2}}) = 0$

TABLE 6.2 Some Binary Logic Configurations Created Using All-Optical Logic Gates

Logic Gate	A	B	Bias Light	Output Light
0	I_{b_1}	I_{b_2}	$I_y(I_y > \phi(I_{b_{1+2}})_H)$	$\tau(I_b)$
A × B	I_{b_1}	I_{b_2}	$I_y(\phi(I_b)_L < I_y < \phi(I_{b_{1+2}})_H)$	$\tau(I_b)$
A + B	I_{b_1}	I_{b_2}	$I_y(I_y < \phi(I_b)_L)$	$\tau(I_b)$
$\overline{\text{A} \times \text{B}}$	I_{b_1}	I_{b_2}	$I_y(\phi(I_y)_L < I_y < \phi(I_y)_H)$	$\tau(I_y)$
$\overline{\text{A} + \text{B}}$	I_{b_1}	I_{b_2}	$I_y(I_y < \phi(I_y)_L)$	$\tau(I_y)$
A ⊕ B	I_{b_1}	$I_y(\phi(I_b)_L < I_y < \phi(I_y)_L)$	Null	$\tau(I_b) + \tau(I_y)$
$\overline{\text{A} \oplus \text{B}}$	I_{b_1}	I_{b_2}	$I_y(\phi(I_b)_L < I_y < \phi(I_y)_L)$	$\tau(I_b) + \tau(I_y)$
1	I_{b_1}	I_{b_2}	$I_y(I_y < \phi(I_y)_H)$	$\tau(I_y)$

Note: The basic logic symbols are × – AND, + – OR, and ⊕ – Exclusive OR.
Source: Data from Zhang, T. et al., *Opt. Eng.*, 39(2), 527–553, 2000.

According to the above discussion, it is possible to generate a variety of logic functions so long as the three beams are appropriately assigned to be the input and bias light sources and a suitable transmitted beam is chosen as the output. Table 6.2 summarizes several common binary logic operations that can be generated using this principle [Zhang et al., 2000].

6.4 Applications of Bacteriorhodopsin Films

When absorbing light energy, the photochromic protein molecules of the bR film undergo a reproducible and repeatable photocycle characterized by several distinct and measurable intermediate states. The light-transmission properties exhibited by the thin bR film can be modified by varying the wavelength and intensity of the light source. Many of these applications described in the literature exploit the light-transmission properties of the bR film to create reusable, erasable photographic film for optical image processing. Several optical image processing applications are described below including gray-level image subtraction, nonlinear logarithmic filter, programmable spatial filter, real-time defect enhancement, holographic associative memory, and optically addressed direct-view display.

6.4.1 Gray-Level Image Subtraction

The block diagram of a simple optical system for incoherent gray-image subtraction using the light-transmission properties of bR film is illustrated in Figure 6.13 [Gu et al., 1996]. Image subtraction is a common strategy used in automated-surveillance and product-inspection systems, as well as in communication systems for bandwidth-compression techniques that are based on the interframe coding. The proposed optical system is a simple, low-cost, and precise method of high-resolution image subtraction using the mutual suppression properties of the thin bR film. Gu et al. [1996] claim that the technique does not have drawbacks such as low space-bandwidth product and coherent noise problems that are common with other gray-level image subtraction methods.

In the reported experiment, incoherent image subtraction is demonstrated by using two transparent letters (E and L) on a black background. When illuminated by blue and yellow beams of light, the bR film will modulate the transmission characteristics in a complementary fashion. The image subtraction is performed on a chemically enhanced bR film, with an *M* state lifetime of 25 s, under different intensities of yellow light (~568 nm). The letter E is projected onto the bR film using a blue (~412 nm) light source and the letter L is projected on the same film using the yellow (~568 nm) light. The letters are adjusted so that the left and bottom sides of the projected images are spatially superimposed on the bR film. When the yellow illumination for L increases, the absorption for the blue E becomes stronger. The results of the published experiments demonstrate that the parts of E, superimposed with L on the bR film, are faint because of the suppressed light transmission. Furthermore, when the illumination intensity for L is increased beyond the threshold intensity, the regions of E that overlap L are competely removed. In this approach, if a bR film with long *M*-state lifetime properties is used, then the device will have good memory properties. The bR film can also be made very large with high resolution.

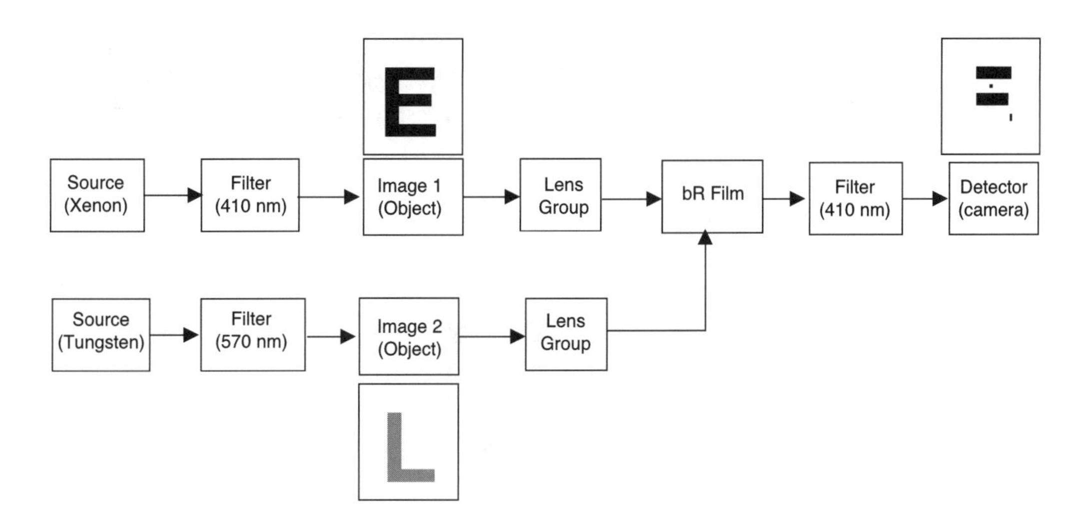

FIGURE 6.13 Block diagram of the experimental setup used by Gu et al. [1996] to perform gray-level image subtraction.

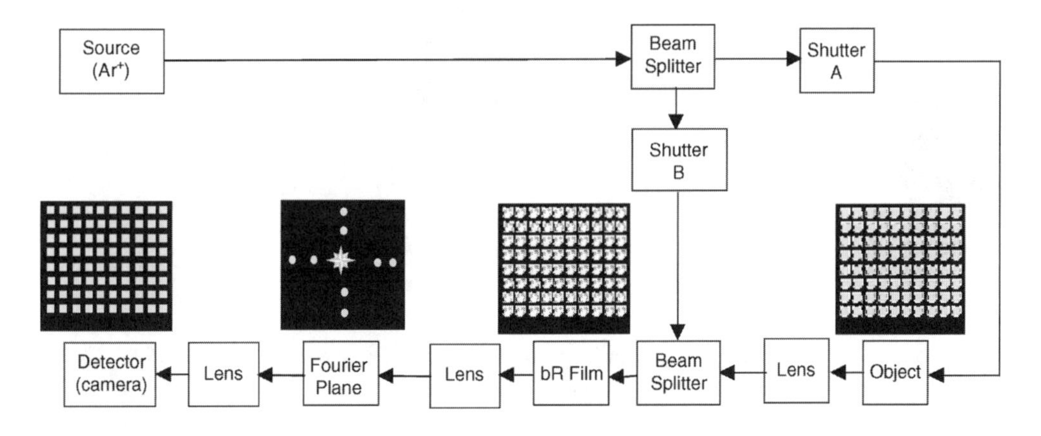

FIGURE 6.14 Simplified block diagram of the optical setup used by Downie [1995] to perform logarithmic transformation operations on images with multiplicative noise. Shutter A is opened during the WRITE process, and shutter B is opened during the READ and image-processing stages. The argon (Ar^+) laser wavelength is 514.5 nm.

6.4.2 Nonlinear Logarithmic Filter

Optical correlation and optical restoration applications often involve spatial images that have multiplicative noise characteristics. One technique used to remove undesirable noise is to transform the original noisy image with a logarithmic function. The logarithmic amplitude-transmission characteristic of the bR film permits the conversion of multiplicative noise to additive noise, which may then be linearly filtered in the Fourier plane of the transformed image.

Downie [1995] demonstrated the principle of an optical nonlinear logarithmic filter and presented experimental results for a variety of different noise conditions including deterministic and speckle. A block diagram of the optical apparatus used to perform logarithmic image transformation and subsequent Fourier-plane spatial filtering operation is shown in Figure 6.14. Shutter A is opened during the WRITE operation to record the image on the bR film, while shutter B is opened only during the READ operation, when image processing tasks are performed. The bR film displayed a logarithmic transmission response for WRITE intensities spanning a dynamic range greater than 2 orders of magnitude [Downie, 1995].

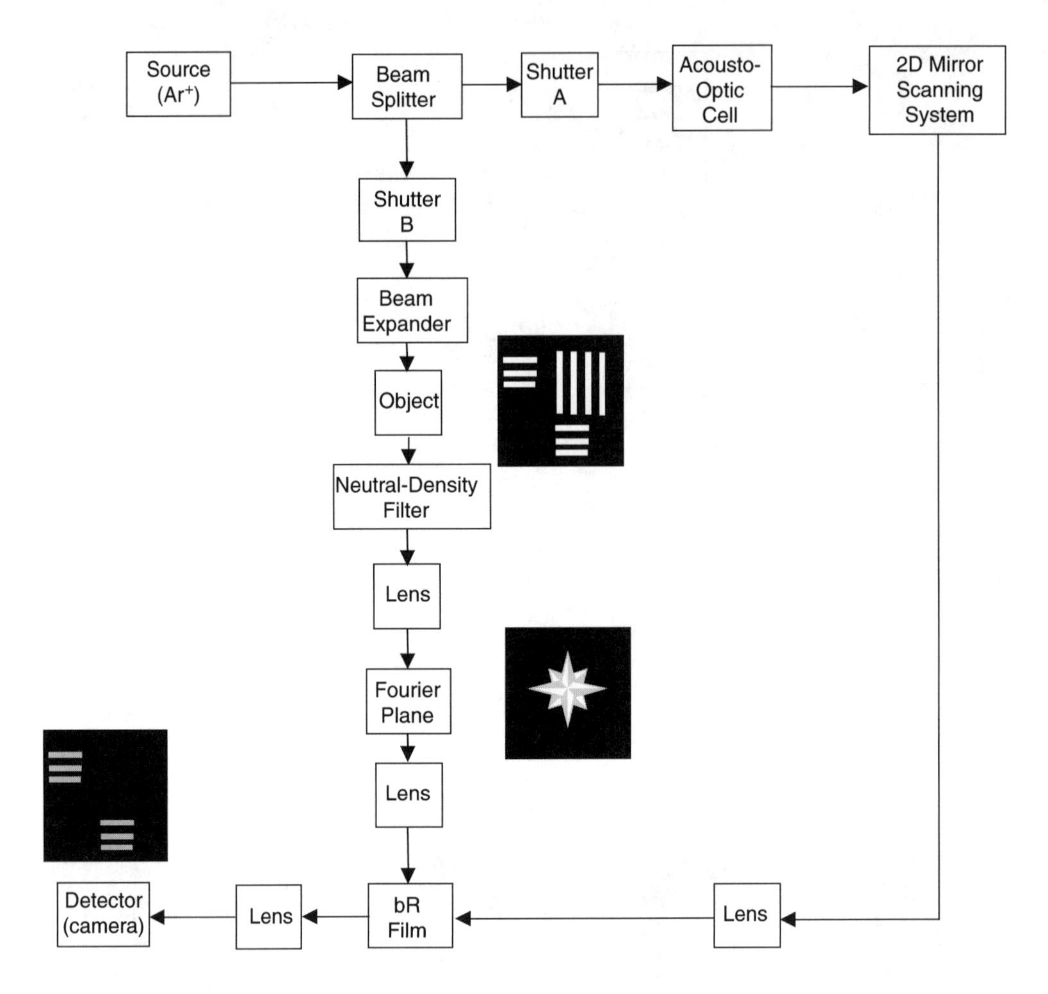

FIGURE 6.15 Block diagram of the programmable spatial filter system described by Storrs et al. [1996]. The planar mirrors have been eliminated in the above diagram.

6.4.3 Programmable Spatial Filter

Fourier spatial-filtering systems are often inflexible because standard photographic film is used as the optical medium to perform the desired filter function. A programmable spatial-filtering system using bR film, as an optically addressed SLM illuminated by a single wavelength, is described by Storrs et al. [1996] (see Figure 6.15). In essence, the bR film serves as a reprogrammable optically addressed SLM. The approach uses a single wavelength of light (~514 nm) and a computer-controlled mirror-based flying spot scanner to write the desired filter function onto the bR film. The relatively long retention time of the bR film (~10 s) provides sufficient opportunity for the scanning system to write a complete filter pattern to the bR film.

Upon illumination the bR film undergoes local changes in the refractive index. In general, the spatial refractive index distribution will modify the system output. The Fourier filter functions written to the bR film were binary, and the transmitted object beam therefore experienced a constant phase shift [Storrs et al., 1996]. As a consequence, the effect of variations in the refractive index were neglected in the published experiments, and additional analysis is required to assess the effects of the resultant phase modulations if continuous functions are to be written to the bR film.

6.4.4 Real-Time Defect Enhancement

Microscope scanning and sophisticated digital-image-processing algorithms are commonly used to detect defects in fine periodic structures such as the photomask used for creating LCD panels or semiconductor wafers. Optical techniques offer a variety of advantages for high-speed, repetitive visual inspection applications because these systems perform image-processing tasks in parallel. In a typical optical-defect detection system a spatial filter is used to enhance faulty patterns by suppressing bright spots diffracted from periodic features and transmitting the weak response from the defects at the Fourier plane. Holographic spatial filters based on traditional photographic films are commonly used in this application. The production of these photographic filters on traditional celluloid film is offline and very time-consuming.

To increase the speed of defect detection, Okamoto et al. [1997] proposed writing holographically generated gratings in bR films. The primary advantage is that the bR film is a reversible optical recording material that can be rewritten more than 10^6 times without significant degradation to the film. In the proposed approach (see Figure 6.16) a straight and equispaced grating that was written in the bR film is simultaneously read by the optical Fourier transform of the object pattern. If the intensity of the READ beam is low compared with that of the WRITE beams, the READ beam is simply diffracted by the grating. However, when the intensity of the READ beam is higher, the beam erases the grating due to saturation and therefore produces a measurable decrease in the diffraction efficiency. In this way the periodic components are suppressed if one uses the Fourier pattern of the periodic patterns to be inspected as READ beams. Okamoto et al. [1997] present experimental results that show enhancement of defects as small as 10 μm in a photomask for liquid crystal display with a pixel pitch of 150 μm.

In practical applications many objects that have similar periodic structures but exhibit different defects must be inspected. For the first object the system requires a finite response time of the bR film of ~1 s for erasure of the grating by the Fourier pattern of the periodic structure at its position. When the second and following objects are inspected, however, the common Fourier pattern has already erased the grating, and only the defect patterns of these objects are imaged in real time. Hence, the slow response of the bR

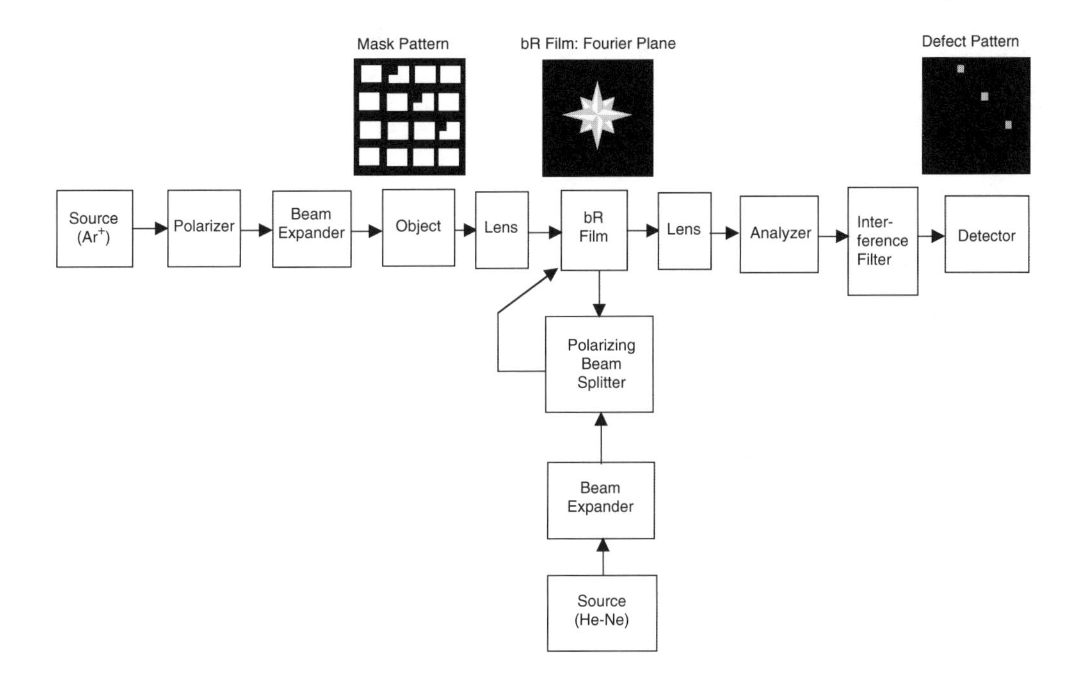

FIGURE 6.16 Simplified block diagram of the defect-enhancement system proposed by Okamoto et al. [1997]. The planar mirrors have been eliminated from the block diagram.

film is rather an advantage to this system. The system is also applicable to moving objects because of the shift invariance of the intensity profile of the Fourier-transformed patterns.

6.4.5 Holographic Associative Memory

Associative memories operate in a manner different from the serial memories found in conventional computers. These devices take an input image, or block of data, and scan the entire memory for a stored image that matches the input. If an exact match cannot be found, then the associative memory device will find the closest match. Finally, the device will return the image found in memory that satisfies the matching criteria, or it will return the address of the image to permit full access of the data. Some forms of associative memory will return a binary bit indicating whether or not the input data are present in the memory.

The bR film can act as the photoactive element in Fourier transform holographic (FTH) associative memory systems. One proposed optical design for achieving associative memory capabilities is described by Birge [1992] and summarized by the block diagram given in Figure 6.17. The input image enters the optical system through the beam splitter at the upper left and illuminates the SLM that is operating in threshold mode. The light reaching the SLM is the superposition of all images stored in the multiplexed holograms. Each image is weighted by the inner product between the recorded patterns from the previous and current iterations. The pinhole array is designed so that the 500-μm-diameter pinholes can be aligned precisely with the optical axes of the multiple FTH reference images that are stored on the two bR films.

The FTH reference images enter from the middle left, and a separate Fourier transform SLM creates a high-frequency image on the bR film in an effort to enhance the autocorrelation peak. The pinholes

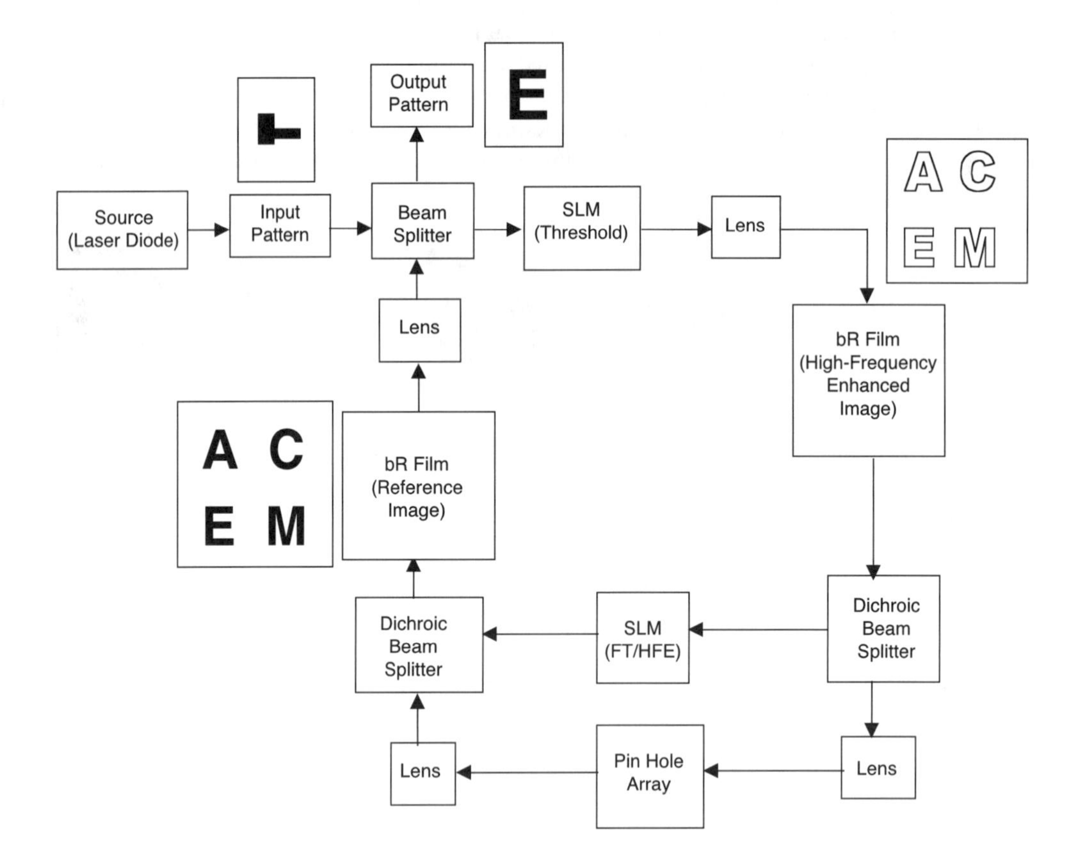

FIGURE 6.17 Simplified block diagram of the Fourier transform hologram (FTH) associative memory system proposed by Birge [1992]. The READ/WRITE reference planes use thin bR films to provide real-time storage of holograms.

eliminate ghost holography, but the optical process loses the spatial invariance of the image reconstruction. Thus, the proper registration of the input image is required for proper associative output. The output image is the full reconstruction of the image stored on the Fourier transform hologram that has the highest correlation with the input images; that is, it produces the largest autocorrelation flux through its aligned pinhole. In this way a partial input image can generate a complete output image.

Thresholding SLM is critical to this application because the response level must be dynamically adjusted to compensate for variable autocorrelation fluxes. The quality of the output image is proportional to the diffraction efficiency of the holographic thin films and the image integration time. By adjusting the threshold level to remove ghost images and by using a charged-coupled detector to integrate the output signal, adequate image quality can be obtained with diffraction efficiencies of 3% [Birge, 1992]. The ability to rapidly change the holographic reference patterns by means of a single optical input while maintaining both feedback and thresholding increases the utility of the associative memory and, in conjunction with solid-state hardware, opens up new possiblities for high-speed pattern-recognition architectures. The defraction limited performance of the bR films, plus the high WRITE/ERASE speeds associated with the excellent quantum efficiencies of the polymer films, is the key element in the potential of bR associative memories. Note that the thresholding SLM and the two holographic memory planes can all be constructed based on bR thin films.

6.4.6 Optically Addressed Direct-View Display

An optically addressed direct-view display with a photoactive layer containing bR in a polymeric matrix has been described by Sanio et al. [1999]. The bR display consists of a glass substrate carrying a dielectric layer with wavelength-selective transmission and reflection, on top of which a thin polymeric film containing the bR film, with enhanced lifetime of the *M* state, as a photoactive medium is deposited. An optional protective layer, either from a polymer or from glass, prevents mechanical damage of the bR layer. The wavelength-selective dielectric layer has two functions. First, it protects the observer in front of the display from the transmitted laser light that is used for writing information to the display. Second, the dielectric layer protects the bR film from the reflected light. During operation, the laser light transmitted by the bR layer is reflected, and, if not adequately protected, it will further bleach the bR layer during the second pass.

A block diagram for the optical apparatus used to demonstrate the direct-view display is shown in Figure 6.18. The expanded laser beam from a laser source (530 nm) passes through a test pattern and

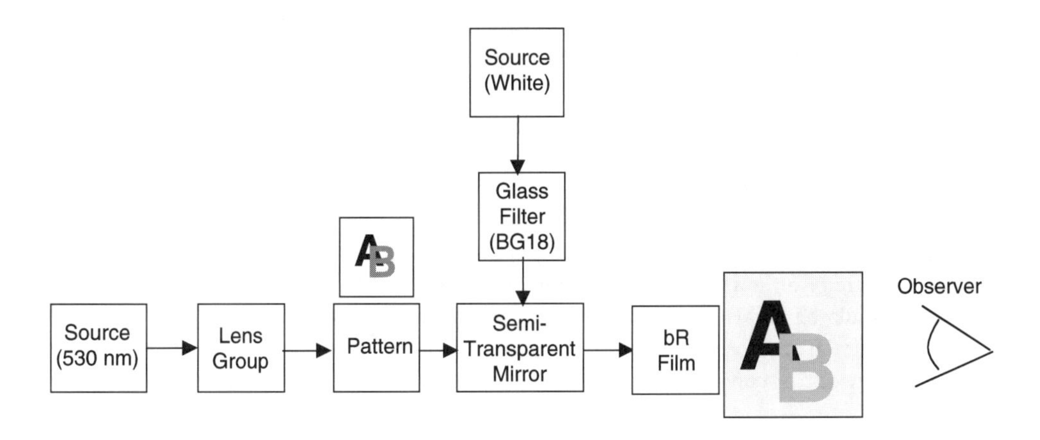

FIGURE 6.18 Block diagram of the optical system proposed by Sanio et al. [1999] to demonstrate how thin bR film can be used for direct-view display. The expanded laser beam (530 nm) is used for writing the information into the bR film. Light from a tungsten lamp, filtered through a BG18 glass filter, is used as observation light and is coupled to the light path by means of a semitransparent mirror.

projects the information onto the rear side of the bR display. Back illumination of the bR display with a suitably filtered white-light source coupled onto the light path by a semitransparent mirror allows the observer to see the displayed information directly with the eye.

The imaged information appears to the eye as an intensity-modulated image in one color (saturation) that is accompanied by a change in the dominant wavelength (hue). The combination of both saturation and hue substantially enhances the ability to discriminate the displayed information from the background. The observer will typically see a bright yellow image on a dark purple background. The bR display shows a high-intensity contrast ratio of 70:1, accompanied by a shift in the color of the visual impression. In an application the image would be recorded on the bR film and periodically be refreshed by means of a laser scanner. The thermal decay time of the bR layer used in the experimental display unit was more than 50 s at room temperature. The contrast decays to 50% of the initial value within 90 s because of the thermal relaxation of the bR film from the M to the B state. Thus, to ensure a minimal contrast ratio of 20:1 the display must be refreshed every 30 s.

6.5 Conclusions

This chapter described several optical sensors and transducers that utilized protein molecules to detect, encode, manipulate, or retrieve information. The bio-molecules acted as signal waveguides, switches, transistors, digital logic gates, and spatial light modulators. Several practical engineering applications were presented in order to illustrate how protein-based devices could be integrated with electronic and optical technologies. The examples are not comprehensive, and numerous alternative applications have been described in the literature. Research continues in this field with the eventual goal of being able to develop integrated optical systems with significantly reduced feature size and virtually nonexistent gate-propagation delays. Furthermore, it is hoped that biomolecular electronics will provide system designers with new hybrid technologies and computing architectures that can perform information-processing tasks faster and more energy efficiently.

Defining Terms

analyte: A biochemical substance that is detected by the sensor or biosensor.
biosensor: A sensor that utilizes biological materials such enzymes, antibodies, and hormones to detect specific chemicals.
chromophore: A chemical group that gives rise to color in a molecule.
enzyme: Complex proteins that are produced by living cells and catalyze specific biochemical reactions at body temperatures.
light transmittance: Fraction of radiant energy that, having entered a layer of absorbing matter, reaches its more distant boundary.
Moore's law: The number of transistors that can be packed into a microchip doubles every 18 to 24 months.
protein: A naturally occurring chain of amino acids that are essential constituents of all living cells and are synthesized from raw materials by plants but assimilated as separate amino acids by animals.
spatial light modulator (SLM): A device that controls the spatial distribution of the intensity, phase, and polarization of transmitted light as a function of electrical signals or a secondary light source.
transducer: A device that converts one form of energy to another.

References

Aston, C., Biological warefare canaries, *IEEE Spectrum*, 38(10), 35–40, 2001.
Birge, R. R., Protein-based optical computing and memories, *IEEE Computer*, November, 56–67, 1992.
Birge, R.R., Protein-based computers, *Sci. Am.*, March, 90–95, 1995.

Birge, R. R., Gillespie, N. B., Izaguirre, E. W., Kusnetzow, A., Lawrence, A. F., Singh, D., Song, Q. W., Schmidt, E., Stuart, J. A., Seetharaman, S., and Wise, K. J., Biomolecular electronics: protein-based associative processors and volumetric memories, *J. Phys. Chem. B*, 103, 10746–10766, 1999.

Chen, X., Zhang, X., Chen, K., and Li, Q., Optical wavelet-matched filtering with bacteriorhodopsin films, *Appl. Opt.*, 36(32), 8413–8416, 1997.

Choi, H. G., Jung, W. C., Min, J., Lee, W. H., and Choi, J. W., Color image detection by biomolecular photoreceptor using bacteriorhodopsin-based complex LB films, *Biosensors Bioelectron.*, 16, 925–935, 2001.

Collier, C. P., Wong, E. W., Belohradsky, M., Raymo, F. M., Stoddart, J. F., Kuekes, P. J., Williams, R. S., and Heath, J. R., Electronically configurable molecular-based logic gates, *Science*, 285, 391–394, 1999.

Downie, J. D., Nonlinear coherent optical image processing using logarithmic transmittance of bacteriorhodopsin films, *Appl. Opt.*, 34(23), 5210–5217, 1995.

Golden, J. P., Saaski, E. W., Shriver-Lake, L. C., Anderson, G. P., and Liger F. S., Portable multichannel fiber optic biosensor for field detection, *Opt. Eng.*, 36(4), 1008–1013, 1997.

Gu, L. Q., Zhang, C. P., Niu, A. N., Li, J., Zhang, G. Y., Wang, Y. M., Tong, M. R., Pan, J. L., Song, Q. W., Parsons, B., and Birge, R. R., Bacteriorhodopsin-based photonic logic gate and its applications to grey level image subtraction, *Opt. Comm.*, 131, 25–30, 1996.

Ivnitski, D., Abdel-Hamid, I., Atanasov, P., and Wilkins, E., Biosensors for detection of pathogenic bacteria, *Biosensors Bioelectron.*, 14, 599–624, 1999.

Kelly, D., Grace, K. M., Song, X., Swanson, B. I., Frayer, D., Mendes, S. B., and Peyghambarian, N., Integrated optical biosensor for detection of multivalent proteins, *Opt. Lett.*, 24(23), 1723–1725, 1999.

Knopf, G. K., Bassi, A. S., Singh, S., Fiorilli, M., and Jauda, L., Optoelectronic biosensor for remote monitoring of toxins, in *OptoMechatronic Systems*, Cho, H. S. and Knopf, G. K., Eds., Proc. Soc. Photo-Opt. Instrum. Eng., Vol. 4190, pp. 9–19, 2000.

Kolodner, P., Lukashev, E. P., Ching, Y. C., and Druzhko, A. B., Electric-field and photochemical effects in D85N mutant bacteriorhodopsin substituted with 4-keta-retinal, *Thin Solid Films*, 302, 231–234, 1997.

Millerd, J. E., Brock, N. J., Brown, M. S., and DeBarber, P. A., Real-time resonant holography using bacteriorhodopsin thin films, *Opt. Lett.*, 20(6), 626–628, 1995.

Min, J., Choi, H. G., Oh, B. K., Lee, W. H., Paek, S. H., and Choi, J. W., Visual information processing using bacteriorhodopsin-based complex LB films, *Biosensors Bioelectron.*, 16, 917–923, 2001.

Oesterhelt, D. and Stoeckenius, W., Rhodopsin-like protein from the purple membrane of *Halobacterium halobium, Nat. New Biol.*, 233, 149–152, 1971.

Okamoto, T., Yamaguchi, I., and Yamagata, K., Real-time enhancement of defects in periodic patterns by use of bacteriorhodopsin film, *Opt. Lett.*, 22(5), 337–339, 1997.

Rao, D. V. G. L. N., Aranda, F. J., Narayana Rao, D., Chen, Z., Akkara, J. A., Kaplan, D. L., and Nakashima, M., All-optical logic gates with bacteriorhodopsin films, *Opt. Comm.*, 127, 193–199, 1996.

Rayfield, G. W., Photodiodes based on bacteriorhodopsin, in *Molecular and Biomolecular Electronics*, Birge, R. R., Ed., American Chemical Society, Washington, D.C., 1994, pp. 561–575.

Sandstrom, K. J. M. and Turner, A. P. F., Biosensors in air monitoring, *J. Environ. Monit.*, 1, 293–298, 1999.

Sanio, M., Settele, U., Anderle, K., and Hampp, N., Optically addressed direct-view display based on bacteriorhodopsin, *Opt. Lett.*, 24(6), 379–381, 1999.

Shelton, D. P., Bacteriorhodopsin optoelectronic synapses, *Opt. Lett.*, 22(22), 1728–1730, 1997.

Song, Q. W., Zhang, C. Blumer, R., Gross, R. B., Chen, Z., and Birge, R. R., Chemically enhanced bacteriorhodopsin thin-film spatial light modulator, *Opt. Lett.*, 18(16), 1373–1375, 1993.

Storrs, M., Mehrl, D. J., and Walkup, J. F., Programmable spatial filtering with bacteriorhodopsin, *Appl. Opt.*, 35(23), 4632–4636, 1996.

Takei, H., Lewis, A., Chen, Z., and Nebenzahl, I., Implementing receptive fields with excitatory and inhibitory optoelectrical responses of bacteriorhodopsin films, *Appl. Opt.*, 30(4), 500–509, 1991.

Timucin, D. A. and Downie, J. D., Phenomenological theory of photochromic media: optical data storage and processing with bacteriorhodopsin films, *J. Opt. Soc. Am. A*, 14(12), 3285–3299, 1997.

Werner, O., Fischer, B., and Lewis, A., Strong self-defocusing effect and four-wave mixing in bacteriorhodopsin films, *Opt. Lett.*, 17(4), 241–243, 1992.

Zhang, T., Zhang, C., Fu, G., Li, Y., Gu, L., Zhang, G., Song, Q. W., Parsons, B., and Birge, R. R., All-optical logic gates using bacteriorhodopsin films, *Opt. Eng.*, 39(2), 527–553, 2000.

For Further Information

Information on biological transducers is included in several professional society journals and conference proceedings. A variety of the articles describing interesting applications of bacteriorhodopsin are found in *Journal of Physical Chemistry, Biosensors and Bioelectronics, Applied Optics, Optics Communication, Optics Letters, Journal of the Optical Society of America*, and *Optical Engineering*.

Two excellent tutorial articles on bacteriorhodopsin films and biomolecular computing are [Birge, 1995] and [Birge et al., 1999]. A number of edited books have been published on the topic of biomolecular computing. One of the most informative is Birge, R. R., *Molecular and Biomolecular Electronics*, American Chemical Society, Washington, D.C., 1994.

A number of textbooks and tutorial articles have been written that provide background on biosensors. These include:

Cass, A. E. G., Ed., *Biosensors: A Practical Approac*h, IRL Press, Oxford University, 1990

Eggins, B. R., *Biosensors: An Introduction*, John Wiley & Sons, New York, 1996

Turner, A. P. F., Karube, I., and Wilson, G. S., Eds., *Biosensors: Fundamentals and Applications*, Oxford University Press, Toronto, 1987

Mulchandani, A. and Bassi, A. S., Principles and applications of biosensors in bioprocess monitoring and control, *CRC Crit. Rev. Biotechnol.,* 15(2), 105–124, 1995

7

Fundamentals of Machine Vision and Their Importance for Real Mechatronic Applications

Roy Davies
Royal Holloway
University of London
Egham, Surrey, U.K.

7.1 Introduction

Vision is one of the five human senses, and it is the one that carries the richest information content. Humans rely on vision for almost all aspects of living, including particularly locomotion, searching for sustenance, and manufacturing myriad edifices, devices, and food products. Thus, it is natural to try to endow man-made machines with vision, with a view to adding to their utility and capabilities. Indeed, it is commonplace to envisage robots as machines that are intelligent and able to move around and act like humans. The necessary transition from biological vision to machine and robot vision hides the fact that vision is a complex ability that has evolved over millions of years to be highly powerful, and apparently effortless and instantaneous, in operation. However, it was discovered long ago that vision is actually quite difficult to engineer and proceeds in two stages that are often confused—seeing and perceiving. The eyes see and pass on basic picture information to the human brain, while the brain itself carries out the perception or detailed understanding of the scene being viewed; this perception is in turn mediated by complex processes of comparison with the huge database of information about the real world that is stored in the brain. By definition, biological and machine vision systems can only carry out processes that are possible, and the scientific study of what is possible is called *computer vision.* It is, then, the application of computer vision to real-world tasks and problems that is the domain of machine vision.

0-8493-1162-4/03/$0.00+$1.50

Robot vision is a subset of machine vision and arises when the machine in question takes the form commonly known as a robot. (It will be of interest that the word *robot* originated from the Polish word *robotnik* meaning "workman".)

There are many aspects of computer vision, and among the earliest that were tackled involved **image processing**—the conversion of one image into another. Converting a gray image into a binary (black and white) image is one of the simplest image-processing tasks, as each pixel is converted in turn from gray to black or white, independently of the other pixels in the image. By contrast, many other types of image-processing operations—such as noise suppression, image enhancement, and edge detection—are only able to determine the final intensity for any given pixel after examining a number of surrounding pixel intensity values [Davies, 1997].

While image processing is useful for improving images for human examination, as when radiographs are enhanced or blurred images are restored, computer analysis of images often requires that image-processing techniques be used to locate image features such as edges or corners of objects. In such cases, the analysis would start from the edge or corner image; then, the coordinates of the edges or corners would be recorded, and the locations of lines, polygons, circles, or other shapes would be deduced abstractly. Thus, the data would have become abstract in the sense that they are no longer recorded in image format. While it might appear that the image-processing phase is soon superseded, in fact many machine vision tasks make continual reference to the original image or other derived images, and the image representation remains a vital one; in fact, image processing is solidly integrated into the whole methodology of machine vision. This makes it well worthwhile to study it in some depth, as this chapter does.

Technically, the area where the image representation is dominant is described as **low-level vision**, while the area where it gradually gives over to a more abstract representation is described as **intermediate-level vision**, the latter being concerned with the extraction of long-range structures within images. *High-level vision* can be regarded as abstract processing and reasoning about the image and involves such aspects as goal seeking, navigation, path planning, and memory access. It is beyond the scope of this article to discuss this topic in great depth. Three-dimensional vision involves trying to make sense of the three-dimensional world from the two-dimensional images that are obtained from camera inputs. This topic is best known as **stereovision** and requires two cameras to provide sufficient information to obtain detailed depth maps of a scene. A further important topic is that of **active vision**, in which the vision system focuses on certain aspects of the input images and attempts to find out more about them, if necessary, by rotating the cameras and zooming in on relevant features. It is especially relevant for autonomous mobile robots that may have to guide themselves along corridors and roads so as to avoid obstacles.

Mechatronics is about the control of machines, and in the present context *control* means vision. In some cases the machines that have to be controlled will be mobile robots; in other cases they may be stationary robots such as those used for assembling and spraying cars. Other mechatronic machines include stationary machines that might not be called robots, though they perform robot functions such as inserting components into printed circuit boards or inspecting and rejecting faulty components at appropriate points on production lines. In all such mechatronic applications vision is a powerful tool, and this article will aim to explain basic vision design methodology. It should be noted that vision can be applied not only in the obvious cases where CCD cameras respond to visible light. Vision may also be applied when the input images are obtained from other modalities such as ultrasonic, microwave, infrared, ultraviolet, or x-ray radiation, or magnetic resonance. This makes vision extremely powerful and considerably widens the scope of this chapter.

The Fundamentals section contains a preliminary review of low-level vision, intermediate-level vision, and topics that can generally be described as shape-analysis techniques. Section 7.3 will then go on to cover three-dimensional vision, motion, active vision, and mechatronics and robotics and finally give a simplified discussion of navigation and path planning for mobile robots. Several short sections covering parameters of importance in applied vision, economic aspects of vision applications, summary, defining terms, references, and further information appear at the end of the chapter. A chapter of this length cannot be comprehensive and necessarily focuses on a limited number of examples. In particular, the chapter concentrates on mobile robots; this is to some extent counterbalanced by a short study of the problems of automated visual inspection. These examples should provide the basis for further study both of general machine vision and of mechatronic or robot applications in a variety of areas.

7.2 Fundamentals

7.2.1 Low-Level Vision

As indicated in the Introduction, low-level vision is concerned with basic operations, such as thresholding, noise suppression, image enhancement, and feature detection, that in some way prepare the image for higher level structural analysis—whether by human or by computer. In general, low-level vision is achieved by local image-processing operations, i.e., operations that convert images into other images that have been modified locally to facilitate abstract analysis.

One of the main categories of operation in this class is the pixel–pixel operation, in which the output pixel intensity depends only on the input intensity of the corresponding pixel. This contrasts with the many other image-processing operations (called window–pixel operations or simply window operations) for which the output intensity is a function of the input intensities of all the pixels within a window around the given pixel; in a good proportion of cases a 3×3 window is sufficiently powerful, though larger windows are sometimes advantageous.

Examples of pixel–pixel operations are thresholding, contrast stretching, and intensity inversion, all of which are useful and widely used, though they are also limited in what they can achieve. For instance, thresholding is used to convert the initial gray-scale image into a binary image (Figure 7.1(a) and (b)): in this case each pixel becomes black if its intensity is lower than a preset threshold; otherwise, it becomes white (though in some cases it is more convenient to interchange the two output intensities). While thresholding can be used to segment dark objects from light backgrounds, if there is any variation in the background illumination, this sort of segmentation can give misleading results. However, if the threshold that is used is permitted to vary over the image, this approach to segmentation can often be made to work and is then extremely useful. Conditions under which suitable dynamically varying thresholds can be found include cases both where the background illumination can be modeled and where a small window is used to examine the pixel intensity values around any given pixel and a threshold is then estimated with the aid of a suitable averaging process. This technique can be highly effective and works well when, for example, printed text is being interpreted under conditions of uneven lighting [Davies, 1997].

7.2.1.1 Use of Convolution Masks

While thresholding is a useful technique, either in its basic or in its dynamically varying form, there are many cases where it is difficult to apply. In such cases edge detection forms a better basis for rigorously segmenting objects from their backgrounds. Edge detection is commonly carried out by window operations, and specifically by convolutions. Indeed, convolutions form one of the most widely used categories of window operation.

A convolution is a linear operation that is independent of the location of the window in the image, and each local output value is given by the expression $\Sigma_i w_i p_i$, where p_i is the intensity of the pixel i in the given window and w_i is the weight assigned to that pixel. It is normally most convenient to describe such an operation in terms of a convolution mask, in which each coefficient represents the weighting factor for the corresponding pixel in the window, as in the case of the following mask, which represents averaging of all the pixels in a 3×3 window to help suppress noise:

$$M = \frac{1}{9}\begin{bmatrix} 1 & 1 & 1 \\ 1 & 1 & 1 \\ 1 & 1 & 1 \end{bmatrix} \tag{7.1}$$

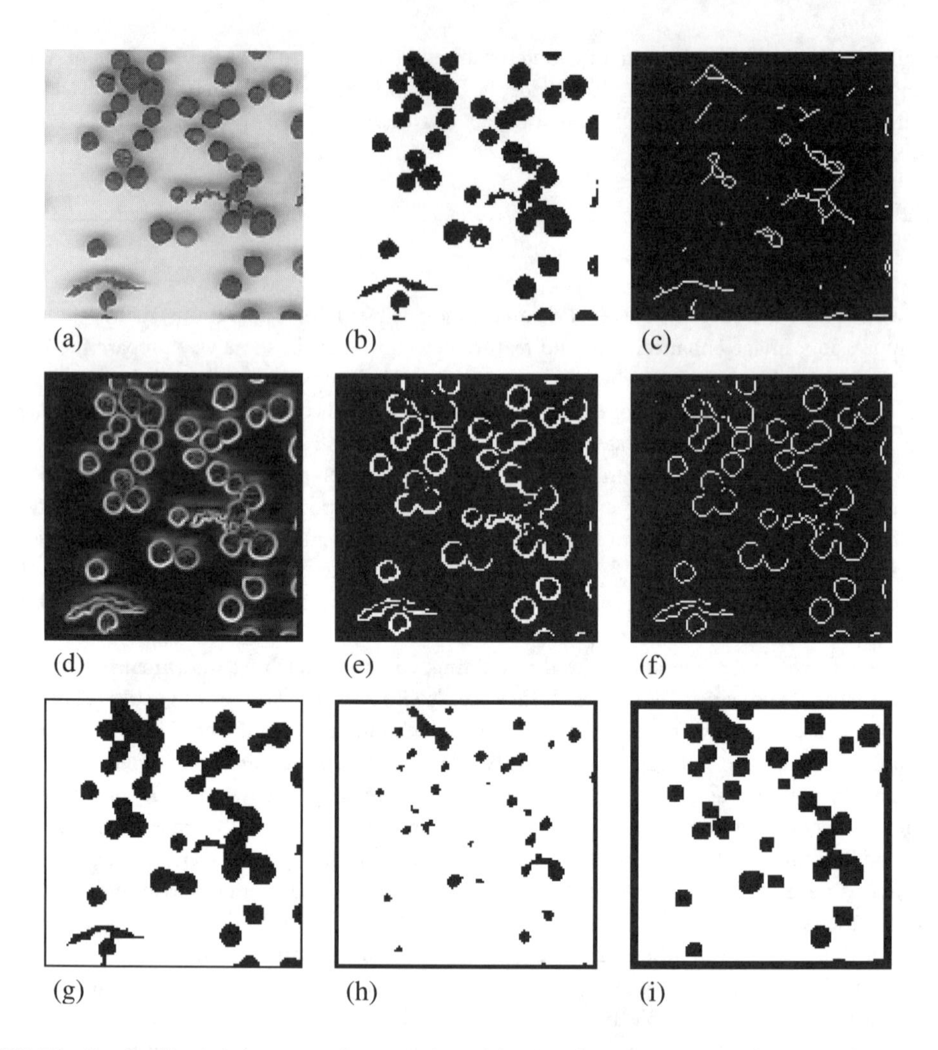

FIGURE 7.1 Result of basic image processing operations: (a) original 128 × 128 gray-scale image of some peppercorns; (b) result of thresholding (a); (c) thinned version of the inverse of (b); (d) effect of applying Sobel edge enhancement operator to (a); (e) result of thresholding (d) to detect edges; (f) thinned version of (e); (g) effect of closing (b) using a 3 × 3 window; (h) effect of eroding (b) using a 5 × 5 window; (i) effect of dilating (h) using a 5 × 5 window. Note that in (c) the skeletons of the small round peppercorns are mostly very short lines or in some cases dots; however, any small hole in the original results in a loop in the skeleton, while touching peppercorns give a variety of oddly shaped skeletons. It will be clear that skeletons do not provide an especially useful representation in this instance. In (e) the detected edges are wide in some places and peter out in others, while in (f) thick edges are thinned to unit width. In (g) closing has eliminated the small holes in the peppercorns, but has also joined some together. In (h) erosion has reduced the peppercorns in size and in many cases has separated those that were touching; it has also eliminated the narrow twiglets, which are not restored in (i) by dilation.

There are many operations for detecting horizontal and vertical edges in an image, but the following Sobel convolution masks are among the most widely known and used [Davies, 1997]:

$$S_x = \begin{bmatrix} -1 & 0 & 1 \\ -2 & 0 & 2 \\ -1 & 0 & 1 \end{bmatrix} \qquad S_y = \begin{bmatrix} 1 & 2 & 1 \\ 0 & 0 & 0 \\ -1 & -2 & -1 \end{bmatrix} \tag{7.2}$$

Strictly speaking, the two Sobel operators are linear and do not themselves detect edges but rather produce edge signals, g_x and g_y, whose contrast is greatest at edge locations. Detection is the process of deciding where the edges actually are, and this process is typically carried out by thresholding the edge-signal images (Figure 7.1(a), (d), and (e)). Note that edges of arbitrary orientation may be detected by estimating the vector edge-signal magnitude g and then thresholding that instead. To find g the following formula must be applied:

$$g = \left(g_x^2 + g_y^2\right)^{1/2} \quad (7.3)$$

The vectorial analysis leading to this equation also permits edge orientation θ to be calculated using the following formula:

$$\theta = \arctan(g_y/g_x) \quad (7.4)$$

This formula is extremely valuable for certain intermediate-level vision operations, as we shall see in Section 7.2.2.2.

Convolution operations are useful for a number of other feature-detection operations, including small-hole or spot detectors and corner detectors, as indicated by the following convolution masks:

$$H = \begin{bmatrix} -1 & -1 & -1 \\ -1 & 8 & -1 \\ -1 & -1 & -1 \end{bmatrix} \qquad C_1 = \begin{bmatrix} -4 & -4 & -4 \\ -4 & 5 & 5 \\ -4 & 5 & 5 \end{bmatrix} \quad (7.5)$$

Notice that these masks resemble the features they are targeted to detect. Ultimately, this is because feature detection needs to couple selectivity with high sensitivity, and matched filters (often used in radar) are appropriate in this sort of situation. However, the masks used in image processing generally have to be zero sum in order to minimize the effects of variations in background illumination; this applies to all the main types of feature detector, including edge, hole, and corner detectors, as will be seen by examining the masks discussed above. In addition, it should be noted that most types of features are anisotropic, so a number of masks are required to ensure that high sensitivity and discriminability are achieved. For example, in a 3 × 3 window eight convolution masks such as C_1 are needed for corner detection. For more advanced corner detectors, see Davies [1997].

7.2.1.2 Use of Nonlinear Operations

The advantage of convolution operations is that they are simple to apply, involve relatively little computation, and are quite powerful. However, apart from the final decision-making operation they are completely linear, and in the end this restricts their use for practical applications. To overcome this problem, nonlinear operations of various sorts need to be applied. Typical among these is the median filter, which is used for eliminating impulse noise.

The median filter operates by examining the intensity values appearing in each window, placing them in order of increasing intensity, and then selecting the central median value as the output intensity value. Notice that the progressive elimination of pairs of extreme values in the distribution will eventually yield the median value; this explains why the median filter is excellent at eliminating the outliers that are responsible for impulse noise. Interestingly, the median filter is capable of not only eliminating impulse noise but also of not causing the blurring that is characteristic of the mean (averaging) filter mentioned above [Davies, 1997].

In fact, there is a range of rank-order filters that, like the median filter, start by ordering the local intensity values. Other notable examples are the minimum and maximum filters, which, respectively, output the lowest and highest local intensity values. In Section 7.2.1.4 we shall see that these are also members of the powerful set of filters known as morphological operations.

7.2.1.3 Use of Logical Operations

Another very powerful set of operations is the set of logical operations that are applied to binary images. Here we assume that the images contain dark objects on a light background and that after thresholding the objects are marked with logical 1s against a background of logical 0s.

One important operation is to separate the image into connected components, which are labeled in sequence throughout the image. The situation is demonstrated in Figure 7.2(a) and (b). Note that there is a problem in that a simple forward raster scan through the image leads to label clashes that have to be resolved before a consistent labeling of the connected components is arrived at. This can be achieved either by repeatedly rescanning the image, sweeping inconsistent labels away and propagating the correct labels, or else by abstract processing of a label-adjacency table and then carrying out one final scan to correct all the inconsistent or erroneous labelings. Space does not permit a full discussion of these techniques, but they are described, for example, in Davies [1997]. Note that once connected components have been labeled correctly, finding the number of objects in the image and the area of each of them is a trivial operation.

In some applications it is useful to examine the edges of the binary picture objects. This can be achieved by eliminating any logical 1s that are completely surrounded by 0s according to the following algorithm:

```
for all pixels in image do
    begin
        if pixel is in object and is completely surrounded by object pixels
            then output : = 0
        else if pixel is in object
            then output : = 1
        else//pixel is in background
            output : = 0
    end
```

There are many ways of rewriting such algorithms, e.g., in terms of logical **or** operations, but space does not permit us to go into further details here. Instead, we examine an alternative representation that is often useful for shape analysis. In this case the object is represented not by its boundary but by its skeleton, which is taken to reflect the internal structure and shape of the object (see Figure 7.1(c) and (f)). To form a skeleton the outer layers of pixels around the object are peeled off one by one until a unit-width structure running medially along the limbs of the object remains. During thinning (skeletonization), care must be taken to prevent the thinner parts of the shape from being eroded altogether, and various rules must be invoked to ensure that the skeleton does not become disconnected. These rules are quite intricate and cannot be explained fully here; a fairly full discussion of the topic appears in Davies [1997]. In principle, the final skeleton will contain a number of limbs equal to those existing on the original shape, though sometimes noise points on the boundary create so-called noise spurs, and removing these is not a trivial process. Likewise, any holes in the original shape will lead to loops in the final skeleton, which can be viewed as a unit-width transform of the original shape (see Figure 7.1(c)). One of the aims of thinning algorithm design is to ensure that the skeleton limbs are not shortened and thus provide good measures of the size of the original object.

It will now be clear that thinning provides a simple way of analyzing shapes and characterizing them in terms of the numbers of branches and loops and their dimensions. Apart from general shape characterization, the skeleton approach is especially well adapted to measuring characters for optical character recognition, to analyzing road patterns appearing in aerial views of the ground, and also to analyzing plant and tree structures. More generally, the skeleton approach is complementary to the boundary-pattern-analysis approach, which will be outlined in Section 7.2.2.1.

Related to the skeleton is the **distance function** of the shape [Rosenfeld and Pfaltz, 1968]. The end result of applying a distance function is that each pixel is given a value that represents its closest distance to the background, i.e., the distance to the nearest logical 0 in the input binary image (see Figure 7.2(c)). Distance functions are valuable in providing techniques for producing skeletons that are not significantly

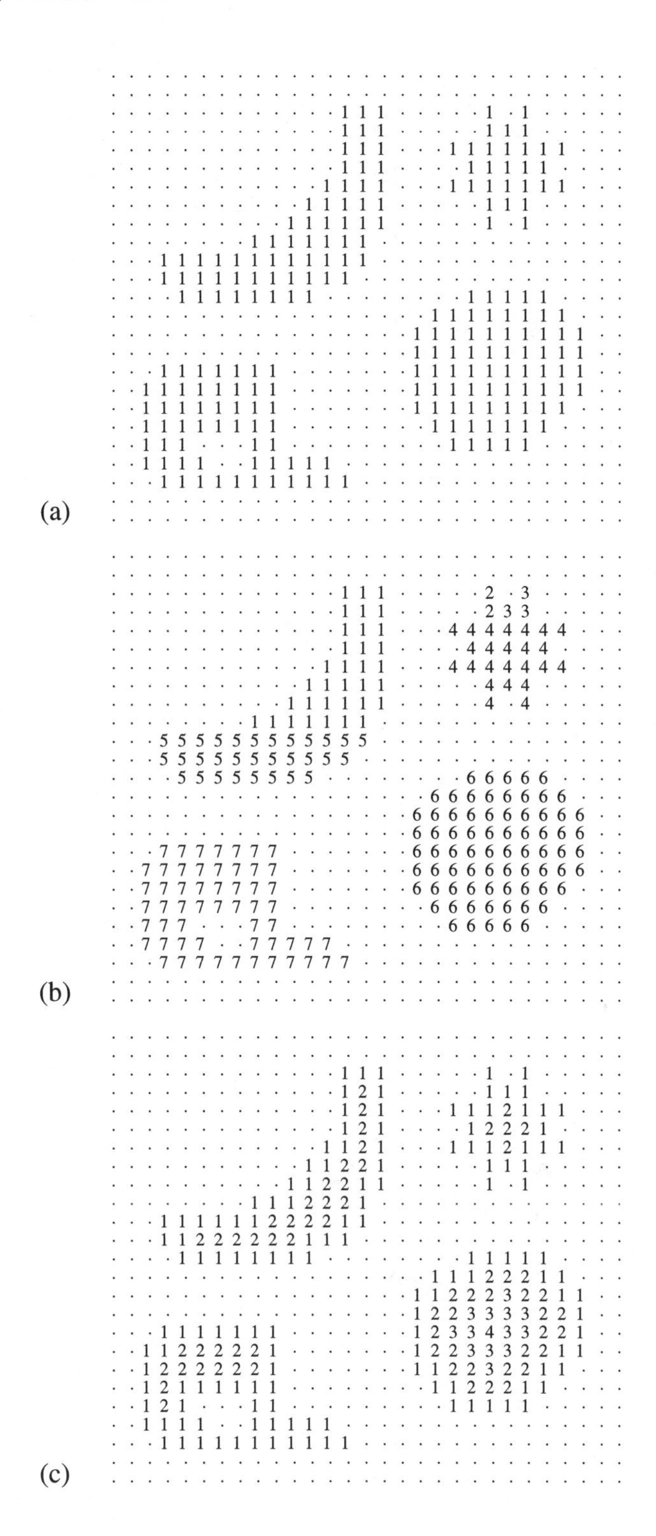

FIGURE 7.2 Numerical operations on binary picture objects: (a) initial binary image containing decorative cake shapes; (b) the result of a single-pass labeling algorithm; (c) distance functions of all the shapes. Repeated scans, or improved algorithms (see text), are able to improve on the results of (b), leading in this case to proper labels in the range 1 to 4.

biased by the successive stripping operations used during thinning. They are also useful for annotating skeletons to give a more complete dimensional analysis of the original shape (the annotated skeleton can be used to almost perfectly reconstruct the original shape, demonstrating that it embodies an accurate representation of it). Distance functions can also be valuable in quite different applications. In Section 7.3.7 we shall see a very useful application of them for robot path planning.

7.2.1.4 Use of Morphological Operations

One further type of local operation is important for shape analysis and recognition, i.e., the set of morphological operations. These are easiest to explain in terms of binary images, though the methodology has been extended to process gray-scale operations. The basic technique is that of eroding or dilating objects in various directions, as dictated by directional structuring elements. First, we can imagine a shape eroded isotropically through the distance of a single pixel. While this operation is clearly similar to the stripping operation that takes place during thinning, it is actually simpler because it does not matter whether or not the shape will be split into two or more parts by the process. By applying N single-pixel erosion operations, it is clear that small objects will be eliminated, and this will help sort the objects in the original image; it can also be imagined that a single isotropic erosion through a distance of N pixels has been performed (Figure 7.1(b) and (h)). Indeed, we can say that the structuring element for erosion is isotropic and has radius of N pixels. A similar dilation process can be applied, either through N single-pixel operations or by using a single process involving an isotropic structuring element having a radius of N pixels (Figure 7.1(h) and (i)).

As indicated earlier, erosion and dilation operations can be directional or isotropic. If several dilation operations are applied in different directions, producing several output images, these can be combined with logical **or** operations; ultimately, if enough such operations are applied and combined in this way, we arrive at an isotropic dilation operation. The same concepts apply for erosion, but in that case the combining operation must be the logical **and**. These ideas show the generality of the approach, but for reasons of space we must now concentrate on isotropic operations.

Let us imagine that an isotropic structuring element of suitable radius has been applied to an image eliminating the small objects, thereby sorting the objects in the image. At this stage the large objects will be reduced in size, and it is convenient to restore them to their original size. This is achieved by applying a dilation operation using the same structuring element. The combined erosion–dilation operation is called an *opening* operation. This is because it has the effect not only of eliminating small objects but also of changing the shapes of certain objects; in fact, if any of the large objects are narrow in certain places, these regions will be opened up and gaps instituted. For example, a dumbbell shape will be replaced by two ball shapes (Figure 7.1(b), (h), and (i)). Notice that any objects containing holes may disappear after opening, whereas those not containing holes may be almost unchanged in shape.

It is also useful to consider the effect of a dilation operation followed by an erosion operation using the same structuring element. This operation is called *closing*, as it has the effect of closing gaps between objects or between different parts of the same object (Figure 7.1(b) and (g)). Thus, a C shape may well become an O shape after a closing operation is applied. It should be noted that opening and closing operations can have rather gross effects on object shapes, though in favorable cases the dilation and erosion elements of these operations almost cancel out. However, the cancellation is seldom exact, and careful studies are required to determine the extent of such changes.

While this discussion has proceeded intuitively, it has largely hidden the fact that this is a highly mathematical subject area that involves operations on sets, upon which many theorems can be, and have been, built [Haralick et al., 1987]. In what follows we attempt to give the flavor of the situation without dwelling unnecessarily on mathematical detail.

First, consider a structuring element B and its application for dilation (denoted by the symbol $\oplus$) and erosion (denoted by $\ominus$). If we also denote closing by the symbol $\bullet$ and opening by $\circ$, we can write the closed and opened forms of shape A respectively by:

$$A \bullet B = (A \oplus B) \ominus B \tag{7.6}$$

$$A \circ B = (A \ominus B) \oplus B \tag{7.7}$$

Ideally, these two shapes would be identical to each other and to A (assuming that we are not required to eliminate A or combine it with another shape), but, as stated above, the shapes may not be identical. Suffice it to say that the mathematics shows that the following restrictions always apply:

$$A \oplus B \supseteq A \bullet B \supseteq A \supseteq A \circ B \supseteq A \ominus B \tag{7.8}$$

However, there is one assumption inherent in the proof: that B contains the identity element. To understand this we first need to explain what is meant by the identity element I. Suppose we are applying a restructuring element B to a region A in order to dilate it. Then I is the value of B when no change in the size or shape of region A occurs during the operation:

$$A \oplus I = A \tag{7.9}$$

The significance of this is that if B contains the identity element, dilation by B will not be able to remove anything from A, and the whole of A will be part of $A \oplus B$. Structuring elements not containing the identity element have interesting properties such as the ability to shift objects rather than merely making them larger.

It is difficult to take the analysis much further in the limited space available here. Instead, we consider a particular type of computation that is important for inspection applications. If we apply a closing operation, local defects in objects, such as small holes, indentations, and cracks, will be eliminated. However, subtracting the original shape will show exactly where these defects are:

$$C = A \bullet B - A \tag{7.10}$$

Similarly, applying an opening operation will eliminate point defects, minor prominences, and hairs, and these can again be revealed by a suitable subtraction operation:

$$D = A - A \circ B \tag{7.11}$$

Another useful inspection application of **morphology** is the application of erosion to separate touching objects, so that they can be counted reliably.

Finally, it is worth remarking that the generalization of morphology to gray-scale images can be achieved by replacing the binary restructuring element by a window of the same size and shape, in which local maximum or minimum operations are applied; local maximum corresponds to dilation of bright regions and local minimum corresponds to erosion of bright regions, or vice versa when the objects being considered are dark on a light background. There are other more complex morphological analogs of dilation and erosion, but again they cannot be considered in detail here [Haralick and Shapiro, 1992].

7.2.2 Intermediate-Level Vision

In the previous subsection we considered how low-level processing could be achieved with the use of local operations. In this subsection we show how the information gleaned during low-level processing can be used to extract global image structures. This is the domain of intermediate-level vision. In particular, features detected by low-level **template matching** operations are grouped together to demonstrate the presence of structures that reflect the existence of objects in the scene [Davies, 1997].

7.2.2.1 Boundary Pattern Analysis

One of the most basic intermediate-level vision techniques is that of tracking around the boundaries of objects, following edge segments that are detected by local operators. Once a connected boundary has been identified, the centroid of the object can be computed, and the boundary can be mapped into a one-dimensional polar (r, θ) representation called a *centroidal profile* (see Figure 7.3(a)). An examination of this profile allows for straightforward identification of circles, as these have constant values of r over the complete

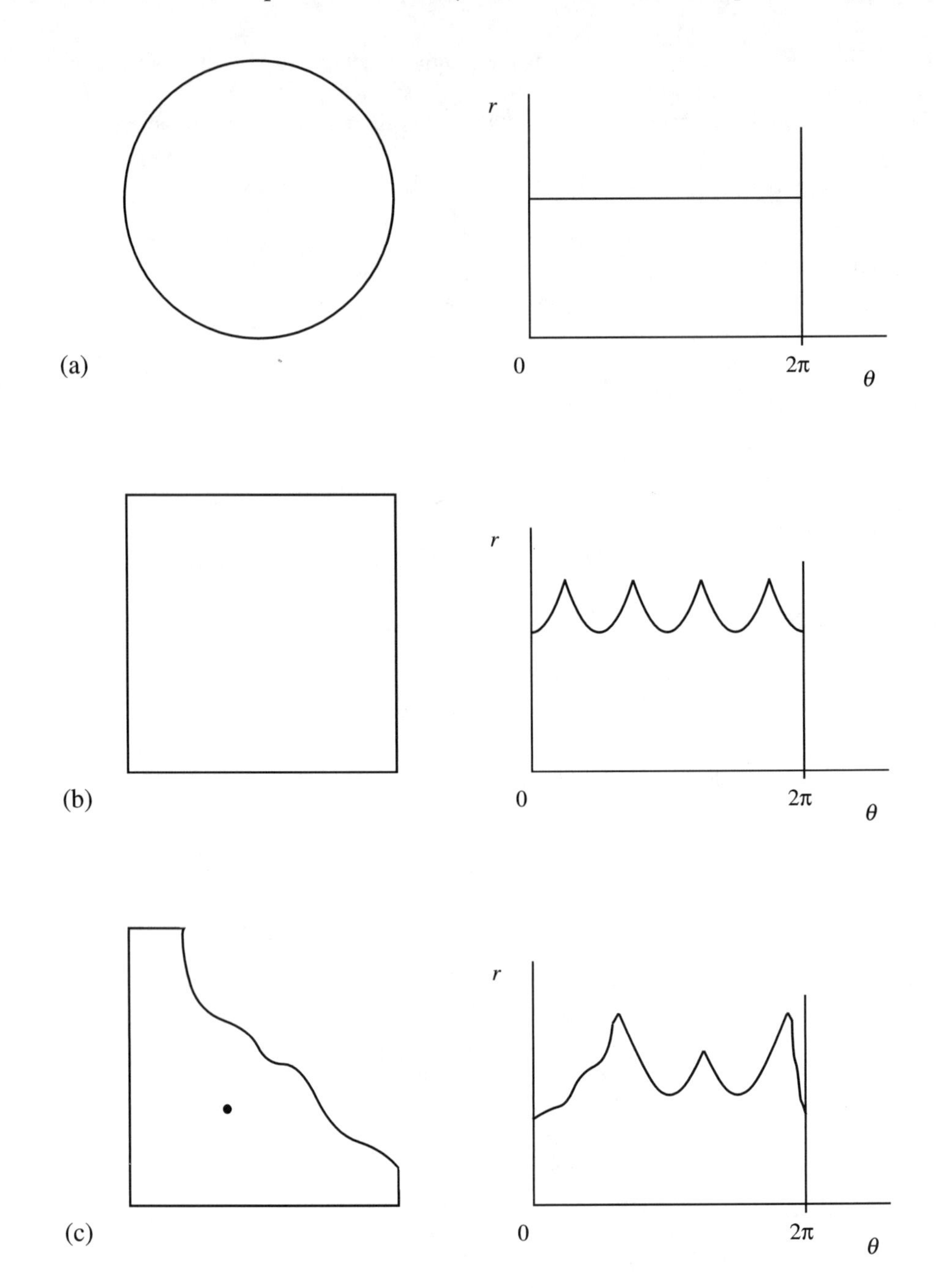

FIGURE 7.3 Centroidal profile and its problems: (a) circle; (b) square; and (c) broken square, together with their centroidal profiles. In (c), it is difficult to get any useful information from the centroidal profile.

range of θ. Square objects are slightly more complex to identify because they have four corners that are quite clear from the profile, but the straight sides of the square map into sec θ curves, which can only be confirmed to represent straight lines by careful analysis (see Figure 7.3(b)). However, measurement of the size and orientation of the square is a simple process. Other shapes, such as rectangles and ellipses, are only a little more difficult to recognize, but it is not profitable to explore the situation further here [Davies, 1997].

Unfortunately, once objects are distorted by breakage or partial occlusion, or even by one object merely touching another, the situation becomes considerably more complex. Not only do the centroidal profiles change shape in obvious ways to match the shapes of the objects, but they also become distorted because the centroid—which is the reference point from which all boundary measurements are made—becomes shifted (see Figure 7.3(c)). At this point the centroidal profiles become difficult to interpret, and the simplicity of the technique is lost; it must be regarded as nonrobust. For this reason recourse must be made to techniques, such as the Hough transform, that are intrinsically robust.

7.2.2.2 The Hough Transform Approach

The **Hough transform** [Hough, 1962] is more robust than many more basic techniques because it concentrates on searching for evidence about the existence of objects and ignores any data that do not support this evidence. For instance, when searching for circles, it aims to accumulate evidence about them by building up votes at potential circle center positions. To this end it examines each edge segment in the image and works out where the center of a circle of radius R would be if that edge segment were part of the circle, and it accumulates that vote at a location in a separate image space called a *parameter space*. When all such votes have been included in the parameter space, the locations of any peaks are noted and taken as possible circle centers. Significant peaks are more likely to correspond to circle center locations than to random accumulations of data, but taking any one as a circle amounts to a hypothesis; in principle, such hypotheses need to be checked by reference to other data in the original image.

The Hough transform calculates the position of candidate center locations by moving a distance R along the edge normal direction from any given edge segment (see Figure 7.4). Thus, it is important to use an edge detector that is capable of giving accurate edge-orientation information (see Section 7.2.1.1). When the value of R is unknown, several values of R can be tried, and the solutions corresponding to the highest peaks in parameter space are the ones most likely to correspond to circle centers and to correct values of R.

The Hough transform approach can be used for locating other shapes such as ellipses or even general shapes. The necessary methodology is covered in texts such as Davies [1997]. Here, there is only space to cover one other application of the Hough transform—that used for detecting straight lines in digital images. This case is especially useful in the context of mobile robots (see Section 7.3.4).

To detect straight lines, all the edge segments in the image are located; then, an extended line is constructed through each edge segment E_i with the same orientation θ_i as E_i, and its distance ρ_i from the origin is calculated; next, a vote is accumulated in an abstract parameter space with coordinates (ρ_i, θ_i); finally, peaks are sought in this parameter space, and the peak coordinates are taken as those of likely lines (or hypotheses of lines) in the original image space. Again, in principle all hypotheses should

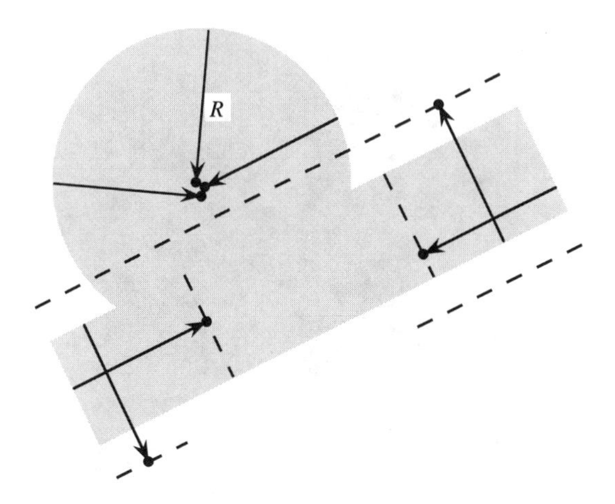

FIGURE 7.4 Hough transform for circle. A circle is partly occluded by a rectangular object and is shown together with the Hough transform. Note that the straight sides of the rectangle lead to straight lines of votes indicated by the five dotted lines, which give low ridges rather than peaks; thus, the peak arising from the partial circle is readily detected.

be checked by reference to the original image data, but those corresponding to the highest peaks are most likely to represent valid lines in the image.

7.2.2.3 The Graph-Matching Approach

The Hough transform is particularly suited to the robust location of objects from their edge features. When objects are to be located from sparse, widely separated point features such as corners or small holes, it has been common to use an alternative approach called *graph matching.* This involves matching the graph joining the point features in the real image against the graph representing the ideal object template. In fact, it is necessary to match subgraphs in each case because (a) some points may be obliterated from the image by damage or occlusion, and (b) other points may appear in the image because of noise or irrelevant background, or indeed objects other than the ones being extracted. The maximal clique graph-matching approach [Bolles and Cain, 1982] works by searching for the set of correspondences between the two graphs that form the largest completely consistent subset; that is, all features of one subset correspond to all features of the other subset. This is checked by ensuring that all distances between feature points are identical in the two subsets. This is a highly rigorous technique: any inconsistency suggests that the evidence being collated is unreliable and so it does not represent a totally valid hypothesis; thus, only smaller pairs of subsets can be matched together exactly. Figure 7.5 shows some planar brackets that have been identified and accurately located by this means.

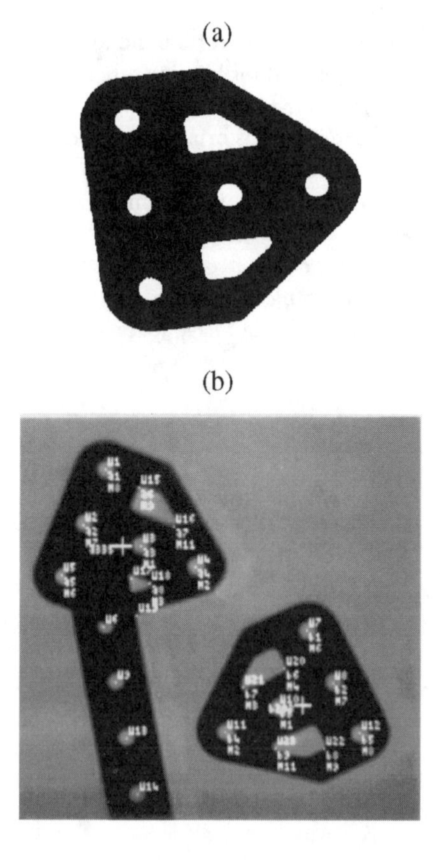

FIGURE 7.5 Object recognition by graph matching: (a) planar bracket which forms a template for the recognition process; (b) two brackets of the specified type that have been located by graph matching. The method is not confused by the holes of a different type of bracket that also appears in the image. (Thanks are due to Dr. Simon Barker for permission to reproduce this figure from his Ph.D. thesis, Royal Holloway, University of London, 1989.)

While this approach is highly effective and about as robust as the Hough transform approach, it is computationally costly; the total computation increases approximately exponentially with the number of image and template features. As a result it works well with objects containing up to five or six features, but for objects with more than 10 or 12 features, alternative methods are generally sought.

Some time ago it was found [Davies, 1992] that the Hough transform approach could also be used for matching point features, with a marked reduction in the amount of computation, by estimating the position of a reference point on any object from each pair of feature points found on it; by recording votes at each such locations, peaks in an image-like parameter space could again provide hypotheses for the locations of the specific type of object being sought. For further details of this and the maximal clique technique, see, for example, Davies [1997].

7.3 Three-Dimensional Vision and Applications

7.3.1 Three-Dimensional Vision

In the earlier sections of this chapter we considered two-dimensional image-processing techniques such as those that would be used for a camera looking straight down on a number of flat objects lying on a horizontal worktable or conveyor. Under these circumstances shape analysis is the dominant class of techniques, with dimensional measurement also playing a crucial role. However, in real mechatronic applications it is rather rare for objects to be flat or to be viewed in this way. Indeed, robot limbs, manufactured products, vehicles, buildings, factories, and outdoor scenes are far from flat structures; consequently, vision systems must be able to cope with three-dimensional objects and scenes. Furthermore, the relative placements of any objects are often complex, and it is rarely possible or desirable to view them from directly overhead.

The fact that objects will be observed from a general viewpoint means that the number of degrees of freedom increases from three (corresponding to translation and orientation in a two-dimensional plane) to six (corresponding to three translation parameters and three orientation parameters). This makes analysis of three-dimensional scenes far more complex; in addition, it will generally be much more computation intensive to analyze this type of situation. This is clear because quantization of each of the parameters into 200 levels immediately changes the number of available configurations *per object* from 8 million in two-dimensional to 64 million million in three-dimensional objects. However, that is not the end of the story, for it is very possible that objects will occlude, overlap, and touch each other, giving at worst misleading visual interpretations and at best limited information. This makes the analysis considerably more complex. In fact, it is well known that the human eye is subject to a plethora of optical illusions; these arise from the hypotheses made by the eye–brain system in order to make sense of the incoming visual information, and in certain circumstances they can be erroneous. A typical case is when the dimensions of objects in the foreground are erroneously estimated because of the effects of **perspective** in the background. We shall consider the effects of perspective in more detail in a later section.

In spite of these complexities, it is evident that the human eye is able to interpret three-dimensional scenes extremely quickly, and this hides the fact that there is any particular difficulty about the process. However, humans have two eyes and perform stereo vision, and it may seem obvious that this provides the key to efficient visual processing. In this respect, two points should be borne in mind: First, with an inter-eye baseline of just a few centimeters, the stereo effect is minimal above 100 m or so, and humans manage to see well using even a single eye; second, stereo vision adds only depth information and does not actually provide mechanisms for immediate recognition of objects. This indicates that vision must be even more complex than might be imagined *a priori*, and in this context the fact that the human eye–brain system contains upwards of 10^{10} neurons, each of which is connected to some 10^4 to 10^5 others, should not be forgotten. But what else helps the eye to interpret visual scenes? The answer lies in the considerable number of cues besides depth. These include shape information from contours, shading, color, texture, focus, and motion. While we are not consciously aware of all the processing involved in assessing these cues, this does not mean that it is not happening. In any case, the important problem

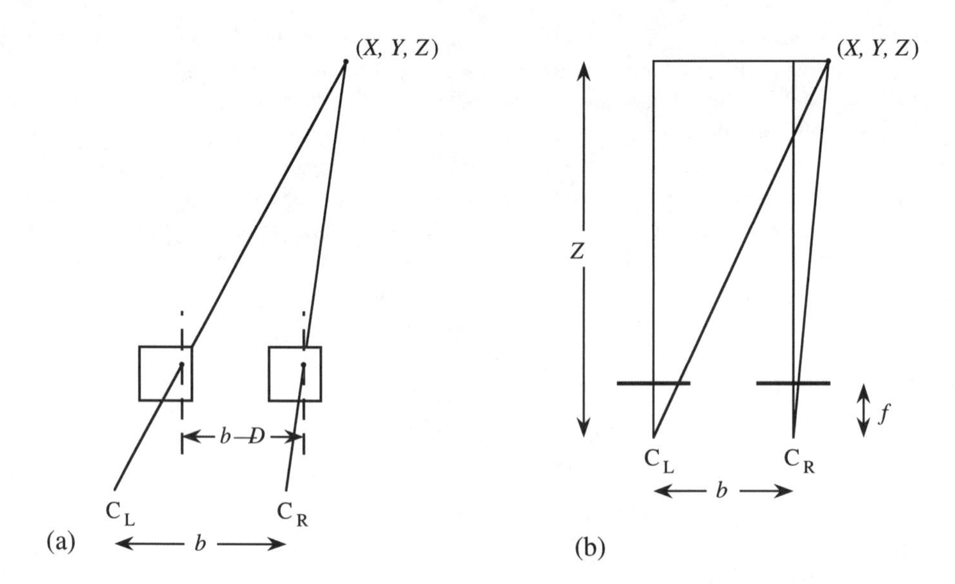

FIGURE 7.6 Depth from binocular vision: (a) a point *X*, *Y*, *Z* in the scene observed by two cameras with centers of projection C_L and C_R; (b) plan view of a showing sets of similar triangles that are used for calculating depth from disparity (see text). While the disparity *D* cannot itself be marked in the diagram, the distance $b - D$ is indicated in (a).

here is how to emulate these processes in machine vision systems. In a short article of this sort, it will be possible devote any attention to only a few aspects, but fortunately quite a lot can be achieved with relatively few techniques [Davies, 1997].

7.3.1.1 Stereo Vision

As indicated above, stereo vision based on two eyes is an important feature of the human visual system and one that should be invaluable when designing a machine vision system. While human eyes can be turned inwards to scrutinize nearby objects more carefully, this makes the geometry unnecessarily complicated for machine vision, and we shall consider only dual cameras with zero vergence (parallel optical axes). Figure 7.6(a) shows the basic configuration of such a system when a distant object is viewed at position (*X*, *Y*, *Z*) relative to an origin on the midpoint of the line bisecting the centers of projection of the two lenses. (With no loss of generality we imagine the image being formed on a focal plane at distance *f* in front of, rather than behind, each lens.)

The sense of depth and the depth information arise because the images in the two eyes are not identical. Indeed, the difference in relative position of corresponding feature points in the two images is called the *disparity D*, and it allows depth *Z* in the scene to be calculated using the following simple formula:

$$Z = fb/D \tag{7.12}$$

where *b* is the length of the baseline between the two centers of projection. To verify this formula, it is only necessary to write down the equations obtained by considering two pairs of similar triangles in Figure 7.6(b):

$$x_L/f = (X + b/2)/Z \tag{7.13}$$

$$x_R/f = (X - b/2)/Z \tag{7.14}$$

Subtracting, we immediately find that:

$$D = x_L - x_R = fb/Z \tag{7.15}$$

and depth Eq. 7.12 follows.

While this equation is very simple, considerable difficulty arises when trying to identify corresponding feature points in the two input images. If the two images were identical, there would be little problem, but then no depth information would be available. In fact, increasing b enhances the depth information and makes depth measurement more accurate, but it also has the effect of making the images even more different and makes it more difficult to locate correspondences. This is known as the *correspondence problem*. It is exacerbated by the fact that as the images arise from different views of the scene, some points are visible in one view but may be hidden in another. This is caused by occlusion and may be due either to objects in front of the object being viewed or to self-occlusion when the object has curved boundaries or several facets. Whatever the reason, the algorithm that is used to match features between the images must permit tolerance for the fact that the two views will not be identical; however, this gives rise to the possibility of false matches and ambiguities, and additional processing will be required to overcome this problem.

Nevertheless, a very useful technique is available to cut down the search problem and to eliminate many of the false matches that can arise. This is the *epipolar line* approach. It involves considering the line in space along which a given feature point in, say, the left image must lie. When this line is viewed in the right image, it takes the form of the so-called epipolar line (Figure 7.7); corresponding feature points must lie on this line and cannot lie elsewhere. As stated above, this clearly provides the capability for eliminating most of the search and most of the false matches. Any remaining false matches will have to be eliminated by considering the relative placements of the features in both images and by identifying those that are capable of constituting parts of reasonable surfaces (mismatches are liable to be characterized by unusual or totally impossible depths in the scene).

As stereo vision requires the identification of point features, it can be applied when easily discernible salient features, such as corners, can be identified; however, it is also applicable when surfaces have a textured appearance, as textures contain a plethora of relevant features. (On the other hand, the large numbers of features available in textured regions could lead in some cases to problems of identification and ambiguity.) Another important point is that plain surfaces, such as sheets of paper or car door panels, may not have identifiable point features, and stereo vision cannot be used for depth measurement in such cases.

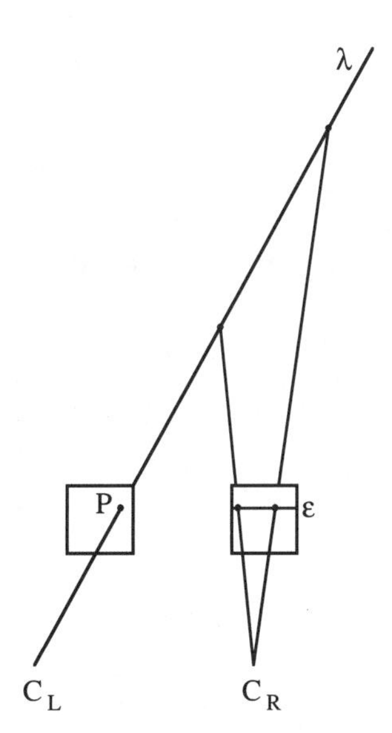

FIGURE 7.7 Epipolar line. Point P is observed by the left-hand camera, and the line λ on which the original object point lies is deduced. λ projects to a line ε in the right-hand camera image; ε is called the epipolar line of the point P.

Nevertheless, it is possible to use a related approach—that of *structured lighting*—to obtain depth maps of plain surfaces, and this method is widely used in machine vision during factory assembly and inspection tasks. Structured lighting involves the projection of patterns of light onto object surfaces, from a known direction, and thus provides the features that can be observed by a camera viewing from a different direction. Thus, one of the stereo cameras has been replaced by its inverse—a projector. The geometry of depth perception and measurement for the two approaches is identical, but stereo vision is applicable when point features are naturally present, while structured lighting is applicable when point features have to be generated artificially. Many configurations and techniques exist for structured lighting, but they are not universally applicable outside the laboratory or factory environment; further discussion of this specialized approach will not be undertaken here [Davies, 1997].

7.3.1.2 Shape from Shading

While stereo vision is clearly very important for providing depth maps, it is also possible to discern depth by monocular vision. In particular, the shading on plain three-dimensional objects, such as fields, hills, sand dunes, or (closer to hand) walls, fan blades, or egg surfaces, readily permits their shapes to be ascertained by the human eye. These objects and surfaces can all be modeled as being approximately Lambertian matte surfaces that reflect light in all directions, the amount of light available being approximately proportional to the cosine of the angle of incidence i of the incoming light. By measuring the reflected light intensity, it is possible to obtain estimates of $\cos i$, such that (1) contours of constant i can be formed, and (2) estimates of the local surface normal directions can be obtained by suitable processing. The type of processing required is nontrivial and involves calculations of mutually consistent orientations for all parts of a curved surface. While this sort of processing is possible, it is quite computation intensive, somewhat difficult to manage as it involves tedious setting up of boundary conditions, and in the end provides orientation maps that then have to be converted to depth maps. At that point the methodology is effectively equivalent to stereo vision: the technique is advantageous to the eye in providing further confirmatory information on the scene, but it has limited value for machine vision. However, an extension known as photometric stereo is somewhat more useful. This involves applying light from three light sources in turn and using the resulting images to estimate the direction cosines of the local surface normal directions; in this case the calculations are simpler, less susceptible to ambiguity, and amenable to reasonably rapid computation for mechatronic applications [Horn, 1986].

7.3.1.3 Effects of Perspective

Interpretation of three-dimensional scenes may also be tackled by examining the effects of perspective. In particular, perspective makes more distant objects appear smaller; it also distorts object shapes, especially when they are nearby and are subject to the phenomenon known as foreshortening. Clearly, these two effects are related and arise because the rays arriving at the camera from the various objects act as if they pass through the center of projection of the camera on their way to the image plane (other rays are focused by the camera lens so that they contribute to the same images).

The mathematics of perspective projection is somewhat tedious, but various key points arise: Straight lines project into straight lines in the image; parallel straight lines project into straight lines passing through a point known as the **vanishing point,** which is situated at a point in the image plane corresponding to infinite distance in the scene (Figure 7.8); in general, equal distances do not project into equal distances; equal angles do not project into equal angles, and right angles do not project into right angles; circles do not project into circles, but into ellipses (see Figure 7.9); ellipses project into other ellipses. Thus, there is considerable complexity in the overall situation.

One simplifying effect is that objects that are imaged under *weak-perspective projection* (WPP), i.e., objects whose differential depths ΔZ are much less than their actual depths in the scene, appear the same as Z varies, except that their sizes in the image are inversely proportional to Z. Under WPP a cuboid will still appear to have parallel edges, but its size will appear inversely proportional to depth in the scene. This is certainly not the case for *full-perspective projection* (FPP), which arises when ΔZ is comparable to Z.

FIGURE 7.8 Vanishing point formation. In this case, parallel lines in space appear to converge toward a vanishing point to the right of the image; although the vanishing point V is only marginally outside this image, in many other cases it lies a considerable distance away.

FIGURE 7.9 Parts for assembly. These parts contain many circles that appear as ellipses; the latter can be identified by specially designed Hough transforms and then used to help recognize and locate the parts in the scene [Davies, 1997].

The distinction between WPP and FPP is particularly important when it comes to recognizing three-dimensional objects from a small number of feature points, which we shall assume have been identified [Davies, 1997]. If three points on an object have been identified, then the object can be matched and orientated in space—*except* that under WPP there will be a single ambiguity because the object can be flipped through an angle around a carefully chosen axis, which will give an identical image. Adding a fourth point coplanar with the other three will eliminate the problem under FPP, but the ambiguity will remain under WPP. On the other hand, if the fourth point is not coplanar with the rest, WPP will eliminate the ambiguity.

Curiously, under FPP, a total of six points are required to eliminate the ambiguity, assuming that no more than three of the points are in the same plane. The situation is quite complex, as more parameters are needed to interpret the image under FPP, though under FPP more information is actually available to help with the interpretation, albeit with greater mathematical complexity. With all this complexity, it is difficult to see how real-time image interpretation could be carried out in the three-dimensional case. However, if we have prior knowledge about the scene and can search for important cues such as parallel lines, the situation should be eased considerably. Fortunately, parallel lines are a frequently occurring phenomenon in the modern world, and their presence can be confirmed via the presence of vanishing points.

7.3.1.4 Vanishing-Point Detection

We next consider how vanishing points (VPs) can be detected. Generally, the process is undertaken in two stages: (1) locate all the straight lines in the image; (2) find those lines that pass through common points and deduce which must be VPs. Stage 1 is in principle a straightforward application of the Hough transform, though how straightforward depends on whether some of the lines are texture edges, which are difficult to locate accurately and consistently. Stage 2 is, in effect, another Hough transform in which peaks are located in a parameter space congruent to the image space, where all points on lines are accumulated. In fact, points on extended lines, as well as those on the lines themselves, must be accumulated. Unfortunately, complications arise as VPs may be outside the confines of the initial image space (Figure 7.8). Extending the image space can help, but there is no guarantee that this will work; in addition, accuracy of location may be poor when the lines leading to a VP converge at a small angle; finally, the density of votes accumulated in the VP parameter space may, even with the application of suitable point-spread functions, be so low that sensitivity of detection will be negligible. This means that special techniques are necessary to guarantee success.

Perhaps the most important technique for detecting VPs is to use the surface of a Gaussian sphere as the parameter space [Magee and Aggarwal, 1984]. Here, a unit sphere S is constructed around the center of projection of the camera, and S is used as a parameter space, instead of an additional image plane. Thus, VPs appear at convenient positions at finite distances: infinite distances are ruled out (see Figure 7.10). Notice that there is a one-to-one correspondence between each point on the extended image plane and each point on the front half of the Gaussian sphere; the back half is not used. Nevertheless, using the Gaussian sphere gives rise to problems when analyzing the vote patterns, as these are not deposited uniformly over the sphere.

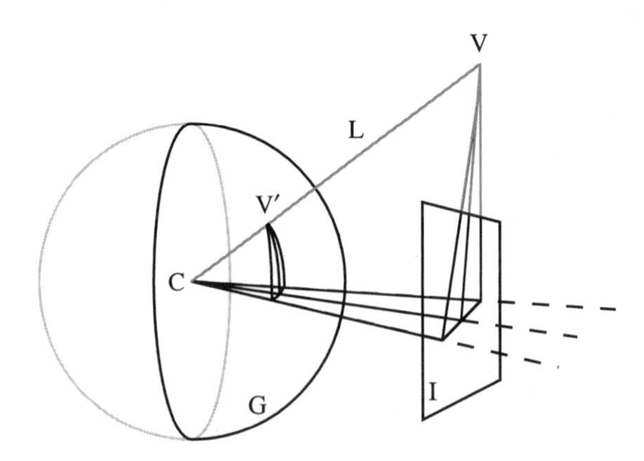

FIGURE 7.10 Use of the Gaussian sphere for vanishing point detection. In this case the vanishing point V is shown well above the input image I. To be sure of detecting such points, the surface of the Gaussian sphere, G, is used as the parameter space for accumulating vanishing point votes. Also marked in the figure are the projection line L from the center of projection, C, to V and V′, which is the position of the vanishing point on G.

In addition, large numbers of irrelevant votes are cast when lines do not pass through actual VPs, as they do not originate from lines that are parallel to other lines in the scene. It would be better if all the votes that were cast at isolated peaks corresponded to likely VPs. To achieve this, *pairs* of lines are taken, and the VPs that would arise if each pair of lines were actually parallel lines in the scene are recorded as votes in the parameter space. If this is implemented for all pairs of lines (or for all pairs of edge segments), all necessary peaks will be recorded in the parameter space, and the amount of unnecessary clutter will be minimal. However, this is not achieved without cost, as the computational load will be proportional to the number of pairs of edge segments or, alternatively, the number of pairs of lines. In fact, if there are N lines, the number of pairs is $\binom{N}{2} = \frac{1}{2}N(N-1)$, which is proportional to N^2 if N is large.

After the Gaussian sphere has been used to locate the peaks corresponding to the VPs, the latter can be recorded in the image space and extended if necessary, though this may prove impossible if one or more of the VPs is situated at infinity. However, the important thing is to deduce which of the initial lines or line segments correspond to the VPs and which correspond to isolated lines or mere clutter. Once this has been achieved, there is considerably greater certainty in the interpretation of the images. By way of confirmation, for a moving robot there should be some correspondence between the VPs seen in successive images.

7.3.1.5 Invariance

One more variable that is important for analyzing three-dimensional images is that of **invariance** [Mundy and Zisserman, 1992]. A particular case is the *cross-ratio*, which shows whether a set of points on a straight line in one image could correspond to a different perspective view of the same set of points in another image. This could be useful when analyzing a sequence of images. Such situations occur during egomotion, e.g., when an automatic vehicle is guiding itself down a road along which objects such as telegraph poles appear at intervals. Another use of such a measure is to indicate whether sets of points on the two sides of a building or corridor could correspond to crossings with sets of parallel lines.

The reason why a cross-ratio could be useful is that in the absence of perspective distortions ratios of distances between points would remain the same at different viewing orientations, whereas when perspective is important, it is intuitively necessary to take ratios of ratios for constancy; cross-ratios are carefully conceived ratios of ratios that are guaranteed to maintain invariance under perspective projection.

7.3.2 Motion

Motion analysis has an obvious application in mechatronics and particularly in the area of vehicle guidance and tracking. Perhaps the most obvious way of studying motion is to make difference images for all pairs of frames in a sequence of images. However, this approach leads to certain difficulties. In particular, if a uniform rectangular object moves in a direction parallel to two of its sides, the amplitude will be zero at these locations in the difference image. Similarly, the body of the uniform rectangular object will not appear in the difference image. Indeed, the difference image will only be nonzero at the front and back of the object. In a more general case, the component of any motion parallel to any edge will not affect the difference image. Only components of the motion perpendicular to any edges will affect the difference image. Furthermore, the difference will be nonzero only if the foreground object contrasts with its background.

These facts may be summarized in the following equation giving the amplitudes within the difference image:

$$\frac{\partial I}{\partial t} = -\nabla I \cdot \mathbf{v} \tag{7.16}$$

where $\mathbf{v} = \mathbf{v}(x, y)$ is the local velocity, and $\boldsymbol{I}$ is the intensity of the input image. Clearly, the final difference signal depends on the motion and the intensity gradient vectors and their relative orientations. This leads to an ambiguity—called the *aperture problem*—in determining the velocity vector using a small aperture to detect the local motion.

In spite of the ambiguities inherent in the approach, the so-called **optical flow** field $\mathbf{v}(x, y)$ is widely used for the purpose of analyzing motion. The ambiguities must be eliminated by iterative or other procedures that are capable of resolving them by communicating information over a larger area until a consistent global solution is arrived at. In some cases the problem is ill conditioned and it is difficult to arrive at globally accurate solutions [Davies, 1997]. One way around this is to use point features such as corners to back up the edge information that is essentially assumed in the foregoing approach. Space precludes a more detailed discussion of the methodology, and in what follows we shall assume that an unambiguous optical flow field has been arrived at in any given case.

When an optical flow field has been computed, interpretation is by no means easy, as a number of different situations may arise: The camera itself may be moving, one or more objects in the scene may be moving, or the camera and some objects in the scene may be moving. If all the objects in the scene are moving at the same speed relative to and toward the camera, there will be a single **focus of expansion** (FOE), with all the velocity vectors in the optical flow field appearing to diverge away from the FOE (see Figure 7.11). An equivalent focus of contraction (FOC) can also occur. In fact, an image can contain several FOEs and FOCs, when objects are moving at different velocities relative to the camera; one example of this is included in Figure 7.11.

Unfortunately, the situation is generally far more complex when objects are rotating as well as translating, or if the camera is rotating (unless the only motion is a single rotation). In the general case complex iterative techniques are required to analyze the full situation and cannot be discussed in detail here. However, when robot or vehicle guidance is being undertaken, the motion will be largely one of translation, and the location of a single FOE will be of potential value for vehicle control.

7.3.3 Active Vision

The term *active vision* has a number of meanings, and it is important to differentiate carefully between them. First, there is a connotation that comes from radar, active radar being different from passive radar in that the aircraft being sensed deliberately returns a special signal that is able to identify it. In vision, this is not a connotation that is normally used. Instead, active vision is taken to mean one of two things: Either (1) moving the camera around to focus on and attend to a particular item of current interest in

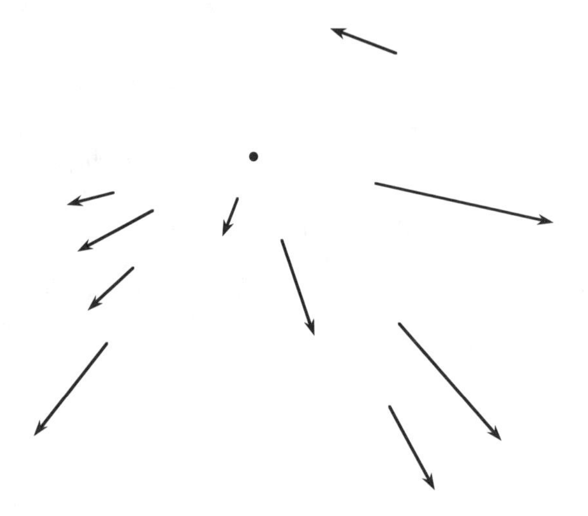

FIGURE 7.11 Focus of expansion. The short arrows in this diagram show the image velocities of points in the scene. With one exception, they appear to diverge from a focus-of-expansion central black dot. The pattern is similar to that for an airplane flying at low level over the ground, so that all but one of the points observed have the same velocity relative to the pilot; the exception relates to another aircraft moving at a different relative velocity. In general, each rigid body in the scene will have its own relative velocity and separate focus of expansion. Thus, it is not surprising when, as here, one or more arrows have exceptional directions.

the environment or (2) attending to a particular item or region in the image and interacting dynamically with it until an acceptable interpretation is arrived at. In either case, there is a focus of attention, and the vision system spends its time actively pursuing and analyzing what is present in the scene, rather than blandly interpreting the scene as a whole and attempting to recognize everything. In active vision, it is up to a higher level processor to identify items of interest and to ask—and answer—strategic questions about them.

While the two modes of active vision outlined above may seem very similar, they have evolved to be quite distinct. In the first one the activity takes place largely in terms of what the camera may be made to look at. In the second, the activity takes place at the processing level, and active agents such as snakes are used to analyze different parts of the received image. Thus, a snake may initially be looped loosely around the whole region of an object and will then gradually move inwards until it has sensed exactly where the object is. This is achieved by interacting intelligently with the object region and interrogating it until a viable solution is arrived at. In one sense both schemes are carrying out interrogation, and in that sense active vision is being undertaken in each case. However, the emphasis of each is different: what is important in the case of the moving camera is that recording and memorizing the whole of a scene in all directions around a robot station is necessarily a costly process and one for which a complete interpretation is unlikely to be fully utilized. It is wasteful to gather information that will later not be used; better indeed to concentrate on finding the part of the scene that is currently most useful and to narrow down its scope and enhance its quality so that really useful answers can be found [Yazdi and King, 1998]. This approach mimics the human interrogation of scenes, which involves making saccades from one focus of attention to another, thereby obtaining a progressive understanding of what is happening in the environment.

The active vision approach is clearly useful for practical control of real robot limbs, but it is clearly even more vital for mobile robots that have to guide themselves around the factory, hospital, office-block, or (in the case of robot vehicles) along the highway; in such cases it is relevant to update only those parts of the scene that give information that is needed for the main task—that of safe guidance and navigation.

7.3.4 Mechatronic Applications

Mechatronics is the study of mechanisms and the electronic systems needed to provide them with intelligent control. It has applications in a wide variety of areas including industry, commerce, transport, space, and even health service. For instance, robots can be used in principle—and in many cases in practice—for mowing lawns, clearing rubbish, driving vehicles, effecting bomb disposal, carrying out reconnaissance, performing delicate operations, and executing automatic assembly tasks. A related task is that of automated inspection for quality control, though in that case the robots take the form of trap doors or air jets that divert defective items into reject bins [Davies, 1997]. It is clear that there are several categories of mechanisms and robots. First, there are the mobile robots, which include robot vacuum cleaners and other autonomous vehicles. Then there are the stationary robots with arms that can spray cars on production lines or assemble television sets. Finally, there are robots that do not have discernible arms but that insert components into printed circuit boards or inspect food products or brake hubs and reject faulty items.

Figure 7.12 shows a printed circuit board using surface-mount and board-grid-array technology, which requires reasonably accurate placement of components so that the soldering process can proceed satisfactorily. Figure 7.13 shows a case of tombstoning, where inaccurate placement has resulted in one joint on a capacitor not taking, with the outcome that the component has risen like a tombstone. This is readily detected by machine vision if (1) the camera is set to view obliquely and (2) the object is partially backlit. Alternatively, it will be evident from an overhead view that the component is not in its proper position, though exact diagnosis could be uncertain. This example illustrates the importance of careful illumination in machine vision.

By definition, all mechatronic systems control the processes they are part of, and to achieve this they need sensors of various sorts. Sensors can be tactile, ultrasonic, visual, or involve one of a number of

FIGURE 7.12 Printed circuit board after assembly.

FIGURE 7.13 Printed circuit board with a case of "tombstoning." Here the board is viewed obliquely and the components are partially back-lit.

other modalities, some of which provide images that can be handled in the same way as normal visual images; instances include ultrasonic, infrared, thermographic, x-ray, and ultraviolet images. In this chapter we are concerned primarily with these sorts of visual image, and there will be some mechatronic applications that do not involve this general class of input. However, there are far too many mechatronic applications that use visual images for control for us to deal with them all here. Thus, we shall concentrate on autonomous mobile robots and automated inspection as two important ends of the mechatronic scale.

7.3.4.1 Autonomous Mobile Robots

Mobile robots are poised to invade human living space at an ever-increasing rate. The mechanical technology is already in place, and the sensing systems required to guide them have been evolving for many years and are sufficiently mature for serious, though perhaps not ambitious, application. For example, robot lawnmowers, robot vacuum cleaners, robot trash gatherers, and robot weed-sprayers have been produced, but more ambitious autonomous vehicles are held back largely because of the risks involved when humans inhabit the same areas—and in case of accidents, litigation could pose serious problems. Indeed, attempting to advance too quickly could actually set the industry back.

In addition to safety, there are other practical issues such as speed of operation and implementation cost. As yet, the problems of applying mobile robots have not all been solved, especially in sight of these practical issues. Setting aside the question of the end use of the robot, one of the main features of robots still to be engineered is how they are to guide themselves safely in territory that will be populated by both humans and other robots. Autonomous guidance involves sensing, scene interpretation, and navigation. On a limited scale, a cleaning robot could proceed until it bumps into an object and then take evading action; in this case tactile sensing is required. However, ultrasonic devices permit proximity sensing and appreciation that other objects that have to be avoided are nearby. Here the need for planning is beginning to be felt [Kanesalingam et al., 1998]. However, serious planning and navigation are probably only possible when vision sensing is used, as the latter is remotely instituted and is rich in information content, and flow rate.

While vision sensing provides an extremely rich flow of information that is potentially exceptionally valuable for autonomous control and path planning, the interpretation task is considerable. In addition, it must be borne in mind that mobile robots will be moving at walking speed indoors and at much greater speeds outdoors, so a significant **real-time processing** problem arises. However, it seems that stereovision offers limited advantage in that (a) the additional information requires additional processing and (b) it does not offer substantial advantages in three-dimensional scene interpretation at distances of more than a few meters. On the other hand, stereo can help the vision system to cue into the complex information present in the incoming images, so if stereo is not to be used, reliable cues have to be found in monocular images. As noted in Section 7.3.1.4, a widely used means by which such cues can be obtained is by VP detection. VPs are all too evident to humans in the sort of environment in which they live and work. In particular, factories, office blocks, hospitals, universities, and outdoor city scenes all tend to contain many structures composed not only of straight lines and planes but also of rectangular blocks and towers (see Figure 7.8).

Vanishing points are useful in several ways. They confirm that various lines in the scene arise from parallel lines in the environment; they help to identify where the ground plane is; they permit local scale to be deduced (for example, objects on the ground plane have width that is referable to, and a known fraction of, the local width of the ground plane); and they permit an estimate to be made of distance along the ground plane by measuring the distance from the relevant image point to the VP. Thus, they are useful for initiating the process of recognizing and measuring objects, determining their positions and orientations, and helping with the task of navigation. Location of VPs has already been covered in Section 7.3.1.4. In Section 7.3.5 we take the process of navigation and path planning further.

7.3.4.2 Inspection

Typical inspection tasks involve the following sequence of actions: acquiring images of the products on the product line, locating individual products and application of a region of interest around each product, conducting close scrutiny and measurement of each product, and accepting or rejecting each product. Product location is likely to be the most computation intensive of these processes, as it involves unrestricted search, often through many images, for the object in question, and it is not usually known in advance what its orientation will be. Thus, it can be quite exacting to devise suitable algorithms for this purpose, as they have to operate in real time at whatever rate products advance along the line [Davies, 2001].

FIGURE 7.14 Texture arising from close-packed objects. This image shows a set of close-packed beans on a conveyor. Attention is on the overall flow rather than on the individual beans. At this scale the picture is still marginally better described as a collection of beans rather than as a texture.

In many cases, such as biscuit or cake manufacture, product flow rates can exceed 20 items per second, though if peas or other small vegetables or ball bearings have to be inspected, flow rates can exceed 200 or even 1000 per second. Once a product has been located, a golden template can in principle be placed over it to verify that it is acceptable for packaging and sale. However, golden templates have restricted value, as many products—including almost all food products—have wide variability and similarly wide acceptance levels. Thus, analytical measures, such as shape and size, must be applied to determine whether such products are acceptable, while states of underfit and overfit are important parameters for mechanical parts. In some cases, artificial neural networks can be used to help with the judgments of acceptability.

Some inspection tasks do not fit into the above category, for example, when powder, grain, paper or sheet metal is passed in an unbroken stream along a conveyor. In that case there is no need for the object location stage, as every part of the conveyor should be covered, and the attention is not on any one grain or product item (see Figure 7.14). Nevertheless, although the purpose of the inspection system is then to identify any foreign bodies in the product flow, or any marks on the paper, the inspection system will have to be sensitive to the possibility of a breakage in the flow of product and to log it separately from a normal defect or foreign body.

When continuous flow of product occurs, it will not always be possible to test for uniformity by applying simple intensity thresholding procedures, such as identifying insects among grain by their relative darkness. Instead it will be important to characterize the normal variability of the texture composed by the many small particles in a powder or grainy flow. This is not a trivial matter, as the shapes of the individual texels can be important for recognition of texture as well as variations in intensity. However, standard techniques are available for achieving this with a high degree of effectiveness [Davies, 1997]. Note, however, that some care is required to select an appropriate technique in any given case, particularly if real-time computation is required.

Finally, while the most obvious purpose of inspection is to reject defective products, a further purpose is to provide feedback to stations further back down the line to ensure that uniformity (e.g., in the sizes of biscuits) can be controlled by adjusting the temperature of the mixture, as this affects its spread. As indicated above, yet another purpose is to log the situation on the product line, noting any problems that arise and their frequency of occurrence.

7.3.5 Navigation for Autonomous Mobile Robots

When a mobile robot is proceeding down a corridor, the field of view will usually contain at least four lines that converge to a VP situated in the middle of the corridor (see Figure 7.15(a)). In many cases this VP can be used for navigation purposes. For example, the robot can steer so that the VP is in the very center of the field of view, thereby giving maximum clearance with the corridor walls. It can also ensure that other objects do not come between it and the VP. To simplify the latter process, the floor region in front of the robot should be checked for continuity. Such actions can be taken by examining the image space itself. Indeed, both navigation and obstacle detection can be carried out by examining the image space, without appeal to special navigation models that map out the entire working area.

While this may be sufficient for guiding a vacuum cleaner, more advanced mobile robots will have to make mapping, path planning, and navigational modeling part of a detailed high-level analysis of the

(a)

FIGURE 7.15 Navigational problems posed by obstacles: (a) view of a corridor with four lines converging to a vanishing point and a narrowed entrance to be negotiated; (b) view of a path with various pillars and bollards to be negotiated; (c) plan view of (b) showing what can be deduced by examination of the ground plane in (b): only walls, W, that can be seen from the viewpoint marked Δ are drawn; however, the pillars, P, bollards, B, and litter-bin, L, which are easily recognizable, are shown in full.

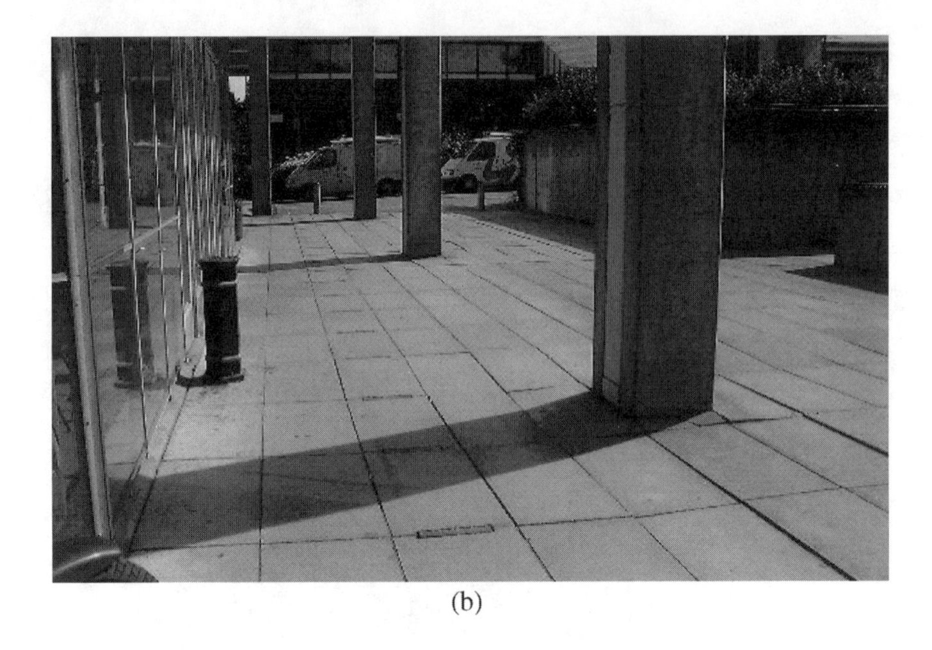

(b)

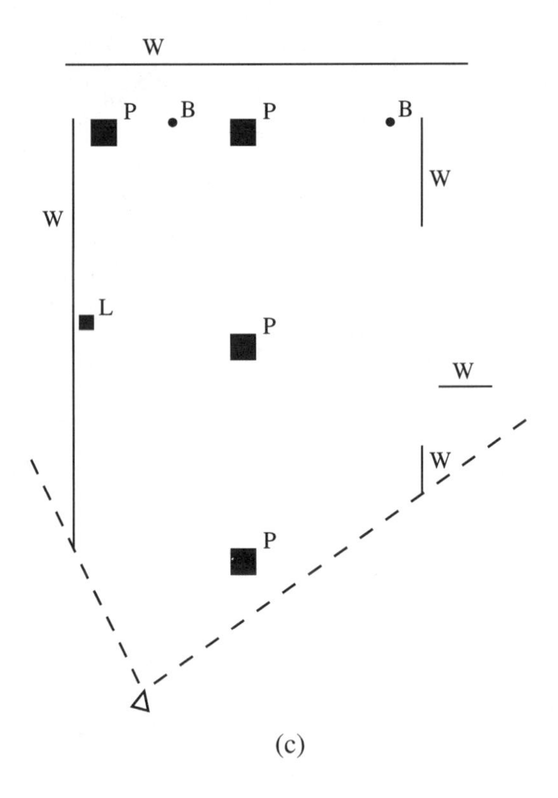

(c)

FIGURE 7.15 (*continued*)

situation [Kortenkamp et al., 1998]. This will apply if the corridor has several low obstacles such as chairs, waste paper bins, display tables, etc. or if a path has obstacles such as bollards or pillars (see Figure 7.15(b) and (c)), but will *a fortiori* be necessary if the robot is running a maze (as in the Robot Micromouse contests). In such cases, the robot will only have limited knowledge and visibility of the working area,

TABLE 7.1 Overall Algorithm for Constructing Plan View of Ground Plane

Perform edge detection; see Section 7.2.1.1.
Locate straight lines using the Hough transform; see Section 7.2.2.2.
Find VPs using a further Hough transform; see Section 7.3.1.4.
Eliminate all VPs except the primary VP, which lies in the general direction of motion.
Identify which lines passing through the primary VP lie within the ground plane.[a]
Mark other lines as not lying within the ground plane.
Find objects not related to lines passing through the primary VP.
Segment these objects and delimit their boundaries; mark them as not being on the ground plane.
Find shadows and delimit their boundaries; mark them as possibly being on the ground plane.[b]
Check object and shadow interpretations for consistency between successive images.[c]
Annotate all feature points in the ground plane with their deduced real-world X and Z coordinates.[d]
Check coordinates for consistency between successive images.
Mark significant inconsistencies as due to objects and artefacts such as shadows.
Mark insignificant inconsistencies as irrelevant noise to be ignored.[e]
Produce final map of X, Z coordinates.

[a]An initial line of attack is to take the ground plane as having a fair degree of homogeneity, in which case it will have the same general characteristics as the ground immediately in front of the robot. This should lead to identification of a number of ground plane lines. Note also that in a corridor any lines lying above the primary VP cannot be within the ground plane.

[b]Identifying shadows unambiguously is difficult without additional information about the scene illumination. However, dark areas within the ground plane can tentatively be placed in this category. In any case the purpose of this step is to move toward an understanding of the scene. This means that the later steps in which consistency between successive images is considered are of greater importance.

[c]One example is that of four points lying close together on a flat surface such as the ground plane; these will maintain their spatial relationship in the image during camera motion, the relative positions changing only slowly from frame to frame. However, if one of the points is actually on a separate surface, such as a pillar, it will move in a totally different way, exhibiting a marked stereo motion effect. This inconsistency will allow the disparate nature of the points to be discerned. In this context it is important to realize that scene interpretation is apparently easy for a human but has to be undertaken by a robot using "difficult" and quite extensive deductive processes.

[d]See Eqs. (7.18) and (7.19).

[e]It will often be sufficient to define inconsistent objects as noise if they occupy fewer than 3 to 5 pixels. They can often by identified and eliminated by standard median filtering operations; see Section 7.2.1.2.

Note: This table gives the basic steps for constructing a plan view of the ground plane. For simplicity each input image is treated separately until the last few stages, though more sophisticated implementations would adopt a more holistic strategy.

and the natural representation for thinking about the situation and path planning is a plan view model of the working area. This plan will gradually be augmented as more and more data come to light. The following section indicates how a (partial) plan view can be calculated for each individual frame obtained by the camera. The overall plan construction algorithm is given in Table 7.1.

7.3.6 Methodology for Constructing Plan View of Ground Plane

This section explains a simple strategy for constructing the plan view of the ground plane starting with a single view of a scene in which (1) the vanishing point V has been determined, and (2) significant feature points on the boundary of the ground plane have been identified. In fact, there is only space enough here to show how distance can be deduced along the ground plane in a line λ defined by the ground plane and a vertical plane containing the center of projection C of the camera and the vanishing point V. In addition, the simplifying assumption is made that the camera has been aligned with its optical axis parallel to the ground plane and calibrated so that its focal length f and the height H of its axis above the ground plane are both known. In that case, Figure 7.16 shows that the angle of declination α of a

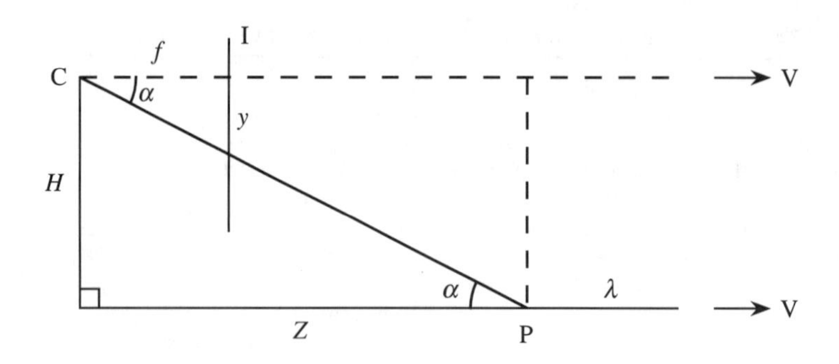

FIGURE 7.16 Geometry for constructing plan view of ground plane. This diagram is a vertical cross-section containing the center of projection C of the camera and the vanishing point V, though the latter can only be indicated by arrows. The cross-section also includes a line λ within the ground plane. The projection line from a general point P on λ to the image plane I has a declination angle α. f is the focal length of the camera lens, and H is the height of the optical axis of the camera above the ground plane. Note that the optical axis of the camera is assumed to be horizontal and parallel to λ.

general feature point P on λ is given by:

$$\tan \alpha = H/Z = y/f \tag{7.17}$$

where Z is the distance of P along λ (Figure 7.16), and y is the distance of the projection of P below the vanishing line in the image plane. This leads to the following value for Z:

$$Z = Hf/y \tag{7.18}$$

Similar considerations lead to the lateral distance from λ being given by:

$$X = Hx/y \tag{7.19}$$

where x is the lateral distance in the image plane.

Overall, we have now related image coordinates (x, y) and world (plan view) coordinates (X, Z). It should be noticed that both X and Z vary rapidly with y when y is small because of the distorting (foreshortening) effects of perspective geometry. Thus, as might be expected X and Z are more difficult to measure accurately in the distance. More general cases are best dealt with using homogeneous coordinates that lead to linear transformations using 4×4 matrices [Davies, 1997].

7.3.7 Other Factors Involved in Mobile Robot Navigation

Overall, we have moved from a simple image view that can successfully indicate a correct way forward to one in which experience of the situation is gradually built up. Notice that the first representation is restricted to providing *ad hoc* help, while the second representation is the natural one for storing knowledge and arriving at globally optimal solutions. It is actually difficult to know which representation humans rely on. While it is apparently the former, the latter is definitely invoked whenever detailed analysis of what is happening (e.g., in a maze) has to be undertaken. However, introspection is not a good means of finding out how humans actually carry out this task. In any case, in a machine vision context, we merely need to see how to set about programming a software system to emulate the human.

It is profitable to explore a little further how the navigation task can be undertaken in the abstract, typified by a maze-running robot. Suppose first that the robot moves over the maze in a search mode, exploring the environment and mapping the allowed and forbidden areas. When a complete map has been built up, some purposeful activity can be imposed. How should the robot go from its current position C to a goal position G in the minimum time? We can run the maze abstractly by applying an algorithm to form a distance function that starts at G and propagates around the maze, constrained only

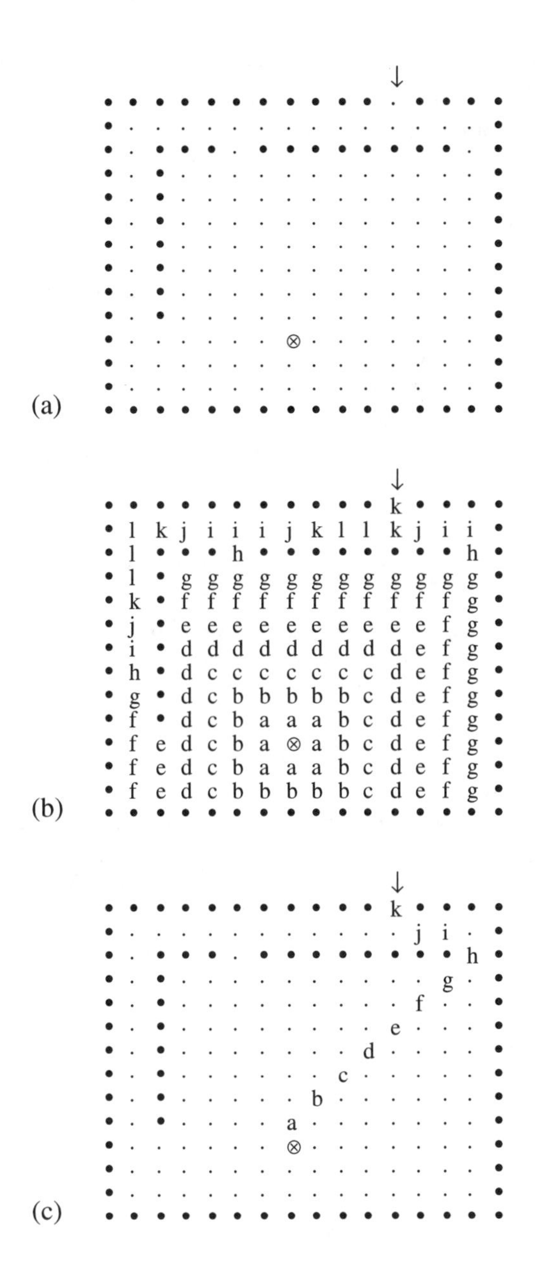

FIGURE 7.17 Use of distance function for path planning: (a) rudimentary maze; (b) distance function of the maze, with origin at the goal marked; (c) optimum path deduced by tracking from the entrance to the maze marked with an arrow along directions of maximum gradient. For convenience of display, the distance function is coded in letters, starting with a = 1. Improved forms of distance function would lead to more accurate path optimization.

by the positions of the walls (see Figure 7.17). When the distance function arrives at C, the distance value is noted and the propagation algorithm is terminated. To find the optimum path, it is now only necessary to proceed downhill from C in the direction of the locally greatest gradient, until arriving at G [Kanesalingam et al., 1998]. The reason for using a distance function is to ensure that the map contains information on which of several alternative routes is the shortest; if it were known that only one route would be possible, simple propagation operations of the type used in connected components analysis would be adequate for locating it.

So far we have only considered mobile robots capable of rolling or walking down corridors or paths. However, much of what has been talked about also applies to vehicles driving down city roads, because perspective lines are quite common in such situations. However, it will be difficult to apply these ideas to country roads that meander along largely unpredictable paths. In such cases, the concept of a VP has to be replaced by a sequence of VPs that move along a vanishing line, typically the horizon line [Billingsley and Schoenfisch, 1995]. While this complication does not invalidate the general technique, the straight-line-detection approach described above becomes inappropriate, and more suitable methods must be devised. The image-based model can probably still be used for instantaneous steering, especially if a Kalman filter is used to maintain continuity. However, practical accident-free country driving will require considerably more realistic scene interpretation than analysis based simply on where the current VP might be. After all, images are rich in cues that the human is able to discern and use by virtue of the huge database of relevant information stored in the human mind. Clearly, robot vision capabilities do not yet match up to this task with the required reliability and speed, though excellent efforts are being made, and the next decade will undoubtedly see a revolution in what can be achieved.

7.4 Parameters of Importance in Applied Vision

In various sections of this chapter certain parameters have been mentioned as important in the design of real-time vision algorithms. Robustness and reliability have repeatedly come up, as there is no sense in producing practical applications—especially those involving robots and all the safety issues that arise with them—that are not robust and reliable. Next, there are factors such as accuracy and sensitivity, which are central to the design of feature detectors and upon which later intermediate-level-vision algorithms strongly depend [Davies, 2000]. Then there are the speed factors involved when images typically containing 100,000 or more pixels have to be accessed and processed several times per second in order to provide real-time control of robots or other mechatronic devices. Such speed factors require fast processors, and this potentially raises system costs—sometimes so much that practical implementation is jeopardized. Optimizing these various important parameters is bound to be difficult, and indeed it is usually a question of trading one against another. A good deal of scientific skill and knowledge is needed to achieve this, and even now it is not always known how to perform all possible visually guided tasks effectively and at sufficiently low cost. Though substantial progress is being made on these problems, much of the progress has been recent, and there is much more work to be done in this area before robots achieve their early promise and, for example, achieve widespread domestic use.

7.5 Economic Factors

Any mechanical system can be expensive, but the costs are largely calculable, and the rate of reduction for quantity production is a known phenomenon. However, the costs of mechatronic systems also depend on the costs of computer control systems, which have been seen to plummet markedly in recent years. On the other hand, the cost of including significant intelligence in the computer systems corresponds to the sophistication of the software rather than the hardware, and this depends on the development costs, which can sometimes be considerable for ambitious systems. On the vision side, costs used to be dominated by the special dedicated hardware required for real-time implementation. However, broadly speaking, the days of discrete logic are over, and the 1990s saw the use of DSP chips bring the price of real-time hardware down to manageable levels (e.g., under $15,000 for typical visual inspection systems). By the year 2000, PC computer chips, such as the Pentium III, that utilize embedded operating systems have made vision software more transportable, and typical inspection systems are dropping further in price.

However, it has to be borne in mind that inspection systems can be highly constrained, and they are in general not ambitious in their visual capabilities. On the other hand, once robots get out into the real world, the vision systems they use have to be able to cope with drastically more variegated input data. In general they cannot yet do this with anything like the sophistication of the human—regardless of price. Nevertheless, robots are starting to make an impact in areas where human operators were formerly required; thus, we are starting to see robots used for collecting trash, mowing lawns, driving vehicles, etc. Such systems are not quite out of the laboratory, though modest versions are in some cases available on the market. In the future, costs will come down, but in a manner depending strongly on tradeoffs between sophistication and mass production. Prices are not currently predictable, but it is clear that the coming decades will see vast growth in the market for vision-based robots. This will be further accelerated by the growing capabilities of cheap computers.

7.6 Summary

This chapter has studied a number of areas of machine vision and mechatronics. First, some basic but nevertheless crucial low-level-vision concepts were reviewed. These included noise suppression, thresholding, detection of features such as edges and corners, and the highly topical methods of morphology. The study of feature detection led naturally to boundary pattern analysis, and the twin approaches of intermediate level vision—Hough transforms and graph matching—which are traditionally associated with, respectively, edge detection and point features such as corners and holes. By the end of the analysis it was evident that the Hough transform approach is also applicable to point pattern matching, and that this is advantageous because it can save computation when the graph-matching approach would grow at faster than polynomial rates with an increase in the number of image features.

The fundamental section of the chapter containing the topics described above was necessarily restricted to analysis of two-dimensional images and was unable to deal with the rich information content of three-dimensional scenes. However, the section of the chapter dealing with three-dimensional vision showed how this restriction could be overcome by the use of two cameras to obtain two images upon which stereo matching could be carried out. Such approaches ultimately yield depth maps of the scene. These do not automatically provide scene interpretation, but they do present the right sort of information that would allow this to be achieved. Because of lack of space, complete analysis of the situation could not be undertaken in this chapter, but it was found that cues of various sorts could facilitate scene interpretation and analysis. Important among these cues is recognition of perspective by identification of vanishing points, while invariants also play a key role. Furthermore, when the camera or objects in the scene are moving, additional cues can be obtained in the form of foci of expansion and contraction, though again a full analysis was beyond the scope of the chapter.

A further important aspect of vision was that of *active vision,* wherein the vision system actively focuses on certain aspects of the input images and attempts to find out more about them. Active vision is especially relevant for autonomous mobile robots, which may have to guide themselves around buildings and avoid obstacles, the focus of attention necessarily changing as time goes on.

The remainder of the chapter concentrated on the use of vision for mechatronic applications. It classified the different types of machines into mobile robots, stationary robots with arms that could be used for tasks such as assembling and spraying cars, and stationary machines that would not normally be called robots. The last named would typically perform functions such as inserting components into printed circuit boards or inspecting and rejecting faulty components at appropriate points on production lines. In all such mechatronic applications, vision is a powerful tool, and this chapter aimed at explaining the basic vision design methodology. It also broached the question of navigation and path planning for mobile robots and other vehicles, using techniques such as distance functions that were originally developed for low-level vision.

While space prevented a more extensive discussion of modalities of vision other than those corresponding to visible light, vision is increasingly being applied using input images obtained with ultrasonic, microwave, infrared, ultraviolet, or x-ray radiation, and techniques such as magnetic resonance. This makes vision extremely powerful and considerably widens the scope of the techniques described here.

At the present time vision is developing rapidly as more sophisticated algorithms are developed and as faster computer hardware that will help these algorithms to be implemented in real time becomes available. This progress serves to remove the speed limitations that would otherwise prevent complete control of guided vehicles and other mechatronic processes from being used in anger. Hence, there is great scope for robots to be brought relatively quickly into the everyday world in the form of robot cleaners, trash collectors, paint sprayers, window cleaners, lawn mowers, car drivers, and a host of other useful devices. Suffice it to qualify the rate of progress by saying that over the coming decade vision-guided robots may well be limited more by what is permissible legally and with regard to insurance premiums than from a purely technical point of view.

Defining Terms

active vision: The use of cameras that can be moved to focus on particular objects in a scene, as dictated by the vision software; active vision may also refer to vision software that focuses on particular parts of a scene stored in the computer memory and that uses agents such as snakes to iteratively analyze them until a satisfactory segmentation or identification is attained.

distance function: Map of an object in which each pixel has a numerical value corresponding to its distance from the nearest background pixel.

focus of expansion: The point in an image from which a number of points in the image appear to diverge when the camera is moving though a scene in the general direction of the corresponding objects; a corresponding *focus of contraction* arises when the camera is moving away from the objects in the scene.

Hough transform: Type of global image transform, used for intermediate-level vision, that operates by accumulating votes in a parameter space and then searching for peaks in this space; each peak results in the identification of a relevant (e.g., geometric) pattern in the image.

image processing: Processing of an image so as to construct another image that typically corresponds locally to the original image; the new image is often a cleaned-up or enhanced version of the original image, or one in which various parts are highlighted by feature detection.

intermediate-level vision: That part of vision whose function is to group low-level features into known types of global geometric shape such as lines, circles, and ellipses.

invariance: Mathematical property of an object or feature that remains constant over a sequence of images in spite of changes of view or perspective and that thus aids identification and tracking.

low-level vision: That part of vision concerned with identifying local image features and in some cases modifying them; this topic includes noise suppression, edge detection, and other feature detection tasks.

morphology: The processing of images by set operations that make use of local structuring elements; typically, structuring elements are used to increase or decrease the sizes of objects in one or more directions, though combinations of such operations are also used. The effect of morphology is to analyze shapes and to select or eliminate features of various sizes and shapes. Also called *mathematical morphology*, the approach has been extended to gray-scale processing of images.

optical flow: The apparent flow of points in an image when the corresponding objects or the camera is in relative motion; the optical flow field is the vector field image in which each pixel carries a vector representing the apparent motion in the image space.

perspective: The changes in view of a scene that result when it is projected by a camera onto an image plane; typically, parallel lines become nonparallel in the image, while circles appear as ellipses.

real-time processing: Processing operations that must proceed at the (real) rate at which objects arrive on the scene, as when objects on a conveyor must be inspected, or when a car is being driven along a road on which obstacles must be negotiated.

stereo vision: The use of two cameras to obtain pairs of images that differ slightly from each other, the differences being used to estimate depth at each point in the scene.

template matching: Comparison of all parts of an image with a template, using processes such as convolution, differencing, or correlation.

vanishing point: The point in the image plane where a number of perspective lines meet, when these lines correspond to parallel straight lines in the original scene.

References

Billingsley, J. and Schoenfisch, M., Vision-guidance of agricultural vehicles, *Autonomous Robots* 2(1), 65–76, 1995.

Bolles, R. C. and Cain, R. A., Recognizing and locating partially visible objects: the local-feature-focus method, *Int. J. Robot. Res.* 1(3), 57–82, 1982.

Davies, E. R., Locating objects from their point features using an optimized Hough-like accumulation technique, *Pattern Recogn. Lett.* 13(2), 113–121, 1992.

Davies, E. R., Low-level vision requirements, *Electron. Commun. Eng. J.* 12(5), 197–210, 2000.

Davies, E. R., Some problems in food and cereals inspection and methods for their solution, *Proc. Int. Conf. Quality Control by Artificial Vision – 2001*, Le Creusot, France 21–23 May, 2001, pp. 35–46.

Haralick, R. M., Sternberg, S. R., and Zhuang, X., Image analysis using mathematical morphology, *IEEE Trans. Pattern Anal. Mach. Intell.*, 9(4), 532–550, 1987.

Hough, P. V. C., Method and means for recognising complex patterns, U.S. Patent 3069654, 1962.

Kanesalingam, C., Smith, M. C. B., and Dodds, S. A., An efficient algorithm for environmental mapping and path planning for an autonomous mobile robot, *Proc. 29th Int. Symp. Robotics*, Birmingham, 1998, pp. 133–136.

Kortenkamp, D., Bonasso, R. P., and Murphy, R., Eds. *Artificial Intelligence and Mobile Robots*, AAAI Press/MIT Press, Menlo Park, CA, 1998.

Magee, M. J. and Aggarwal, J. K., Determining vanishing points from perspective images, *Computer Vision Graph. Image Process.*, 26(2), 256–267, 1984.

Mundy, J. L. and Zisserman, A., Eds., *Geometric Invariance in Computer Vision*, MIT Press, Cambridge, MA, 1992.

Rosenfeld, A. and Pfaltz, J. L., Distance functions on digital pictures, *Pattern Recogn.*, 1, 33–61, 1968.

Yazdi, H. R. and King, T. G., Application of "Vision in the Loop" for inspection of lace fabric, in Davies, E. R. and Ip, H. H. S., Eds., *Real Time Imaging*, 4(5), 317–332, 1998.

Further Information

Ballard, D. H. and Brown, C. M., *Computer Vision*, Prentice-Hall, Englewood Cliffs, NJ, 1982. This early book on computer vision is a well-known classic and is still worth referring to.

Batchelor, B. G., Hill, D. A., and Hodgson, D. C., Eds., *Automated Visual Inspection*, North-Holland, Amsterdam, 1985. This early book on inspection contains a wealth of information on illumination systems and other practical details.

Billingsley, J., Ed., *Mechatronics and Machine Vision*, Research Studies Press, Baldock, England, 2000. This volume is aimed primarily at mechatronics; about half the chapters have useful references to robot and computer vision.

Blake, A. and Yuille, A., Eds., *Active Vision*, MIT Press, Cambridge, MA, 1992. This volume studies many aspects of active vision which are especially relevant for mobile robots.

Davies, E. R., *Machine Vision: Theory, Algorithms, Practicalities*, 2nd ed., Academic Press, London, 1997. This book starts at an elementary level and covers in considerable depth most of the machine vision topics mentioned in this chapter; it emphasizes the practicalities of vision algorithms.

Davies, E. R., *Image Processing for the Food Industry*, World Scientific, Singapore, 2000. This recent research monograph shows how image processing has been applied in the food industry for food inspection, food processing, and agriculture.

Davies, E. R. and Atiquzzaman, M., Eds., Special Issue on Projection-Based Transforms, *Image Vision Comput.*, 16(9–10), 1998. This recent special issue is valuable in showing how the Hough transform approach has developed over the past decade or so.

Davies, E. R. and Ip, H. H. S., Eds., Special Issue on Real-Time Visual Monitoring and Inspection, *Real-Time Imaging*, 4(5), 1998. This recent special issue contains applications of vision that are highly relevant to mechatronics.

Haralick, R. M. and Shapiro, L. G., *Computer and Robot Vision*, Vol. 1, Addison-Wesley, Reading, MA, 1992. This volume is the first of a weighty pair covering the subject area is widely useful for its in-depth mathematical coverage of computer and robot vision; it is especially valuable for its detailed study of mathematical morphology.

Horn, B. K. P., *Robot Vision*, MIT Press, Cambridge, MA, 1986. This is a good early reference on robot vision that is still useful for its rigor and set problems.

Sangwine, S. J. and Horne, R. E. N., *The Colour Image Processing Handbook*, Chapman & Hall, London, 1998. This is a useful source of information on color processing.

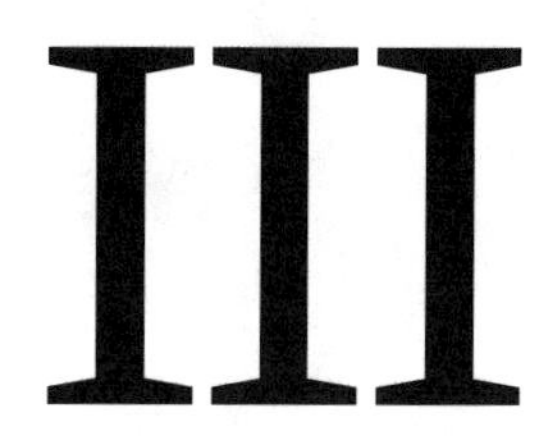

Optical Information Processing and Recognition

8

Volume Holographic Imaging

George Barbastathis
Massachusetts Institute of Technology
Cambridge, Massachusetts

Overview

Volume holographic imaging is emerging as one of the major applications of holography for a wide range of applications including microscopy and spectroscopy for bioimaging, remote sensing, and metrology. The purpose of this chapter is to provide a basic qualitative understanding of volume holographic imaging as well as some quantitative tools for imaging system design. For this purpose we first describe existing imaging systems and classify them according to the share of labor between optical elements (lenses, interferometers, etc.) and digital electronic processing toward the process of image formation. Our main argument is that this hybrid mode of imaging provides the richest information to the user about the object that is being imaged. Volume holographic imaging is introduced as a "superset" of imaging systems, as volume holographic elements provide the richest possible set of optical transformations for processing the optical signals received from the object. The function of volume holographic imagers is described in terms of their Bragg selectivity properties and is contrasted with thin optical elements such as lenses and thin holograms. Some examples of how volume holograms can be used for depth-selective imaging and spectral imaging are also described. The chapter concludes with a discussion of specific applications in distortion compensation and surface metrology, an overview of qualitative resources for volume holographic imaging from the literature, and a discussion of future directions for volume holographic imaging.

8.1 Introduction

Imaging remains one of the primary applications of optical science. An imaging system receives emitted or scattered radiation from the environment and transforms it in order to recover the structure of an object of interest within the scene. Therefore, imaging may be defined as the solution to an inverse problem of light propagation. In turn, image information can be used for tracking, metrology, etc. on opto-mechanical systems that may be in the context of industrial, biological, etc. applications. The purpose of this chapter is to describe a new imaging technique, volume holographic imaging (VHI), which was invented in 1999 and is currently experiencing rapid growth in popularity because it returns

0-8493-1162-4/03/$0.00+$1.50

more information in real time than any other imaging method. VHI refers to a class of imaging techniques that use a volume hologram in at least one location within the optical path. The volume hologram acts as a "smart lens," which processes the optical field to extract spatial information in three dimensions (lateral as well as longitudinal) and spectral information.

Until the development of CCD cameras in the early to mid-1970s the almost universal purpose of imaging systems had been the production of geometrically exact projections of the three-dimensional world onto two-dimensional sensor arrays, permanent (such as photographic film) or instantaneous (such as an observer's retina). Remarkable exceptions such as computed tomography and radio astronomy, known earlier, had limited or no use in the domain of optical frequencies. However, the gradual increase in availability of digital cameras, ample digital computing power to process the images, and digital networks to distribute them produced a revolutionary shift in optical imaging design. The necessity to produce "images" on the detector plane that are physically analogous to the imaged objects became secondary, as long as the detector captured sufficient information to allow the recovery of the objects after appropriate processing operations. This paradigm shift had two apparent additional benefits. First, it simplified the design of optical elements, as several geometric imaging imperfections (e.g., defocus) could, to some extent, be corrected digitally. Second, and most important, digital processing allows the user to recover more object information from the sensor data. For example, by processing several two-dimensional intensity patterns one may recover surface topography data about opaque objects (2 + 1/2D images) or volumetric structure data (three-dimensional images), and even spectral data in addition (four-dimensional images), and semitransparent objects. Examples of this principle in action are triangulation-based imaging systems, the confocal microscope, and coherence imaging systems.

The generality of digital processing is undoubtedly the principal reason for the increased power of imaging systems. It would be desirable to have optical elements available with equally general behavior to maximize design flexibility. Lenses, for example, are rather limited to low-power polynomial surface shapes due to manufacturing constraints present even in modern techniques such as injection molding and servo-controlled grinders. Diffractive optimal elements achieve a much broader range of responses, but the transformations they can effect are still limited to two dimensions. The most general optical elements available are volume holograms, which can be thought of as self-aligned three-dimensional stacks of diffractive elements operating on the incident field coherently as it propagates through. In the remainder of this chapter, we first discuss the background of digital and hybrid imaging in Section 8.2, then we introduce volume holographic imaging qualitatively (Section 8.3) and quantitatively (Section 8.4), and conclude with some example applications of volume holographic imaging in Section 8.5 and assistance to the reader in finding more resources about the topic (Section 8.6).

8.2 Digital and Hybrid Imaging

A generic imaging system is shown in Figure 8.1. *Field-transforming elements* are the first to receive radiation from the object; they can be, for instance, lenses, prisms, holograms, etc., and they operate directly on the optical field. The field produced by this stage is converted to electrical signal by the intensity detector. The electrical signal is then fed to the *intensity-transforming elements*, which may apply very general algorithms, e.g., iterative and statistical processing, to produce the final image.

As the respective names suggest, there is a fundamental difference in the nature of transformations that can be affected by field- and intensity-transforming imaging elements. Field-transforming elements operate directly on the electromagnetic field that composes the optical wave. For coherent fields, this property allows field-transforming elements to modify the amplitude as well as the phase of the (deterministic) input wave. For example, a spherical lens imposes a quadratic phase modulation in the paraxial approximation. In the case of partially coherent fields, field-transforming elements operate rather on the random process that represents one realization of the optical field (different in every experiment); as a result, these elements are capable of modifying all the moments of the random field process.

At optical frequencies, the detection step (i.e., the conversion of the electromagnetic field to an electrical signal, such as a voltage) is fundamentally limited to return the intensity of the field. For coherent fields,

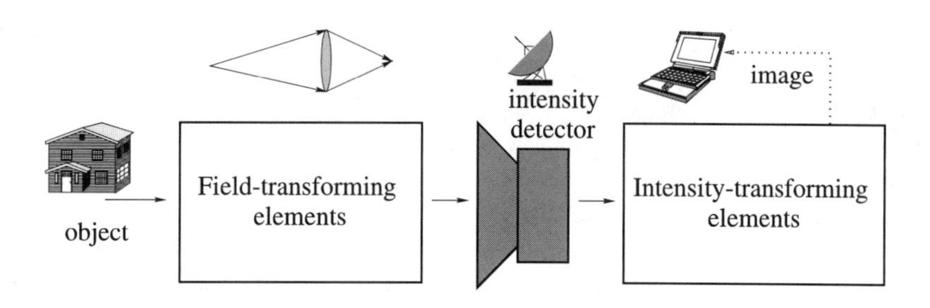

FIGURE 8.1 General classification of elements composing an imaging system.

the intensity is proportional to the modulus of the complex amplitude; that is, phase information has been lost. For partially coherent fields, the only information available past the detector is the first moment (intensity) of the coherence function. In both cases, the intensity is a measure of the average energy (normalized over the unit area) carried by the optical field, after canceling out field fluctuations. It is well known that the intensity-detection step is not by any means destructive for other types of information carried by optical fields. For example, interferometry exploits the detectors' quadratic nonlinearity law to map the phase of the field onto the intensity for coherent fields and the phase of the second-order correlation function onto the intensity for partially coherent fields.

Therefore, the designer has considerable freedom in selecting the location of the detection step in an optical imaging system and, subsequently, the division of labor between the field- and intensity-transforming parts of the system. In practice, some configurations work better than others, depending on the application. The three examples below illustrate this point.

The confocal microscope, invented by Minsky[1] and popularized by Wilson and Sheppard,[2,3] is an example of an imaging system that is primarily based on field transformations. It operates by the lowest-dimensional measurements possible, i.e., point measurements, and constructs a three-dimensional image by scanning the volume of the specimen and obtaining the emitted intensity values one point at a time. The geometry of the optical system is such that light emitted locally from a very small portion of the object is only allowed to reach the detector. The rest of the light is rejected by the detector pinhole. The proportional light contribution to a single measurement as a function of object coordinates is equivalent to the three-dimensional point-spread function (PSF) of the system; it can be calculated accurately under various aberration conditions using Fourier optics.[4,5] Clearly, the function of partitioning the source is performed in the field domain by elements such as the objective lenses and the detector pinhole. The role of intensity transformations is confined to storing the scanned data in a three-dimensional matrix for subsequent display, a role that is relatively minor as far as the imaging operation is concerned. Confocal microscopy has been implemented in many different variants for improved light efficiency or resolution, e.g., differential interference,[6] fluorescence,[7] two-photon,[8] etc.; it has been spectacularly successful, primarily in various applications of biological and biomedical imaging.

Coherence imaging is an example of computational imaging that relies on global, rather than local, measurements. It is based on a fundamental result, derived independently by van Cittert[9] and Zernicke,[10] that states that the degree of statistical correlation of the optical field in the far zone, expressed as a complex function over the exit pupil of the imaging system, is the Fourier transform of the object-intensity distribution. Therefore, the object can be recovered by measuring the coherence function through interferometry and then inverse-Fourier-transforming the result. The application of the van Cittert–Zernicke theorem in the radiofrequency spectral region is the basis of radio astronomy,[11] which yields by far the most accurate images of the most remote cosmic objects. The most common formulation of the theorem relates the mutual coherence in a plane at infinity to a two-dimensional source intensity distribution, but extensions to three-dimensional sources have been derived by various authors.[12–15] The far-field version of the extended van Cittert–Zernicke theorem was recently implemented experimentally.[16–18] A full generalization of the theorem has also been developed and experimentally implemented to allow Fresnel zone reconstruction in projective coordinates.[19] In most coherence imaging systems, the "work balance" of the imaging operation is shifted almost exclusively to intensity transforms (from the intensity one derives the correlation function and then Fourier-transforms digitally to retrieve the source itself). The role of field transforms is limited to "folding" the field in order to generate a self-interference pattern that maps the field auto-correlation onto the intensity detector.[19–21]

Extended depth-of-field imaging with the use of cubic phase masks[22,23] is an example of a system where the roles of field and intensity transforms are more balanced than in the two systems described previously. The cubic phase mask "scrambles" the optical field, thereby forming a strongly blurred intensity image in the focal plane. However, application of the inverse deblurring digital transform leads to recovery of the object(s) in the input field with relative insensitivity to their depth. Thus, a clever combination of field and intensity transforms leads to a significant advantage in improving the perceived quality of the recovered digital images.

Optical coherence tomography[24] is an interesting hybrid between a confocal microscopy and coherence imaging. It should be classified as a field-transforming technique, because it uses field decorrelation due to decoherence to achieve depth slicing (in the confocal microscope, the slicing function is performed by the pinhole). An extension, spectroscopic optical coherence tomography,[25] exploits the phase information in the recorded fringes to additionally extract the spectral composition of the object through a Fourier transform. This version assigns increased importance to the intensity-transforming operation, thereby increasing the amount of information that it returns about the object.

Summarizing the above descriptions, their marked design differences also indicate the diversity in their domain of application. Confocal microscopy and optical coherence tomography are excellent for imaging thick specimens such as tissue and other semitransparent materials. In these cases the pure version of nonconfocal (global) coherence imaging would have limited fringe visibility due to increased quasi-constant background radiation. On the other hand, for objects that can be modeled as superposition of discrete, mutually incoherent radiators distributed in three-dimensional space, or for opaque objects, coherence imaging overperforms other techniques due to the reduced cost of the required optical elements and the reduced scanning time. Cubic-phase-mask-based imaging is appropriate for digital-photography kinds of applications, where defocus-free two-dimensional projections of three-dimensional objects are sufficient, but it can be combined with scanning to return 2 + 1/2D data as well.[23]

8.3 Description of Volume Holographic Imaging

Figure 8.2 is a generic VHI system. The object is either illuminated by a light source (e.g., sunlight or a pump laser) as shown in the figure, or it may be self-luminous. Light scattered or emitted by the object is first transformed by an objective lens and then illuminates the volume hologram. The role of the objective is to form an intermediate image that serves as input to the volume hologram. The volume

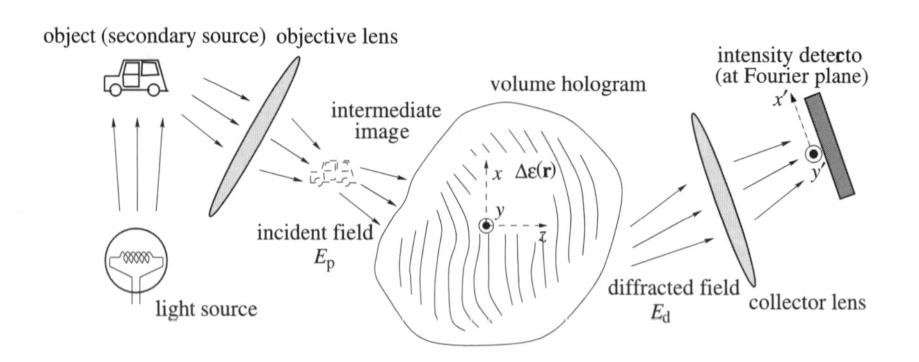

FIGURE 8.2 Volume holographic imaging (VHI) system.

hologram itself is modeled as a three-dimensional modulation $\Delta\varepsilon(\boldsymbol{r})$ of the dielectric index within a finite region of space. The light entering the hologram is diffracted by $\Delta\varepsilon(\boldsymbol{r})$ with efficiency η, defined as

$$\eta = \frac{\text{Power diffracted by the volume hologram}}{\text{Power incident to the volume hologram}}. \tag{8.1}$$

We assume that diffraction occurs in the Bragg regime. The diffracted field is Fourier-transformed by the collector lens, and the result is sampled and measured by an intensity-detector array (such as a CCD or CMOS camera).

Intuitively, we expect that a fraction of the illumination incident upon the volume hologram is Bragg-matched and diffracted toward the Fourier-transforming lens. The remainder of the incident illumination is Bragg-mismatched and as result is transmitted through the volume hologram undiffracted. Therefore, the volume hologram acts as a filter that admits the Bragg-matched portion of the object and rejects the rest. When appropriately designed, this "Bragg imaging filter" can exhibit very rich behavior, spanning the three spatial dimensions and the spectral dimension of the object.

To keep the discussion simple, we consider the specific case of a transmission geometry volume hologram, described in Figure 8.3. The volume hologram is created by interfering a spherical wave and a plane wave, as shown in Figure 8.3(a). The spherical wave originates at the coordinate origin. The plane wave is off-axis and its wave vector lies on the xz plane. As in most common holographic systems, the two beams are assumed to be at the same wavelength, λ, and mutually coherent. The volume hologram results from exposure of a photosensitive material to the interference of these two beams.

First, assume that the object is a simple point source. The intermediate image is also approximately a point source, which we refer to as a "probe," located somewhere in the vicinity of the reference point source. Assuming the wavelength of the probe source is the same as that of the reference and signal beams, volume diffraction theory shows that:

1. If the probe point source is displaced in the y direction relative to the reference point source, the image formed by the volume hologram is also displaced by a proportional amount. Most common imaging systems would be expected to operate this way.
2. If the probe point source is displaced in the x direction relative to the reference point source, the image disappears (i.e., the detector plane remains dark).
3. If the probe point source is displaced in the z direction relative to the reference point source, a defocused and faint image is formed on the detector plane. "Faint" here means that the fraction of energy of the defocused probe transmitted to the detector plane is much smaller than the fraction that would have been transmitted if the probe had been at the origin.

Now consider an extended, monochromatic, spatially incoherent object and intermediate image, as in Figure 8.3(f). According to the above description, the volume hologram acts as a "Bragg slit" in this case. Because of Bragg selectivity, the volume hologram transmits light originating from the vicinity of the y axis and rejects light originating anywhere else. For the same reason the volume hologram affords depth selectivity (like the pinhole of a confocal microscope). The width of the slit is determined by the recording geometry and the thickness of the volume hologram. For example, in the transmission recording geometry of Figure 8.3(a), where the reference beam originates at distance z_0 away from the hologram, the plane wave propagates at angle θ with respect to the optical axis z (assuming $\theta \ll 1$ radian), and the hologram thickness is L (assuming $L \ll z_0$), the width of the slit is found to be

$$\Delta x \approx \frac{\lambda z_0}{L\theta}. \tag{8.2}$$

The imaging function becomes richer if the object is polychromatic. In addition to its Bragg slit function, the volume hologram, like all diffractive elements, then exhibits dispersive behavior. In this particular

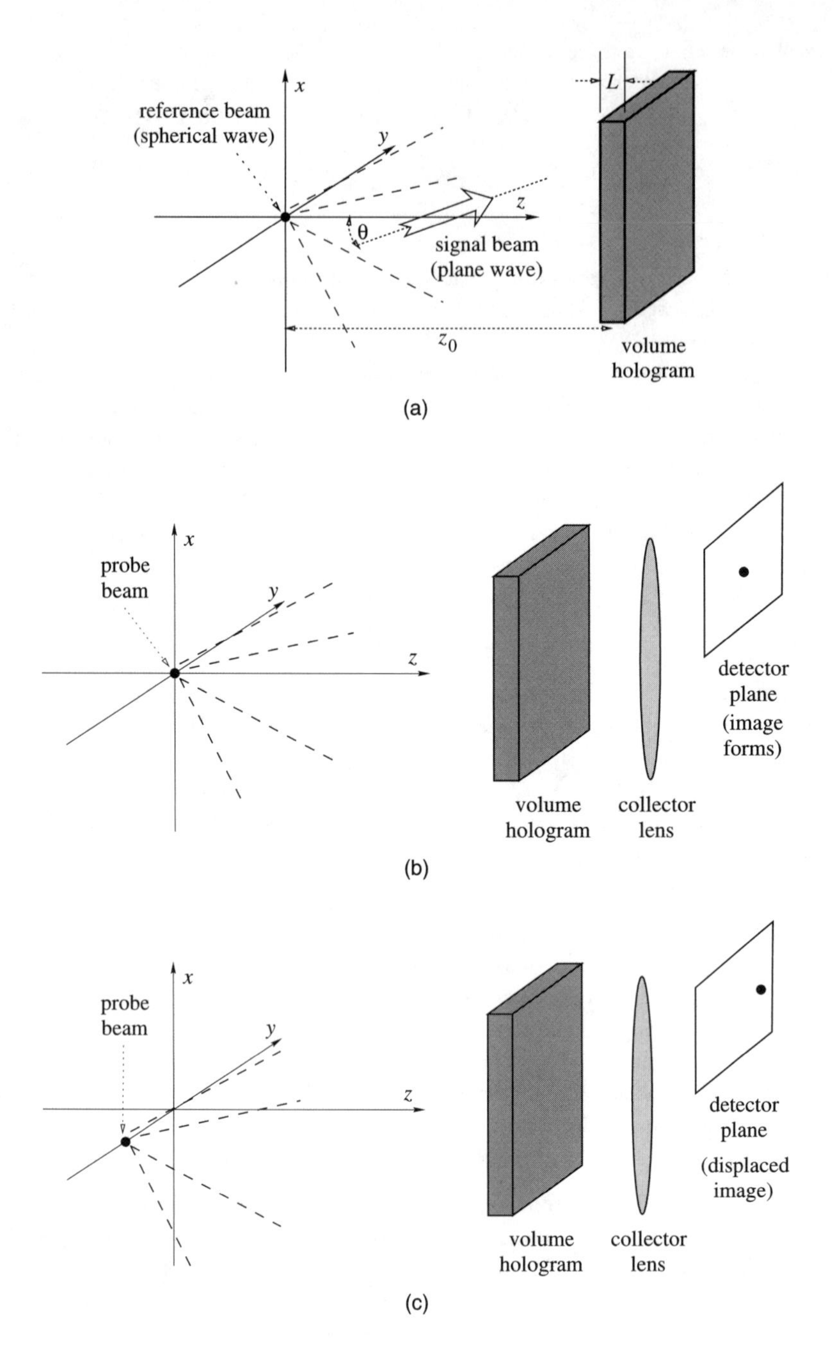

FIGURE 8.3 (a) Recoding of a transmission-geometry volume hologram with a spherical wave and an off-axis plane wave with its wave vector on the *xz* plane; (b) imaging of a probe point source that replicates the location and wavelength of the reference point source using the volume hologram recorded in part (a); (c) imaging of a probe point source at the same wavelength but displaced in the *y* direction relative to the reference point source; (d) imaging of a probe point source at the same wavelength but displaced in the *x* direction relative to the reference point source; (e) imaging of a probe point source at the same wavelength but displaced in the *z* direction relative to the reference point source; (f) bragg slitting: imaging of an extended monochromatic, spatially incoherent object using a volume hologram recoded as in (a); (g) joining Bragg slits from different colors to form rainbow slices: imaging of an extended polychromatic object using a volume hologram recorded as in (a); (h) multiplex imaging of several slices using a volume hologram formed by multiple exposures.

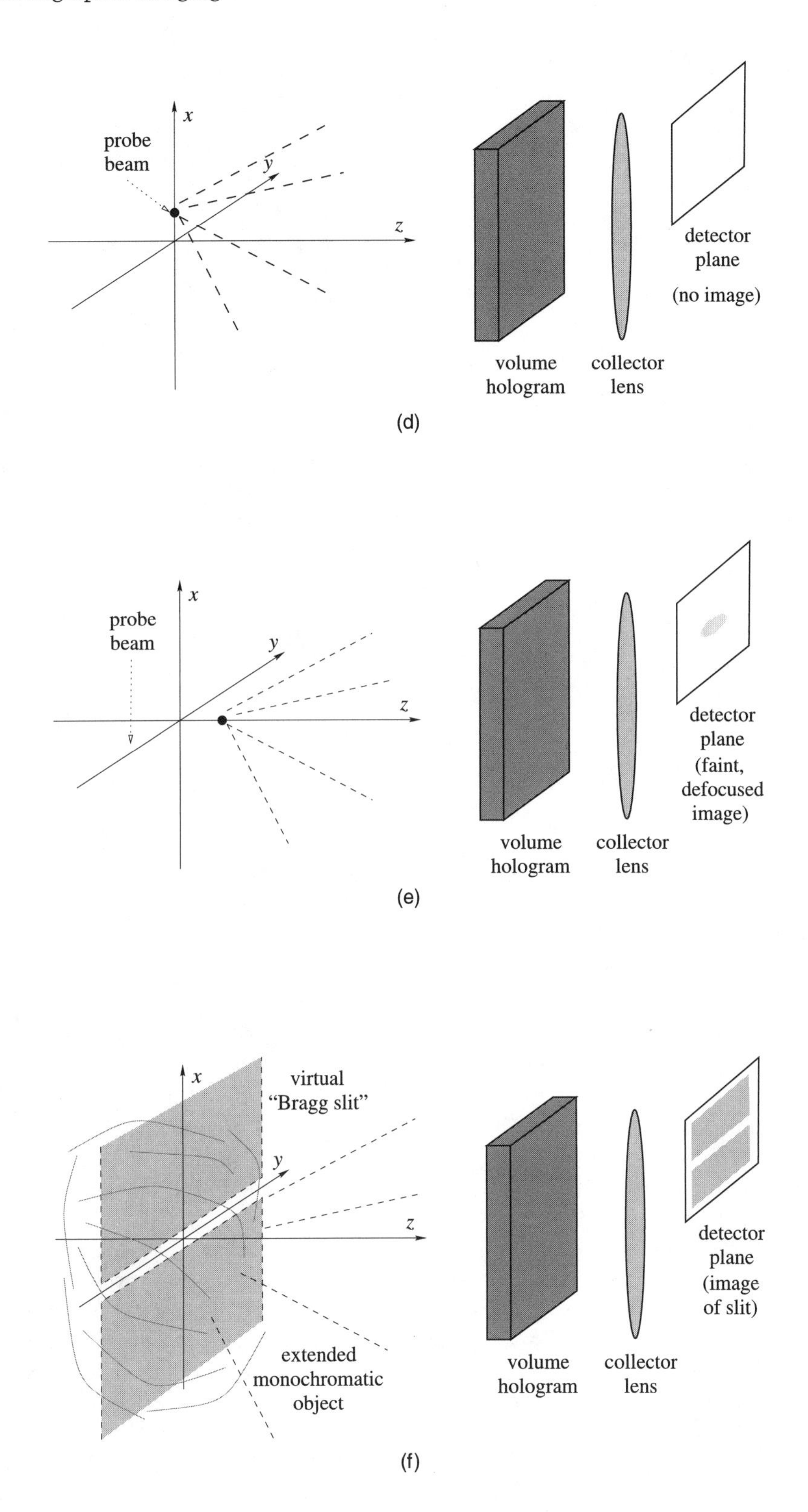

FIGURE 8.3 *(continued)*

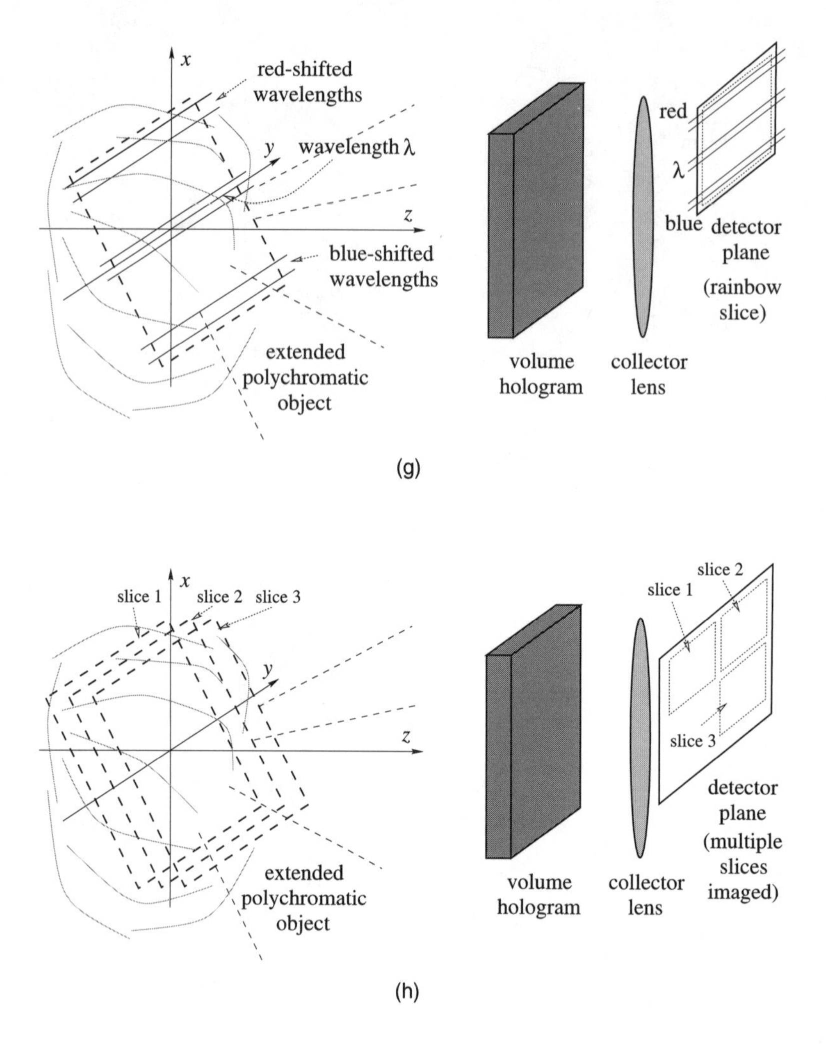

FIGURE 8.3 (*continued*)

case, dispersion causes the hologram to image simultaneously multiple Bragg slits, each at a different color and parallel to the original slit at wavelength λ, but displaced along the z axis. Light from all these slits finds itself in focus at the detector plane, thus forming a "rainbow image" of an entire slice through the object, as shown in Figure 8.3(g).

To further exploit the capabilities of volume holograms, we recall that in general it is possible to "multiplex" (superimpose) several volume gratings within the same volume by successive exposures. In the imaging context, suppose that we multiplex several gratings similar to the grating described in Figure 8.3(a) but with spherical reference waves originating at different locations and plane signal waves at different orientations. When the multiplexed volume hologram is illuminated by an extended polychromatic source, each grating forms a separate image of a rainbow slice, as described earlier. By spacing appropriately the angles of propagation of the plane signal waves, we can ensure that the rainbow images are formed on nonoverlapping areas on the detector plane, as shown in Figure 8.3(h). This device is now performing true four-dimensional imaging; it is separating the spatial and spectral components of the object illumination so that they can be measured independently by the detector array. Assuming the photon count is sufficiently high and the number of detector pixels is sufficient, this "spatio-spectral slicing" operation can be performed in real time, without need for mechanical scanning.

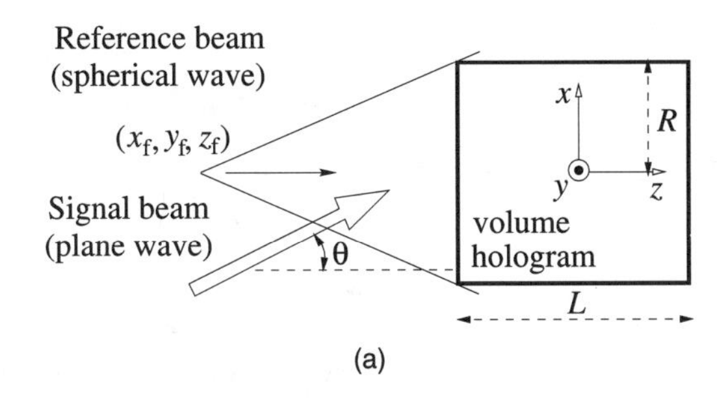

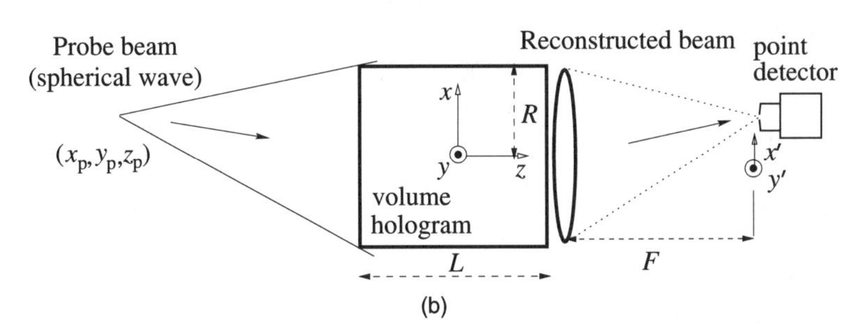

FIGURE 8.4 Notation for transmission geometry: (a) recording; (b) imaging.

8.4 Quantifying Volume Holographic Imaging

Consider the transmission geometry of recording and imaging, shown again in Figure 8.4 with some additional notation. The volume hologram is created by interfering a spherical wave and a plane wave, as shown in Figure 8.4(a). The spherical wave originates at $\mathbf{r}_f = (x_f, y_f, z_f)$. The plane wave is propagating at angle θ with respect to the optical axis. The photosensitive region where the hologram is recorded is a cylinder of height L (in the $\hat{\mathbf{z}}$ direction) and radius R.

The probe point source is located at $\mathbf{r}_p = (x_p, y_p, z_p)$ in front of the volume hologram, as in Figure 8.4(b). Using the first Born approximation to volume diffraction and the paraxial approximation, it can be shown (see Reference 26, pp. 38–42) that the intensity-normalized, Fourier-transformed intensity as a function of detector coordinates (x', y') is

$$E_d(x', y') = \frac{\sqrt{\eta}}{V}\left|\int_{-L/2}^{L/2} F(2\pi AR^2,\ 2\pi BR)e^{iC}dz\right|^2, \tag{8.3}$$

where

- V is the volume of the hologram.
- $F(u, v)$ is the function that describes the field distribution near the focal region of a quadratic lens (see Reference 27, Section 8.8.1).
- For the monochromatic case, the coefficients, $A, B = \sqrt{B_x^2 + B_x^2}, C$ are given, in terms of the recording, imaging, and detector geometries, as

$$A(z) = \frac{1}{\lambda(z - z_f)} - \frac{1}{\lambda(z - z_p)} \tag{8.4}$$

$$B_x(z) = \frac{x_p}{\lambda(z - z_p)} + \frac{x_f}{\lambda(z - z_f)} - \frac{x'}{\lambda F} + \frac{u}{\lambda}, \tag{8.5}$$

$$B_y(z) = -\frac{y_p}{\lambda(z-z_p)} + \frac{y_f}{\lambda(z-z_f)} - \frac{y'}{\lambda F}, \tag{8.6}$$

$$C(z) = \frac{x_p^2 + y_p^2}{\lambda(z-z_p)} - \frac{x_f^2 + y_f^2}{\lambda(z-z_f)} + \left(\frac{x'^2 + y'^2}{\lambda F^2} - \frac{u^2}{\lambda}\right) z. \tag{8.7}$$

The simplest way to think about the integral in Equation 8.3 is as a superposition of lenses, coming to focus approximately at the detector plane. At Bragg match, i.e., when $A = 0$, $B = 0$, and $C = 0$, the lenses all come to focus precisely at the same point, and they are all in phase. Therefore, strong diffraction is obtained. If any of these conditions are violated, however, the contributions are mismatched and the amount of light received at the detector plane drops. This happens when $x_p \neq x_f$ or $z_p \neq z_f$, for example. On the other hand, if $y_p \neq y_f$, it can be seen that all coefficients are approximately equal to zero if $y' = -Fy_f/z_f$. This behavior explains the slit-forming operation of the volume hologram. For complete analysis of the spatial as well as the spectral behavior of VHI, see Reference 28.

8.5 Applications

1. *Precomposition of aberrations and multicolor readout*: If the aberrations in the optical path are fixed and known *a priori*, then the hologram can be prerecorded with a reference beam passing through the same aberrated optical path. We have done preliminary experiments that support this hypothesis for the simple case of a reflective flat surface covered by a thin turbulent layer.[29] Obviously more challenging is the case of semitransparent objects occluded by turbulence (e.g., thick human tissue). Assuming the spectral properties of the object are *a priori* known in addition to the precompensated aberrations, we propose to use polychromatic illumination and exploit the spectral imaging properties of volume holograms[28] to improve the three-dimensional spatial resolution of the object of interest. The proposed method is shown in Figure 8.5. An alternative but equivalent technique is the use of wide-band, pulsed illumination with the spectrum conjugate-matched to that of the object of interest. Essentially, this proposed imaging method is a fundamentally new type of optical coherence tomography (OCT) enhanced by the spectral matched-filtering action of the hologram. It can also be thought of as a form of spectral imaging diversity conditioned on our *a priori* knowledge of the object spectrum.
2. *Objects constrained on surface manifolds*: Assume an object constrained on a surface manifold of dimension d, $2 \leq d < 3$. Because the rank of the volume holographic imaging transform is 3,

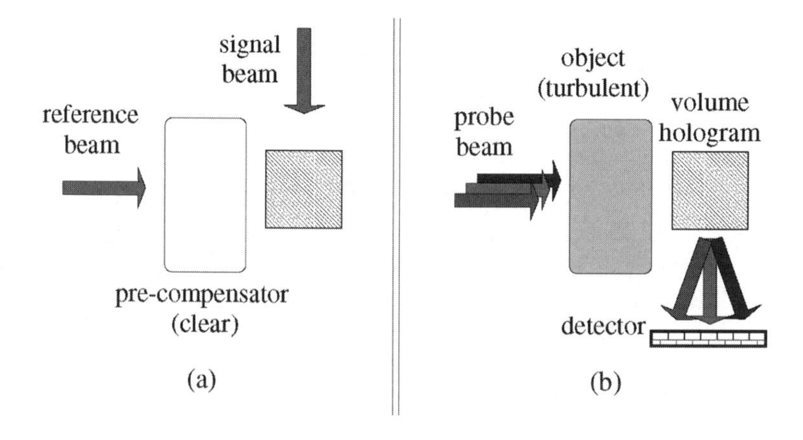

FIGURE 8.5 Imaging through turbulence using a precompensated volume hologram and polychromatic probe beam: (a) recording; (b) imaging.

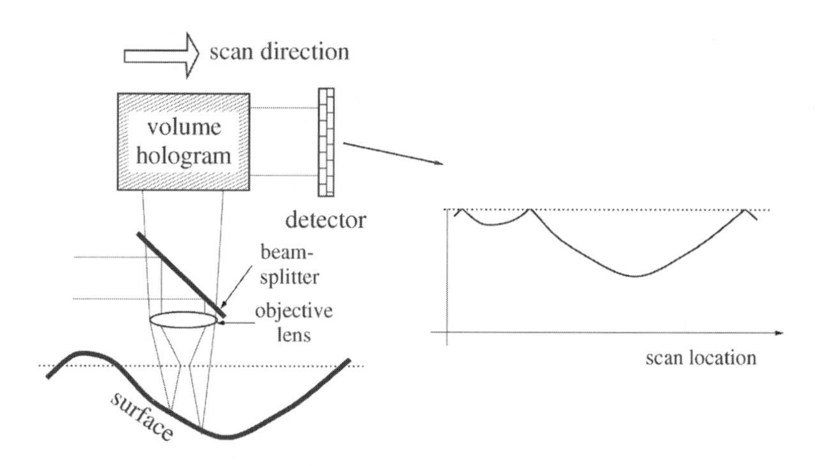

FIGURE 8.6 Mapping height to detected intensity in a scanning volume holographic confocal microscope using active illumination, opacity, and constant reflectivity constraints. For the surface shown, the additional constraint of continuous surface gradient is necessary to resolve the surface within a factor of ±1.

remapping the volume holographic transform onto the object surface leads to one-shot sampling denser than would be possible with thin optical elements. Unlike other analog super-resolution methods (e.g., apodization), the method proposed herein does not suffer from loss of power or increased side lobes, simply because the dense mapping is inherently consistent with volume diffraction. Active illumination improves system performance even further. An example is confocal microscopy using a volume holographic filter on reflective surfaces; by keeping the microscope at a constant height we obtain a map of the object height onto the detected intensity at each scan point (see Figure 8.6). The determination of a topographic "height map" on the surface is well posed if the detector dynamic range is sufficient.

8.6 Further Resources

Experimental demonstrations of VHI have been performed in the context of a confocal microscope where the volume hologram performs the function of a "Bragg pinhole"[30] as well as a real-time four-dimensional imager.[31] The limited diffraction efficiency η of volume holograms poses a major concern for VHI systems. Holograms, however, are known to act as matched filters.[32] It has been shown[33,34] that the matched filtering nature of volume holograms as imaging elements is superior to other filtering elements (e.g., pinholes or slits) in an information-theoretic sense. Efforts are currently underway to strengthen this property by design of volume holographic imaging elements that are more elaborate than described here.

8.7 Conclusions

Volume holographic imaging is unusual in that it allows simultaneous depth and spectral data to be acquired simultaneously, i.e., it allows real-time four-dimension imaging to be performed. This property opens new possibilities for quantifying complex phenomena with spatial and spectral dimensions (e.g., chemical reactions on surfaces or in volumes accessible to the imaging system). Perhaps more importantly, volume holographic imaging is a new paradigm for next-generation imaging systems, which rely on sophisticated transformations of the optical field directly, before the application of digital processing. The former improve the stability of the imaging problem and set an upper bound on the amount of information that the system can recover (because they are applied directly on the optical field before any phase information is lost), whereas the latter perform the task of actually extracting as much information as the bound allows.

The task of systematically optimizing the information throughput of optical systems is by no means complete, but it is a fascinating endeavor that will be in increasing demand in the next decade as existing measurement and metrology schemes run out of steam trying to meet the requirements from instrument users.

References

1. Minsky, M., Microscopy apparatus, U.S. Patent 3,013,467, 1961.
2. Wilson, T. and Sheppard, C.J.R., *Theory and Practice of Scanning Optical Microscopy*, Academic Press, New York, 1984.
3. Wilson, T., Ed, *Confocal Microscopy*, Academic Press, New York, 1990.
4. Sheppard, C. J. R. and Choudhury, A., Image formation in the scanning microscope, *Opt. Acta*, 24, 1051–1073, 1977.
5. Sheppard, C. J. R. and Cogswell, C.J., Three-dimensional image formation in confocal microscopy, *J. Microscopy*, 159(2), 179–194, 1990.
6. Cogswell, C. J. and Sheppard, C. J. R., Confocal differential interference contrast (DIC) microscopy: including a theoretical analysis of conventional and confocal DIC imaging, *J. Microscopy*, 165(1), 81–101, 1992.
7. Cox, I. J., Sheppard, C. J. R., and Wilson, T., Super-resolution by confocal fluorescence microscopy, *Optik*, 60, 391–396, 1982.
8. So, P.T.C., French, T., Yu, W.M., Berland, K.M., Dong, C.Y., and Gratton, E., Time-resolved fluorescence microscopy using two-photon excitation, *Bioimaging*, 3(2), 49–63, 1995.
9. Van Cittert, P.H., *Physica*, 1, 201, 1934.
10. Zernike, F., *Proc. Phys. Soc.*, 61, 158, 1948.
11. Schooneveld, V., Image formation from coherence functions in astronomy, in Reidel, D., Ed., *IAU Colloquium*, Vol. 49, Grøningen, 1978.
12. Devaney, A. J., The inversion problem for random sources, *J. Math. Phys.*, 20, 1687–1691, 1979.
13. Carter, W. H. and Wolf, E., Correlation theory of wavefields generated by fluctuating, three-dimensional, primary, scalar sources. I. General theory, *Opt. Acta*, 28, 227–244, 1981.
14. LaHaie, I. J., Inverse source problem for three-dimensional partially coherent sources and fields, *J. Opt. Soc. Am. A*, 2, 35–45, 1985.
15. Zarubin, A. M., Three-dimensional generalization of the van Cittert–Zernike theorem to wave and particle scattering, *Opt. Commun.*, 100(5–6), 491–507, 1992.
16. Rosen, J. and Yariv, A., Three-dimensional imaging of random radiation sources, *Opt. Lett.*, 21(14), 1011–1013, 1996.
17. Rosen, J. and Yariv, A., General theorem of spatial coherence: application to three-dimensional imaging, *J. Opt. Soc. Am. A*, 13(10), 2091–2095, 1996.
18. Rosen, J. and Yariv, A., Reconstruction of longitudinal distributed incoherent sources, *Opt. Lett.*, 21(22), 1803–1805, 1996.
19. Marks, D., Stack, R., and Brady, D.J., Three-dimensional coherence imaging in the Fresnel domain, *Appl. Opt.*, 38(8), 1332–1342, 1999.
20. Itoh, K. and Ohtsuka, Y., Fourier-transform spectral imaging: retrieval of source information from three-dimensional spatial coherence, *J. Opt. Soc. Am. A*, 3(1), 94–100, 1986.
21. Marks, D.L., Stack, R.A., Brady, D.J, Munson, Jr., D.C., and Brady, R.B., Visible cone-beam tomography with a lensless interferometric camera, *Science*, 284(5423), 2164–2166, 1999.
22. Dowski, E.R. and Cathey, W.T., Extended depth of field through wave-front coding, *Appl. Opt.*, 34(11), 1859–1866, 1994.
23. Marks, D.L., Stack, R., Brady, D., and van der Gracht, J., Three-dimensional tomography using a cubic-phase plate extended depth-of-field system, *Opt. Lett.*, 24(11), 253–255, 1999.
24. Huang, D., Swanson, E.A., Lin, C.P., Schuman, J.S., Stinson, W.G., Chang, W., Hee, M.R., Flotte, T., Gregory, K., Puliafito, C.A., and Fujimoto, J.G., Optical coherence tomography, *Science*, 254(5035), 1178–1181, 1991.

25. Morgner, U., Drexler, W., Kartner, F.X., Li, X.D., Pitris, C., Ippen, E.P., and Fujimoto, J.G., Spectroscopic optical coherence tomography, *Opt. Lett.*, 25(2), 111–113, 2000.
26. Coufal, H., Psaltis, D., and Sincerbox, G., Eds., *Holographic Data Storage*, Springer-Verlag, New York, 2000.
27. Born, M. and Wolf, E., *Principles of Optics*, 7th ed., Pergamon Press, Elmsford, NY, 1998.
28. Barbastathis, G. and Brady, O.J., Multidimensional tomographic imaging using volume holography, *Proc. IEEE*, 87(12), 2098–2120, 1999.
29. Balberg, M., Barbastathis, G., Fantini, S., and Brady, D.J., Confocal imaging through scattering media with a volume holographic filter, *Proc. SPIE*, 3919, 69–74, 2000.
30. Barbastathis, G., Balberg, M., and Brady, D.J., Confocal microscopy with a volume holographic filter, *Opt. Lett.*, 24(12), 811–813, 1999.
31. Liu, W., Psaltis, D., and Barbastathis, G., Real time spectral imaging in three spatial dimensions, *Opt. Lett.*, in press.
32. VanderLugt, A., Signal detection by complex spatial filtering, *IEEE Trans. Inf. Th.*, 10(2), 139–145, 1964.
33. Barbastathis, G. and Sinha, A., Information content of volume holographic images, *Trends Biotechnol.*, 19(10), 383–392, 2001.
34. Barbastathis, G., Sinha, A., and Neifeld, M.A., Information-theoretic treatment of axial imaging, in *OSA Topical Meeting on Integrated Computational Imaging Systems*, Albuquerque, NM, 2001, pp. 18–20.

9

Pattern Recognition

Carl G. Looney
University of Nevada
Reno, Nevada

9.1 Introduction

9.1.1 Classification and Recognition

The conceptualization of things, or objects, as belonging to classes is at the core of all knowledge. We must caution, however, that classes are in the mind of the beholder and that there are often many possible attributes and ways to classify a set of objects. Commonly agreed upon classifications for many objects used routinely are a part of human cultures. We examine here some modern nonlinear methods for classifying and recognizing objects that include clustering, improved clustering, a newer type of fuzzy clustering, probabilistic and fuzzy neural networks, radial basis function neural networks, radial basis functional link nets, ellipsoidal basis function neural networks, ellipsoidal basis functional link nets, and fuzzy ellipsoidal classifiers.

Pattern recognition involves two processes: (1) **classification,** where a sample from a population of objects is partitioned into groups called **classes;** and (2) **recognition,** where a given unknown object from the same population is recognized as belonging to one of the established classes. Recognition is sometimes broken into recognition and **identification,** which means that a particular individual object is recognized. The terms *classification* and *recognition* are sometimes used interchangeably in the literature.

0-8493-1162-4/03/$0.00+$1.50

A classification process examines a sample of objects that represents a population of such objects and partitions it into subsets (the classes) according to similarity of the objects within classes and dissimilarity of the objects between classes. This type of process is called **self-organization, unsupervised learning,** or **clustering** of the sample into **clusters** (classes or subclasses). On the other hand, once a process has clustered the sample into classes and each object is assigned a class **label**—an index value (or codeword) that designates the particular class—a **recognizer** can be trained to assign a class label to any unknown object from the same population (pattern recognition). The training process is called **supervised learning** or **training** of the recognizer. A trained recognizer can perform pattern recognition online.

9.1.2 Features, Vectors, and Prototypes

The objects to be recognized as belonging to one class or another have certain properties that we use to distinguish between classes. These properties (attributes) are the observables, where the observation provides a value for each of a set of properties. A fixed set of properties is used for a particular population and the set of their values for an object determines whether it belongs to a class or not. The individual properties are called **features** of the population. Suppose that there are N features for a population to be used for recognition. These N features are ordered into an N-tuple so that a set of observed values for an object forms a vector, called a **feature vector**. Thus, the feature vectors represent the objects in a population. Pattern recognition is done on feature vectors.

For example, suppose a type of beetle in a certain geographical region has three subtypes and that one is a voracious eater that will destroy the crops. One subtype has a gray-green back when mature, while the other two have light green and dark green backs when mature. Further, suppose that the gray-green beetles have shorter length and are wider than the light and dark green beetles and that the dark green ones have longer antennae. Let us choose these properties as the features and assign quantities to them. For example, for the first feature we could assign 1 to gray-green, 2 to light green, and 3 to dark green. We can also assign the width and length as the second and third features so that the feature vector for each observed beetle is then the 3-tuple (n, w, l).

In the real world the situation is usually more complex. Suppose that the younger beetles are not well differentiated as to color, so it is not a clear case of gray-green, light green, or dark green. Also, the younger beetles are smaller so that the width and height are both smaller numbers. One way to attain greater accuracy is to let $x = 0$ represent gray-green, $x = 0.5$ represent light green, and $x = 1.0$ represent dark green. The observer could then assign a value such as 0.3 to color that is slightly more light green than gray-green. To account for age in the width and length measurements the ratio $r = w/l$ of width to length represents a shape parameter that is more independent of age. If we now observe young beetles in the spring to obtain feature vectors $\{(x, r)\}$, we can use these measurements to recognize the subtype to determine whether or not eradication methods are necessary to save the crop, provided that we know the typical feature vector for each subtype.

Figure 9.1 shows a set of vectors in the plane that represents a sample of beetles. The typical vector of each subtype sample is a **prototype** vector that represents that subtype, or class, of beetles. Given a feature vector (x, r), we test it against each prototype to determine the one it most resembles. A metric is used such as Euclidean distance or maximum component magnitude distance so that the closest prototype to (x, r) determines the subtype of (x, r).

Figure 9.2 shows the steps in the development of a PR system. Let us sample a given population of objects and determine a set of features to distinguish its classes. Then we draw a representative random sample for classification and obtain the observations for each feature of each sampled object to obtain a sample of feature vectors. This sample represents the population of interest. Now we **classify** the feature vectors and assign a class label to each one. The labels can be satisfactorily assigned by humans at times but is usually done by computational algorithms. Self-organizing learning is often better in cases where the feature values are noisy or the classes are not well known. In many cases the number K of classes is not known, but in others it is known, such as $K = 2$ for a *defective* or *nondefective*

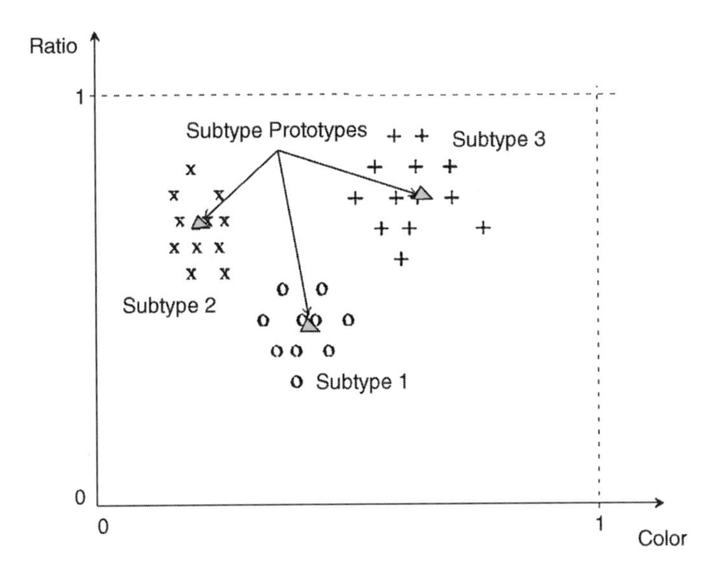

FIGURE 9.1 Feature vectors of observations on beetles.

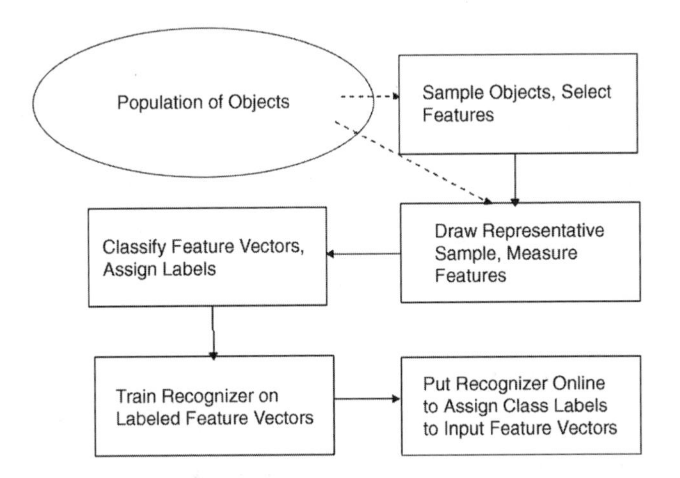

FIGURE 9.2 Development of a PR system.

product or for *malignant* or *benign* tumors, or $K = 3$ for *strong edge, weak edge* or *no edge* in image edge detection.

After classification and labeling of the sample feature vectors, the next step is to train a process via supervised learning to read any input feature vector from the same population and to output an accurate class label for it, that is, to *recognize* it. Recognition may require preprocessing raw data to extract special features, for example, taking the fast Fourier transform of a satellite image to obtain the power in certain frequency bands as texture feature values for either crop recognition or ocean wind speed from wave frequency patterns.

Feature selection is very important for classification and recognition [Looney, 1997]. If two features are strongly correlated, then they differentiate the same feature vectors and are redundant. Redundancy can be detected by performing a correlation of the features over the sample. We could eliminate one of each pair of correlated feature vectors to obtain a minimal set. An important principle is: *For every pair of classes, there must be at least one feature that separates them* (although it is preferable to have more than one such separating feature for each pair of classes).

9.2 Classification

9.2.1 Clustering with the *k*-Means Algorithm

Clustering is a self-organizing process that partitions an exemplar set of feature vectors into clusters (subsets) that represent classes. The ***k*-means** clustering algorithms are the simplest. These were developed by Forgy [1965] and MacQueen [1967]; we present the one by Forgy. Given a set of Q unlabeled feature vectors $\{\mathbf{x}^{(q)}: q = 1,\ldots,Q\}$, where each has N features, $x^{(q)} = (x_1^{(q)},\ldots,x_N^{(q)})$, we want to classify them into K clusters, where K is input by the user. The basic k-means algorithm is described below, where the **center,** or **prototype,** that represents a cluster is the average vector of that cluster.

Forgy's *k*-means Algorithm

Step 1: Randomly order the Q-feature vectors and input K (the number of classes).

Step 2: Select the first K of the Q-feature vectors as **seeds** (initial prototypes of classes).

Step 3: Assign each of the Q-feature vectors to the nearest prototype to form K classes (use the index $c[q] = k$ to designate that $\mathbf{x}^{(q)}$ belongs to Class k and count cluster sizes with $s[k]$, which is incremented every time a vector is assigned to Class k).

Step 4: Average the feature vectors in each class to find K new centers. To average all vectors in Class k, use the following technique, starting with $a[n][k] = 0.0$, where $a[n][k]$ is the value of component n of the Class k average vector and $x[n][q]$ is the nth component of $\mathbf{x}^{(q)}$.

```
for k = 1 to K do                              //For each Class k:
    for n = 1 to N do a[n][k] = 0.0;           //initialize averages
    for q = 1 to Q do                          //For given Class k, find
        if (c[q] = = k) then                   //all vectors in Cluster k
            for n = 1 to N do                  //and for each component n,
                a[n][k] = a[n][k] + x[n][q];   //sum that component
    if (s[k] > 1) then
        a[n][k] = a[n][k]/s[k];                //Average Cluster k
```

Step 5: *if ((not first pass) and (no class has changed)) then exit, else go to Step 3 above*

Many algorithms [Pena et al., 1999] use the k-means to start but apply some adjustment to aid with the seeding. They showed that random drawing of many more seeds worked better than the use of the first K-feature vectors. Selim and Ismail [1984] showed that the algorithm converges to a local minimum in the sum-squared error

$$E = \sum\nolimits_{(k=1,K)} \sum\nolimits_{\{c[q]=k\}} \{\sum\nolimits_{(n=1,N)} (x[n][q] - a[n][k])^2\} \tag{9.1}$$

However, examples [Looney, 2002] showed that there is no guarantee of optimality of the clustering, which depends on K and the K initial seeds.

MacQueen's adjusted algorithm recomputes a new center every time a feature is assigned to a cluster. Snarey et al. [1997] proposed a maximum method for selecting a subset of the feature vectors for seeds. Kaufman and Rousseeuw [1990] used medians of clusters for the prototypes and then optimized a sum of squared distances (errors) in each cluster.

9.2.2 The Number *K* of Classes and Clustering Validity

The number of classes, K, may not be known, so we must reorder the Q-feature vectors and repeat the clustering several times to find the best clustering for estimated K, $K + 1$, $K - 1$, etc., to find a better clustering. But how do we know if a clustering is better? We can use a **clustering validity measure** as

```
XXXXXXXXXXX      XX XX     X X          XXX
XXXXXXXXXXX      X XX  X    X  X  X XX    X X XX
XXXXXXXXXXX       X  XX    XX XXX X     XX X X
XXXXXXXXXXX      X XX X     X X    X      X XX
XXXXXXXXXXX        X          X   XX X
XXXXXXXXXXX
```

No Clustering Structure   Clustering Structure

FIGURE 9.3 Structure in feature vector data.

described below. Validity measures in general are known to be problematic [Dubes and Jain, 1998], but it is more problematic when K is unknown to not use one and so we use the best one (see below).

A clustering is good if the clusters are relatively **compact** (packed closely about the center) and relatively **well separated** (no two centers are too close). Let σ_k^2 be the mean-square error (variance) of the kth cluster (fix k in Eq. 9.1 above). If these are relatively small for all k, then the clusters are compact, which is desirable. Let $D_{\min}$ be the minimum distance between all pairs of cluster *centers*, where a center is a single prototype for a cluster. It is desirable for this distance to be larger, rather than smaller, in which case the clusters are *well separated*. The **Xie-Beni** (XB) **clustering validity measure** is:

$$XB = (\sigma_1^2 + \cdots + \sigma_k^2)/D_{\min} \tag{9.2}$$

The smaller this measure is, the better is the clustering [Xie and Beni, 1991]. Thus, we move K in the direction that decreases XB until we get a minimum value for XB and accept the corresponding K. Figure 9.3 shows two sets of vectors, of which one has a moderately strong clustering structure.

There remains a problem with this algorithm. If two seeds are close together, then the result will be two clusters close together that should be merged into a single one, while a seed far from any other seeds can yield a large cluster that should be broken into two or more clusters. Ways to improve this have been found. For example, Chen and Wang [1999] used the *equalized universe* method, whereby a fine grid of initial seeds was used and a bell-shaped fuzzy-set membership function was centered on each with 25% overlap. But the number of such functions grows exponentially with the number of dimensions. The method of Chiu [1994] is a more efficient variant of the *mountain method* of Yager and Filev [1994], which sums Gaussians centered on the feature vectors to build a combined mountain function, which in turn is a variant of *Parzen windows* (summed Gaussians). However, these methods are all rather computationally expensive.

9.2.3 An Improved *k*-Means Algorithm

A simpler way to prevent bad clustering due to inadequate seeding is to modify the basic k-means algorithm to obtain the *improved k-means algorithm* [Looney, 2002]. We start with a large number of uniformly distributed seeds in the bounded N-dimensional feature space, but we reduce them considerably by eliminating those that are too close to another one. The test threshold is the average distance between seed vectors, which may be computed from a sample of seeds. Thus, we obtain a smaller number K of initial seeds that are uniformly distributed. Next, we assign each of the Q-feature vectors to a seed by minimum distance and then eliminate the empty clusters and any clusters of a size less than p (p is given by the user). Iteration now converges quickly to a lot of relatively small clusters. The closest ones are merged until the XB measure stops decreasing or until there are only two clusters remaining. The improved algorithm is given below.

The Improved *k*-Means Algorithm

Step 1: Draw a large number K of uniform random seed vectors $\mathbf{z}^{(k)}$ as initial centers.
Step 2: Eliminate all seed vectors that are too close to other seed vector and reduce K as needed.

```
for k1 = 1 to K – 1 do                    //For all except the last seed
  for k2 = k1 + 1 to K do                 //pair it with a different seed
     if (distance(k1,k2) < ϵ) then         //If these two seeds are too close
        for k = k2 to K – 1 do            //then eliminate one by closing
           z(k) = z(k+1);                 //up the indices over it
        K = K – 1;                        //and reducing no. clusters
```

Step 3: Assign each of the Q-feature vectors $\mathbf{x}^{(q)}$ to the nearest random seed vector by the assignment $c[q] = k$ and increment the size $s[k]$ of Class k

```
for k = 1 to K do s[k] = 0;               //Initialize cluster sizes to 0
Dmin = 99999.9;                           //Initialize large min. distance
for q = 1 to Q do                         //For every feature vector q
  for k = 1 to K do                       //and every center seed k
     d = distance(k,q);                   //compute distance between them
     if (d < Dmin) then                   //If distance is smallest
        Dmin = d;                         //then save it and also save
        kmin = k;                         //its index
  c[q] = kmin;                            //Assign vector to nearest center
  s[kmin] = s[kmin] + 1;                  //and increase size of that cluster
```

Step 4: Eliminate all clusters that have fewer than p vectors and reduce K as needed, where $s[k]$ is the size of Cluster k and $c[q] = k$ means feature vector q belongs to Class k:

```
for k1 = 1 to K do                        //For each cluster:
  if (s[k1] < p) then                     //if cluster size is too small
    for k2 = k1 + 1 to K do               //then eliminate that cluster
      for q = 1 to Q do                   //by finding its vectors and then
         if (c[q] = k2) then              //(if vector q is in Class k2)
           c[q] = k2 – 1;                 //re-indexing them accordingly
      s[k2 – 1] = s[k2];                  //Also re-index sizes of clusters
      K = K – 1;                          //and reduce no. clusters
```

Step 5: Assign features to cluster centers, average cluster centers, and stop if no centers changed or else repeat this step.

Step 6: Compute the XB measure:

```
if ((this is not the first pass) and (XB increases)) then
  stop                                    //accept clustering of the previous iteration
else
  merge()                                 //merge two clusters with closest centers
  go to Step 5.
```

9.2.4 An Improved Fuzzy k-Means Algorithm

The next problem in clustering to be considered is the fact that outliers in a cluster can unduly affect the center, or prototype, when it is obtained as the cluster average (by averaging each component). Medians can be used in place of averages, although they may throw away good points as well as outliers. We use a type of fuzzy averaging here that puts the center prototype among the more densely situated points by using a weighted average. In the following algorithm we start with the improved k-means average of each cluster as the prototype, but then we center a Gaussian on that average prototype and compute fuzzy weights for a weighted average. We iterate this process until the fuzzy average does not change any further. We could also use medians or α-trimmed means [Bednar and Watt, 1984], where we throw away the α greatest and α least for each component and average the remainder component-wise.

A Weighted Fuzzy k-Means Averaging Algorithm

Step 1: Draw a large number K of uniformly distributed random seed vectors in the feature space.

Step 2: Eliminate any seed vectors that are too close to other seed vectors and reduce K accordingly (see Step 2 of the improved k-means algorithm above).

Step 3: Assign each of the Q-feature vectors $\mathbf{x}^{(q)}$ to the nearest random seed vector.

Step 4: Eliminate all seed vectors that are centers of empty clusters or have fewer than p vectors, and reduce K accordingly (see Step 4 of the improved k-means algorithm above).

Step 5: Compute the **weighted fuzzy average** of each class as the new class prototype with current K (see the description of *weighted fuzzy average* below).

Step 6: Assign each of the Q-feature vectors to the class with the nearest weighted fuzzy average.

Step 7: If (first pass) or (any weighted fuzzy average has changed) then go to Step 5.

Step 8: Compute the XB measure (use Equation 9.1):

if ((not the first pass) and (XB increases)) then

stop *//accept clustering of previous iteration*

Step 9: Merge the two clusters whose prototypes are closest and use the average of their two prototypes as a new prototype (seed) and reduce K accordingly.

Step 10: Go to Step 5 (to find new class prototypes and reassignment, etc.)

The **weighted fuzzy average** (WFA) of the vectors in a cluster is done component-wise, so we explain it here for a single dimension. Let $\{x_1, \ldots, x_P\}$ be a set of P real numbers. To find its weighted fuzzy average we initially take the sample mean $\mu^{(0)}$ and variance σ^2 to start the process. We center a Gaussian over the current approximate WFA $\mu^{(r)}$ and iterate as follows:

$$w_p^{(r)} = \exp[-(x_p - \mu^{(r)})/2\sigma^2] / \sum_{(m=1,P)} \exp[-(x_m - \mu^{(r)})/2\sigma^2] \quad (9.3)$$

$$\mu^{(r+1)} = \sum_{(p=1,P)} w_p^{(r)} x_p, \quad r = 0, 1, 2, \ldots \quad (9.4)$$

The denominator in Eq. (9.3) standardizes the weights so they all sum to unity. We compute σ^2 on each of three or four iterations and then leave it fixed. After about five iterations the approximate WFA is sufficiently close to the true WFA. Schneider and Craig [1992] used a *weighted fuzzy expected value* for histogram adjustment, but it was based on a decaying exponential. We use Gaussians that are canonical fuzzy-set membership functions. Figure 9.4 below shows an example of five points (circles) that compares the mean, median, and the WFA.

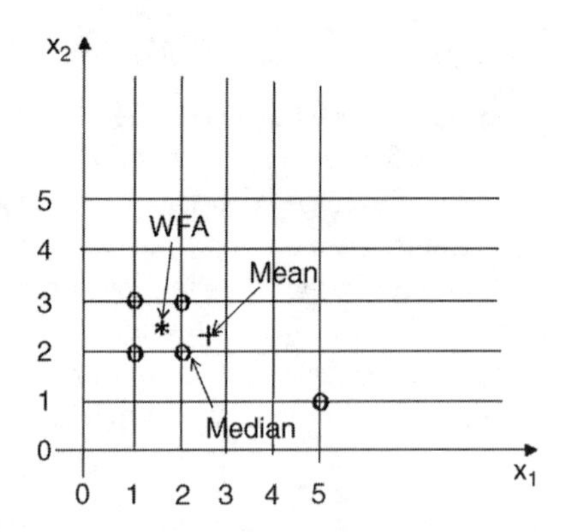

FIGURE 9.4 A mean, median, and WFA of five points.

We mention here that Bezdek's *fuzzy c-means* algorithm [Bezdek, 1973] computes weights for a weighted averaging of the vectors in a cluster. The solution for the weights optimizes a Lagrangian expression of the summed-squared error of the weighted average with respect to the centers, suitably constrained with Lagrangian multipliers. The solution weights are computed as

$$w_{qk} = \{1/\|x^{(q)} - z^{(k)}\|^2\}^{1/(p-1)} / \sum_{r=1,K} \{1/\|x^{(q)} - z^{(r)}\|^2\}^{1/(p-1)} \tag{9.5}$$

for $p > 1$ (p is usually 2 to 3 and remains fixed over all prototypes $z^{(k)}$. The advantage was considered to be that every feature vector had a fuzzy weight for membership in each cluster (a new concept).

The performance of the fuzzy *c*-means clustering, however, can be very inadequate. It appears that smaller clusters should have a different value of p than larger clusters, i.e., a different weighting function, which is the case for our weighted fuzzy average with a different & value to fit each cluster that also evaluates the fuzzy-set membership of any feature vector in each class. The improved k-means performs well and is included in our weighted fuzzy clustering. We could also use medians or α-trimmed means [Bednar and Watt, 1984] as centers.

9.3 Recognition

Here we assume that we have a set of **exemplar** feature vectors (a sample of feature vectors that represent all classes of the population and that have been labeled by a classification process that assigns each feature vector a label to represent the class to which the vector belongs). The following methods train a recognition process via supervised learning for online recognition of an incoming stream of feature vectors from the population of interest. Of course, the process can learn only the labels that it is trained on in the supervised learning mode, so if the feature vectors are mislabeled or contain too much noise, then the trained process is inaccurate. However, a recognizer that is well trained on accurate data from representative exemplars can perform very well when put online.

9.3.1 Probabilistic Neural Networks

A probabilistic neural network (PNN) has three layers of nodes [Specht, 1988; 1990] (see [Anagnostopoulis et al., 2001] for an application to human face recognition). Figure 9.5 displays the architecture of a PNN that recognizes $K = 2$ classes, but it can be extended to any number K of classes. The input layer (on the left) contains N nodes: one for each of the N input features of a feature vector. These are fan-out nodes that branch at each feature input node to all nodes in the hidden (or middle) layer so that each hidden node receives the complete input feature vector $\mathbf{x}$. The hidden nodes are collected into groups, one group for each of the K classes, as shown in Figure 9.5. Each hidden node in the group for Class k corresponds

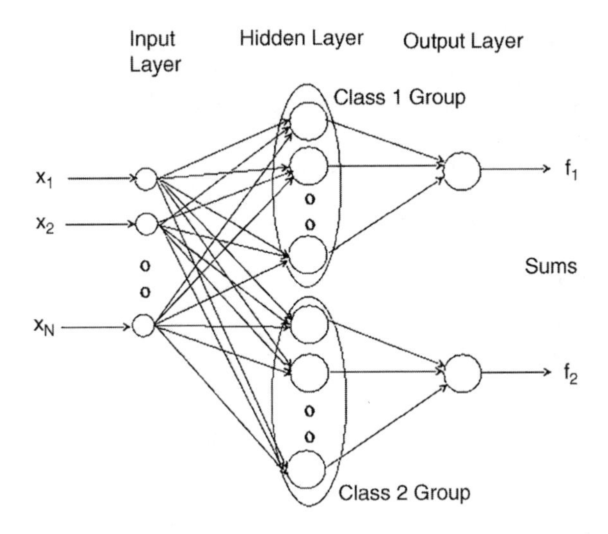

FIGURE 9.5 A two-class PNN.

to a Gaussian function centered on its associated feature vector in the kth class (there is a Gaussian for each exemplar feature vector). All of the Gaussians in a class group feed their functional values to the same output layer node for that class, so there are K output nodes (one for each class).

At the output node for Class k, all of the Gaussian values for Class k are summed and the total sum is scaled to force the integral of the sum to be unity so that the sum forms a probability density function. Here we temporarily use special notation for clarity. Let there be P exemplar feature vectors $\{\boldsymbol{x}^{(p)}: p = 1,\ldots,P\}$ labeled as Class 1, and let there be Q exemplar feature vectors $\{\boldsymbol{y}^{(q)}: q = 1,\ldots,Q\}$ labeled as Class 2 (see Figure 9.5). In the hidden layer there are P nodes in the group for Class 1 and Q nodes in the group for Class 2. The equations for each Gaussian centered on the respective Class 1 and Class 2 points $\boldsymbol{x}^{(p)}$ and $\boldsymbol{y}^{(q)}$ (feature vectors) are:

$$g_1(\boldsymbol{x}) = [1/\sqrt{(2\pi\sigma^2)}]\exp\{-\|\boldsymbol{x} - \boldsymbol{x}^{(p)}\|^2/(2\sigma^2)\} \tag{9.6}$$

$$g_2(\boldsymbol{y}) = [1/\sqrt{(2\pi\sigma^2)}]\exp\{-\|\boldsymbol{y} - \boldsymbol{y}^{(q)}\|^2/(2\sigma^2)\} \tag{9.7}$$

The σ values can be taken to be one half the average distance between the feature vectors in the same group, or at each exemplar it can be one half the distance from the exemplar to its nearest other exemplar vector. The kth output node sums the values received from the hidden nodes in the kth group, called *mixed Gaussians* or *Parzen windows*. The sums are defined by:

$$f_1(\boldsymbol{x}) = [1/\sqrt{(2\pi\sigma^2)^P}](1/P)\sum\nolimits_{(p=1,P)} \exp\{-\|\boldsymbol{x} - \boldsymbol{x}^{(p)}\|^2/(2\sigma^2)\} \tag{9.8}$$

$$f_2(\boldsymbol{y}) = [1/\sqrt{(2\pi\sigma^2)^Q}](1/Q)\sum\nolimits_{(q=1,Q)} \exp\{-\|\boldsymbol{y} - \boldsymbol{y}^{(q)}\|^2/(2\sigma^2)\} \tag{9.9}$$

Any input vector is put through both functions, and the maximum value (**maximum *a posteriori*,** or MAP value) of f_1 and f_2 decides the class. For $K > 2$ classes the process is analogous. There is no iteration or computation of weights. For a large number of Gaussians in a sum, the error can be significant (see [Wedding and Cios, 1998]). Thus, the feature vectors in each class may be reduced by thinning those that are too close to another one.

9.3.2 Fuzzy Neural Networks

Our **fuzzy neural networks** (FNNs) are similar to the PNNs. Let there be K classes, and let $\boldsymbol{x}$ be any feature vector from the population of interest to be recognized. The Class k exemplar feature vectors are denoted by $\boldsymbol{x}^{(qk)}$ for $qk = 1, \ldots, Qk$. We replace the functions of Eqs. (9.8) and (9.9) with the following more simply scaled sums.

$$f_1(\boldsymbol{x}) = (1/Q_1)\sum_{(p=1,P)} \exp\{-\|\boldsymbol{x} - \boldsymbol{x}^{(q1)}\|^2/(2\sigma^2)\}$$

$$\vdots \qquad\qquad\qquad \vdots \tag{9.10}$$

$$f_K(\boldsymbol{x}) = (1/Q_K)\sum_{(q=1,Q)} \exp\{-\|\boldsymbol{x} - \boldsymbol{x}^{(qK)}\|^2/(2\sigma^2)\}$$

These functions are the K fuzzy-set membership functions whose functional values are the relative fuzzy truths of memberships in the K respective classes. Thus, $\boldsymbol{x}$ belongs to the class with the highest fuzzy value. When there is a clear winner, then $\boldsymbol{x}$ belongs to a single class, but otherwise it may belong to more than one class with the given relative fuzzy truths. No training of weights is required. The exemplar vectors may be thinned as previously done.

9.3.3 Radial Basis Function Neural Networks

These neural networks (NNs) have three layers: (1) the input layer of N nodes, where the respective N features in a feature vector are input; (2) the hidden (middle) layer, where at each node the input feature vector is put through a Gaussian **radial basis function** (RBF) that is centered on a corresponding exemplar vector $\boldsymbol{v}$, as shown in Figure 9.6; and (3) the output layer, where the inputs from all of the hidden nodes are combined at each output node in a weighted average. The weights $\{u_{mj}\}$ used in the weighted average are adjusted during the training so as to map each exemplar feature vector into its correct output codeword to approximately equal the **target** vector (desired output codeword that is a label). Figure 9.7 presents the architecture of a **radial basis function neural network** (RBFNN).

Again we let $\{\boldsymbol{x}^{(q)}: q = 1, \ldots, Q\}$ be a set of Q exemplar feature vectors for training, where each has N components.

$$x^{(q)} = (x_1^{(q)}, \ldots, x_N^{(q)}) \tag{9.11}$$

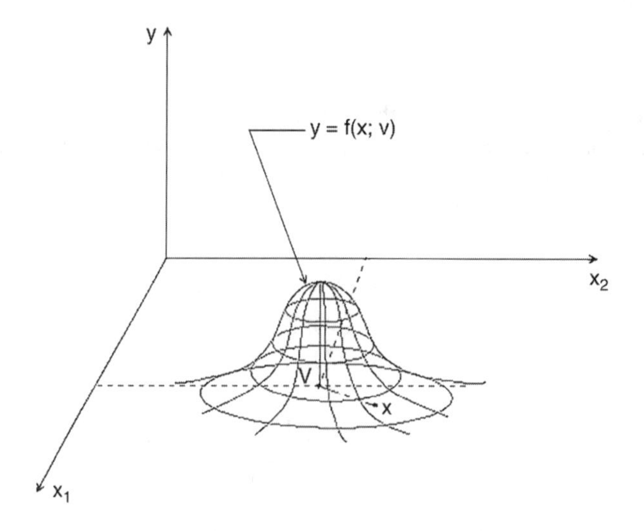

FIGURE 9.6 A radial basis function centered on v.

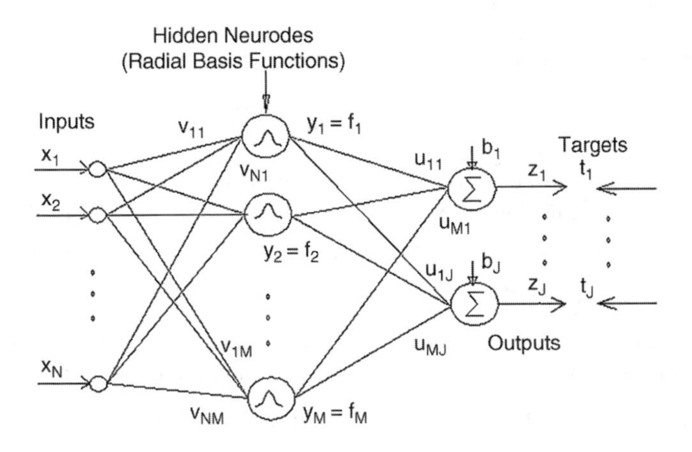

FIGURE 9.7 The radial basis function neural network architecture.

Each qth exemplar feature vector $\boldsymbol{x}^{(q)}$ corresponds to a hidden node as the center for a Gaussian (RBF). Each jth output node receives all of the values $\{y_{mj}\}_{m=1,M}$ from the M hidden-node Gaussians and takes a weighted average of them to obtain the final output value from the jth output node. All y_m values are less than or equal to 1, and all weights w_{mj} are less than or equal to 1, so the weighted sum is less than or equal to M. Thus, we divide the sum by M to keep the output less than or equal to 1 in magnitude.

Each actual output $z_j^{(q)}$ from the jth output node (for the input $\boldsymbol{x}^{(q)}$) is forced to match the target componnent label $t_j^{(q)}$ by adjusting the weights $\{w_{mj}\}$ (for all m). A bias b_j is added onto the sum and is also adjusted to help approximate the target value $t_j^{(q)}$, as shown in Figure 9.7.

All of the exemplar feature vectors are usually used as centers of RBFs. The output from the mth hidden node for any input vector $\boldsymbol{x}^{(q)}$ has the following form:

$$y_m^{(q)} = g_m(\boldsymbol{x}^{(q)}) = \exp[-(\boldsymbol{x}^{(q)} - \boldsymbol{x}^{(m)})^2/(2\sigma^2)] \tag{9.12}$$

The output value from each of the J nodes in the output layer for input $\boldsymbol{x}^{(q)}$ is

$$z_j^{(q)} = (1/M)\sum\nolimits_{m=1,M} w_{mj}\, y_m^{(q)} + b_j \tag{9.13}$$

where b_j is a bias to account for translation of the output values.

The training starts with an initial set of weights that is drawn randomly such that each weight is between –0.5 and 0.5. The basic idea is to adjust the weights $\{w_{mj}\}$ to minimize the *total sum-squared error* (TSSE) E between the actual outputs and the target output labels. E is defined over all exemplar feature vectors $\{\boldsymbol{x}^{(q)}: q = 1, \ldots, Q\}$ as inputs and over all J output values as:

$$E = \sum\nolimits_{q=1,Q}\sum\nolimits_{j=1,J}(t_j^{(q)} - z_j^{(q)})^2 \tag{9.14}$$

We could set the derivative $\partial E/\partial w_{mj}$ of E with respect to each weight equal to 0, and if we have the same number of equations and unknowns, we could solve the set of linear equations in the same number of unknowns. Usually these numbers are different, but even if they are not, the equations may be ill conditioned (two equations are approximately a multiple of each other), and the convergence may not be stable. We could compute a reduced basis of vectors for which the system is not ill conditioned, but this appears not to be as fruitful as the usual method of *steepest descent*.

Starting with a set of random weights $\{w_{mj}\}$ between –0.5 and 0.5 from a uniform random number generator, the convergence is quick. The iterative procedure is

$$w_{mj}^{(i+1)} = w_{mj}^{(i)} - \alpha(\partial E / \partial w_{mj}) \tag{9.15}$$

where the superscripts $i+1$ and i designate the iteration numbers. This is the familiar steepest descent method, where we start at an initial point and then move in the direction of steepest descent of E. Because E is a strictly convex function of the w_{mj}, it has a single minimum that is the global minimum.

The RBFNN Training Algorithm

Step 1. Read in all exemplar feature vectors in $\{\boldsymbol{x}^{(q)}: q = 1, \ldots, Q\}$ and their associated target vectors (code words or labels) in $\{\boldsymbol{t}^{(q)}: q = 1, \ldots, Q\}$, set iteration number $i = 0$.

Step 2. Compute the initial value $s = (1/2)[1/M]^{1/N}$ for the Gaussians.

Step 3. Input an integer value M for the number of exemplar feature vectors to use as centers (M may be equal to Q if Q is not too large).

Step 4. Select initial weights $\{w_{mj}\}$ and biases $\{b_j\}$ randomly between –0.5 and 0.5.

Step 5. Compute $\{y_m^{(q)}\}$ for all m for every input feature vector $\boldsymbol{x}^{(q)}$ (use Eq. (9.12)).

Step 6. Compute the outputs $\{z_j^{(q)}\}$ for all j and q (use Eq. (9.13)).

Step 7. Compute the TSSE value E_0 (use Eq. (9.14)).

Step 8. Iterate over the set of all M hidden nodes and all J output nodes:

for j = 1 to J do *//For all output nodes and for all*

for m = 1 to M do *//hidden nodes, adjust weights*

$$w_{mj}^{(i+1)} = (w_{mj}^{(i)} - \alpha(\partial E/\partial w_{mj})) = w_{mj}^{(i)} + (\alpha/M)\sum\nolimits_{q=1,Q}(t_j^{(q)} - z_j^{(q)})(y_j^{(q)})$$

$$b_j^{(i+1)} = b_j^{(i)} + (\gamma/M)\sum\nolimits_{q=1,Q}(t_j^{(q)} - z_j^{(q)})$$ *//for (i+1)st iteration*

i = i + 1; *//Update iteration number*

Step 9. Compute a new value for E (use Eq. (9.14)).

Step 10. *If ($|E_0 - E| < \epsilon$) then stop;* *//Stop if error change negligible*

else

$E_0 = E$; *//or else save current error*

go to Step 8; *//and repeat adjustments*

The *step sizes* α and γ are also called *learning rates*. These are numbers that usually start low (about 0.4). We then apply our *en route* method on α so that if $E < E_0$, then the step was a success and we increase the learning rate by $\alpha = 1.24\alpha$ to speed up the convergence, or else we decrease it by $\alpha = 0.96\alpha$. There are various strategies, but it is possible to increase α rapidly (a **greedy algorithm** that takes chances to converge very quickly) to several hundreds and then decrease it as needed as the minimum is approached.

9.3.4 Radial Basis Functional Link Nets

If we start with the RBFNN and add extra lines from the input nodes to the output nodes and also weight these with weights $\{u_{nj}\}$, then we have two sets of values to average at the output nodes: (1) the x_n from the input nodes weighted by u_{nj}, and (2) the y_m from the hidden nodes weighted by w_{mj}. Thus, the outputs

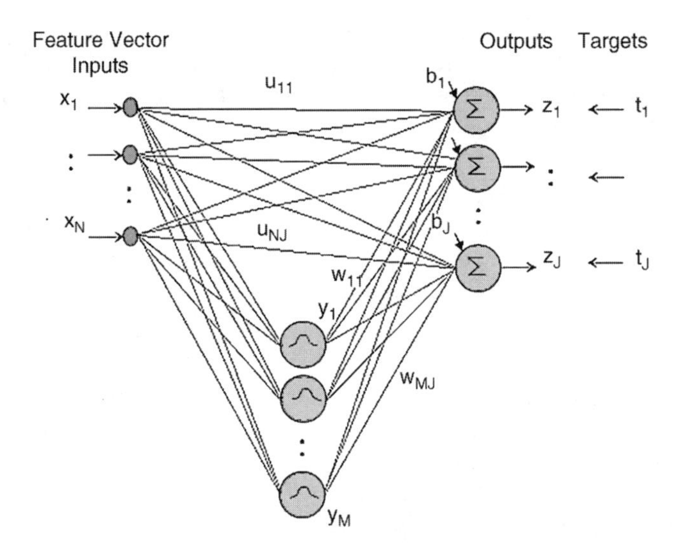

FIGURE 9.8 The radial basis functional link net architecture.

from the nodes at the output layer now become

$$z_j = [1/(M+N)]\{\sum_{m=1,M} w_{mj} y_m^{(q)} + \sum_{n=1,J} u_{nj} x_n^{(q)} + b_j\} \tag{9.16}$$

We call this network a **radial basis functional link net** (RBFLN) [Looney, 2001a] (Looney calls it a *random vector quantization functional link net* (RBQFLN) [1997]). The name originated with the functional link net of Pao et al. [1994]. The RBFLN is more powerful than the RBFNN because it includes the linear combinations of the inputs as well as the nonlinear part from the hidden nodes and the biases. In this case, we also adjust the new weights $\{u_{nj}\}$, in addition to adjusting the previous weights $\{w_{mj}\}$ and biases $\{b_j\}$, all by steepest descent iteration.

Figure 9.8 shows the radial basis functional link-net architecture. These networks perform better than RBFNNs, which in turn perform better than the so-called *multiple-layered perceptrons* (MLPs), also known as *feed forward neural networks*, with the so-called *back-propagation algorithm*, which is also steepest descent [Looney, 1997], but that is problematic due to multiple local minima.

To adapt the RBFNN to that of the RBFLN we must also randomly initialize the second set of weights $\{u_{nj}\}$ in addition to $\{w_{mj}\}$ in Step 4 above and train them on the second set of weights as well as the first. The iterative Step 8 in the previous RBFN algorithm changes as given below.

The RBFLN Algorithm (only Step 8 need be shown here)

Step 8. Iterate weight and bias adjustments over the set of all output, hidden and input nodes:

for j = 1 to J do //For each outut node: for each

for m = 1 to M do //hidden node, update weights

$$w_{mj}^{i+1} = (w_{mj}^{(i)} - \alpha(\partial E / \partial w_{mj})) = [a/(M+N)]\sum_{q=1,Q}(t_j^{(q)} - z_j^{(q)})(y_m^{(q)})$$

for n = 1 to N do //and for each input node, update weights

$$u_{nj}^{i+1} = (u_{nj}^{(i)} - \beta(\partial E / \partial u_{nj})) = [\beta/(M+N)\sum_{q=1,Q}(t_j^{(q)} - z_j^{(q)})(x_n^{(q)})$$

$b_j = b_j + (\gamma/M)\sum_{q=1,Q}(t_j^{(q)} - z_j^{(q)})$ //Update bias at each output node

The advantage of using the radial basis functional link net (RBFLN) over the RBFNN is that it adds a linear input–output functional component to the outputs so that any linear relationships do not need to be approximated by the nonlinear part coming through the hidden nodes. Looney has shown [2001a] over many simulation runs that it learns with significantly fewer iterations and learns better (fewer mistakes). There appears to be no good reason to use the RBFNN instead of the RBFLN.

9.3.5 A Simplified Approach to RBFNNs and RBFLNs

A promising modification to the RBFNN and the RBFLN to make them more efficient is the reduction in the number of Gaussians. This also prevents extraneous error from building up in the summing of the Gaussians at the output nodes. It involves the thinning of the exemplar vectors in each class, especially for classes with large numbers of exemplars.

For each Class k in turn, we examine the set of all (labeled) exemplars $\{\mathbf{x}^{(q)}: q = 1, \ldots, Q\}$ to find those that belong to Class k via $c[q] = k$. We save these in a new data structure of vectors and then perform a search with each one to see if there are any other exemplars in this class that are too close, according to a threshold T_k for Class k. We take T_k to be a constant α times the average distance between feature vectors in that class, where $0.4 < \alpha < 0.8$ (a higher value yields fewer exemplars in Class k). We eliminate any exemplar vectors that are too close to the current one and reindex. Then we select the next exemplar remaining in the class and search for exemplars in the class that are too close to it. We continue this process until every exemplar vector in this class has been checked (this is the same thinning process that we used in the improved k-means algorithm).

After such processing, we use the remaining exemplar vectors for each class as centers for Gaussians as before. At this point we have a smaller subset of Gaussians that is sparser but distributed over the classes so that the processing is more efficient and the extraneous errors are smaller. Figure 9.9 shows a thinned set of five mixed Gaussians that cover a class of exemplar vectors (the class need not be a circular region and usually is not).

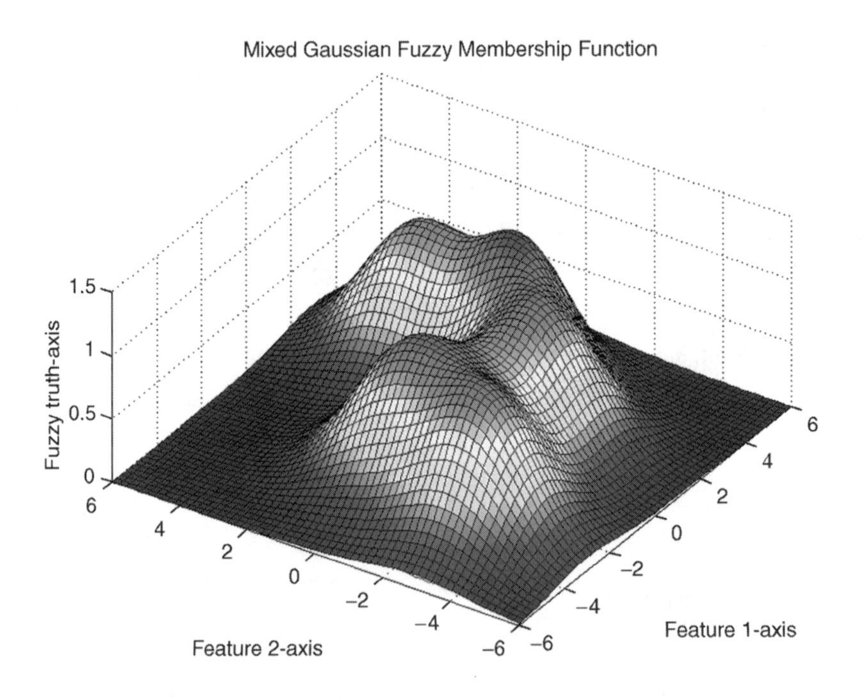

FIGURE 9.9 Thinned multiple Gaussians covering a cluster.

9.3.6 Using Ellipsoidal Basis Functions

In the algorithms given above we have used radial (circular) Gaussians as the radial basis functions. It is also possible to use *ellipsoidal basis functions* that have been examined in recent research by Abe and Thawonmas [1997] and Looney [2001a; 2002b]. To use these, we must first perform classification to get labeled feature vectors, separate them by class to obtain a data structure for each class of these exemplar vectors, and compute the respective covariance and inverse covariance matrices C_k and C_k^{-1} for each kth class. Then we can use the full multivariate Gaussian function for the resulting ellipsoidal (nonradial) basis functions, given by:

$$g_k(\boldsymbol{x}) = \exp\{-(1/2)(\boldsymbol{x} - \boldsymbol{z}^{(k)})^t C^{-1}(\boldsymbol{x} - \boldsymbol{z}^{(k)})\} \tag{9.17}$$

where k designates the kth class and $\boldsymbol{z}^{(k)}$ is the prototype (center) for that class.

Classes often do not fit inside circles, so we use the data in each class to shape an ellipsoid to contain the class exemplar vectors. The *generalized Gaussian* is defined by Eq. (9.17) above. The lines of equal functional value of this Gaussian form not circles, but ellipses instead. We can use ellipsoidal basis functions for training RBFNNs and RBFLNs, in which case they become, respectively, **ellipsoidal basis function NNs** (EBFNNs) and **ellipsoidal basis functional-link nets** (EBFLNs). In this case we can use the mean, median, the α-trimmed mean [Bednar and Watt, 1984], or the weighted fuzzy value as the single prototype of each class. We can multiply the covariance matrix C by a factor $\theta > 1.0$ (which multiplies the covariance matrix C^{-1} by $1/\theta$) to expand the ellipses slightly for better results ($\theta = 1.2$ is an empirical value that improves performance for ellipsoidal Gaussians).

9.4 Fuzzy Classifiers

9.4.1 Using a Fuzzy Ellipsoidal Classifier

Given the exemplar feature vectors $\{\boldsymbol{x}^{(q)}: q = 1, \ldots, Q\}$, we do the following steps to design a **fuzzy classifier** [Looney, 2001b] (see also Maturino-Lozoya et al. [2000]), which is actually a recognizer that is similar to a fuzzy neural network. It uses a fuzzy-set membership function for each class to provide the fuzzy truth that an input feature vector belongs to that class. The exemplar feature vectors must first be clustered and labeled. Here we use a **fuzzy ellipsoidal classifier,** so that this classifier is similar to the EBFNN, except that we have no weights to train.

A Fuzzy Ellipsoidal Classifier

Step 1. Obtain the Q-labeled exemplar feature vectors and separate the K classes of vectors.

Step 2. For each class, compute the mean, α-trimmed mean, median, or weighted fuzzy average, $\boldsymbol{c}^{(k)}$.

Step 3. For each Class k compute the covariance matrix C_k and its inverse C_k^{-1}.

Step 4. For each Class k define the ellipsoidal Gaussian fuzzy-set membership function:

$$g_k(\boldsymbol{x}) = \exp\{-[1/(2\theta)](\boldsymbol{x} - \boldsymbol{z}^{(k)})^t C^{-1}(\boldsymbol{x} - \boldsymbol{z}^{(k)})\} \tag{9.18}$$

where $\theta > 1.0$ is discussed above and $\boldsymbol{z}^{(k)}$ is the center of Class k.

Step 5. Implement the following algorithm for online processing of the input vector stream $\{\boldsymbol{x}\}$

```
Repeat                          //Iterate over input vectors
    read x                      //Read next feature vector to recognize
    for k = 1 to K do           //For each Class k compute the
        f[k] = g_k(x);          //fuzzy membership function on x
```

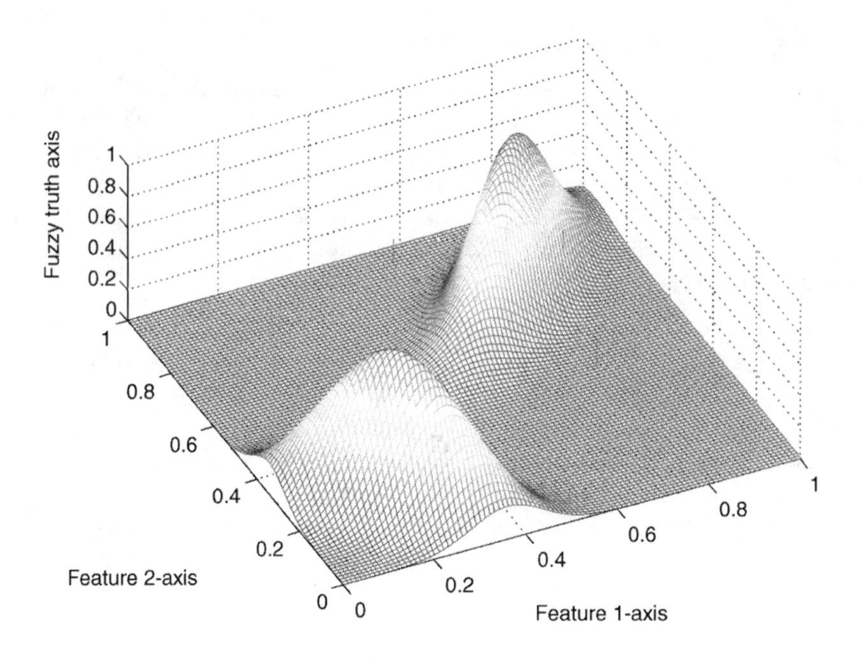

FIGURE 9.10 Ellipsoidal Gaussians covering two classes.

```
        fmax = 0.0;                    //Find the maximum fuzzy membership value
        for k = 1 to K do              //over all K classes and record
            if (f[k] > fmax) then
                fmax = f[k];           //the maximum value and the
                kmax = k;              //index for the maximum value
        output(fmax, kmax);            //Output the index kmax of winner
    until process is stopped;
```

Figure 9.10 shows two ellipsoidal Gaussians that cover two classes in the plane. These classes are well separated on synthetic data for the purpose of illustration, but classes may be closer together on real-world data. The maximum value *f[kmax]* of the fuzzy-set membership functions determines the winner, which is the Class *kmax* that $\boldsymbol{x}$ is recognized as belonging to.

9.4.2 Computing the Covariance Matrix

The covariance matrix must be computed for each class. We do this with the following algorithm, where $c[q] = k$ means that feature vector $\boldsymbol{x}^{(q)}$ belongs to the kth class, $\{cv[i][j]: 1 \le i, j \le N\}$ is the computed covariance matrix, and $c[n][k]$ is the nth component of the center for the kth class.

The Covariance Matrix Algorithm

```
Step 1: for k = 1 to K do                          //For each Class k
            for n = 1 to N do                      //and each component
                read x[n][k] for Class k;          //input class exemplars
            for n = 1 to N do                      //Similarly for the prototypes
                read prototype vector z[n][k] for Class k;
```

```
Step 2: for k = 1 to K do                          //For each class, process to get C
     for n1 = 1 to N do                             //For each first component
          for n2 = n1 to N do                       //and second component
               sum = 0.0;                           //initialize sum, check each feature
               for q = 1 to Q do                    //vector to determine if it belongs to Class k
                    if (c[q] = = k) then            //and if so, process it in covariance value
                         sum = sum + (x[n1][q] – z[n1][k])*(x[n2][q] – z[n2][k]);
               if (count [k] > 1) then              //Average if more than 1 vector
                         cv[n1][n2][k] = sum/(count[k] – 1);
               else                                 //else covariance is 0
                         cv[n1][n2][k] = 0.0;
               if (n1! = n2) then                   //If not diagonal element then
                         cv[n2][n1] = cv[n1][n2];   //fill in matrix by symmetry
```

9.4.3 Computing the Inverse Covariance Matrix

To compute the inverse C^{-1} of the covariance matrix C of the kth class, we form an extended matrix by putting C and the $N \times N$ identity matrix I together in $[C \mid I]$, which is $N \times (2N)$ in size. We first do the upper triangularization on C to zero out all entries in C below the diagonal and then perform lower triangularization to zero out everything above the diagonal of C. Then, we normalize the diagonal elements so that the part that was C becomes the identity matrix I. For every operation that we do to C, we do to I also, so that by transforming C into I we also transform I into C^{-1} at the same time; that is, $[C \mid I]$ is transformed to $[I \mid C^{-1}]$. In the algorithm below, we denote the extended matrix $[C \mid I]$ for the kth class by $cx[n][col][k]$, where n is the row and col is the column.

The Inverse Covariance Matrix Algorithm

Step 1: Upper triangularization of extended matrix

```
     for k = 1 to K do                              //For each kth class
          for col = 0 to N do                       //Eliminate below each column
               for n1 = col to N – 1 do             //row entries (make zero)
                    if (cx[n1 +1][col][k] ! = 0) then        //If element not 0 pivot on it,
                         t = cx[n1 +1][col][k]/cx[col][col][k];  //get elimination factor
                         for n2 = col to 2N do      //and apply to column elements
                              cx[n1 + 1][n2][k] = cx[n1 + 1][n2[k] – t*cx[col][n2][k]
```

Step 2: Lower triangularization of extended matrix

```
     for k = 1 to K do                              //For each class
          for n = 1 to N do                         //go through extended matrix
               col = N – n;                         //and get pivot element
```

```
    for m = 0 to col do                          //for elimination
      n_1 = col – m – 1;
      if (cx[n_1][col][k] ! = 0.0 then           //If element not 0, then use as pivot
          t = cx[n_1][col][k]/cx[col][col][k];   //Compute elimination factor
          for n_2 = col to 2N do                 //and apply to column elements
            cx[n_1][n_2][k] = cx[n_1][n_2][k] – t*cx[col][b_2][k];
```

Step 3: Normalize diagonal elements in extended matrix

```
for k = 1 to K do                                //For each class
  for n_1 = 1 to N do                            //go through rows and divide
    t = 1.0/cx[n_1][n_1][k];                     //by diagonal element
    for n_2 = n_1 to 2N do                       //across all columns from diagonal
      cx[n_1][n_2][k] = t*cx[n_1][n_2][k];       //element to the right
```

Step 4: Retrieve inverse covariance matrix from extended matrix

```
for k = 1 to K do                                //For each class retrieve inverse
  for n_1 = 1 to N do                            //from [I | C^-1], so get all rows
    for n_2 = N + 1 to 2N do                     //and all columns on right
      ci[n_1][n_2][k] = cx[n_1][n_2 +N][k];      //This N × N matrix is C^-1
```

We can check C to see if it is invertible (nonsingular) before we try to invert it. C is symmetrical with nonnegative diagonal elements. A practical check can be made on the upper triangularized matrix $C^{[T]}$ by testing the diagonal elements to ensure that none are approximately 0 (recall that the determinant of a triangular matrix is the product of the diagonal elements). If all exemplar feature vectors in a class have the same value for a given feature, then a diagonal element (a component variance) will be zero and C (and its upper triangularization) will be singular. This should never occur on real world data where the features have been selected for their separation properties.

9.5 An Application: Edge Recognition in Images

9.5.1 Image Edge Detection

Edges are defined as locations in an image where there is a significant variation in the gray level or intensity of color of pixels in some direction across a small number of pixels [Efford, 2000]. They are one of the most important visual clues for interpreting images [Gose et al., 1996]. The process of **edge detection** reduces an image to show only its edges, which appear as the outlines of objects within the image that can be used in subsequent image analysis operations for feature detection and object recognition. Although there are many different methods for edge detection, such as Sobel 3×3 filtering, Prewit 3×3 filtering, Laplacian of Gaussian filtering, moment-based operators, the Shen and Castan operator, and the Canny and Deriche operator, some common problems of these methods are a large volume of computation and too much sensitivity to noise and anisotropy.

Russo [1992, 1993] and Russo and Ramponi [1992] designed fuzzy rules for edge detection. Such rules can smooth while sharpening edges, but this method requires a rather large rule set [Looney, 2000]. Here we describe a special fuzzy classifier for edge detection that does not require training [Liang and Looney, 2001]. Our fuzzy classifier uses the two classes of *edge* and *background*. Its advantages are easy modeling, efficient computation, low sensitivity to noise, and isotropy (it detects edges in all directions).

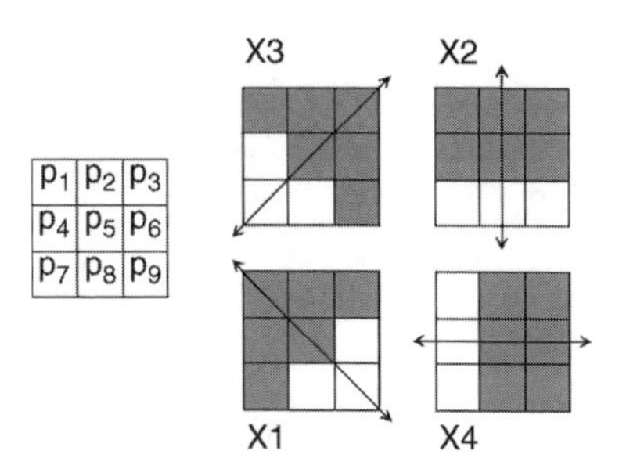

FIGURE 9.11 The edge classes.

9.5.2 Fuzzy Edge Detection

Figure 9.11 shows the 3 × 3 neighborhood of the center pixel p_5 and also the four directions of gray-level change. The magnitudes of the gray-level changes in these directions are defined in the horizontal, vertical, and two diagonal directions and are designated by, respectively, X1, X2, X3, X4. The four features are calculated by

$$X1 = |p_1 - p_5| + |p_9 - p_5|, \quad X2 = |p_2 - p_5| + |p_8 - p_5| \qquad (9.19a,b)$$

$$X3 = |p_3 - p_5| + |p_7 - p_5|, \quad X4 = |p_4 - p_5| + |p_6 - p_5| \qquad (9.19c,d)$$

For each image pixel (not on the outer boundary of the image), we compose the four-dimensional feature vector $\boldsymbol{x} = (X1, X2, X3, X4)$ that contains the gray-level difference magnitudes in the four directions of its 3 × 3 neighborhoods. X1 is diagonal (upper left to lower right); X2 is vertical (downward along the center column); X3 is diagonal (upper right to lower left); and X4 is horizontal (rightward along the center row).

9.5.3 Pixel Classes and Their Feature Vectors

Four classes of edges and a fifth background class are differentiated in our **competitive fuzzy classifier**. Four edge situations are used for each class, which are those shown in Figure 9.11, their opposites determined by 180% rotation and those with the dark and light pixels reversed. The five prototypical feature vectors are designated by $\boldsymbol{x}_0, \ldots, \boldsymbol{x}_4$ and each represents four neighborhood situations.

We use the linguistic variable *Low* to substitute for 0, and *High* to substitute for 255 in the directional difference magnitudes (see Figure 9.11). Letting L denote *Low* and H denote *High*, the feature vectors become:

Background pixel, Class 0: $\boldsymbol{x}_0 = \{L, L, L, L\}$

Edge pixel, Class 1: $\boldsymbol{x}_1 = \{L, H, H, H\}$ Edge pixel, Class 2: $\boldsymbol{x}_2 = \{H, L, H, H\}$

Edge pixel, Class 3: $\boldsymbol{x}_3 = \{H, H, L, H\}$ Edge pixel, Class 4: $\boldsymbol{x}_4 = \{H, H, H, L\}$

In practice, the values of *Low* and *High* can be defined by the user for each particular image to achieve a desirable result, for example, $L = 5$ and $H = 20$ gray levels. With these values defined, every interior pixel of the input image can be classified under one of the above classes by its feature vector of directional difference magnitudes on its 3 × 3 neighborhood.

9.5.4 The Fuzzy Classifier Architecture

Our *competitive fuzzy classifier* is made up of our basic fuzzy classifier and some competitive rules that allow a competition between pixels for designation as an edge (see Chen and Chi [1998] and Chung and Lee [1994] for other competitions). Our fuzzy classifier classifies each pixel in an image as a white background pixel, or else one of the four classes of edge pixels shown above, by providing fuzzy truths

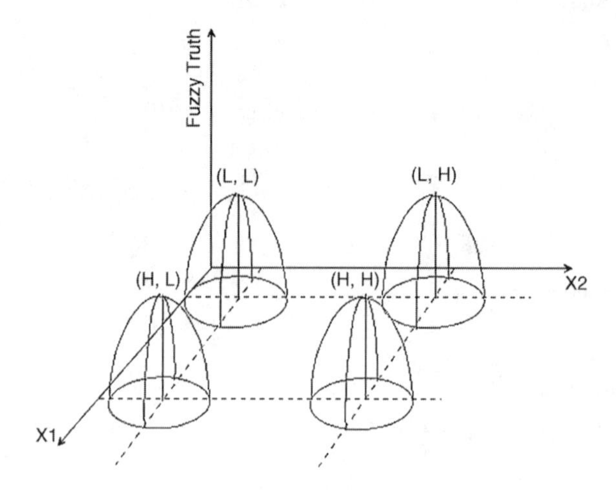

FIGURE 9.12 A two-dimensional depiction of the fuzzy edge classifier.

of those classes based on the feature vector of that pixel. Then our competitive rules [Liang and Looney, 2001] are applied according to the output of the original fuzzy classifier. Only the pixels that are classified as edge (directional) pixels and that subsequently win in the local competition are mapped to black edges in the new output image. This creates a black line drawing on a white background. If there are adjacent edge pixels in the same direction, only the one with the greatest fuzzy value is the winner.

On the four-dimensional feature space, we define the **fuzzy membership functions** for the five classes with *extended Epanechnikov* [Looney, 2001b] functions defined by:

$$\text{Background Class 0: } f(\boldsymbol{x}) = \max\{0, 1 - \|\boldsymbol{x} - \boldsymbol{x}_0\|^2/\beta^2\} \tag{9.20a}$$

$$\text{Edge Class 1: } f(\boldsymbol{x}) = \max\{0, 1 - \|\boldsymbol{x} - \boldsymbol{x}_1\|^2/\beta^2\} \tag{9.20b}$$

$$\text{Edge Class 2: } f(\boldsymbol{x}) = \max\{0, 1 - \|\boldsymbol{x} - \boldsymbol{x}_2\|^2/\beta^2\} \tag{9.20c}$$

$$\text{Edge Class 3: } f(\boldsymbol{x}) = \max\{0, 1 - \|\boldsymbol{x} - \boldsymbol{x}_3\|^2/\beta^2\} \tag{9.20d}$$

$$\text{Edge Class 4: } f(\boldsymbol{x}) = \max\{0, 1 - \|\boldsymbol{x} - \boldsymbol{x}_4\|^2/\beta^2\} \tag{9.20e}$$

where $\boldsymbol{x}$ is any input feature vector for a pixel and $\boldsymbol{x}_0, \ldots, \boldsymbol{x}_4$ were defined above. Thus, the quality of the edge detection, as measured by the fuzzy truth of its memberships in the fuzzy classes, depends on the parameters L, H, and β, and thus on the particular image.

Figure 9.12 provides a two-dimensional portrayal of two features vs. fuzzy truths for easy visualization. The upside-down cups in the figure are the extended Epanechnikov functions [Looney, 2001b] that are shown here with small diameters for clarity (in applications they overlap). Each input feature vector $\boldsymbol{x}$ falls into one or more of these fuzzy-set membership functions. Given any input feature vector $\boldsymbol{x}$, the maximum fuzzy truth-value of the three shown fuzzy-set membership functions evaluated on X determines the class of the pixel.

9.5.5 Competitive Edge Rules

Before an edge pixel is changed to either white or black in the output image, a competition with its neighbor edge pixels is conducted. For the edge pixel neighbors that belong to the same class of edge, only the one with the largest feature (difference magnitude in the direction associated with that class) is considered an edge and turned to black. Thus, the edges in the output image will be thin instead of thick. Rules for the competition are given below.

IF (white class wins) THEN (change to white).

IF (edge Class 1 wins) THEN compete X3 with neighbor pixels in direction 3.

IF (win) THEN (change to black) ELSE (change to white).

IF (edge Class 2 wins) THEN compete X4 with neighbor pixels in direction 4.

IF (win) THEN (change to black) ELSE (change to white).

IF (edge Class 3 wins) THEN compete X1 with neighbor pixels in direction 1.

IF (win) THEN (change to black) ELSE (change to white).

IF (edge Class 4 wins) THEN compete X2 with neighbor pixels in direction 2.

IF (win) THEN (change to black) ELSE (change to white).

9.5.6 The Algorithm

Figure 9.13 is a flowchart that shows the algorithm in detail for operating on PGM files of grayscale images, which could also be intensity data from color images. The third block from the top processes each pixel not on the image boundary to compute its fuzzy truth of membership in each of the five

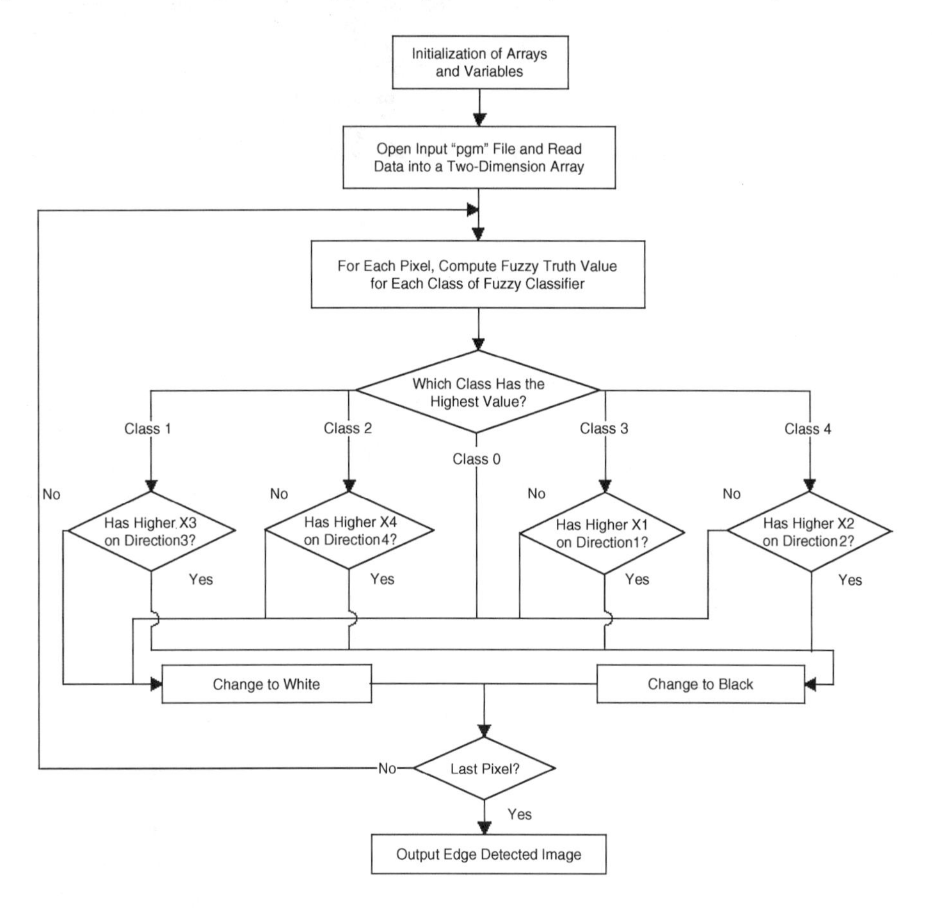

FIGURE 9.13 The algorithm flowchart.

classes. The winning class determines the rule to be used in the competitive comparisons to determine whether a pixel is converted to white (background) or black (edge).

9.5.7 Edge Detection Results

All of our results are obtained by using a 3×3 neighborhood of the center pixel and the fuzzy-set membership functions and rules established above. The threshold parameters (L and H) are adjusted to achieve good results.

Figure 9.14 shows the original image *building.pgm*. The results of using the public-domain software *XView* that is included with the *Linux* operating system on the image are shown in Figure 9.15. The results of using the Canny edge detector with respective low and high edge sensitivity are shown in Figures 9.16 and 9.17. The competitive fuzzy-edge detection results for the respective high and low sensitivity to edges are shown in Figures 9.18 and 9.19, and they were obtained in less than 1/20 of the computing time required for the Canny edge detector.

FIGURE 9.14 The original *building.pgm* image.

FIGURE 9.15 The *XView* edges.

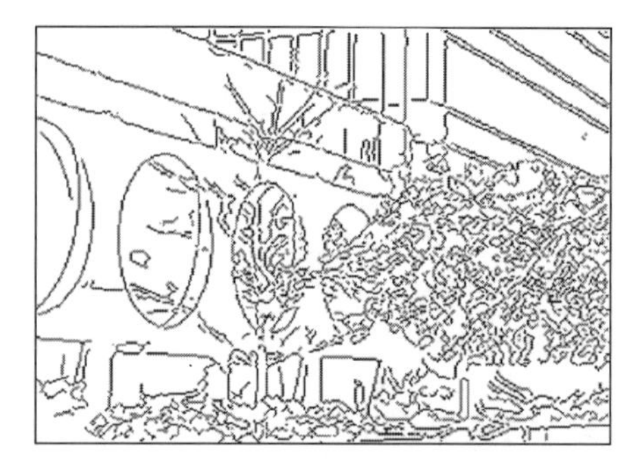

FIGURE 9.16 The Canny edges, low sensitivity.

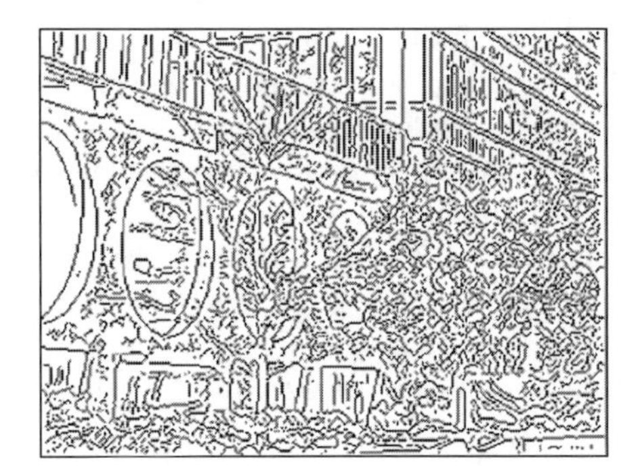

FIGURE 9.17 The Canny edges, high sensitivity.

FIGURE 9.18 Competitive fuzzy edges, higher sensitivity.

FIGURE 9.19 Competitive fuzzy edges, lower sensitivity.

9.6 Summary

We have discussed classification and recognition, although the literature at times uses them interchangeably. For classification, the problems with the k-means algorithm for clustering were listed and two improved algorithms were provided, of which the second was our weighted fuzzy k-means algorithm. For recognition, we have provided the following powerful algorithms: (1) probabilistic neural networks; (2) fuzzy neural networks; (3) radial basis function neural networks; (4) radial basis functional link nets; (5) ellipsoidal-basis function neural networks; (6) ellipsoidal radial basis functional link nets; and (7) fuzzy ellipsoidal classifiers.

For the networks that use Gaussian radial basis functions, we have suggested a thinning algorithm that can be used to reduce the number of mixed Gaussians to provide efficiency and also to alleviate the build-up of error from extraneous noise. By choosing a combination of the clustering and recognition algorithms presented here, one can attack most classification and pattern recognition problems successfully. However, it is possible to modify any of these for special purposes.

As an application, we have given an example of fuzzy edge detection and an improvement called *competitive fuzzy edge recognition* for yielding thinner and cleaner lines. This latter method can be combined with some preprocessing, such as frequency filtering, to remove the highest frequencies of noise or despeckling. We have put the neighborhood directional difference magnitudes into the competitive fuzzy classifier to classify a pixel as being of type *edge* or *background* and then thinned the edges. The result is a line drawing of thin black lines on a white background, whereas the usual methods yield thick lines, or, in the case of the Canny edge detector, noise is converted to a plethora of mostly useless edge traces.

Defining Terms

α-trimmed mean: A representative value for a set of values determined by omitting the α smallest and the α largest values and averaging the remainder.

artificial intelligence: Any machine process that emulates human reasoning or decision making by means of deduction/abduction, computational methods (statistical/mathematical), fuzzy methods, genetic algorithms, or other means.

center: The typical (prototypical) representative feature vector for a class of feature vectors that represent objects in a population of such objects.

classes: Groups of objects such that those within a group are similar according to some criteria and those in different groups are dissimilar according to the same criteria.

classification: A set of classes determined for a population; a research area of study; a process of determining a set of classes (see *classify*).

classify: The process of partitioning the objects of a population into groups (*classes*) that are similar within each group and dissimilar between groups.

clustering: The process of partitioning a population of objects into groups to form classes, a type of classification.

clustering validity measure: A measure of a clustering (classification) to determine how good the clustering is, used in a relative sense to compare multiple clusterings.

clusters: Groups of objects that form a partition of a population of objects.

compact: The property of a class of feature vectors having a small mean square error (variance) relative to the distances to the centers of other classes.

competitive fuzzy classifier: A fuzzy classifier that implements a competition between objects that are classified by a fuzzy classifier and there is a single winner of the competition for some purpose (such as determining which of neighboring edge pixels is the edge).

edge detection: A process of finding (detecting or recognizing) the pixels in an image that are edges.

ellipsoidal basis function: A multivariate (ellipsoidal) Gaussian function centered on a vector.

ellipsoidal basis functional link net: A radial basis functional link net where one or more radial basis functions are replaced by ellipsoidal basis functions.

ellipsoidal basis function neural network: A neural network similar to a radial basis function neural network except that the ellipsoidal basis functions replace one or more radial basis functions.

exemplar: A feature vector that is labeled with a particular class codeword (it contains extra components for the label).

feature: A property of a population of objects that can be observed (measured, detected) as a real number value for any selected object.

feature vector: A vector of observed feature values for a fixed set of features for a given population of objects.

fuzzy classifier: A system such that a feature vector is input and the outputs are the fuzzy truths of the memberships of the feature vector in the various classes, a fuzzy recognizer.

fuzzy ellipsoidal classifier: Fuzzy classifier where one or more of the fuzzy-set membership functions are ellipsoidal.

fuzzy neural network: A neural network with *N* input nodes corresponding to the components of a feature vector such that either the inputs are feature-wise fuzzy truths, or such that the final outputs represent the respective fuzzy truths of memberships in the various classes (or in some cases, both fuzzy truth inputs and outputs).

fuzzy-set membership function: A function on a real or vector domain whose values are between 0 and 1 and whose value represents the fuzzy truth of a property of the argument.

hidden layer: A set of nodes in a network that neither receive inputs from the outside nor send output values to the outside, nodes that connect only to other nodes in the network.

identification: The process of recognizing a unique individual from a population of objects.

input layer: A set of nodes in a network each of which receives the input of a component of some data structure (such as a vector) from a source external to the network.

***k*-means:** An algorithm that iterates: (1) the assignment of feature vectors to class centers to form new classes and (2) the determination of new class centers by averaging the vectors in each class, where the seeds (initial centers) are randomly selected feature vectors.

label: An integer or code word that represents the class to which the feature vector for an object belongs, assigned by humans or assigned by a clustering process.

maximum *a posteriori*: Also designated as MAP, the maximum of the probabilities of membership of an input feature vector in the various classes that is computed from the input feature vector and the prior probability distributions (in this case, mixed Gaussians).

output layer: A set of nodes in a network that receive input values from nodes within the network and send output values to destinations or sinks outside of the network.

pattern recognition: The process of recognizing an object as belonging to a particular class of the population by working on a set of measurement values for the object features (attributes).

probabilistic neural network: A network that accepts feature vectors as inputs and outputs the probabilities of belonging to particular classes.

prototype: A center or other representative vector for a class or cluster that represents the class.

radial basis function: A circular Gaussian centered on a vector; that is, a function whose value is the same for all vectors of equal distance to the center.

radial basis functional link net: A radial basis function neural network that has extra lines from the input nodes to the output nodes and an extra set of weights on these lines (thus a linear part is added to the model for efficiency and accuracy).

radial basis function neural network: A neural network with an input, a hidden, and an output layer of nodes where: (1) each input node accepts a component of a feature vector; (2) each hidden node receives all input components and outputs the value of the Gaussian radial basis function centered on its associated exemplar feature vector; and (3) the output nodes sum the weighted radial basis function values from the hidden nodes (the training on exemplars adjusts the weights by steepest descent until each input feature vector yields a correct approximate output target).

recognition: The same process as pattern recognition.

recognizer: An online process that performs recognition of feature vectors from a particular population on which it has been trained.

seed: An initially given center (prototype) on which to form a class as the first step in a classification (clustering) process.

self-organization: A system that interacts with an environment and adjusts itself in some fashion to perform one or more tasks more optimally relative to the interaction; a form of machine learning.

supervised learning: A process whereby a system is trained on examples to perform a task, for example, recognition.

target: The desired output number or vector that the actual computed outputs are to approach (approximately), usually by iterations.

training: A process of presenting examples and the correct responses to a system until the system parameters are adjusted to put out the correct responses to inputs.

unsupervised learning: A process whereby a local system interacts with an external system and processes the response data to adjust its parameters to more optimally interact with the system for some purpose; a form of machine learning (see *self-organization*).

weighted fuzzy average: The weighted average of a set of values obtained by an iterative process of taking weighted averages of N values, where the weights are initially $1/N$ and are computed as the values of a radial fuzzy set membership function centered on the current average (done component-wise for vectors), designated as *WFA*.

well separated: A property of a set of clusters (classes) whereby the centers of the clusters are located relatively far apart when compared to their variances.

Xie-Bene clustering validity measure: A measure computed for a particular clustering of a set of feature vectors that takes the sum of the cluster variances and divides by the minimum distance between cluster centers.

References

Abe, S. and Thawonmas, R., A fuzzy classifier with ellipsoidal regions, *IEEE Trans. Fuzzy Sys.*, 6(2), 358–368, 1997.

Anagnostopoulos, C., Anagnostopoulos, J., Vergados, D., Kayafas, E., Loumos, V., and Stassinopoulos, G., A neural network and fuzzy logic system for face detection on RGB images, *Proc. ISCA Int. Conf. Computers Their Applications*, 2001, pp. 233–236.

Bednar, J. B. and Watt, T. L., Alpha-trimmed means and their relation to median filters, *IEEE Trans. Acoustics*, Speech and Signal Processing, 32(1), 145–153, 1984.

Bezdek, J. C., Fuzzy Mathematics in Pattern Classification, Ph.D. thesis, Center for Applied Mathematics, Cornell University, Ithaca, NY, 1973.

Chen, K. and Chi, H., A method of combining multiple probabilistic classifiers through soft competition on different feature sets, *Neurocomputing*, 20, 227–252, 1998.

Chen, M. S. and Wang, S. W., Fuzzy clustering analysis for optimizing fuzzy membership functions, *Fuzzy Sets Systems*, 103, 239–254, 1999.

Chiu, S. L., Fuzzy model identification based on cluster estimation, *J. Intelligent Fuzzy Sys.*, 2, 3, 1994.

Chung, F. L. and Lee, T., Fuzzy competive learning, *Neural Networks*, 7(3), 539–551, 1994.

Dubes, R. C. and Jain, A. K., *Algorithms for Clustering Data*, Prentice-Hall, Englewood Cliffs, NJ, 1998.

Efford, N., *Digital Image Processing*, Addison-Wesley, Reading, MA, 2000.

Forgy, E., Cluster analysis of multivariate data: efficiency versus interpretability of classifications, *Biometrics*, 21, 768–776, 1965.

Gose, E., Johnsonbaug, R., and Jost, S., *Pattern Recognition and Image Analysis*, Prentice-Hall, Upper Saddle River, NJ, 1996.

Kaufman, L. and Rousseeuw, P., *Finding Groups in Data: an Introduction to Cluster Analysis*, John Wiley & Sons, NY, 1990.

Liang, L. and Looney, C., Competitive Fuzzy Edge Detection, Technical Report, Computer Science Dept., University of Nevada, Reno, 2001.

Liang, L., Basallo, E., and Looney, C., Image edge detection with fuzzy classifier, *Proc. ISCA 14th Int. Conf. CAINE-01,* Las Vegas, 2001, pp. 279–284.

Looney, C. G., Interactive clustering and merging with a new fuzzy expected value, *Pattern Recognition Lett.*, 35, 187–197, 2002.

Looney, C. G., Radial basis functional link nets and fuzzy reasoning, *Neurocomputing,* 2001a.

Looney, C. G., A Fuzzy Classifier Network with Ellipsoidal Epanechnikovs, Technical Report, Computer Science Dept., University of Nevada, Reno, 2001b.

Looney, C. G., Nonlinear rule-based convolution for refocusing, *Real-Time Imaging*, 6, 29–37, 2000.

Looney, C., *Pattern Recognition Using Neural Networks*, Oxford University Press, New York, 1997.

MacQueen, J. B., Some methods for classification and analysis of multivariate observations, *Proc. 5th Berkeley Symp. Probability Statistics*, University of California Press, Berkeley, 1967, pp. 281–297.

Maturino-Lozoya, H., Munoz-Rodriguez, D., Jaimes-Romero, F., and Tawfik, H., Handoff algorithms based on fuzzy classifiers, *IEEE Trans. Vehicular Technol.*, 49(6), 2286–2294, 2000.

Pao, Y. H., Park, G. H., and Sobajic, D. J. Learning and generalization characteristics of the random vector functional link net, *Neurocomputing*, 6, 163–180, 1994.

Pena, J. M., Lozano, J. A., and Larranago, P., An empirical comparison of four initialization methods for the *k*-means algorithm, *Pattern Recognition Lett.*, 20, 1027–1040, 1999.

Russo, F., A new class of fuzzy operators for image processing, *IEEE Int. Conf. Neural Networks*, 1993, pp. 815–820.

Russo, F., A user-friendly research tool for image processing with fuzzy rules, *Proc. First IEEE Int. Conf. Fuzzy Sys.*, San Diego, 1992, pp. 561–568.

Russo, F. and Ramponi, G., Fuzzy operator for sharpening of noisy images, *IEEE Electron. Lett.*, 28, 1715–1717, 1992.

Schneider, M. and Craig, M., On the use of fuzzy sets in histogram equalization, *Fuzzy Sets Systems,* 45, 271–278, 1992.

Selim, S. Z. and Ismail, M. A., *k*-means type algorithms: a generalized convergence theorem and characterization of local optimality, *IEEE Trans. Pattern Analysis Machine Intelligence*, 6, 81–87, 1984.

Snarey, M., Terrett, N. K., Willet, P., and Wilton, D. J., Comparison of algorithms for dissimilarity-based compound selection, *J. Mol. Graphics Modeling*, 15, 3782–3785, 1997.

Specht, D. F., Probabilistic neural networks for classification, mapping or associative *memory, Proc. IEEE Int. Conf. Neural Networks*, 1, 525–532, 1988.

Specht, D. F., Probabilistic neural networks, *Neural Networks*, 1(3), 109–118, 1990.

Wedding, D. K., II and Cios, K. J., Certainty factors versus Parzen windows as reliability measures in RBF networks, *Neurocomputing*, 19, 151–165, 1998.

Xie, X. L. and Beni, G., A validity measure for fuzzy clustering, *IEEE Trans. Pattern Analysis Machine Intelligence*, 13(8), 841–847, 1991.

Yager, R. R. and D. P. Filev, Approximate clustering via the mountain method, *IEEE Trans. Sys., Man Cybernetics,* 24, 1279–1284, 1991.

Further Information

For information on statistical pattern recognition, see Duda, R., Hart, P., and Stork, D., *Pattern Classification*, Second Edition, Wiley-Interscience, New York, 2001.

Bayesian (belief) networks can be applied to pattern recognition. For these networks, see Jensen, F., *Bayesian Networks and Decision Graphs*, Springer, New York, 2001.

For PR fundamentals (statistical, syntatic, graphical, neural network and other), see Looney, C. G., *Pattern Recognition Using Neural Networks,* Oxford University Press, New York, 1997; Chen, C. H., Pau, L. F., and Wang, P. S., Eds., *Handbook of Pattern Recognition & Computer Vision*, World Scientific, Singapore, 1999.

Theory and applications can be found in the following journals: *Pattern Recognition* (Elsevier), *Pattern Recognition Letters* (Elsevier), and *Transactions on Pattern Analysis and Machine Intelligence* (IEEE).

10

Real-Time Feature Extraction

Farrokh Janabi-Sharifi
Ryerson University
Toronto, Ontario, Canada

10.1 Introduction

The development of robust vision is still a major issue for real-time industrial and service applications. Industrial vision sensors are generally subject to considerable noise and produce a large amount of data. Therefore, a selected number of **features** are used to measure some properties of the task. A feature is any scene property that can be mapped onto the image plane such as corner position, edge length, and centroid. Feature extraction has been historically associated with pattern recognition and refers to the process of mapping to reduce the dimensionality of the patterns. Feature extraction also improves the generalization ability and computational requirements of pattern classification. Therefore, feature extraction has received significant attention during the last two decades. However, the focus of past approaches was on off-line pattern recognition, and little effort has been spent on real-time applications of feature extraction. Considerable changes of spatial–temporal conditions in real-time applications make the task of robust feature extraction quite challenging. The main requirements of robust visual measurements, i.e., speed, accuracy, and reliability, depend on the image processing and, specifically, the feature extraction method used. To achieve a robust and effective feature-extraction process, the methods used could be tailored for the application under study. In this chapter, we will introduce some generalities on real-time feature extraction but the focus of the chapter will be on **visual servoing** application.

Sophisticated techniques exist to properly process image data and to remove noise. However, these techniques are often computationally too expensive to meet real-time calculation requirements. For example, a typical visual servoing system must operate with the sample rates of 50 to 100 Hz, indicating a calculation rate of less than 10 to 20 ms for image processing [Wilson et al., 2000]. Many previous approaches to visual servoing have assumed over-simplified environments for ease of extraction, e.g., by using artificial targets [Feddema et al., 1991]. In less structured environments, vision systems usually use sharp contrast markings in the image such as corners, holes, and circles. The focus and challenge of

0-8493-1162-4/03/$0.00+$1.50

many real-time vision applications such as visual servoing are to use simple, computationally feasible, yet robust feature-extraction techniques to retrieve the necessary information. The common feature-extraction task in visual servoing is to determine the image location of features such as holes and corners. This is due to the availability of these features in many industrial parts and because of the ease and robustness of their extraction. Window-based methods have been used in visual servoing systems to provide computational simplicity, reduced requirements for special image-processing hardware, and ease of reconfiguration for different applications. A few reviews of the feature-extraction methods based on specialized hardware using temporal and geometric constraints are also available in the literature (e.g., O. Faugeras' book *Three-Dimensional Computer Vision,* MIT Press, 1993).

10.2 Background

Feature extraction and measurement can be formulated as a mapping $\boldsymbol{F}$ from d-dimensional image space (input space) to an e-dimensional image feature parameter space (output), i.e., $\boldsymbol{F}: \Re^d \rightarrow \Re^e$, where usually $d \geq e$, such that some criterion, **C**, is optimized. In digital images, input space usually consists of image intensity, $I(x, y)$, and coordinates of the pixel of image (x, y). In many applications, including visual servoing, output space will be the image plane coordinates of the feature $[x^i, y^i]^T$. A large number of feature-extraction approaches are available in the pattern-recognition literature. The approaches differ from each other in the characteristics of $\boldsymbol{F}$ and **C** being used. Choice of a proper feature-extraction technique depends on the available information, *a priori* knowledge about the image input space, and task requirements. In visual servoing, some information about the object and environment is usually available. This information could be obtained in advance, e.g., from the CAD models of the system.

The most important task requirement for feature extraction in visual servoing is robustness, which implies reliable, accurate, and fast measurements of the image feature parameters despite relatively significant noises and image changes due to the relative motion between the vision sensor and object. In order to meet the robustness requirement, feature extraction methods for visual servoing have relied on window-based techniques with simple extraction methods such as binary centroid computation, correlation matching, or 1D gradient [Arbter et. al., 2000]. Further robustness will be provided by the integration of feature selection and planning methods to enable dynamic feature switching.

This chapter will describe window-based techniques for visual servoing. These methods allow feature extraction of multiple features at frame rate without requiring any special hardware, unless otherwise specified. In many visual servoing systems, a Kalman filter (KF) is used for partial or full relative pose estimation between the object and the camera. For pose estimation robustness, feature-extraction methods must provide an unbiased estimate of the location of the true feature point description in the image plane such that the estimate error can be approximated as Gaussian noise with a measurable covariance. Two feature types that meet this requirement are hole and corner features [Smith, 1989]. These features could also be easily extracted and are popular feature types in visual servoing. Therefore, the focus of this chapter will be on the extraction of hole and corner features.

The important steps involved in window-based feature extraction are: (1) pixel classification, (2) window placement, (3) pixel representation, (4) feature description (or feature measurement and estimation), and (5) window tracking and adjustment. In the next sections each of the above steps will be described. Next, the effects of parameters influencing the robustness of feature extraction will be discussed with respect to error analysis.

10.3 Pixel Classification

Pixels are classified into spatial sets according to their scalar or vector properties such as intensity, range, or speed. For instance, optical flow techniques could be used to classify moving pixels from nonmoving ones. However, the intensity is the simplest and fastest classification characteristic. For digital images, dynamic pixel numbers could be used. A popular example is binary classification,

$$I(x_i, y_j) = \begin{cases} 1 & \text{if } f(x_i, y_j) \le I_T \\ 0 & \text{otherwise} \end{cases} \tag{10.1}$$

where $I(x_i,y_j)$ is the segmented intensity value associated with pixel (i,j) with the coordinates (x_i,y_j), I_T is a threshold, and $f(x_i,y_j)$ is the pixel intensity. Many visual servoing systems use this simple binary segmentation method. If 100% intensity amplitude corresponds to cases in which a pixel is completely covered by the image, 50% peak amplitude should ideally be the optimal threshold level. In practice, however, a different level might be selected to compensate for poor radiometric conditions. The selection of threshold value affects the accuracy and, hence, robustness of image processing. Adaptive and optimal thresholding methods have also been proposed.

In adaptive thresholding, the threshold value varies spatially or temporally as a function of local image characteristics. These characteristics could include information about the object and prediction of the local (windowed) feature characteristics. That is:

$$I_T = I_T(t_k, f, f_w, \hat{C}_\varepsilon) \tag{10.2}$$

where t_k is the sampling time; f and f_w denote the whole and windowed images, respectively; and $\hat{C}_\varepsilon$ is the estimated feature characteristic. An example of a feature characteristic is the predicted hole area in visual servoing. If the measured area exceeds the prediction bound, a new threshold value could be selected. In adaptive thresholding, the threshold could be determined independently in each window. If the threshold cannot be determined in some windows, it can be interpolated from thresholds in neighboring windows. Each window is then processed with respect to its own threshold. Some suboptimal thresholding methods have also been proposed, for instance, at step t_k, $(I_T)_B^k$ and $(I_T)_O^k$, as the mean background and object gray level, respectively, are calculated. The segmentation into background and objects at step t_k is defined by the threshold value in the previous step t_{k-1}, i.e.,

$$(I_T)_B^k = \frac{\sum_{(i,j)\in B} f(x_i, y_j)}{N_B} \qquad (I_T)_O^k = \frac{\sum_{(i,j)\in O} f(x_i, y_j)}{N_O} \tag{10.3}$$

where N_B and N_O are the number of background and object features, respectively. The value of threshold will be updated as:

$$(I_T)^{k+1} = \frac{(I_T)_B^k + (I_T)_O^k}{2} \tag{10.4}$$

and $(I_T)^{k+1}$ provides an updated background-object segmentation. If $(I_T)^{k+1} = (I_T)^k$, the algorithm will stop. However, both of the above approaches are time consuming and therefore might not fully meet the speed requirement of robust visual servoing. Despite the proposals for optimal and adaptive thresholding [Sahoo et. al., 1988], the problem of appropriate threshold selection has not yet been resolved.

Corke [1996] suggests a threshold of at least $3\sigma_c$ above the background gray level and at least $3\sigma_c$ below the object gray level (or vice versa) to overcome the camera-noise effect and poor radiometric conditions, where σ_c^2 is the gray-level variance and is an almost linear function of pixels intensity. It is possible to obtain values of σ_c^2 for different gray levels by experiments for a particular vision system and environment. Next, the maximum value of variance could be selected and used for threshold selection. For example, Corke [1996], with noise measurements for a Pulnix camera, indicated a maximum variance of 4 gray levels^2. Certain features are also more sensitive to thresholding than others. For instance, experiments with typical corner and hole features have indicated that thresholding affects the accuracy of corner feature extraction considerably, while the effects on the accuracy of hole feature extraction were minimal [Wilson et. al., 2000].

The experiments indicated that corner feature estimations would tend to move towards (away from) the corner orientation with low (high) threshold settings.

Methods for color classification have also been proposed in the literature due to its distinction power. The main limitation of color classification is its processing time, which might not meet the real-time requirement of a machine vision application such as visual servoing. Supervised segmentation is usually applied to analyze the color distribution of the scenes using **HS** histograms and to choose distinctive colors for real-time segmentation. A typical example is given in Arbter et al. [2000] that uses color markers for visual servoing. The advantage is that no initialization for a marker position is required, and each marker is easily distinguished from others, simplifying feature- and control-switching. The disadvantages are (pre- and post-) processing time, resources required, and the need for marking the objects.

10.4 Window Placement

Windows could be placed using the initial information about the possible type, size, and location of the features in the image plane. Further tracking and adjustments will be discussed later. Many approaches to visual servoing assume a constant window size. However, a planned and adaptive windowing would be more appropriate to meet the robustness requirement of visual servoing. In this section, it is assumed that initial information about the image features is available and methodologies are provided to determine the center and size of the windows for corner and hole features. The method of Janabi-Sharifi and Wilson [1997] can be extended to other types of features and will be used further in the window-adjustment and windowing-constraint definition for feature selection and planning.

The center coordinates C_{x^i} and C_{y^i} and size of a window W_x and W_y in x and y directions for a hole feature (see Figure 10.1) can be easily determined from the image coordinates, x_k^i and y_k^i, of the perimeter pixels of the feature, where k is the perimeter pixel index. That is:

$$C_{x^i} = \frac{1}{2}(\max_k(x_k^i) + \min_k(x_k^i)), \quad C_{y^i} = \frac{1}{2}(\max_k(y_k^i) + \min_k(y_k^i)), \tag{10.5}$$

with

$$W_x = \max_k(x_k^i) - \min_k(x_k^i) + 2\delta_{W_x}, \quad W_y = \max_k(y_k^i) - \min_k(y_y^i) + 2\delta_{W_y}, \tag{10.6}$$

where δ_{W_x} and δ_{W_y} are the window boundaries' clearance in the x and y directions (Figure 10.1).

Both double-windowing and single-windowing techniques are possible for a corner feature [Janabi-Sharifi and Wilson, 1997]. In this section, only the single-windowing technique will be presented. Similar to a hole feature, the window center will be located on the feature corner estimation. For window sizing

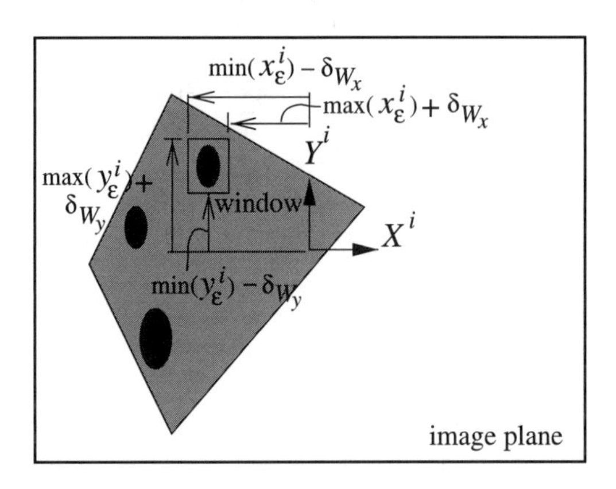

FIGURE 10.1 Windowing of a hole feature.

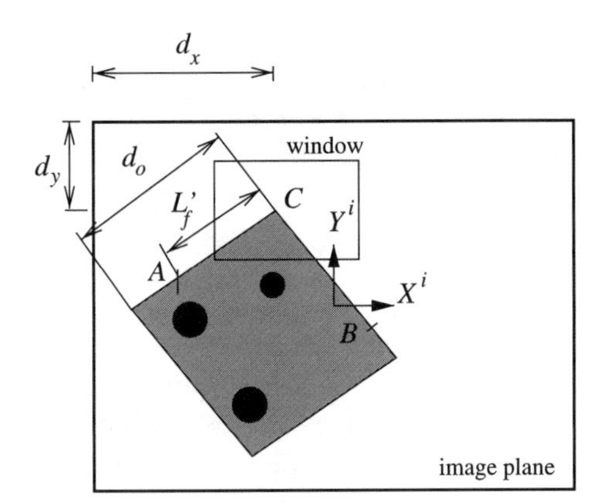

FIGURE 10.2 Windowing of a corner feature.

(Figure 10.2) first a secure distance is calculated by $L'_f = d_o / C_f$ where $C_f > 1$ is a clearance factor to reduce the likelihood of a window interference with other features. Let A and B be points on the image plane along two edges of the corner with distance L'_f from the corner. Then Δ can be defined as:

$$\Delta \equiv \max\left\{\left|x_C^i - x_A^i\right|, \left|x_C^i - x_B^i\right|, \left|y_C^i - y_B^i\right|, \left|y_C^i - y_B^i\right|\right\} \tag{10.7}$$

In addition, a clearance is provided from image boundaries by defining

$$d_x \equiv \frac{1}{2} N_x P_x - \left|x_C^i\right|, \qquad d_y \equiv \frac{1}{2} N_y P_y - \left|y_C^i\right|, \tag{10.8}$$

where $N_x, N_y, P_x,$ and P_y are the number of pixels and inter-pixel spacings in the x and y directions, respectively. The size of a corner feature is then calculated from:

$$W_x = \min\{2\Delta,\ 2d_x,\ (S_{\max})_x\}, \quad W_y = \min\{2\Delta,\ 2d_y,\ (S_{\max})_y\} \tag{10.9}$$

where $(S_{\max})_x$ and $(S_{\max})_y$ are the maximum allowable window sizes along the x and y directions, respectively. Window sizes should be greater than the minimum window size determined by the feature extraction method and hardware.

10.5 Feature Pixel Representation

Within each window segmented image pixels of interest must be represented. Two common representation schemes are regions and edges (or boundaries). For regional representation pixels with similar characteristics are grouped together. Examples include hole- or area-based marks representation. In edge representation, pixels with a discontinuous characteristic, such as intensity, are identified as edges that could be represented by fitted curves. These two schemes are interrelated and could be used together if required by the task. For instance, the edges of a hole region could be identified and used for outlier rejection and adaptive window sizing.

The most common problems of pixel representation are image noise and unsuitable information in an image. Different factors such as inappropriate thresholding, poor radiometric conditions (e.g., non-uniform illumination, changes in contrast, and reflectivity), and camera noise factors (e.g., photoelectron noise, sensor receiver shot, and thermal noise) all contribute to the image noise. Other severe factors such as feature occlusion might also occur. Some problems such as feature occlusion can be handled by feature selection and a planning mechanism, which will be discussed in subsequent chapters. Also, the real-time Hough transform has been proposed to make feature measurements robust against partial occlusions, low contrast, or poor lighting [Arbter et al., 2000]. Hough transform will be discussed in the feature description section. Examples of image noises include isolated pixels, discontinuous edges, etc. In environments with high contrast and an appropriate threshold, camera noise has minor effects. The probability of camera noise affecting the binary image is inversely proportional to the edge gradient; therefore, it is recommended to provide edge gradient $p_e > 3\sigma_c$ gray level/pixel [Corke, 1996]. This is easily met by quality binary images.

Three basic region-growing methods used in visual servoing are region merging, region splitting, and split-and-merge region growing. The problem with region growing is that the obtained image often contains too many regions or too few regions due to nonoptimal parameter settings. Therefore, some post-processing will be required, making region-based segmentation computationally nonfeasible. Region-growing techniques are more suitable for very noisy images where borders are extremely difficult to detect. For edge-based representation, edge-detection operators and filters (e.g., median or high-pass filters) are applied to remove noise and to single out edge pixels. This will eliminate whole window-scanning effort and will increase the accuracy and speed of feature extraction.

Linking techniques and connectivity analysis are applied to define meaningful groupings for pixels. The output will be fitted to boundary or region representations. Further details on the connectivity and linking techniques can be found in the literature (e.g., *Digital Picture Processing* by A. Rosenfeld and A.C. Kak, Academic Press, 1982). Linking techniques rely on some measurements for common properties of pixels to link them and are classified into *global* and *local* linking analyses. In a local analysis, gradient characteristics such as the strength and direction of each edge pixel in a small neighborhood are analyzed to determine if they satisfy certain conditions. Local linking methods are not computationally feasible for the real-time requirement of visual servoing and will not be discussed further. In global analysis, proximity to an approximate border location or its shape is used to link pixels together. Of particular interest is the detection of straight edges present in many industrial objects. The least-squares method lends itself to this application. Here, distances of pixels from a straight line can be evaluated and the best line, which minimizes the distance error function, can be found to represent the edge. This approach will be explored in edge representation and in the description scheme. Once the pixels of interest, representing features such as edge pixels or hole pixels, are extracted, one can apply measurement techniques to describe a feature by a scalar or a vector.

10.6 Feature Description and Measurement

In feature description and measurement, the segmented features must be expressed in some mathematical form. Following pixel representation, the feature description methods can be classified into region- and edge-based (or contour-based) descriptions. Region-based description contains simple descriptors of the shape such as area, projections, direction (or orientation), and moments. In edge-based description, edge (or contour) pixels could be expressed by their coordinates, sequence, and orientation (**chain code**, **crack code**), or by geometrical properties of the surrounded region (such as boundary length, curvature, and **signature**). There are many approaches to feature description. However, many of them are extremely noise-sensitive and usually do not meet requirements of visual servoing for robustness (such as speed and accuracy). Ideal descriptors for real-time vision applications such as visual servoing are those that meet the following requirements:

- Time and space computational efficiency, i.e., they are easy to compute in a short time
- Insensitivity to small changes in the represented pixels, e.g., due to nonideal segmentation and noise effects
- Invariance with respect to (edge or region) pixel magnitude, e.g., the number of pixels considered for feature description
- Calculation robustness irrespective of the relative view-pose (between the camera and feature)

Some of these requirements are interrelated, and improving one might degrade another. Among the descriptors used feature moments largely meet the above requirements and could be used for both binary and gray-level descriptions. If R is the region of interest and (i, j) represents the pixels of the region (e.g., a hole), then the moment of order $(p + q)$ is given by:

$$m_{pq} = \sum_i \sum_j x_i^p y_j^q I(x_i, y_j), \quad (i, j) \in R \tag{10.10}$$

which depends on scaling, translation, orientation, and thresholding. Moments define the object pixels distribution with m_{00} as the total area of region (such as hole) and the centroid of the region defined by:

$$S = m_{00} = \sum_i \sum_j I(x_i, y_j), \quad (i, j) \in R \tag{10.11}$$

$$x_c = \frac{m_{10}}{m_{00}}, \quad y_c = \frac{m_{01}}{m_{00}} \tag{10.12}$$

In binary segmentation, S will be equal to N, the number of pixels representing the region (e.g., a hole). The moments may also be normalized by using N. That is:

$$\overline{m}_{pq} = \frac{1}{N} \sum_i \sum_j x_i^p y_j^q I(x_i, y_j) \tag{10.13}$$

Translation-invariant moments can be calculated by the central moments, defined as:

$$\mu_{pq} = \sum_i \sum_j (x_i - x_c)^p (y_j - y_c)^q I(x_i, y_j), \quad (i, j) \in R \tag{10.14}$$

Similarly, scale-invariant and orientation-invariant moments could be defined. Many issues and analyses related to moments can be found in Savini [1988]. Moment characteristics have been also used for shape description and feature measurements, even if the region is represented by its boundary. A closed boundary R_b (perimeter of the area) can be characterized by N_b pixels and the normalized contour central moment will then be:

$$\overline{\mu}_{pq} = \frac{1}{N_b} \sum_i \sum_j (x_i - x_c)^p (y_j - y_c)^q I(x_i, y_j), \quad (i, j) \in R_b \tag{10.15}$$

The advantage of contour moments is that they are computationally less demanding. Also, in object classification they have proved to be more accurate than area-based moments. The introduced contour central moment can be used for calculating the direction (or orientation) of a region. In this chapter, the direction and orientation terms for an image region will be used interchangeably. Let a region be represented by an equivalent ellipse with its axes along the principal axes (Figure 10.3). Direction of a closed elongated region is then defined as the angle between the elongated side with the positive image x axis, or the direction of major axis of the equivalent ellipse. Direction is undefined for circular images. If the moments are known, the direction θ of a region will be calculated by:

$$\theta = \frac{1}{2} \tan^{-1} \left(\frac{2\overline{\mu}_{11}}{\overline{\mu}_{20} - \overline{\mu}_{02}} \right) \tag{10.16}$$

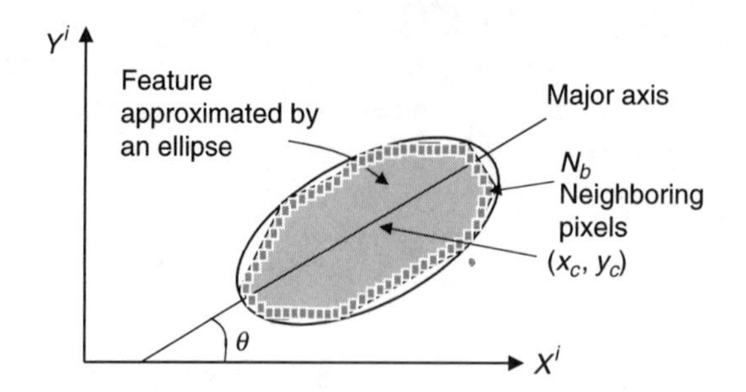

FIGURE 10.3 Direction and center of region-based feature.

Direction is independent of all linear transformations that do not include rotation. The centroid of a convex region can also be calculated from the boundary pixels by:

$$x_c = \frac{1}{N_b} \sum_{i=1}^{N_b} x_i, \qquad y_c = \frac{1}{N_b} \sum_{j=1}^{N_b} y_j . \tag{10.17}$$

In this section we will look at two popular feature descriptors for point-feature measurements in visual servoing, i.e., corner and hole feature description. Next, we will discuss other descriptors such as Hough transforms.

10.6.1 Hole Feature Description

Holes are bounded **contiguous pixels** (regions) that are not contiguous with an image's limits and have a distinct intensity level. If a region (e.g., a surface) has no holes, it is called simply a contiguous region and the one with holes is denoted by multiple contiguous region. Holes could be classified into circular (elliptical), polygonal, and irregular holes. Four hole feature measurements are common in visual servoing and include (1) area, (2) centroid, (3) direction, and (4) perimeter. In the measurements, the window must capture the hole with good resolution. Otherwise, window adjustments will be carried out to meet this requirement.

Area, centroid, and direction can be calculated from Eqs. (10.11), (10.12), and (10.16), respectively. These parameters could be used for pose estimation (e.g., in a Kalman-filter-based approach to position-based visual servoing) and outlier rejection (characterized by poor extraction or identification). A significant shift in these parameter values during servoing will indicate poor feature extraction and could be used for feature switching. Also, area and centroid measurements will be used for feature selection and planning. Perimeter measurement could easily be done after edge detection and might be utilized for feature selection, outlier rejection, and window adjustment.

10.6.2 Corner Feature Description

Corners are the points where two or more straight edges of the object intersect. Corners in the objects can be located using corner detectors such as Moravec or ZH (Zuniga–Haralick) operators. The inputs to these images are gray-level images and the output will be the images with values proportional to the likelihood of a pixel being the corner point. For example, if the image function $f(i, j)$ in the neighborhood of a pixel (i, j) is expressed by a cubic polynomial with coefficients $c_k (k = 1, 2, \ldots, n)$, the ZH operator will be given by:

$$\mathrm{ZH}(i, j) = \frac{-2(c_2^2 c_6 - c_3^2 c_4 - c_2 c_3 c_5)}{(c_2^2 + c_3^2)^{3/2}} . \tag{10.18}$$

However, these operators operate on whole images and are computationally expensive to meet the robustness requirement of visual servoing. Further details on corner detectors are available in the literature (e.g., *Image Processing Analysis and Machine Vision*, by M. Sonka, V. Holavac, and R. Boyle, Brooks/Cole, 1998).

A fast approach would be to detect edges (or contour curvatures) and to apply optimized estimation techniques to find the corner point location. Of particular interest is the description of a corner with straight edges. Both single- and two-window approaches are possible [Madhusudan, 1990]. In the two-window approach, two windows are placed on each edge of the corner using the information about the location of edges and corner orientation. This information can be obtained from the model of the object or from an estimation of the pose of the object (using KF, for example). Least-square lines are fit to each edge pixels, and the intersection of the lines will be calculated to describe the corner feature. However, it is not always possible to place two windows along the edges of a corner, e.g., when the corner angle is too small. Also, because windows are not usually placed near the corner and because line fits are determined away from the corner, considerable errors and noise sensitivity will be experienced in corner feature extraction. The single-window approach does not have the above disadvantages. Also, it is computationally faster to process a single window than double windows. However, the problem with the single-window approach is in dividing the window into two areas for processing. Therefore, the focus of this section will be on the single-window approach.

There are several methods for corner feature description using a single window in visual servoing. Moment-based methods have been described in Madhusudan [1990]. The method of moments does not provide a limited variance with a zero mean, and it depends on the corner angle and orientation. Two other methods have also been proposed to resolve the above problems, but they have met with limited success. These approaches attempt to address the problem of separating the edge pixels into two sets on both sides of the corner. The first approach uses the line between the moment-based estimation of the corner and the mid-point of the line connecting window-edge points (the intersection of a detected edge with a window boundary) to split the area into two sub-areas. Edge detection and least-squares line fitting is applied to each set of edge pixels on both sides of the split-line and a new corner feature estimate is made. The second method uses two pivot points to divide the window into two subareas for edge detection and least-squares fit. The extreme column (or row) of the object is taken and two pivot points are determined by taking the last pixels of that column (or row). These two approaches outperform the moment-based approach for corner feature description. However, the performance of both approaches depends on the corner angle and orientation. Experiments have indicated that these methods produce inaccurate results for corner angles greater than 90°. In particular, the second method is very noise sensitive.

Recently, Wong [1996] proposed a maximum-distance (MD) approach. This approach outperforms previous methods in terms of accuracy and computational efficiency. Let (x_1, y_1) and (x_2, y_2) be the coordinates of the window-edge points (Figure 10.4). An initial estimation of the corner location (x_c, y_c) is made by searching the pixel that has the largest vertical or horizontal distance from the window-edge points. In the exceptional case when two window-edge points have the same x or y coordinates, only the horizontal or vertical distance will be used for searching the initial estimate of the corner pixel. A line L_1 is drawn from pixel (x_c, y_c) to the mid-point pixel of L line connecting (x_1, y_1) and (x_2, y_2). This point is denoted by (x_m, y_m). Line L_1 divides the window into two areas, A_1 and A_2, each surrounded by L_1, L, and an edge of the corner. Edge pixels are detected for each area, and the least-squares method is applied to describe the location of the corner, as follows. The number of edge pixels for area A_k, denoted by N_k, and the normalized moment for each area, denoted by $(\overline{m}_{pq})_k$, can be calculated from Eqs. (10.11) and (10.13), respectively. The least-squares solution to each line intercept C_k will be calculated using:

$$C_k = -\sin\theta_k(\overline{m}_{10})_k - \cos\theta_k(\overline{m}_{01})_k, \quad k = 1, 2 \tag{10.19}$$

which leads to the coordinates of the intersection point, i.e., corner description as:

$$\hat{x}_c = \frac{C_1\cos\theta_2 - C_2\cos\theta_1}{\sin(\theta_2 - \theta_1)}, \quad \hat{y}_c = \frac{C_1\sin\theta_2 - C_2\sin\theta_1}{\sin(\theta_2 - \theta_1)} \tag{10.20}$$

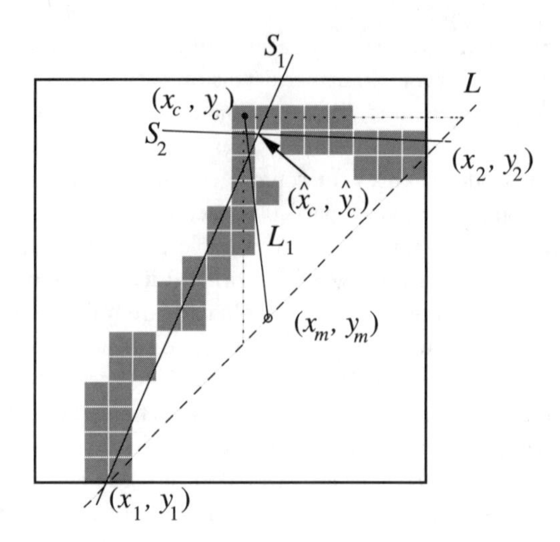

FIGURE 10.4 Corner description using the MD approach.

where θ_k is calculated from:

$$\theta_k = \frac{1}{2}\tan^{-1}\left(\frac{2(\overline{m}_{11} - \overline{m}_{10}\overline{m}_{01})}{(\overline{m}_{20} - \overline{m}_{02}) - \overline{m}_{10}^2 + \overline{m}_{01}^2}\right), \quad k = 1, 2 . \tag{10.21}$$

Experiments with the MD approach have revealed this to be a noise-sensitive, orientation-dependent approach as well. However, its performance for different corner angles is superior to the previous approaches.

10.6.3 Hough Transform

Hough transform can be applied to feature description if the shape and size of the feature is known. The main advantage of this transform is its robustness to segmentation (e.g., thresholding) results, i.e., the results will not be very sensitive to noise or imperfect data. Examples of such image imperfection include partially occluded features and sudden changes in the contrast. Many geometrical features such as edges, curves, and holes could be extracted by the Hough transform. It works well if the geometry of the feature is known *a priori*. The main disadvantage is related to its demanding computation time and space. However, its inherent parallel character creates good potential for its real-time implementation, e.g., for visual servoing. There has been some progress in improving Hough transforms for real-time applications [Arbter et al., 2000]. There are also reviews of Hough transforms in the literature (see Further Information). For feature extraction, each geometric feature could be expressed by $g(x, y, \boldsymbol{d})$, where $\boldsymbol{d}$ is the vector of parameters to be determined (Figure 10.5). For this purpose, parameter space within the limits of $\boldsymbol{d}$ is quantized and an n-dimensional accumulator array is formed, with n being the dimension of parameter space. For each image pixel (x, y) the accumulator cell $A(\boldsymbol{d})$ is increased by an adaptation rule of:

$$A(\boldsymbol{d})_{k+1} = A(\boldsymbol{d})_k + \Delta A \qquad \text{if } g(x, y, \boldsymbol{d}) = 0. \tag{10.22}$$

Local maxima in $A(\boldsymbol{d})$ correspond to feature parameters that are present in the image. Therefore, feature extraction reduces to maxima detection in accumulator space. The per-pixel computational load is $O(m^{n-1})$, where m is the number of quantized values per parameter, and n is 1, 2, and 4 for lines, circles, and ellipses, respectively.

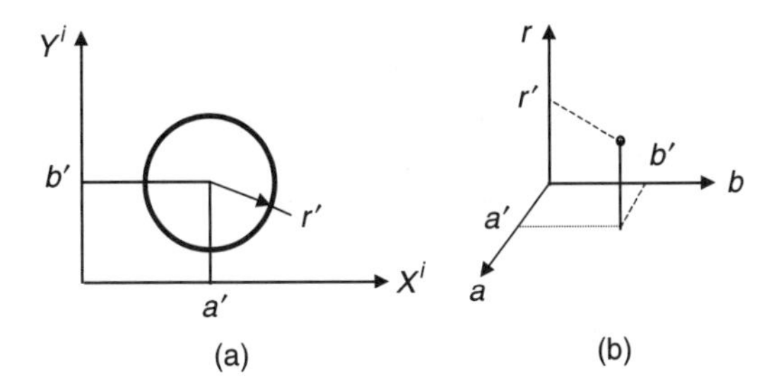

FIGURE 10.5 Hough transform in $d = (a, b, r)$ space: (a) a circle in image space; (b) (a, b, r) parameter space.

10.6.4 Window-Based Tracking Methods

With these methods the features are first located once across the entire image and then feature localization and description overlap with window tracking. Two types of tracking methods have been proposed: edge-based and area-based techniques. Edge-based methods (e.g., in the X-vision system of Yale CS-RR-1078) are not robust and are susceptible to mistracking by occluding edges. In area-based methods, first features are chosen based on certain measures, such as confidence measures [Papanikolopoulos, 1992]. The average value of each measure is calculated for a window of a specific size in different areas such as a uniform-intensity area, across an edge, and around corners. The threshold value for confidence is set between the average value of confidence measure both along an edge and around a corner. This way the windows with uniform intensity are disregarded. Next, tracking is based on matching the feature patterns in consecutive image frames with the assumption that the changes of the small region around the features in a frame sequence are not significant (temporal consistency assumption). Assume a window image was acquired at time $(k - 1)$ with a feature point $\mathbf{f}_{k-1}$ located at (x_{k-1}, y_{k-1}). Ideally, the best $\boldsymbol{r} = (u, v)^T$ is found, such that $\mathbf{f}_k = (x_{k-1} + u, y_{k-1} + v)^T$, which is the point in which the feature moves in frame k. Assuming a temporal consistency for windows of size N in the neighborhood of feature point $\mathbf{f}$, the similarity measure for the window pixels will be:

$$e(\boldsymbol{r}) = \sum_{m,n \in N} g[I_{k-1}(x_{k-1} + m, y_{k-1} + n) - I_k(x_{k-1} + m + u, y_{k-1} + n + v)] \tag{10.23}$$

where m and n are the indices for the pixels in the neighborhood N, $I_{k-1}(.,.)$ and $I_k(.,.)$ represent the intensity values in images k and $k - 1$, and u and v are bounded displacements. $g(.)$ is a scalar function that could be $g(x) = |x|$ denoted by the sum of absolute differences (SAD) or $g(x) = x^2$ for the sum of squared differences (SSD). The goal of the search will then be to find an $\boldsymbol{r}$ to minimize $e(r)$. Papanikolopoulos has reported an implementation of the SSD technique [1992]. Because an exhaustive search in a high-resolution image would be time consuming, optimization techniques are usually adopted for search in a low-resolution approximation. Because the location of the feature with respect to the window remains constant, the calculation of $\boldsymbol{r}$ will describe features and will also indicate the window-displacement vector for the next frame. This method works well for only small motions. Also, because the method relies on exact gray-level matching, small changes in radiometric conditions might lead to tracking failure. Due to the volume of the computations involved, special signal processors might be required with this method for frame-rate visual servoing.

10.7 Window Tracking and Adjustment

Initial window placement was discussed in Section 10.4. This section will discuss the issues related to the tracking and further adjusting of windows during servoing.

During servoing the relative pose of the object with respect to the camera and, hence, the locations of the image features will change. In window-based techniques, it is important to track the motion of the features and windows associated with the features of robust visual servoing. This is a problem of feature-window tracking. Many approaches assume a dynamic model of the target motion (such as the constant-velocity model) and apply filters such as the Kalman filter [Ficocelli and Janabi-Sharifi, 2001; Smith, 1989] to predict the next-time step feature (and hence window) location on the image plane, based on the current feature point measurements. Other filters and models such as tracking filters and the ARMAX model [Papanikolopoulos, 1992] have also been applied. In the window-based tracking method for feature description, e.g., SSD-based feature matching, the search result $\boldsymbol{r}$ will provide an estimate for the next search window location.

When the relative speed (between the camera and object) is high or when the object gets closer to the camera, the inter-frame motion of the image feature might result in a dislocated or oversize feature image with respect to its window. The last occurrence is also possible with large features. Another undesired case is having multiple features per window. This usually happens when the objects gets significantly away from the camera. These situations will lead to incorrect image feature descriptions and hence poor servoing. These issues are related to the windowing problem and can be handled by adjusting the size and location of the window(s). Ideally, only one feature must be present per window at its center at good distance from the window boundaries.

The problem of multiple features (per window) could be solved by separating the feature of interest from other features (Figure 10.6). A simple solution would be to (1) locate the feature of interest and (2) mark the feature pixels. The first task could be accomplished by doing a complete search of special descriptors of the feature of interest and matching them with the features present in the image. Special descriptors are invariant to partial occlusion and could be matched using an SSD-type objective function, assuming a temporal consistency. For example, if there are two corners per window, the corner angle and direction viewed in the image will not change considerably during inter-frame motion and could be used for identifying the corner feature of interest. A crude distance-based method has also been proposed for elliptical holes [Wong, 1996]. By this method, the hole that is closest to the center of the image is marked as the feature of interest. This is because windows are usually centered on each hole, and it is very likely that the closer feature (to the center) will be the feature of interest. The second task could be done by recursive marking of the feature bounded by the window. Subsequent adjustments should be applied to capture the feature of interest completely.

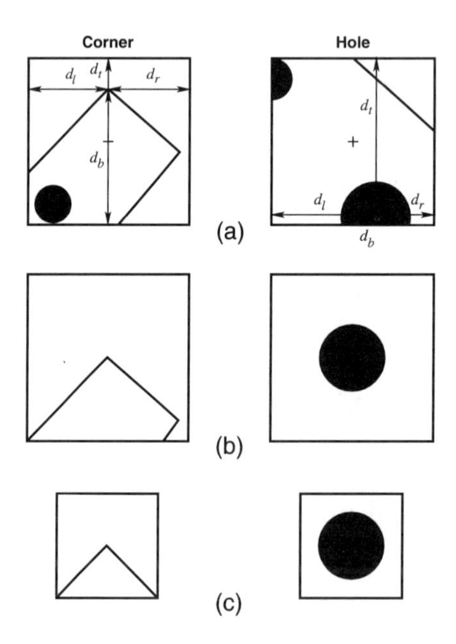

FIGURE 10.6 Separation of multiple features, and incomplete feature recovery in a window.

The problem of a partially captured feature might often occur in visual servoing due to the fast motions of the object and the presence of oversized features (Figure 10.6). This problem could be avoided by a high sampling rate or by the implementation of recovery strategies. First the feature inclusion test is required. This can be done by checking the window border pixels against the feature pixels. For a hole feature, none of its pixels must lie along the border, and for the corner feature both edges must be present in the image. If the test is negative, the search should be established in the neighborhood of the feature to locate the missing pixels. The window-border test and the information obtained in the previous stage are used to guide the search. If C_{x^i} and C_{y^i} denote the image coordinates of a window with sizes W_x and W_y, the window could move to a new location $C_{x^i}^{new}$ and $C_{y^i}^{new}$ such that

$$C_{x^i}^{new} = C_{x^i} + \Delta C_x, \qquad C_{y^i}^{new} = C_{y^i} + \Delta C_y \tag{10.24}$$

and

$$\Delta C_x = \begin{cases} 0.5(d_l - d_r) & \text{if } (d_l + d_r) > 1 \text{ pixel} \\ -0.5\Delta W_x & \text{otherwise} \end{cases} \quad \text{and} \quad \Delta C_y = \begin{cases} 0.5(d_b - d_t) & \text{if } (d_b + d_t) > 1 \text{ pixel} \\ -0.5\Delta W_y & \text{otherwise} \end{cases} \tag{10.25}$$

for centering the feature in the window [Wong, 1996]. Here d_r, d_l, d_t, and d_b are the minimum distances of the feature perimeter from the right, left, top, and bottom window borders, respectively. Also, ΔW_x and ΔW_y are the window size corrections in the x and y directions of the image, respectively. The window enlargement will be required when the window boundaries are very close to the feature perimeter (less than a pixel).

Window sizing will be required to capture the feature, to exclude multiple feature inclusion per window, to minimize the measurement error, and to decrease the image-processing effort. A tradeoff must be made among these multidirectional requirements. For example, decreasing the size of the window would improve image-processing speed but increase the risk of capturing the feature incompletely. Error of the feature point measurement originates predominantly from corner features and the method used for their extraction. This will be discussed in the next section, and it will be shown that larger windows provide better results when the MD method, for example, is used. Assuming that the features are approximately centered, the window size corrections will follow the method outlined by Janabi-Sharifi and Wilson [1997]:

$$\Delta W_x = 2\delta_w - 2\min\{d_r, d_l\}, \qquad \Delta W_y = 2\delta_w - 2\min\{d_t, d_b\}, \tag{10.26}$$

where δ_w is the window border clearance requirement, implied by the extraction method and servoing condition. For practical purposes, its recommended lower and upper values are two and five pixels, respectively.

10.8 Error Analysis

There are many sources for the errors in feature measurements, including classification (quantization) and description methods, relative motion, defocusing, camera noise, radiometric conditions, and electromagnetic noise. Camera noise with well-selected threshold and good radiometric conditions should not be a problem. Heuristic rules for threshold level and edge gradient to reduce the effects of poor radiometric conditions and camera noise were provided in Section 10.3. Electromagnetic noise can be greatly reduced by using differential circuits and good bandwidth cables. In this section, the focus will be on the effects of quantization, extraction methods, and defocusing.

The quantization error depends on pixel size (spatial quantization) and threshold level (amplitude quantization). For a row of pixels, such as edge, it has been shown that under certain conditions the variance of errors in the centroid is independent of threshold level. In addition, Madhusudan [1990] has concluded that decreasing the pixel density for minimizing the quantization errors due to relative velocity between the camera and the object is in contrast to increasing that for minimizing the static quantization error. The results agree with the analytical results of Corke [1996] that the centroid estimate is unbiased and independent of threshold level. Furthermore, the variance of the centroid estimate is shown to decrease with the diameter of the hole. Ho [1983] derived the following approximation:

$$\sigma_{x_c}^2 = \frac{1}{9\pi^2}\left(\frac{4}{d} + \frac{1}{d^3}\right) \tag{10.27}$$

for a disk of diameter d. Therefore, it is not necessary to have a sharp image for a good estimate of the centroid of a region because the centroid estimate is unbiased. However, the average length (or width) measurement of a binary region during servoing depends on the threshold level. An unbiased estimate of width is possible only when the threshold is midway between the object and background intensities. Therefore, the width of the features should not be used. Instead, distances between the centroids can be used, if necessary.

Analysis and experiments have been performed to measure the accuracy of corner and hole feature extraction methods introduced [Smith, 1989; Madhusudan, 1990; Wong, 1996]. It has been shown that hole feature measurements exhibit robustness to signal-to-noise-ratio (SNR) variations and are angle independent. Corner feature measurements are highly sensitive to decreases in image SNR, and the accuracy of the measurements also depends on the corner direction and angle [Smith, 1989].

For hole feature extraction, the error of area and its variance increase as the size of the hole increases. However, according to Jarnik's theorem, the area error cannot be greater than N_b. The area error percentage will decrease with the increase in the size of a hole, e.g., going from 38 to 18% when the radius of the hole image increases from three to seven pixels [Madhusudan, 1990]. If p represents the **normalized perimeter** of a hole image, the variance of area error has been shown to be $0.075p$ while the variance of direction (orientation) error has been obtained as $0.00774\ p^2$. The error of centroid is independent of hole interior pixels, and its variance decreases as the size of the hole increases. For example, the centroid error variance will go from about 0.15 pixels^2 to 0.003 pixels^2 as the radius of the hole image changes from two to seven pixels. Like other hole parameters, centroid errors are mainly due to the quantization effects and exhibit a zero-mean Gaussian distribution.

For corner feature extraction by the two-window method, the errors are zero-mean with a variance of 0.36 pixels^2 for a corner of 90°. The moment-based corner-extraction method with a single window indicates an error distribution with non-zero mean and large variance under similar conditions. The errors of two other single-window methods, proposed by Madusudan [1990], however, are zero-mean with almost 70% variance decrease in comparison with a moment-based method. The corner angle change, from 45 to 135°, indicates a variance change of from 0.375 to 0.07 pixels^2 with the minimum at 90°. Also, the variance of the error has been shown to increase with the increase in window size due to the inclusion of the pixels further away from the corner and an increase of quantization effects. Finally, Wong tested the MD approach [1996] and concluded that the corner estimate is orientation dependent with maxima near orientation angles of multiples of 45°. The error distribution is zero-mean with its variance changing from 0.0533 to 0.0161 pixels^2 in the x direction when switching from a 16 × 16-pixel window size to a 32 × 32-pixel-size one. Therefore, with the MD method, using a larger window will provide more accurate corner estimations.

Defocusing of image features is common in visual servoing due to the relative motion between the camera and the object. Either cameras with servo-controlled focus or a large depth-of-field must be used. Cameras with servo-controlled focuses are expensive and bulky, have limited adjustment rates, and need to get information about the distance to the object during the motion. Even with automated focusing, because of the differences in relative distance to the camera, all the features might not be in focus. The second solution is possible if a small aperture is used. However, to avoid motion blur the exposure time must be short. Consequently, the sensor must be sensitive, or the ambient illumination should be high. In summary, defocusing cannot be avoided with many industrial settings. By simulation and experiments with holes and corner features under different threshold levels, defocusing has virtually no effect on hole feature measurement accuracy. However, the corner feature point estimation tends to shift towards the corner direction in a low threshold setting and toward the opposite direction with a high threshold setting. Also, defocusing might cause overlapping of the **circles of confusion** (or **blur circles**) belonging to two or more closely located object features. Therefore, from the point of view of defocusing hole features with sufficient distance from each other are preferred.

10.9 Summary

Robustness of real-time vision-based systems, such as visual servoing, largely depends on the methods used for feature extraction. Primarily three characteristics of robust real-time vision, i.e., speed, accuracy, and reliability, have been taken into consideration, and window-based methodologies for robust feature extraction have been introduced. Feature measurements for two popular features, i.e., hole and corner features, and visual servoing applications have been highlighted.

Major steps in window-based feature extraction have been reviewed, and the effects of different parameters on the robustness of feature extraction have been studied. These steps include pixel classification, window placement, pixel representation, feature description and measurement, and window tracking and adjustment. In pixel classification, methods and guidelines for binary and adaptive thresholding have been given. For window placement, techniques for defining the location and size of the windows for both corner and hole features have been provided. With respect to pixel representation, common problems have been discussed, and an explanation of global and local linking methods has been given with an emphasis on the least-squares method. In the feature description section, requirements from the perspective of visual servoing have been enumerated, and the feature moments methods have been described, paying special attention to techniques for hole and corner feature descriptions. Also, Hough transform and window-based tracking methods for feature description have been discussed. For window tracking and adjustment, methods have been explored to address the issue of window location prediction and problems of multiple and partially occluded features. Finally, errors due to quantization, extraction methods, and defocusing have been discussed and practical conclusions have been presented.

Defining Terms

chain code: Represents an edge by direction vectors that link the centers of the edge pixels.

circle of confusion or blur circle: If the image of a point source falls inside of a circle of diameter a or smaller, that circle is called a circle of confusion or blur circle.

contiguous pixels: Pixels that have at least one path between them.

crack code: Represents an edge by a series of vertical and horizontal line segments that follow the cracks between pixels around the boundary of the pixel set.

feature: Any scene property that can be mapped onto and measured in an image plane. Any structural feature that can be extracted from an image is called an image feature that usually corresponds to the projection of a physical feature of objects onto the image plane. The image features can be divided into region-based features such as planes, areas, holes, and edge segment-based features such as corners and edges.

hue and saturation (HS): Hue refers to the perceived color (dominant wavelength), e.g., orange; saturation is a measure of its dilution by white light, resulting in light orange or dark orange, for example.

normalized perimeter: Estimated by moving along the boundary and counting any diagonal move as $\sqrt{2}$ linear resolution and any move along x or y axis of image as one resolution.

signature: A sequence of normal contour distances. For each point on the border, the shortest distance to an opposite border is calculated in a direction perpendicular to the border tangent at that point.

visual servoing: Also called vision-guided servoing; the use of vision in the feedback loop of the lowest level of a (usually robotic) system control with fast image processing to provide reactive behavior. The task of visual servoing for robotic manipulators (or robotic visual servoing, RVS) is to control the pose of the robot's end-effector relative to either a world coordinate frame or an object being manipulated, using real-time visual features extracted from the image. A camera can be fixed or mounted at the end-point (eye-in-hand configuration).

Acknowledgment

This work was supported by the Natural Sciences and Engineering Research Council of Canada (NSERC) through Research Grant #203060–98.

References

Arbter, K., Hirzinger, G., Langwald, J., Wei, G.Q., and Wunsch, P., Proven techniques for robust visual servo control, in *Robust Vision for Vision-Based Control of Motion,* Vincze, M. and Hager, G. D., Eds., IEEE Press, New York, 2000.

Corke, P. I., *Visual Control of Robots: High Performance Visual Servoing,* Research Studies, Ltd., Somerset, England, 1996.

Feddema, J. T., Lee, C. S. G., and Mitchell, O. R., Weighted selection of image features for resolved-rate visual feedback control, *IEEE Trans. Robot. Automat.*, 7(1), 31–47, 1991.

Ficocelli, M. and Janabi-Sharifi, F., Adaptive filtering for pose estimation in visual servoing, *IEEE/RSJ Int. Conf. Intelligent Robots Systems, IROS 2001* Maui, Hawaii, 2001, pp. 19–24.

Ho, C. S., Precision of digital vision systems, *IEEE Trans. Pattern. Anal. Machine Intell.,* 5(6), 593–601, 1983.

Janabi-Sharifi, F. and Wilson, W. J., Automatic selection of image features for visual servoing, *IEEE Trans. Robot. Automat.,* 13(6), 890–903, 1997.

Madhusudan, C., Error Analysis of the Kalman Filtering Approach to Relative Position Estimation Using Noisy Vision Measurements, Master's thesis, Dept. of Electrical and Computer Engineering, University of Waterloo, Waterloo, Canada, 1990.

Papanikolopoulos, N. P., Controlled Active Vision, Ph.D. dissertation, Dept. of Electrical and Computer Engineering, Carnegie Mellon University, Pittsburgh, PA, 1992.

Sahoo, P. K., Soltani, S., and Wong, A. K. C., A survey of thresholding techniques, *Comp. Vision Graph. Image Processing,* 41, 233–260, 1988.

Savini, M., Moments in image analysis, *Alta Frequenza,* 57(2), 145–152, 1988.

Smith, R. A., Relative Position Sensing Using Kalman Filtering of Vision Data, Master's thesis, Dept. of Electrical and Computer Engineering, University of Waterloo, Waterloo, Canada, 1989.

Wilson, W. J., Williams Hulls, C. C., and Janabi-Sharifi, F., Robust image processing and position-based visual servoing, in *Robust Vision for Vision-Based Control of Motion*, Vincze, M. and Hager, G. D., Eds., IEEE Press, New York, 2000, pp. 163–201.

Wong, J. S. H., Close Range Object Operations Using a Robot Visual Servoing System, M.A.Sc. dissertation, Electrical Engineering, University of Waterloo, Waterloo, Canada, 1996.

For Further Information

IEEE Transactions on Pattern Analysis and Machine Intelligence also provides advancements in feature extraction, image processing, and machine vision techniques. For example, a review of many feature extraction techniques is available in Vol. PAMI-3, No. 6, "Evaluation of projection algorithms," by G. Biswas, A. K. Jain, and R. C. Dubes. A review of Hough transforms is also reported in Vol. 16, pp. 329–341, 1994, in an article by J. Princen, J. Illingworth, and J. Kittler, "A Framework for Analyzing and Optimizing Hough Transform Performance."

Computer Vision, Graphics, and Image Processing contains recent reports on advances in image processing and machine vision techniques, for example, "Survey of the Hough Transform," by J. Illingworth, and J. Kittler, 44(1), pp. 87–116, 1988.

Digital Picture Processing, by A. Rosenfeld and A. C. Kak (Academic Press, 1982), provides details and examples for different feature-extraction techniques; in particular, sections on the representation aspect of feature extraction are comprehensive. *Image Processing: Analysis and Machine Vision* by M. Sonka, V. Holavac, and R. Boyle (Brooks/Cole, Pacific Grove, CA, 1998) also contains descriptive sections on classification and representation aspects of feature extraction. A chapter of this book on three-dimensional vision, geometry, and radiometry is also relevant to modeling aspects of visual servoing.

Three-Dimensional Computer Vision, by O. Faugeras (MIT Press, Cambridge, MA, 1993), provides a good review of feature extraction methods based on specialized hardware using temporal and geometric constraints.

Handbook of Pattern Recognition and Computer Vision, edited by C. H. Chen, L. F. Pau, and P. S. P. Wang (World Scientic, Singapore, 1999), contains a descriptive chapter on feature extraction and pattern recognition. In particular, a chapter called "Color in Computer Vision" explores in detail the applications of color in computer vision.

Robot Vision, by B. K. Horn (McGraw-Hill, New York, 1986), is a comprehensive introductory book for the application of machine vision in robotics.

Robust Vision for Vision-Based Control of Motion, edited by M. Vincze and G. D. Hager (IEEE Press, New York, 2000), provides recent advances in the development of robust vision for visual-servo-controlled systems. The articles span issues including object modeling, feature extraction, feature selection, sensor data fusion, and visual tracking. Also, *Visual Control of Robots: High Performance Visual Servoing,* by P. I. Corke (Research Studies, Ltd., Somerset, England, 1996), encompasses both theoretical and practical aspects of visual servoing of robotic manipulators.

11

Real-Time Image Recognition

Francesca Odone
University of Genova
Genova, Italy

Alessandro Verri
University of Genova
Genova, Italy

11.1 Introduction

Image recognition covers a wide range of diverse application areas (for example, security, surveillance, image-guided surgery, quality control, entertainment, autonomous navigation, and image retrieval by content). A fundamental requirement of an image-recognition system is the ability to automatically adjust to the changing environment or to be easily reusable for different tasks. For example, an effective object-detection system is expected to be able to operate in rather different domains and for a relatively broad range of different objects.

One way of ensuring this flexibility is provided by adaptive techniques based on the *learning from examples* paradigm. According to this paradigm, the solution to a certain problem can be learned from a set of examples. In the training stage this set of *positive* and *negative* examples—e.g., for object detection, images either containing or not containing the object—determines a discriminant function that will be used in the test stage to decide whether or not the same object is present in a new image. In this scenario, positive examples, typically collected by hand, are images or portions of images containing the object of interest, possibly under different poses. Typically, positive examples are collected manually in the training stage. By contrast negative examples, though comparatively much less expensive to collect, are somewhat ill defined and difficult to characterize. In the case of object detection, for example, negative examples are all the images or image portions *not* containing the object. Problems for which only positive examples are easily identifiable are sometimes referred to as novelty-detection problems. In the central part of this chapter, we will concentrate on these problems, arguing that this is a fairly typical example of the sort of problems image-recognition systems are required to solve.

In essence, a typical trainable computer vision system consists of three modules: preprocessing, representation, and learning modules. In the system described later in this chapter preprocessing is kept to a minimum and more attention is devoted to the representation and learning modules. The architecture

0-8493-1162-4/03/$0.00+$1.50

of trainable systems is well suited for real-time implementation. Typically, the test stage reduces to the computation of a function value that leads to a certain decision.

The plan of the chapter is as follows. Following a discussion of some of the recent work on image recognition we describe a trainable system developed in our lab for addressing problems of object detection and recognition. To this end we first present the notions of statistical learning and the similarity measure, which constitute the backbone of the system, and then illustrate the experimental results obtained in two different applications. Finally, in the concluding section we discuss the real-time issue and some open problems within the presented framework.

11.2 Previous and Current Work

The diversity of the computer vision methods that form the basis of image-recognition systems and the vast range of possible applications make it difficult to provide even a simple list of all the significant contributions. The most important advances can be tracked by referring to the specialized literature given at the end of this chapter. In this section we restrict our attention to object detection and recognition from images.

11.2.1 Object Detection

Object detection refers to the problem of finding instances of different exemplars of the same class, or category, in images. Face detection in static images is possibly the most important instance of the problem of object detection. Example-based strategies can be found in Leung et al. [1995], Osuna et al. [1997], Wiskott et al. [1997], Rowley et al. [1998], Rikert et al. [1999], and Schneiderman and Kanade [2000]. All these approaches attempt to detect faces in cluttered scenes and show some robustness to nonfrontal views.

Another active area of research is people detection, mainly tackled in image sequences [Wren, 1995; Intille et al., 1997; Haritaoglu et al., 1998; Bobick et al., 1999] but with more emphasis on the construction of *ad hoc* models and heavy use of tracking strategies.

A third area of interesting research is car detection. Bregler and Malik [1996] describe a real-time system based on mixtures of experts of second-order Gaussian features identifying different classes of cars. Beymer et al. [1997] propose a traffic-monitoring system consisting of a car detection module that locates corner features in highway sequences and groups features from the same car through temporal integration.

The object-detection system originally proposed by Papageorgiou et al. [1998] and later extended and refined by Papageorgiou and Poggio [2000] provides a very good example of a prototype system able to adapt to several object-detection tasks. Essentially the same system has been trained on different object data sets to address problems of face, people, and car detection from static images. Each object class is represented in terms of an overcomplete dictionary of Haar wavelet coefficients (corresponding to local, oriented, multiscale intensity differences between adjacent regions). A polynomial support vector machine (SVM) classifier is then trained on a training set of positive and negative examples. The system achieved very good performances on several data sets, and a real-time application of a person detection system as part of a driver assistance system has been reported.

11.2.2 Object Recognition

While object detection refers to the problem of finding instances of different exemplars of the same class, the task of object recognition is to distinguish between exemplars of the same class. The main difficulty of this problem is that objects belonging to the same class might be very similar. Face identification and recognition are classic problems of computer vision. Among the various proposed approaches we mention methods based on eigenfaces [Sirovitch and Kirby, 1987; Turk and Pentland, 1991], linear discriminant analysis [Belhumeur et al., 1997], hidden Markov Models [Nefian and Hayes, 1999], elastic graph matching [Wiskott et al., 1997], and, more recently, SVMs [Guodong et al., 2000; Jonsson et al., 2000].

One of the most important problems for object identification is pose invariance. The learning-from-examples paradigm fits nicely with a view-based approach to object recognition in which a three-dimensional (3D) object is essentially modeled through a collection of its different views. This idea, which seems to be one of the strategies employed by our brain for performing the same task [Bülthoff and Edelman, 1992; Logothetis et al., 1995], has been explored in depth in the last few years [Murase and Nayar, 1995; Sung and Poggio, 1998].

Pontil and Verri [1998] propose a simple appearance-based system for 3D object identification. The system was assessed on the COIL (Columbia Object Image Library) database, consisting of 7200 images of 100 objects (72 views for each of the 100 objects). The objects were positioned in the center of a turntable and observed from a fixed viewpoint. For each object the turntable was rotated a few degrees per image. In the reported experiments the color images were transformed into gray-level images of reduced size (32×32 pixels) by averaging the original images over 4×4 pixel patches. The training set used in each experiment consists of 36 images for each object. The remaining 36 images for each object were used for testing. Given a subset of 32 of the 100 objects, a linear SVM was trained on each pair of objects using the gray values at each pixel as components of a 32×32 input vector. Recognition, without pose estimation, was then performed on a one-to-one basis. The filtering effect of the averaging preprocessing step was sufficient to induce very high recognition rates even in the presence of large amounts of random noise added to the test set and for slight shift of the object location in the image.

A major limitation of the above system was the global structure of the classifier, which was ill suited to deal with occlusions and clutter. In order to circumvent this problem, one could resort to strategies based on object components. Brunelli and Poggio [1993], for example, performed face recognition by independently matching templates of three facial regions (eyes, nose and mouth). A similar approach with an additional alignment stage was proposed by Beymer [1993]. More recent work on this topic can be found, for example, in Mohan et al. [2001].

11.2.3 Event Recognition

Extending the analysis from static images to image sequences, image-recognition systems can address problems of event detection and recognition. In the case of event recognition, most of the systems that can be found in the literature for detecting and recognizing people's actions and gestures are based on a rather sophisticated algorithmic front-end and often on a fairly complex event representation and recognition scheme. Bregler [1997], for example, based many levels of representation on mixture models, EM, and recursive Kalman and Markov estimations to learn and recognize human dynamics. Pfinder [Wren et al., 1995] adopts a maximum *a posteriori* probability approach for detecting and tracking the human body using simple two-dimensional models. It incorporates *a priori* knowledge about people to bootstrap and recover from errors and employs several domain-specific assumptions to make the vision task-tractable.

W^4S [Haritaoglu et al., 1998] is a real-time visual surveillance system for detecting and tracking people while monitoring their activities in an outdoor environment. The system learns and models background scene statistically to detect foreground objects, if the background is not completely stationary, and distinguishes people from other objects using shape and periodic-motion cues. W^4S integrates real-time stereo computation into an intensity-based detection and tracking system without using color cues. KidRooms [Bobick et al., 1999] is a tracking system based on the idea of *closed-world* regions [Intille et al., 1997]. A closed world is a region of space and time in which the specific context of what is in the region is assumed to be known. Visual routines for tracking, for example, can be selected differently based on the knowledge of which other objects are in the closed world. The system tracks in real-time domains, where object motions are not smooth or rigid and multiple objects are interacting.

Instead of assuming a prior model of the observed events, it is interesting to explore the possibility of building a trainable dynamic event recognition system. Pittore et al. [2000] describe an adaptive system for recognizing visual dynamic events from examples. During training a supervisor identifies and labels the events of interest. These events are used to train the system that, at run time, automatically detects and classifies new events. The system core consisted of an SVM for regression [Vapnik, 1998], used for representing dynamic events of different durations in a feature vector of fixed length, and an SVM for classification [Vapnik, 1995] for the recognition stage. The system has been used to test recognition of people's movements within indoor scenes and has produced very interesting percentages of success.

11.3 Statistical Learning Fundamentals

In this section we briefly review the main issues of statistical learning relevant to the novelty detection technique proposed by Campbell and Bennett [2001] and that forms the basis of the system described later in the chapter. We first emphasize the similarity between various learning problems and corresponding techniques and then discuss in some detail the problem of novelty detection and the strong links between the solution proposed for this problem by Campbell and Bennett [2001] and SVMs for classification. For more details and background on the subject of statistical learning theory and its connection with regularization theory and functional analysis, see Vapnik [1995, 1998], Evgeniou et al. [2000], and Cucker and Smale [2001].

11.3.1 Preliminaries

In statistical learning, many problems—including classification and regression—are equivalent to minimizing functionals of the form

$$I_\lambda[f] = V(f(\mathbf{x}_i), y_i) + \lambda \|f\|^2,$$

where $(\mathbf{x}_i, y_i)$ for $i = 1, \ldots, N$ are the training examples, V is a certain function measuring the cost, or *loss*, of approximating y_i with $f(\mathbf{x}_i)$, λ is a regularization parameter, and the sum is for $i = 1, \ldots, N$. The y_i are labels (function values) corresponding to the point $\mathbf{x}_i$ in the case of classification (regression). The norm of f is computed in an appropriate Hilbert space, by a process called reproducing kernel Hilbert space [Wahba, 1990; Evgeniou et al., 2000], defined in terms of a certain function $K(\circ, \circ)$, named *kernel*, satisfying some mathematical constraints [Vapnik, 1995]. The norm can be regarded as a term controlling the smoothness of the minimizer. For a vast class of loss functions, the minimizer f_λ can be written as

$$f_\lambda(\mathbf{x}) = \sum_{i=1}^{N} \alpha_i K(\mathbf{x}, \mathbf{x}_i),$$

[Girosi et al., 1995], where the coefficients α_i are determined by the adopted learning technique. Intuitively, for small λ the solution f_λ will follow very closely the training data without being very smooth, while for large λ the solution f_λ will be very smooth, though not necessarily close to each data point. The goal (and problem) of statistical learning theory is to find the best trade-off between these two extreme cases. SVMs and regularization networks [Evgeniou et al., 2000] are among the most popular learning methods that can be cast in this framework.

For a fixed learning strategy, such as the SVM method for classification, the class of a new data point is then obtained by setting a threshold on the value of $f_\lambda(\mathbf{x})$, so that the point is a positive example if

$$\sum_{i=1}^{N} \alpha_i K(\mathbf{x}, \mathbf{x}_i) + b > 0, \tag{11.1}$$

and negative otherwise. The b in Eq. (11.1) is a constant term computed in the training stage.

11.3.2 Novelty Detection

We now discuss a technique proposed by Campbell and Bennett [2001] for novelty detection sharing strong theoretical and computational similarities with the method of SVMs [Vapnik, 1995]. Novelty detection can be thought of as a classification problem in which only one class, the class of *positive* examples, is well defined, and the classification task is to identify whether a new data point is a positive

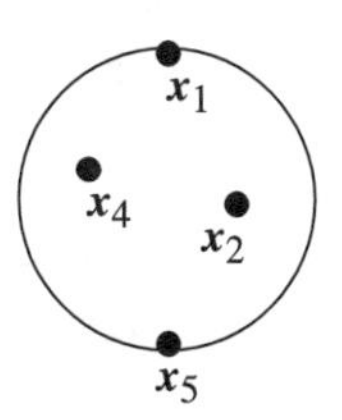

FIGURE 11.1 The smallest sphere enclosing the positive examples $\mathbf{x}_1,\ldots,\mathbf{x}_5$. For the current value of C the minimum of problem (2) is reached for $\xi_3 \neq 0$, effectively considering $\mathbf{x}_3$ as an outlier.

or a *negative* example. Given $\mathbf{x}_i$ for $i = 1, \ldots, N$ training examples, the idea is to find the sphere in the feature space of minimum radius in which most of the training data $\mathbf{x}_i$ lie (see Figure 11.1). The possible presence of outliers in the training set (e.g., negative examples labeled as positive) is countered by using slack variables. Each slack variable ξ_i enforces a penalty term equal to the square of the distance of $\mathbf{x}_i$ from the sphere surface. This approach was first suggested by Vapnik [1995] and interpreted and used as a novelty detector by Tax and Duin [1999] and Tax et al. [1999]. In the classification stage a new data point is considered positive if it lies sufficiently close to the sphere center.

In the linear case, the primal minimization problem can be written as

$$\begin{aligned} &\min_{\mathrm{R,p},\xi} \quad \mathrm{R}^2 + \mathrm{C}\sum_{i=1}^{N}\xi_i \\ &\text{subject to} \quad \|\mathbf{x}_i - \mathbf{p}\|^2 \leq \mathrm{R}^2 + \xi_i \\ &\text{and} \quad \xi_i \geq 0 \quad \text{for} \quad i = 1, \ldots, N, \end{aligned} \tag{11.2}$$

with $\mathbf{x}_i$ for $i = 1, \ldots, N$ the input data, R the sphere radius, $\mathbf{p}$ the sphere center, $\boldsymbol{\xi} = (\xi_1, \ldots, \xi_N)$ the slack variable vector, C a regularization parameter controlling the relative weight between the sphere radius and the ouliers, and the sum is for $i = 0, \ldots, N$.

The solution to Eq. (11.2) can be found by introducing the Lagrange multipliers $\alpha_i \geq 0$ and $\beta_i \geq 0$ for the two sets of constraints. Setting the partial derivatives equal to zero with respect to R and to each component of $\mathbf{p}$ and $\boldsymbol{\xi}$ we obtain

$$\begin{aligned} &\mathbf{p} = \sum_{i=1}^{N}\alpha_i\,\mathbf{x}_i \\ &\sum_{i=1}^{N}\alpha_i = 1 \\ &\mathrm{C} = \alpha_i + \beta_i \quad \text{for} \quad i = 1, \ldots, N. \end{aligned} \tag{11.3}$$

Plugging these constraints into the primal problem, after some simple algebraic manipulations, the original minimization problem becomes a QP maximization problem with respect to the Lagrange multiplier vector $\boldsymbol{\alpha} = (\alpha_1, \ldots, \alpha_N)$ or

$$\begin{aligned} &\max_{\boldsymbol{\alpha}} \; -\sum_{i=1}^{N}\sum_{j=1}^{N}\alpha_i\,\alpha_j\,\mathbf{x}_i \circ \mathbf{x}_j + \sum_{i=1}^{N}\alpha_i\,\mathbf{x}_i \circ \mathbf{x}_i \\ &\text{subject to} \quad \sum_{i=1}^{N}\alpha_i = 1 \\ &\text{and} \quad 0 \leq \alpha_i \leq \mathrm{C} \text{ for } i = 1, \ldots, N. \end{aligned} \tag{11.4}$$

The constraints on α_i in Eq. (11.4) define the *feasible* region for the QP problem. As in the case of SVMs for binary classification, the objective function is quadratic, the Hessian function is positive semi-definite, and the inequality constraints are box constraints. The three main differences are the form of the linear term in the objective function, the fact that all the coefficients are positive, and the equality constraints (here the Lagrange multipliers sum to 1 rather than to 0). As in the case of SVMs, the training points for which $\alpha_i \neq 0$ are the *support vectors* for this learning problem.

From Eq. (11.3) we see that the sphere center $\mathbf{p}$ is found as a weighted sum of the examples, while the radius R can be determined from the Kuhn–Tucker conditions associated with any support vector $\mathbf{x}_i$ having $\alpha_i \neq C$. In the classification stage a new point $\mathbf{x}$ is a *positive* or *negative* example, depending on its distance d from $\mathbf{p}$, where d is computed as

$$d^2 = \|\mathbf{x} - \mathbf{p}\|^2 = \mathbf{x} \circ \mathbf{x} + \mathbf{p} \circ \mathbf{p} - 2\,\mathbf{x} \circ \mathbf{p}.$$

In full analogy to the SVM case, because both the solution of the dual maximization problem and the classification stage require only the computation of the inner product between data points and sphere center, one can introduce a positive definite function K [Vapnik, 1995] that effectively defines an inner product in some new space, called *feature space*, and solve the new QP problem

$$\max_{\boldsymbol{\alpha}} - \sum_{i=1}^{N}\sum_{j=1}^{N} \alpha_i\, \alpha_j K(\mathbf{x}_i, \mathbf{x}_j) + \sum_{i=1}^{N} \alpha_i K(\mathbf{x}_i, \mathbf{x}_j)$$

$$\text{subject to} \quad \sum_{i=1}^{N} \alpha_i = 1 \tag{11.5}$$

$$\text{and} \quad 0 \leq \alpha_i \leq C \text{ for } i = 1, \ldots, N.$$

Recall that, given a set $X \subseteq \mathbb{R}^N$ a function K: $\mathrm{X} \times \mathrm{X} \rightarrow \mathbb{R}$ is *positive definite* if for all integers N and all $\mathbf{x}_1, \ldots, \mathbf{x}_N \in \mathrm{X}$, and $\alpha_1, \ldots, \alpha_N \in \mathbb{R}$,

$$\sum_{i=1}^{N}\sum_{j=1}^{N} K(\mathbf{x}_i, \mathbf{x}_j) \geq 0$$

The fact that a kernel must be *positive definite* ensures that K *behaves* like an inner product in some suitably defined feature space. Intuitively, one can think of a mapping φ from the input to the feature space implicitly defined by K. In this feature space the inner product between the points $\varphi(\mathbf{x}_i)$ and $\varphi(\mathbf{x}_j)$ is given by

$$\varphi(\mathbf{x}_i) \circ \varphi(\mathbf{x}_i) = K(\mathbf{x}_i, \mathbf{x}_j).$$

The sphere center is now a point in feature space, $\varphi(\mathbf{p})$. If the mapping $\varphi(\circ)$ is unknown, $\varphi(\mathbf{p})$ cannot be computed explicitly, but the distance D between $\varphi(\mathbf{p})$ and a point $\varphi(\mathbf{x})$ can still be determined as

$$D^2 = K(\mathbf{x}, \mathbf{x}) + \sum_{i=1}^{N}\sum_{j=1}^{N} \alpha_i \alpha_j K(\mathbf{x}_i, \mathbf{x}_j) - 2\sum_{i=1}^{N} \alpha_i K(\mathbf{x}_i, \mathbf{x}).$$

In addition, we note that if $K(\mathbf{x}, \mathbf{x})$ is constant, the dependence of D on $\mathbf{x}$ is only in the last term of the summation on the right-hand side. Consequently, D is sufficiently small if for some fixed threshold τ

$$\sum_{i=1}^{N} \alpha_i K(\mathbf{x}_i, \mathbf{x}) \geq \tau \tag{11.6}$$

The striking similarity between Eqs. (11.1) and (11.6) emphasizes the link between these two learning strategies. We now discuss a similarity measure, which we will use to design a kernel for novelty detection in computer vision problems.

11.4 Measuring the Similarity between Images

In this section we first describe a similarity measure for images inspired by the notion of Hausdorff distance. Then, we determine conditions under which this measure defines a legitimate kernel function. We start by reviewing the notion of Hausdorff distance.

11.4.1 Hausdorff Distance

Given two finite point sets A and B, both regarded as subsets of some Euclidean n-dimensional linear space, the directed Hausdorff distance h can be written as

$$h(A, B) = \max_{a \in A} \min_{b \in B} \|a - b\|.$$

In words, $h(A, B)$ is given by the distance of point A furthest away from set B. The directed Hausdorff distance, h, though not symmetric (and thus not a true distance), is very useful to measure the degree of mismatch of one set with respect to another. To obtain a distance in the mathematical sense, symmetry can be restored by taking the maximum between $h(A, B)$ and $h(B, A)$. This leads to the definition of *Hausdorff distance H*, that is,

$$H(A, B) = \max \{h(A, B), h(A, B)\}$$

A way to gain an intuitive understanding of Hausdorff measures is to think in terms of set inclusion. Let B_ρ be the set obtained by replacing each point of B with a disk of radius ρ and taking the union of all of these disks. The following simple proposition then holds:

P1. The directed Hausdorff distance $h(A, B) \leq \rho$ if and only if $A \subseteq B_\rho$.

Proposition *P1* follows from the fact that, in order for every point of A to be within distance ρ from some points of B, A must be contained in B_ρ (see Figure 11.2). We now see how these notions can be useful for the construction of similarity functions between images.

11.4.1.1 Hausdorff Similarity Measure

The idea of using Hausdorff measures for comparing images is not new. Huttenlocher et al. [1993], for example, used the Hausdorff distance to compare edge, i.e., binary, images. Here we want to extend this use to the gray-value image structure thought of as a 3D point set.

Suppose we have two gray-level images, I and J, for which we want to compute the degree of similarity. Ideally, we would like to use this measure as a basis for deciding whether the two images contain the same object, possibly represented in two slightly different views, or under different illumination conditions. In order to allow for gray-level changes within a fixed interval or small local transformations (for example, small-scale variations or affine transformations), the following function may be evaluated [Odone et al., 2001]:

$$k(I, J) = \Sigma_p \vartheta(\varepsilon - \min_{q \in N_p} |I[p] - J[q]|) \quad (11.7)$$

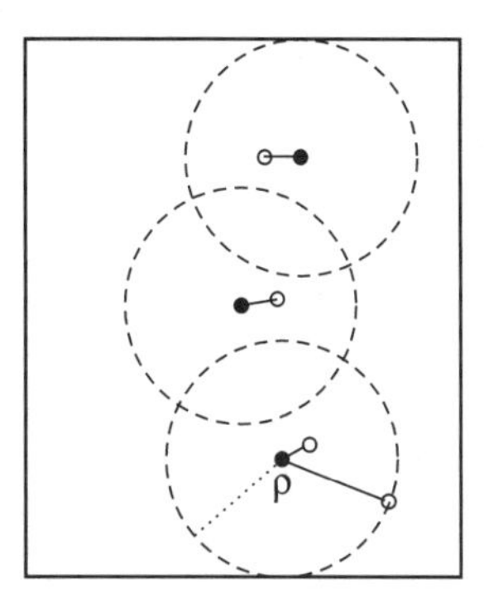

FIGURE 11.2 Interpreting the directed Hausdorff distance *h(A,B)* with set inclusion. The open dots are the points of set A and the filled dots the points of set B. Because all points of A are contained in the union of the disks of radius ρ centered in the points of B, we have $h(A, B) \leq \rho$.

where ϑ is the unit step function. The function k in (11.7) counts the number of pixels p in I that are within a distance ε (on the grey levels) from at least one pixel q of J in neighborhood N_p of p. Unless N_p coincides with p, the function k is not symmetric, but symmetry can be restored, for example, by taking the average

$$K(I, J) = 1/2(k(I, J) + k(J, I)) \tag{11.8}$$

Equation (11.8), interpreted in terms of set dilation and inclusion, leads to an efficient implementation [Odone et al., 2001] that can be summarized in three steps:

1. Expand the two images I and J into 3D binary matrices $\boldsymbol{I}$ and $\boldsymbol{J}$, the third dimension being the gray value. If (i, j) denotes a pixel location and g its gray level, we can write

$$\boldsymbol{I}(i, j, g) = 1, \text{ if } I(i, j) = g; \quad \boldsymbol{I}(i, j, g) = 0 \text{ otherwise.}$$

2. Dilate both matrices by growing their nonzero entries by a fixed amount ε (in the gray-value dimension), and ε_r and ε_c (sizes of the neighborhood N_p in the space dimensions). Let $\boldsymbol{M}_I$ and $\boldsymbol{M}_J$ be the resulting 3D dilated binary matrices. This dilation varies according to the degrees of similarity required and the transformations allowed.
3. Compute the size of the intersections between $\boldsymbol{I}$ and $\boldsymbol{M}_J$ and $\boldsymbol{J}$ and $\boldsymbol{M}_I$, and take the average of the two values obtaining $K(I, J)$.

If the dilation is isotropic in the spatial and gray-level dimensions, the similarity measure k in Eq. (11.7) is closely related to the directed Hausdorff distance h, because computing k is equivalent to fixing a maximum distance ρ allowed between the two sets and to determining how much of the first set lies within that distance from the second set. In general, if we have $k(I, J) = m$, we can say that a subset of I cardinality m is within a distance ρ from J. Odone et al. [2001] showed that this similarity measure has several properties, such as tolerance to small changes affecting images due to geometric transformations, viewing conditions, and noise, as well as robustness to occlusions.

We now discuss the issue of whether or not the proposed similarity measure can be viewed as a kernel function.

11.4.2 Kernel Functions

To prove that a function K can be used as a *kernel* is often a difficult problem. A sufficient condition is provided by Mercer's theorem [Vapnik, 1995], which is equivalent to requiring positive definiteness for K. Intuitively, this means that K computes an inner product between point pairs in some suitably defined feature space.

Looking directly into the QP optimization problem (11.5), we see that a function K can be used as a kernel, if the objective function is convex, or if the associated Hessian matrix $K_{ij} = K(\mathbf{x}_i, \mathbf{x}_j)$ in the dual maximization problem is such that

$$\sum_{i=1}^{N}\sum_{j=1}^{N} \alpha_i \alpha_j K_{ij} \geq 0 \tag{11.9}$$

for all α_i in the *feasible* region.

11.4.2.1 Hausdorff Kernel

We start with novelty detection. We note that by construction $K_{ij} \geq 0$ for all image pairs. Because the feasible region is

$$0 \leq \alpha_i \leq C \text{ for all } i = 1,\ldots,N \text{ with } \sum_{i=1}^{N} \alpha_i = 1,$$

we immediately see that inequality (11.9) is always satisfied in the feasible region and conclude that the Hausdorff similarity measure can always be used as a kernel for novelty detection.

In view of extending the use of the Hausdorff similarity measure to SVMs for classification, it is of interest to consider the problem of what happens to the inequality (11.9) when the feasible region is (see [Vapnik, 1995])

$$-C \leq \alpha_i \leq C \text{ for all } i = 1, \ldots, N \text{ with } \sum_{i=1}^{N} \alpha_i = 0. \tag{11.10}$$

We start by representing our data (say, $M \times M$ pixel images with G gray values) as M^2G-dimensional vectors. In what follows we say that the triple (i', j', g') is *close* to (i, j, g) if $\mathbf{M}_I(i', j', g') = 1$ because $\mathbf{I}(i, j, g) = 1$.

1. An image is represented by an M^2G-dimensional vector $\mathbf{b}$ such that $b_k = 1$ if $I(i, j) = g$; $b_k = 0$, otherwise,

   where $k = 1, \ldots, M^2G$ is in one-to-one correspondence with the triple (i, j, g), for $i, j = 1, \ldots, M$ and $g = 1, \ldots, G$. Note that for each i and j there is one and only one gray value g for which $b_k = 1$, hence only M^2 of the M^2G components of $\mathbf{b}$ are nonzero.

2. A given dilation of an image is obtained by introducing a suitable vector $\mathbf{c}$. We let $c_k = 1$ if $b_k = 1$, and $c_{k'} = 1$ if k' corresponds to a triple (i', j', g') *close* to (i, j, g).
3. Finally, if $\mathbf{B}_i = (\mathbf{b}_i, \mathbf{c}_i)$ and $\mathbf{B}_j = (\mathbf{b}_j, \mathbf{c}_j)$ denote the vectors corresponding to images I_i and I_j, respectively, the intersection leading to the Hausdorff similarity measure in Eq. (11.8) can be written as

$$K(I_i, I_j) = 1/2(\mathbf{b}_i \circ \mathbf{c}_j + \mathbf{b}_j \circ \mathbf{c}_i). \tag{11.11}$$

The use of ordinary dot products in Eq. (11.11) suggests the possibility that K is an inner product in the linear span of the space of the vectors $\mathbf{B}_i$. Indeed, for the null dilation we have $\mathbf{c}_i = \mathbf{b}_i$, $\mathbf{c}_j = \mathbf{b}_j$, and equation (11.11) reads

$$1/2(\mathbf{b}_i \circ \mathbf{b}_j + \mathbf{b}_j \circ \mathbf{b}_i) = \mathbf{b}_i \circ \mathbf{b}_j$$

thus K, in Eq. (11.11), reduces to the standard dot product in a M^2G-dimensional Euclidean space. In general, for *any* nonzero dilation, K in Eq. (11.11) is symmetric, linear, and

$$K(I_i, I_j) \geq 0$$

for all images I_i and corresponding vectors $\mathbf{B}_i$, but K *does not* satisfy inequality (11.9) for all possible choices of α_i.

However, it is possible to show that through a modification of the dilation step and, correspondingly, of the $\mathbf{c}$-vector definition, the function K in Eq. (11.11) becomes an inner product. For a fixed dilation—say, $\varepsilon = \varepsilon_r = \varepsilon_c$—for simplicity's sake, we write $1/\Delta = (2\varepsilon + 1)^3 - 1$, and modify the dilation step by defining

$$c_k = \sum b_k + \Delta b_{k'}, \tag{11.12}$$

where the sum is over all k' corresponding to triples (i', j', g') close to (i, j, g). With this new definition of $\mathbf{c}_i$, the function K in Eq. (11.11) can be regarded as a nonstandard inner product defined as

$$1/2(\mathbf{b}_i \circ \mathbf{c}_j + \mathbf{b}_j \circ \mathbf{c}_i) = \langle \mathbf{B}_i, \mathbf{B}_j \rangle$$

in the linear span of the vectors $\mathbf{B}_i$. Indeed, this product is now symmetric and linear, and, as can easily be shown by square completion, satisfies the nonnegativity condition for all vectors in the linear span of the $\mathbf{B}_i$. Two questions arise. First, how much smaller are the off-diagonal elements in the corresponding Hessian matrix due to the presence of the Δ factor in the dilation step of Eq. (11.12). On average, one might expect that with sufficiently smooth images about $(2\varepsilon + 1)^2$ components of $\mathbf{b}$ (roughly, the size of the spatial neighborhood) will contribute to c_k. From this we conclude that for each $c_k \neq 0$ we are likely to have

$$c_k \simeq b_k + 1/2\varepsilon.$$

This means that, with respect to the original definition, the nonzero components of $\mathbf{c}$ can now be greater or smaller than 1, depending on whether $b_k = 0$ or 1. To what extent this reduces the offdiagonal elements in the Hessian matrix remains to be seen and will ultimately depend on the specific image structure of the training set.

A second question is related to the fact that, in many practical cases, the image spatial dimensions and the number of gray levels are so much larger than the dilation constants that even the original definition of the Hausdorff similarity measure gives rise to Hessian matrices that are diagonally dominant. This is consistent with the empirical observation that the Hessian matrix obtained with null dilation in all three dimensions is typically almost indistinguishable from a multiple of the identity matrix.

In summary, the Hausdorff similarity measure can *always* be used as a kernel for *novelty detection* and, *after the correction of above*, as a kernel for binary *classification* with SVMs.

11.5 Experiments

In this section we present results obtained on several image data sets, acquired from multiple views and for different applications: face images for face recognition and images of 3D objects for 3D object modeling and detection from image sequences. The two problems are closely related but distinct. As we will see, in the latter the notion of object is blurred with the notion of scene (the background being as important as the foreground), while in the former this is not the case.

11.5.1 System Structure

The system consists of a learning module based on the novelty detector, previously described, equipped with the Hausdorff kernel discussed in the previous section. In training the system is shown positive examples only. The novelty detector has been implemented by suitably adjusting the decomposition technique developed for SVMs for classification described by Osuna et al. [1997] to the QP problem in Eq. (11.5). In the discussed applications we make minimal use of preprocessing. In particular, we do not compute accurate registration because our similarity measure takes care of spatial misalignments. Also, because our mapping-in feature space allows for some degree of deformation in both the gray-levels and the spatial dimensions, the effects of small illumination and pose changes are attenuated. Finally, we exploit full 3D information on the object by acquiring a training set that includes frontal and lateral views. All the low-resolution images used have been resized to 58 × 72 pixels; therefore, we work in a space of 4176 dimensions.

11.5.2 Face Identification

We run experiments on two different data sets. In the first set of images, we acquired both training and test data in the same session. We collected four sets of images (frontal and rotated views), one for each of four subjects, for a total of 353 images, samples of which are shown in Figure 11.3. To test the system we have used 188 images of ten different subjects, including test images of the four people used to acquire the training images (see the examples in Figure 11.4). All the images have been acquired in the same location and thus have a similar background. No background elimination is performed because the face occupies a substantial part of the image — about 3/4 of the total image area, implying that even images of different people have, on average, 1/4 of the pixels that match.

The performance of the Hausdorff kernel has been compared with a linear kernel on the same set of examples. The choice of the linear kernel is due to the fact that it can be proved to be equivalent to

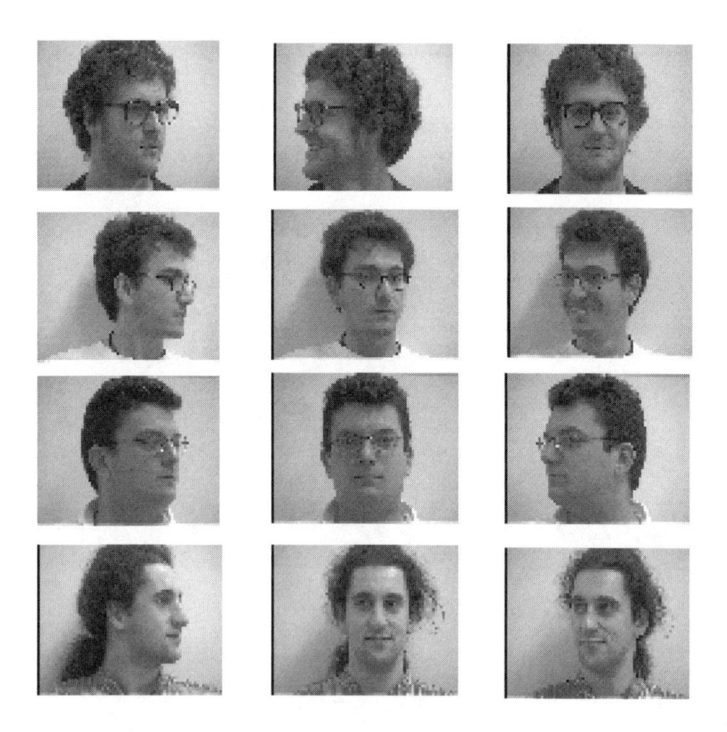

FIGURE 11.3 Examples of training images of the four subjects (first data set).

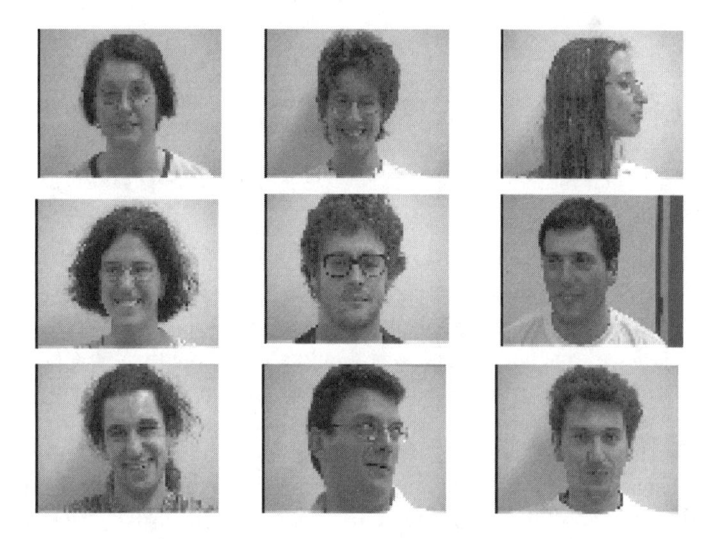

FIGURE 11.4 Examples of test images (first data set).

correlation techniques (sum of squared differences and cross correlation) widely used for the similarity evaluation between gray-level images.

The results of this comparison are shown as receiver operating characteristic (ROC) curves. Each point of an ROC curve represents a pair of false alarms and the hit rate of the system for a different rejection threshold τ is shown in Eq. (11.6). The system efficiency can be evaluated by the growth rate of its ROC curve: for a given false-alarm rate, the better system will be the one with the higher hit rate. The overall performance of a system, for example, can be measured by the area under the curve.

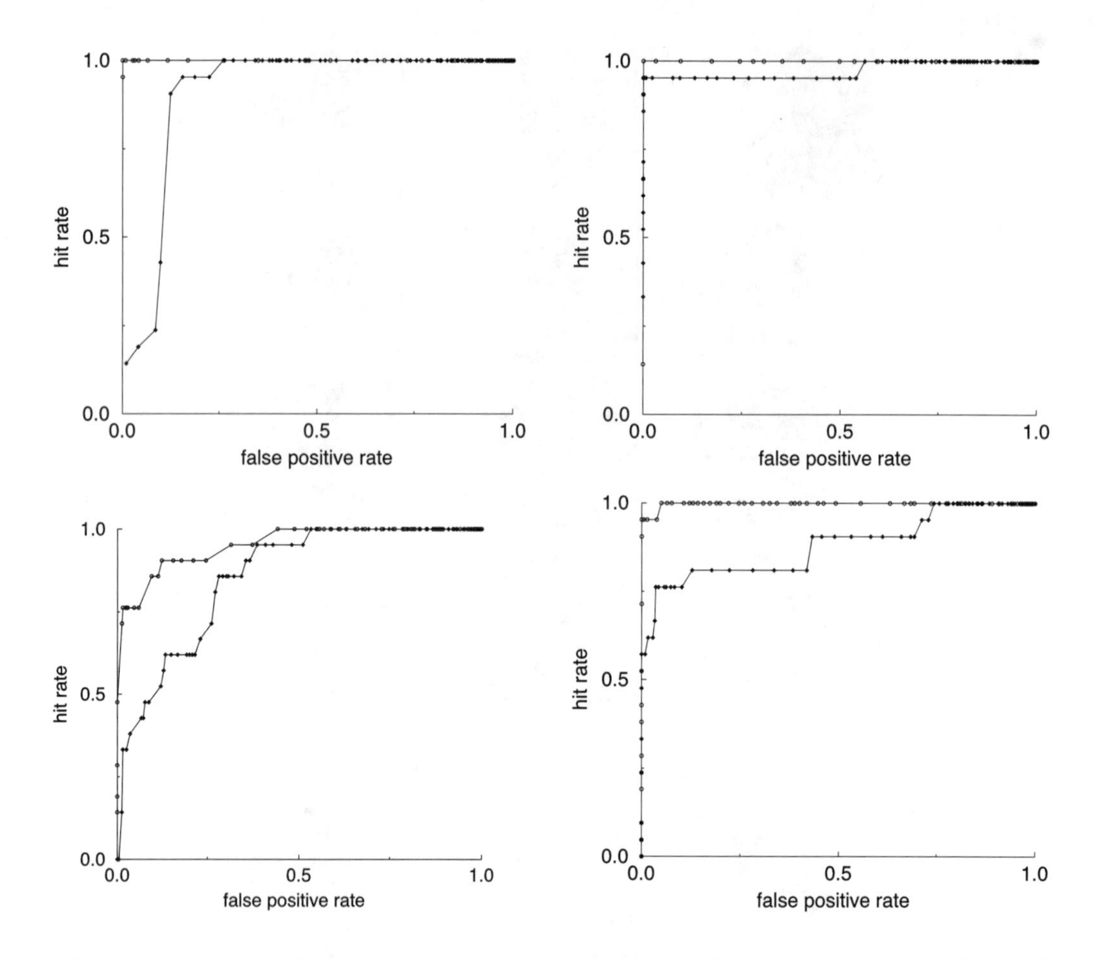

FIGURE 11.5 ROC curves for the four training sets. In all four cases the linear kernel curves are well below the Hausdorff kernel curves.

Figure 11.5 shows the ROC curves of a linear and Hausdorff kernels for all the four face-identification problems. The curve obtained with the Hausdorff kernel is always above the one of the linear kernel, showing superior performance. The linear kernel does not appear to be suitable for the task: in all four cases, to obtain a hit rate of 90% with the linear kernel one should accept more than 50% false positives. Instead, the Hausdorff kernel ROC curve increases rapidly and shows good properties of sensitivity and specificity.

In a second series of experiments on the same training sets, we estimate the system performance in the leave-one-out mode. Figure 11.6 shows samples from two training sets. Even if the set contains a number of images that are difficult to classify, only 19% were found to lie outside the sphere of minimum radius, but all of them were found to lie within a distance of less than 3% of the estimated radius.

Other sets of face images were acquired on different days and under unconstrained illumination (see examples in Figure 11.7). In this case, as shown in the scale change between Figures 11.3 and 11.7, a background elimination stage from the training data was necessary. To this end we performed a semi-automatic preprocessing of the training data, exploiting the spatio-temporal continuity between adjacent images of the training set. Figure 11.8 shows an example of the reduced images (bottom row) obtained by manually selecting a rectangular patch in the first image of the sequence in the top row and then tracking it automatically through the rest of the sequence through the Hausdorff similarity measure. The tracking procedure consisted in selecting in each new frame the rectangular patch most similar, in the Hausdorff sense, to the rectangular patch selected in the previous frame.

Figure 11.9 shows some of the images that were used to test our system, the performance of which, for the four identification tasks, is described by the ROC curves of Figure 11.10. The four different training

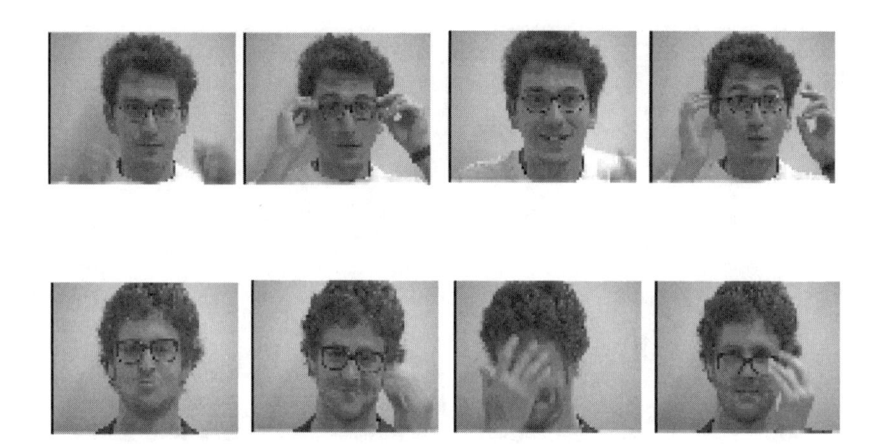

FIGURE 11.6 Examples of data sets used for the leave-one-out experiments.

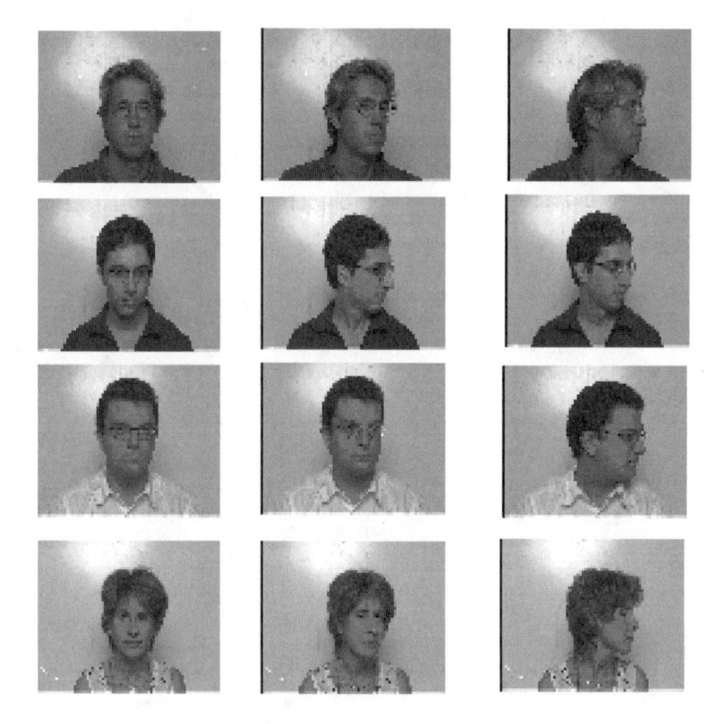

FIGURE 11.7 Examples of training images of the four subjects (second data set).

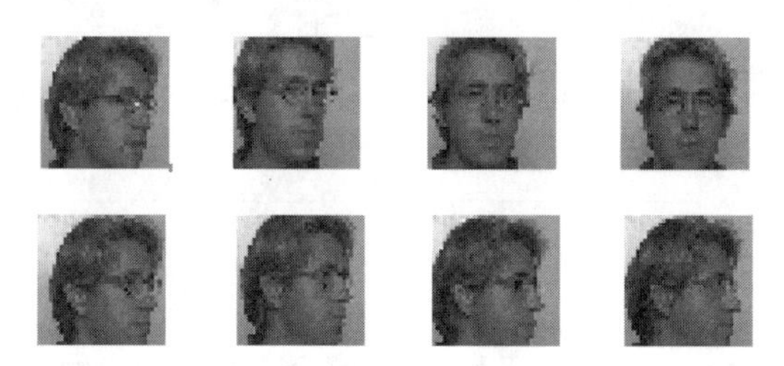

FIGURE 11.8 Background elimination: examples of reduced images (bottom row) for one training sequence (top row).

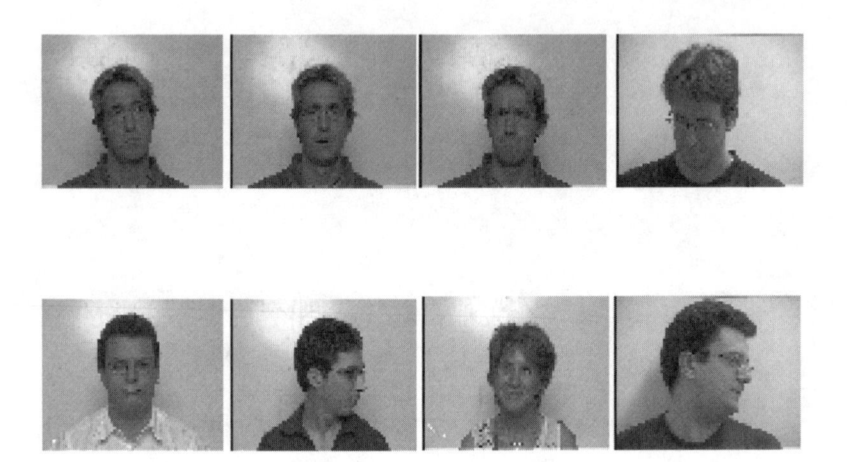

FIGURE 11.9 Examples of test images (second data set).

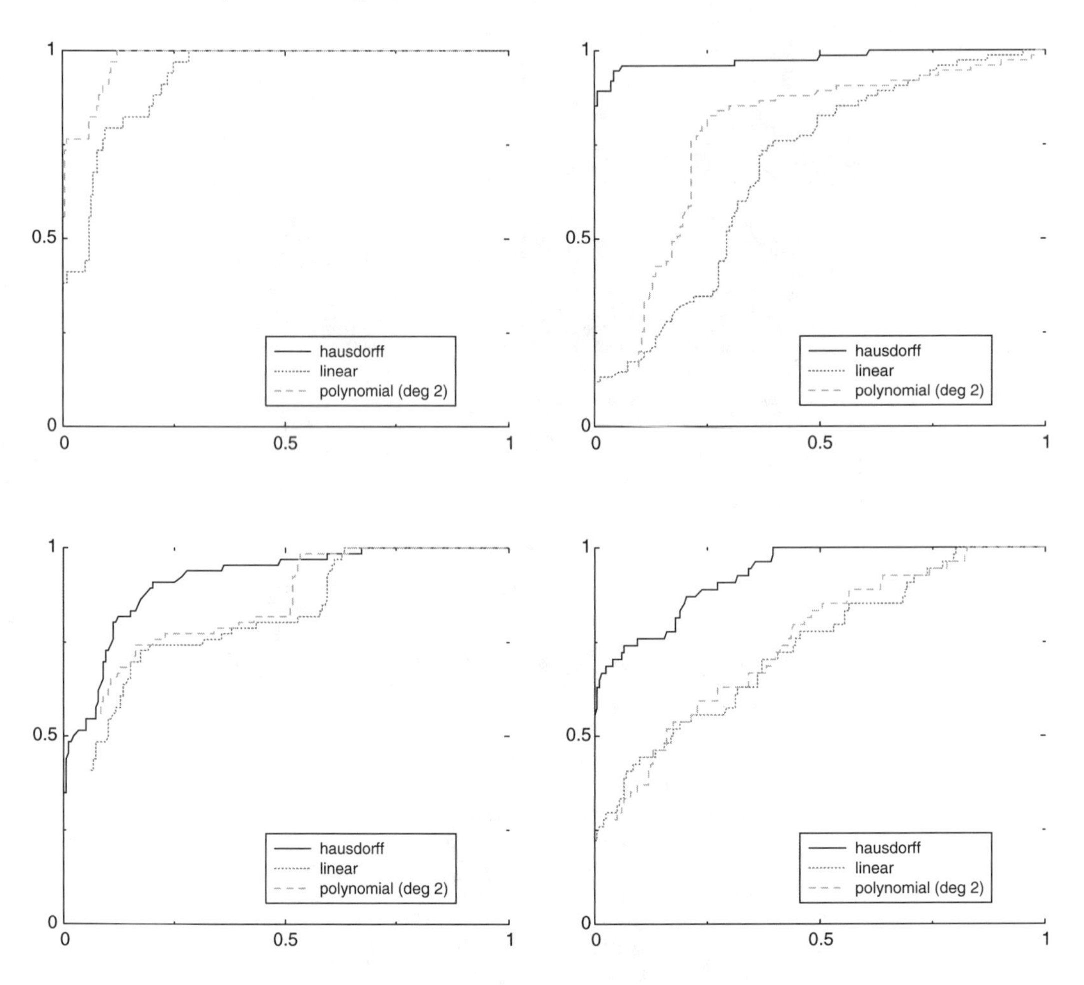

FIGURE 11.10 ROC curves for the four subjects of the second data set. Comparison between the linear and polynomial (degree 2) kernels and the Hausdorff kernel. Top left: training 126 images, test 102 (positive) and 205 (negative) images. Top right: training 89 images, test 288 (positive) and 345 (negative) images. Bottom left: training 40 images, test 149 (positive) and 343 (negative) images. Bottom right: training 45 images, test 122 (positive) and 212 (negative) images.

sets were of 126, 89, 40, and 45 images, respectively (corresponding to the curves in clockwise order, from the top left). See the figure caption for the test-set size. With this second class of data sets, in most cases the results of the linear kernel are too poor to represent a good comparison, so we also experimented with polynomial kernels of various degrees. In the ROCs of Figure 11.10 we included the results obtained with a polynomial of degree 2, as it produced the best results. One of the reasons of the rather poor performance of classic polynomial kernels may be related to the fact that the data we use are very weakly correlated, i.e., images have not been accurately registered with respect to a common reference. During the acquisition the subject was free to move and, in different images, the features (eyes, for instance, in the case of faces) can be found in different positions.

11.5.3 Three-Dimensional Object Recognition

The second type of data represents 3D rigid objects found in San Lorenzo Cathedral. The objects are six marble statues, and the ones we used for these experiments are all located in the same chapel (Cappella di S. Giovanni Battista); thus, all were acquired under similar illumination conditions, which results in noticeable similarities in the brightness pattern of all the images (see Figures 11.11 and 11.12). In this

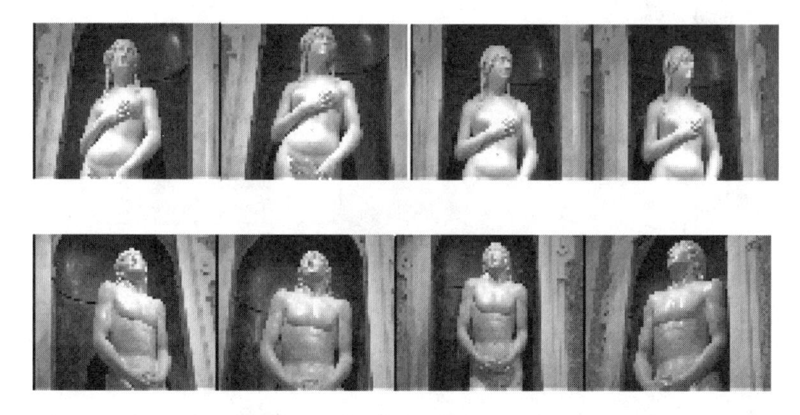

FIGURE 11.11 Samples of the training sets for statue A (above) and statue B (below).

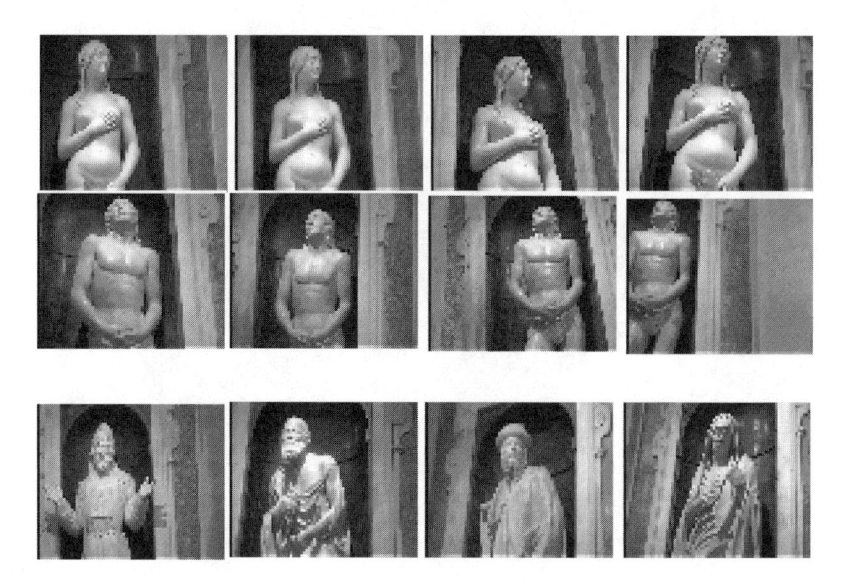

FIGURE 11.12 Samples of the test set: statue A in the top row, statue B in the middle row, and other statues in the bottom row.

case no segmentation is advisable, as the background itself is representative of the object. Here we can safely assume that the subjects will not be moved from their usual position.

Figure 11.11 shows samples of the training sets used to acquire the 3D appearance of two of the six statues; in what follows we will refer to these two statues as statue A and statue B. Figure 11.12 illustrates samples of positive (two top rows) and negative (bottom row) examples with respect to both statues A and B, used for preliminary tests of the system, which produced the ROC curves of Figures 11.13 and 11.14,

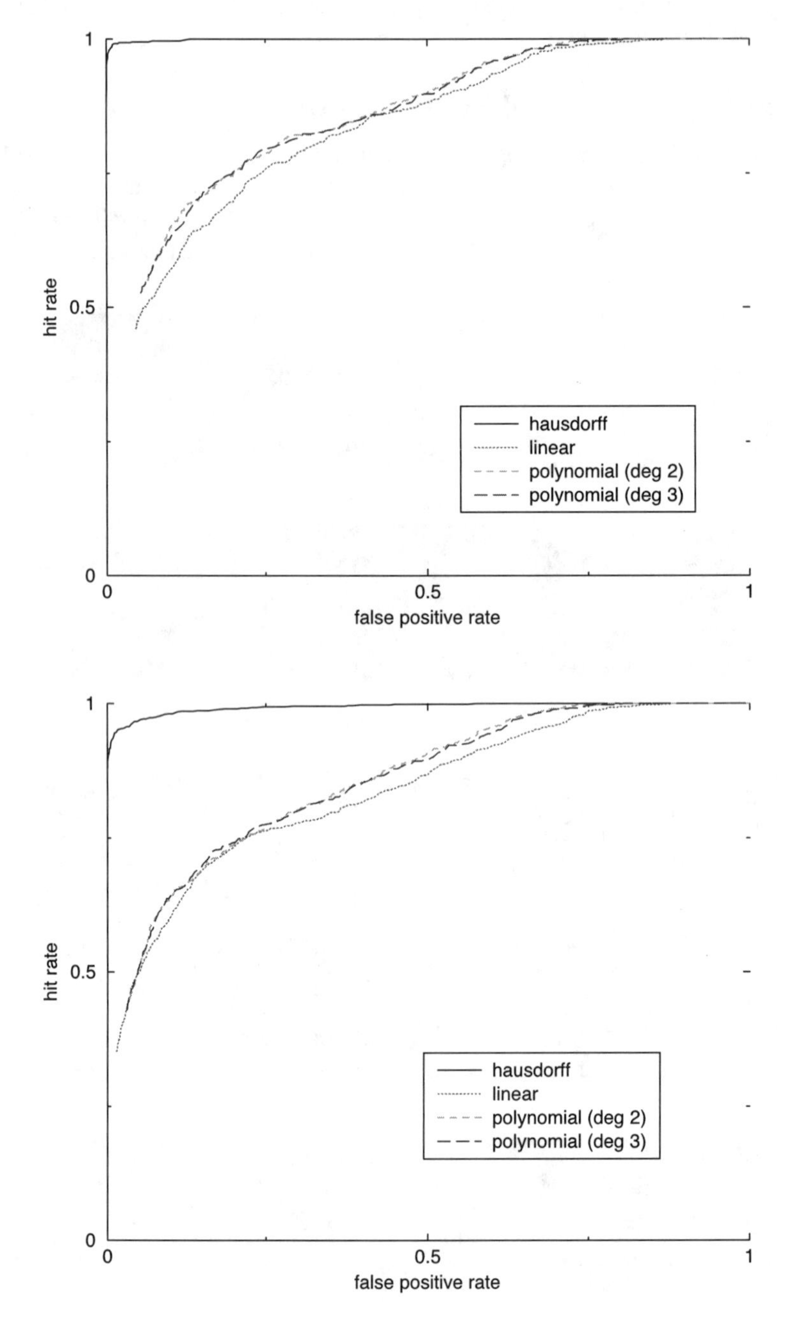

FIGURE 11.13 ROC curves showing the effect of decreasing the size of the training set for statue A, from 314 (top) down to 96 elements (bottom). The size of the test set is 846 (positives) and 3593 (negatives).

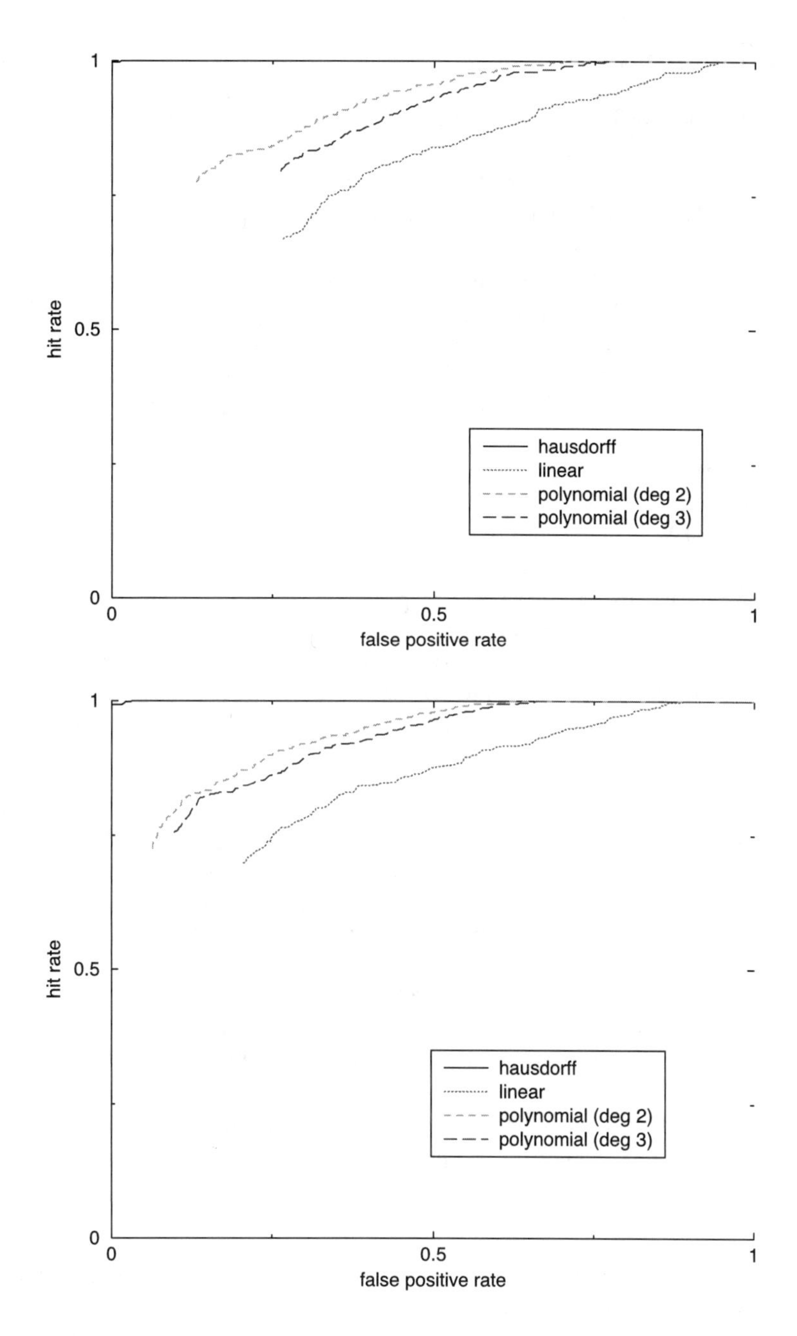

FIGURE 11.14 ROC curves showing the effect of decreasing the size of the training set for statue B, from 226 (top) down to 92 elements (bottom). The size of the test set is 577 (positives) and 3545 (negatives).

respectively. Notice that some of the negative examples look similar to the positive ones, even to a human observer, especially at the image sizes (58×72) used in the representation or training stage.

From the point of view of representation, we have trained the system with increasingly small training sets to check for graceful degradations of the results. In Figures 11.13 and 11.14 the top curves have been obtained with bigger training sets (314 images for statue A and 226 images for statue B) than the bottom ones (97 images for statue A and 95 images for statue B). Notice that the results with the Hausdorff kernel are still very good (in three of the four graphs the Hausdorff kernel curve almost goes through the top left corner).

11.6 Conclusions

In this concluding section we first discuss the real-time aspect of implementing image-recognition systems and then mention the main unresolved problems in this framework. Real-time implementation of trainable image-recognition systems depends on two main issues. The first concerns scale. If the apparent size of the object to be detected is unknown, the search must be performed across different scales. In this case, a spatial multiresolution implementation is often the simplest way to reduce the computation time, possibly avoiding the need for special-purpose hardware. The second issue concerns the location of the object in the image. Clearly, if some prior knowledge about the object location in the image is available, the detection stage can be performed on certain image regions only. Note that for both these problems a parallel architecture is well suited for the optimal reduction of computing time.

A preliminary version of a real-time face-detection and identification system based on the modules described in this chapter is currently under development in our lab. In this case the location of both the scale and the image is determined fairly accurately in the training stage because the system has to detect faces of people walking across the lab main door. In this experimental setting all faces have nearly equal size and appear in well-defined portions of the image.

Trainable image-recognition systems seem to provide effective and reusable solutions to many computer vision problems. In this framework a primary unresolved problem concerns the development of kernels able to deal with different kinds of objects and invariant to a number of geometric and photometric transformations. From the real-time viewpoint, it is important to develop efficient implementation of kernels and kernel methods. Finally, the investigation of procedures able to learn from partially labeled data sets will make it possible to realize systems that could be trained with much less effort.

11.7 Summary

This chapter examines the structure and methods underlying the construction of efficient image-recognition systems. From the very beginning we argue in favor of adaptive systems, or systems that can be *trained* instead of *programmed* to perform a given image-recognition task. We first discuss the philosophy of the approach and focus on some of the methods recently proposed in the area of object detection and recognition. Then, we review the fundamentals of statistical learning, a mathematically well-founded framework for dealing with the problem of supervised learning—or *learning from examples*—and we describe a method for measuring the similarity between images, allowing for an effective representation of visual information. Finally, we present some experimental results of a trainable system developed in our lab and based on this similarity measure applied to two different object-detection problems and discuss some of the open issues.

References

Belhumeur, P., Hespanha, P., and Kriegman, D., Eigenfaces vs. Fisherfaces: recognition using class specific linear projection, *IEEE Trans. PAMI*, **19**, 711–720, 1997.

Beymer, D. J., Face Recognition under Varying Poses, AI Memo 1461, MIT, 1993.

Beymer, D. J., McLauchlan, P., Coifman, B., and Malik, J., A real time computer vision system for measuring traffic parameters, in *Proc. IEEE Conf. CVPR*, Puerto Rico, 1997, pp. 495–501.

Bobick, A., Intille, S., Davis, J., Baird, F., Pinhanez, C., Campbell, L., Ivanov, Y., Schütte, A., and Wilson, A., The KidsRoom: a perceptually-based interactive and immersive story environment, *PRESENCE: Teleoperators and Virtual Environments*, **8**, 367–391, 1999.

Bregler, C., Learning and recognizing human dynamics in video sequences, in *Proc. IEEE Conf. CVPR*, Puerto Rico, 1997.

Bregler, C. and Malik, J., Learning appearance based models: mixtures of second moments experts, in *Advances in Neural Information Processing Systems*, MIT Press, Cambridge, MA, 1996.

Brunelli, R. and Poggio, T., Face recognition: features versus templates, *IEEE Trans. Pattern Analysis Machine Intelligence*, **15**, 1042–1052, 1993.

Bülthoff, H. H. and Edelman, S., Psychophysical support for a two-dimensional view interpolation theory of object recognition, *Proc. Natl. Acad. Sci. U.S.A.*, **89**, 60–64, 1992.

Campbell, C. and Bennett, K. P., A linear programming approach to novelty detection, in *Advances in Neural Information Processing Systems*, 2001.

Cucker, F. and Smale, S., On the mathematical foundations of learning, *Bull. Am. Math. Soc.*, **39**, 1–49, 2002.

Evgeniou, T., Pontil, M., and Poggio, T., Regularization networks and support vector machines, *Advances Computational Math.*, **13**, 1–50, 2000.

Girosi, F., Jones, M., and Poggio, T., Regularization theory and neural networks architectures, *Neural Computation*, **7**, 219 –269, 1995.

Guodong, G., Li, S., and Kapluk, C., Face recognition by support vector machines, in *Proc. IEEE Intl. Conf. Automatic Face Gesture Recognition*, 2000, pp. 196–201.

Haritaoglu, I., Harwood, D., and Davis, L., W^4S: a real time system for detecting and tracking people in 2.5 D, *Proc. Eur. Conf. Computer Vision*, Freiburg, Germany, 1998.

Huttenlocher, D. P., Klanderman, G. A., and Rucklidge, W. J., Comparing images using the Hausdorff distance, *IEEE Trans. Pattern Analysis Machine Intelligence*, **9**, 850–863, 1993.

Intille, S., Davis, J., and Bobick, A., Real-time closed-world tracking, in *Proc. IEEE Conf. CVPR*, Puerto Rico, 1997.

Jonsson, K., Matas, J., Kittler, J., and Li, Y., Learning support vectors for face verification and recognition, *Proc. IEEE Intl. Conf. Automatic Face Gesture Recognition*, 2000.

Leung, T. K., Burl, M. C., and Perona, P., Finding faces in cluttered scenes using random labeled graph matching, in *Proc. Intl. Conf. Comput. Vision*, Cambridge, MA, 1995.

Logothetis, N., Pauls, J., and Poggio, T., Shape representation in the inferior temporal cortex of monkeys, *Current Biol.*, **5**, 552–563, 1995.

Mohan, A., Papageorgiou, C., and Poggio, T., Example-based object detection in images by components, *IEEE Trans. Pattern Analysis Machine Intelligence*, **23**, 349–361, 2001.

Murase, H. and Nayar, S. K., Visual learning and recognition of three-dimensional object from appearance, *Int. J. Comput. Vision*, **14**, 5–24, 1995.

Nefian, A. V. and Hayes, M. H., An embedded HMM-based approach for face detection and recognition, in *Proc. IEEE ICSSP*, 1999.

Odone, F., Trucco, E., and Verri, A., General purpose matching of grey level arbitrary images, *LNCS*, **2059**, 573–582, 2001.

Osuna, E., Freund, R., and Girosi, F., Training support vector machines: an application to face detection, *Proc. IEEE Conf. CVPR*, Puerto Rico, 1997, pp. 130–136.

Papageorgiou, C. and Poggio, T., A trainable system for object detection, *Int. J. Comput. Vision*, **38**, 15–33, 2000.

Papageorgiou, C., Oren, M., and Poggio, T., A general framework for object detection, in *Proc. Int. Conf. Comput. Vision*, Bombay, India, 1998, pp. 555–562.

Pittore, M., Campani, M., and Verri, A., Learning to recognize visual dynamic events from examples, *Int. J. Comput. Vision*, **38**, 35–44, 2000.

Pontil, M. and Verri, A. Object recognition with support vector machines, *IEEE Trans. Pattern Analysis Mach. Intell.*, **20**, 637–646, 1998.

Rikert, T. K., Jones, M. J., and Viola, P., A cluster-based statistical model for object detection, in *Proc. IEEE Conf. CVPR*, Fort Collins, CO, 1999.

Rowley, H. A., Baluja, S., and Kanade, T., Neural network-based face detection, in *IEEE Trans. Pattern Analysis Machine Intelligence*, **20**, 23–38, 1998.

Schneiderman, H. and Kanade, T., A statistical method for three-dimensional object detection applied to faces and cars, in *Proc. IEEE Conf. Comput. Vision Pattern Recognition*, 2000.

Sirovitch, L. and Kirby, M., Low-dimensional procedure for the characterization of human faces, *JOSA A*, **2**, 519–524, 1987.

Sung, K. K. and Poggio, T., Example-based learning for view-based human face detection, *IEEE PAMI*, **20**, 39–51, 1998.

Tax, D. and Duin, R., Data domain description by support vectors, in *Proc. ESANN99*, 1999, pp. 251–256.

Tax, D., Ypma, A., and Duin, R., 1999 Support vector data description applied to machine vibration analysis, in *Proc. 5th Annual Conference of the Advanced School for Computing and Imaging*, 1999.

Turk, M. and Pentland, A., Face recognition using eigenfaces, in *Proc. IEEE Conf. CVPR*, 1991.

Vapnik, V. N., *The Nature of Statistical Learning Theory*, Springer-Verlag, Berlin, 1995.

Vapnik, V. N., *Statistical Learning Theory*, Wiley, New York, 1998.

Wahba, G., *Splines Models for Observational Data*, Series in Applied Mathematics, Vol. **59**, SIAM, Philadelphia, PA, 1990.

Wiskott, L., Fellous, J. M., Kruger, M., and von der Malsburg, C., A statistical method for three-dimensional object detection applied to faces and cars, in *Proc. IEEE Int. Conf. on Image Processing*, Vienna, Austria, 1997.

Wren, C., Azarbayejani, A., Darrell, T., and Pentland, A., Pfinder: real-time tracking of the human body, *IEEE Trans. PAMI*, **19**, 780–785, 1995.

For Further Information

For a thorough and technical introduction to statistical learning the interested reader can consult Vapnik [1995], Evgeniou et al. [2000], and Cucker and Smale [2001]. The literature of image recognition is vast. Reviewing the proceedings of the major conferences (International Conference of Computer Vision, International Conference of Computer Vision and Pattern Recognition, European Conference of Computer Vision) is the simplest way to catch up with recent advances. Several workshops—for example, the International Workshop on Face and Gesture Recognition or the Workshop on Machine Vision Applications—cover more specific topics. A good starting point is the references above and the references therein.

Finally, for software, demos, data sets, and more, search for the up-to-date URL of the Computer Vision homepage on your favorite search engine.

12

Optical Pattern Recognition

Katsunori Matsuoka
National Institute of Advanced Industrial Science and Technology
Human Stress Signal Research Center
Osaka, Japan

12.1 Introduction

Laser technology brought us a variety of optical information technology in many application fields such as optical computing, optical pattern recognition, optical encryption, optical networks, etc. Optical information processing has a great ability to perform fast operations on two-dimensional information, such as images, by taking advantage of parallel processing.

In pattern recognition, the correlation of images is a necessary operation that involves searching the target image in a whole input image. This correlation can be carried out by the optical system in a short time, while a computer requires considerable time for it because of its sequential operation. This advantage of optics is useful and necessary for pattern-recognition systems to realize real-time processing in the real world.

Many studies on optical pattern recognition have been conducted. In the 1960s, an optical spatial frequency filtering was shown to have the ability to perform matched filtering for extracting a target signal from the input [VanderLugt, 1964]. Matched spatial filtering is optimal in the sense that it provides the maximum signal-to-noise ratio in the output. But it is sometimes too sensitive for small changes in the target. To perform distortion-invariant pattern recognition, Hester and Casasent [1980] proposed the idea of a synthetic discriminant function (SDF), and several advanced versions of SDFs were also proposed. To provide rotation-invariant recognition, the circular harmonic expansion filter was proposed [Hsu and Arsenault, 1982]. Horner and Gianino [1984] proposed another type of filter called the phase-only filter, designed to improve sensitivity and light efficiency; it uses only the phase signal with uniform amplitude in the frequency domain.

0-8493-1162-4/03/$0.00+$1.50

In the development of optical pattern-recognition systems, a three-dimensional disk-based optical correlator was developed [Curtis and Psaltis, 1994] that has the storage capacity of 150,000 images on a photopolymer disk and can perform 6000 correlations per second. A compact real-time multichannel joint-transform correlator was reported by Rajbenbach et al. [1992].

A pattern-recognition system based on the single-correlation system is simple to implement and can successfully achieve high performance when the correlation filter is designed properly. But it sometimes produces false signals caused by background or nonuniform illumination because of its poor discrimination of images. In practical use, false detection restricts the applicability of optical pattern recognition to real-world processing.

To achieve robust processing with an optical system, we have proposed an **optical multiple-correlation method.** It can achieve high discrimination of target patterns; for example, it can discriminate 10^{30} kinds of patterns when 100 multiple correlations are used in the system. This discrimination ability makes it possible to detect targets even if an input image includes many unknown images such as background or noise.

In this chapter, optical pattern recognition based on the optical correlation is described. First, the method for carrying out optical pattern recognition and its basic theory are explained. Second, the limitation of optical pattern recognition with a single correlation is described. The multiple-correlation method is introduced to achieve high performance of pattern recognition. Third, an implementation of an optical system executing the multiple correlation method is shown, and its application to the robot vision system is demonstrated. Finally, the performance of an optical system executing the multiple-correlation method is evaluated.

The research work on the multiple-correlation method and its implementation of optical systems were carried out under the joint research project of "Tera-Optical-Information-Processing Technology" sponsored by the Japan Science and Technology Corporation.

12.2 Optical Pattern Recognition with a Single Correlation

12.2.1 Basic Operation of the Optical Pattern-Recognition System

In optical pattern recognition, the correlation between an input image and a filter image is the fundamental operation for discriminating an input image. Correlation can be described as

$$O(x, y) = \iint f(x', y')\, i(x + x', y + y')\, dx'dy', \tag{12.1}$$

where $O(x, y)$, $f(x, y)$, and $i(x, y)$ denote a correlation result, a filter image, and an input image, respectively, and x and y denote the coordinates on the image plane.

The correlation pattern between the same two images usually has a large correlation peak in the center of the correlation pattern, while the correlation pattern between two different images has a smaller correlation peak than that between the same two images. Then, an input image can be discriminated with the correlation peak value.

For example, a correlation pattern between character patterns of "A" and "B" can be obtained as Figure 12.1. The correlation between the same two patterns of "A" has a large correlation peak, and the correlation peak between "A" and "B" is small. Then, an input image can be recognized as "A" when the correlation peak is larger than the threshold level, while an input image is recognized as not "A" when the correlation peak is smaller than the threshold level.

The ability of some optical systems to achieve a fast operation of correlation between two images is superior to that of the digital computer. Figure 12.2 shows an optical correlation system that is known as the VanderLugt correlator [VanderLugt, 1964]. The Fourier transform of the filter image is recorded on the photo-film at the P_2 plane by a holographic process. When an input image on a photo-film is located on the P_1 plane and is illuminated by a coherent light such as a laser light, the Fourier transform of the input image appears on the P_2 plane and is modulated with the Fourier-transformed signal of the

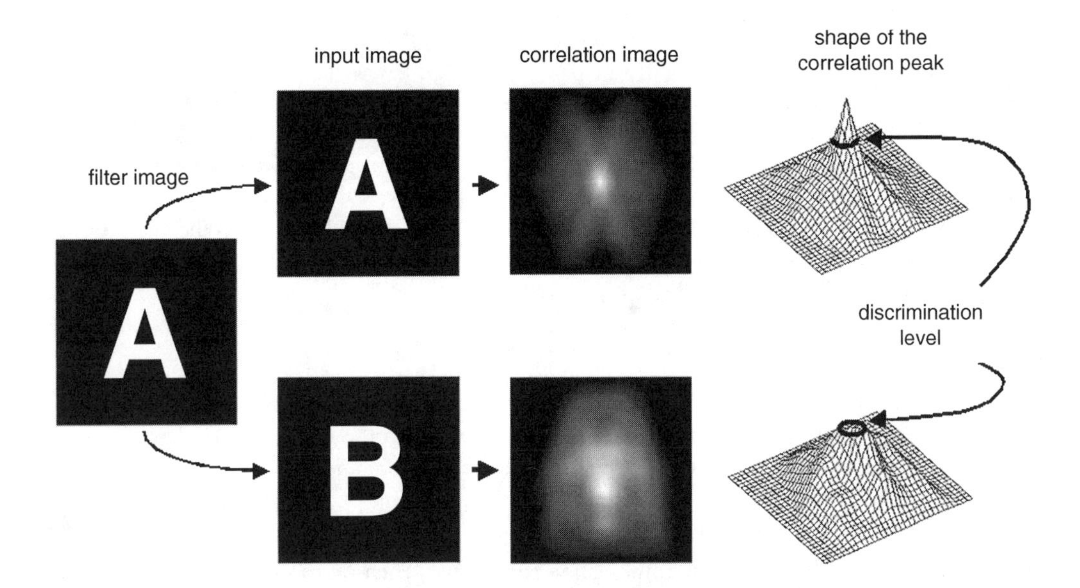

FIGURE 12.1 Correlation patterns between character images of "A" and "B."

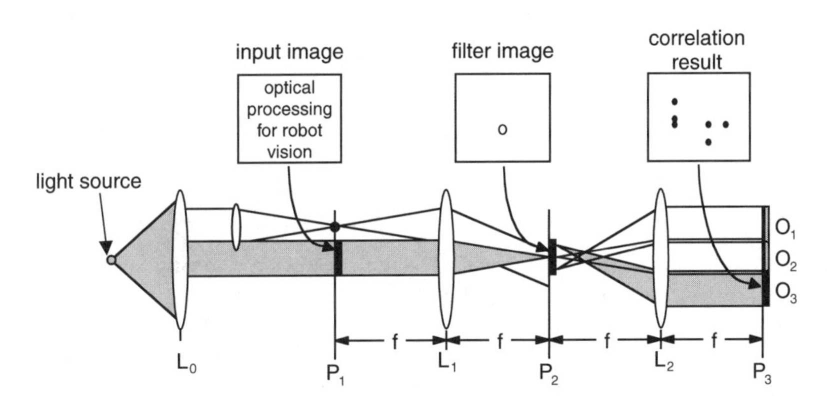

FIGURE 12.2 Optical correlation system (VanderLugt type).

filter image recorded on the P_2 plane. Then, the modulated light is transferred through lens L_2 and the Fourier-transformed pattern of light distribution after the P_2 plane appears on the P_3 plane. Then, the correlation pattern between the input and the filter can be obtained on position O_3 of the P_3 plane.

The system illustrated in Figure 12.2 requires only three lenses (L_0, L_1, and L_2) with focal length f. The correlation result can be obtained during light propagation from the light source to the O_3 plane, which is performed in a very short time. All correlation peaks that denote the positions of target images in the input image can be obtained simultaneously as the bright spots on the result, while the computer usually searches each position of the target sequentially.

Figure 12.2 diagrams the detection of character "o" in the input image. All positions of character "o" can be found as the spots on the correlation result simultaneously.

12.2.2 Design of a Single Optical Correlation Filter

In optical pattern recognition, the function of the correlation filter defines the performance of recognition. To get high performance, many superior methods for filer design have been proposed and were reviewed by Kumar [1992]. Some well-known design methods for the correlation filter are reviewed here.

The most familiar filter is the matched filter, which is the same as the pattern to be detected. For example, character pattern "A" is the matched filter for detecting character "A." The matched filter provides the largest signal-to-noise ratio at the correlation peak, but its pattern discrimination is poor compared with other filters.

To obtain the desired function of pattern discrimination, a filter design using the synthetic discriminant function (SDF) has been proposed [Hester and Casasent, 1980; Casasent, 1984]. It enables us to control the correlation peak value obtained with each of the typical input images, known as **training images**, at the design stage.

Let us consider the filter design for n training images by using the SDF method. When each training image is represented by a vector $\mathbf{x}_i$ ($i = 0, \ldots, n - 1$), the **correlation value** c_i between a training image $\mathbf{x}_i$ and the filter image $\mathbf{f}$ can be described as

$$\mathbf{c} = \mathbf{X}^t\mathbf{f}, \tag{12.2}$$

where the vector $\mathbf{c}$ consists of the correlation value c_i, the column vector of matrix $\mathbf{X}$ denotes each training image $\mathbf{x}_i$, and superscript t denotes the transpose of matrix. When the filter $\mathbf{f}$ can be described as the linear combination of training images $\{\mathbf{x}_i\}$, then that filter can be obtained as

$$\mathbf{f} = \mathbf{X}\mathbf{a}, \tag{12.3}$$

where the vector $\mathbf{a}$ denotes the linear-combination coefficients. Then we can obtain the filter designed by the SDF method from Eqs. (12.2) and (12.3) as

$$\mathbf{f} = \mathbf{X}(\mathbf{X}^t\mathbf{X})^{-1}\mathbf{c}. \tag{12.4}$$

The SDF filter defined uniquely as Eq. (12.4) will produce a correlation peak with a predefined value by $\mathbf{c}$ for each input image $\mathbf{x}_i$.

To enhance the sensitivity or the noise reduction of the designed correlation filter, we can introduce a function $\mathbf{W}$ to Eq. (12.3) as

$$\mathbf{f} = \mathbf{W}\mathbf{X}\mathbf{a}. \tag{12.5}$$

The matrix $\mathbf{W}$ affects the training images as **spatial frequency filtering** such as high-frequency-enhancement filtering or high-frequency-cut filtering. From Eqs. (12.2) and (12.5), we obtain the correlation filter as

$$\mathbf{f} = \mathbf{W}\mathbf{X}(\mathbf{X}^t\mathbf{W}\mathbf{X})^{-1}\mathbf{c}. \tag{12.6}$$

The correlation filter described in Eq. (12.6) is the generalized SDF filter. When the Fourier transform of $\mathbf{W}$ is selected to be equal to the inverse of the average of power spectrums of all training images, the filter of Eq. (12.6) is called the minimum average correlation energy (MACE) filter [Mahalanobis et al., 1987]. By adding the constraints for suppressing noise to the MACE filter, the filter of Eq. (12.6) represents the minimum noise and correlation energy (MINACE) filter [Ravichandran and Casasent, 1992]. A typical constraint used in the MINACE filter is to limit the upper level of the inverse of the averaged power spectrum according to the signal-to-noise ratio of the input image. The MACE filter and the MINACE filter usually generate a sharp correlation peak and reduced side lobes of a correlation peak, but it is more sensitive to noise than the SDF filter.

The performance of pattern recognition based on the correlation can be designed properly by these design methods for the correlation filter.

12.3 Optical Pattern Recognition with Multiple Correlations

12.3.1 Limitations of Pattern Recognition with a Single Correlation

Optical correlation systems are able to perform fast correlation for the pattern recognition. However, a single correlation is limited to pattern recognition in real-world processing.

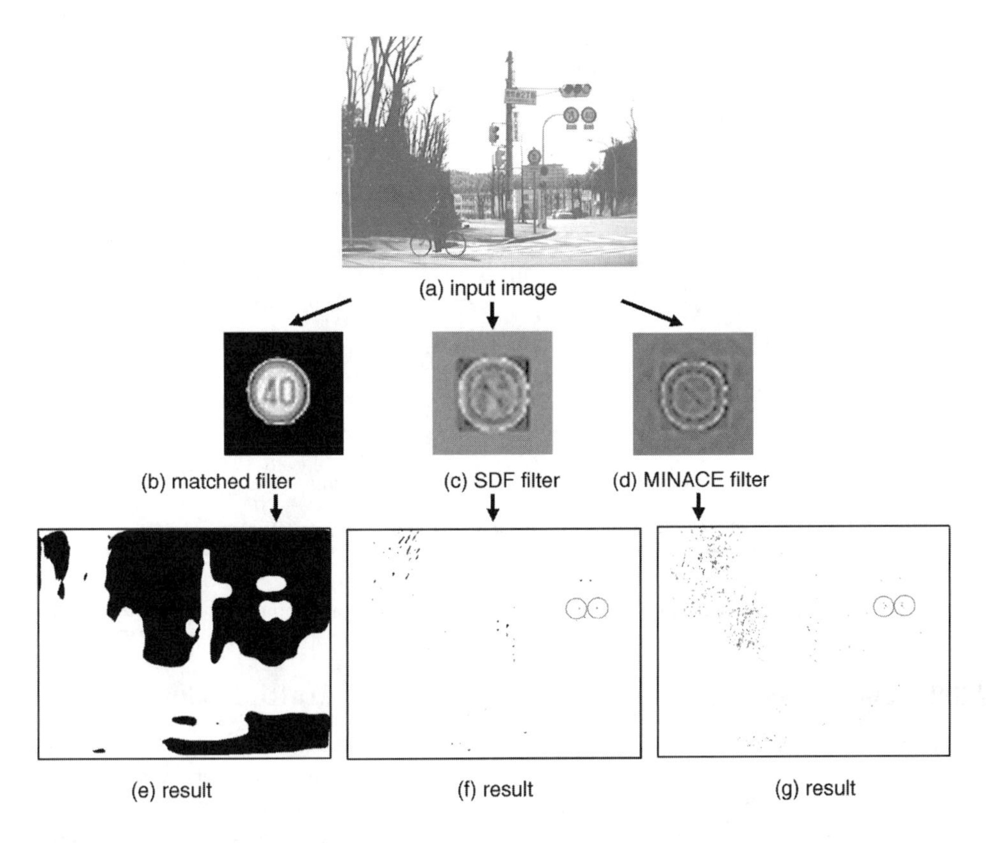

FIGURE 12.3 Results of road-sign detection by using a single correlation with conventional correlation filters. In the resulting images, dark spots denote the obtained signals and circled spots indicate correct signals. (From Matsuoka, K., *Jpn. J. Opt.*, 25, 265–271, 1996. With permission.)

For real-world image processing, a robust processing must be realized against the following unknown factors:

1. Locations and number of targets are unknown *a priori.*
2. Contrast or brightness of an input image is not uniform.
3. The shape of target changes according to the view angle or the view distance.
4. Many unknown images such as noise or background are included in a real-world scene.

A conventional optical system based on a single correlation cannot achieve robustness against all these unknown factors. Therefore, in many cases a single-correlation process produces many false results caused by the above-mentioned unknown factors.

For example, Figure 12.3 shows the results of detecting road signs from a real-world scene by using a single correlation with well-known correlation filters—matched filter, SDF filter, and MINACE filter. Dark spots denote the signals detected by a single correlation and circled spots in the figures denote correct detections. The matched filter shown in Figure 12.3(b) has positive values in the road sign area with zero background, and bright areas in the input are detected in the result of Figure 12.3(e). Each filter shown in Figures 12.3(c) and 12.3(d) has positive and negative values in the road-sign area with zero background; in this state the filters can suppress the detection of the uniformly bright area. But many false signals caused by an unknown background appear in the results.

The limitation of the single-correlation method can be observed when it is applied to pattern recognition.

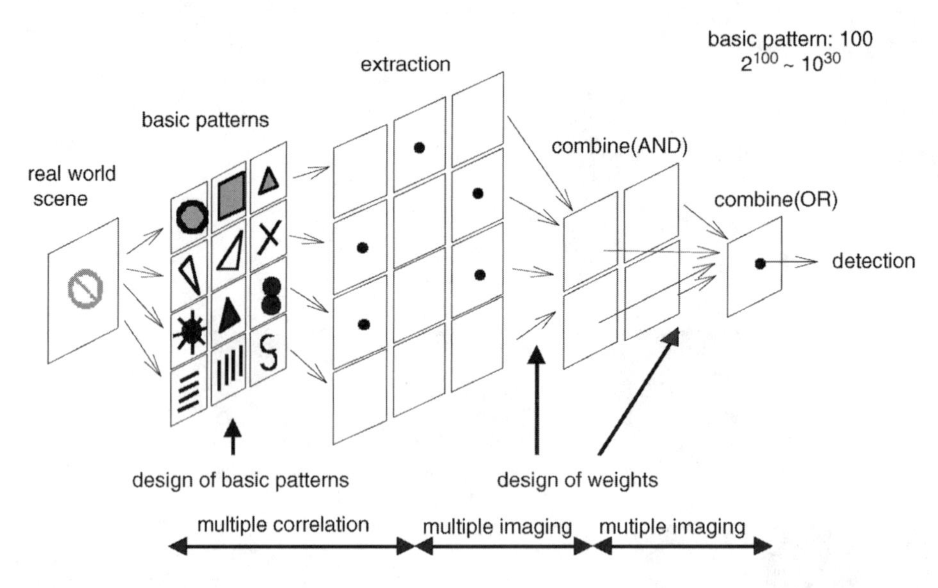

FIGURE 12.4 Basic concept of pattern recognition based on multiple correlations. (From Matsuoka, K., Optical pattern recognition system using "multiple correlators," SPIE's Int. Tech. Group Newsletter, 2000. With permission.)

12.3.2 The Concept of Pattern Recognition with Multiple Correlations

To realize a high performance of pattern recognition with an optical correlation system, some pattern discrimination methods based on multiple correlations have been proposed [Braunecker, 1979; Matsuoka, 1996]. The multiple-correlation method has great potential for discriminating many patterns (N patterns) with a small number ($\log_2 N$) of filters. The optical pattern recognition system based on multiple correlations that we are now developing is described below.

The basic concept of pattern recognition based on multiple correlations is illustrated in Figure 12.4. It consists of two processes—multiple correlations and combining the multiple-correlation results. The multiple-correlation process detects which basic patterns are included in an input image. The combination of extracted basic patterns identifies an input image. Then N number of input images can be identified with the $\log_2 N$ number of basic patterns, when the basic patterns can be designed properly.

This means that if we use only 100 basic patterns, 2^{100} images ($\sim 10^{30}$ images) can be identified by the multiple-correlation system. This ability for high discrimination makes it possible to detect target patterns without false signals, even if an input image contains unknown background.

The multiple correlation method is carried out as shown in Figure 12.5. Three correlation filters were designed as the basic patterns for detecting human faces, as shown in Figure 12.5. The design rule of multiple correlation filters is described in the next section. First, three correlation results of an input image with three correlation filters are obtained individually. Then, the results are binarized with the threshold to get the signals detecting the basic patterns individually. Finally, all correlation results are combined as an AND operation. In the final result, the signals of face positions can be obtained without false signals, while intermediate correlation images have false signals. This signifies that the multiple correlation method increases noise tolerance by suppressing false signals with other intermediate correlation results.

Figure 12.6 shows a comparison between results obtained with the multiple-correlation method and with the SDF filter. The target is a road sign. Three basic patterns shown in Figure 12.6(d) are designed to detect circle-type road signs, and they are used as the multiple-correlation filters. Figure 12.6(a) is an input image, (b) is the result obtained by the multiple-correlation method, and (c) is the result obtained by the SDF filter. The threshold for each result to discriminate the signal was individually defined so as to get both signals of road signs, and the detection signals are shown as dark spots in the results. Figure 12.6 shows that the multiple-correlation method is superior to the single-correlation method with respect to the amount of false signals.

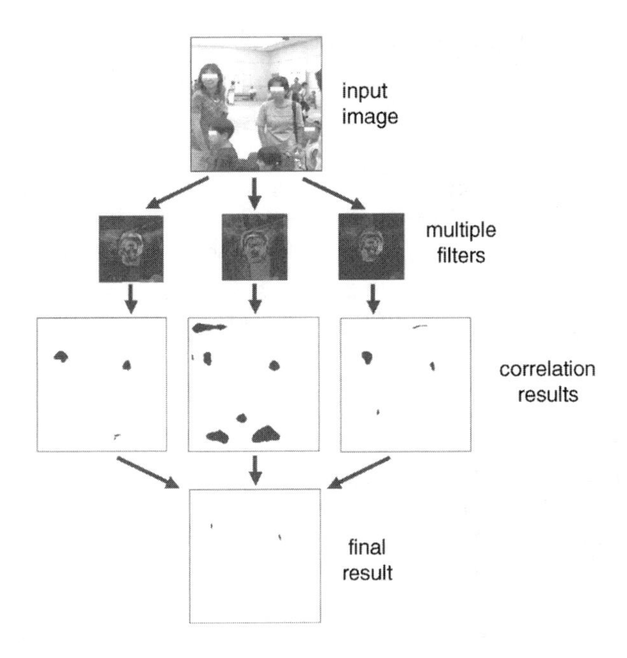

FIGURE 12.5 Process flow of the multiple-correlation method.

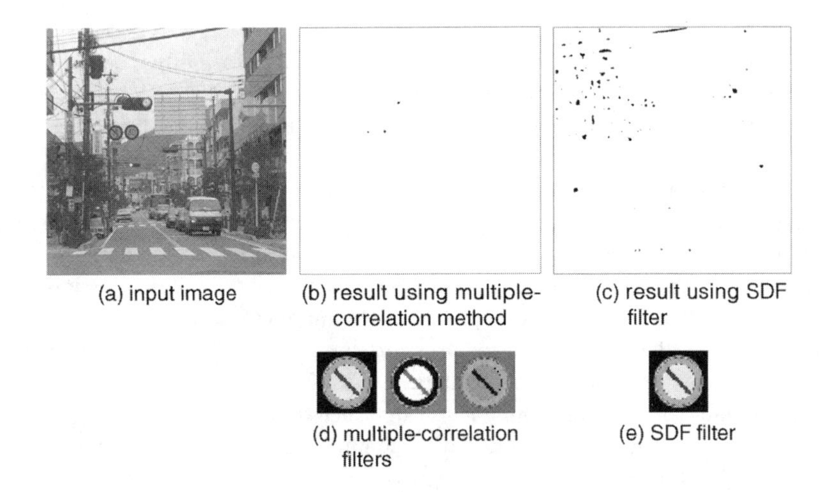

FIGURE 12.6 Comparison between results obtained with the multiple-correlation method and with the SDF filter. The target is a road sign.

12.3.3 Design of Multiple-Correlation Filters

The performance of pattern recognition by using the multiple-correlation method depends on the set of basic patterns to be used as the multiple-correlation filters. The purpose of the basic patterns is to find a minimum and necessary set of patterns that describe all training images properly. We applied the **principal component analysis** to find the set of basic patterns.

Principal component analysis finds component images that are commonly included in the images to be analyzed, and all component images extracted by the principal component analysis become **orthogonal** to each other. Now let us consider the principal component analysis for the training images. When the component image $\mathbf{p}$ can be described as the linear combination of training images, the vector $\mathbf{p}$ is expressed as

$$\mathbf{p} = \mathbf{X}\mathbf{w}, \tag{12.7}$$

where the vector $\mathbf{w}$ is the unit vector (that is, $\mathbf{w}^t\mathbf{w} = 1$), and it expresses the coefficient of the linear combination. The column vector of matrix $\mathbf{x}$ denotes each training image. The principal component analysis leads the component with the largest power. It is described as

$$\mathbf{p}^t\mathbf{p} = \mathbf{w}^t\mathbf{X}^t\mathbf{X}\mathbf{w} \text{ -> maximize.} \tag{12.8}$$

The matrix $\mathbf{X}^t\mathbf{X}$ denotes the correlation matrix of training images. It is well known that the vector $\mathbf{w}$ maximizing the inner product $\mathbf{p}^t\mathbf{p}$ is the eigenvector of the correlation matrix $\mathbf{X}^t\mathbf{X}$, which has the largest eigenvalue λ_0. Then, we can obtain the principal component $\mathbf{p}_0$ described as Eq. (12.7), with the eigenvector $\mathbf{w}_0$ corresponding to the eigenvalue λ_0. Therefore, the principal component $\mathbf{p}_0$ denotes the largest component that is commonly included in the training images.

The first principal component $\mathbf{p}_0$ is obtained by Eq. (12.7) for the largest eigenvalue λ_0, the second principal component $\mathbf{p}_1$ is obtained for the second largest eigenvalue λ_1, and so on, and the eigenvalues of the correlation matrix $\mathbf{X}^t\mathbf{X}$ are described as λ_0, λ_1, λ_2, and so on. The inner product of the principal components $\mathbf{p}_i$ and $\mathbf{p}_j$ are described as

$$\mathbf{p}_i^t\mathbf{p}_j = \mathbf{w}_i^t\mathbf{X}^t\mathbf{X}\mathbf{w}_j = \lambda_j\mathbf{w}_i^t\mathbf{w}_j = \begin{cases} \lambda_j & \text{for } i = j, \\ & \text{for } i \neq j, \end{cases} \tag{12.9}$$

because the eigenvectors of the correlation matrix are orthogonal to each other. Equation (12.9) shows that the principal components are orthogonal to each other. The value $\lambda_j/\Sigma\lambda_i$ is called the contribution rate for the principal component $\mathbf{p}_j$, and a large contribution rate signifies that the principal component contributes strongly to all training images. As a matter of course, the accumulated contribution rate for all principal components becomes one.

Next, let us consider the extraction of the necessary common components from the training images for the filter design. To perform robust pattern recognition and to reduce the noise it is necessary to design the correlation filter to be sensitive to the common components of all training images and not to be sensitive to the weak or individual components of the training images. We can find the common component from the training images by using principal component analysis; the principal component with a large contribution rate is a common component and that with a small contribution rate is a weak component. The number of common components can be determined by selecting the principal components in the order of contribution rates until their accumulated contribution rate exceeds the predetermined value, which is usually 0.8 to 0.95 in our experiments. Then, we can properly describe all of the training images with the limited number of principal components.

The selected principal components represent the common pattern in the training image, but they are not suitable for basic patterns in the multiple correlation system because their correlation values with the training images cannot be designed. Thus, we have developed a design process for the basic pattern, as shown in Figure 12.7. The design process is based on two stages. In the first stage, common components are extracted from the training images by using principal component analysis. Two sets of components are extracted, one for training images describing true images (target images) and another for training images describing false images (background). In the second stage, basic patterns are synthesized with these two sets of common components. In the synthesizing stage, each basic pattern is designed to generate a predetermined correlation value for each training image by using the same method as the SDF calculation. The number of basic patterns is equal to the number of common components extracted by the principal component analysis. In this process, the basic patterns to be used in the multiple correlation system can be designed. These basic patterns are used as the multiple-correlation filters.

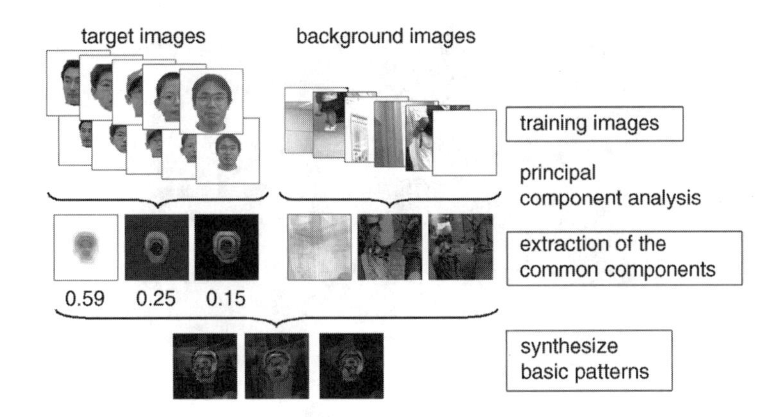

FIGURE 12.7 Design method for multiple-correlation filters.

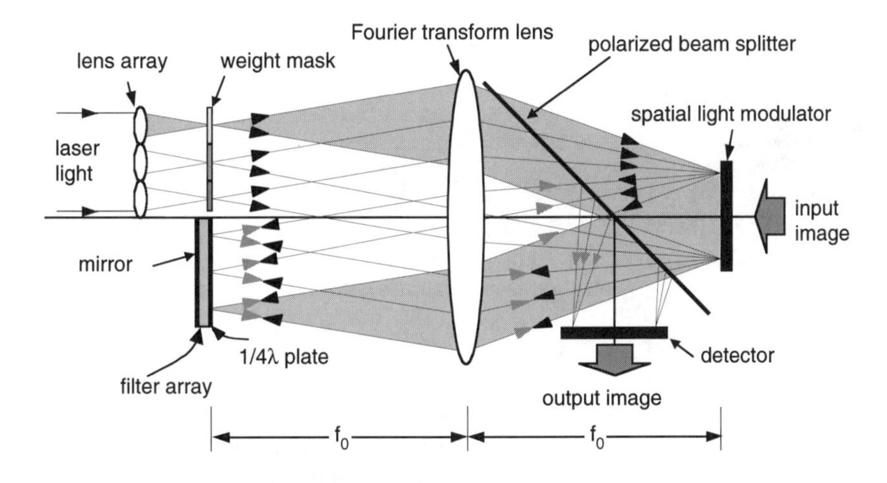

FIGURE 12.8 An all-optical system executing multiple correlations. (Matsuoka, K., Optical pattern recognition system using "multiple correlators," SPIE's Int. Tech. Group Newsletter, 2000. With permission.)

12.4 Implementation of Optical Multiple-Correlation Systems

12.4.1 Optical Multiple-Correlation System

The multiple-correlation method described in Figure 12.4 can be implemented with the optical system illustrated in Figure 12.8.

In Figure 12.8, a real-world image is focused onto the spatial light modulator (SLM) and is read out by each of the collimated beams generated by a lens-array and by the Fourier transform lens. Then, multiple Fourier images of an input image appear at the filter array plane through the Fourier transform lens. An array of the Fourier transforms of basic patterns is recorded as the multiple-correlation filters on the filter array plane. Each of the multiple Fourier images of the input is modulated with multiple-correlation filters individually, and their inverse Fourier transforms are formed on the detector plane through the Fourier transform lens. Therefore, the image on the detector plane is the superimposed one of each correlation image between the input and each of the basic patterns.

The light coming from or going toward the spatial light modulator can be separated with the polarized beam splitter by rotating the polarization of the light with the quarter-wavelength plate at the filter array plane.

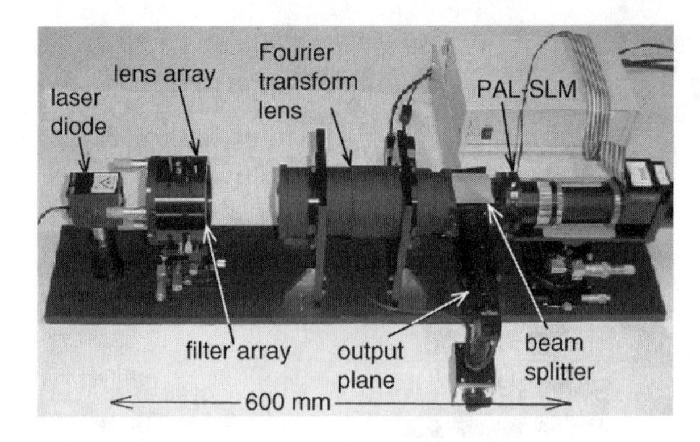

FIGURE 12.9 A prototype system of the optical multiple-correlation system.

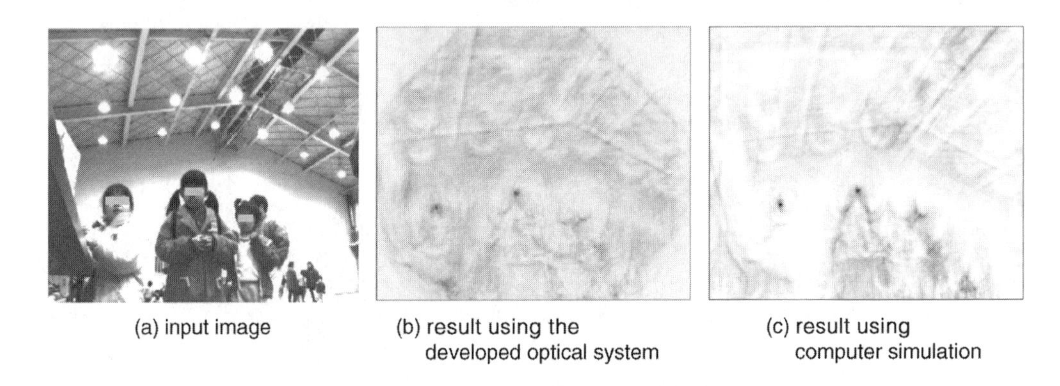

(a) input image (b) result using the developed optical system (c) result using computer simulation

FIGURE 12.10 Result of face finding by using the optical multiple-correlation system.

The operation of multiple correlations is carried out in the path from the spatial light modulator to the filter array plane, and the operation of combining the correlation results is carried out in the path from the filter array plane to the detector plane. The final result of the multiple correlations is obtained on the detector.

Figure 12.9 shows the prototype system of the multiple-correlation system we have developed. Its size is about 1000 mm in length with an 80-mm diameter. This system is designed to perform 100 multiple correlations simultaneously by using a 10×10 array of filters on the filter array plane and a lens array, each having a 3-mm diameter. In this system, the following main optical elements are used to perform the correlation: a Fourier transform lens with a focal length of 80 mm by COSINA Co., Ltd., and a spatial light modulator (PAL-SLM) by Hamamatsu Photonics K.K.

The designed filter array for multiple correlations was fabricated on a silica-glass by lithography technology using the electric-beam drawing system (JBX-5000LS, JEOL) in the laboratory of Osaka Science and Technology Center, which has a line resolution of 30 nm and a drawing area of 100 mm.

Figure 12.10 shows the multiple-correlation result of human face detection using the developed optical system. The result using the optical system is shown in (b), and the result using computer simulation is shown in (c). Both results represent a similar pattern and indicate that the optical system works properly.

12.4.2 Application to the Robot Vision System

The optical multiple-correlation system was applied to the robot vision system shown in Figure 12.11, and its performance was evaluated. The robot vision system consists of a CCD camera on the rotation stage, the optical multiple-correlation system shown in Figure 12.9, and a computer that controls the angle of the CCD camera by using the rotation stage.

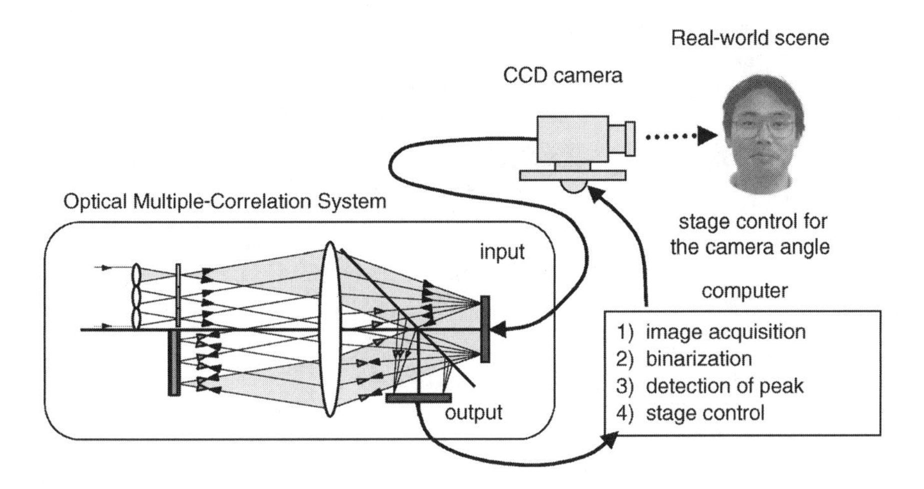

FIGURE 12.11 Robot vision system using the optical multiple-correlation system. This system gazes at a human face to start the communication.

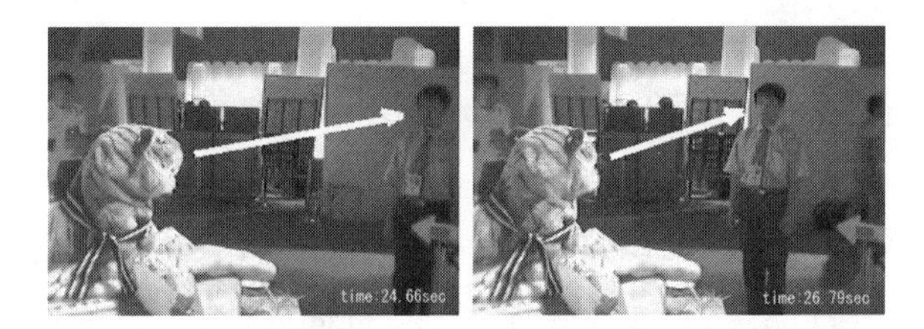

FIGURE 12.12 The developed robot vision system named "Gazing Tiger" and its operation. It gazes at a human face by detecting the position of the face using the optical multiple-correlation system.

The CCD camera functions as an eye for the robot. It acquires an image of the real-world scene around the robot and sends the acquired image to the optical multiple-correlation system. The optical multiple-correlation system is designed to detect human faces. To detect several kinds of human faces and different angles of faces, the multiple-correlation filters were designed with 15 different kinds of facial images as training images. Then, the optical multiple-correlation system generates signals on the locations of human faces in the input image, and the signals are transferred to the computer through the CCD camera in the optical system. The computer finds the strongest signal and moves the CCD camera to look at the position of the signal. Then, the system can always examine the human face that produces the strongest signal in the optical multiple-correlation system.

The tiger shown in Figure 12.12 shows the view of a developed robot vision system called "Gazing Tiger." The CCD camera is located in the mouth of the tiger, and the optical system is behind the tiger. Figure 12.12 shows that the tiger's head moves according to the movement of the human face. It shows that the developed system works properly. In the developed system, it took about 80 msec for the processing of optical multiple correlations, which is strongly dependent on the response time of the spatial light modulator we used.

This robot vision system was demonstrated at the robot festival in Osaka in 2001 over a 10-day period, and it worked without incident even in crowds. The developed optical system is clearly stable enough for real-world processing.

The development of this robot vision system was the result of collaborative efforts by the Osaka Science and Technology Center, Kanebo, Ltd.; COSINA Co., Ltd.; and the National Institute of Advanced Industrial Science and Technology of Japan.

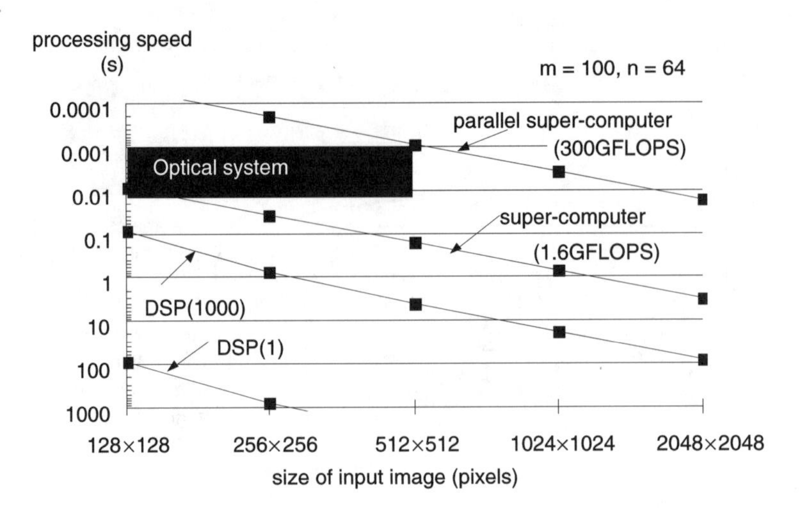

FIGURE 12.13 Comparison of the processing speed of the optical multiple-correlation system and a computer system in the execution of 100 multiple correlations with a filter size of 64 × 64 pixels. (From Matsuoka, K., *Jpn. J. Opt.*, 25, 265–271, 1996. With permission.)

12.4.3 Performance of the Optical Multiple-Correlation System

In this section, we consider the performance of the optical multiple-correlation system by comparing it with that of a computer system. When the size of an input image is 512 × 512 pixels and the size of each correlation filter is 64 × 64 pixels, a super computer with 300 GFLOPS (floating point operations per second) can perform about 100 multiple correlations per millisecond, as shown in Figure 12.13. On the other hand, an optical system performs the same number of multiple correlations in about 1 msec to 10 msec, when the spatial light modulator operates at a speed of 1 msec to 10 msec, a speed that is available in light valves containing ferro-electric liquid crystals.

Figure 12.13 shows that the performance of an optical multiple-correlation system is equivalent to that of a super computer in terms of its multiple-correlation operation speed. The size of the optical multiple-correlation system is now about 1 m in length, but it can be designed in a smaller size. We are now designing a 17-cm-long version that can execute 14 correlations simultaneously.

In terms of operation speed and extent, the optical multiple-correlation system has the potential to become useful as a pattern recognition system, but cost is the main obstacle to bringing this about. It is our hope that the spatial light modulator and the Fourier transform lens will become less expensive.

12.5 Conclusion

Optical multiple-correlation systems achieve high pattern-discrimination performance and can operate at high speeds. We have developed a prototype system of an optical multiple-correlation system for evaluating its ability. We applied it to the robot vision system for detecting a variety of human faces in the real world.

Optical multiple-correlation systems are expected to find application in face-identification security systems; car navigation systems that will inform drivers about road-sign information; thermal control systems in large public spaces, such as train stations, that will monitor people's movements; and water-purity monitoring systems that will automatically count the number of microbes contained in a given pool of water. In the future, it might be possible to create a robot or machine with an intelligent vision system by using optical pattern recognition technology.

Defining Terms

correlation value: The inner product of two images, the center value of the correlation of two images.
optical multiple-correlation method: Method of optical pattern recognition by using multiple correlations.
orthogonal: When two images are orthogonal, the correlation value of two images is zero.
principal component analysis: One kind of component analysis for multivariant data.
spatial frequency filtering: Filtering in the spatial frequency domain.
training images: Images to be used at the design process of the correlation filter, which represents the typical input images.

References

Braunecker, B., Hauck, R. W., and Lohmann, A. W., Optical character recognition based on nonredundant correlation measurements, *Appl. Opt.*, 18, 2746–2753, 1979.

Casasent, D., Unified synthetic discriminant function computational formulation, *Appl. Opt.*, 23, 1620–1627, 1984.

Curtis, K. and Psaltis, D., Three-dimensional disk-based optical correlator, *Opt. Eng.*, 33, 4051–4054, 1994.

Hester, D. L. and Casasent, D., Multivariant technique for multiclass pattern recognition, *Appl. Opt.*, 19, 1758–1761, 1980.

Horner, J. L. and Gianino, P. D., Phase-only matched filtering, *Appl. Opt.*, 23, 812–816,1984.

Hsu, Y. N. and Arsenault, H. H., Optical character recognition using circular harmonic expansion, *Appl. Opt.*, 21, 4016–4019, 1982.

Kumar, B. V. K. V., Tutorial survey of composite filter designs for optical correlators, *Appl. Opt.*, 31, 4773–4801, 1992.

Mahalanobis, A., Kumar, B. V. K. V., and Casasent, D., Minimum average correlation energy filters, *Appl. Opt.*, 26, 3633–3640, 1987.

Matsuoka, K., Intelligent optical vision system, *Jpn. J. Opt.*, 25, 265–271, 1996.

Rajbenbach, H. et al., Compact photorefractive correlator for robotic applications, *Appl. Opt.*, 31, 5666–5674, 1992.

Ravichandran, G. and Casasent, D., Minimum noise and correlation energy optical correlation filter, *Appl. Opt.*, 31, 1823–1833, 1992.

VanderLugt, A., Signal detection by complex spatial filtering, *IEEE Trans. Inf. Theory IT*, 10, 139–145, 1964.

IV

Opto-Mechatronic Systems Control

13

Real-Time Control of Opto-Mechatronic Systems

Hyungsuck Cho
Korea Advanced Institute of Science and Technology
Taejeon, South Korea

13.1 Introduction

The ultimate goal of modern control in most engineered devices, products, machines, processes, and systems, is to ensure high functionality, high performance quality of products, high productivity, and complete autonomy, regardless of changes in operating environments [Cho, 2001]. To achieve this the design of such control systems must often overcome a serious challenge to increased precision, adaptation, and embedded intelligence. The opto-mechatronic systems control is no exception; it deals with the control of a system integrated with optical and mechatronic elements.

According to the Introduction in this handbook, opto-mechatronic systems can be classified into three categories. The shortcomings of this classification are restated here:

1. *Opto-mechatronically fused systems:* in these systems, optical and mechatronic elements are not separable in the sense that if optical or mechatronic elements are removed from the system they constitute the system cannot function properly.
2. *Mechatronically embedded optical systems:* many optical systems require positioning or servoing optical elements/devices to manipulate and align beams and to control the polarization of beams.
3. *Optically embedded mechatronic systems:* in these systems, the optical element is embedded into a mechatronic system. The optical element is separable from the system; the system can function, but at a reduced level of performance.

0-8493-1162-4/03/$0.00+$1.50

The regulation of opto-mechatronic systems described above encompasses a wide range of problems, from well-known simple time-invariant systems to uncertain time-varying nonlinear systems. For example, consider the control of a simple mirror that steers and scans light beams in sequence in a desired direction. When the motor of the steering mirror is modeled and friction and backlash of the motor dynamics are neglected, the control problem is a simple linear, time-invariant system. In this case, a simple controller may be used to actuate the mirror. But when product control of manufacturing processes is considered, the dynamics is not as simple as that of the mirror steering system; they are quite complicated, uncertain, time-varying, highly nonlinear systems, and unless appropriate control with optical elements or sensors is applied in an effective way, their quality often varies from product to product.

Another very interesting category of problems connected with the control of opto-mechatronic systems is the feedback control of adaptive optics imaging systems. The control problem is concerned with enhancing the resolution of optical telescopes corrupted by a turbulent atmosphere. The control system corrects for the disturbance-induced degradation by means of controlling one or more deformable mirrors. This problem is characteristics of distributed-parameter system control.

As with other control problems a prerequisite for a design approach to the control of opto-mechatronic systems is to obtain a dynamic model that accurately describes system properties. In some cases, due to the intractable properties of opto-mechatronic processes, however, no clear-cut method to model their dynamic model has been available. This has limited the number of research works in this area, which demonstrated various feedback-controller design techniques based on the conventional approach of utilizing a mathematical model. The use of simple PID, H^{∞}, time-delay, adaptive, fault-tolerant, and optimal controllers is the main focus of such efforts. In recent years, a new control concept has evolved called "intelligent control," which utilizes artificial intelligence (neural networks, fuzzy logic, expert systems, and GA). The power of this method lies in the fact that it can be effectively designed even with systems that are difficult to design and whose mathematical model is not easily attainable. Because of this favorable property of the controller-design approach a variety of applications has come out in various opto-mechatronic systems.

The purpose of this chapter is to provide basic control concepts and methodologies needed for the control of opto-mechatronic systems. Because system dynamics covers a wide spectrum of characteristics, several controllers will be introduced in some detail to help the reader understand design concepts when opto-mechatronic control problems are encountered. Section 13.2 discusses the characteristics and classification of opto-mechatronic control systems with illustrations of some actual control problems. Section 13.3 introduces and reviews the principles and design methods popularly used in various practical systems. Section 13.4 illustrates the actual control systems associated with opto-mechatronic characteristics. These include a microassembly of microelectromechanical (MEM) devices, laser generation control, and a welding quality-control system based on optical information feedback. Finally, conclusions are drawn regarding system performance and the prospects of future control applications to opto-mechatronic systems.

13.2 Characteristics and Classification of Opto-Mechatronic Control Systems

From the point of view of previous discussions opto-mechatronic control systems can be classified into two different categories. The first one is according to system type while the other is according to dynamic system characteristics involved with the system to be controlled—actuators and sensors.

Figure 13.1 depicts the three different classes of opto-mechatronics. The figure shows the typical control structure of (a) an opto-mechatronically fused system, (b) an optical system, and (c) a mechatronic system. In Figures 13.1(a) and 13.1(b) the sensor in the feedback loop indicates any type of sensor, whether optical or nonoptical. However, in the case of the mechatronic system control in 13.1(c), the sensor must be of the optical type, unless the system to be controlled contains any optical elements. A practical control system that belongs to category (a) is the atomic force microscope (AFM), where the probe tip (cantilever tip) is always kept constant at a certain distance from the surface of objects to be

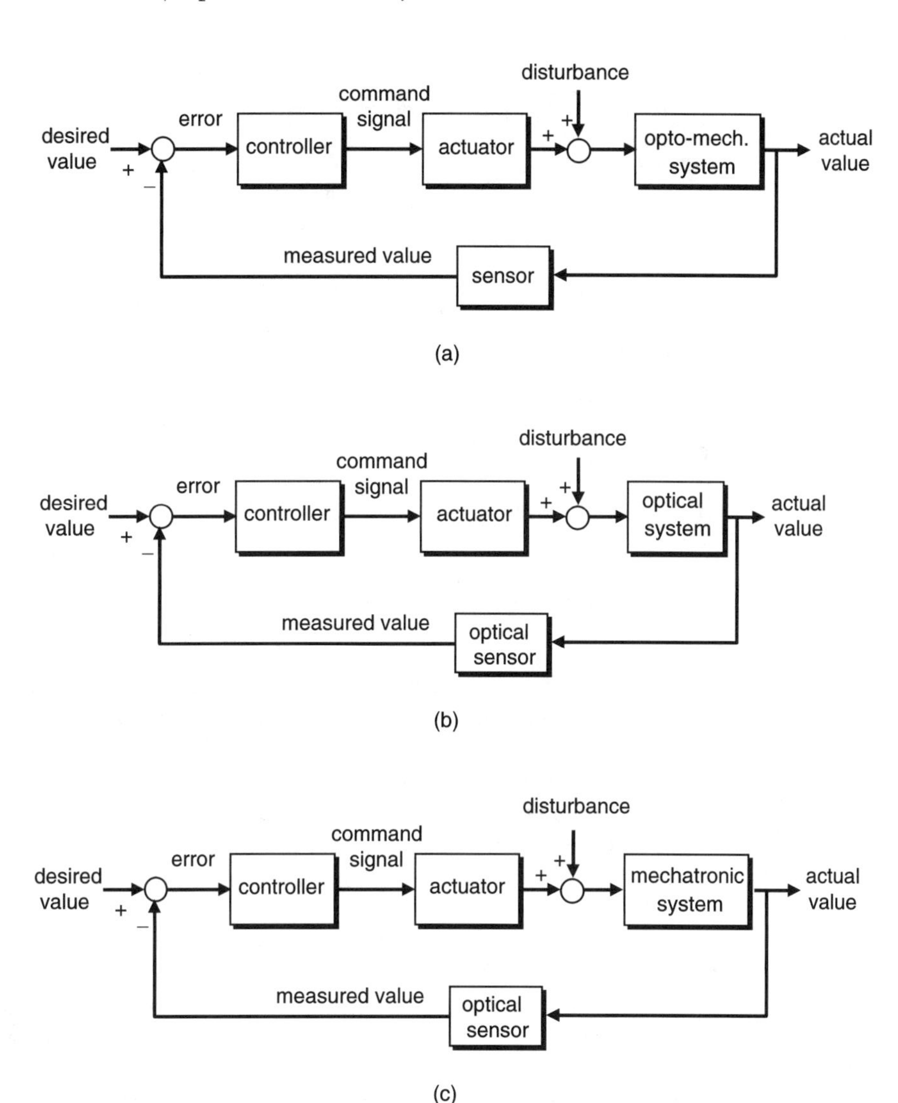

FIGURE 13.1 Schematic diagram of various opto-mechatronic control: (a) control of opto-mechatronically fused systems; (b) control of optically embedded mechatronic systems; (c) control of mechatronically embedded systems.

measured [Wong and Welland, 1993; Hsu and Fu, 1999]. The optical/visual tracking system belongs to 13.1(b), where a servoing mechanism is actuated to track a desired target object at any instant. The mobile robot navigation system belongs to 13.1(c). It houses a control system that it uses to navigate through environments with the aid of an optical/visual sensor. In addition, many manufacturing systems and laser-based material-processing systems belong to this class. Also typical of this category are unmanned-vehicle and gun-turret systems that utilize optical/visual sensing.

The opto-mechatronic control system is classified according to dynamic characteristics, as shown in Figure 13.2: (1) linear/nonlinear, (2) time-invariant/time-varying, (3) lumped/distributed, (4) known/uncertain, (5) deterministic/stochastic, and (6) single/multivariable.

According to the characteristics many controller design methodologies have been developed and utilized effectively in the past. The control system that is being widely used in various optical measurement systems adopts the principle of autofocusing of the lens system. In order to control the focus, servo motors and piezoactuators are mostly used as actuators. The dynamics of servo motors can often be linearized to

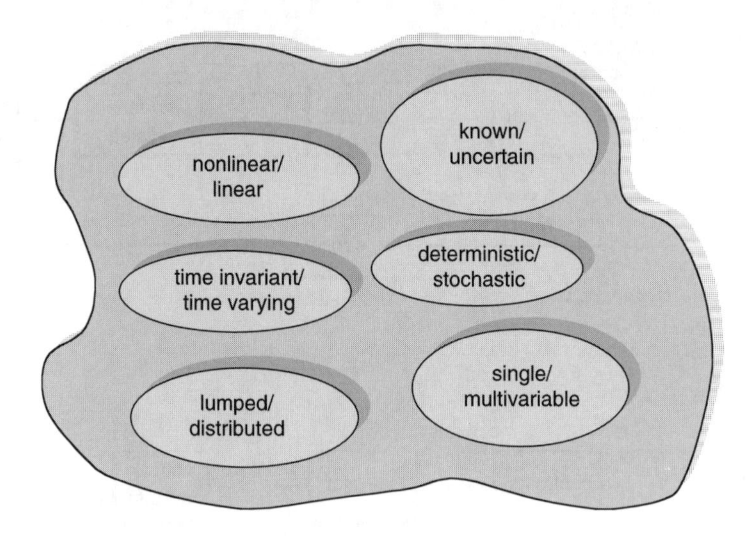

FIGURE 13.2 Dynamic characteristics of opto-mechatronic systems.

make the control system design simpler. However, piezoactuators are characterized by nonlinear dynamic properties such as hysterisis and dead zone, which make it difficult to design appropriate controllers needed for precision positioning or tracking. The time-varying and nonlinear properties can be found in many manufacturing-related opto-mechatronic systems. These include optical sensing-based machining processes, textile fiber fabrication processes, laser machining, and laser welding. In laser welding the gap between the shield-gas nozzle surrounding the optics and the workpiece is often subject to variations in the gap, which include defocusing of the laser spot at the workpiece generated by the optical system. This variation in light intensity influences welding process dynamics and thus necessitates precise correction of the defocusing problem [Haran et al., 1996].

The control problem that is characterized by uncertain nonlinear and time-varying properties is the visual feedback control via hand-in-eye of 6 degree-of-freedom robotic manipulators. This problem concerns subsystems. In this control system, the robot dynamics represents the nonlinear, uncertain, and time-varying properties. The uncertainty comes from the fact that the measurement information is undefined due to the uncertainty of the end effectors' positioning [Papanikolopolous et al., 1993].

The control of adaptive optical systems belongs to a distributed-parameter system. A typical system includes ground-based, optical telescopes that are subject to turbulent atmospheric conditions. The purpose of the control system here is to enhance the resolution of the telescope by means of one or more adaptive mirrors. The system measures the characteristics of the place of the arriving wave front and corrects for the degradations caused by deformable mirrors [Plemmons and Pauca, 2000].

13.2.1 Types of Control Problems

The control of opto-mechatronic systems deals with many different types of objectives depending on system performance requirements. Many requirements differ from system to system, but typical requirements may be listed here for each system.

13.2.1.1 Opto-Mechatronically Fused Systems

The control of these systems is depicted in the block diagram in Figure 13.1(a). Control systems of this type can be found in a number of practical systems; typical ones include deformable mirror control, atomic force microscope (AFM) laser wavelength control, and sensory feedback optical system control.

Confocal measuring system control: A piezoelectric actuator is used to drive the optical head that automatically searches the focal position on a tested object surface [Zhang and Cai, 1997]. The principle behind this method is applied to the control of the atomic force microscope (AFM)

[Veeco Co., 2000]. In the noncontact mode of the AFM, the probe initially moves up to the surface of a tested object, thereby preventing the probe from being touched. The position of the probe is controlled at a certain level close to the surface by piezoelectric actuators. Measurement of the gap is carried out by an optical photo diode sensor. With this configuration, the control objective is to actuate the PZT so as to follow the profile of the object, whose movement represents the surface profile to be obtained.

Adaptive mirror control: The adaptation of deformable mirrors is essential to compensate for blurring caused by air turbulence in telescope systems. A closed-loop feedback system is constructed to control the deformation of the mirrors based on measurements of the incoming wave front captured by a wave-front sensor.

Contact force control: The purpose of this system's construction is to accurately achieve contact force control. As shown in Figure 13.3, it consists of a contact tip, a force sensor, an optical displacement sensor, and a static electromagnetic actuator [Shim and Cho, 1999]. This actuator provides the contact tip with micromotion with the aid of an optical sensor so that the force can be as low as possible, when the tip makes contact with the environment.

Sensory-information-based optical system control: Optical system control is based on external sensory feedback information. A number of systems belonging to this category can be found in the literature. A typical example of this control method is the zoom-and-focus control system. The objective of this system is to control a zoom lens so that the sharp eye's image appears in the same size irrespective of back-and-forth head movement against the camera. In this system an ultrasonic distance measurement device is utilized to control the zoom and focus of a camera [Ebisawa, et al., 1996].

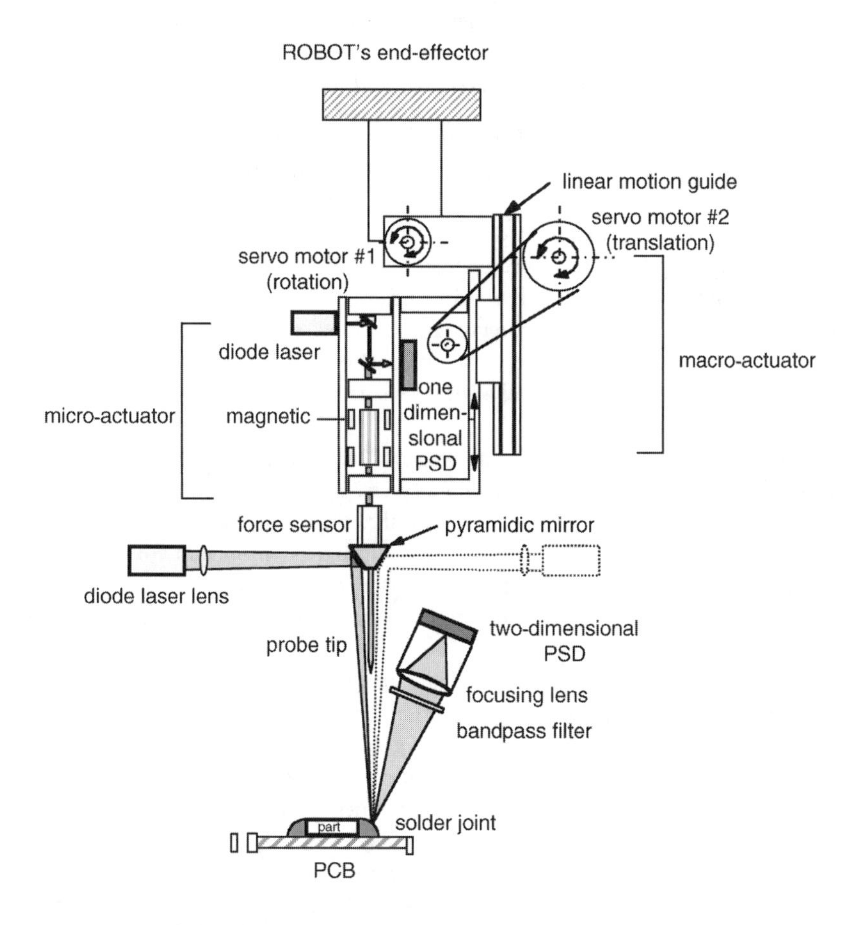

FIGURE 13.3 A sensor-guided probing robot with an optical force sensor.

13.2.1.2 Mechatronically Embedded Optical Systems

Figure 13.1(b) represents this control system, which typically includes beam steering control, optical system laser focus control, laser stability control, iris control, and optical switching.

Optical scanner system: To generate a sequential motion of optical elements such as optical-based sensors and light source, scanning action is needed. In this case, servomechanisms provide such motion by controlling the position of the sensors or the beam direction. Polygon mirror, galvanometer, and pen-tilt mechanisms are typical devices in this system.

Laser focus control: One important issue in laser-based measurement systems and in-process monitoring of laser material processing is defocusing of the laser spot at a workpiece. In the case of laser processing, to bring the workpiece into focus, measurement of optical radiation emitted from the process is made by an optical fiber, which collects the radiation. This information is fed to a controller to adjust a motorized stage accordingly.

Laser-generation control system: The control objective of this system is to achieve stability of laser light by controlling the pulse energy and width of the laser.

Optical data-transmission or -switching control system: In the mirror-based switching control system, mirrors are controlled to connect the light path to any port by using servo controllers. Precision positioning accuracy needs to be achieved by some actuating means [Robinson, 2001].

13.2.1.3 Optically Embedded Mechatronic Systems

This control system is schematically shown in Figure 13.1(c) and typically includes control of weld pools in arc welding, visual feedback control, tracking motion control, positioning control, track following control of flexible disk drive, home appliances control, and robotic systems control.

Optical sensor-based position control: The control of many positioning systems requires sensory information detected by optical sensors. The servo system controls the position of positioning mechanisms based on this feedback information. A number of systems belong to this class. In integrated-circuit devices, precision control of stage-aligning wafers for microlithography processes is required to possess an unprecedented level of performance; positioning reproducibility is less than 20 nm. To fulfill this, a three-dimensional (x–y–θ) servo control is adopted, often by a dual servo composed of a microservo and a global servo. The microservo is actuated by microactuators such as PZT and electrostatic actuators. The measurement of the (x–y–θ) positioning is made by three-dimensional optical sensors such as plane mirrors and angle measuring interferometers.

Optical sensor-based tracking control: The objective of this control is to make the servomechanism follow the instantaneous motion of moving objects, based on optical sensory information.

Track following control of flexible disk drive: A typical example system is track following the control of high-capacity flexible disk drives, which is ultimately important for magnetic heads that read and write data. The control is provided by an actuator called a voice coil motor (VCM), by an optical tracking sensor, and sometimes by a magnetic scale for backwards-compatibility mode.

Visual motion feedback control: Frequently, feedback control action can be achieved via visual sensory information. The action includes object pattern recognition, target positioning and tracking, environment recognition for mobile robot navigation, and part assembly.

Optical information-based process control: Visual/optical information that represents the current controlled state is very useful in the control of machines, processes, and systems. The information obtained by optical/visual sensors is utilized for adjusting operating control inputs to achieve certain desired objectives associated with their performance and output quality. Control of machining systems based on optical measurements of machined workpiece geometry and surface is a typical example of this category.

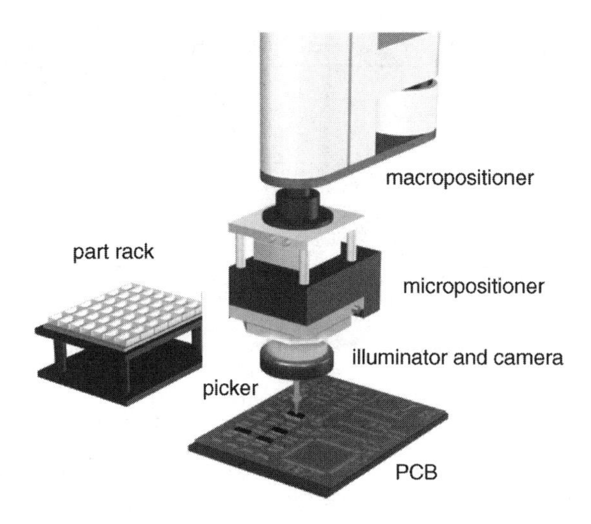

FIGURE 13.4 Vision-guided micropositioning system.

13.2.1.4 Dual Servo Control of Opto-Mechatronic Systems

This control system is composed of a global servomechanism and a microservomechanism. The global positioning system provides initial, relatively coarse positioning, whereas the microservomechanism produces precision positioning. As shown in Figures 13.3 and 13.4, this type of control system can be found in a variety of positioning or tracking systems such as the lithography wafer stage [Lee and Kim, 1997], the optical-based contact force system (Figure 13.3) [Shim and Cho, 1999], the robotic manipulator with a microprecision positioning device (Figure 13.4) [Chen and Hollis, 1999], the track searching and following control system for optical/magnetic disk drive systems [Jeong et al., 1999], etc. Table 13.1 illustrates various opto-mechatronic control systems frequently utilized in practical applications and their control components and characteristics.

13.3 Controller Design Methodologies

A general feedback control system as shown in Figure 13.5 consists of a controller, an actuator, a sensor, and a feedback element that feeds the measured system signal to the controller. The role of the controller is to adjust its command signal depending on error characteristics. Therefore, performance of the controller significantly affects the overall performance of the control. Equally important is the performance of the actuator and sensor to be used for control. Unless these are suitably designed and selected, the control performance could not be guaranteed, although the controller is designed to best reflect the system characteristics.

The greatest difficulty in system controller design is the fact that an accurate model of the process dynamics often does not exist. The lack of a physical model makes the design of a system controller difficult and renders conventional control methodologies virtually impractical to use. In this situation, there are two widely accepted methods of designing system controllers. One is to approximate the exact mathematical model dynamics by making certain assumptions about system mechanisms and phenomena. The system model thus approximately obtained can be utilized for the design of conventional controllers, called a "model-based control," which include all the model-based control schemes such as PID (proportional-integral-derivative), adaptive, optimal, predictive, robust, and time-delay controls, etc.

The advantage of the approach using model dynamics is that the analytical design method is possible because it enables us to investigate the effects of the design parameters. The disadvantage is that control performance may not be satisfactory when compared with the desirable performance of the ideal case because the controller is designed based on an approximate model. Furthermore, when changes in the system characteristics occur over time, the designed controller may be further deteriorated.

TABLE 13.1 Control Components and Their Characteristics in Various Opto-Mechatronic Control Systems

System Type	Control System	Control Components	Component Physical Phenomena	Control System Characteristics
Opto-mechatronically fused system	AFM	•Cantilever probe •Optical sensor •Piezoactuator	•Flexibility hysteresis •Linearity	•Nonlinear •Uncertain
	Optical switch	•Mirror array •Laser input source •Piezoactuator	•Hysteresis •Friction	•Nonlinear
	Optical disk storage	•Beam focusing system •Voice coil motor •Optical sensor	•Nonlinear •Friction	•Nonlinear •Time-varying
	Lithography stage	•Directed beam •Position table •Optical sensor	•Nonlinear •Friction	•Nonlinear
	Automatic camera	•Lens •Servo motor •Optical sensor	•Friction •Backlash	•Nonlinear
Mechatronically embedded optical system	Map building mobile robot	•Mobile robot with positioning mechanism •Servo motor •Camera/optical sensor	•Friction •Backlash	•Nonlinear •Time varying •Uncertain
	Vision system with pan-tilt mechanism	•Positioning system •Servo motor •Camera/optical sensor	•Friction •Backlash	•Nonlinear
	Galvanometer	•Mirror positing system •Servo motor •Encoder	•Friction •Backlash	•Nonlinear
Optically embedded mechatronic system	Optical-based coordinates measuring machine (CMM)	•Optical probe positioning table •Servo motor piezoactuator •Encoder	•Inertia of table •Friction backlash •Hysteresis	•Nonlinear
	Laser welding	•Beam positioning •Servo motor •Optical sensor •Weld torch nozzle	•Nonlinear •Nozzle damage	•Nonlinear •Uncertain

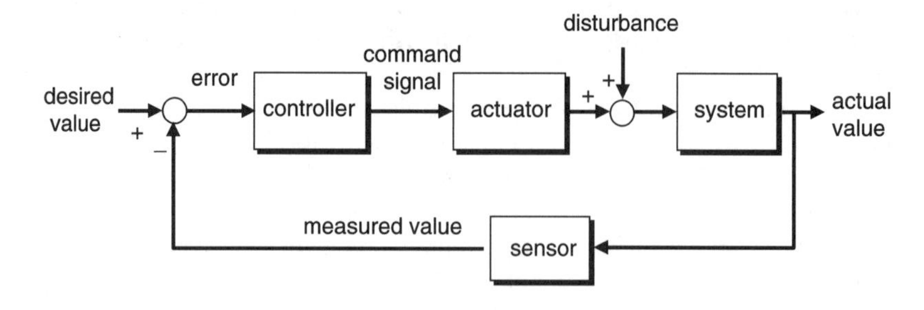

FIGURE 13.5 A general block diagram of a manufacturing process control system.

The other widely accepted approach, called intelligent control, is based on artificial intelligence that uses heuristics of human operators rather than a mathematically based algorithm. In this case, human operators design the controller, making use of their own knowledge and past experience of the control action based on observation of dynamic characteristics. The control actions of a human operator are generated from the result of inference of rules that he uses to formulate his knowledge. Accordingly, the performance of the control largely depends on how broad and deep is the operator's knowledge of dynamic characteristics of the process and how he can construct the appropriate rule base utilizing his knowledge and experience. Reliable control performance may not be guaranteed by just the human operator's observations and experience, when system characteristics become uncertain and time-varying in nature.

13.3.1 Model-Based Control vs. Intelligent Control

The system model approximating process dynamics can be writtenas

$$\begin{aligned} \dot{\mathbf{x}} &= \mathbf{f}(\mathbf{x}, \mathbf{u}, \boldsymbol{\alpha}(t), \mathbf{d}(t), t) \\ \mathbf{y} &= \mathbf{c}(\mathbf{x}, \mathbf{d}(t)) \end{aligned} \tag{13.1}$$

where $\boldsymbol{\alpha}(t)$ and $\mathbf{d}(t)$ denote, respectively, time-varying properties of the system parameters and disturbances. That is, if the system dynamics is exactly known, the problem becomes relatively easy, the difficulty being only a matter of the nonlinearity of the functional form **f**. When systems are subject to parameter variations and disturbances, the $\boldsymbol{\alpha}(t)$ and $\mathbf{d}(t)$ contained in Eq. (13.1) are uncertain in nature. In this situation, the controller design needs to take these properties into consideration. When the function **f** is uncertain, incorrect, or unavailable to represent a system's dynamics, we must resort to some other means to design a controller. The approach in this case is based on artificial intelligence, which uses heuristics of human operators or artificial neural network models rather than a mathematically based algorithm.

13.3.2 Methods for Controller Design

The objective of PID controllers is to regulate a process output at a constant set point. The regulation objective is mainly obtained by the integral operation for any unmeasured constant external disturbance. The output of the PID controller is split into three terms:

$$u(t) = K_p e(t) + K_d \frac{de(t)}{dt} + K_i \int_o^t e(\tau)\,d\tau \tag{13.2}$$

where $e(t)$ is the error of the state of a system from its desired value $r(t)$ ($e(t) = x(t) - r(t)$). The K_p, K_d, and K_i denote, respectively, the proportional, derivative, and integral gain. If it is denoted in a Laplace domain, then $u(t)$ is represented by

$$U(s) = K_p E(s) + K_d E(s) + K_I E(s)/s \tag{13.3}$$

where s is the Laplace operator and $U(s)$ and $E(s)$ are the Laplace-transformed variables of $u(t)$ and $e(t)$, respectively.

The principle behind this controller is very intuitive. In proportion to the error magnitude the proportional action tends to reduce the deviation with the control action. Derivative action tends to stabilize the control system and reduce overshoots by anticipatory control. Integral action is generally necessary to regulate the process output to a predefined set point. The steady-state control input required to zero the steady-state error is automatically obtained by this integral action. Several techniques are available to design the PID controller, ranging from the classical- and state-space methods to the intelligent-design method. For example, the classical method includes Zigler–Nicols tuning, the root-locus technique, the frequency-domain method, the method of optimal tuning based on certain criteria, etc. Because PID tuning is somehow fundamental to the controller design, the details of the techniques will not be introduced here but can be found in the literature.

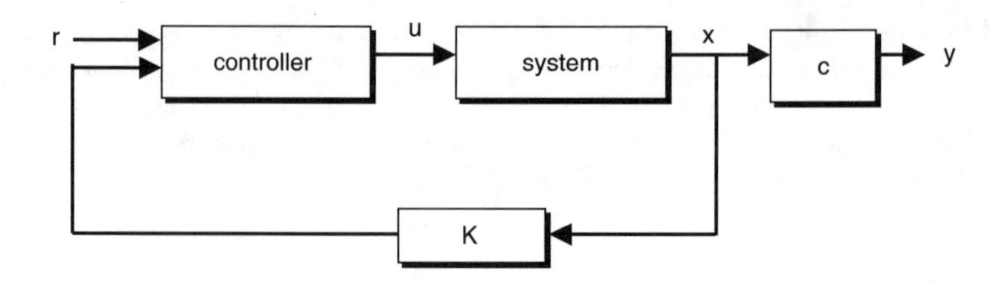

FIGURE 13.6 A general structure of optimal regulator control.

13.3.3 Linear Optimal Control

The optimal control problem is usually concerned with many different types of optimality to minimize time, error of regulation or tracking, energy, and terminal value at the final control point, etc. Here we shall consider the optimal control problem in brief. The problem begins with the linear system equation given in the following state form:

$$\dot{\mathbf{x}} = \mathbf{Ax} + \mathbf{Bu} \tag{13.4}$$

where $\dot{\mathbf{x}}$ is the n state vector for an nth-order lumped-parameter object, $\mathbf{u}$ is the control vector, $\mathbf{A}$ is the $n \times n$ system matrix, and $\mathbf{B}$ is the $m \times 1$ control input vector. The control objective here is to determine the gain matrix of the optimal control vector shown in Figure 13.6.

$$\mathbf{u} = -\mathbf{K}(t)\mathbf{x}, \quad t \in [t_0, t_f], \tag{13.5}$$

so as to minimize the performance index defined by

$$J = h(\mathbf{x}(t_f)) + \int_{t_0}^{t_f} \mathbf{f}(\mathbf{x}, \mathbf{u})dt \tag{13.6}$$

where t_0, the time at the initial point, is prespecifed and t_f, the time at the final point, may be known or unknown beforehand [Lewis and Syrmos, 1995]. With linear quadratic regulator (LQR) problems, the J is given by

$$J = \mathbf{x}(t_f^{\mathrm{T}})\,\mathbf{H}\mathbf{x}(t_f) + \int_{t_0}^{t_f} [\mathbf{x}^{\mathrm{T}}\mathbf{Q}\mathbf{x} + \mathbf{u}^{\mathrm{T}}\mathbf{R}\mathbf{u}]\,dt \tag{13.7}$$

where $\mathbf{Q}$ is a positive-definite (or positive semidefinite) real symmetric matrix and $\mathbf{R}$ is a positive-definite real symmetric matrix. The optical gain matrix $\mathbf{K}$ is given by

$$\mathrm{K}(t) = \mathbf{R}^{-1}\mathbf{B}^{\mathrm{T}}\mathbf{P} \tag{13.8}$$

where the positive-definite matrix satisfies the following Riccati differential equation:

$$\dot{\mathbf{P}} = \mathbf{A}^{\mathrm{T}}\mathbf{P} + \mathbf{PA} - \mathbf{PR}^{-1}\mathbf{B}^{\mathrm{T}}\mathbf{P} + \mathbf{Q}$$

and

$$\mathrm{K}(t_f) = \mathbf{H} \tag{13.9}$$

where $\dot{\mathbf{P}}$ denotes the time derivative of $\mathbf{P}$.

When $t_f \to \infty$ and $t_0 \to 0$ the equation reduces to

$$\mathbf{A}^{\mathrm{T}}\mathbf{P} + \mathbf{PA} - \mathbf{PR}^{-1}\mathbf{B}^{\mathrm{T}}\mathbf{P} + \mathbf{Q} = 0 \tag{13.10}$$

Substituting the result of Eq. (13.10) yields the optimal gain matrix. It can be shown that the resulting closed loop system **A** – **BK** is a stable one.

13.3.4 Adaptive Control

When a system is made or tuned at least infrequently, a fixed control is tuned just once. This simplifies the system operation and is justified if system characteristics are invariant. However, if the process dynamics has uncertainty or is changed appreciably, satisfactory control performance cannot be guaranteed. In this case, the controller must adapt to these process characteristics so as to eliminate the unfavorable effects. The adaptive control system is one that is able to adjust itself to provide consistent control results in the presence of uncertainty or changes of the system. Typically, adaptive control is used where constant performance is desired in spite of high nonlinearities in the process dynamics and long-term changes in the physical process characteristics; that is, where the elements being controlled experience changes in connection with operating environments or time.

The block diagrams of three control systems are shown in Figure 13.7. Figure 13.7(a) shows an example of direct adaptive control. Here, each control signal u is generated based on the actual y and desired

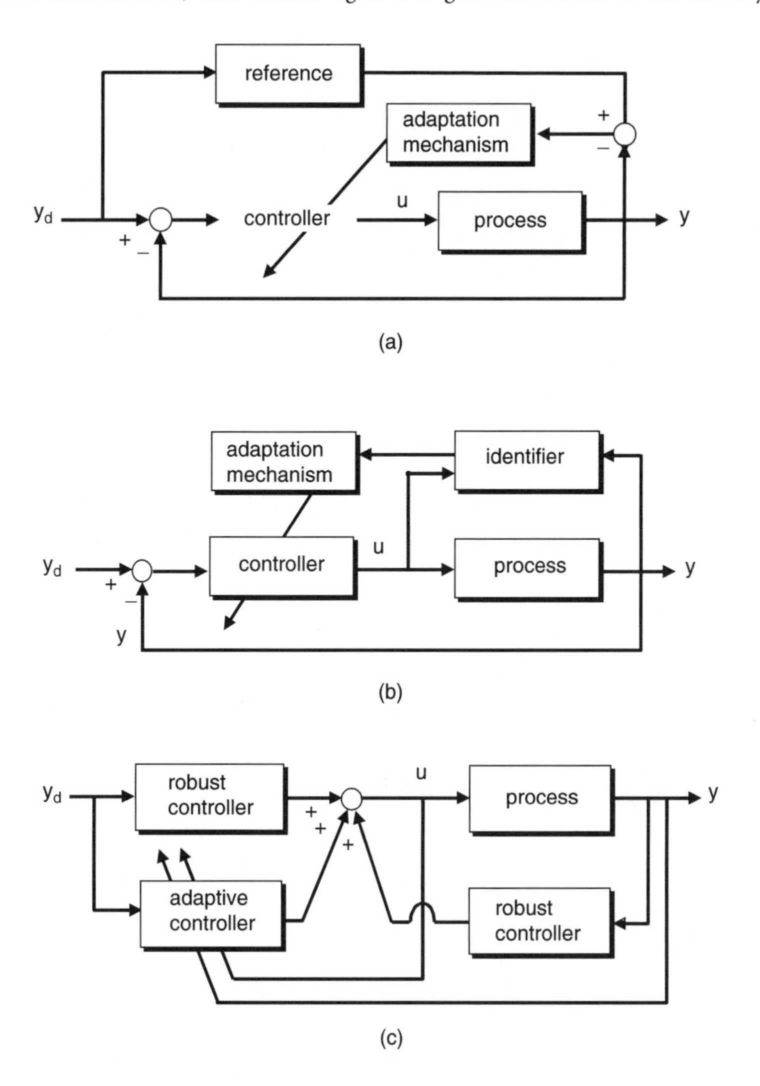

FIGURE 13.7 Adaptive control systems of various architectures: (a) direct adaptive control architecture; (b) indirect adaptive control architecture; (c) adaptive nonlinear-robust control architectures.

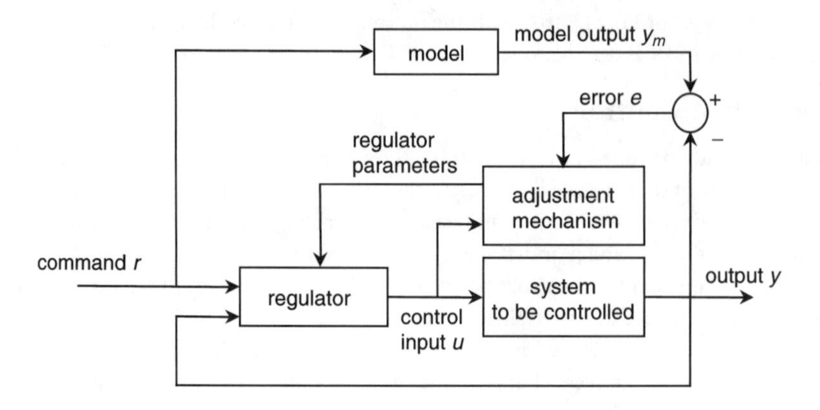

FIGURE 13.8 Typical structure of a model reference adaptive control (MRAC).

output y_d, internal state of the controller, and adaptation mechanism to tune the pertinent controller parameters. The adaptation is carried out at each time-step based on the error e between the measured process output and the reference output y_r. This is direct adaptive controlling, which does not need to perform online process identification. In the indirect adaptive approach shown in Figure 13.7(b), an identifier updates the model of the process as quickly as possible based on the input and measured outputs. Based on the estimation of the model, the adaptation mechanism adjusts the controller parameters online. The last block diagram in Figure 13.7(c) shows the adaptive-nonlinear-robust control architecture. The integrated controller generates the control signal from a combination of three separate controllers: A robust controller that assumes linear nominal models is designed so that the unmodeled dynamics, disturbance, and parameter variations are accounted for, while a feed-forward nonlinear controller is included to account for process nonlinearities. The last is the adaptive controller in the outer loop. It provides adaptation toward parameter variation.

The model reference and self-tuning methods can be shown to be in some sense equivalent, even if the approach appears to be totally different. In this section, we will introduce the MRAC design methodology. As shown in Figure 13.8, the MRAC method attempts to adjust the controller in such a way that the response of the closed-loop system $\mathbf{x}$ in Eq. (13.4) to a reference input r is identical to that of the reference model $\mathbf{x}_m$.

$$\dot{\mathbf{x}}_m = \mathbf{A}_m \mathbf{x}_m + \mathbf{B}_m \mathbf{u}_m \tag{13.11}$$

where $\mathbf{A}_m$ and $\mathbf{B}_m$ are the system and input matrices for the reference model, respectively, and $\mathbf{u}_m$ denotes the desired input ($\mathbf{u}_m = \mathbf{r}$).

The plant input $\mathbf{u}$ in Eq. (13.11) can be represented by summation of a linear input signal and adaptation signal as follows:

$$\mathbf{u} = \mathbf{u}_l + \mathbf{u}_a \tag{13.12}$$

where $\mathbf{u}_l$ and $\mathbf{u}_a$ are the linear input signal and the adaptation signal, respectively, and they are given in the form of

$$\begin{aligned} \mathbf{u}_l &= -\mathbf{K}_x \mathbf{x} + \mathbf{K}_m \mathbf{x}_m + \mathbf{K}_u \mathbf{u}_m \\ \mathbf{u}_\alpha &= \mathbf{D}\mathbf{K}_x(\mathbf{e},t)\mathbf{x} + \mathbf{D}\mathbf{K}_u(\mathbf{e},t)\mathbf{u}_m. \end{aligned} \tag{13.13}$$

In the above, $\mathbf{e}(t)$ is an error signal related to the difference between the outputs of the plant and the reference model ($\mathbf{e} = \mathbf{x}_m - \mathbf{x}$). When the Popov–Landau approach [Landau, 1987] is used, the adaptive loop will have a compensator to generate the signal

$$\boldsymbol{\sigma} = \mathbf{Le} \tag{13.14}$$

where **L** denotes the linear compensator, which is computed from the condition that the linear part of the equivalent feedback system corresponding to the Popov–Landau design method is a strictly positive real transfer function. Then, the parameter adaptation law for $\Delta\mathbf{K}_x(\mathbf{e}, t)$ and $\Delta\mathbf{K}_u(\mathbf{e}, t)$ in Eq. (13.13) can be obtained as

$$\begin{aligned} \Delta\mathbf{K}_x(\mathbf{e},t) &= \Delta\mathbf{K}_x(\boldsymbol{\sigma},t) = \int_o^t \phi_1(\boldsymbol{\sigma},t,\tau)d\tau + \phi_2(\boldsymbol{\sigma},t) + \Delta\mathbf{K}_x(0) \\ \Delta\mathbf{K}_u(\mathbf{e},t) &= \Delta\mathbf{K}_u(\boldsymbol{\sigma},t) = \int_o^t \psi_1(\boldsymbol{\sigma},t,\tau)d\tau + \psi_2(\boldsymbol{\sigma},t) + \Delta\mathbf{K}_u(0). \end{aligned} \tag{13.15}$$

where ϕ_1 and φ_2 denote a nonlinear time-varying relation between $\Delta\mathbf{K}_x(\sigma, t)$ and $\Delta\mathbf{K}_u(\sigma, t)$ and the value of σ for $0 \le \tau \le t$, respectively. ϕ_1 and φ_2 denote a nonlinear time-varying relation for the value having the property $\phi_1(0, t) = 0$ and $\varphi_2(0, t) = 0$ for all t.

13.3.5 Fuzzy Rule-Based Control Method

The modern control theories presented thus far are based on the assumption that a mathematical model is available for the design of the controllers. In real industrial situations, however, highly precise and detailed modeling of the relevant process is often difficult to achieve due to the complexity of the process as well as imperfections in describing the physical phenomena. Taking this fact into consideration, many control engineers tried to find a controller that emulates human reasoning and action in various circumstances. A fuzzy rule-based controller is one such effort that utilizes the fuzzy-set theory [Cho et al., 2000]. In the fuzzy-set theory, the concept of graded membership provides a basis for describing and manipulating the imprecision and vagueness of natural phenomena. Thus, the notion of the fuzzy-set theory helps to implement an operator's experiences and heuristics on a controller. For example, as shown in Figure 13.9, the operator's imprecise expression "positive small" can be expressed as fuzzy set "PS."

A general structure of a fuzzy rule-based control is shown in Figure 13.10. As shown in the figure, the fuzzy reasoning procedure includes four modules: (a) fuzzification, (b) a rule base, (c) an inference mechanism, and (d) a defuzzification module.

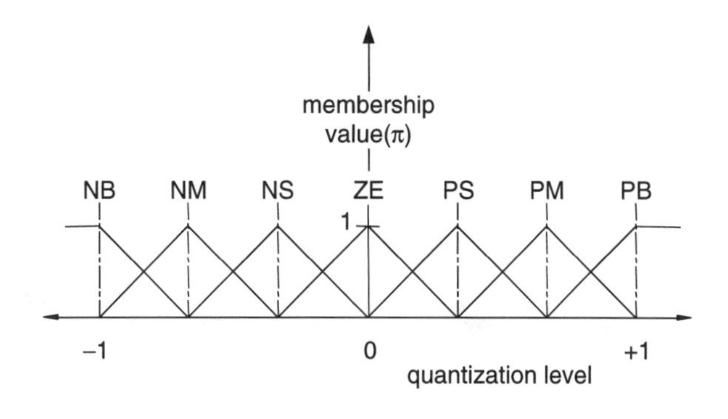

FIGURE 13.9 Membership function of the linguistic values.

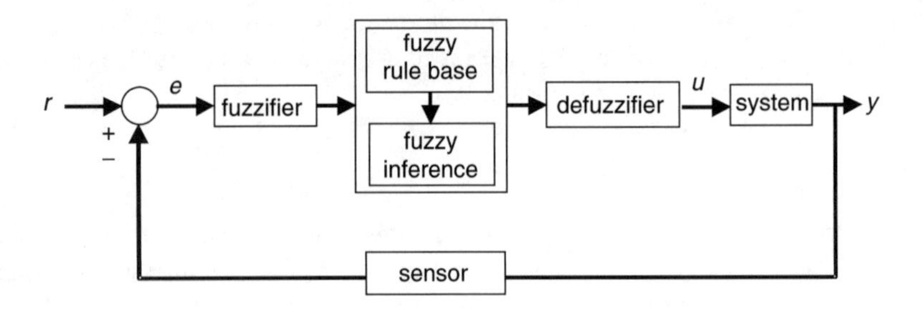

FIGURE 13.10 A general structure of a fuzzy rule-based control.

13.3.5.1 Fuzzification

Fuzzification is defined as mapping from an observed input space to fuzzy sets in a certain universe of discourse. Because all the manipulations in the fuzzy controller are based on fuzzy-set theory, fuzzification is necessary in the earlier stages. Defining the fuzzification operator $F(\cdot)$, the fuzzification procedure can be expressed by

$$\begin{aligned} \tilde{e}(k) &= F\{e(k)\} \\ \tilde{c}(k) &= F\{c(k)\} \\ \Delta\tilde{u}(k) &= F\{\Delta u(k)\} \end{aligned} \tag{13.16}$$

where $e(k)$, $c(k)$, and $\Delta u(k)$ are, respectively, the error, change in error, and additional control input at the kth step, and $\tilde{e}(k)$, $\tilde{c}(k)$, and $\Delta\tilde{u}(k)$ represent the corresponding fuzzy variables. Fuzzification includes scale mapping from the range of input variables to the corresponding universe of discourse.

13.3.5.2 Rule Base

The collection of fuzzy-learning control rules that are expressed in fuzzy conditional statements forms the rule base of the fuzzy controller. A set of linguistic rules based on collected expert knowledge can be described by the following sets of linguistic statements:

$$\begin{aligned} &\text{If} \quad \tilde{e} = PS \quad \text{and if} \quad \tilde{c} = NS, \quad \text{then} \quad \Delta\tilde{u} = ZE \\ &\text{If} \quad \tilde{e} = NB \quad \text{and if} \quad \tilde{c} = PM, \quad \text{then} \quad \Delta\tilde{u} = NS \\ &\qquad \vdots \\ &\text{If} \quad \tilde{e} = PB \quad \text{and if} \quad \tilde{c} = NM, \quad \text{then} \quad \Delta\tilde{u} = PS \end{aligned}$$

where the linguistic values are assigned by NB, negative big; PB, positive big; NM, negative medium; PM, positive medium; NS, negative small; PS, positive small; ZO, zero.

13.3.5.3 Inference Mechanism

The function of the fuzzy control algorithm is to infer the controller input to apply from the error and change in error. If, at a particular sampling instance, the linguistic value of $\tilde{e}$ and $\tilde{c}$ are E' and C', respectively, then the linguistic values of the control output, $\Delta U'$, can be inferred from the inputs by using Zadeh's compensational rule of inference [Zadeh, 1972]:

$$\Delta U' = E' \circ C' \circ R \tag{13.17}$$

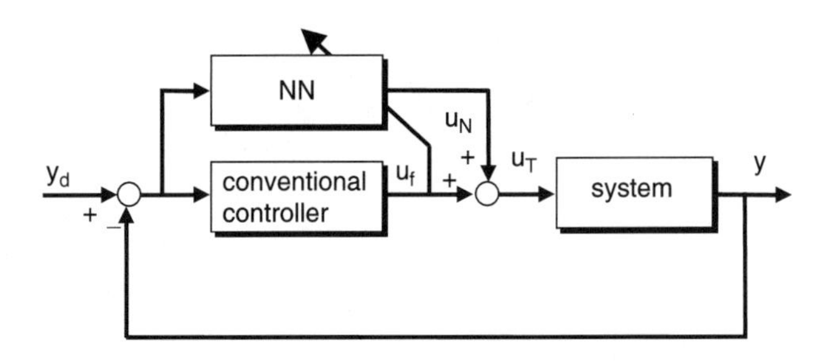

FIGURE 13.11 A neural controller combined with a conventional feedback controller.

13.3.5.4 Defuzzification Module

To generate an appropriate input to the actuator, it is necessary to calculate the crisp value of the control input from the corresponding fuzzy values. Therefore, one of the defuzzification methods must be used to find crisp value Δu that corresponds to the fuzzy set $\Delta\tilde{u}$

$$\Delta u = DF(\Delta\tilde{u}), \tag{13.18}$$

where $DF(\cdot)$ is the defuzzification operator. In general, the center-of-gravity method [Lee et al., 1990] is widely used as a defuzzification scheme.

13.3.5.5 Neural Network-Based Control

Here, we will introduce a combined controller form that consists of a neural network part and a conventional feedback part. In this control structure, shown in Figure 13.11, the conventional controller is used to provide the network controller with stability of system response, which may be needed at an initial learning stage. This control structure is called *feedback error learning*, and it means that learning is provided by a neural network [Kawato, 1987]. The network is used here to learn the inverse model of the process, that is, the relationship that relates the desired output y_d to the corresponding control action u. To achieve this, the neural network is trained to minimize the squared error function E defined by

$$E = \frac{1}{2}(u_T - u_N)^2. \tag{13.19}$$

In the above, the resultant control effort u_T consists of u_N, the control part generated by the network, and u_f, that of the feedback signal. The basic idea of this control scheme is that the u_N takes a major role in control as the feedback signal gets increasingly smaller. Eventually, the u_N represents a steady-state controller when u_f is faded out. The controller design problem then becomes focused on how we determine the u_N as time goes on. Here, we will use a multilayer feedback for that purpose [Zurada, 1992]. A generic structure of the network is illustrated in Figure 13.12. Now, with the network output error signal given by Eq. (13.19), the weights of the network are trained iteratively such that the error can be made approaching zero as the time step goes on. If the error back-propagation algorithm is adapted, the weights of the network are updated in an iterative manner. The algorithm starts with individual weight adjustment Δw_{lk} of the weight in the output layer l as follows:

$$\Delta w_{lk} = -\eta \frac{\partial E}{\partial w_{lk}} \tag{13.20}$$

where η is the learning rate concerned with the speed of convergence of the error, w_{lk} is the weight linking the lth node in the output layer and kth node in the hidden layer. It is noted that the gradient of error with respect to weights, $\partial E/\partial w_{lk}$, is a function of $\partial u_N/\partial w_{lk}$, and u_f.

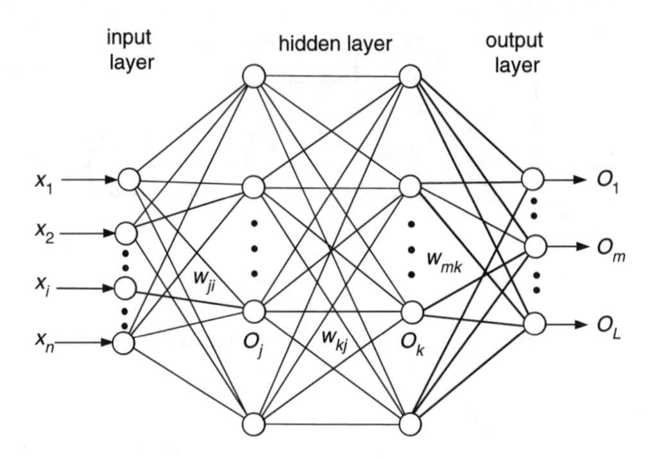

FIGURE 13.12 Topology of a multilayer perceptron.

The final update equation is given by

$$\begin{aligned} w_{lk}(t+1) &= w_{lk}(t) + \eta\delta_l(t)o_l(t) + \alpha\Delta w_{lk}(t) \\ o_l(t) &= (u_T - u_N)f'(net_l(t)) \end{aligned} \tag{13.21}$$

where f is the activation function, f' denotes the derivative of $(\cdot)$, and t and α are, respectively, the number of training iterations and the momentum rate to avoid the oscillating phenomena, and o_1 is the output of the αth node. In the above, $net_l(t)$ is given by

$$net_l(t) = \sum_{k=0}^{L} w_{lk}(t)o_k(t) \quad (l = 1,\ 2,\ \ldots,\ L) \tag{13.22}$$

where L is the total number of neurons in the last layer. Hidden-layer weights are adjusted according to

$$\begin{aligned} \delta_k(t) &= f'(net_k(t))\sum_{l=0}^{L}\delta_l(t)\cdot w_{lk}(t) \quad (k = 1, 2, \ldots, K) \\ w_{kj}(t+1) &= w_{kj}(t) + \eta\cdot\delta_k(t)\cdot o_j(t) \quad (j = 1, 2, \ldots, J) \end{aligned} \tag{13.23}$$

where j is the number of neurons of the jth hidden layer and w_{kj} are the weights linking the kth node of the last hidden layer and the jth node of the jth hidden layer. In this application, learning and momentum gains η and α are normally chosen to be within the range of 0.7 and 0.5, respectively.

13.4 Online Control of Opto-Mechatronic Systems

13.4.1 Microassembly of MEM Devices [Zhou et al., 2001]

As MEM devices or hybrid MEM devices such as actuators, sensors, optical devices, miniature drug pumps, printer nozzles, etc. are becoming commercialized, the needs for assembling and packaging them become apparent. To date, however, economical methods for assembling and packaging such services of any complexity are not available due to the uncertainty involved with the dimension of fabricated MEM devices and assembly tools. The solution to this problem is the integration of data from multiple sensors, which can provide a means of coping with uncertainty and complexity in a microdomain.

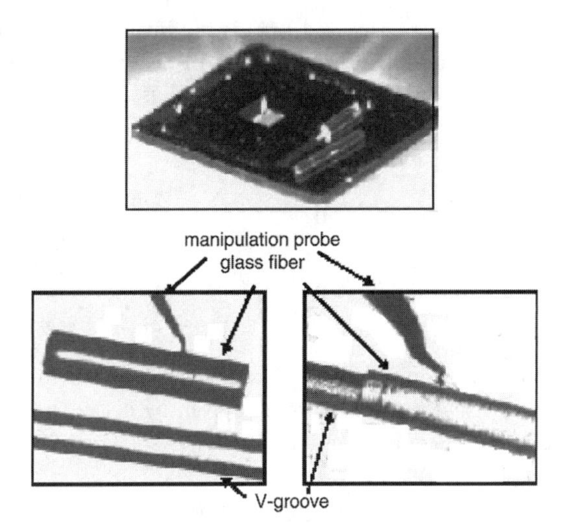

FIGURE 13.13 Manipulating a 308-μm-diameter glass fiber into a 270-μm-wide V-groove for constructing an electron column for a miniature scanning electron micro-scope. (From Zhou, Y. et al., *J. Intelligent Robotic Sys.*, 28, 259–276, 2000. With permission.)

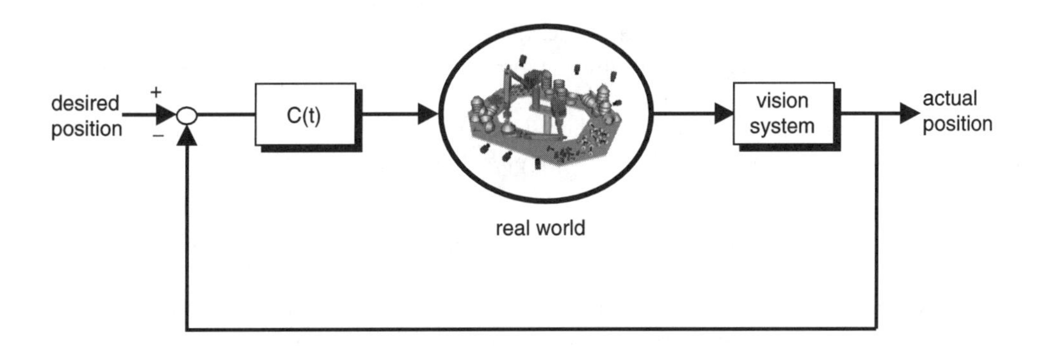

FIGURE 13.14 A visual servoing control loop.

One interesting and feasible method based on the integration of visual feedback with force sensing has been introduced for this kind of microassembly [Zhou et al., 2000]. Currently, microdevices requiring complex manipulation are assembled by hand using an optical microscope and probes or small tweezers, which is basically a form of teleoperated micromanipulation. Figure 13.13 shows the manipulation of one such device. Here, a 308-μm-diameter glass fiber is inserted into a 270-μm-wide V-groove for constructing an electron column for a miniature scanning electron microscope (SEM).

13.4.1.1 Visual Servoing Formulation in a Microdomain

Figure 13.14 indicates a visual feedback control system to manipulate microparts to be assembled at a desired location. The objective of this control system is to control the end-effector motion of a manipulator in order to place the image plane coordinates of features on the target at some desired position.

To model the visual feedback system we need to relate the two coordinate systems—the sensor space and the task space. In formulating this, Jacobian mapping from task space to sensor space is used. The Jacobian for a camera-lens system is of the form

$$\dot{\mathbf{x}}_s = \mathbf{J}_v(\phi)\dot{\mathbf{X}}_T \tag{13.24}$$

where $\dot{\mathbf{x}}_s$ is a velocity vector in sensor space and $\mathbf{J}_v(\phi)$ is the image Jacobian matrix. In the above, the $\mathbf{J}_v(\phi)$ is a function of the extrinsic and intrinsic parameters of the vision sensor as well as the number of features tracked and their locations on the image plane and $\dot{\mathbf{X}}_T$ is a velocity vector in the task space. For a microscope that is allowed to translate and rotate, $\mathbf{J}_v(\phi)$ is of the form

$$J_v = \begin{bmatrix} -\frac{m}{s_x} & 0 & 0 & 0 & -\frac{Z_c m}{s_x} & \frac{y_s s_y}{s_x} \\ 0 & -\frac{m}{s_y} & 0 & \frac{Z_c m}{s_y} & 0 & -\frac{x_x s_x}{s_y} \end{bmatrix} \tag{13.25}$$

where s_x and s_y are the pixel dimensions of the CCD camera and m is the total linear magnification.

The state equation for the visual servoing system is given by discretizing Eq. (13.24) into a digital domain and writing the discretized equation as

$$\mathbf{x}(k+1) = \mathbf{x}(k) + \Delta T J_v(k)\mathbf{u}(k) \tag{13.26}$$

$$\mathbf{u}(k) = [\dot{X}_T \dot{Y}_T \dot{Z}_T \omega_{X_T}\ \omega_{Y_T}\ \omega_{Z_T}]^T \tag{13.27}$$

where $\mathbf{x}(k) \in \mathbb{R}^{2M}$ (M is the number of features being tracked); ΔT is the sampling period of the vision system; $\mathbf{u}(k)$ is a velocity vector in task manipulator end-effector velocity; and k is the kth time step. The Jacobian is written as $J_v(k)$ in order to emphasize its time-varying nature due to the changing feature coordinates on the image plane. The intrinsic parameters of the camera-lens system are constant.

The desired image-plane coordinates could be constant or changing with time. The control strategy used to achieve the control objective is based on the minimization of an objective function that places a cost on errors in feature positions and a cost on providing control energy

$$\mathbf{J}(k+1) = [\mathbf{x}(k+1) - x_D(k+1)]^T\ \mathbf{Q}[\mathbf{x}(k+1) - \mathbf{x}_D(k+1)] + \mathbf{u}^T(k)\mathbf{R}\mathbf{u}(k)\,. \tag{13.28}$$

where $\mathbf{Q}$ is the weighting matrix for the state and $\mathbf{R}$ is that for the control input. The expression is minimized with respect to the current control input $\mathbf{u}(k)$. According to the optimal control input presented in Eqs. (13.5) and (13.8) in Section 13.2, the result yields the following expression

$$\mathbf{u}(k) = -(T\,\mathbf{J}_v^T(k)\,\mathbf{Q}T\,\mathbf{J}_v(\mathbf{R})^{-1}\,\Delta T\,\mathbf{J}_v^T(k)\,\mathbf{Q}[\mathbf{x}(k) - \mathbf{x}_D(K+1)]\,. \tag{13.29}$$

The weighting matrices $\mathbf{Q}$ and $\mathbf{R}$ allow the user to place more or less emphasis on the feature error and the control input.

The measurement of the motion of the features on the image plane needs to be done continuously and quickly. The method used to measure this motion is based on an optical flow technique called sum of squares (SSD). The method assumes that the intensities around a feature point remain constant as that point moves across the image plane. The displacement of a point $P_a = (x_s, y_s)$ at the next time increment to $P_{a'} = (x_s + \Delta x, y_s + \Delta y)$, is determined by finding the displacement $\Delta x = (\Delta x, \Delta y)$,, which minimizes the SSD measure $e(P_a, \Delta x)$. A pyramidal search scheme is used to reduce the search space. Nelson [1993] provides a more complete description of the algorithm and its implementation.

13.4.1.2 Optical Force Sensing

Although visual feedback control can achieve high positioning accuracy, the accuracy is not enough to carry out microassembly, which requires nanometric repeatability. To improve the repeatability of the assembly system, an optical microforce sensing technique is adopted. The principle of the method is

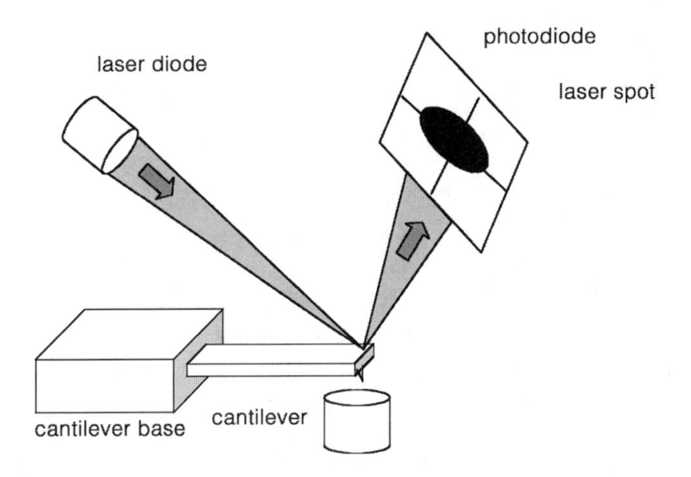

FIGURE 13.15 Optical beam deflection technique.

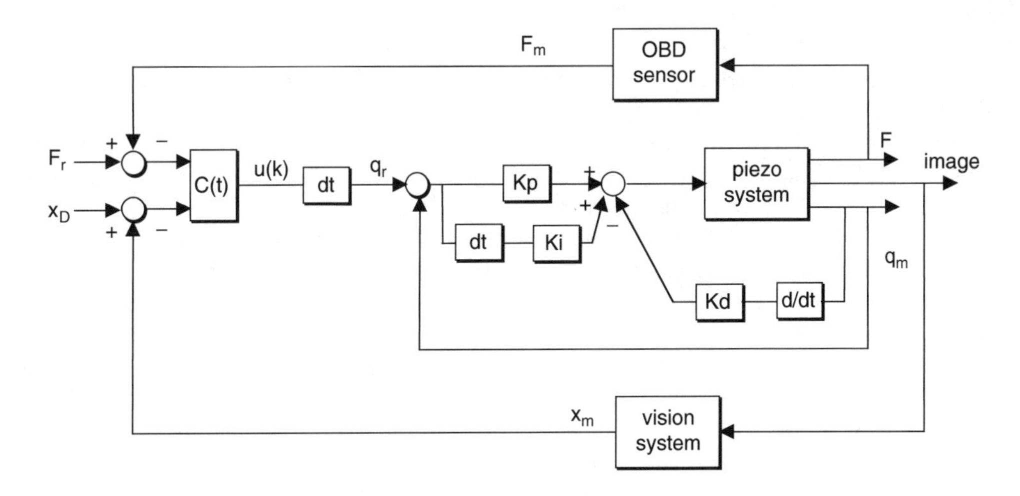

FIGURE 13.16 Force and vision in the feedback loop.

shown in Figure 13.15. The sensor consists of a laser diode, a Si or Si_3Ni_4 cantilever, and a photodiode. Light from a laser diode is focused onto the tip of a Si or Si_3Ni_4 cantilever. When the cantilever probe is deflected due to contact between the objects to be assembled, the laser beam is deflected by the cantilever onto a quadratic photodiode. The four voltages of output from the photodiode are used to measure changes in the deflection of the beam, which can be easily obtained from a simple beam theory.

13.4.1.3 Force and Vision Feedback

In an effort to achieve this a control strategy utilizing two-sensor information is shown by block diagram in Figure 13.16. The two inputs to the system are the desired state of visual features observed by the optical microscope, $\mathbf{x}_D$, and the desired contact force for the cantilever with the surface, $\mathbf{F}_{Pr}$. The desired feature state, $\mathbf{x}_D$, represents visual features expected upon contact. When the error in visual feature states is great, indicating the surfaces are "far" from one another, the controller $\mathbf{C}(k)$ uses pure vision feedback to control the system in accordance with Eq. (13.30). When $|x_d - x_m|$ falls below a threshold indicating a surface is being approached, $\mathbf{C}(k)$ switches to pure force control to initiate contact and to maintain the desired contact force $\mathbf{F}_r$. The control law that switches between and vision control is given by

$$\dot{x}_{ref_v} = -(J_v^T(k)QJ_v(k)+L)^{-1} J_v^T(k)\,Q[x(k)-x_D(k+1)] \tag{13.30}$$

$$\dot{x}_{ref_f} = \mathbf{G}_F(\mathbf{F}_r(k)-\mathbf{F}_m(k))$$

$$\mathbf{u}(k) = \mathbf{S}\dot{x}_{ref_v} + (1-\mathbf{S})\dot{x}_{ref_f}$$

$$\text{if } ((|F_m| > F_{Th}) \vee |x_D - x_m| < \varepsilon),\ \mathbf{S}=0.0,$$

$$\text{else } \mathbf{S}=1.0.$$

F_{Th} and ε provide thresholds to determine when the system should be operating in a force-control or visual-servoing mode. The output of the controller $u(k)$ is a velocity command that is integrated and sent to the piezoactuator as a position set point. The piezoactuator control system accepts this command as a reference input and uses an internal PID loop to servo the system to the desired position with an inner-loop control frequency of 4 kHz. Positional feedback q_m is provided to the piezoactuator by a calibrated capacitance sensor.

13.4.1.4 Performance Test Results

A series of tests was conducted with the microassembly workstation, which is centered around a Wentworth MP950 Integrated Circuit Probe Station with a Mitutoyo FS60 optical microscope. The nanopositioner has a 21-bit positioning resolution. Along the z axis, which has a range of motion of 15 μm, this translates into a resolution beyond the 0.05-nm calibration range Queensgate quotes as their limit. Visual feedback is used to guide the z axis nanopositioner using a Navitar TenX motorized zoom/focus lens with a 10× objective and a 5× zoom. This results in a total magnification for this system of 50×.

The optical beam deflection sensor was built using an AFM head supplied by Digital Instruments. Si cantilevers 450 μm long were used and the stiffness of the cantilever, calculated according to simple beam theory, was estimated to be 0.22 N/m.

The step response of the visual servoing system with four different objective lenses was tested using the vision system, microscope, and brushless DC motors. Overall results are shown in Table 13.2. Images of a miniature scanning electron microscope electrostatic column that was servoed by the positioning stage of the probe station. Settling time for each of the four trials was quite similar, approximately 0.1 s. Optimal performance was achieved by tuning the values of the diagonal terms in the control gain matrix Q in Eq. (13.28). Relatively small adjustments in Q between magnifications were required in order to achieve optimal performance. Critically damped response was easily achieved. More detailed experimental results for this system have been recently published.

To test the performance of the OBD sensor, impact results were performed in which the pizoactuator was commanded to impact the Si cantilever. A/D output from the photodiodes in the OBD sensor was monitored and the piezoactuator's motion was commanded through communication between the computer and the piezocontroller. Various gains k were chosen for the proportional force control law

$$u(k) = G_F(F_r(k) - F_m(k))\,. \tag{13.31}$$

TABLE 13.2 Visual Servoing Results

TM	NA	Gain	R (μm)	Δ (μm)	Step (pix)	Step (μm)	Rep. (μm)
20×	0.055	90	5.0	220.8	50	218.75	4.37
100×	0.28	75	1.0	8.6	50	43.21	0.86
200×	0.42	100	0.7	3.3	50	21.875	0.43
500×	0.42	75	0.7	2.2	50	8.75	0.17

Note: TM: total magnification; NA: numerical aperture; Gain: experimentally tuned value of diagonal elements of Q; R: resolution; Δ: depth of field; Step: commanded step size in pixels and microns; Rep.: estimated repeatability of positioning system based on controller deadband.

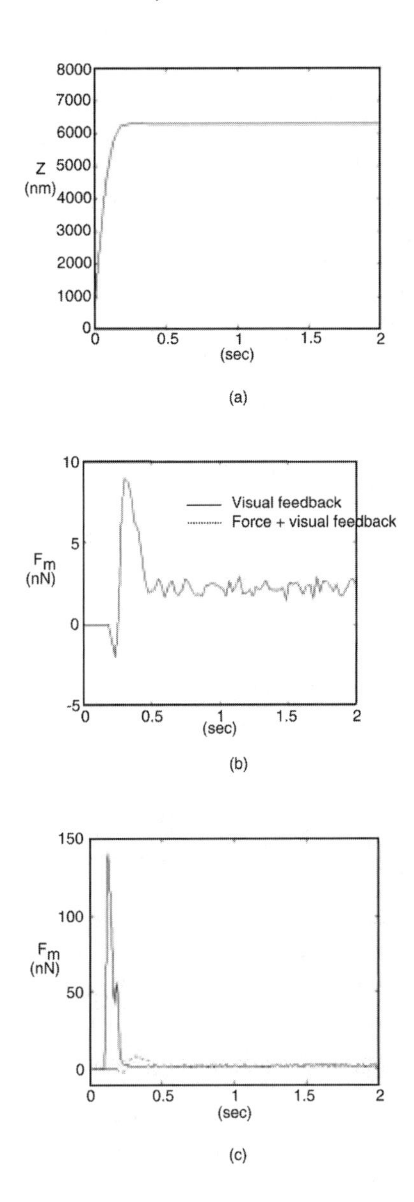

FIGURE 13.17 Results from impact trial that integrates force and vision data, and a combined plot comparing the initial impact force resulting from force feedback alone and force and vision feedback: (a) approach velocity; (b) controlled force with combined force and visual feedback; (c) comparison of the two results. (From Zhou, Y. et al., *J. Intelligent Robotic Sys.*, 28, 259–276, 2000. With permission.)

Although detailed results have not been shown here, the position and force are found to show a relatively small impact force of 9 nN and very little position overshoot. The drawback to this controller is that the slow approach velocity generated approximately 7 m/sec.

In the final trial, force and vision data are integrated according to the control law given in Equation 13.29. Results from an experimental trial are shown in Figure 13.17. A fast approach velocity of approximately 80 m/s, an adhesive force pulling the cantilever toward the surface, a small impact velocity of 9 nN, and a quick convergence to the desired 2 nN contact force are all observed. The lower plot combines the measured force from the trial in which force control alone was used with a high gain to increase the approach velocity, and from this trial that combines force and vision feedback. It is apparent that a much lower impact force results; however, there is a cost in the slower setting time of a 2-nN contact force.

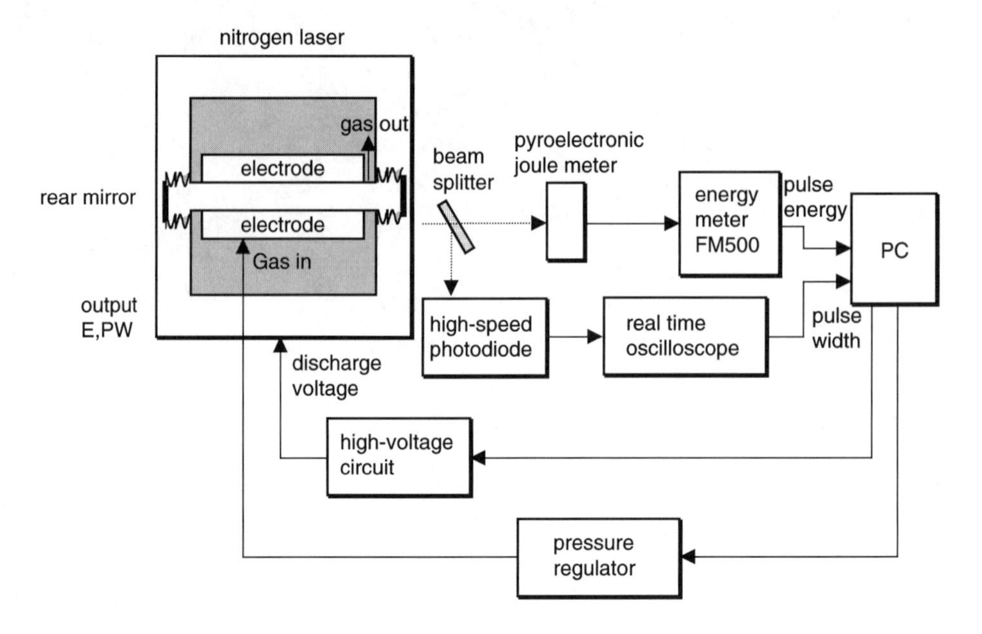

FIGURE 13.18 Measurement and control of a nitrogen laser generation system. (From Tam, S. et al., *Opt. Eng.*, 40(2), 237–244, 2001. With permission.)

13.4.2 Control of a Laser Generation System

Maintaining the dynamic stability of the output characteristics is ultimately important for any laser used in practical applications. For example, the temporal characteristics of the laser pulse are vital in shadowgraphy, micromachining, Schlieren photography, etc. Similarly, in an excimer laser, the laser duration control is essential in determining the amount of ablation of the cornea tissues.

The schematic diagram of a pulsed nitrogen laser control system is shown in Figure 13.18 [Tam et al., 2001]. The most important variables involved with laser performance are pulse energy and pulse width, which are dependent on various variables such as pulse charging driving voltage and nitrogen pressure. Therefore, the control effort to obtain the target values for energy and pulse width can be made by controlling instantaneous values of discharge voltage and nitrogen pressure. The discharge circuit for experimental lasers is of the conventional Blumlein design modified by the on-demand pulse-charging technique. The discharge voltage operates from 9.0 to 11.1 kV and the nitrogen pressure is varied from 250 mbar to 1.0 bar. The pressure of the nitrogen gas that flows into the laser is controlled by an electro-vacuum regulator.

The output beam from the N_2 laser is split into two paths for simultaneous measurements of pulse energy and pulse width. The pulse energy is measured with a pyroelectric detector connected to an energy meter. The digital signal bearing the energy content of the pulse is sent to the PC controller via an RS232 link. The pulse width is measured with a photodiode. Because the laser generator is often subject to parameter variation and exhibits nonlinearity in its dynamic characteristics, a simple conventional controller such as PID requires fine-tuning whenever operating points change. The fuzzy controller is schematically shown in a block diagram in Figure 13.19. The desired inputs are the pulse energy (E) and pulse width (PW), and these are the variables to be controlled for stability. The control variables are nitrogen pressure (P) and discharge voltage (V). Here, two membership value functions are constructed for the controller, one based on the error signal of the energy and the other based on that of the pulse width. Then, the fuzzy logic controller can be formulated utilizing the fuzzy if–then rules and algorithms discussed in Eqs. (13.16) through (13.18). A detailed discussion of the rule construction is provided in the reference section.

TABLE 13.3 Control Results (E_d = 0.25 mJ, PW_d = 1.90 ns)

	Energy (≠J)			Pulse Width (ns)		
Controller Type	Accuracy (%)	rms Deviation (%)	Max. Drift (%)	Accuracy (%)	rms Deviation (%)	Max. Drift (%)
Free-running laser	—	1.82	6.97	—	4.58	6.63
PID control	100.00	1.46	1.21	100.00	4.46	3.03
Fuzzy logic control	99.98	1.02	2.28	99.92	4.24	3.88

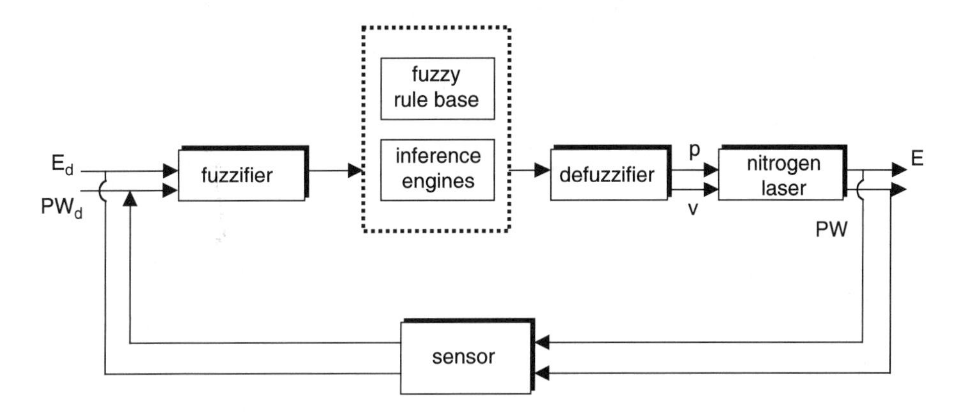

FIGURE 13.19 Schematic of a fuzzy logic control of the laser system. (From Tam, S. et al., *Opt. Eng.*, 40(2), 237–244, 2001. With permission.)

13.4.2.1 Controller Performance

The fuzzy controller thus designed was tested for a specified operating point and condition. The membership value function for both the energy control and pulse width control consists of a Z function, three triangular functions and an S function, and 10 if–then fuzzy rules. The open-loop response of the nitrogen laser was used as the baseline reference. The temporal responses were obtained for 4000 laser pulses, at a normalized discharge voltage of 220 and a normalized nitrogen pressure of 130, and at a firing rate of 5 s/pulse. The desired pulse energy and pulse width are 0.25 mJ and 1.97 ns, respectively.

In Table 13.3, the overall results are compared for three different cases: open-loop scheme, two-channel decoupled PID controller, and the fuzzy logic controller. This table shows that the two-channel PID controller performs better when compared to the open-loop system. The drawback of this controller is that for every input setting the user must fine-tune the control parameters extensively to achieve substantial output stability. In addition, the same values of the control parameters cannot be reused for different desired output values. The fuzzy logic controller gives overall improvements when compared to the open-loop system. The rms instability for the pulse energy improves from 1.82 to 1.02%, and that for the pulse width also improves from 4.58 to 4.24%. The drift in pulse energy improves from 6.97 to 2.28% and that for the pulse width improves from 6.63 to 3.88%.

13.4.3 Laser Material Processing: Laser Grooving Process

13.4.3.1 Process Description

In many industrial processes lasers have been used for machining metals, ceramics, glass, plastics, composites, and heat-treating surfaces of mechanical parts. In any laser machining process *dimensional accuracy* and *surface quality* are the important quality variables that are major issues in producing products. As shown in Figure 13.20, in laser drilling surface quality and dimensional accuracy are dependent on hole taper. In laser through-cutting, dimensional accuracy is related to kerf taper, and surface quality is related to surface roughness, dross formation, and thermal material damage (heat-affected

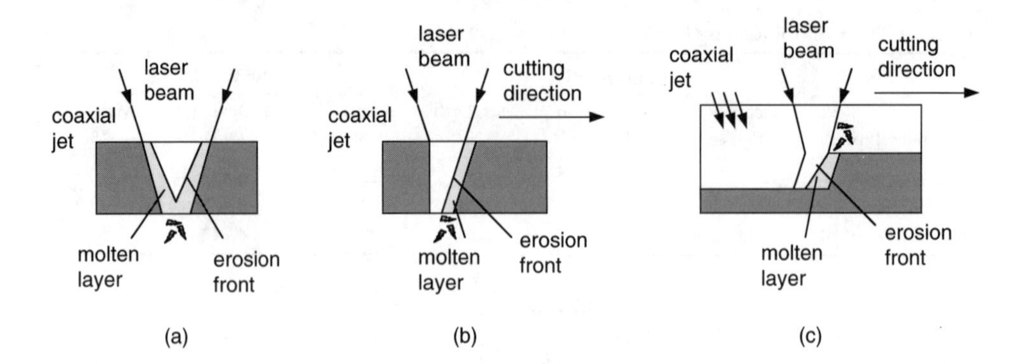

FIGURE 13.20 Various laser machining processes: (a) drilling; (b) through-cutting; (c) grooving. (From Chryssolouris, G. et al., *ASME Trans. Eng. Industry*, 113, 268–275, 1991. With permission.)

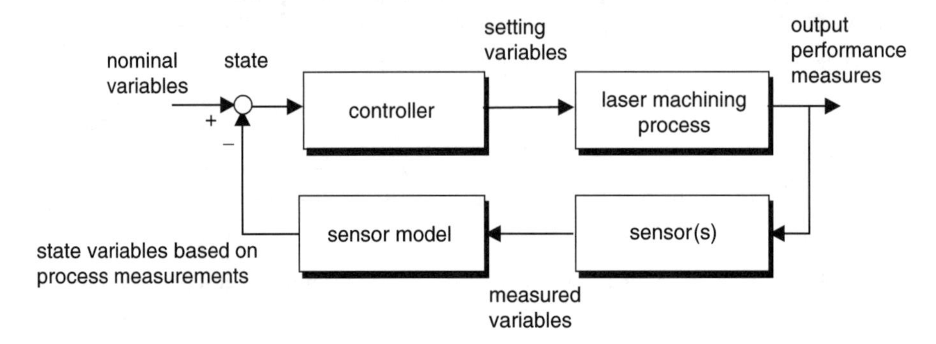

FIGURE 13.21 A control system model for a laser grooving process. (From Chryssolouris, G. et al., *ASME Trans. Eng. Industry*, 113, 268–275, 1991. With permission.)

zone) due to micro-crack formation, changes in crystalline structure, matrix recession, and polymer charring. In laser grooving (as well as in three-dimensional laser machining), dimensional accuracy is largely dependent on the precision of the workpiece positioning system and the shape of the groove surface (taper angle). The surface quality can be related to the measurement of the heat-affected zone. Therefore, a laser machining process must achieve certain requirements in surface quality and dimensional accuracy, and one or more output variables (depth of cut, heat-affected zone, and hole/groove/kerf taper) can be defined for reference into the control system.

13.4.3.2 Groove Sensing and Control System Design

Figure 13.21 shows a control system for a laser grooving process that requires a sensor model to detect the groove depth [Chryssolouris et al., 1991]. The direct depth measurement is not easy and requires an estimator to instantaneously assess the depth. For this purpose, the acoustic-sensing technique can be integrated into the control scheme for laser grooving, where acoustic emission from the process is measured by a sensor. The sensor model converts the measured variable (resonant frequency) into a state variable (groove depth). If f_o is denoted by the resonant frequency of the acoustic wave within the groove, then the sensor model is described by

$$f_0 = \frac{A_1 B}{\pi(A_0 + A_1 D)} \tag{13.32}$$

where $A_0 = \pi/4 \cdot w^2$, $A_1 = \pi/4 \cdot w \cdot b$, $B = 0.88$ for laser grooving, and D is the groove depth. Here, w is the width of the groove and b is the jet expansion. A PID controller is implemented to regulate the groove depth, as shown in Figure 13.22. A comparison between the estimated and nominal groove depth yields an error, which becomes the input to a PID controller. The controller relates the error to a change in the setting

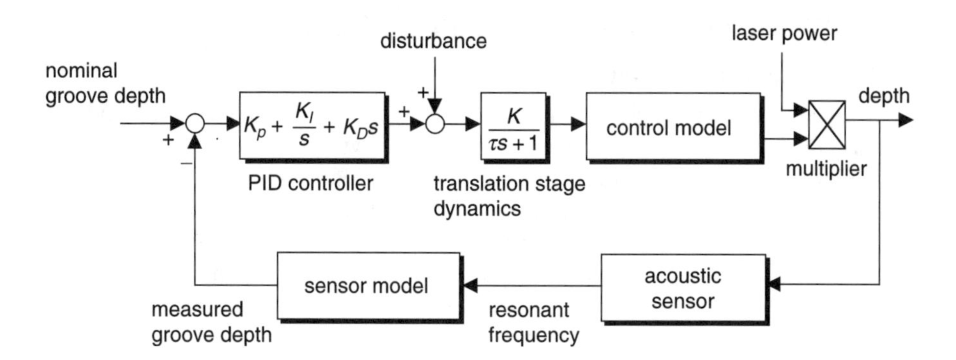

FIGURE 13.22 A PID controller for a groove depth control system. (From Chryssolouris, G. et al., *ASME Trans. Eng. Industry*, 113, 268–275, 1991. With permission.)

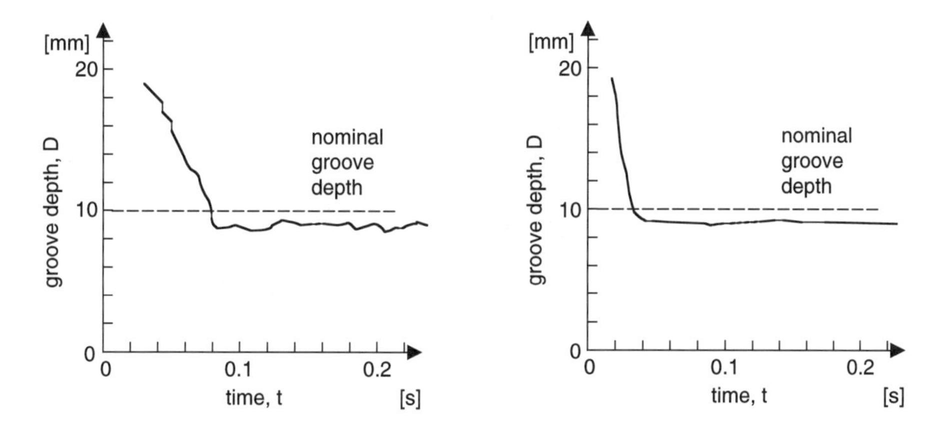

FIGURE 13.23 Groove responses with and without closed loop control. (From Chryssolouris, G. et al., *ASME Trans. Eng. Industry*, 113, 268–275, 1991. With permission.)

variable (scanning velocity). The improvements in groove-depth fluctuations due to disturbances to the system and the dynamic response of the control system in selecting a setting variable value corresponding to a desired groove depth can be analyzed by performing simulations on the proposed control scheme.

The laser groove process was dynamically modeled as a first-order system with the scanning velocity $v(t)$ as input. The workpiece translation stage was assumed to have a first-order behavior. The process model used was an analytical solution derived from heat-transfer considerations, which provided a relationship between the groove depth and the setting variable (scanning velocity) through heat-transfer analysis both at the groove surface and in the interior of the workpiece. Laser power affected system response through changes in gain for the process model.

$$D = \frac{2aP}{\pi^{1/2}\rho V d(c_p(T_{vap} - T_\infty) + L)} \tag{13.33}$$

where D is the groove depth, and the material properties for acrylic are: density ρ, thermal conductivity k, specific heat c_p, latent heat of vaporization L, and vaporization temperature T_{vap}.

In Figure 13.23, the system responses to a step input ($D = 10$ mm) are shown with and without closed-loop control. In this case a step increase in velocity was made as an external disturbance. The steady-state open-loop response shows groove-depth variations of 4% of nominal value, while the closed-loop response shows much less sensitivity to the disturbance. However, in the case of the open-loop system the time required to converge the desired value is rather long compared with that of the closed-loop system.

13.4.4 Arc-Welding Process Control with Optical Information Feedback

13.4.4.1 The Process Description

In GMA molding processes an electric arc is generated by the flow of an electric current and maintained between a consumable wise electrode and the weldment, as shown in Figure 13.24(a). The consumable electrode is automatically fed by a wise feeding device and provides additional filler metal. In manual operation skilled workers control the weld quality by adjusting the weld parameters such as the electrode feeding, weaving, and moving based on their viewing of the weld pool and its surrounding area. In an automated environment this motion has now been replaced mostly by welding robots that are able to

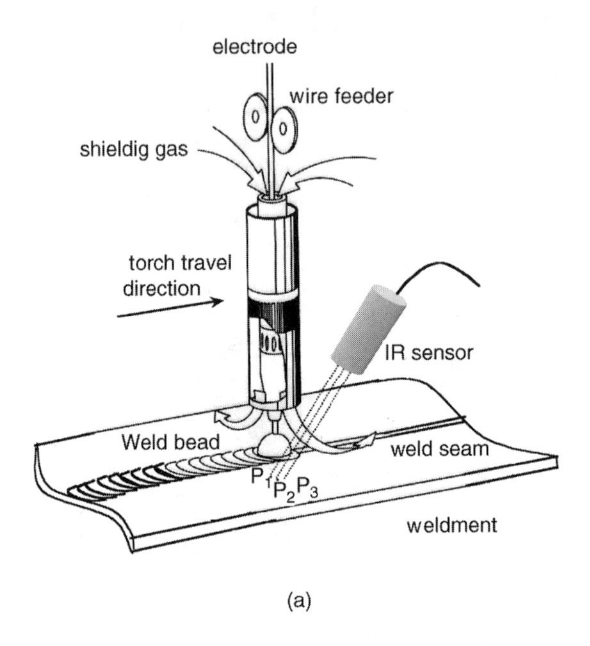

(a)

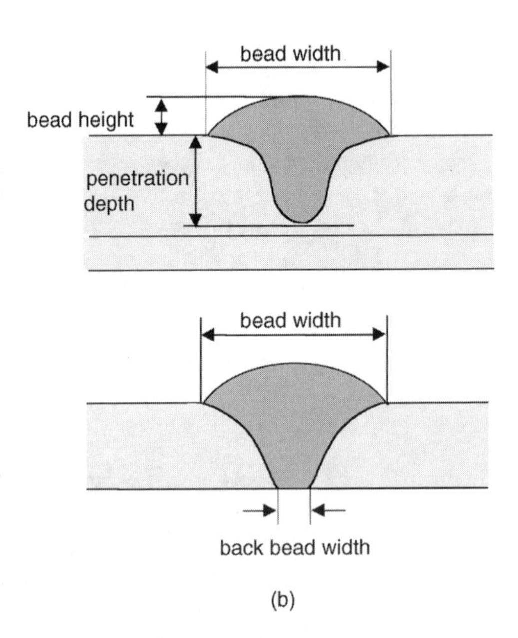

(b)

FIGURE 13.24 GMA welding process: (a) schematic description of GMA welding process; (b) weld bead geometry.

seam-track and pool geometry control. The geometry of the resulting weld bead in the case of the pull penetration is represented by the top bead width, the depth of penetration, and the back bead width, as shown in Figure 13.24(b). The objective of automated welding is to produce welds of high mechanical strength. The strength indicative of quality is usually represented by the geometry of the weld pool as well as its width and penetration.

A great deal of effort has been expended to measure the weld-pool geometry and control it online. From the point of view of monitoring and control, direct weld-pool measurement is essential. However, the geometrical variables are not readily measurable during welding due to its hostile environment. In fact, the pool appearing outside does not represent the actual melted zone and the penetration cannot be measured online. As an alternative, pervasive efforts have been made in taking indirect measurements. One such effort is to use surface temperatures near the torch area, since surface temperature distribution is indicative of the weld-pool geometry [Boo, 1991]. Figure 13.24(a) shows a feedback control system for monitoring and controlling a welding process. The IR sensor, which is an infrared noncontacting sensing system, measures point temperatures on the top surface of the element.

13.4.4.2 The Neural Network-Based Controller

The temperature measurement is utilized to estimate an instantaneous weld-pool size, which is otherwise unattainable. The control system to regulate the weld-pool size is shown in Figure 13.25 [Lim and Cho, 1993]. In the figure, PS indicates the weld-pool size and T_i denotes the temperature of the ith location. PS_d denotes a desired pool size, and PS_N indicates the estimated size by a neural network. The error is defined by

$$e = PS_d - PS_N. \tag{13.34}$$

The system is basically a feedback-error learning system adopting two neural networks, one for estimation of the weld-pool size and one for a feedback forward controller. The network structure is a multilayer perceptron, and the error back-propagation method is utilized for training them. This architecture is the feedback-error learning architecture that essentially utilizes the inverse dynamics of the welding process. The total input u_T is given by

$$u_T(t) = u_N(t) + u_f(t) \tag{13.35}$$

where u_N is the network-generated control signal and u_f is the feedback control signal. In fact, u_f can be any of the conventional controllers. In the neural network estimator, the network receives several inputs, including heat-input power, torch travel speed, surface temperatures $\{T_i(k), T_i(k-1), T_i(k-2)$, where $i = 1, 2, \ldots, 5\}$ and temperature differences between the sensing points. Here the argument k indicates the kth time step. As the outputs, the network generates the actual weld-pool size, top bead width, and penetration plus half back width.

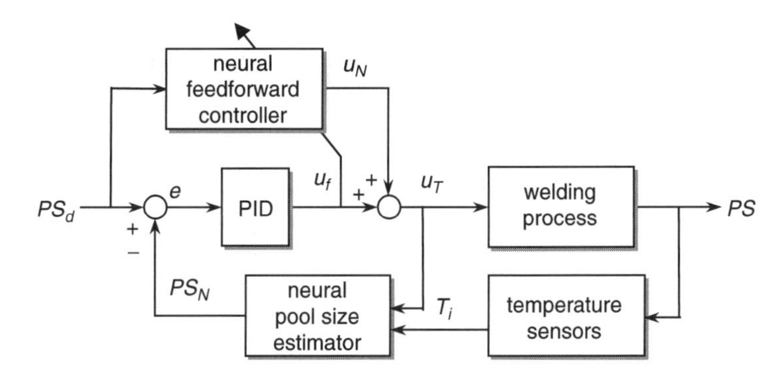

FIGURE 13.25 A neural network-based estimation and control system for the GMA welding process.

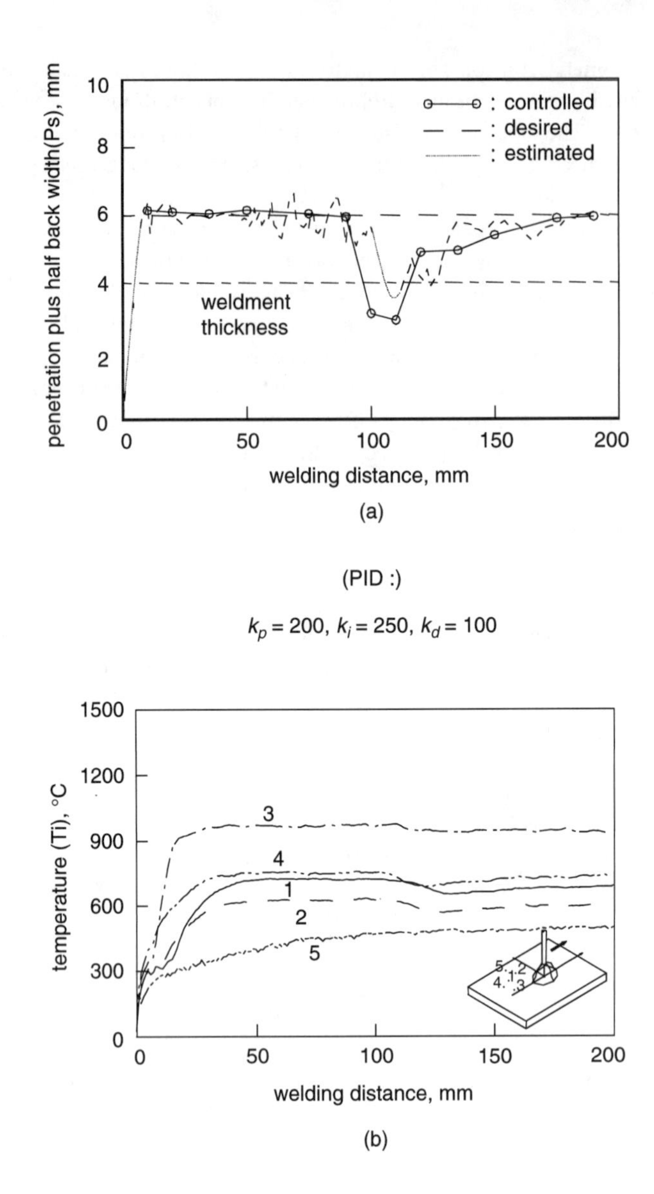

FIGURE 13.26 Neural control of a weld-pool size with an weld pool estimation: (a) controlled weld-pool size; (b) measured temperature.

To validate the capability of this control architecture, a series of welding experiments is performed taking into consideration external disturbances. Two different network architectures are considered: 30 × 40 × 40 × 1 for the estimator and 30 × 50 × 50 × 1 for the controller. As an external disturbance, torch travel speed is increased from 4 mm/sec to 6 mm/sec, while the other input parameters are kept unchanged. In Figure 13.26(a), the results of the welding control in the presence of the disturbance are shown together with the responses of surface temperatures at five locations ($i = 1, 2, \ldots, 5$), shown in Figure 13.26(b). It can be seen that the pool size obtained by this controller converges to a desired value, 6 mm, although the response slowly resumes the desired value. The controlled value at the transition, however, exhibits small fluctuations due to that of the estimation. The estimation results, denoted by a dotted line, indicate that the neural network estimates the actual pool size with satisfactory accuracy, which in this case is the controlled one.

13.5 Conclusions

Most engineered devices, products, machines, processes, and systems are expected to continue to evolve toward increased precision, downsizing, adaptation, and embedded intelligence. This can be achieved in a more accelerated way if and only if optical elements are integrated with mechatronic elements together in a systematic way to achieve such objectives. One technology needed to pursue this goal is "control system technology," which in this chapter on control problems related to opto-mechatronic systems have been addressed, and these control systems have been classified in accordance with how the optical and mechatronic elements are combined together to form an integrated system. In addition, characteristics of each system have been discussed and analyzed in view of its control objective. The control system design for these systems is normally involved with actuators, sensors, signal processing, instrumentation and measurement, control-related software, artificial intelligence techniques, and others. Because other chapters discuss some of these techniques, only certain controller design methodologies have been reviewed in brief. To provide insight into designing a control system appropriate for given systems some practical control systems have been illustrated in detail, including their control objectives, system configurations, control algorithms, and control performance.

Because achieving the above-mentioned evolution will require evolutionary concepts, sophisticated mechanisms, intelligent software, and very high performance, future control system developments will need to keep in step with these to make the evolution possible and successful.

Defining Terms

intelligent control: A control technique whose design method does not need any mathematical model but utilizes artificial intelligence techniques.

distributed-parameter system: A system whose controlled or control variables are spatially distributed.

model-based control: A control technique whose design methodology is based upon mathematical models of the systems to be controlled.

opto-mechatronic systems control: Control techniques for the systems whose components are coupled with optical and mechatronic elements in their structure and functions.

References

Boo, K. S., A Study on Analysis and Control of Gas Metal Arc Welding Process, Ph.D. thesis, Korea Advanced Institute of Science and Technology, Taejeon, Korea, 1991.

Chen, M. and Hollis, R., Vision-Guided Precision Assembly, http://www-2.cs.cmu.edu/afs/cs/project/-msl/www/tia/tia_desc.html, 1999.

Cho, H. S., Park, Y. J., and Park, H. J., Techniques and applications of accurate hydraulically-operated processes and machines, in *Machines, Electromechanical Systems and Mechatronic Systems: Techniques and Applications*, Leondes, C. T., Ed., Gorden & Breach, New York, 2000, pp. 171–229.

Cho, H. S., Neural network applications to manufacturing processes: monitoring and control, in *Computational Intelligence in Manufacturing Handbook*, Wang, J. and Kusiak, A., Eds., CRC Press, Boca Raton, FL, 2001, pp. 12-1–12-33.

Chryssolouris, G., Sheng, P., and Alvensleben, F., Process control of laser grooving using acoustic sensing, *ASME Trans. Eng. Industry*, 113, 268–275, 1991.

Ebisawa, Y., Ohtani, M., and Sugioka, A., Proposal of a zoom and focus control method using an ultrasonic distance-meter for video-based eye-gaze detection under free-head conditions, *18th Annu. Conf. IEEE Eng. Medicine Biol. Soc.*, Vol. 2, 1996, pp. 523–525, Amsterdam, the Netherlands.

Haran, F. M., Hand, D. P., Peters, C., and Jones, J. D. C., Real-time focus control in laser welding, *Measurement Sci. Technol.*, 7, 1095–1098, 1996.

Hsu, S. and Fu, L., Robust output high-gain feedback controllers for the atomic force microscope under high data sampling rate, *Proc. of the 1999 IEEE Int. Conf. on Control Applications*, 1999, pp. 1626–161, Kohala Coast-Island, Hawaii.

Jeong, H. M., Choi, J. J., Kim, K. Y., Lee, K. B., Jeon, J. U., and Pak, Y. E., Milli-scale mirror actuator with bulk micromachined vertical combs, *Proc. of the Transducer*, 1999, pp. 1006–1009, Sendai, Japan.

Kawato, M., Furukawa, K., and Suzuki, R., A hierarchical neural network model for control and learning of voluntary movement, *Biological Cybernetics*, 57, 169–185, 1987.

Landau, Y. D., *Adaptive Control*, Marcel Dekker, New York, 1987.

Lee, C. and Kim, S., An ultraprecision stage for alignment of wafers in advanced microlithography, *Precision Eng.*, 21, 113–122, 1997.

Lee, C. C., Fuzzy logic in control systems: fuzzy logic controller, *IEEE Trans. SMC*, 20, 404–435, 1990.

Lewis, F. L. and Syrmos, V. L., *Optimal Control*, John Wiley & Sons, New York, 1995.

Lim, T. G. and Cho, H. S., Estimation of weld pool sizes in gma welding process using neural networks, *Proc. Inst. Mechanical Engineers*, 207, 15–26, 1993.

Nelson, B., Papanikolopoulos, P., and Khosla, P. K., Visual servoing for robotic assembly, in *Visual Servoing*, Hashimoto, K., Ed., World Scientific Series in Robotics and Automation, Vol. 7, World Scientific, River Edge, NJ, 1993, pp. 139–164.

Papanikolopoulos, N. P., Khosla, P. K., and Kanade, T., Visual tracking of a moving target by a camera mounted on a robot: a combination of control and vision, *IEEE Trans. Robotics Automation*, 9(1), 14–35, 1993.

Plemmons, R. J. and Pauca, V. P., Some computational problems arising in adaptive optics imaging systems, *J. Comput. Appl. Math.*, 123, 467–487, 2000.

Robinson, S. D., MEMS Technology—Micromachines Enabling the All Optical Network, *Electron. Components Technol. Conf.*, 2001, pp. 423–428, Orlando, FL.

Shim, J. H. and Cho, H. S., A new macro/micro robotic probing system for the in-circuit test of PCBs, *Mechatronics*, 9, 589–613, 1999.

Tam, S., Tan, A., Neo, W., Foong, S., Chan, C., Ho, A. T. S., Chua, H., and Lee, S., Fuzzy logic control of a nitrogen laser, *Opt. Eng.*, 40(2), 237–244, 2001.

Veeco Co., contact AF, http://www.tmmicro.com/tech/modes/contact.htm,.

Wong, T. M. H. and Welland, M. E., A digital control system for scanning tunneling microscopy and atomic force microscopy, *Measurement Sci. Technol.*, 4, 270–280, 1993.

Zadeh, L. A., A rationale for fuzzy control, *ASME J. DSMC*, 94, 3–4, 1972.

Zhang, J. and Cai, L., An autofocusing measurement system with a piezoelectric translator, *IEEE/ASME Trans. Mechatronics*, 2(3), 213–216, 1997.

Zhou, Y., Nelson, B. J., and Vikramaditya, B., Integrating optical force sensing with visual servoing for microassembly, *J. Intelligent Robotic Sys.*, 28, 259–276, 2000.

Zurada, J. M., *Introduction to Artificial Neural Systems*, West Publishing, St. Paul, MN, 1992.

14

Feature Selection and Planning for Visual Servoing

Farrokh Janabi-Sharifi
Ryerson University
Toronto, Ontario, Canada

14.1 Introduction

Feature selection and planning play important roles in providing robust vision for **visual servoing**. In visual servoing a vision sensor is used to verify and update the task object location by processing the image for a selected number of **features**. A feature is any scene property that can be mapped onto the image plane such as corner position, edge length, and centroid. The main requirements of robust visual measurements for real-time applications, i.e., speed, accuracy, and reliability, depend on the features selected and used for servoing. At the same time the above overlapping requirements present a significant challenge to the design of visual servoing systems.

The concept of feature selection in visual servoing is different from that in object recognition. The rest of this chapter will focus on feature selection in the context of visual servoing. The features selected will substantially affect vision robustness and hence visual servoing performance [Janabi-Sharifi and Wilson, 1997]. Therefore, feature selection constitutes an important aspect of visual servoing, and its goal is to find the best features that provide speed, accuracy, and reliability of image measurements and characteristics of vision robustness. Examples of task failures with poor feature selection can be found in Smith and Papanikolopoulos [1996]. Two fundamental issues in feature selection are the *number* and *quality* of the features [Feddema et al., 1991]. Because processing of a large feature set is computationally infeasible, the focus will be on the selection of a minimum but information-intensive and reliable feature set.

The minimum number of features required depends on the degrees of freedom (DOF) of the visual servo control loop and the control method used. Depending on the definition of the control input, three

0-8493-1162-4/03/$0.00+$1.50

classes of visual servoing control architectures can be distinguished: image-based [Papanikolopoulos, 1992], position-based [Wilson et al., 2000], and hybrid (or 2.5D) visual servoing [Malis et al., 1999], all described in Chapter 15. For example, four features are required for a unique *pose* (position and orientation) estimate in general full six DOF motion [Yuan, 1989], or eight features in 2.5D visual servoing [Malis et al., 1999]. Many image-based control methods use three to four features [Feddema et al., 1991]. In position-based visual servoing use of more than six features for full six DOF pose estimation has been shown not to significantly improve the performance of pose estimation in extended Kalman filtering [Wang, 1992].

The quality of features will depend not only on the feature types, feature extraction and image-processing techniques, and control methodology, but also the relative status and geometry of the vision-robot system and surrounding objects as well as radiometric conditions with respect to the task object. Because these relative conditions change during servoing, the features best suited for each instance of servoing might also vary. For instance, during servoing some of the features might move out of field of view or get occluded, leading to task failure. Many visual servoing systems preselect a *fixed* feature set for tracking in an *ad hoc* fashion. Unfortunately, this is not a reliable method and leads to task failure. Hence, automatic feature selection and planning is a primary research area for visual servoing, but little research effort has been devoted to it.

One of the earliest works to consider feature selection for image-based visual servoing is reported in Feddema et al. [1991], where each feature is assigned a weight that is based on a set of image recognition and control constraints. Another approach for image-based servoing has also been reported in Papanikolopoulos [1992] that is based on an SSD optical-flow technique, which is sensitive to large rotations and small changes in lighting. Wunsch and Hirzinger [1997] also report a feature-selection method that is based on only feature occlusion and good numerical conditioning, ignoring other system constraints such as field of view or resolution constraints. These works are either highly coupled to image-based visual servoing or are nongeneral. Recently, a more comprehensive approach [Janabi-Sharifi and Wilson, 1997] has been reported to address automatic feature selection for position-based visual servoing. However, the method can be extended to other control schemes. The quality of features is assessed by geometric, feature-extraction, and pose-estimation constraints. The proposed method has also been successfully integrated into the vision-based grasp-planning method (see Further Information). The feature-selection method presented in this chapter is based on previous work by Janabi-Sharifi and Wilson [1997] with a recent extension [Ficocelli and Janabi-Sharifi, 2001] to include radiometric synthesis and constraints. Also, some constraints from image-based techniques have been derived and added for the sake of completeness. The next section will summarize the feature-selection criteria.

The objective of feature planning is to assign a subset of optimal features selected for each node of the endpoint (or camera) trajectory relative to the task object. The goal is to overcome the shortcomings of previous approaches, which assume a fixed set of features for the whole trajectory. A major problem with feature planning is the dimensionality of the combined feature- and robot-configuration (or task) space. Some strategies will be introduced to alleviate the above problem. Further discussion on feature planning will be given in Section 14.3.

14.2 Feature Selection Criteria

The quality of the features depends on a blend of criteria that will be applied (as constraints) to reduce the dimensionality of feature selection. These constraints can be classified into:

1. Task constraints, including geometric and radiometric constraints
2. Feature-extraction, pose-estimation, and control constraints

In the remainder of this section we will introduce the above constraints and mathematically formulate measures for each constraint. This will provide a systematic method of synthesizing the task features and will allow the removal of the *ad hoc* analysis and selection of features for the robust performance of a

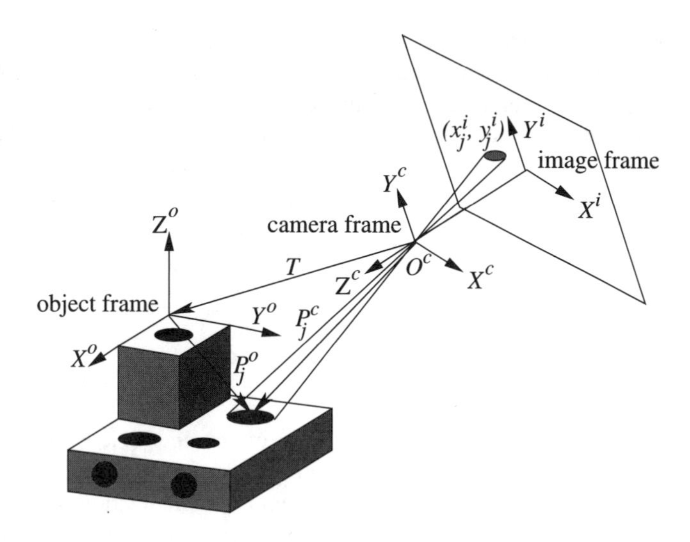

FIGURE 14.1 Projection of an object feature onto an image plane.

visual-servoing task. This will eliminate possible human errors and inconsistencies, resulting in a more autonomous robotic system.

14.2.1 Task Constraints

Task constraints include both *geometric* and *radiometric* constraints. Geometric constraints are established by the geometry of the workspace including the object, vision sensor, and obstacles. Radiometric constraints are imposed by illumination conditions and the contrast of the features with their surroundings. The projection model of the camera is shown in Figure 14.1. Here $P_j^c = (X_j^c, Y_j^c, Z_j^c)^T$ and $P_j^o = (X_j^o, Y_j^o, Z_j^o)^T$ are the coordinate vectors of the *j*th object feature center in the camera and object frames, respectively; $T = (X, Y, Z)^T$ denotes the relative position vector of the object frame with respect to the camera frame. We will also denote $\Theta = (\phi, \alpha, \psi)^T$ as the relative orientation (or viewing direction) vector with roll, pitch, and yaw parameters, respectively. The point $\tau : (T, \Theta)$ will also be used as the view point or a relative trajectory node. The coordinates of the projection of this feature centered on the image plane will be x_j^i, y_j^i. An illumination direction of $\boldsymbol{L}$ will be assumed in the camera frame. The **required edge list** *RE* will be used extensively for corner feature analysis and is defined as the set of all possible combinations of its edges taken two at a time. That is $RE = \{RE_1, RE_2, \ldots, RE_n\}$ with $RE_i = \{e_1^i, e_2^i\}$, where e_j^i denotes the *j*th edge associated with RE_i. The binomial coefficient *n* will be $n = \frac{k!}{(k-2)!2!}$ for *k* edges of a corner.

To model the illumination of the object features we assume that the light reflected from a surface I_r will be only a portion of the object irradiance E_o, and be expressed as

$$I_r = \bar{\phi} E_o = (s\phi_d + (1-s)\phi_s)\frac{\Phi_P}{4\pi R_d^2}(N \cdot L) = \left(s\frac{\bar{\ell}}{\pi} + (1-s)\frac{DGF_r}{N \cdot \Theta}\right)\frac{\Phi_P}{4\pi R_d^2}(N \cdot L) \tag{14.1}$$

where $\bar{\phi}$ accounts for the portion of the light reflected, Φ_p is the light power, R_d is the distance between the point light source and the object surface point, and $\boldsymbol{N}$ and $\boldsymbol{L}$ are the surface normal and the illumination-direction unit vectors, respectively. Also, using the model of Foley et al. [1997], the reflected light can be modeled as a combination of the diffuse component (ϕ_d) and the specular component (ϕ_s), where $0 < s < 1$ accounts for the percentage of diffuse reflection. We assume a Lambertian surface with equal brightness in all directions. The reflection of light from a Lambertian surface depends on the illumination direction $\boldsymbol{L}$ and surface normal $\boldsymbol{N}$; therefore, the bidirectional reflection function of a Lambertian surface is $\phi_d = \frac{\bar{\ell}}{\pi}$, where $\bar{\ell}$ varies for different materials. In Eq. (14.1), the specular reflection has been modeled using the Torrance and Sparrow model, where F_r is the Fresnel reflectance term that accounts for the fraction of incident illumination that is reflected. Also, D is the microfacet distribution function that accounts for the amount of surface that causes perfect reflection.

The commonly used distribution function is $D = \exp(-(\beta^{-1}\ln(\sqrt{2})\arccos(\mathbf{N}\cdot\mathbf{H}))^2$, where $H = (\overline{\Theta}+L)/|\overline{\Theta}+L|$. Here $\boldsymbol{\Theta}$ serves as the unit view vector and β as the angle between $\boldsymbol{H}$ and $\boldsymbol{N}$. In Eq. (14.1), G accounts for roughness of the surface, which may cause the reflected illumination to be blocked and not reach the viewer, $G = \min((1, \frac{2(NH)(N\Theta)}{\Theta H}, \frac{2(NH)(NL)}{\Theta H})$. Because a fraction of the light reflected will actually reach the vision sensor and affect the pixels, the resulting pixel irradiance can be calculated from

$$E_P = L_c I_r = \left(t_\ell \frac{4}{\pi}\left(\frac{a}{z_\ell}\right)^2 \cos^4 \alpha_r \right) I_r, \tag{14.2}$$

where L_c is the lens effect and is calculated from the camera optics model. Here α is the camera aperture, z_ℓ is the distance between the lens and image plane approximated by the focal length, α_r is the angle of the ray entering the lens with respect to the optical axis, and t_ℓ is the fraction of light passing through the lens. Finally a gray-scale value is assigned to a given pixel, resulting in the pixel dynamic number, DN,

$$DN = \left(\frac{2^r - 1}{E_{\max} - E_{\min}}\right) E_P + \left(\frac{2^r - 1}{E_{\max} - E_{\min}}\right) E_{\min} \tag{14.3}$$

The first coefficient is the gain, determined by the minimum illumination intensity, $E_{\min}$, which can be measured by the camera sensor, and the maximum illumination intensity, $E_{\max}$, which saturates the sensor over the entire discrete range of the illumination resolution, r. The second term is the camera offset, defined as the minimum gray-scale that will excite the CCD sensor due to noise in the camera. The introduced camera, object, and illumination models will be used for the measure development related to the constraints.

Task constraints (with the exception of contrast sensitivity constraint) are usually hard in the sense that the features that do not satisfy any of these constraints, i.e., nonfeasible features, are rejected. In the following subsections we will provide an analysis and measure formulation for each task constraint.

14.2.1.1 Geometric Constraints

The geometric constraints include visibility, resolution, field of view, depth of field, and windowing constraints [Janabi-Sharifi and Wilson, 1997]. The corresponding measures have also been summarized for each feature ε in Table 14.1.

14.2.1.1.1 Visibility

A feature must be completely visible from a selected sensor position. Among the previous approaches to visibility detection [Tsai and Tarabanis, 1990; Janabi-Sharifi and Wilson, 1997], the method provided in the latter report is simple and detects nonoccluded portions of an edge, necessary for geometric quality measurements. Two criteria have been considered for the formulation of the visibility measure: accuracy and robustness to occlusions. The closer the nonoccluded portions of an edge to the corner, the higher the accuracy of corner point measurement. The robustness criterion requires large nonoccluded portions of an edge to remain nonoccluded for a longer servoing time with high probability. Let $(\alpha_m^{ij} b_m^{ij})$ be the mth nonoccluded segment of the edge e_j^i associated with corner feature ε; ν_t, the visibility threshold, i.e., the minimum extractable length of the edge; C_ε, the vertex of the corner feature; and $\|\cdot\|_2$ the Euclidean distance (Figure 14.2). Then, the measure for the edge will be defined as:

$$J_{f_1}(e_j^i, T) = \frac{1}{\max\limits_{e_k^i \in RE_i} \left\| e_k^j \right\|_2} \cdot \sum_{(a_m^{ij} b_m^{ij}) \in e_j^i} \begin{cases} \left\| a_m^{ij} b_m^{ij} \right\|_2 \left(1 - \dfrac{\min(\left\| a_m^{ij} C_\varepsilon \right\|_2, \left\| b_m^{ij} C_\varepsilon \right\|_2}{\left\| e_j^i \right\|_2} \right) & \text{if } \left\| a_m^{ij} b_m^{ij} \right\|_2 \geq \nu_t \\ 0 & \text{otherwise} \end{cases} \tag{14.4}$$

TABLE 14.1 Geometric Constraints and Measures

Criterion	Measure
Visibility	$J_{f_1}(\varepsilon,T)=\begin{cases}1 & \text{if } \varepsilon \text{ fully visible}\\ 0 & \text{otherwise}\end{cases}$ or $J_{f_1}(\varepsilon,T)=\max_{RE_i\in RE}\prod_{e_j^i\in RE_i} J_{f_1}(e_j^i,T)$ and $J_{f_1}(e_j^i,T)$ from Eq. (14.4)
Resolution	$J_{f_2}(RE_i,T)=\max_{RE_i\in RE} J_{f_2}(RE_i,T)$ and $J_{f_2}(RE_i,T)=\begin{cases}1 & \text{if } \delta>\delta_f \;\; \forall e\in RE_i\\ 0 & \text{otherwise}\end{cases}$
Field of view	$J_{f_3}(\varepsilon,T,\Theta)=\max_{RE_i\in RE} J_{f_3}(RE_i,T,\Theta)$ and $J_{f_3}(RE_i,T,\Theta)=\begin{cases}1 & \text{if Eq.(14.6) holds } \forall e\in RE_i\\ 0 & \text{otherwise}\end{cases}$
Depth of field	$J_{f_4}(\varepsilon,T,\Theta)=\max_{RE_i\in RE} J_{f_4}(RE_i,T,\Theta)$ $J_{f_4}(RE_i,T,\Theta)=\begin{cases}1 & \text{if Eq.(14.8) holds} \forall \text{ vertices of} \forall e\in RE_i\\ 0 & \text{otherwise}\end{cases}$
Windowing	$J_{f_5}(\varepsilon,T,\Theta)=\begin{cases}1 & \text{if Eq.(14.9) is satisfied}\\ 0 & \text{otherwise}\end{cases}$

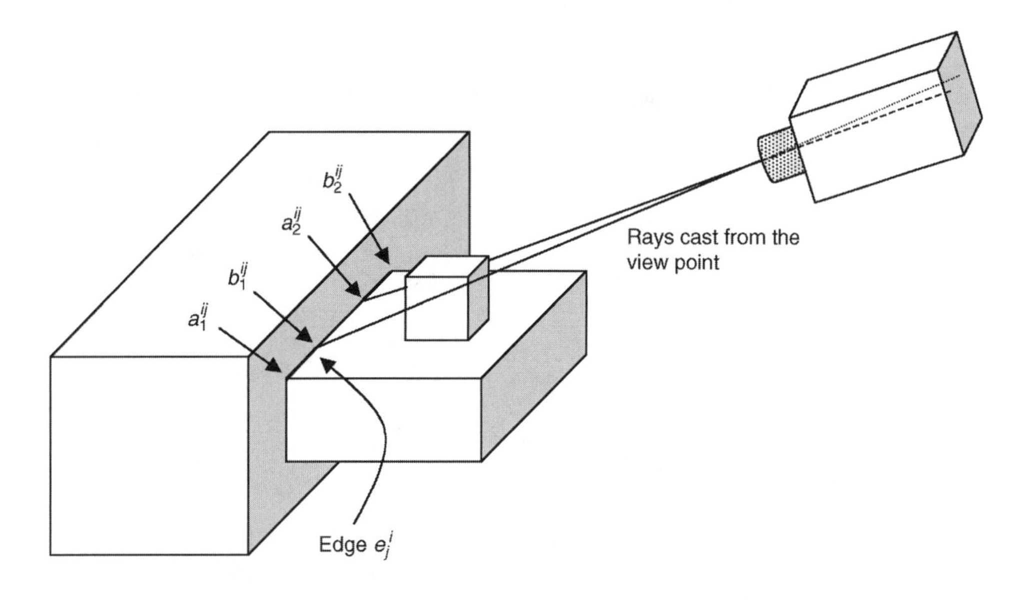

FIGURE 14.2 Identification of visible segments of an edge for visibility constraint.

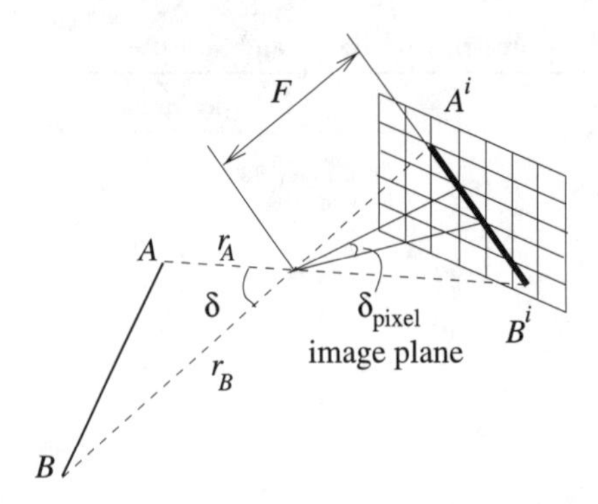

FIGURE 14.3 Image of a line segment.

If a specific RE_i has an occluded edge, the corresponding measure will clearly be zero, and in the case of complete visibility it will be 1. The visibility measure for a feature is then defined as in Table 14.1.

14.2.1.1.2 Resolution

This implies that the object feature is to be observed at a minimum image resolution. The method described by Janabi-Sharifi and Wilson [1997] provides a simple way of calculating the resolution quality of a feature. Based on the proposition in Janabi-Sharif and Wilson [1997], we assume that the focal axis direction coincides with the bisector of the view angle (Figure 14.3). The proposition implies that if the resolution constraint is satisfied for this specific orientation, it will be satisfied for other orientations. With the minimum resolution occurring along the diagonal of each pixel, resolution can be defined as a length l that has at least angle δ_f corresponding to the diagonal of N_δ pixels. The number of pixels $N_\delta \frac{L_f}{RC}$, where L_f is the feature length and RC is the minimum required resolution in (length unit/pixels). Therefore, if

$$\delta = \arccos\frac{r_A^2 + r_B^2 - L_f^2}{2r_A r_B} > \delta_f = 2\frac{L_f}{RC}\arctan\left(\frac{\sqrt{P_x^2 + P_y^2}}{2F}\right) \tag{14.5}$$

the resolution constraint will be satisfied. Thus, we can define resolution measure accordingly (Table 14.1).

14.2.1.1.3 Field of View

This is required in order for the feature image to lie completely inside the image plane for a given configuration of the camera and object. The field-of-view constraint puts a lower bound on the distance between the feature center and the view point, while the resolution constraint provides the upper bound. If a point feature ε lies at $P_\varepsilon^c = (X_\varepsilon^c, Y_\varepsilon^c, Z_\varepsilon^c)^T$, using Figure 14.1, it can be shown that the field-of-view constraint will be satisfied if:

$$\left|\frac{X_\varepsilon^c}{Z_\varepsilon^c}\right| \le \frac{N_x P_x}{2F}, \left|\frac{Y_\varepsilon^c}{Z_\varepsilon^c}\right| \le \frac{N_y P_y}{2F} \tag{14.6}$$

For high accuracy at close distances the vertices of the hole feature or the end of visible portions of the edges for a corner feature should be checked. Consequently, the field of view measure can be formulated as in Table 14.1.

14.2.1.1.4 Depth of Field (Focus)

This constraint requires that the diameter of a feature's blur circle be less than the minimum pixel dimension. A feature point at distance D_f is in focus if it lies between maximum distance D_1 and minimum distance D_2, i.e.,

$$D_1 = \frac{D_f \alpha F}{\alpha F - c(D_f - F)} \quad \text{and} \quad D_2 = \frac{D_f \alpha F}{\alpha F + c(D_f - F)} \tag{14.7}$$

where D_1 and D_2 have been defined in terms of F, a (the diameter of the lens aperture), and c (the minimum of P_x and P_y), as

$$D_2 \leq D_f \leq D_1. \tag{14.8}$$

Because $(D_1 - D_2)$ (depth of field) increases as the D level is elevated, the focus constraint provides a lower boundary for the distance of the camera from the target features. Accuracy issues related to defocused features are discussed in Chapter 10. From a practical point of view, unfocused features might be desirable for certain feature-extraction conditions. Also, in many assembly operations, fine motions cannot guarantee perfectly focused features in close operations. Therefore, a lower weight is usually given to the depth-of-field measure defined in Table 14.1.

14.2.1.1.5 Windowing

The windowing constraint is to assure the feasibility of a target feature for windowing and its robustness for tracking induced motions. This implies that the target-feature window size will not exceed the window size range, the window will have enough clearance from the image boundaries, and it will not overlap with other windows or portions of other features. A method for calculating the feasible window size is provided in Janabi-Sharifi and Wilson [1997] for both single- and double-windowing of features and is also described in Chapter 10. In summary, a window size must be bounded by

$$(S_{\min})_x \leq W_x \leq (S_{\max})_x \quad \text{and} \quad (S_{\min})_y \leq W_y \leq (S_{\max})_y \tag{14.9}$$

with $(S_{\min})_x$ and $(S_{\min})_y$ as the minimum allowable window sizes along the X^i and Y^i directions, respectively. The measure associated with the windowing constraint is given in Table 14.1.

14.2.1.2 Radiometric Constraints

Radiometric constraints include light visibility, feature position, contrast, and contrast-sensitivity constraints. The illumination model introduced in the previous subsection will be used. The corresponding measures [Ficocelli and Janabi-Sharifi, 2001] have also been summarized for each feature ε in Table 14.1.

14.2.1.2.1 Light Visibility

This constraint requires that a feature be well illuminated by a light source. If part of the feature is covered by a shadow, the shadow may be mistaken as part of the feature. Therefore, shadows must be avoided whenever possible. The ray-casting technique can be used to determine if m points of an edge are visible from a light point. This information can be stored in a *light buffer*. If the ray reaches the given edge point without intersecting another object, then the light buffer (M_i) is given a value of 1; otherwise, a value of 0 will be assigned. In many cases the illumination source is not a point but rather a linear light source such as a tube. In this case, the light source can be approximated by a set of point sources. Therefore, the measure for a corner edge with illumination direction $\boldsymbol{L}$ can be devised as:

$$J_{f_6}(e_j^i, L) = \frac{1}{m}\sum_{l=1}^{m} M_l \tag{14.10}$$

where e_j^i is the jth edge belonging to RE_i, and M_l is the light buffer value, for the lth edge point. The measure for a corner feature can be calculated by evaluating the combined edges in each $\boldsymbol{RE}$ associated with the corner. The hole-feature measure is similar to the corner-feature measure except that each point on the hole must be visible from at least one light source. The light visibility measure is given in Table 14.2.

TABLE 14.2 Radiomentric Constraints and Measures

Criterion	Measure
Light visibility	$J_{f_6}(\varepsilon, L) = \begin{cases} 1 & \text{if } J_{f_6}(e_j^i, L) = 001 \ \forall e \in \varepsilon \text{ and } \varepsilon \text{ a hole feature} \\ \max\limits_{RE_i \in RE} \prod\limits_{e_j^i \in RE_i} J_{f_6}(e_j^i, L) & \text{if corner feature} \\ 0 & \text{otherwise} \end{cases}$ and $J_{f_6}(e_j^i, L)$ from Eq. (14.10)
Feature position	$J_{f_7}(\varepsilon, \Theta) = \begin{cases} 1 & \text{if } J_{f_7}(e_j^i, \Theta) = 1 \ \forall e \in \varepsilon \text{ and } \varepsilon \text{ a hole feature} \\ \max\limits_{RE_i \in RE} \prod\limits_{e_j^i \in RE_i} J_{f_7}(e_j^i, \Theta) & \text{if corner feature} \\ 0 & \text{otherwise} \end{cases}$ and $J_{f_7}(e_j^i, \boldsymbol{\Theta})$ from Eq. (14.11)
Contrast	$J_{f_8}(\varepsilon, L, \boldsymbol{\Theta}) = \begin{cases} 1 & \text{if } J_{f_8}(e_j^i, L, \boldsymbol{\Theta}) = 1 \ \forall e \in \varepsilon \text{ and } \varepsilon \text{ a hole feature} \\ \max\limits_{RE_i \in RE} \prod\limits_{e_j^i \in RE_i} J_{f_8}(e_j^i, L, \boldsymbol{\Theta}) & \text{if corner feature} \\ 0 & \text{otherwise} \end{cases}$ $J_{f_8}(e_j^i, L, \boldsymbol{\Theta})$ from Eq. (14.12)
Contrast sensitivity	$J_{f_9}(\varepsilon, L, \boldsymbol{\Theta}) = \begin{cases} J_{f_9}(\varepsilon, L, Q) & \text{if } \varepsilon \text{ hole feature} \\ \max\limits_{RE_i \in RE} \prod\limits_{e_j^i \in RE_i} J_{f_9}(e_j^i, L, \boldsymbol{\Theta}) & \text{if corner feature} \\ 0 & \text{otherwise} \end{cases}$ and $J_{f_9}(e_j^i, L, \boldsymbol{\Theta})$ from Eq. (14.14)

14.2.1.2.2 Feature Position

This constraint implies desirability of edge features formed by the known object surfaces with known radiometric properties. This is because in images edges often occur in the intersection of two faces or between the object boundary and an unknown background. However, those created by two intersecting surfaces are more reliable than those created between boundary and background. This is because the background may not be known, or unknown objects may be moving in the environment as the object and camera move. As with the light-visibility constraint, the ray-casting technique is utilized. A set of m rays can be cast from the camera focal point toward either side of the edge. In other words, the edge can be approximated by m points, where a ray will be cast slightly to the right and to the left of the given edge point. Therefore, each edge point is composed of two rays, M^l and M^r. If the ray intersects one of the two surfaces that make up the edge, a value of 1 is stored in a *ray buffer*; otherwise, if the ray intersects an unknown background, a value of 0 is stored in the ray buffer. The measure for an edge making a corner is given by:

$$J_{f_7}(e_j^i, \boldsymbol{\Theta}) = \frac{1}{m}\left(\sum_{k=1}^{m} M_k^l . M_k^r\right), \tag{14.11}$$

where M_k^l and M_k^r are, respectively, the ray buffer values for the left and right ray cast to the kth edge point. The measure for each *RE* is then calculated for the target corner. For a hole feature, the entire hole edge should be created by known surfaces of the object and not by an unknown background. Therefore, the feature position measure could be defined as shown in Table 14.2.

14.2.1.2.3 Contrast

The contrast constraint imposes a requirement for a large intensity gradient between neighboring pixels for ease and accuracy of extraction. For an edge to be recognized in an image, the contrast across the edge must surpass the threshold value C_t defined by the user. The accuracy requirement also implies that as many edge points as possible must be detected to achieve accurate measurements. However, there is always a trade-off between time efficiency and accuracy. One can use the ray-casting method to calculate the illumination-intensity value on either side of a given edge. For each point along the edge, a contrast value is computed from $C_k = |DN_k^\ell - DN_k^r|$, where DN_k^ℓ and DN_k^r are the pixel-illumination value (or dynamic number) to the left and right of the kth edge point, respectively, calculated from the illumination model. Therefore, the contrast measure for an edge can be formulated as:

$$J_{f_8}(e_j^i, L, \Theta) = \frac{1}{m}\sum_{k=1}^{m} \operatorname{sgn}(C_k - C_t), \tag{14.12}$$

where

$$\operatorname{sgn}(x) = \begin{cases} 1 & \text{if } x \geq 0 \\ 0 & \text{otherwise} \end{cases} \tag{14.13}$$

The measure for the feature is then calculated by evaluating the edges of the feature as shown in Table 14.2.

14.2.1.2.4 Contrast Sensitivity

For accuracy and robustness of measurements, features with contrast values insensitive to and well above the contrast threshold should be chosen. This is because sensor and illumination noise or even small variations in the object pose might cause the points with contrast values near the threshold to pass to the other side of the threshold. This would result in false detection and degraded accuracy of the measured feature point. The variance of an edge point σ^2 is inversely related to contrast C^2 and proportional to the noise variance in the image σ_ς^2, i.e., $\sigma^2 = k\frac{\sigma_\varsigma^2}{C^2}$. The contrast-sensitivity constraint is a measure of how the contrast varies along the edge with respect to the threshold. For an edge, contrast sensitivity can be measured as:

$$J_{f_9}(e_j^i, L, \Theta) = \frac{1 - e^{-B_{CS}\frac{M_{CS}^i}{|DN_{\max} - C_t|}}}{(1 - e^{-B_{CS}})}, \qquad (14.14)$$

with $DN_{\max}$ as the maximum digital number produced by the sensor. M_{cs}^i denotes an average measure for the closeness of a given edge contrast of the given edge (e_j^i) to the threshold value C_t, i.e.,

$$M_{CS}^i = \frac{1}{n_T}\sum_{k=1}^{n_T}|C_k - C_t| \qquad (14.15)$$

where n_T is the number of contrast values that are above the threshold value, and C_k is the edge-point-contrast value above the threshold. B_{cs} is a sensitivity constant that affects the speed at which the measure approaches 1 (Figure 14.4). The selection of B_{cs} depends on a noise model of the sensor and fluctuations of the illumination source. That is, for ideal sensors with almost no noise, a large value of B_{cs} must be chosen in order to reach saturation quickly, and vice versa. This is because with an ideal sensor the contrast near the threshold would be good enough for a satisfactory edge detection and large differences between contrast values and the threshold do not lead to great improvements in the detectability and

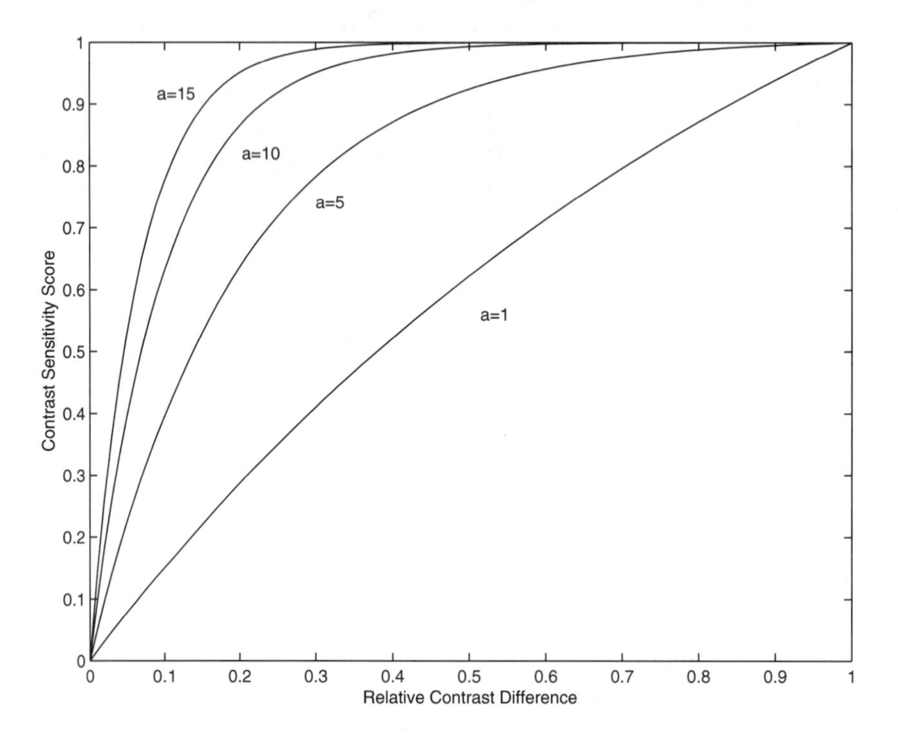

FIGURE 14.4 Contrast-sensitivity measure with different sensitivity constants ($a = B_{CS}$, relative contrast difference = $\frac{M_{CS}^k}{DN_{\max} - C_t}$).

robustness of the edge-point measurements. The contrast-sensitivity measures for both hole and corner features have been formulated accordingly, as shown in Table 14.2.

14.2.2 Feature Extraction, Pose Estimation, and Control Constraints

These constraints impose conditions for improving the speed, ease, and robustness of the feature-extraction process, enhancing controllability of an image-based system, and enriching the observability, accuracy, robustness, and sensitivity of relative pose estimations. These constraints are usually defined over a set of features and hence will be denoted by feature-set constraints (with the corresponding measures denoted by J_{s_i}) as opposed to feature constraints such as task constraints (with denoted measures of J_{f_i}). Contrary to previous measures, the measures introduced here will be soft, implying that the features of sets that do not satisfy these constraints might not be necessarily rejected. In order to analyze different numbers of features at a time, the notion of a ***q*-combination set**, q-*COMB* ϑ, is introduced. Let ϑ be a set of m nonrepeated features ε and $q \leq m$ a positive integer number. A q-*COMB* ϑ is the set of all possible combinations, q-*COMB* ϑ_k, of m features taken q at a time, where the kth combination is $q\text{-}COMB\ \vartheta_k \triangleq \{\varepsilon_1^k, \varepsilon_2^k, \ldots, \varepsilon_q^k\}$. As discussed in the introduction, e.g., for position-based servoing, $4 \leq q \geq 6$.

Feature extraction, pose estimation, and control constraints include type, size, optical angle, set nonambiguity, set apartness, set noncollinearity, set noncoplanarity, set angle, set sensitivity, set uniqueness, set observability, set controllability, set secondary sensitivity, and set durability [Janabi-Sharifi and Wilson, 1997]. The following subsections will provide an analysis of each constraint, with the corresponding measures given in Table 14.3.

14.2.2.1 Type of Features

Some features are easy to identify and measure. Because hole and corner features are common in visual servoing, the discussion in this chapter will be limited to these features. A hole feature is easier to identify: Its center can be extracted with higher accuracy; measurements associated with a hole feature are less sensitive to the orientation of the hole on the object [Madhusudan, 1990]; its center can be extracted with higher accuracy; measurement errors associated with a hole feature are less sensitive to the orientation of the hole on the object; a mathematical expression can be used to update the measurement covariance matrix R in Kalman filtering for a hole feature; and, finally, the errors in the centroid of a circular hole can be approximated by zero mean Gaussian noise. On the other hand, because the projection of two edges associated with a corner feature is sufficient to measure the location of its corner, corner features are more robust to the changes in the range of camera with respect to the object, especially in close ranges. A trade-off has been made by the formulation of a feature type measure shown in Table 14.3, where d_t is a threshold, determined experimentally, and D_f is the distance of focal point to the center of the feature [Janabi-Sharifi and Wilson, 1997].

14.2.2.2 Size of Features

Feature size affects feature-extraction robustness and cost. The use of large holes minimizes the variance of errors in their centroids and reduces measurement errors, thereby increasing feature-extraction accuracy [Madhusudan, 1990]. The cost of extraction would be higher, though. Another disadvantage would be the possibility of large holes moving out of the field of view and the violation of the windowing constraint. Similarly, for the corners large edges associated with each corner are preferred. A trade-off has been made by formulating a measure of size as shown in Table 14.3, where r_f denotes the radius of the feature-circumscribing circle, which can be calculated easily. Also, $A(x)$ maps x to a value between 0 and 1 (e.g., using a sigmoid function with low slope such as 0.1) and ζ is a bell-shaped function in the form of $\exp(-\frac{1}{2\sigma^2}(y-\mu)^2)$ with $\mu = 1.5$ and $\sigma = 0.25$. Note that in the measure the factors of reliability $0 \leq \lambda_1 \leq 1$ and time efficiency $0 \leq \lambda_2 \leq 1$ for the feature-extraction method used for feature ε have been included. If only simple features such as holes and corners are used, it could be assumed that $\lambda_1 = \lambda_2 = 1$ for simplicity. This measure almost encompasses the robustness and computational expense measures in Feddema et al. [1991].

TABLE 14.3 Feature-Extraction, Pose-Estimation, and Control Constraints and Measures

Criterion	Measure
Type	$J_{f_{10}}(\varepsilon,\boldsymbol{T})=\begin{cases}1 & \text{if hole and } D_f > d_t \text{ or corner and } D_f \le a\\ 0.5 & \text{otherwise}\end{cases}$
Size	$J_{f_{11}}(\varepsilon,\boldsymbol{T},\Theta)=\begin{cases}\lambda_1\lambda_2 & \text{if } \zeta(\frac{A(D_f)}{A(r_f)}) \ge 0.75\\ 0.5\lambda_1\lambda_2 & \text{otherwise}\end{cases}$
Optical angle	$J_{f_{12}}(\varepsilon,\boldsymbol{\Theta})=\max_{RE_i\in RE} J_{f_8}(RE_i,\boldsymbol{\Theta})$
	$J_{f_{12}}(RE_i,\Theta)=\begin{cases}1 & \text{if } -1\le \mathbf{k}\cdot\mathbf{n}_\varepsilon \le o_t \text{ fc}\\ 0 & \text{otherwise}\end{cases}$
Set nonambiguity	$J_{s_1}(\vartheta,\boldsymbol{T},\Theta)=\prod_{k=1}^{n} J_{s_1}(\varepsilon_1^k,\varepsilon_2^k,\boldsymbol{T},\Theta)$
	$J_{s_1}(\varepsilon_1^k,\varepsilon_2^k,\boldsymbol{T},\Theta)=\begin{cases}1 & \text{if } \lvert x_1^i-x_2^i\rvert>\rho_t \text{ or } \lvert y_1^i-y_2^i\rvert>\rho_t\\ 0 & \text{otherwise}\end{cases}$
Set apartness	$J_{s_2}(\vartheta)=\begin{cases}\sigma_s=\frac{1}{q}\sum_{k=1}^{q}\sigma_k & \text{if } \sigma_s>\sigma_t\\ 0 & \text{otherwise}\end{cases}$
Set noncollinearity	$J_{s_3}(\vartheta)=\frac{1}{n}\sum_{k=1}^{n} J_{s_3}(\varepsilon_1^k,\varepsilon_2^k,\varepsilon_3^k)$
	$J_{s_3}(\varepsilon_1^k,\varepsilon_2^k,\varepsilon_3^k)=\begin{cases}1 & \text{if Eq. (14.16) holds}\\ 0 & \text{otherwise}\end{cases}$
Set noncoplanarity	$J_{s_4}(\vartheta)=\frac{1}{m}\sum_{k=1}^{m} J_{s_4}(\varepsilon_1^k,\varepsilon_2^k,\varepsilon_3^k,\varepsilon_4^k)$
	$J_{s_4}(\varepsilon_1^k,\varepsilon_2^k,\varepsilon_3^k,\varepsilon_4^k)=\begin{cases}1 & \text{if Eq. (14.17) holds}\\ 0 & \text{otherwise}\end{cases}$
Set angle	$J_{s_5}(\vartheta)=1-\frac{\sum_{j=1}^{q}\left\lvert \kappa_{j,j+1}-\frac{360^\circ}{q}\right\rvert}{360}$
Set sensitivity	$J_{s_6}(\vartheta,\boldsymbol{T},\Theta)=\frac{1}{q}\sum_{k=1}^{q} J_{s_6}(\varepsilon_k,\boldsymbol{T},\Theta)$
	$J_{s_6}(\varepsilon,\boldsymbol{T},\Theta)=\begin{cases}\dfrac{\frac{1}{Z_\varepsilon^c}+R_\varepsilon^c}{\frac{1}{Z_{ct}+\max_{Z_\varepsilon} R_{ct}}} & \text{if } R_\varepsilon^c<R_{ct} \text{ and } Z_\varepsilon^c>Z_{ct}\\ 0 & \text{otherwise}\end{cases}$
Set uniqueness	$J_{s_7}(\vartheta)=\frac{1}{Nq(m-1)}\sum_{i=1}^{q}\sum_{j=1}^{N} M_{ij}$
Set observability	$J_{s_8}(\vartheta,\boldsymbol{T},\Theta)=1-\frac{E^o}{E^o_{\max}}$
Set controllability	$J_{s_9}(\vartheta,\boldsymbol{T},\Theta)=1-\frac{c(J_i)}{c_{\max}}$
Set secondary sensitivity	$J_{s_{10}}(\vartheta,\boldsymbol{T},\Theta)=1-\frac{\hat{s}_2(J_i)}{s_{\max}}$
Set durability	$J_{s_{11}}(\vartheta)=\sum_{i=k}^{\ell}\gamma_\vartheta+\sum_{j=1}^{q}\sum_{i=k'}^{\ell'}\gamma_{\varepsilon_j} \quad \forall\varepsilon_j\in\vartheta$

14.2.2.3 Optical Angle

The features with unit surface normals $\boldsymbol{n}_\varepsilon$ almost perpendicular to the unit optical axis vector $\boldsymbol{k}$ lead to poor pose estimations by EKF [Wang, 1992]. Therefore, the product $\boldsymbol{k}\cdot\boldsymbol{n}_\varepsilon$ must lie between −1 and a threshold o_t that is a small negative number close to zero. The corresponding measure could be formulated accordingly, as shown in Table 14.3. The measure application to a hole feature is straightforward. For a corner feature, the surfaces formed by each RE_i of RE are checked.

14.2.2.4 Set Nonambiguity

This constraint implies that each feature in the set should have enough clearance from other features of the set in the image plane. The set-nonambiguity measure in Table 14.3 assigns the score to a feature set ϑ based on its feature clearance, where ε_1^k, $\varepsilon_2^k \in$ 2-*COMB* ϑ_k of 2-*COMB* ϑ and $n = \frac{m!}{(m-2)!2!}$ for m features in the set. Also p_t is the nonambiguity threshold. For instance, a triangulation method was applied in a relatively high-speed scenario of about 20 cm/s and servoing at a frame rate of 61 Hz, and it was found that $p_t = 0.06$ (cm) (10 pixels) [Wang, 1992].

14.2.2.5 Set Apartness

For improved feature extraction and pose estimation, set features are required to be sufficiently apart [Yuan, 1989]. Let $\sigma_k \stackrel{\Delta}{=} \min_j \|\varepsilon_k,\varepsilon_j\|_2$ with $\varepsilon_j \in \vartheta$, $j \neq k$, where $\|\varepsilon_k,\varepsilon_j\|_2$ denotes the Euclidean distance between the centers of the features in a set ϑ and σ_t is the apartness threshold. Then, the measure of apartness could be formulated as in Table 14.3, using average set apartness σ_s.

14.2.2.6 Set Noncolinearity

The collinear features set will lead to singularity of the control matrix **B** and tracking failure in image-based servoing [Papanikolopoulos, 1992]. In position-based servoing, it has been shown that noncollinear features provide improved observability, accuracy, and robustness in the pose-estimation process [Janabi-Sharifi and Wilson, 1997]. If three features are considered at a time, i.e., $\{\varepsilon_1^k, \varepsilon_2^k, \varepsilon_3^k\} =$ 3-*COMB* $\vartheta_k \in$ 3-*COMB* ϑ, noncollinearity of these feature points will be guaranteed when the following numerically robust condition is satisfied:

$$\begin{vmatrix} X_3^o - X_1^o & X_2^o - X_1^o \\ Y_3^o - Y_1^o & Y_2^o - Y_1^o \end{vmatrix} \geq \bar{\varepsilon} \quad \text{or} \quad \begin{vmatrix} X_3^o - X_1^o & X_2^o - X_1^o \\ Z_3^o - Z_1^o & Z_2^o - Z_1^o \end{vmatrix} \geq \bar{\varepsilon} \quad \text{or} \quad \begin{vmatrix} Y_3^o - Y_1^o & Y_2^o - Y_1^o \\ Z_3^o - Z_1^o & Z_2^o - Z_1^o \end{vmatrix} \geq \bar{\varepsilon} \tag{14.16}$$

where $\bar{\varepsilon}$ is a small scalar and $|\cdot|$ denotes the absolute value of a matrix determinant. The above principle can then be used to obtain a measure of noncollinearity (Table 14.3).

14.2.2.7 Set Noncoplanarity

Noncoplanar features outperform coplanar features in terms of accuracy and robustness in both direct solution of photogrammetric equations [Yuan, 1989] and Kalman filter-based pose estimation [Wang, 1992; Janabi-Sharifi, 1995]. The following condition provides the necessary and sufficient condition of noncoplanarity for four feature points $\varepsilon_k (k = 1, \ldots, 4)$, with the homogeneous coordinates $[X_k Y_k Z_k w_k]^T$ defined in any coordinate frame:

$$\begin{vmatrix} X_1 & Y_1 & Z_1 & w_1 \\ X_2 & Y_2 & Z_2 & w_2 \\ X_3 & Y_3 & Z_3 & w_3 \\ X_4 & Y_4 & Z_4 & w_4 \end{vmatrix} \geq \bar{\varepsilon} \tag{14.17}$$

where $w_i = 1$ $(i = 1, 2, 3, 4)$ for simplicity. The measure of set noncoplanarity can then be formulated using Eq. (14.17), as shown in Table 14.3.

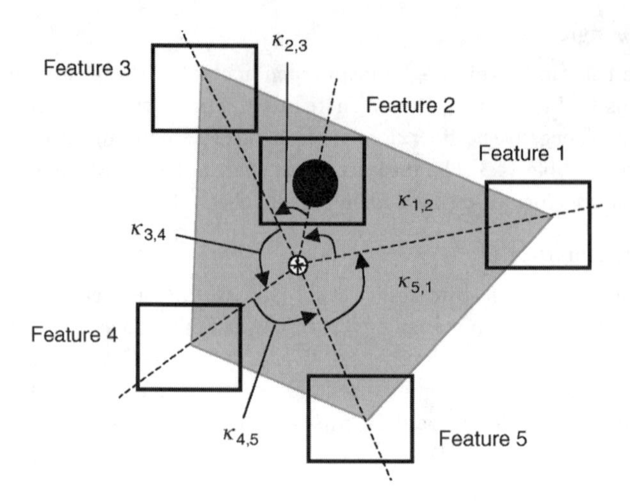

FIGURE 14.5 Central angles of five features used for set-angles constraint.

14.2.2.8 Set Angle

If q features in a set are distributed at an almost equal central angle of $\frac{360^\circ}{q}$, the oscillations in the state prediction error covariance matrix of the EKF algorithm will be minimized and, as a result, pose estimation improved. The central angle of two sequent features ε_i and ε_{i+1}, $\kappa_{i,i+1}$, is the smallest angle subtended at the center of features (or at the image plane center in image plane), as shown in Figure 14.5. In the measure formulated for the set angle (Table 14.3), the features are assumed to be ordered according to their central angle with respect to the center of the features in the preprocessing step, $\kappa_{q,q+1} \equiv \kappa_{q,1}$, and angles are measured in degrees. Obviously, the closer the central angle to $\frac{360^\circ}{q}$, the higher the measure value is.

14.2.2.9 Set Sensitivity

Robust and accurate servoing requires substantially visible motions of the feature images due to the relative pose changes. A feature-sensitivity function could be defined as the Frobenius norm of the feature image Jacobian matrix J_i [Janabi-Sharifi, 1995], i.e.,

$$s_1(J_i) = \| J_i \|_F = \sqrt{\sum_{k=1}^{2} \sum_{m=1}^{6} |(J_i)_{km}|^2} \tag{14.18}$$

In position-based control, relatively high values associated with sensitivity are desired in order to accurately sense the object's motion. Nevertheless, $\|J_i\|$ should not be too large to cause camera sensitivity to noise. In an image-based approach, due to the image Jacobian singularity and inverse update issues, the objective is slightly different. These issues will be discussed under the set-controllability constraint. The sensitivity of a feature ε is a function of $R_\varepsilon^c \equiv \sqrt{X_\varepsilon^{c^2} + Y_\varepsilon^{c^2}}$, and $\theta_\varepsilon \equiv \arctan \frac{Y_\varepsilon^c}{X_\varepsilon^c}$. Sensitivity will increase as the value of R_ε^c increases or the value of Z_ε^c decreases. At the neighborhood of a critical depth Z_{ct}, the slope of sensitivity S increases, due to the R_ε^c increase, and becomes very sharp. Therefore, to provide high sensitivity the depth of a feature Z_ε^c should be chosen so that it is as low as possible, yet Z_ε^c should be greater than Z_{ct}. Also, the radius R_ε^c must be chosen to be as high as possible. Janabi-Sharifi [1995] showed that this not only increases the sensitivity but also yields higher accuracy in *(Z, pitch, yaw)* parameter estimations. However, very high values of R_ε^c may lead to poor field of view and, with a small reduction in the Z_ε^c parameter, features may move out of the field of view. In order to avoid frequent switching of the features and hence improve durability, R_ε^c values should not be very close to an upper bound, i.e., R_{ct}. One can calculate R_{ct} from $R_{ct} = C_r\, Z_\varepsilon^c \tan(\frac{\alpha_f}{2})$ where $\alpha_f = 2\arctan(\frac{\min(N_x P_x, N_y P_y)}{2F})$

and C_R is a clearance factor that should be less than 1. The above points have been taken into consideration in the formulation of a sensitivity measure for position-based control (Table 14.3).

14.2.2.10 Set Uniqueness

A unique feature has an identifiable characteristic that makes it easy to be identified among other features. This is interesting from the point of view of control because a unique feature could be likely identified quickly if it is lost during servoing. If m is the total number of **feasible features** (after the application of task constraints), N the total number of trajectory nodes, and M_{ij} the number of features dissimilar to feature i in image $j\,(0 \le M_{ij} \le m-1)$, the measure of uniqueness could be defined as in Table 14.3 and varies between 0 and 1. This measure is a modified version of the uniqueness measure produced by Feddema et al. [1991]. Note that if the images could be known in advance, a less conservative approach would be to take m as the total number of features in images.

14.2.2.11 Set Observability

Set observability is defined as the ability to estimate the relative pose of the target object with respect to the camera using a particular estimation method at a given view point [Feddema et al., 1991]. The estimation error function can be defined as:

$$E^o = \frac{1}{2}(Z_k - \hat{Z}_k)^T (Z_k - \hat{Z}_k) \tag{14.19}$$

where Z_k and $\hat{Z}_k$ are the measured (or real) and estimated image feature point coordinates, respectively. Usually, $Z_k = [x_1^i, y_1^i, \ldots, x_q^i, y_q^i]^T$, the image coordinates of the feature points. $\hat{Z}_k$ could be calculated using estimated relative pose and photogrammetric equations. Therefore, this error could be measured and would indicate a measure for the observability of the part's pose with respect to the camera. Then, the set observability measure could be defined as in Table 14.3. The measure has been normalized using maximum acceptable error $E^o_{\max}$ to limit the range of the measure to the range of [0, 1]. Also, the measure can be averaged over N view points or nodes:

$$J_{S_8}(\vartheta) = \frac{1}{N}\sum_{j=1}^{N} J_{S_9}(\vartheta, T, \Theta) \tag{14.20}$$

14.2.2.12 Set Controllability

In image-based control, the inverse of the image Jacobian matrix, J_i^{-1}, is used to calculate the desired change in the camera position when the distance between the actual and desired feature locations is short. In this sense the relative pose is said to be controllable if J_i^{-1} exists and is nonsingular [Feddema et al., 1991]. The controllability of the image-based control can be measured by inspecting the condition of J_i, defined as:

$$c(J_i) = \|J_i\| \, \|J_i^{-1}\| \tag{14.21}$$

where $\|\cdot\|$ could be any norm of the matrix. In the above equation, J_i^{-1} could be replaced with pseudo-inverse J_i^+ in the case of nonsquare J_i matrix, e.g., when more than three features are used. Small values of $c(J_i)$ imply that the Jacobian matrix does not approach image singularity. Also, the norm of J_i is a measure of the change in image features for a unit change in the relative pose. If $\|J_i\|$ becomes too large, the camera will be sensitive to noise. On the other hand, if $\|J_i\|$ becomes too small, the object motion might not be sensed accurately. Similarly, $\|J_i^{-1}\|$ is a measure of the accuracy of detecting changes in the object's pose. From the perspective of image-based control, $\|J_i^{-1}\|$ should also be small. Therefore, the condition of J_i needs to be minimized for improving both the sensitivity and the controlability of an image-based control system. For position-based control, this constraint, opposed to the sensitivity constraint, will reduce $\|J_i\|$, resulting in less camera sensitivity to noise. Using the largest acceptable value of (J_i), denoted by $c_{\max}$, a corrected condition could be defined as:

$$c(\hat{J}_i) = \begin{cases} c(J_i) & \text{if } J_i^{-1} \text{ exists and } c(J_i) \leq c_{\max} \\ c_{\max} & \text{if } J_i^{-1} \text{ exists and } c(J_i) > c_{\max} \\ c_{\max} & \text{otherwise} \end{cases} \tag{14.22}$$

Then the measure of controlability could be defined as shown in Table 14.3. Again, the measure could be averaged over the N nodes of the trajectory.

A similar measure has been introduced under the measure of the observability of robot motion [Sharma and Hutchinson, 1994]. The introduced measure is motivated by the manipulability measure and is defined as:

$$w_{im} = \sqrt{\det(J_i J_i^T)} \tag{14.23}$$

The observability measure in Eq. (14.23) is also related to the resolvability concept in Nelson and Khosla [1994]. In singular configurations w_{im} becomes zero, and when there are no redundant features (i.e., square Jacobian matrix), the measure reduces to $|\det(J_i)|$. The measure of w_{im} can be used to define a similar controlability measure, as shown in Table 14.3.

14.2.2.13 Set Secondary Sensitivity

In image-based control the image Jacobian matrix J_i is required to be updated for real-time control. To reduce the computational time the image Jacobian elements should be almost constant or slowly time varying. Therefore, we must minimize the sensitivity of J_i to the changes of image feature points. The secondary sensitivity function can then be defined as [Feddema, et al., 1991]:

$$s_2(J_i) = \sum_{i=1}^{q} \sum_{k=1}^{2q} \sum_{m=1}^{6} \left\{ \left| \frac{\partial (J_i)_{km}}{\partial X_i^c} \cdot \frac{X_i^c}{(J_i)_{km}} \right| + \left| \frac{\partial (J_i)_{km}}{\partial Y_i^c} \cdot \frac{Y_i^c}{(J_i)_{km}} \right| + \left| \frac{\partial (J_i)_{km}}{\partial Z_i^c} \cdot \frac{Z_i^c}{(J_i)_{km}} \right| \right\} \tag{14.24}$$

where each $(J_i)_{km}$ is an element of the image Jacobian matrix. The function can be corrected using the maximum acceptable value of $s_2(J_i)$, i.e., $s_{\max}$, as follows:

$$\hat{s}_2(J_i) = \begin{cases} s_2(J_i) & \text{if } s_2(J_i) \leq s_{\max} \\ s_{\max} & \text{otherwise.} \end{cases} \tag{14.25}$$

The secondary sensitivity measure can then be normalized using $s_{\max}$ to have a value between 0 and 1, as shown in Table 14.3. Like the previous constraint, the measure could be averaged over the N nodes of the trajectory.

14.2.2.14 Set Durability

Frequent switching of the feature sets is not desirable from the point of view of servoing accuracy and speed (Figure 14.6). The measure for the durability constraint of a set ϑ with q features [Janabi-Sharifi, 1995] is given in Table 14.3, where γ_ϑ is the set durability factor (usually taken to be 1), k is the smallest node number such that the set ϑ is valid for all nodes $i = k, k+1, \ldots, \ell$, including the node under study. Similarly, γ_ε is the feature durability factor (usually $\gamma_\varepsilon \ll \gamma_\vartheta$ such as $1/q$), and the feature ε of ϑ is valid for all nodes $i = k', k'+1, \ldots, \ell'$. Note that the durability constraint is similar to the feature-set completeness measure in Feddema et al. [1991]. However, the completeness measure assigns a score based only on the presence of a member of the set in the image. The notion of durability is stronger in the sense

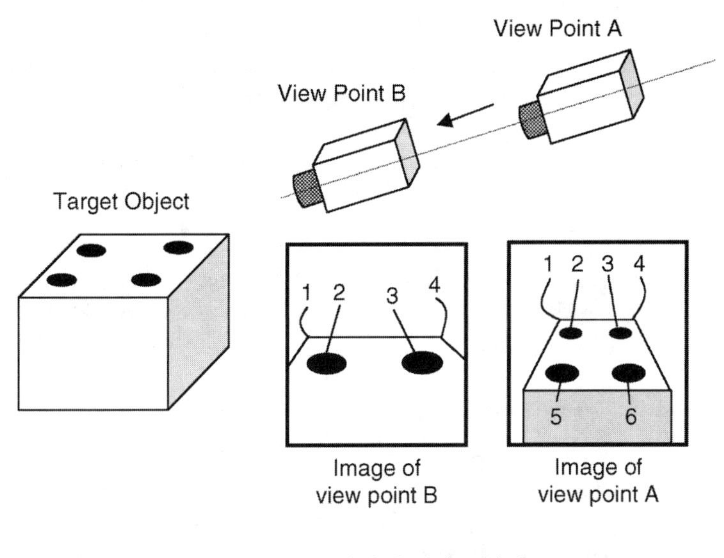

FIGURE 14.6 Durable features for a given relative trajectory.

that it rewards the features that are not only in an image, but also feasible from geometric and radiometric perspectives.

14.3 Feature Selection and Planning

Vision-based control and planning pose an important problem of moving robots among physical and sensory constraints. Physical constraints, such as obstacles for the links, and kinematic constraints (e.g., robot joint limits) are dealt with by conventional methods for robot-motion planning, and consequently an admissible trajectory is generated. However, visual constraints, such as image singularity and camera occlusion avoidance, need to be incorporated into a robot-motion plan for the effective, continuous, and robust integration of visual information into control actions. The objective of feature selection and planning is to integrate vision constraints into robot-motion planning by selecting and assigning optimal feature sets to each node of the generated trajectory. In visual servoing, this trajectory is usually a relative Cartesian trajectory between the camera (eye-in-hand) and the object. The selected features could also be used for feature-based trajectory planning in image-based servoing. The main issue in feature planning is computational complexity, mainly due to the high dimensionality of feature space in many workpieces. For instance, for a set of n features, dimension of q-$COMB\vartheta$ will be high, i.e., it is possible to form $\sum_q \frac{n!}{(n-q)!q!}$ feature sets each with q nonrepeated features. Dimensionality reduction can be achieved by partitioning the whole selection problem into the following subproblems. Here we assume that a relative trajectory has already been generated by an automatic trajectory planner or by teaching in the off-line phase of operation. Each node of the trajectory will be denoted by $\tau{:}(T,\Theta)$. The following strategy [Janabi-Sharifi, 1995] could be used for feature selection and planning.

14.3.1 Features Representation

This is done by selecting and representing the candidate features, $\gamma = \{\varepsilon_1, \varepsilon_2, \ldots, \varepsilon_n\}$, that are allowed by the problem-formulation, feature-extraction, and pose-estimation processes. For instance, in many visual servoing implementations, hole and corner features are used. The representation could be achieved using the solid object modeler as in the AFS (automatic feature selection) package [Janabi-Sharifi, 1995].

14.3.2 Feasible Features Selection

The set of feasible features $\Gamma \subseteq \gamma$ for a node $\tau{:}(T,\Theta)$ of a trajectory could be found by testing the set of candidate features γ using the feature constraint measures J_{f_i} $(i = 1, 2, \ldots, 12)$ 12. Features that do not meet any of these measures (i.e., nonfeasible features) are removed in the early stages of calculation, before they increase the dimensionality of computation. The dimension of Γ is denoted by m.

14.3.3 Feasible Feature Sets Formation

All $q\text{-}COMB\Gamma_k$ sets from Γ are formed, where q depends on the servoing strategy. For instance, in position-based strategy q = 4, 5, 6. Each of these sets will be denoted by ϑ_k. Obviously there will be $s = \sum_{q=q_1}^{q_2} \frac{m!}{(m-q)!q!}$ feature sets with q_1 and q_2 as the lower and upper allowable numbers of features in a set.

14.3.4 Admissible Feature Sets Selection

This is done by applying the feature-set constraints J_{s_i} $(i = 1, 2, \ldots, 10)$ (as the secondary filter) to feasible feature sets. The result will be a set of admissible feature sets Λ_α associated with a node $\tau{:}(T,\Theta)$ of a trajectory. Note that $J_{s_4} - J_{s_7}$ are not associated with the hard constraints, and 0 values for these measures of a feature set will not exclude that set from further considerations. Also, the measures $J_{s_8} - J_{s_{10}}$ are more relevant to image-based visual servoing, while the set-sensitivity measure J_{s_6} is related primarily to position-based visual servoing.

14.3.5 Optimal Feature Set Selection

First, the durability constraint is imposed over the admissible feature sets in $\Lambda_a(T,\boldsymbol{\Theta})$. Next, the overall optimality measure will be calculated and the optimal feature set χ for a node will be obtained by

$$J_{op}(\chi,T,\Theta) = \max_{\vartheta_k \in \Lambda_a}\left(\sum_{j=1}^{11} J_{s_j}(\vartheta_k,T,\Theta) + \frac{1}{q_k}\sum_{i=1}^{q_k}\sum_{j=7}^{12} J_{f_j}(\varepsilon_i^k,T,\Theta)\right) \tag{14.26}$$

The main challenge remaining with feature selection and planning algorithm is its time complexity. For certain applications, this might be problematic. In the next section, complexity issues and possible relevant solutions to this issue will be discussed.

14.4 Computational Efficiency and Real-Time Feature Selection

The time complexity of the feature selection and planning algorithm can be shown to be $O(N(n_f + n_a + n))$ for N nodes of the trajectory, n_f feasible feature sets, and n_a admissible feature sets [Janabi-Sharifi and Wilson, 1997]. This could result in a relatively long running time, making it inappropriate for reactive motions. In such online operations, there is no access to a relative trajectory, and a point-to-point motion is carried out. Here, the system must plan for the next point of motion according to predictions made, e.g., by Kalman filter. In this section, techniques for improving the computational efficiency of feature planning will be discussed.

14.4.1 Parallel Processing Method

Many parts of an algorithm can be processed in parallel, resulting in a shorter processing time. Partitioning steps (2) to (4) among N processors (one for each node of the trajectory) is possible and would decrease the running time by $O(N)$ for the above stages. Also, it is possible to further partition the task of step (2) among 12 parallel processors (one for each J_{f_i}) and that of step (4) among 11 parallel processors (one for each J_{s_i}), reducing the execution time to $1/12 \times 1/11$ of that with sequential processing. Further parallel processing is possible by allocating a processor to each feasible feature set during the selection of the admissible feature set. This will further reduce the time by $O(n_f)$. Hence, the full parallel-processing task will require $11Nn_f$ parallel processors (assuming $n_f \geq 2$). In point-to-point motion, since $N = 1$, at most only $11n_f$ processors will be necessary.

14.4.2 Loci and Reduced Constraints Method

Some of the constraints are not crucial for the robustness of visual servoing and so can be dropped from the feature planning algorithm. These constraints include optical angle, depth of field, set noncoplanarity, set apartness, set angle, sensitivity, set uniqueness, and durability. Some of the constraints are also specific to the control method used and could be excluded when another servoing method is used. For instance, set controlability, set observability, and set secondary sensitivity would not be applied in the position-based servoing method. Also, the loci of some constraints such as visibility can be determined in advance [Janabi-Sharifi, 1995]. Therefore, during the online phase of operation, the satisfaction of the constraints can be verified easily by checking the loci of the corresponding constraints.

14.4.3 Space of Admissible Feature Sets Method

First, the **feasible features space** of an object P viewed from all possible poses X of sensor C (denoted by $Fspace_C(P)$) is calculated. That is:

$$Fspace_C(P) = \left\{ \varepsilon \in \Gamma_X^P \middle| X \in Cspace_C^P \right\} \tag{14.27}$$

where Γ_X^P is the set of feasible features of object P viewed from X, and $Cspace_C^P$ denotes the configuration space of C with respect to the frame of object P. Similarly, the **space of admissible feature sets** $AFspace_C(P)$ can be calculated. If the operational zones of $AFspace_C(P)$ can be calculated, real-time feature selection can be achieved in a very short time. However, $AFspace_C(P)$ is high-dimensional, e.g., at least seven-dimensional for the general relative motion of the end-effector with respect to the object in position-based control. Further assumptions might be required to reduce the dimensionality of feature selection and planning. For example, one might assume that the relative orientation of the camera is fixed by assuming that the optical axis passes through the center of the features. This assumption applies to many industrial inspection and assembly cases. In order to further reduce the computational cost, the three-dimensional relative position of the end-effector (or the camera) could be divided into patches, e.g., by defining the concentric viewing spheres $VWS(r_o)$, each with radius r_o, around the object, as defined by Janabi-Sharifi [1995]:

$$VWS(r_o) = \left\{ (i, j) \middle| \; 0° \leq i \leq 180° \text{ and } 0° \leq j \leq 360° \sin\left(\frac{\pi}{180°} i\right) \right\} \tag{14.28}$$

It is easy to show that there are $2\sum_{i=0}^{90}(360° \sin(\frac{\pi}{180°} i)) - 359$ view positions or patches on each VWS. After the discretization process, the set of admissible (or optimal) feature sets associated with each patch can be calculated using a feature-planning algorithm. This completes the $AFspace_C(P)$ calculation algorithm. During the execution, the relative pose of the sensor can be determined from the pose estimations obtained, e.g., from EKF, and the admissible (or optimal) feature set associated with that patch can be read from $AFspace_C(P)$ map accordingly.

14.5 Example

The correctness of the developed measures has been examined through simulations and experiments. A simulation environment can be generated to test the quality of the features by applying the introduced measures before experiments or any visual servoing task. Both MATLAB™ and an AFS package [Janabi-Sharifi, 1995] have been used for simulations. Models of the camera, lighting, objects, environment, and robots are usually available. Models of EG&G Reticon MC9000 CCD camera and a five-degrees-of-freedom

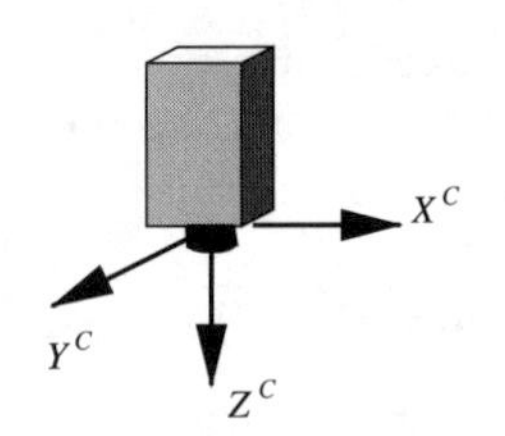

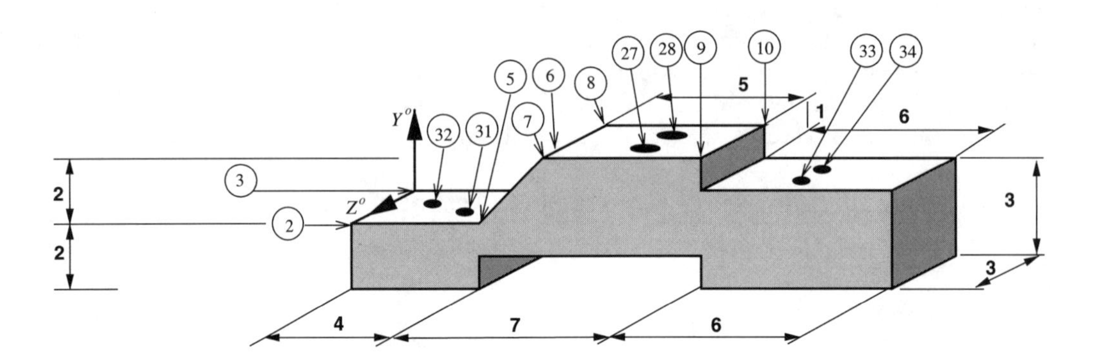

FIGURE 14.7 Testing object model with some hole and corner features identified.

CRS Plus SRS-M1A robot have been considered. The object model is shown in Figure 14.7. Overall system parameters have been summarized in Table 14.4.

For simulations an additive noise has been considered to represent more realistic situations. The object and the camera (mounted on the robot) had relative motion through a relative trajectory. For comparison, only the results for one node of the trajectory will be presented, such that τ: $[-4, 50, -1.5, 0°, 0°, -90°]^T$, which is the pose of object frame with respect to camera frame. All length and distance units will be centimeters by default unless otherwise mentioned. An incandescent light source is located approximately at $[20, 30, -150, 0°, 0°, 0°]^T$ with respect to the camera frame. The objective is to find the best 4 to 6 out of 36 features that will provide the best performance for extraction, pose estimation, and position-based (visual) control of the camera with respect to the object. Tables 14.5, 14.6, and 14.7 summarize the results of feature and feature set measures applied to the object features. Note that features 27 and 28, despite their good properties, are too close to the image boundaries and are removed from further analysis by applying the windowing constraint.

In evaluating the set-uniqueness measure, only the 18 upper features of the object are taken into consideration (i.e., $m = 18$, which included 6 holes and 12 corner features). Only the results of evaluations for a few feature sets are shown in Tables 14.6 and 14.7. To evaluate the set observability it is assumed that the estimation error is proportional to the summation of each feature-extraction error. For hole features with an image size of 4 pixels at the given view point, a variance of 0.1 pixels^2 is interpolated, resulting in a maximum error of 0.94 pixels. For corner features, a window size of 16×16 is assumed and the overall variance is calculated as 0.1417 pixels^2 leading to a maximum error of 1.11 pixels. The maximum acceptable error is calculated to be 2 pixels, equivalent to a projection of 2.8 mm in the Cartesian space. Also, because image-based control is not used, set controllability and set secondary sensitivity are assumed to be 1 for all the sets. For one node of the trajectory considered, the durability measure is set to 1 for all the feature sets. Finally, the overall score for feature sets is evaluated (Table 14.7), and the optimal feature set is chosen to be (5, 6, 7, 8, 31, 32). The obtained result has been verified experimentally as well.

The experimental setup consists of the aforementioned camera mounted at the endpoint of the same robot. For real-time visual servoing, a distributed computing architecture is used. A special-purpose

TABLE 14.4 System Parameters for Simulations and Experiments

Camera and Lighting	
Focal length, F	1.723 cm
Lens aperture, a	1.077 cm
Interpixel spacings, P_x, P_y	0.006 cm
Number of pixels, N_x, N_y	128
Light power, Φ_P	7.2 W
Percentage of diffuse reflection, s (object and table)	0.95, 0.6
Diffuse coefficient, $\overline{\ell}$ (object and table)	1.34, 1.0
Lens transmission, t_ℓ	0.95
Lens distance to image plane, z_ℓ	1.730 cm
Camera illumination intensity, E_{max}, E_{min}	26 nJ/cm^2, 40 pJ/cm^2
Measures, Parameters, and Thresholds	
Visibility threshold, v_t	1
Contrast sensitivity, B_{CS}	6.5
Contrast threshold, C_t	25
Type distance threshold, d_t	15 cm
Optical angle threshold, o_t	–0.1
Set nonambiguity threshold, p_t	0.05 cm
Set apartness threshold, σ_t	0.5 cm
Set noncollinearity and noncoplanarity threshold, $\bar{\varepsilon}$	0.01
Minimum required resolution, RC	1.2 cm/pixel
Window clearance factor, C_f	1.1
Window boundaries clearance, δ_{w_x}, δ_{w_y}	0.03 cm
Minimum window size, $(S_{min})_x$, $(S_{min})_y$	0.090 cm
Maximum window size, $(S_{max})_x$, $(S_{max})_y$	0.288 cm
Critical depth, Z_{ct}	7 cm
Sensitivity clearance factor, C_R	0.9

TABLE 14.5 Results of Feature-Measure Evaluation for Some Features of a Sample Object

Feature No.	J_{f_1}	J_{f_2}	J_{f_3}	J_{f_4}	J_{f_5}	J_{f_6}	J_{f_7}	J_{f_8}	J_{f_9}	$J_{f_{10}}$	$J_{f_{11}}$	$J_{f_{12}}$	$\sum J_{f_i}$
3	1	1	1	1	1	1	0.20	0.92	0.91	0.5	1.0	1	10.03
5	1	1	1	1	1	1	0.75	0.83	0.69	0.5	0.5	1	10.27
6	1	1	1	1	1	1	0.75	0.89	0.88	0.5	0.5	1	10.52
7	1	1	1	1	1	1	0.80	0.81	0.76	0.5	0.5	1	10.37
8	1	1	1	1	1	1	0.80	0.95	0.85	0.5	0.5	1	10.60
15	0	—	—	—	—	—	—	—	—	—	—	—	—
27	1	1	1	1	0	—	—	—	—	—	—	—	—
28	1	1	1	1	0	—	—	—	—	—	—	—	—
31	1	1	1	1	1	1	1	1	0.89	1	0.5	1	11.39
32	1	1	1	1	1	1	1	1	0.90	1	0.5	1	11.40
33	1	1	0	—	—	—	—	—	—	—	—	—	—

TABLE 14.6 Results of Feature-Set Measure Evaluation for Some Feature Sets of a Sample Object

Feature Set	J_{s_1}	J_{s_2}	J_{s_3}	J_{s_4}	J_{s_5}	J_{s_6}	J_{s_7}	J_{s_8}	J_{s_9}	$J_{s_{10}}$	$J_{s_{11}}$
3, 5, 6, 7, 8, 31	1	1.82	1	0.87	0.78	1.35	0.41	0.46	1	1	1
3, 5, 6, 7, 31, 32	1	1.65	0.95	0.67	0.58	1.49	0.47	0.47	1	1	1
5, 6, 7, 8, 31, 32	1	1.82	1	0.87	0.77	1.29	0.47	0.47	1	1	1
3, 5, 7, 8, 31, 32	1	1.73	0.95	0.93	0.61	1.46	0.47	0.47	1	1	1

TABLE 14.7 Optimum Feature Set Evaluation

Feature Set	$\frac{1}{6}\sum_{i=1}^{6}\sum_{j=7}^{12} J_{f_j}$	$\sum_{i=1}^{11} J_{s_i}$	Σ
3, 5, 6, 7, 8, 31	4.53	10.69	15.22
3, 5, 6, 7, 31, 32	4.66	10.28	14.94
5, 6, 7, 8, 31, 32	4.76	10.69	**15.45**
3, 5, 7, 8, 31, 32	4.67	10.62	15.29

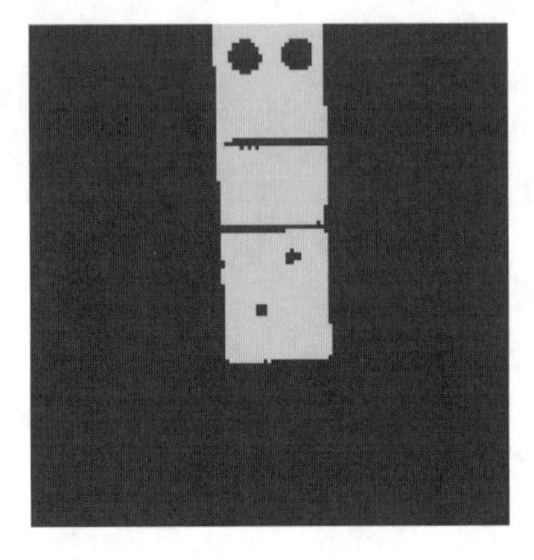

FIGURE 14.8 Robot tracking an object at specified node with features image.

image preprocessor receives the estimated window size and locations for each feature with its type and performs feature extraction. To handle intensive computations related to Kalman filtering, inverse Jacobian matrix, and trajectory control, a parallel network of RISC processors has been implemented using a Northern Digital transputer board mounted in an IBM AT 486 PC that manages the communication between the transputer network, robot controller, image preprocessor, and a supervisory computer. The feature-point measurements, joint-encoder values, and reference-relative pose vector are received from the preprocessor, the robot controller, and the supervisory computer, respectively, and passed to the transputer network. The reference joint position updates, and state predictions (next image feature point locations) are calculated in the transputer network and are sent to the robot controller and preprocessor. The object was placed in a relative pose and the robot was given the relative trajectory. The object moved at a speed of about 5 cm/s. Figure 14.8 shows a snapshot of a robot tracking the object at the specified node τ: $[-4, 50, -1.5, 0°, 0°, -90°]^T$ with the features image. The image clearly indicates the appropriateness of the selected features (5, 6, 7, 8, 31, 32) at this relative pose.

14.6 Summary

Feature selection and planning is an important aspect of many machine vision applications such as visual servoing. In particular, robustness of real-time vision-based systems depends highly on the features selected and used, for example, during servoing. This chapter introduced an algorithm for automatic selection and planning of the features. Also, the measures for various system constraints have been developed. Three characteristics of robust real-time vision, i.e., speed, accuracy, and reliability, have been taken into consideration, and the selection methods for two popular features (holes and corners) have been highlighted. Automatic feature selection has many advantages for real-time vision application. First, by excluding the irrelevant features, a simpler feature-extraction task could be defined and the time required to measure the features reduced. Second, by identifying the optimal features for any situation (e.g., relative pose of vision with respect to the object) the accuracy and reliability of feature extraction, pose estimation, and control will be improved.

The results could be used primarily to visually control the relative pose of the camera (mounted at the endpoint of a robot) with respect to an object. However, they are general enough to be applied to other real-time vision applications. The measures have been imposed by the constraints related to task (geometric-radiometric), feature extraction, pose estimation, and control. Geometric measures include visibility, resolution, field-of-view, depth-of-field, and windowing. Radiometric measures are light visibility, feature position, contrast, and contrast sensitivity. The measures to improve feature extraction, pose estimation, and control include type, size, optical angle, set nonambiguity, set apartness, set noncollinearity, set noncoplanarity, set angle, set sensitivity, set uniqueness, set observability, set controllability, set secondary sensitivity, and set durability. The introduced measures apply to both methods of visual servoing, namely, image-based and position-based visual servoing. Furthermore, computational efficiency and real-time feature selection have been discussed and remedies for enhancing the speed of feature selection and planning have been provided.

Defining Terms

admissible feature sets: Feasible feature sets that satisfy feature-set constraints (mainly feature-extraction, pose-estimation, and control constraints). The admissible feature set space of an object P viewed from all possible poses X of sensor C (denoted by $AFspace_C\ (P)$) is the set of all possible admissible feature sets of object P viewed from X, belonging to the configuration space of C with respect to the frame of object P.

feasible features: Features that satisfy all feature constraints (mainly task constraints). The feasible feature space of an object P viewed from all possible poses X of sensor C (denoted by $Fspace_C\ (P)$) is the set of all possible feasible features of object P viewed from X, belonging to the configuration space of C with respect to the frame of object P.

feature: Any scene property that can be mapped onto and measured in the image plane. Any structural feature that can be extracted from an image is called an image feature and usually corresponds to the projection of a physical feature of objects onto the image plane. Image features can be divided into region-based features, such as planes, areas, holes, and edge segment-based features, such as corners and edges.

q-combination-set: The set of all possible combinations of m features taken q at a time. Let ϑ be a set of m nonrepeated features ε and $q \leq m$ a positive integer number. A $q\text{-}COMB\ \vartheta$ is the set of all possible combinations, $q\text{-}COMB\ \vartheta_k$, of m features taken q at a time, where the kth combination is $q\text{-}COMB\ \vartheta_k \equiv \{\varepsilon_1^k, \varepsilon_2^k, \ldots, \varepsilon_q^k\}$.

required edge list (RE): The set of all possible combinations of a corner edge taken two at a time.

visual servoing (vision-guided servoing): The use of vision in the feedback loop of the lowest level of a (usually robotic) system control with fast image processing to provide reactive behavior. The task of visual servoing for robotic manipulators (or robotic visual servoing, RVS) is to control the pose of the robot's end-effector relative to either a world coordinate frame or an object being manipulated, using real-time visual features extracted from the image. A camera can be fixed or mounted at the endpoint (eye-in-hand configuration).

Acknowledgment

This research was supported by the Natural Sciences and Engineering Research Council of Canada (NSERC) through Research Grant #203060-98.

References

Feddema, J. T., Lee, C. S. G., and Mitchell, O. R., Weighted selection of image features for resolved rate visual feedback control, *IEEE Trans. Robot. Automat.*, 7(1), 31–47, 1991.

Ficocelli, M. and Janabi-Sharifi, F., Radiometric measures for feature selection in visual servoing, *Proc. 27th Annual Conf. IEEE Industrial Electron., IECON 2001*, Denver, CO, 2001, pp. 404–409, Nov.

Foley, J. D., van Dam, A., Feiner S., and Hughes J. F., *Computer Graphics: Principles and Practice*, 2nd ed., Addison-Wesley, Reading, MA, 1997.

Janabi-Sharifi, F., A Supervisory Intelligent Robot Control System for a Relative Pose-Based Strategy, Ph.D. dissertation, Dept. of Electrical and Computer Engineering, University of Waterloo, Waterloo, Canada, 1995.

Janabi-Sharifi, F. and Wilson, W. J., Automatic selection of image features for visual servoing, *IEEE Trans. Robot. Automat.*, 13(6), 890–903, 1997.

Madhusudan, C., Error Analysis of the Kalman Filtering Approach to Relative Position Estimation Using Noisy Vision Measurements, Master's thesis, Dept. of Electrical and Computer Engineering, University of Waterloo, Waterloo, Canada, 1990.

Malis, E., Chaumette, F., and Boudet, S., 2.5D visual servoing, *IEEE Trans. Robot. Automat.*, 15(2), 238–250, 1999.

Nelson, B. and Khosla, P. K., The resolvability ellipsoid for visual servoing, *Proc. 1994 Conf. Comput. Vision Pattern Recog. CPVR94,* 1994, pp. 829–832.

Papanikolopoulos, N. P., Controlled Active Vision, Ph.D. dissertation, Dept. of Electrical and Computer Engineering, Carnegie Mellon University, Pittsburgh, PA, 1992.

Sharma, R. and Hutchinson, S., On the observability of robot motion under active camera control, *Proc. 1994 IEEE Int. Conf. Rob. Automation,* 1994, pp. 162–167.

Smith, C. E. and Papanikolopoulos, N. P., Vision-guided robotic grasping: issues and experiments, *Proc. 1991 IEEE Int. Conf. Robot. Automat.*, Minneapolis, MN, 1996.

Tsai, R. and Tarabanis, K., Occlusion-free sensor placement planning, in *Machine Vision for Three-Dimensional Scenes*, Freeman, H., Ed., Academic Press, San Diego, 1990.

Wang, J., Optimal Estimation of 3D Relative Position and Orientation for Robot Control, M.A.Sc. dissertation, Dept. of Electrical and Computer Engineering, University of Waterloo, Waterloo, Canada, 1992.

Wilson, W. J., Williams Hulls, C. C., and Janabi-Sharifi, F., Robust image processing and position-based visual servoing, in *Robust Vision for Vision-Based Control of Motion*, Vincze, M. and Hager, G. D., Eds., IEEE Press, New York, 2000, pp. 163–201.

Wunsch, P. and Hirzinger, G., Real-time visual tracking of three-dimensional objects with dynamic handling of occlusion, *Proc. IEEE Int. Conf. Robot. Automation,* Albuquerque, NM, 1997, pp. 2868–2873.

Yuan, J. S. C., A general photogrammetric method for determining object position and orientation, *IEEE Trans. Robot. Automat.,* 5(2), 129–142, 1989.

For Further Information

A detailed analysis of feature selection, including formulation of feature-selection measures, is presented in Janabi-Sharifi [1995], Janabi-Sharifi and Wilson [1997], and Ficocelli and Janabi-Sharfi [2001]. The integration of the feature-selection method with grasp planning has been examined in "Automatic Grasp Planning for Visual-Servo Controlled Robotic Manipulators," *IEEE Trans. Syst., Man, Cybern.*, 28(5), 693–711, 1998, by F. Janabi-Sharifi and W. J. Wilson. *Proceedings of IEEE International Conference on Robotics and Automation: Workshop on Integrating Sensors with Mobility and Manipulation* (San Francisco, CA, April 2000) provides articles related to the integration of vision with grasp and motion planning.

IEEE Transactions on Pattern Analysis and Machine Intelligence provides advancements in feature extraction and selection in image processing and machine vision. Also, *Computer Vision, Graphics, and Image Processing* contains recent reports on advances in image processing and machine vision techniques.

Image Processing: Analysis and Machine Vision by M. Sonka, V. Holavac, and R. Boyle (Brooks Cole, Pacific Grove, CA, 1999) also includes descriptive sections on classification and representation aspects of feature extraction. A chapter of this book on three-dimensional vision, geometry, and radiometry is also relevant to modeling aspects of visual servoing.

Illumination and Color in Computer Generated Imagery, by R. Hall (Springer-Verlag, 1989), is a good introductory book for illumination and light modeling of objects. *Handbook of Pattern Recognition and Computer Vision*, edited by C. H. Chen, L. F. Pau, and P. S. P. Wang (World Scientific, Singapore, 2000) contains descriptive chapters on feature extraction and pattern recognition. In particular, a chapter called "Color in Computer Vision" explores in detail the applications of color in computer vision. An article by R. L. Cook and K. E. Torrence contains further details on the reflection mode: "A Reflection Model for Computer Graphics," *ACM Trans. Graphics*, 1(1), 7–24, 1982.

Robust Vision for Vision-Based Control of Motion, edited by M. Vincze and G. D. Hager (IEEE Press, New York, 2000), provides recent advances in the development of robust vision for visual-servo-controlled systems. The articles span issues including object modeling, feature extraction, feature selection, sensor data fusion, and visual tracking. *Robot Vision*, by B. K. Horn (McGraw-Hill, 1986), is a comprehensive introductory book for the application of machine vision in robotics.

A good collection of articles on visual servoing can be found in *IEEE Trans. Robot. Automat.*, 12, 5, 1996. This issue includes an excellent tutorial on visual servoing. A good reference book is *Visual Control of Robots: High Performance Visual Servoing* by P. I. Corke (Research Studies, Ltd., Somerset, England, 1996), encompassing both theoretical and practical aspects related to visual servoing of robotic manipulators.

Proceedings of IEEE International Conference on Robotics and Automation, IEEE Robotics and Automation Magazine, Proceedings of IEEE/RSJ International Conference on Intelligent Robots and Systems, and *IEEE Transactions on Robotics and Automation* document the latest developments in visual servoing, including solutions to issues related to feature selection and planning and image processing.

15

Visual Servoing: Theory and Applications

Farrokh Janabi-Sharifi

15.1 Introduction

Conventional robotic systems are limited to operating in highly structured environments, and considerable efforts must be expended to compensate for their limited accuracy. This is because conventional robotic manipulators operate on an open kinematic chain for placing and operating a tool (or end-effector) with respect to a workpiece. This is done by joint-to-joint kinematic transformations in order to define the *pose* (position and orientation) of the endpoint with respect to a fixed-world coordinate frame. Similarly, the workpiece must be placed accurately with respect to the same coordinate frame. Any uncertainty or error about the pose of the endpoint or workpiece would lead to task failure. Potential sources (such as gear backlashes, bending of the links, joints slippage, poor fixturing) would contribute to errors in the endpoint or workpiece poses. Therefore, considerable effort and cost are expended to overcome the above issues, for example, to design and manufacture special-purpose end-effectors, jigs, and fixtures. Consequently, due to the need for an accurate world model, the cost of changing the robot task would be quite high.

Alternatively, visual servoing provides direct measurements and control of the robot endpoint with respect to the workpiece and hence does not rely on open-loop kinematic calculations. **Visual servoing** or vision-guided servoing is the use of vision in the feedback loop of the lowest level of a (usually robotic) system control with fast image processing to provide reactive behavior. The task of visual servoing for robotic manipulators (or robotic visual servoing, RVS) is to control the pose of the robot's end-effector relative to either a world coordinate frame or an object being manipulated, using real-time visual features extracted from the image. A camera can be fixed or mounted at the endpoint (eye-in-hand configuration). The advantages of robotic visual servoing can be summarized as follows:

1. It will relax the requirement for the *exact* specification of the workpiece pose. Therefore, it will reduce the costs associated with robot teaching and special-purpose fixtures. For example, it will allow operations on moving or randomly placed workpieces.
2. The requirement for *exact* positioning of the endpoint will be relaxed. Therefore, the robot operation will not highly depend on stiffness and mechanical accuracy of the robot structure so

0-8493-1162-4/03/$0.00+$1.50

the robot mechanisms could be built lighter. This will lead to reduced cost of robot manufacture and operation, and decreased robot cycle time.

In summary, task specifications in a visual servoing framework would support robotic cells that are relatively robust to many disturbing effects in unstructured environments and can adapt readily to minor changes in the task or workpiece without needing to be reprogrammed.

One must distinguish visual servoing from traditional vision integration with robotic systems. Traditional vision-based control systems [Shirai and Inoue, 1973] are typically *look-and-move* structures, where visual sensing and manipulation control are combined in an open-loop fashion. Therefore, image processing and robot control are independent with sequential operations. With look-and-move systems, image-processing times are long (in the order of 0.1 to 1 seconds), and the accuracy depends on the accuracy of the vision system and the robot manipulator. In real visual servoing systems, the control feedback loop is closed around real-time image processing and measurements. The visual feedback loop operates at a high sampling rate in the range of 100 Hz for direct control of the robot endpoint with respect to an object and allows the operation of robot inner joint-servo loops at high sampling rates. Therefore, visual servoing systems provide improved accuracy and robustness to disturbing elements of unstructured environments such as kinematic modeling errors and randomly positioned parts. Visual servoing is a truly mechatronic stream combining results from real-time image processing, kinematics, dynamics, control theory, real-time computation, and information technology.

The fundamentals of system modeling and image projection will be summarized in Section 15.2, the basic classes of visual servoing systems will be presented in Section 15.3, simulation and experimental examples will be provided in Section 15.4, and the applications will be given in Section 15.5. Finally, the chapter will be summarized in Section 15.6.

15.2 Background

As shown in Figure 15.1, the coordinate frames that might be used for visual servoing include coordinate frames attached to the base (world frame), endpoint of the robot, camera, and object. For example, the location of an object viewed by the camera would be calculated with respect to the camera frame, or the location of the object might be determined with respect to the base frame. The homogeneous transforms between these frames are $\overline{T}_B^E, \overline{T}_E^C, \overline{T}_C^O$, and $\overline{T}_B^O$. Here $\overline{T}_j^i$ is the homogenous transform from frame i to frame j, specified by a rotation matrix R_j^i and translation vector T_j^i. The camera is usually fixed with respect to the endpoint or the world coordinate frame. Therefore, $\overline{T}_E^C$ or $\overline{T}_B^C$ could be obtained from kinematic calibration tests. A common configuration is eye-in-hand configuration, where the camera is mounted at the endpoint, hence $\overline{T}_E^C = I$, i.e., coordinate frames E and C coincide. This configuration provides better viewing possibilities with less likelihood of viewing obstruction from the moving arm and target.

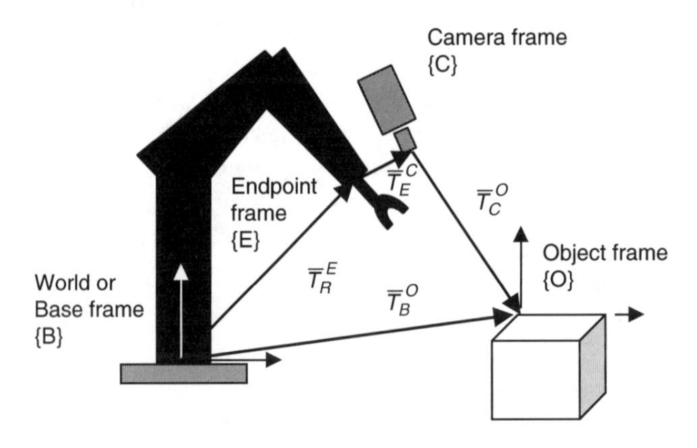

FIGURE 15.1 Relations between different coordinate frames: world, endpoint, camera, and object.

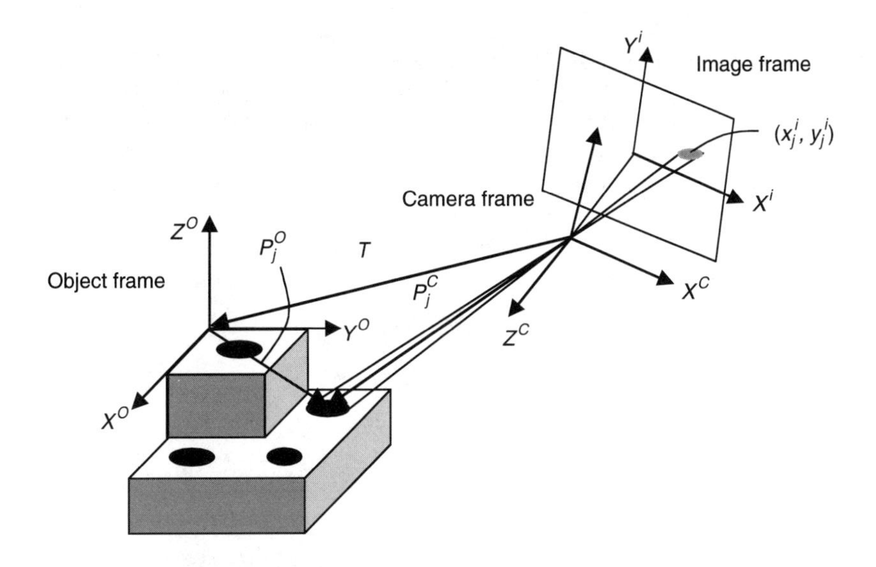

FIGURE 15.2 Projection of an object feature onto the image plane.

For simplicity, in the remainder of this chapter, we will use frames E and C interchangeably, unless otherwise specified.

Let $\boldsymbol{T} = (X, Y, Z)^T$ denote the relative position vector of the object frame with respect to the camera frame (Figure 16.2). We will also denote $\Theta = (\phi, \alpha, \psi)^T$ as the relative orientation (or viewing direction) vector with roll, pitch, and yaw parameters, respectively. The pose $\boldsymbol{W}$ (position and orientation vector) of the object relative to the robot endpoint (or camera) will be then

$$W = (T, \Theta)^T = (X, Y, Z, \phi, \alpha, \psi)^T. \tag{15.1}$$

We can define the relative task space $\tau = SE^3 = \Re^3 \times SO^3$ as the set of all possible positions and orientations that could be attained by the end-effector. In general, however, we prefer to represent the relative pose by a 6D vector $\boldsymbol{W} \in \Re^6$ of rather than by $\boldsymbol{W} \in \tau$.

The camera-projection model is shown in Figure 15.2. Each camera contains the projection of a three-dimensional scene in its two-dimensional (2D) image plane. Because depth information is lost, additional information will be required to estimate the three-dimensional coordinates of a point $\boldsymbol{P}$. This information could be obtained by multiple views of the object or by knowledge of the geometric relationship between a set of **feature** points ε_i on the object. This information is obtained from **image feature parameters.** Although different image feature parameters are used in vision, in this chapter we will use the coordinates of image feature points as image feature parameters. Good visual features depend on many parameters such as the feature's radiometric properties, its visibility from many view points, and its ease of extraction without any ambiguity. Feature selection and planning for visual servoing is an important component of any visual servoing system [Janabi-Sharifi and Wilson, 1997] and received a detailed discussion in another chapter (see Chapter 14). In practice, hole and corner features are readily available in many objects and have proven to serve well for many visual servoing tasks [Feddema et al., 1991; Wilson et al., 2000]. Therefore, the rest of this chapter will focus on the use of hole and corner features.

A set of image feature parameters could be chosen to provide information about 6D relative pose vectors at any instance of servoing. Therefore, image feature vector will be defined as $\boldsymbol{s} = [s_1, s_2, \ldots, s_k]^T$ where each $s_i \in \Re$ is a bounded image feature parameter. Therefore, image feature parameter space (or shortly image space) $\boldsymbol{S}$ is defined as $\boldsymbol{s} \in \boldsymbol{S} \subseteq \Re^k$. The projection will be a mapping denoted by

$$G: \tau \rightarrow \boldsymbol{S}. \tag{15.2}$$

For instance, each pair of s_i could be thought of as the coordinates $[x^i, y^i]^T$ of the projection of a feature point $\boldsymbol{P}$ onto the image plane, and then $\boldsymbol{S} \in \Re^2$. The number of feature points used depends on the servoing strategy, but for 6D relative pose estimation, the image feature coordinates of at least three features will be necessary [Yuan, 1989]. Three projection models used in visual servoing are perspective projection, scaled orthographic projection, and affine projection [Hutchinson et al., 1996]. However, the scaled orthographic projection and affine projection models are approximations of the perspective projection model; therefore, we will adopt the commonly used perspective projection. Here $P_j^c = (X_j^c, Y_j^c, Z_j^c)^T$ and $P_j^o = (X_j^o, Y_j^o, Z_j^o)^T$ are the coordinate vectors of the jth object feature center in the camera and object frames, respectively (Figure 15.2). The feature point can be described in the camera frame using the following transformation:

$$P_j^c = T + R(\phi, \alpha, \psi) P_j^o \tag{15.3}$$

where the rotation matrix is

$$R(\phi,\alpha,\psi) = \begin{bmatrix} \cos\phi\cos\alpha & \cos\phi\sin\alpha\sin\psi - \sin\phi\cos\psi & \cos\phi\sin\alpha\cos\psi + \sin\phi\sin\psi \\ \sin\phi\cos\alpha & \sin\phi\sin\alpha\sin\psi + \cos\phi\cos\psi & \sin\phi\sin\alpha\cos\psi - \cos\phi c\sin\psi \\ -\sin\alpha & \cos\alpha\sin\psi & \cos\alpha\cos\psi \end{bmatrix}. \tag{15.4}$$

The position vector of the feature in the object frame, P_j^o, is usually known from the CAD model of the object. The task of visual servoing is to control the pose $\boldsymbol{W}$ using the visual features extracted from the object image. The coordinates of the projection of this feature center on the image plane will be x_j^i, y_j^i, given by Figure 15.2:

$$\begin{bmatrix} x_j^i \\ y_j^i \end{bmatrix} = -\frac{F}{Z_j^c} \begin{bmatrix} \dfrac{X_j^c}{P_X} \\ \dfrac{Y_j^c}{P_Y} \end{bmatrix} \tag{15.5}$$

where P_X and P_Y are interpixel spacing in X^i and Y^i axes of the image plane, respectively, and F is the focal length. This model assumes that the origin of the image coordinates is located at the principal point and $|Z_j^c| \gg F$. The perspective projection model requires both **intrinsic** and **extrinsic camera parameters**. The intrinsic camera parameters (P_X, P_Y, F) and coordinates of optical axis on the image plane (principal point) O^i are determined from camera calibration tests. Additionally, camera calibration tests will provide radial distortion parameters (r, K_1, K_2), tangential distortion parameters (P_1, P_2) and γ aspect ratio [Ficocelli, 1999]. The extrinsic camera parameters include the pose of the camera with respect to the end-effector or the robot base frame. The extrinsic camera parameters are calculated by inspection of the camera housing and kinematic calibration [Corke, 1996]. Excellent solutions to the camera calibration problem exist in the literature [Tsai and Lenz, 1989] (see also Further Information).

15.3 Servoing Structures

Three major classifications of visual servoing structures are *position-based* visual servoing (PBVS), *image-based* visual servoing (IBVS), and *hybrid* visual servoing (HVS). The basic structures of these systems are shown in Figures 15.3 to 15.5. The first two classifications were initially introduced by Sanderson and Weiss [1980]. In all of the structures the image features are extracted and image feature parameters are measured (i.e., mapping G). Processing of the entire image would be time consuming for real-time visual servoing, so windowing techniques are used to process a number of selected features. Windowing methods not only provide computational speed but also reduce the requirement for special-purpose hardware.

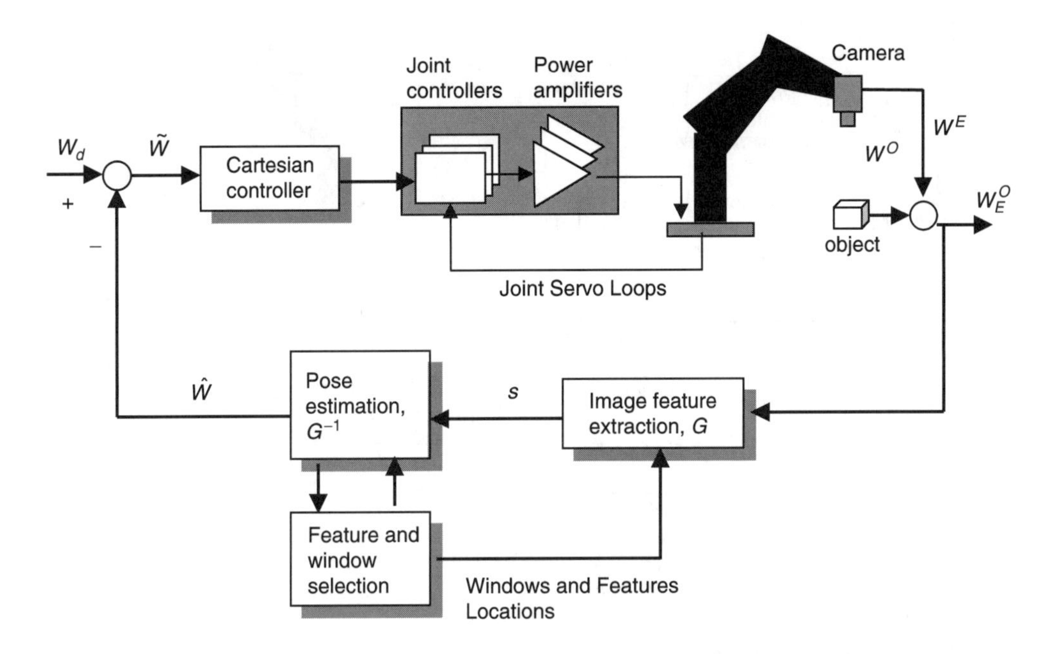

FIGURE 15.3 Structure of position-based visual servoing (PBVS).

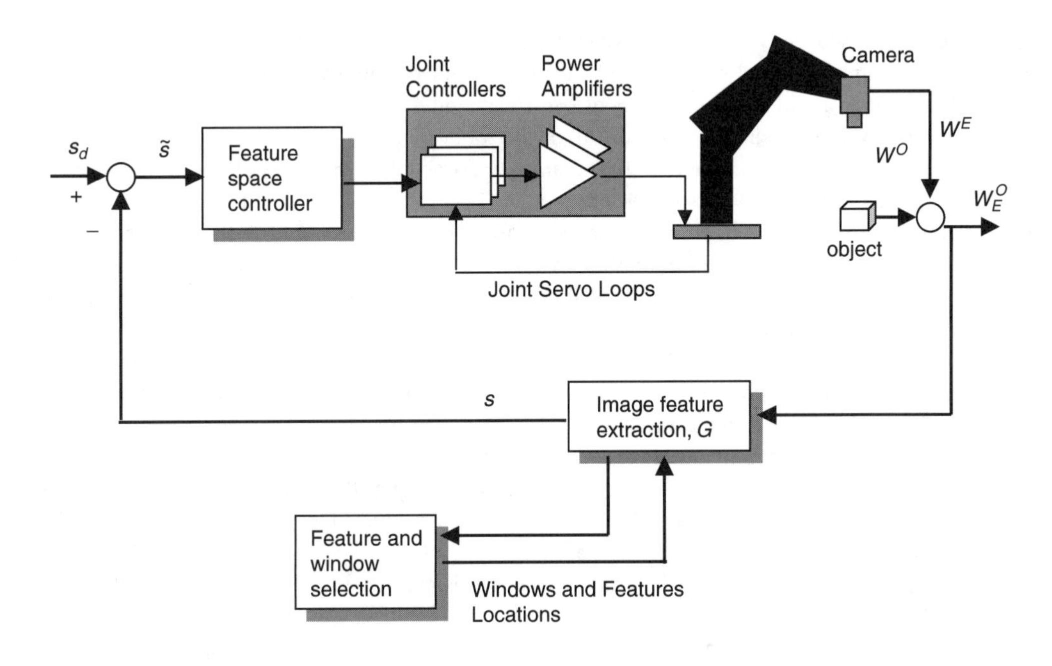

FIGURE 15.4 Structure of image-based visual servoing (IBVS).

Real-time feature extraction and robust image processing are crucial for successful visual servoing and will be discussed in detail in another chapter (see Chapter 10). The feature- and window-selection block in all of the shown structures uses current information about the status of the camera with respect to the object and the models of the camera and environment to prescribe the next time-step features and the locations of the windows associated with those selected features. Feature-selection and planning issues are discussed in another chapter as well (see Chapter 14). In all of the structures, the visual servo controllers

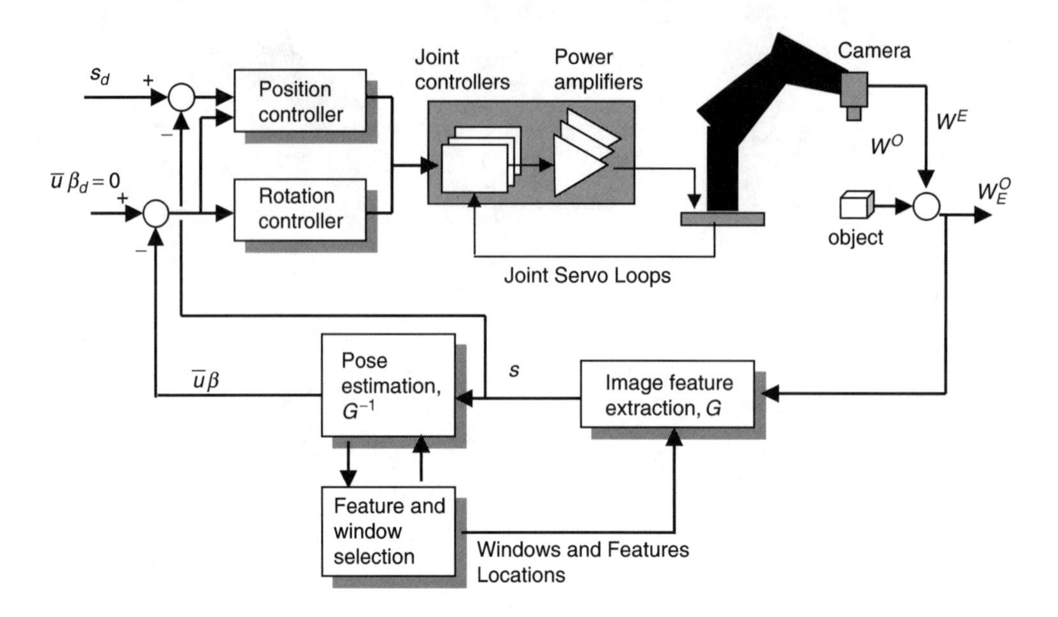

FIGURE 15.5 Hybrid 2-1/2D visual servoing (HVS).

determine set points for the robot joint-servo loops. Because almost any industrial robot has a joint–servo interface, this simplifies visual servo-control integration and portability. Therefore, the internal joint-level feedback loops are inherent to the robot controller, and visual servo-control systems do not need to deal with the complex dynamics and control of the robot joints.

In position-based control (Figure 15.3) the parameters extracted ($\boldsymbol{s}$) are used with the models of camera and object geometry to estimate the relative pose vector ($\hat{\boldsymbol{W}}$) of the object with respect to the end-effector. The estimated pose is compared with the desired relative pose ($\boldsymbol{W}_d$) to calculate the relative pose error ($\tilde{\boldsymbol{W}}$). A Cartesian control law reduces the relative pose error, and the Cartesian control command is transformed to the joint-level commands for the joint-servo loops by appropriate kinematic transformations.

In image-based control (Figure 15.4), the control of the robot is performed directly in the image parameters space. The feature parameters vector extracted ($\boldsymbol{s}$) is compared with the desired feature parameter vector ($\boldsymbol{s}_d$) to determine the feature-space error vector ($\tilde{\boldsymbol{s}}$). This error vector is used by a feature-space control law to generate a Cartesian or joint-level control command.

In hybrid control (Figure 15.5) such as 2-1/2D visual servoing, the pose estimation is partial and determines rotation parameters only. The control input is expressed partially in three-dimensional Cartesian space and in part in two-dimensional image space. An image-based control is used to control the camera translations, while the orientation vector $\bar{u}\beta$ is extracted and used to control the camera rotational degrees of freedom.

Each of the above strategies has its advantages and limitations. Several articles have reported the comparison of the above strategies (see Further Information). In the next sections these methods will be discussed, and simulation results will be provided to show their performance.

15.3.1 Position-Based Visual Servoing (PBVS)

The general structure of a PBVS is shown in Figure 15.3. A PBVS system operates in Cartesian space and allows the direct and natural specification of the desired relative trajectories in the Cartesian space, often used for robotic task specification. Also, by separating the pose-estimation problem from the control-design problem, the control designer can take advantage of well-established robot Cartesian control algorithms. As will be shown in the Examples section, PBVS provides better response to large translational and rotational camera motions than its counterpart IBVS. PBVS is free of the image singularities, local minima, and camera-retreat problems specific to IBVS. Under certain assumptions, the closed-loop stability of PBVS is robust with respect to bounded errors of the intrinsic camera-calibration and object model. However, PBVS is more sensitive to camera and object model errors than IBVS. PBVS provides no mechanism for regulating the features in the image space. A feature-selection and switching mechanism would be necessary [Janabi-Sharifi and Wilson, 1997]. Because the relative pose must be estimated online, feedback and estimation are more time consuming than IBVS, with accuracy depending on the system-calibration parameters.

Pose estimation is a key issue in PBVS. Close-range photogrammetric techniques have been applied to resolve pose estimation in real-time. The disadvantages of these techniques are their complexity and their dependency on the camera and object models. The task is to find (1) the relative pose of the object relative to the endpoint ($\boldsymbol{W}$) using two-dimensional image coordinates of feature points ($\boldsymbol{s}$) and (2) knowledge about the camera intrinsic parameters and the relationship between the observed feature points (usually from the CAD model of the object). It has been shown that at least three feature points are required to solve for the 6D pose vector [Yuan, 1989]. However, to obtain a unique solution at least four features will be needed.

The existing solutions for the pose-estimation problem can be divided into analytic and least-squares solutions. For instance, unique analytical solutions exist for four coplanar, but not collinear, feature points. If intrinsic camera parameters need to be estimated, six or more feature points will be required for a unique solution (for further details see Further Information). Because the general least-squares solution to pose estimation is a nonlinear optimization problem with no known closed-form solution, some researchers have attempted iterative methods. These methods rely on refining the nominal pose estimation based on real-time observations. For instance, Yuan [1989] reports a general iterative solution to pose estimation independent of the number of features or their distributions. To reduce the noise effect, some sort of smoothing or averaging is usually incorporated.

Extended Kalman filtering (EKF) provides an excellent iterative solution to pose estimation. This approach has been successfully examined for 6D control of the robot endpoint using observations of image coordinates of 4 or more features [Wilson et al., 2000; Wang, 1992]. To adapt to the sudden motions of the object an adaptive Kalman filter estimation has also been formulated recently for 6D pose estimation [Ficocelli and Janabi-Sharifi, 2001]. In comparison to many techniques Kalman-filter-based solutions are less sensitive to small measurement noise. An EKF-based approach also has the following advantages. First, it provides an optimal estimation of the relative pose vector by reducing image-parameter noise. Next, EKF-based state estimations improve solution impunity against the uniqueness problem. Finally, the EKF-based approach provides feature-point locations in the image plane for the next time-step. This allows only small window areas to be processed for image parameter measurements and leads to significant reductions in image-processing time. Therefore, this section provides a brief discussion of EKF-based method that involves the following assumptions and conditions.

First, the target velocity is assumed to be constant during each sample period. This is a reasonably valid assumption for small sample periods in real-time visual servoing. Therefore, the state vector $\boldsymbol{W}$ is extended to include relative velocity as well. That is:

$$W = [X, \dot{X}, Y, \dot{Y}, Z, \dot{Z}, \varphi, \dot{\phi}, \alpha, \dot{\alpha}, \psi, \dot{\psi}]^T. \tag{15.6}$$

A discrete dynamic model will then be:

$$W_k = AW_{k-1} + \gamma_k \tag{15.7}$$

with diagonal $\boldsymbol{A}$ matrix defined as follows:

$$\mathrm{A} = \begin{bmatrix} 1 & T & & & \\ 0 & 1 & & & \\ & & \ddots & & \\ & & & 1 & T \\ & & & 0 & 1 \end{bmatrix}_{12\times 12} \tag{15.8}$$

where T is the sample period, k is the sample step, and γ_k denotes the dynamic model disturbance noise vector described by a zero mean Gaussian distribution with covariance Q_k.

The output model will be based on the projection model given by Eqs. (15.3) to (15.5) and defines the image feature locations in terms of the state vector $\boldsymbol{W}_k$ as follows:

$$Z_k = G(W_k) + \nu_k \tag{15.9}$$

with

$$Z_k = [x_1^i, y_1^i, x_2^i, y_2^i, \ldots, x_p^i, y_p^i]^T \tag{15.10}$$

and

$$G(W_k) = -F\left[\frac{X_1^c}{P_X Z_1^c}, \frac{Y_1^c}{P_Y Z_1^c}, \ldots, \frac{X_p^c}{P_X Z_p^c}, \frac{Y_p^c}{P_Y Z_p^c}\right]^T \tag{15.11}$$

for p features. Here, X_j^c, Y_j^c, and Z_j^c are given by Eqs. (15.3) and (15.4). ν_k denotes the image parameter measurement noise that is assumed to be described by a zero mean Gaussian distribution with covariance $\boldsymbol{R}_k$.

The recursive EKF algorithm consists of two major parts, one for prediction and the other for updating, as follows.

Prediction:

$$\hat{W}_{k,k-1} = A\hat{W}_{k-1,k-1} \tag{15.12}$$

$$P_{k,k-1} = AP_{k-1,k-1}A^T + Q_{k-1} \tag{15.13}$$

Linearization:

$$H_k = \left.\frac{\partial G(W)}{\partial W}\right|_{W=\hat{W}_{k,k-1}} \tag{15.14}$$

Kalman gain:

$$K = P_{k,k-1}H_k^T(R_k + H_k P_{k,k-1} H_k^T)^{-1} \tag{15.15}$$

Estimation updates:

$$\hat{W}_{k,k} = \hat{W}_{k,k-1} + K(Z_k - G(\hat{W}_{k,k-1})) \tag{15.16}$$

$$P_{k,k} = P_{k,k-1} - KH_k P_{k,k-1} \tag{15.17}$$

In the above equations, $\hat{W}_{k,k-1}$ denotes the state predictions at time k based on measurements at time $k-1$. Also, $\hat{W}_{k,k}$ is the optimal state estimation at time k based on the measurements at time k. $\boldsymbol{P}_{k,k-1}$ and $\boldsymbol{P}_{k,k}$ are state prediction error and state estimation error covariance matrices, respectively. $\boldsymbol{K}$ is the Kalman gain matrix. As mentioned above, the number of features used (p) should be above four to obtain a unique solution to 6D pose-vector estimation. However, inclusion of more features will improve the performance of an EKF-based estimation, with a concomitant increase in cost for additional computations. It has been shown that the inclusion of more than six features will not improve the performance of EKF estimation significantly [Wang, 1992]. Also, the features need to be noncollinear and noncoplanar to provide good results. Consequently, $4 \leq p \leq 6$ [Janabi-Sharifi and Wilson, 1997]. Further discussion on feature-selection issues is provided in Chapter 14.

The vision-based control system consists of fast inner joint-servo control loops. The slower outer loop is a visual servo control loop. The regulation aim in position-based control is to design a controller that computes joint-angle changes to move the robot such that the relative pose reaches the desired relative pose in an acceptable manner. Therefore, control design for visual servo-control loops requires calculation of the error vector or the control command in joint space to provide the input commands to the joint-servo loops (Figure 15.3). For this purpose first the Euler angles Θ must be converted to the total rotation angles θ with respect to the endpoint frame [Wilson et al., 1996]. Assuming a slow-varying endpoint and object frames within a sample period, the total rotation angles could be obtained from:

$$\theta = \begin{bmatrix} \theta_X \\ \theta_Y \\ \theta_Z \end{bmatrix} \approx \begin{bmatrix} -\alpha\sin\phi + \psi\cos\phi\cos\alpha \\ \alpha\cos\phi + \psi\sin\phi\cos\alpha \\ \phi - \psi\sin\alpha \end{bmatrix} \tag{15.18}$$

The orientation vector θ and relative position vector $\boldsymbol{T}$ are compared with the input reference trajectory point to determine the endpoint relative-pose error in Cartesian space. Also, assuming slow-varying motion of the system without commanding any large abrupt changes, the endpoint relative position and orientation changes, e.g., error equations required for controlling the robot endpoint, are related to the joint changes via:

$$\Delta \boldsymbol{q} \approx \boldsymbol{J}^{-1} \begin{bmatrix} \boldsymbol{R}_B^E & \boldsymbol{0} \\ \boldsymbol{0} & \boldsymbol{R}_B^E \end{bmatrix} \begin{bmatrix} \Delta \boldsymbol{T}_E \\ \Delta \theta_E \end{bmatrix} \tag{15.19}$$

One can regulate the joint level error $\Delta \boldsymbol{q}$ or Cartesian error $\Delta \boldsymbol{W}_E = [\Delta \boldsymbol{T}_E, \Delta\theta_E]^T$. Different control laws have been examined in the literature, e.g., PD control law to regulate $\Delta \boldsymbol{q}$ [Wilson, et al., 1996]. Note that the inner joint–servo control loops of common industrial robots operate at high sample rates, and for smooth tracking performance high sample rates such as 60 Hz are usually expected from the visual servo control loop (i.e., outer loop). Due to the load of computations involved in PBVS, distributed computing architectures have been proposed to provide reasonable sample rates for the visual servo-control loop. For instance, Wilson et al. [1996] have reported a PBVS design and implementation using a transputer-based architecture with a coordinating PC. However, recent advances in microprocessor technology have made high-speed PC-based implementation possible as well [Ficocelli and Janabi-Sharifi, 2001].

15.3.2 Image-Based Visual Servoing (IBVS)

In image-based visual servoing (Figure 15.4) the error signal and control command are calculated in the image space. The task of the control is to minimize the error of the feature-parameter vector, given by $\tilde{s} = s_d - s$. An example of a visual task specified in the image space is shown in Figure 15.6. It shows the initial and desired views of an object with five hole features. The advantage of IBVS is that it does not require full pose estimation and hence is computationally less involved than PBVS. Also, it is claimed

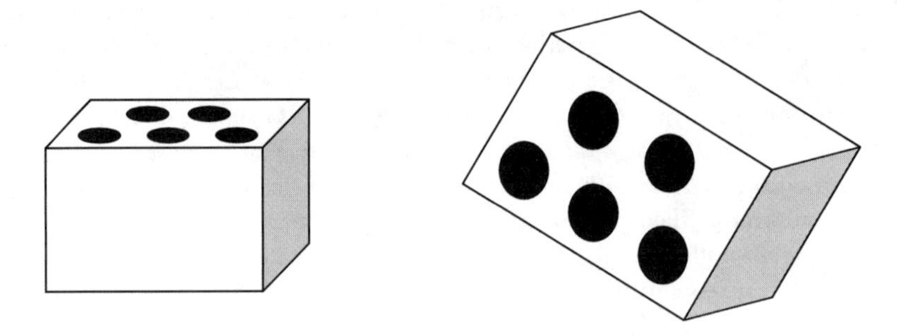

FIGURE 15.6 An example of a task specified in the image space with the initial and desired views.

that the positioning accuracy of IBVS is less sensitive to camera-calibration errors than PBVS. However, IBVS would lead to image singularities that might cause control instabilities. Another issue with IBVS is the camera-retreat problem: For the commanded pure rotations around the optical axis, the camera often moves away from the target in a normal direction and then returns. Moreover, the convergence of IBVS is ensured only in a small neighborhood of the desired camera pose. The domain of this neighborhood is analytically impossible to determine [Chaumette, 1998]. The closed-loop stability is also robust with respect to the errors of the camera-calibration and target model.

The goal of IBVS control is to find appropriate endpoint pose changes (or velocity) required to minimize error in the image space defined by $\tilde{s} = s_d - s$ (or its time rate). Because the robot control is usually in either Cartesian or joint space, the resultant control in image space must be converted into the Cartesian- or joint-space error. The velocity (or differential changes) of the camera $\dot{\boldsymbol{r}}$ (or $\Delta \boldsymbol{r}$) or its relative pose $\dot{W}$ (or $\Delta \boldsymbol{W}$) can be related to the image feature velocities $\dot{\boldsymbol{s}}$ (or $\Delta \boldsymbol{s}$) by a differential Jacobian matrix, $\boldsymbol{J}_i$, called **image Jacobian**. This matrix is also referred to as feature Jacobian matrix, feature sensitivity matrix, interaction matrix, or $\boldsymbol{B}$ matrix. Because we have assumed a hand-eye configuration, the endpoint frame (E) is coincident with the camera frame (C). Therefore, we will use endpoint and camera frames or poses interchangeably. Then, $\boldsymbol{J}_i \in \Re^{k \times m}$ is a linear transformation from the tangent space of τ at W (or r) to the tangent space of $\boldsymbol{S}$ at $\boldsymbol{s}$. Here, k and m are the dimensions of the image and task spaces, respectively. Depending on the choice of endpoint pose representation, there are different derivations for the image Jacobian matrix (see Further Information). For instance, if the relative pose of the object with respect to the endpoint is considered, we will have

$$\Delta s = J_i \Delta W_E \tag{15.20}$$

where $\Delta \boldsymbol{W}_E$ is the differential change of the relative pose of the object defined in the endpoint (or camera) frame. The corresponding image Jacobian matrix is given in Feddema et al. [1991]. It is also possible to relate the motion of a task frame to the motion of the feature points. In hand-eye configurations, it is preferable to define endpoint (or camera) motions with respect to the endpoint (or camera) frame. Therefore, $\Delta \boldsymbol{W}_E = [\Delta \boldsymbol{T}_E, \Delta \boldsymbol{\theta}_E]^T$ in Eq. (15.20) will imply the differential corrective motion of the endpoint with respect to the endpoint frame. The image Jacobian matrix for one feature can then be written as:

$$J_i = \begin{bmatrix} \dfrac{F}{P_X Z^c} & 0 & \dfrac{x^i}{Z^c} & -\dfrac{P_Y}{F} x^i y^i & \left(\dfrac{F}{P_X} + \dfrac{P_X}{F} x^{i^2}\right) & \dfrac{P_Y}{P_X} y^i \\ 0 & \dfrac{F}{P_Y Z^c} & \dfrac{y^i}{Z^c} & -\left(\dfrac{F}{P_Y} + \dfrac{P_Y}{F} y^{i^2}\right) & \dfrac{P_X}{F} x^i y^i & -\dfrac{P_X}{P_Y} x^i \end{bmatrix}. \tag{15.21}$$

Because several features (p) are used for 6D visual servoing, the image Jacobian matrix will have the form of:

$$J_i = \begin{bmatrix} (J_i)_1 \\ (J_i)_2 \\ \vdots \\ (J_i)_p \end{bmatrix} \tag{15.22}$$

where $(J_i)_k$ is a 2×6 Jacobian matrix for each feature given by Eq. (15.21). Therefore, one can use Eqs. (15.20) and (15.22) to calculate endpoint differential motion (or its velocity) for a given change or error expressed in the image space. That is:

$$\Delta W_E = J_i^{-1} \Delta s, \qquad \text{or} \quad \dot{W}_E = J_i^{-1} \dot{s} \tag{15.23}$$

which assumes a nonsingular and square image Jacobian matrix. A control law can then be applied to move the robot endpoint (or camera) toward the desired feature vector. The earliest and easiest control approach is resolved-rate motion control (RRMC), which uses a proportional control $\dot{\tilde{s}} = K_p \tilde{s}$ to warrant exponential convergence $\tilde{s} \to \mathbf{0}$. Therefore:

$$\boldsymbol{u} = K_p J_i^{-1} \tilde{s} \tag{15.24}$$

where $\boldsymbol{K_p}$ is a diagonal gain matrix [Feddema et al., 1991] and $\boldsymbol{u}$ is the endpoint velocity screw. At this stage a Cartesian-space controller can be applied. Because the control output will indicate endpoint changes (or velocity) with respect to the endpoint frame, transformations must be applied to calculate the equivalent joint angles changes. Under the same assumptions as Eqs. (15.19), these calculations are given by:

$$\Delta \boldsymbol{q} \approx J^{-1} \begin{bmatrix} R_B^E & \mathbf{0} \\ \mathbf{0} & R_B^E \end{bmatrix} K_p J_i^{-1} \tilde{s}\,. \tag{15.25}$$

RRMC is a simple and easy-to-implement method with a fast response time; however, it is not prone to singularity and provides highly coupled translational and rotational motions of the camera. In addition to RRMC, other approaches to IBVS exist in the literature such as optimal control techniques, model reference adaptive control (MRAC) methods, etc. (see Corke [1996]). Close examination of Eq. (15.25) reveals several problems with IBVS.

First, the image Jacobian matrix, given by Eqs. (15.21) and (15.22) depends on the depth Z^c of the feature. For a fixed camera, when the end-effector or the object held in the end-effector is tracked as the object, this is not an issue because depth can be estimated using robot forward kinematics and camera-calibration data. For an eye-in-hand configuration, however, the depth or image Jacobian matrix must be estimated during servoing. Conventional estimation techniques can be applied to provide depth estimation; however, they increase computation time. Adaptive techniques for online depth estimation can be applied, with limited success. Some proposals have been made to use an approximation to the image Jacobian matrix using the value of the image Jacobian matrix or the desired features depth computed at the desired camera position. This solution avoids local minima and online updating of the image Jacobian matrix. However, the image trajectories might be unpredictable and would leave image boundaries. Further sources are provided in Further Information.

Second, the inverse of the image Jacobian matrix in Eq. (15.25) might not be square and full-rank (nonsingular). Therefore, pseudo-inverse solutions must be sought. Because p features are used for servoing, the dimension of the image space is $k = 2p$. Two possibilities exist when the image Jacobian matrix is full rank, i.e., $\text{rank}(\boldsymbol{j}_i) = \min(2p, m)$ but $2p \neq m$. If $2p < m$, the pseudo-inverse Jacobian and least-squares solution for $\dot{W}_E$ (or ΔW_E) will be:

$$\dot{W}_E = J_i^+ \dot{s} + (I - J_i^+ J_i)v \tag{15.26}$$

$$J_i^+ = (J_i^T J_i)^{-1} J_i^T \tag{15.27}$$

with v as an arbitrary vector such that $(I - J_i^+ J_i)v$ lies in the null space of J_i and allows endpoint motions without changes in the object features velocity. If $2p > m$, the pseudo-inverse Jacobian and least-squares solution for $\dot{W}_E$ (or ΔW_E) will be:

$$\dot{W}_E = J_i^+ \dot{s} \tag{15.28}$$

$$J_i^+ = (J_i^T J_i)^{-1} J_i^T. \tag{15.29}$$

When the image Jacobian matrix is not full rank, the singularity problem must be dealt with. Small velocities of the image features near image singularities would lead to large endpoint velocities and result in control instabilities and task failure. This problem might be treated by singular value decomposition (SVD) approaches or via damped least-squares inverse solutions.

Finally, IBVS introduces high couplings between translational and rotational motions of the camera, leading to a typical problem of camera retreat. This problem has been resolved by the hybrid 2-1/2D approach, which will be introduced in the next section. A simple strategy for resolving the camera-retreat problem has also been introduced in Corke and Hutchinson [2000].

15.3.3 Hybrid Visual Servoing (HVS)

The advantages of both PBVS and IBVS are combined in recent hybrid approaches to visual servoing. Hybrid methods decouple control of certain degrees of freedom; for example, camera rotational degrees could be controlled by IBVS. These methods generally rely on the decomposition of an image Jacobian matrix. A homography matrix is considered to relate camera configurations corresponding to the initial and desired images. This matrix can be computed by a set of corresponding points in the initial and desired images. It is possible to decompose the homography matrix into rotational and translational components. Among hybrid approaches, 2-1/2D visual servoing is well established from an analytical point of view. Therefore, the rest of this section will be devoted to the analysis and discussion of 2-1/2D visual servoing. Further details on other approaches can be found in the sources listed in Further Information.

In 2-1/2D visual servoing [Malis et al., 1999] the controls of camera rotational and translational degrees of freedom are decoupled. That is, the control input is expressed in part in three-dimensional Cartesian space and in part in two-dimensional image space (Figure 15.5). IBVS is used to control the camera translational degrees of freedom while the orientation vector $\bar{u}\beta$ is extracted and controlled by a Cartesian-type controller. This approach provides several advantages. First, since camera rotation and translation controls are decoupled, the problem of camera retreat is solved. Second, 2-1/2D HVS is free of image singularities and local minima. Third, this method does not require full pose estimation, and it is even possible to release the requirement for the geometric model of the object. Fourth, Cartesian camera motion and image plane trajectory can be controlled simultaneously. Finally, this method can accommodate large translational and rotational camera motions. However, in comparison with PBVS and IBVS, 2-1/2D HVS introduces some disadvantages that will be discussed after the introduction of the methodology, as follows.

For 2-1/2D HVS, first a pose estimation technique (e.g., scaled Euclidean reconstruction algorithm, SER [Malis et al., 1999]) is utilized to recover the relative pose between the current and the desired camera frames, and then the $\bar{u}\beta$ vector is extracted. A rotation matrix R exists between the endpoint frame (or the camera frame) denoted by E and the desired endpoint frame by E*. This matrix must reach identity matrix at the destination (Figure 15.7). In order to avoid workspace singularities, the rotation matrix will be

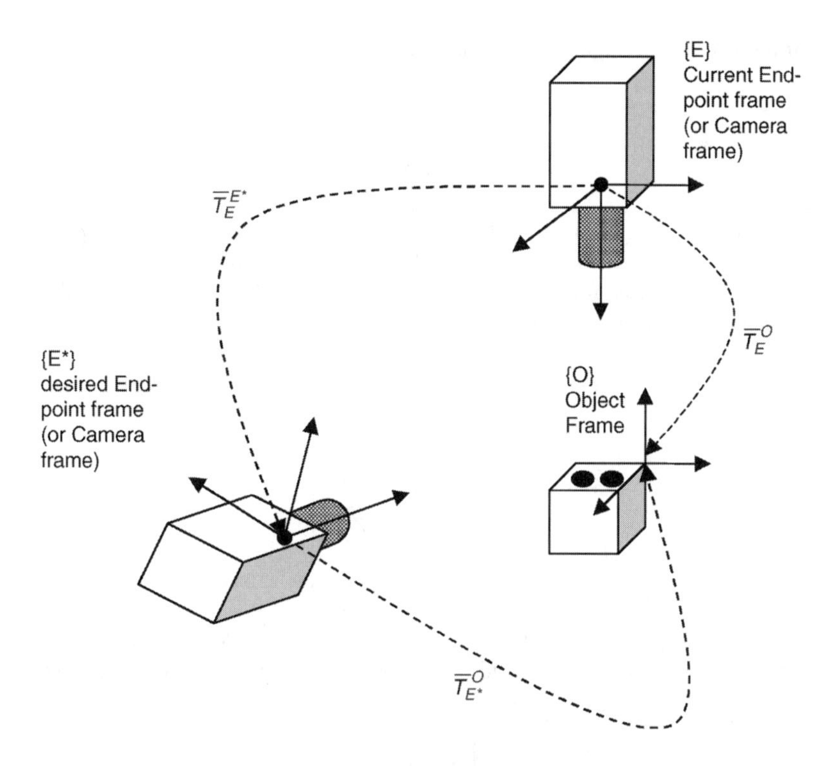

FIGURE 15.7 Modeling of the endpoint (or camera) displacement in a hand-eye configuration for 2-1/2D HVS.

represented by a vector $\overline{u}$ (a unit vector of rotation axis) and a rotation angle β. For a 3 × 3 rotation matrix with elements r_{ij}, the $\overline{u}\beta$ vector could be obtained from:

$$\overline{u}\beta = \frac{1}{2\ \text{sinc}\beta}\begin{bmatrix} r_{32} - r_{23} \\ r_{13} - r_{31} \\ r_{21} - r_{12} \end{bmatrix} \quad \text{for } \beta \neq \pi \tag{15.30}$$

and

$$\text{sinc}\beta = \begin{cases} 1 & \text{if } \beta = 0 \\ \dfrac{\sin\beta}{\beta} & \text{otherwise} \end{cases} \tag{15.31}$$

where β = π can be detected through the estimation of β. Then the axis of rotation $\overline{u}$ will be given by the eigenvector associated with the eigenvalue of 1 of the rotation matrix. The endpoint velocity screw can be related to the time derivative of the $\overline{u}\beta$ vector by a Jacobian matrix $\boldsymbol{L}_\omega$, given by:

$$\boldsymbol{L}_\omega = \boldsymbol{I} - \frac{\beta}{2}S(\overline{\boldsymbol{u}}) + \left(1 - \frac{\text{sinc}(\beta)}{\text{sinc}^2(\frac{\beta}{2})}\right)S^2(\overline{\boldsymbol{u}}) \tag{15.32}$$

where $S(.)$ is the skew-symmetric matrix associated with $\boldsymbol{u}$ and $\boldsymbol{I}$ is an identity matrix. Malis et al. [1999] have shown that $\boldsymbol{L}_\omega$ is singularity free.

Next, a hybrid error vector (Figure 15.5) is defined as:

$$e = \begin{bmatrix} m_e - m_e^* \\ \bar{u}\beta - \mathbf{0} \end{bmatrix} \tag{15.33}$$

where m_e and m_e^* are current and desired extended image parameter vectors, respectively. Consider a reference feature point $\boldsymbol{P}$ with coordinates $[X^E, Y^E, Z^E]^T$ with respect to the endpoint (or camera) frame. A supplementary normalized coordinate could be defined as $z = \log Z^E$ with Z^E as depth. Then, the normalized and extended image vector will be:

$$m_e = \begin{bmatrix} x^i \\ y^i \\ z^i \end{bmatrix} = \begin{bmatrix} \frac{X^E}{Z^E} \\ \frac{Y^E}{Z^E} \\ \log Z^E \end{bmatrix} \tag{15.34}$$

Let $\dot{W}_E = [v, \omega]^T \equiv [\dot{T}_E, \dot{\theta}_E]$ represent the endpoint (or camera) velocity screw with respect to the endpoint frame. The velocity screw $\dot{W}_E$ could be related to the hybrid error velocity $\dot{e}$ by:

$$\dot{e} = \begin{bmatrix} \dot{m}_e \\ \frac{d(\bar{u}\beta)}{dt} \end{bmatrix} = L\dot{W}_E \tag{15.35}$$

where

$$L \equiv \begin{bmatrix} -L_v & L_{(v,\omega)} \\ 0 & L_\omega \end{bmatrix}. \tag{15.36}$$

Here, $\boldsymbol{L}$ is the hybrid Jacobian matrix, with $\boldsymbol{L}_\omega$ given by Eq. (15.32),

$$L_v \equiv \begin{bmatrix} \frac{1}{Z^E} & 0 & -\frac{X^E}{Z^{E^2}} \\ 0 & \frac{1}{Z^E} & -\frac{Y^E}{Z^{E^2}} \\ 0 & 0 & \frac{1}{Z^E} \end{bmatrix} \tag{15.37}$$

and

$$L_{(v,\omega)} \equiv \begin{bmatrix} x^i & -(1+x^{i^2}) & y^i \\ 1+y^{i^2} & -x^i y^i & -x^i \\ -y^i & x^i & 0 \end{bmatrix}. \tag{15.38}$$

Obviously, one can consider differential changes of the endpoint (or camera) pose and error instead of their velocities. Note that Jacobian matrix $\boldsymbol{L}$ is singular only when $Z^E = 0$, or $\frac{1}{Z^E} = 0$, or with $\beta = \pm 2\pi$. These cases are exterior to the task space τ, so the task space is free of image-induced singularities.

Finally, the exponential convergence of $\boldsymbol{e} \to \mathbf{0}$ is achieved by imposing a control law of the form:

$$\dot{e} = -K_p e \tag{15.39}$$

or

$$\dot{W}_E = -K_p L^{-1} e \tag{15.40}$$

which could be simplified to:

$$\dot{W}_E = -K_p \begin{bmatrix} -L_v^{-1} & L_v^{-1} L_{(v,\omega)} \\ 0 & I \end{bmatrix} e \tag{15.41}$$

Note that $L_\omega^{-1} = I_{3\times3}$ because $L_\omega^{-1}\bar{u}\beta = \bar{u}\beta$, as proven by Malis et al. [1999]. Also, because L^{-1} is an upper triangular matrix, the rotational control loop is decoupled from the translational one. Like Eq. (15.25), endpoint screw velocity can be expressed in terms of changes required in joint angles that will be sent to the robot joint-servo loops for the execution.

Despite numerous advantages there are a few problems associated with 2-1/2D HVS. One of the problems is the possibility of features leaving image boundaries. Some approaches have been proposed to treat this problem. Among them is the approach of Morel et al. who use a modified feature vector (see Further Information).

The second problem is related to noise sensitivity and computational expense of partial estimation in HVS. A scaled Euclidean reconstruction (SER) was originally used to estimate camera displacement between the current and desired relative poses [Malis et al., 1999]. This method does not require a geometric three-dimensional model of the object; however, SER uses recursive KF to extract the rotation matrix from the homography matrix, while KF might be sensitive to camera calibration. Also, the estimation of the homography matrix requires more feature points, especially with noncoplanar objects. This is because when the object is noncoplanar, the estimation problem becomes nonlinear and will necessitate at least eight features for homography matrix estimation at the video rate of 25 Hz. Another approach would be to estimate the full pose of the object with respect to the camera by a fast and globally convergent method such as the orthogonal iteration algorithm (OI) of Lu and Hager (see Further Information). Next, a homogeneous transformation could be applied to obtain camera-frame displacement. However, this would require a full three-dimensional model of the object.

Finally, the selection of reference feature point affects the performance of 2-1/2D HVS. A series of experiments would be required to select the best reference point for the improved performance.

15.4 Examples

Simulations and experiments were run to compare the performances of PBVS, IBVS, and HVS. A MATLAB™ environment was created to simulate a five-degrees-of-freedom CRS Plus SRS-M1A robot with an EG&G Reticon MC9000 CCD camera mounted at its endpoint (Figure 15.8). A 16-mm lens was used. Simulation parameters are shown in Table 15.1. As shown in Figure 15.8, an object with five noncoplanar hole features was considered. The initial poses of the object and robot endpoint are also shown in Table 15.1.

A number of tests were run to investigate the effect of different parameters and compare the performances of different visual servoing methods. The control methods used for each strategy are EKF-based pose estimation and PD control for PBVS, RMRC for IBVS, and 2-1/2D for HVS. The control gains were tuned by running a few simulations for stable and fast responses. The results are shown in Figures 15.9 to 15.12. For simulations, an additive noise has been considered to represent more realistic situations. The general relative motion is shown in Figures 15.9 to 15.11 for PBVS, IBVS, and HVS. Initially, the robot endpoint is located above the object, i.e., $W_E^\circ = [0,\ 0,\ 30,\ 0°,\ 0°,\ 0°]^T$ in cm and rad, from Table 15.1.

TABLE 15.1 System Parameters for Simulations and Experiments

Parameter	Value
Focal length, F	1.723 cm
Interpixel spacings, P_x, P_y	0.006 cm/pixel
Number of pixels, N_x, N_y	128
Features coordinates in frame {O}	(5, 5, 0), (5, −5, 0), (−5, 5, 0), (−5, −5, 0), (0, 0, 3) cm
Initial pose of object with respect to the base frame, W_B°	[33.02, −2.64, 0.38, −0.0554, 0.0298, −3.1409] (in cm and rad)
Initial pose of the camera/endpoint with respect to the base frame, $\mathbf{W}_B^E$	[33.02, −2.64, 30.38, −0.0554, 0.0298, −3.1409] (in cm and rad)
Initial relative depth, Z_E^O	30 cm
Output measurement covariance matrix of EKF, $\boldsymbol{R}_k$	diag[0.04, 0.01, 0.04, 0.01, 0.04, 0.01, 0.04, 0.01, 0.04, 0.01] (pixel^2)
Disturbance noise covariance matrix of EKF, $\mathbf{Q}_k$	diag[0, 0.8, 0, 0.8, 0, 0.8, 0, 0.8, 0, 0.001] in (cm/s)^2 and (deg/sec)^2

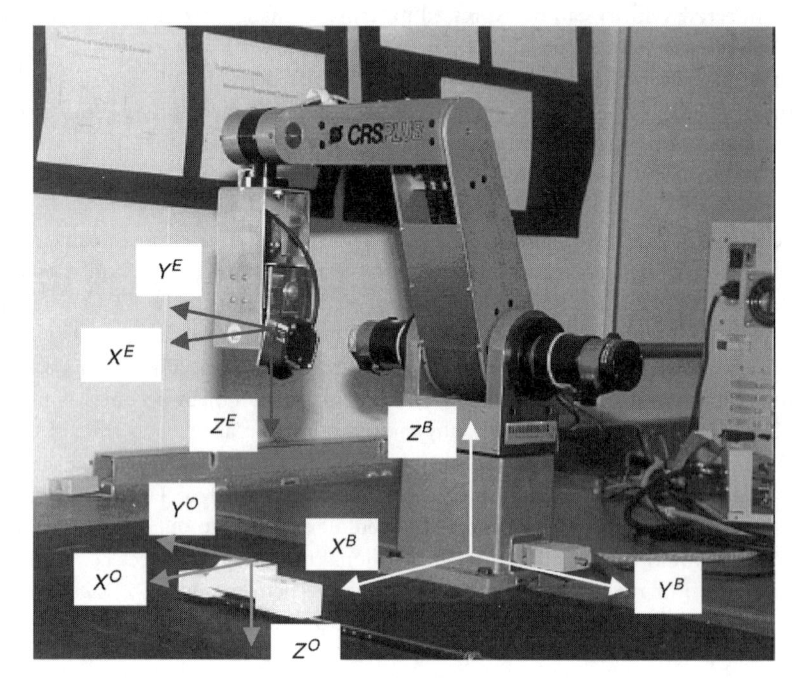

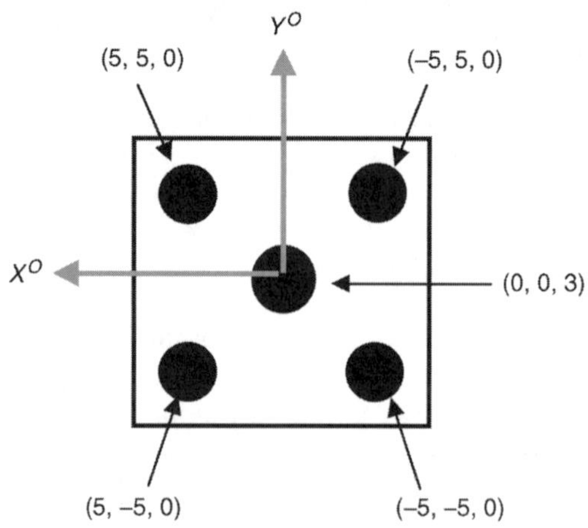

FIGURE 15.8 Simulation environment with object model and its five features.

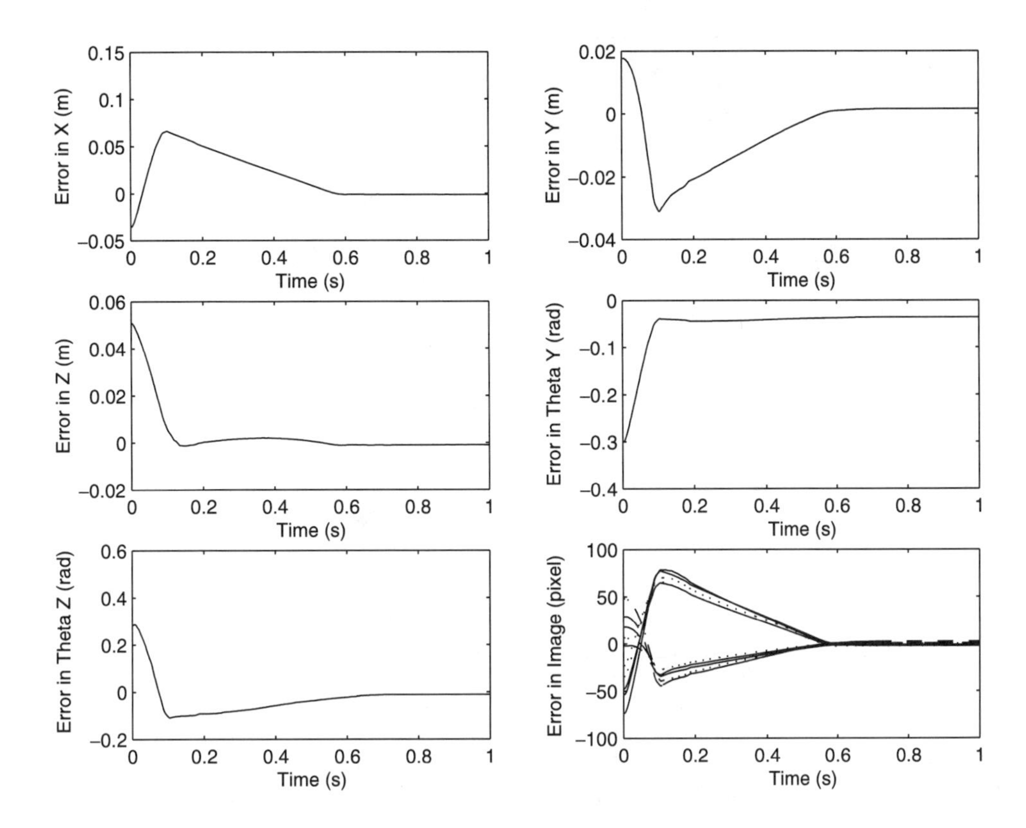

FIGURE 15.9 Simulation results for PBVS: desired relative pose of the object with respect to the endpoint frame: $(-3, -3, 25 \text{ cm}; 0, 0.3, -0.3 \text{ rad})$. The relative pose errors are specified with respect to the endpoint frame {E}.

The figures show the relative pose errors for $(X, Y, Z, \theta_X, \theta_Y, \theta_Z)$, i.e., the current relative pose minus the desired relative pose, in the endpoint frame. The endpoint desired relative motion is indicated in the caption of each figure. That corresponds to (3, –3, –5 cm, 0, 0.3, –0.3 rad) translation and rotation along and around the *X*, *Y*, and *Z* axes of the base frame, respectively, for the stationary object. Because the robot had five degrees of freedom, the commanded rotation about *X* axis was set to zero. Therefore, no error of θ_X is shown. Also, errors in the image plane for the *X* and *Y* positions of five feature points (in pixels) are shown in Figures 15.9 and 15.11. Figure 15.12 shows the camera-retreat problem for IBVS. Only a pure rotation around the optical *Z* axis of the camera was requested, but the camera moved away from the object and then returned. This problem was observed in neither PBVS nor HVS.

Some experiments were run for a different number of features and with different initial conditions and relative poses. The joint couplings in the CRS robot had negative effects on the control responses for three-dimensional relative motions. The tests showed that a reasonably larger number of features tended to improve the system response. The steady-state errors increased for all the methods with large commanded relative motions. This applied particularly to the IBVS that used a nondecoupled image Jacobian matrix leading to coupled translational and rotational motions. When the commanded motions were very close to the object, the steady-state error increased. Moreover, it was observed that PBVS and IBVS had almost the same response speed; however, HVS demonstrated a bit faster response than IBVS and PBVS, mainly due to the incorporation of the fast OI algorithm for estimation, instead of the SER used by Malis et al. [1999]. Choosing different reference points for HVS apparently had minimal effects on system performance. Finally, joint couplings of the CRS robot had considerable effect on the control responses for three-dimensional relative motions.

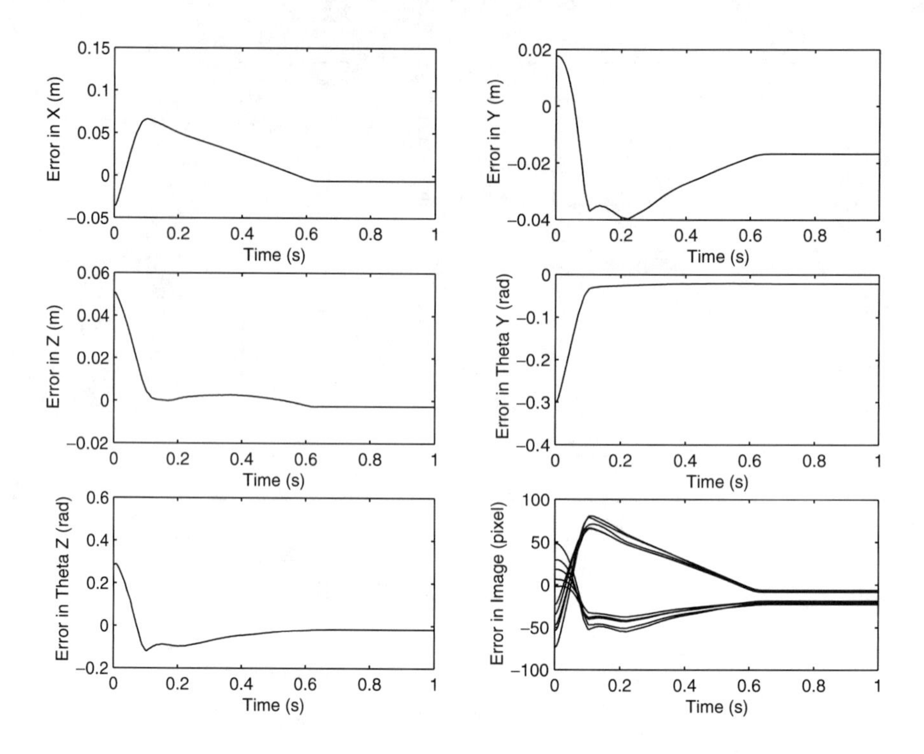

FIGURE 15.10 Simulation results for IBVS: desired relative pose of the object with respect to the endpoint frame: (−3, −3, 25 cm; 0, 0.3, −0.3 rad). The relative pose errors are specified with respect to the endpoint frame {E}.

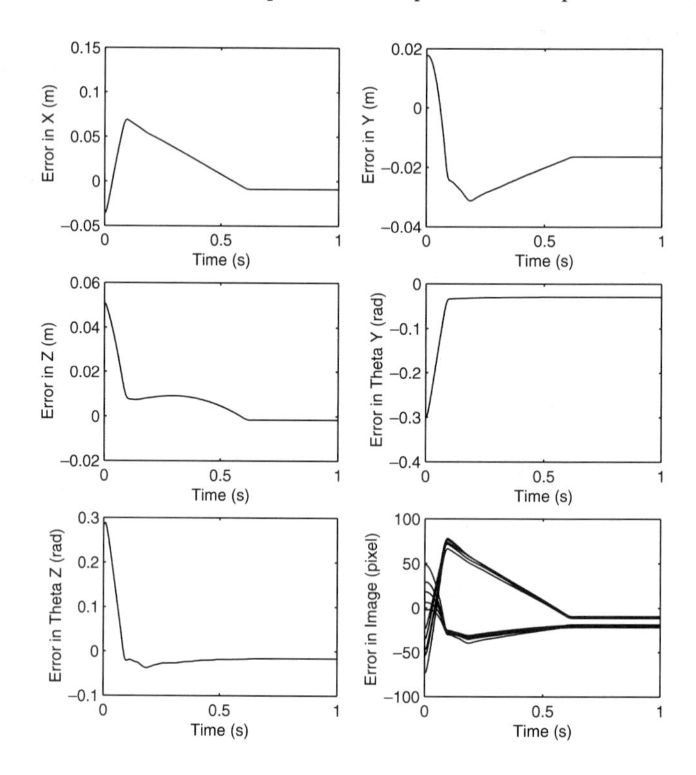

FIGURE 15.11 Simulation results for HVS: desired relative pose of the object with respect to the endpoint frame: (−3, −3, 25 cm; 0, 0.3, −0.3 rad). The relative pose errors are specified with respect to the endpoint frame {E}.

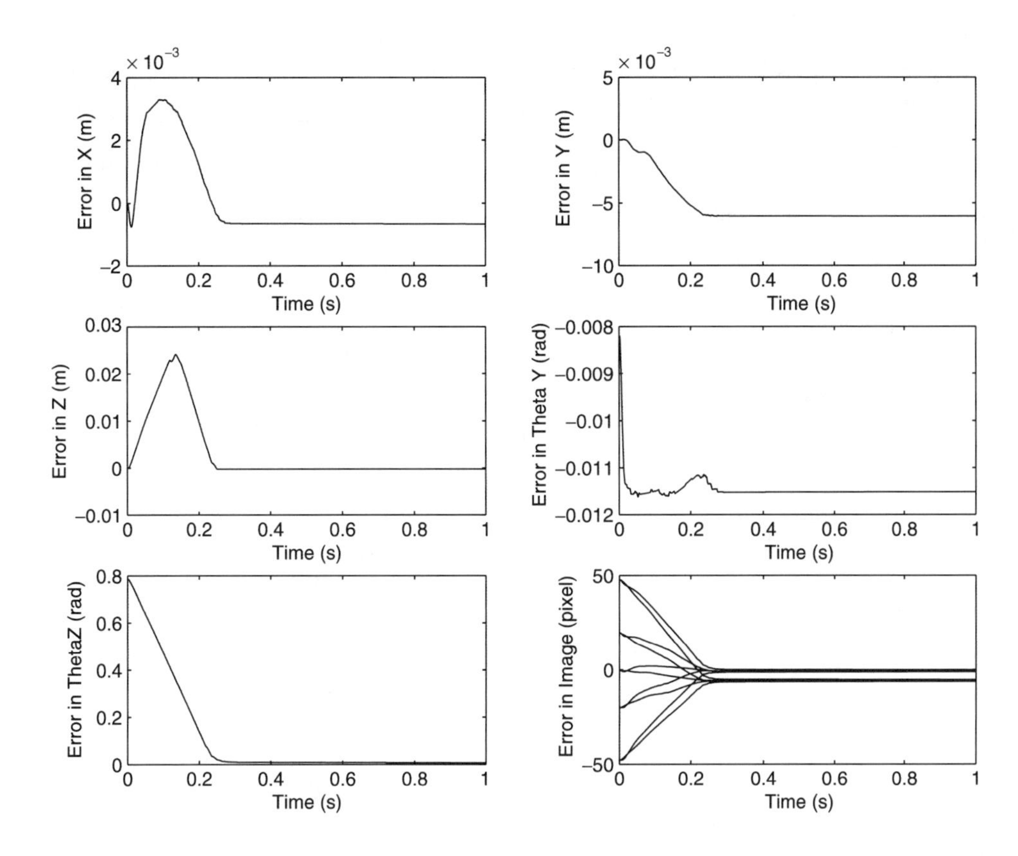

FIGURE 15.12 Simulation results for IBVS: desired relative pose of the object with respect to the endpoint frame: $(0, 0, 0 \text{ cm}; 0, 0, -\frac{\pi}{4} \text{ rad})$. The relative pose errors are specified with respect to the endpoint frame {E}.

15.5 Applications

Many applications of visual servoing have been limited to the laboratory and structured environments. For instance, many visual servoing systems use markers, structured light, and artificial objects with high-contrast features. Recent advances in opto-mechatronics technology have led to significant improvements of visual servoing science and practice. A list of visual servoing achievements and applications can be found in Corke [1996]. With the recent progress, it has been possible to design visual servoing systems operating above 60 Hz, tracking and picking objects moving at 30 cm/s, applying sealants at 40 cm/s, and guiding vehicles moving about 96 km/h. In summary, the current state of visual servoing technology has the potential to support robot operation in more realistic environments than today's structured environments. This is particularly required in many emerging technologies for autonomous systems.

Table 15.2 summarizes some of the demonstrated applications of visual servoing. However, many of these applications would not justify visual servoing applications, mainly due to the effectiveness of existing traditional solutions. Commercial applications of visual servoing will occur with those applications that there are not substitute technologies. A good example is applications that require precise positioning, such as fixtureless assembly, within dynamic and uncertain environments.

Table 15.3 summarizes potential new applications of visual servoing. The main obstacle for the development of new applications is related to the robustness of vision in adapting to different and noisy environments. Research and development is underway to address vision and image-processing robustness. This topic is discussed further in Chapters 10 and 14.

TABLE 15.2 Demonstrated Applications of Visual Guidance and Servoing (Speeds Represent Those of the Objects; Numbers with Hz Denote Bandwidth of Visual Servoing System)

Application	Investigators or Organizations, Date
Bolt insertion, picking moving parts from conveyor	Rosen et al., 1976–1978 (SRI Int.)
Picking parts from fast moving conveyor (30 cm/s)	Zhang et al., 1990
Tracking and grasping a toy train (25 cm/s, 60 Hz)	Allen et al., 1991
Visual-guided motion for following and grasping	Hill and Park, 1979
Three-dimensional vision-based grasping of moving objects (61 Hz)	Janabi-Sharifi and Wilson, 1998
Fruit picking	Harrell et al., 1989
Connector acquisition	Mochizuki et al., 1987
Weld seam tracking	Clocksin et al., 1985
Sealant application (40 cm/s, 4.5 Hz)	Sawano et al., 1983
Rocket-tracking camera with pan/tilt (60 Hz)	Gilbert et al., 1980
Planar micro-positioning (300 Hz)	Webber and Hollis, 1988 (IBM Watson Research Center)
Road vehicle guidance (96 Km/h)	Dickmanns and Graefe, 1988
Aircraft landing guidance	Dickmanns and Schell, 1992
Underwater robot control	Negahdaripour and Fox, 1991
Ping-pong bouncing	Anderson, 1987
Juggling	Rizzi and Koditscek, 1991
Inverted pendulum balancing	Dickmanns and Graefe, 1988
Labyrinth game	Anderson et al., 1991
Catching ball	Bukowski et al., 1991
Catching free-flying polyhedron	Skofteland and Hirzinger, 1991
Part mating (10 Hz)	Geschke, 1981
Aircraft refuelling	Leahy et al., 1990
Mating U.S. space shuttle connector	Cyros, 1988
Telerobotics	Papanikolopoulos and Khosla, 1992
Robot hand-eye coordination	Hashimoto et al., 1989

TABLE 15.3 Potential Applications of Visual Guidance and Servoing

Potential Applications
Fixtureless assembly
Automated machining
PC board inspection and soldering
IC insertion
Remote hazardous material handling (e.g., in nuclear power plants)
Weapons disassembly
Remote mining
Textile manufacturing
Automated television and surveillance camera guidance
Remote surgery
Satellite tracking and grasping
Planetary robotic missions

15.6 Summary

Visual servoing, when compared with conventional techniques, offers many advantages for the control of motion. In particular, visual servoing supports autonomous motion control systems without requiring exact specifications of poses for the object and tracker (e.g., robot endpoint). Also, visual servoing could relax the requirement for an exact object model. Visual servoing integration with robotic environments has significant implications such as fixtureless positioning, reduced robot training, and lower robot manufacturing costs and cycle time. In this chapter, the fundamentals of visual servoing were discussed and emphasis was placed on robotic visual servoing with eye-in-hand configurations. In particular, the emphasis was on the introduction of background theory and well-established methods of visual servoing.

An overview of the background related to visual servoing notations, coordinate transformations, image projection, and object kinematic modeling was given. Also, relevant issues of camera calibration were discussed briefly.

The main structures of visual servoing, namely, position-based, image-based, and hybrid visual servoing structures, were presented. The basic and well-established control method for each structure was given. The advantages and disadvantages of these control structures were compared.

Separation of the pose-estimation problem from the control-design problem, in position-based techniques, allows the control designer to take advantage of well-established robot Cartesian control algorithms. Also, position-based methods permit specification of the relative trajectories in a Cartesian space that provides natural expression of motion for many industrial environments, e.g., tracking and grasping a moving object on a conveyor. The interaction of image-based systems with moving objects, for example, has not been fully established. Position-based visual servoing provides no mechanism for regulating features in image space and, in order to keep the features in the field of view, must rely heavily on feature selection and switching mechanisms. Although both image-based and position-based methods demonstrate difficulties in executing large three-dimensional relative motions, image-based techniques show a response that is inferior to that of position-based methods, mainly due to the highly coupled translational and rotational motions of the camera. Hybrid systems, such as the 2-1/2D method, show superior response in comparison with their counterparts for long-range three-dimensional motions. This is mainly due to the provision of decoupling between translation and rotation of the camera in hybrid systems.

Also, two major issues with image-based methods are the presence of image singularities and camera-retreat problems. These problems do not exist with position-based and hybrid methods. One disadvantage of position-based methods over image-based and hybrid techniques is their sensitivity to camera calibration and object model errors. Furthermore, the required computation time of the position-based method is greater than that in image-based and hybrid methods. Hybrid methods usually rely on the estimation of a homography matrix. This estimation might be computationally expensive and sensitive to camera calibrations. For instance, with noncoplanar objects and conventional SER estimation of a homography matrix, more feature points might be required than those with other visual servoing methods. However, with the recent advances in microprocessor technology, the computation-time should not pose any serious problems. In all of the techniques, the visual control loop must be designed to provide higher bandwidth than that of robot position loops. Otherwise, the system control, like any other discrete feedback system with a delay, might become unstable by increasing the loop gain.

Simulations were done to demonstrate the performance of each servoing structure with the subscribed control strategy. Moreover, the effects of different design parameters were studied and some conclusions were drawn.

Finally, the demonstrated and potential applications of visual servoing techniques were summarized. Future research and development activities related to visual servoing were also highlighted.

Defining Terms

extrinsic camera parameters: Characteristics of the position and orientation of a camera, e.g., the homogeneous transform between the camera and the base frame.

feature: Any scene property that can be mapped onto and measured in the image plane. Any structural feature that can be extracted from an image is called image feature and usually corresponds to the projection of a physical feature of objects onto the image plane. Image features can be divided into region-based features, such as planes, areas, holes, and edge segment-based features, such as corners and edges.

image feature parameter: Any quantity with real value that can be obtained from image features. Examples include coordinates of image points; the length and orientation of lines connecting points in an image, region area, centroid, and moments of projected areas; parameters of lines, curves, or regular regions such as circles and ellipses.

image Jacobian matrix: Relates the velocity (or differential changes) of the camera $\dot{r}$ (or Δr) or its relative pose $\dot{W}$ (or ΔW) to the image feature velocities $\dot{s}$ (or Δs). This matrix is also referred to as feature Jacobian matrix, feature sensitivity matrix, interaction matrix, or $\boldsymbol{B}$ matrix.

intrinsic parameters of the camera: Inner characteristics of the camera and sensor, such as focal length and radial and tangential distortion parameters, and the coordinates of the principal point, where the optical axis intersects the image plane.

visual servoing (vision-guided servoing): The use of vision in the feedback loop of the lowest level of a (usually robotic) system control with fast image processing to provide reactive behavior. The task of visual servoing for robotic manipulators (or robotic visual servoing, RVS) is to control the pose of the robot's end-effector relative to either a world coordinate frame or an object being manipulated, using real-time visual features extracted from the image. The camera can be fixed or mounted at the endpoint (eye-in-hand configuration).

Acknowledgments

This work was supported by the Natural Sciences and Engineering Research Council of Canada (NSERC) through Research Grant #203060-98. I would also like to thank my Ph.D. student Lingfeng Deng for his assistance in the preparation of the simulation results.

References

Chaumette, F., Potential problems of stability and convergence in image-based and position-based visual servoing, *The Confluence of Vision and Control,* Vol. 237 of *Lecture Notes in Control and Information Sciences*, Springer-Verlag, New York, 1998, pp. 66–78.

Corke, P. I., *Visual Control of Robots: High Performance Visual Servoing*, Research Studies, Ltd., Somerset, England, 1996.

Corke, P. I. and Hutchnison, S. A., A new hybrid image-based visual servo-control scheme, *Proc. IEEE Int. Conf. Decision and Control,* 2000, pp. 2521–2526.

Feddema, J. T., Lee, C. S. G., and Mitchell, O. R., Weighted selection of image features for resolved rate visual feedback control, *IEEE Trans. Robot. Automat.*, 7(1), 31–47, 1991.

Ficocelli, M., Camera Calibration: Intrinsic Parameters, Technical Report TR-1999-12-17-01, Robotics and Manufacturing Automation Laboratory, Ryerson University, Toronto, 1999.

Ficocelli, M. and Janabi-Sharifi, F., Adaptive Filtering for Pose Estimation in Visual Servoing, *IEEE/RSJ Int. Conf. on Intelligent Robots and Systems, IROS 2001*, Maui, Hawaii, 2001, pp. 19–24.

Hutchinson, S., Hager, G., and Corke, P. I., A tutorial on visual servoing, *IEEE Trans. Robot. Automat.*, 12(5), 651–670, 1996.

Janabi-Sharifi, F. and Wilson, W. J., Automatic selection of image features for visual servoing, *IEEE Trans. Robot. Automat.*, 13(6), 890–903, 1997.

Malis, E., Chaumette, F., and Boudet, S., 2-1/2D visual servoing, *IEEE Trans. Robot. Automat.*, 15(2), 238–250, 1999.

Sanderson, A. C. and Weiss, L. E., Image-based visual servo control using relational graph error signals, *Proc. IEEE*, 1980, pp. 1074–1077.

Shirai, Y. and Inoue, H., Guiding a robot by visual feedback in assembling tasks, *Pattern Recognition*, 5, 99–108, 1973.

Tsai, R. and Lenz, R., A new technique for fully autonomous and efficient three-dimensional robotic hand/eye calibration, *IEEE Trans. Robot. Automat.*, 5(3), 345–358, 1989.

Wang, J., Optimal Estimation of Three-Dimensional Relative Position and Orientation for Robot Control, M.A.Sc. dissertation, Dept. of Electrical and Computer Engineering, University of Waterloo, Waterloo, Canada, 1992.

Wilson, W. J., Williams Hulls, C. C., and Bell, G. S., Relative end-effector control using cartesian position-based visual servoing, *IEEE Trans. Robot. Automat.*, 12(5), 684–696, 1996.

Wilson, W. J., Williams-Hulls, C. C., and Janabi-Sharifi, F., Robust image processing and position-based visual servoing, in *Robust Vision for Vision-Based Control of Motion*, Vincze, M. and Hager, G. D., Eds., IEEE Press, New York, 2000, pp. 163–201.

Yuan, J. S. C., A general photogrammetric method for determining object position and orientation, *IEEE Trans. Robot. Automat.*, 5(2), 129–142, 1989.

For Further Information

A good collection of articles on visual servoing can be found in *IEEE Trans. Robot. Automat.*, 12(5), 1996. This issue includes an excellent tutorial on visual servoing. A good reference book is *Visual Control of Robots: High Performance Visual Servoing* by P. I. Corke (Research Studies, Ltd., Somerset, England, 1996), encompassing both theoretical and practical aspects related to visual servoing of robotic manipulators. *Robust Vision for Vision-Based Control of Motion*, edited by M. Vincze and G. D. Hager (IEEE Press, New York, 2000) provides recent advances in the development of robust vision for visual servo-controlled systems. The articles span issues including object modeling, feature extraction, feature selection, sensor data fusion, and visual tracking. *Robot Vision*, by B. K. Horn (McGraw-Hill, New York, 1986) is a comprehensive introductory book for the application of machine vision in robotics.

Proceedings of IEEE International Conference on Robotics and Automation, IEEE Robotics and Automation Magazine, Proceedings of IEEE/RSJ International Conference on Intelligent Robots and Systems, IEEE Transactions on Robotics and Automation, and *IEEE Transactions on Systems, Man*, and *Cybernetics* document the latest developments in visual servoing.

Several articles have compared the performances of basic visual servoing methods. Among them are "Potential Problems of Stability and Convergence in Image-Based and Position-Based Visual Servoing," by F. Chaumette, in *The Confluence of Vision and Control*, Vol. 237 of *Lecture Notes in Control and Information Sciences* (Springer-Verlag, New York, 1998, pp. 66–78), and "Stability and Robustness of Visual Servoing Methods," by L. Deng, F. Janabi-Sharifi, and W. J. Wilson, in *Proc. IEEE Int. Conf. Robot. Automat.* (Washington, D.C., May 2002).

Good articles for camera calibration include the one by Tsai and Lenz [1989] and also "Hand-Eye Calibration," by R. Horaud and F. Dornaike, in *International Journal of Robotics Research*, 14(3), 195–210, 1995. Implementation details can also be found in Corke [1996].

Articles for analytical pose estimation include the following. For pose estimation using four coplanar but not collinear feature points, see "Random Sample Consensus: A Paradigm for Model Fitting with Applications to Image Analysis and Automated Cartography," by M. A. Fischler and R. C. Bolles, in *Comm. ACM*, 24, 381–395, 1981. If camera intrinsic parameters need to be estimated, six or more feature points will be required for a unique solution. This is shown in "Decomposition of Transformation Matrices for Robot Vision," by S. Ganapathy, in *Pattern Recog. Lett.* 401–412, 1989. See also "Analysis and Solutions of the Three Point Perspective Pose Estimation Problem," by R. M. Haralick, C. Lee, K. Ottenberg, and M. Nolle, in *Proc. IEEE Conf. Comp. Vision, Pattern. Recog.*, pp. 592–598, 1991, and "An Analytic Solution for the Perspective 4-Point Problem," by R. Horaud, B. Canio, and O. Leboullenx, *Computer Vision Graphics, Image Process.*, no. 1, 33–44, 1989.

The articles for the least-squares solutions in pose estimation include an article by S. Ganapathy [1989], mentioned above, and "Determination of Camera Location from 2-D to 3-D Line and Point Correspondences," by Y. Liu, T. S. Huang, and O. D. Faugeras, in *IEEE Trans. Pat. Anal. Machine Intell.*, no. 1, 28–37, 1990. Also, "Constrained Pose Refinement of Parametric Objects," by R. Goldberg, in *Int. J. Comput. Vision*, no. 2, 181–211, 1994. A fast and globally convergent orthogonal iteration (OI) algorithm is introduced in "Fast and Globally Convergent Pose Estimation from Video Images," by C. P. Lu and G. D. Hager, in *IEEE Trans. Patt. Analysis and Machine Intell.*, 22(6), 610–622, June 2000.

The following references provide solutions to online estimation of depth information and image Jacobian matrix for IBVS. "Manipulator Control with Image-Based Visual Servo," by K. Hashimoto, T. Kimoto, T. Ebine, and H. Kimura, in *Proc. IEEE Int. Conf. Robot. Automat.*, Piscataway, NJ, 1991, pp. 2267–2272 provides an explicit depth estimation based on the feature analysis. An adaptive control-based method for depth estimation is provided in "Controlled Active Vision," by N. P. Papanikolopoulos, Ph.D. dissertation, Dept. of Electrical and Computer Engineering, Carnegie Mellon University, 1992. It is also proposed to use an approximation to the value of the image Jacobian matrix computed at the desired camera position, in "A New Approach to Visual Servoing in Robotics," by B. Espiau, F. Chaumette, and P. Rives, in *IEEE Trans. Robot. Automat.*, 8(3), 313–326, 1992.

There are different image Jacobian derived in the literature. The following sources could be studied for further details: [Feddema et al., 1991], and "Vision Resolvability for Visually Servoed Manipulation," by B. Nelson, and P. K. Khosla, in *Journal of Robotic Systems*, 13(2), 75–93, 1996. Also see "Controlled Active Vision," by N. P. Papanikolopoulos, Ph.D. dissertation, Dept. of Electrical and Computer Engineering, Carnegie Mellon University, 1992.

Other forms of image Jacobian matrices have been derived using geometrical entities, such as spheres and lines, are also available in: "A New Approach to Visual Servoing in Robotics," by B. Espiau et al., mentioned above. Also see Malis et al. [1999] for Jacobian matrix for 2-1/2D servoing.

In addition to Malis et al. [1999], the following papers are good resources for the study of HVS. For instance, the following paper proposes a solution for features leaving field-of-view in 2-1/2D HVS: "Explicit Incorporation of 2-D Constraints in Vision-Based Control of Robot Manipulators," by G. Morel, T. Liebezeit, J. Szewczyk, S. Boudet, and J. Pot, in *Experimental Robotics VI*, P. I. Corke and J. Trevelyan, Eds., Vol. 250, Springer-Verlag, 2000, pp. 99–108. The following article proposes to compute translational velocity instead of rotational one for HVS: "Optimal Motion Control for Image-Based Visual Servoing by Decoupling Translation and Rotation," by K. Deguchi, in *Proc. IEEE Int. Conf. Intel. Robotics and Systems*, 1998, pp. 705–711. Also see Corke and Hutchinson [2000] for a new partitioning of camera degrees of freedom. That includes separating optical axis z and computing z-axis velocity using two new image features.

The following sources are useful for the study of homography matrix and its decomposition used in HVS: *Three-Dimensional Computer Vision*, by O. Faugeras, MIT Press, Cambridge, MA, 1993, and "Motion and Structure from Motion in a Piecewise Planar Environment," in *Int. J. Pattern Recogn. Artificial Intelligence*, 2, 485–508, 1988.

The following sources cover the applications developed using VS techniques: Corke [1996] and "Visual Servoing: A Technology in Search of an Application," by J. T. Feddema, in *Proc. IEEE Int. Conf. Robot. Automat: Workshop on Visual Servoing: Achievements, Applications, and Open Problems*, San Diego, CA, 1994.

16

Optical-Based In-Process Monitoring and Control

Masatake Shiraishi
Ibaraki University
Ibaraki, Japan

16.1 Introduction

The monitoring of the working conditions of a system is becoming an increasingly important issue in its automation. Despite many successful results, this area is still not mature enough to be used for complete unmanned automation. With the advantages of noncontact and the remote sensing approach, an optical sensing or monitoring technique such as light projection, light focusing, light detection, etc. is the most promising method when compared with other approaches. However, the problem with the optical approach is its reliability and repeatability, because it generally provides a high measuring resolution and is easily affected by disturbances. Among the optical techniques a laser beam is traditionally a major tool for sensing and monitoring and has often been used in conjunction with a CCD camera in processes such as machining, assembly, welding, textile fabrication, quality inspection, etc. The results of this method are normally satisfactory when the disturbance is not serious and a certain countermeasure is sometimes implemented to withstand the disturbance. An incandescent light is sometimes used instead of a laser light, but the application is limited because of a low sensitivity and narrow measuring range. The CCD camera itself is also a major tool for detection and often used as a machine vision or vision sensor. This approach uses an image-processing technique that generally is burdened by low-speed detection (recognition). An optical fiber sensor has basically two functions—light projection and light detection—depending on the application. It is often used with a laser beam as a light source.

Under these circumstances reliable process/machine condition monitoring and control in an online fashion is a practical step toward real automation in industry. Such automation must be performed with reliable sensors and an associated control strategy that is able to interpret incoming sensor information and decide on an appropriate control action. Successful implementations of monitoring system states based on sensing techniques depend on two factors: first, the quality of useful information provided by

0-8493-1162-4/03/$0.00+$1.50

the sensors and, second, the techniques used to process the obtained information in order to make decisions for better control strategy. A sensor-fusion approach is one of the solutions for future sensing and monitoring technologies that aims to achieve those goals.

16.2 Fundamentals

16.2.1 Role of In-Process Monitoring and Control [Cho, 2001]

The ultimate goal of in-process monitoring and control is to ensure high quality of output, high-speed reliability, and complete autonomy in process operation. Achieving this goal would solve the problems of quality monitoring and control, which ensure that process outputs conform to target value specifications. The achievement, however, is not an easy task because most processes have the characteristics of a highly nonlinear, uncertain, and complex system, as depicted in Figure 16.1. This may be attributable to the fact that during operating time complex physical phenomena occur, and the input variables operating the processes interact with each other. Furthermore, the performance of operating equipment gradually degrades due to changes in the external environment and changes in their mechanical structures. In short, the processes are exposed to external disturbances and noise and often subjected to parameter variations. Due to these characteristics the output quality often varies from output to output, impairing uniformity. These bring about the degradation of output quality unless effective monitoring and control of the processes are properly performed. In other words, to avoid any occurrence that is unfavorable for output quality maintenance, monitoring and control functions are essential for the process.

Monitoring of the process state requires three major steps that must be carried out online: (1) the process is continuously monitored with a sensor or multiple sensor, (2) the sensor signals are conditioned and preprocessed so that certain features sensitive to the process states can be obtained, and (3) the process states are identified by pattern recognition based on (1) and (2). Feedback control of the process state is comprised of the following steps: (1) measurement of the process state, (2) generation of the control signal based on the dynamic characteristics of the process, and (3) correction of the process operation, observing the resulting process output and comparing the observed with the desired output. In the last step, the observed state must be related to quality.

A general procedure for evaluating the process output from monitoring of process variables and machine condition variables is shown in Figure 16.2. This procedure requires a number of activities that are performed by the sensing element, signal interpretation elements, and the quality evaluation unit. The sensors may be composed of multiples of different types having different principles of measurement or multiples of one type. In using sensors of different types, sensing reliability becomes quite important in synthesizing the information required to estimate the process condition or product quality. The reliability of the sensors may change relative to one another. This necessitates careful development of the

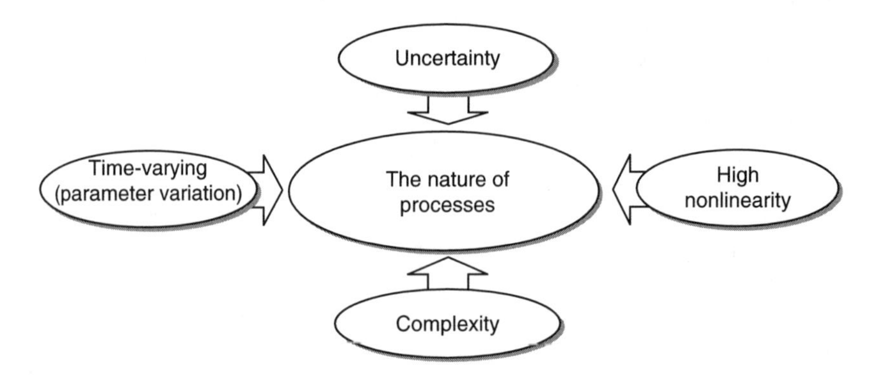

FIGURE 16.1 Process characteristics.

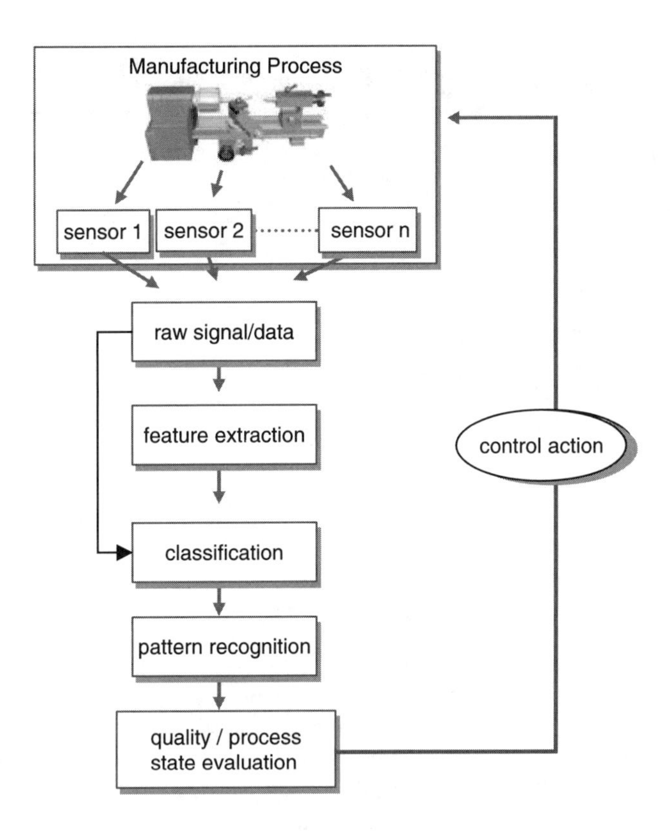

FIGURE 16.2 A general procedure for quality monitoring.

synthesis method. In reality in almost all processes whose quality cannot be measured directly multisensor integration/fusion is vital for characterizing product quality.

With respect to handling types of information, there are two typical methods used to evaluate product quality. One is to make direct use of raw signal, while the other is to use features extracted from the raw signal. In the case of using raw signal, indicated by a dotted line, the amount of data can be a burden on tasks for clustering and pattern recognition. On the other hand, the feature extraction method is very popular because it allows analysis of data in lower dimensional space and provides efficiency and accuracy in monitoring. The feature values used could be of entirely different properties depending on monitoring applications. For example, in most industrial inspection problems adopting machine vision techniques, image features such as area, center of gravity, periphery, and moment of inertia of the object image are frequently used to characterize the shapes of the object under inspection. In some complicated problems the number of features used may be as many as 20 in order to achieve problem solution. By contrast, in some simple problems one single feature may suffice to characterize the object. Monitoring machine conditions frequently employ time or frequency domain features of the sensor signal such as mean variance, kurtosis, crest factor, skewness, and power in a specified frequency band. When the choice of features is appropriately made and their values calculated, the next task is to find the similarity between the feature vector and the quality variables or process conditions, that is, to perform a classification task. Depending on the nature of the problem, the classifier should differ in its discriminating characteristics because there is no universal classifier that can be effectively used for a large class of problems. In fact, the literature indicates that a specific method works for a specific application.

The operating values should be adjusted in order to ensure desired output quality or a desired state variable when process parameters are subject to change and external disturbances are present, which is usually the case in manufacturing processes. When it comes to output quality control, the output quality should be measured by a sensor online. When this happens, there may be two methods of quality control,

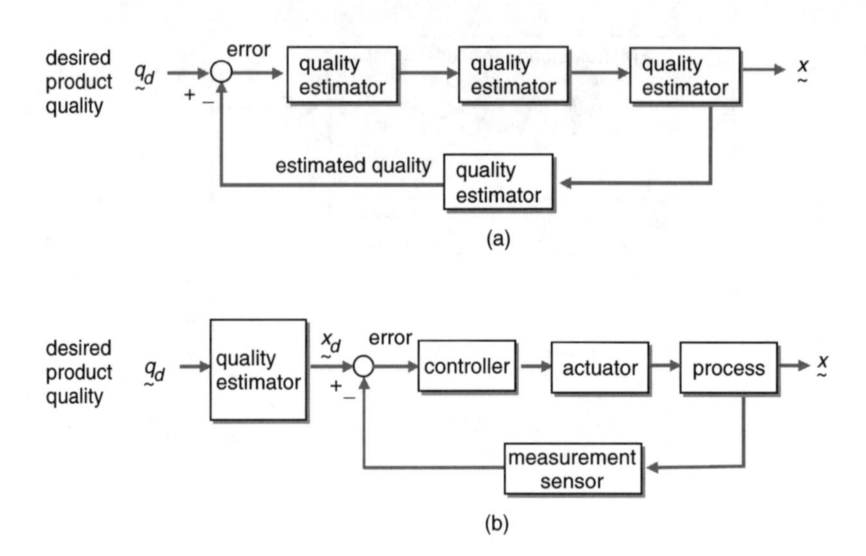

FIGURE 16.3 Comparison of direct (a) and indirect (b) quality-control methodologies.

depending on measurability of the variables: (1) direct method and (2) indirect method. As illustrated in Figure 16.3, the direct method shown in (a) utilizes the quality variables as outputs of the controlled system, while the indirect method shown in (b) uses some process-state variables. This situation is exactly the same as in the case of output quality monitoring. Direct method is desirable because it directly controls output quality. The difficulty, however, is that very few quality variables are measurable in most manufacturing processes. The indirect method uses an inverse quality estimator in order to directly control the process state variables instead of quality variables.

The estimator is derivable from

$$\bar{q} = \bar{f}(\bar{x}) \tag{16.1}$$

and it can be given by

$$\bar{x} = \bar{f}^{-1}(\bar{q}) \tag{16.2}$$

where the $\bar{q}$ vector consists of the quality variables to be controlled, $\bar{x}$ consists of those process-state variables, and $\bar{f}$ represents the functions linking quality variables with process-state variables. It simply transforms the quality variables to be controlled to the corresponding process-state variables. The estimator design should consider several factors with respect to determination of the relevant state variables, selection of sensors, and construction of the functional. State variables must be measurable and selected to give distinct correlation with the quality variables; that is, the $\bar{f}$ function must be determined uniquely and be less complex in its form. In addition, state variables must show high sensitivity to variation of the quality variables. The selection of sensors must be made to meet control requirements such as resolution response speed, linearity, and sampling time.

16.3 Basic Concept of In-Process Monitoring and Control [Shiraishi, 1994]

16.3.1 Selection of Measuring Method

A fundamental selection of in-process monitoring can be done from the standpoint of contact and noncontact methods. Table 16.1 shows a general comparison of two measuring methods. The advantage of the contact method is its reliability in practical use, while the noncontact method—for example, an

TABLE 16.1 Comparison Between Contact and Noncontact Methods

Feature	Contact	Noncontact
Resolution	Not so high	High
Measuring range	Wide	Generally narrow
Operational speed	Not so high	High
Calibration	Easy	A little complicated
Signal processing	Easy	A little complicated
Reliability	High	Relatively low but depends on sensor design
Contact load	Large	No load
Remote sensing	Available with limitation	Available
Mounting space	With limitation	Relatively free

optical approach—offers the capability for remote sensing and performing measurements with high resolution without physical contact. The importance of selecting this measuring method is embodied in the answers to the following questions:

1. What is the required accuracy?
2. What is the required operational speed of the sensor?
3. Will the space allowed be adequate for the sensor, and will the sensor be in the best position for measuring the space?
4. How hostile will the environment be? Will it be removed by suitable countermeasures (especially with the noncontact method)?

Of course, cost is an important factor for selection. However, what must be considered is the result using the advanced techniques of these methods as compared with the traditional method. Figure 16.4 indicates a key point of sensor selection from more practical standpoints. Time delay in measurement is an important factor in practical use because of instability in the control process. The detecting point should be as close to the phenomenon as possible. With the advantages of noncontact and high resolution, an optical method is described throughout this chapter.

16.3.2 Ideal Sensors for Optical-Based In-Process Measurement

By taking the aforementioned discussions into account, the ideal optical sensor for in-process measurement will include following requirements.

1. General features:
   - Long life and stability in characteristics
   - Rugged and simple construction, and easy to use (calibration, maintenance, inspection, etc.)
   - Insensitive to environmental conditions
   - Easy signal processing
   - Inexpensive
2. Multipoint measurements: The in-process technique is largely accomplished by single-point detection, and the phenomenon taking place is representatively evaluated only by that measurement. Therefore, the measured values normally differ when the detection points are changed. To avoid such errors multipoint measurable sensors must be developed.
3. Development of signal processing technique: The sensors developed so far have not been used to their full potential. In applying a laser displacement sensor, for example, the output generally includes information such as surface roughness, waviness, etc. All advanced laser displacement systems, however, detect only a single phenomenon through adequate electrical filters. Other significant information should not be thrown away from measured signals. A multisignal processing technique must be developed from the point of view of both hardware and software.

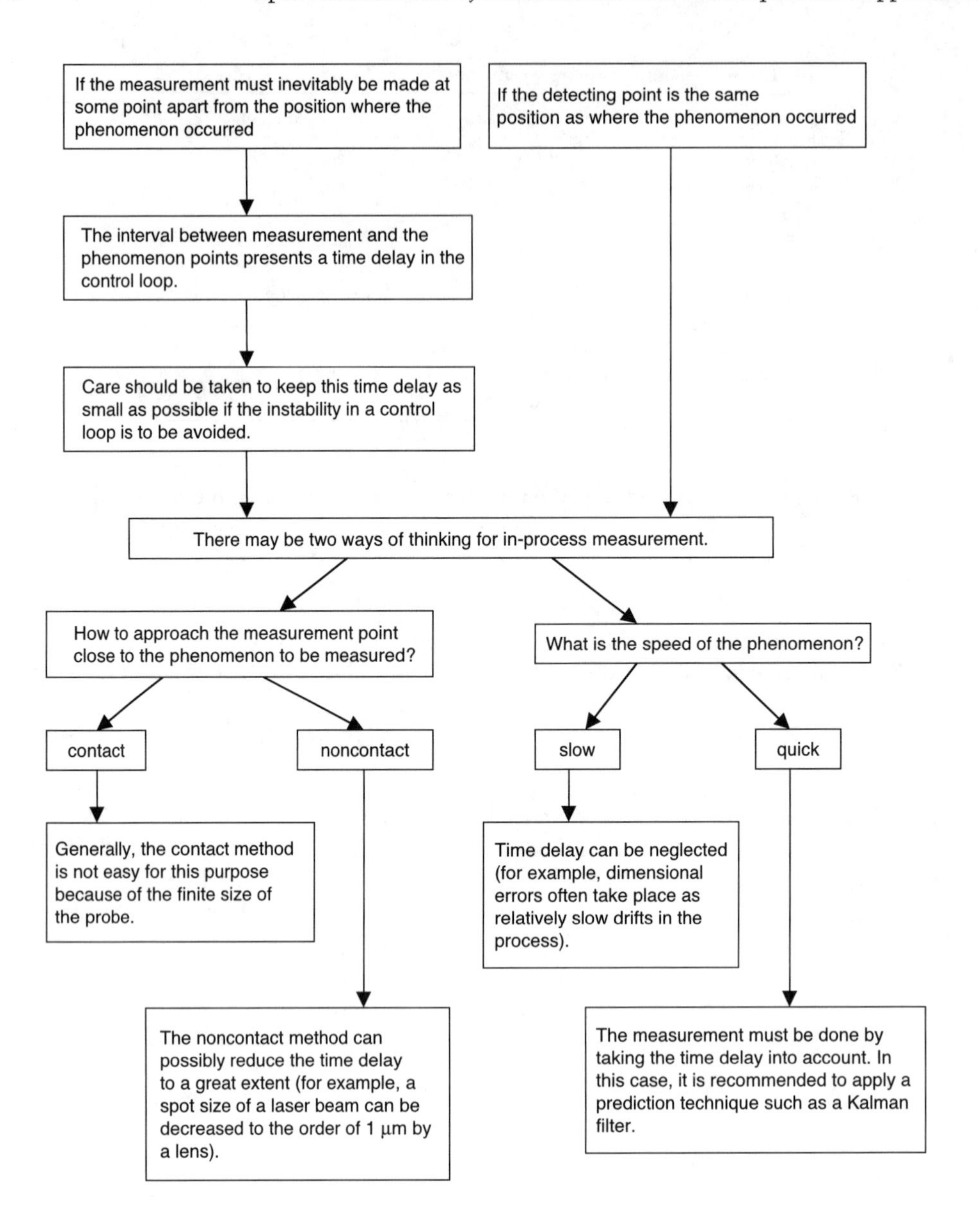

FIGURE 16.4 Flow chart of measurement selection.

4. Smart sensor: Required sensors and transducers must be compact, rugged, and insensitive to environmental conditions other than those they are to measure. In order to satisfy these specifications, "smart sensors" should be more actively developed by using solid-state technology. These types of sensor are integrated hardware/software systems that not only measure several important physical properties but also extract and interpret relevant data, in terms of established criteria and priorities, for online measurement and control of complex processes. A conventional sensing system is composed of a sensing device, a signal converter, a signal transmission element, a receiver, and a separate output element connected to the computer through an interface. A smart sensing system, however, should be integrated in a single-package integrated sensor for the following reasons (Figure 16.5):
   a. A high output signal and sensitivity should be obtained in order to remove various influences under actual working conditions.

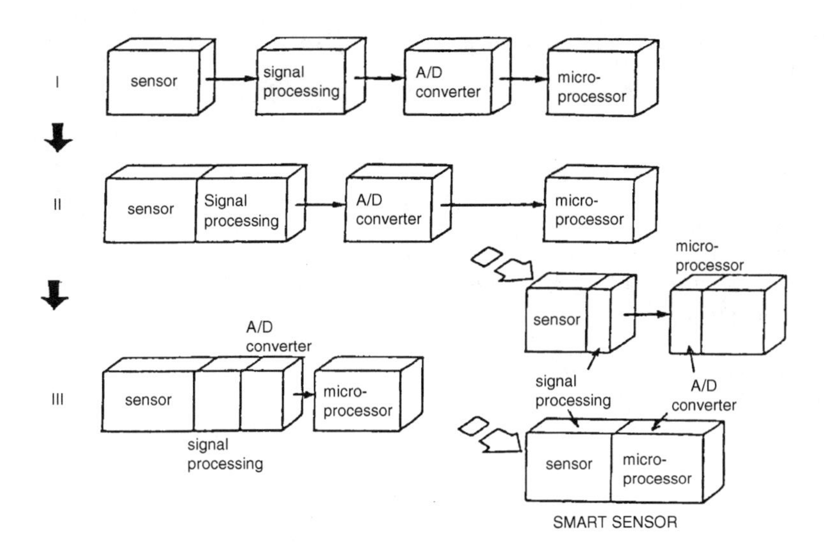

FIGURE 16.5 Construction of smart sensor.

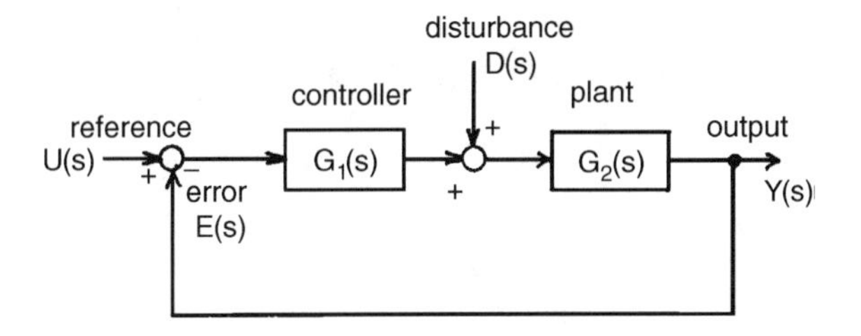

FIGURE 16.6 Basic control block diagram.

b. A high total performance will be expected when the performance and accuracy of each element are adjusted.
c. Input/output terminals and transmission lines should be omitted as much as possible for the compact sensing system.
d. Parallel data processing can be implemented to increase input information.

16.3.3 Control Technology

Successful opto-mechatronic systems also rely heavily on control technology. In a highly advanced precision positioning system with an optical sensing device, for example, positioning accuracy is easily affected by disturbances such as temperature change, pressure change, air fluctuation, etc. An adequate compensation for these influences is usually implemented using the control technology approach.

Figure 16.6 shows a basic control block diagram of mechatronic systems with a disturbance $D(s)$ applied to the plant. The output $Y(s)$ and steady-state error $E(s)$ are represented by

$$Y(s) = \frac{G_1(s)G_2(s)}{1+G_1(s)G_2(s)}U(s) + \frac{G_2(s)}{1+G_1(s)G_2(s)}D(s) \tag{16.3}$$

$$E(s) = \frac{1}{1+G_1(s)G_2(s)}U(s) - \frac{G_2(s)}{1+G_1(s)G_2(s)}D(s) \tag{16.4}$$

where $U(s)$ is the desired value, $G_1(s)$ is the transfer function of the controller, and $G_2(s)$ is the transfer function of the plant.

The objectives of the control are as follows:

1. *Keep the system in stable condition.* The characteristic equation in Eq. (16.3) is given by

$$1 + G_1(s)G_2(s) = 0 \tag{16.5}$$

which is a polynomial function of Laplace transformation s. The stable condition is to assign all poles in Eq. (16.5) in the left half plane (negative) of the coordinate system.

2. *Follow the output $Y(s)$ exactly to a desired value $U(s)$, that is, to eliminate the steady-state error $E(s)$.* This implies that

$$\frac{Y(s)}{U(s)} \rightarrow 1 \tag{16.6}$$

or

$$E(s) \rightarrow 0 \tag{16.7}$$

To satisfy this requirement, $G_1(s)$, i.e., the controller gain, should be as large as possible.

3. *Eliminate the effects of disturbance.* This implies that

$$\frac{Y(s)}{D(s)} \rightarrow 0 \tag{16.8}$$

where the controller gain of $G_1(s)$ should also be as large as possible.

4. *Maintain system robustness even in the presence of parameter changes or disturbances.* A recent control issue of mechatronics is mainly focused on "robust control" such as H_∞ control, disturbance observer, intelligent control, etc.

These requirements are not always satisfied at the same time; therefore, many compensation approaches are being implemented in both classical and modern control technologies.

Example of the Positioning System

To achieve high-performance servo control with nanometer-precision accuracy, feed forward and disturbance compensation by an observer is effectively and successfully implemented in a single axis positioning system [Yaskawa Electric Co., 1997]. The experimental device is schematically depicted in Figure 16.7. A command from the computer drives the DC servo motor, which rotates the screw shaft. The resultant axial force moves the table by the corresponding displacement, which is measured by the laser measuring apparatus with a resolution of 2.5 nm. The difference Δx between the actual travel y and the desired position x is then registered by the computer and fed to the servo motor. In this system, torque variation is one of the crucial disturbances and affects positioning accuracy. Figure 16.8 demonstrates the control system with a feed forward compensation and disturbance observer. The motor angle and angular velocity are governed by the proportional control mode and the proportional-integral control mode, respectively. The dominant closed-loop dynamics of this system is approximated by the second-order lag system with response time τ_r and damping ratio ζ. With the current control mode, response errors due to response delay and load disturbance are inevitable, thus affecting positioning accuracies.

To avoid these inconveniences feed forward compensations for acceleration and velocity are added to the original control scheme. The gains of the feed forward compensations are set at τ_r^2 for acceleration compensation and $2\zeta\tau_r$ for velocity compensation, which makes the positioning system more quick and accurate. The response error can be reduced to less than one tenth of the original amount. Next, systematic errors due to disturbance torque are compensated by adding a disturbance observer to the control loop.

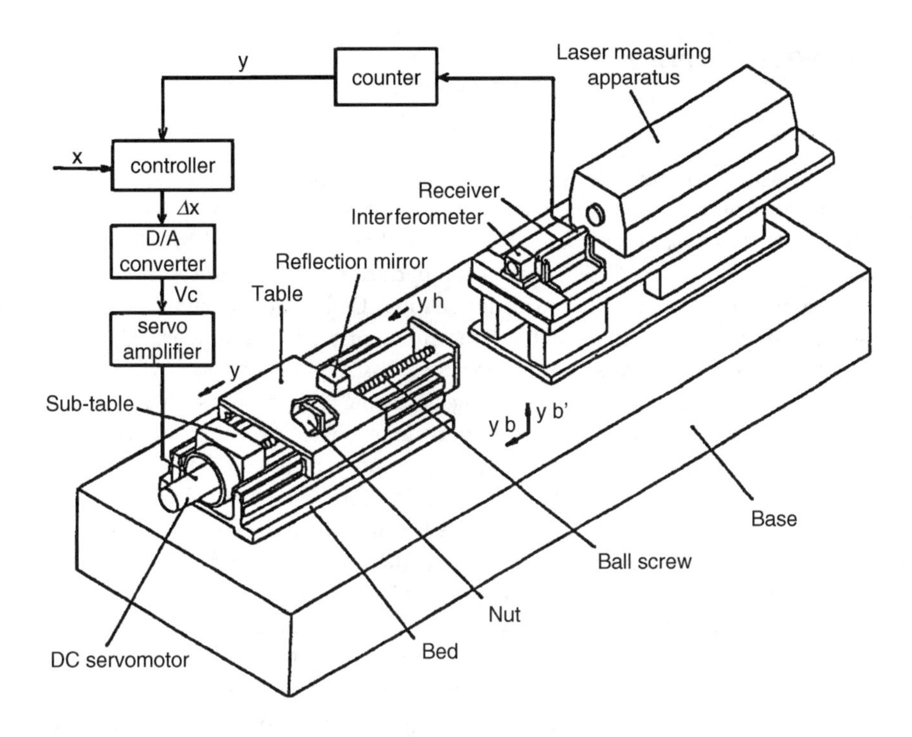

FIGURE 16.7 Experimental positioning device.

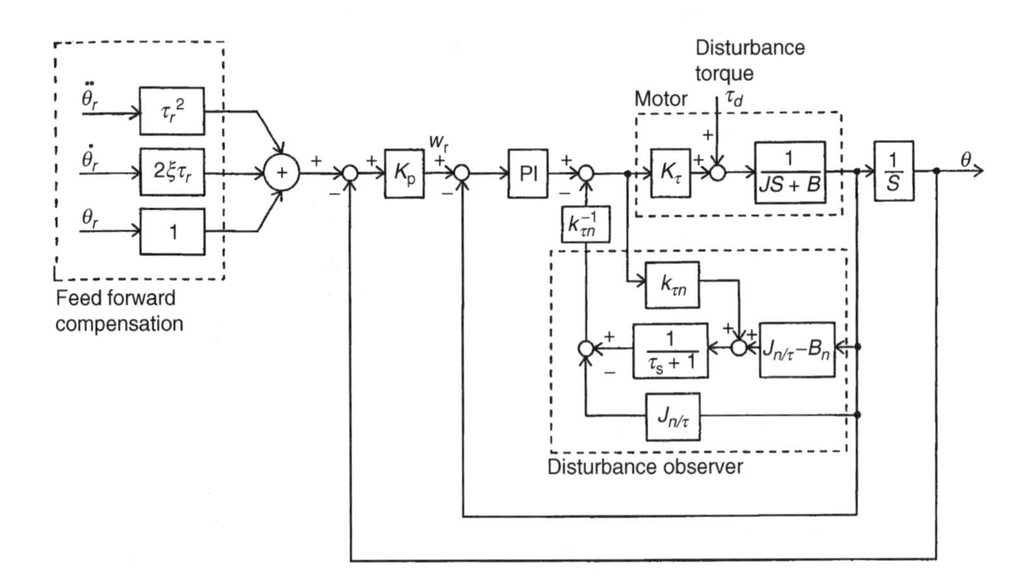

FIGURE 16.8 Servo control system with feed forward compensation and disturbance observer.

By using signals on the motor angular velocity and motor current, the observer estimates the applied load torque. The estimated torque is then subtracted from the motor current reference to achieve load-torque compensation. Compensation by the disturbance observer makes it possible to achieve fine control while the system remains insensitive to unavoidable load changes present in the precise servo control system with nanometer positioning resolution.

16.4 Sensor Fusion

16.4.1 Outline

The metal-cutting process, for example, is a very complicated phenomenon, and the use of single sensor is, therefore, not sufficient for monitoring machining states. There are many kinds of phenomena to be measured, as discussed in Section 16.2, and ideally they should be detected by using a single sensor. However, it is impossible because of the nature of complicated machining process itself, so the problem is to identify the best or most suitable sensor for sensing. Under these circumstances the integrated sensing system based on a sensor fusion will be a promising measuring technique.

Figure 16.9 shows a concept of the sensor fusion system based on multiple sensors in connection with decision-making. This is sometimes called an "intelligent sensor." Object states such as profile, dimension, tilt, etc. are detected by several sensors, and their outputs are adequately processed for better interpretation of object states. Signals obtained from sensors are noise corrupted and basically show stochastic behaviors. Therefore, decision-making using a neural network, GMDH (group method of data handling), Kalman filter, etc. plays an important role in determining or estimating the phenomena. To this end, the development of hardware (sensing technique) and software (data processing and decision-making) should be made simultaneously not only in industry but also in the laboratory.

Figure 16.10 shows the classification of sensor-integration techniques. In the pattern-recognition method, the cutting tool condition, for example, is divided into different classes: sharp, small wear, normal wear, severe wear. Then, one may try to monitor the tool condition by processing the signals from sensors using one of the pattern-recognition methods. Usually by one set of experimental data obtained from different classes some kind of pattern recognition method is trained, and then the trained scheme is used to predict the actual class of tool condition.

In the modeling approach, on the other hand, tool wear is related to signals from sensors. However, the establishment of this relationship using a theoretical method is very difficult; therefore, a new modeling

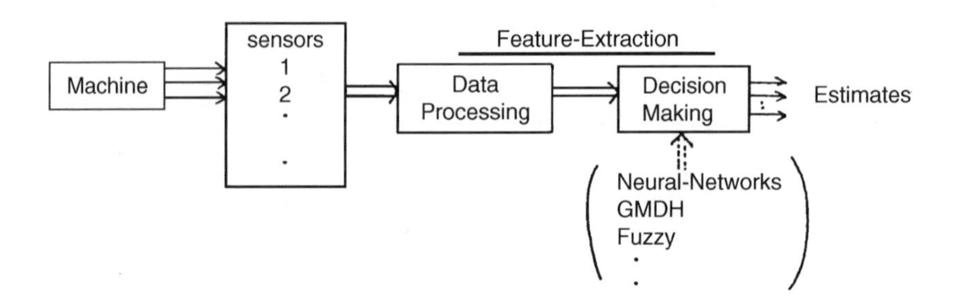

FIGURE 16.9 Structure of sensor fusion system.

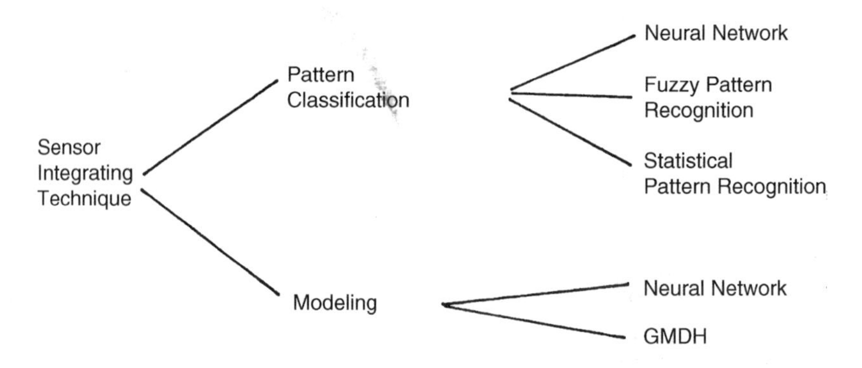

FIGURE 16.10 Sensor integration technique.

technique such as a GMDH or a neural network is used. Like the first approach, these schemes can be trained by the experimental data. Then tool wear conditions are obtainable by processing sensor signals using the trained scheme.

16.4.2 Data-Processing Technique

The problem of a sensor-fusion technique is that it is not clear what kind of data processing is required for the extraction of necessary information from the sensor. Some examples are the raw signal itself, average signal, moving average, variance, covariance, skewness, FFT processed signal, etc. There is no general solution for this selection. However, the fundamental way of thinking about this may be as follows. In practical terms, in applying the sensor fusion technique the representation of signals in discrete form requires a large number of sampled data points in the time domain. Each sampled waveform, considered as a pattern vector X, is of the form:

$$X^T = [x_1, x_2, \ldots, x_n] \tag{16.9}$$

and contains information that is, of course, characteristic of the mechanism by which it is generated. In general a feature selector is used to extract the most useful information from the original signal by mapping a pattern vector into a feature space in the frequency domain by using the FFT. Each of the resulting power spectral components is then defined as $(k=1, 2, \ldots, m)$ in a vector of given dimension, m, Y with each spectral component y_k being a feature.

$$Y = [y_1, y_2, \ldots, y_m] \tag{16.10}$$

When properly chosen, an effective set of features tends to assemble patterns from different sources into well-separated clusters in the feature space. To reduce the amount of data the class mean scatter criterion is used for feature selection. The feature mean for each class $\overline{Y}_i$ is determined as

$$\overline{Y}_i = \frac{1}{m} \sum_{K=1}^{m_i} Y_{ik} \tag{16.11}$$

where m_i is the number of patterns in class C_i. The overall system mean $\overline{Y}$ is then given by

$$\overline{Y} = \sum_{i=1}^{C} P_i \overline{Y}_i \tag{16.12}$$

where P_i = *a priori* probability of membership in class C_i, and C = the number of classes. Then, the scatter within each class is obtained by calculating the frequency domain covariance matrix as

$$R_i = \frac{1}{m_i} \sum_{K=1}^{m_j} (Y_{ik} - \overline{Y}_i)(Y_{ik} - \overline{Y}_i)^T \tag{16.13}$$

giving an overall system covariance matrix of

$$R = \sum_{i=1}^{C} P_i R_i \tag{16.14}$$

The scatter between the individual classes is defined as

$$R_c = \sum_{K=1}^{C} P_i (\overline{Y}_{iK} - \overline{Y}_i)(\overline{Y}_{iK} - \overline{Y}_i)^T \tag{16.15}$$

From which the feature selection criterion is defined as

$$Q = \frac{R_c(j,j)}{R(j,j)} \tag{16.16}$$

where $R_c(j,j)$ and $R(j,j)$ are the jth diagonal elements of the covariance matrices R_c and R, respectively. This mathematical treatment is implemented in a computer.

Because the objective is to minimize scatter between classes, the desired number of features with a maximum of Q values are selected as features. Whatever kind of output signal is derived by data processing, the aforementioned procedure is the fundamental approach to data processing of feature selection and is applied in the decision-making. The number of features used is very important because in addition to reducing the computational efficiency, a higher dimensionality requires more experimental data for training. The number of training sets required is four or more times the number of features.

16.4.3 Decision-Making

There are several kinds of decision-making, and representative techniques are briefly introduced below.

16.4.3.1 Neural Network

A neural network can be thought of as a collection of interconnected parallel processing elements, in which knowledge possessed by the network is represented by the strength of interconnections between processors. The processing element shown in Figure 16.11(a) may have any number of input paths but only one output as interconnection links. The input signals are modified by the interconnection weights W_{ji} and combined to form a single result by their summation. The combined input, s, is then modified by an activation function $f(s)$. This function can be as simple as a threshold function, in which an output is produced only if the combined inputs exceed a given level, or nonlinear continuous function such as the sigmoid or hyperbolic tangent, which generates an output Y_j. The output signal (or signals) may either become the inputs to other processing elements or sent to an external source or sources for interpretation. A neural network consists of several processing elements joined together in the manner described above. The network architecture generally resembles layers of processing elements with full or random connections between successive layers, as demonstrated in Figure 16.11(b). The first layer is an input buffer where the data are presented to the network. The last layer is an output buffer, which holds the response of the network. Layers between the input and output buffers are hidden layers.

There are two distinct phases in the operation of a neural network: learning and recall. In learning a back-propagation algorithm is used by presenting a known set of data to the network. The network calculates an output response using the current set of weights and thresholds. The output signal is then compared with the desired output signal, and an error between them is computed. The error is propagated

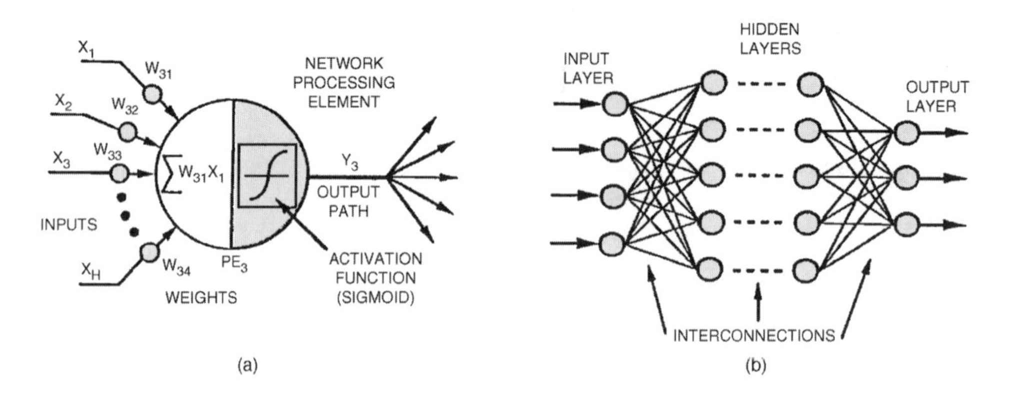

FIGURE 16.11 Structure of a neural network: (a) processing element; (b) structure.

back through the network to adjust the weights and thresholds in order to minimize this error. This process of feeding forward and propagating the error back is repeated for every signal in the training set until the network converges and responds with the desired signal. The training sets include input signal and the desired output signal. The recall process starts by presenting the input layer with input signals that are broadcast to the hidden-layer processors through the connection weights W_{ji}. Signals are multiplied by the weights and summed by the hidden-layer processor. A threshold is also included in this summation. The summed input is passed through the activation function $f(s)$ to yield an output signal Y_j, which propagates through the weights W_{ji} to the output layers. There, each processor receives the weighted output of every element in the hidden layer as before. For a network of multiple hidden layers the operation will be repeated layer by layer until it produces a set of outputs at the output layer.

The neural network may be the best way at present to integrate information from multiple sensors for recognition or monitoring of object states. The advantages of using this technique are the superior learning and noise-suppression abilities as well as its high flexibility.

16.4.3.2 GMDH

The GMDH algorithm is an identification and forecasting method based on the principle of a heuristic self-organization structure. It was developed to solve complex systems that are difficult to formulate when the data sequence is very short or the relationships between data have characteristics of multicollinearity. Figure 16.12 shows the structure of the GMDH. This technique models the input–output relationship of a complex system using a multilayered perceptron-type network structure. Each element in the network implements a nonlinear function of its inputs. The function, i.e., algorithm, implemented is usually a second-order polynomial of the inputs. Because each element generally accepts two inputs, this function implemented by an element in one of the layers is:

$$y = a_0 + a_1 x_i + a_2 x_j + a_3 x_i x_j + a_4 x_i^2 + a_5 x_j^2 \tag{16.17}$$

where x_i and x_j denote input variables such as a cutting force and sensor outputs. As shown in Figure 16.12, it can be seen that a number of self-selection thresholds are used. Their purpose is to filter out, at each layer, those elements that are least useful for predicting the correct output Y. Only those elements whose performance indices exceed the threshold at that layer are allowed to pass to the next layer.

The design procedure is briefly described as follows:

1. By taking the correlation between output y and each input variable x, the inputs that provide large correlation coefficients are chosen as a set of "useful" data.
2. Divide the time series data into training and checking data. For example, the training data are provided with large variances and the checking data with small variances.
3. For the two selected inputs, x_i and x_j, the intermediate variable Z_k is given by

$$Z_k = a_0 + a_1 x_i + a_2 x_j + a_3 x_i^2 + a_4 x_j^2 + a_5 x_i x_j \tag{16.18}$$

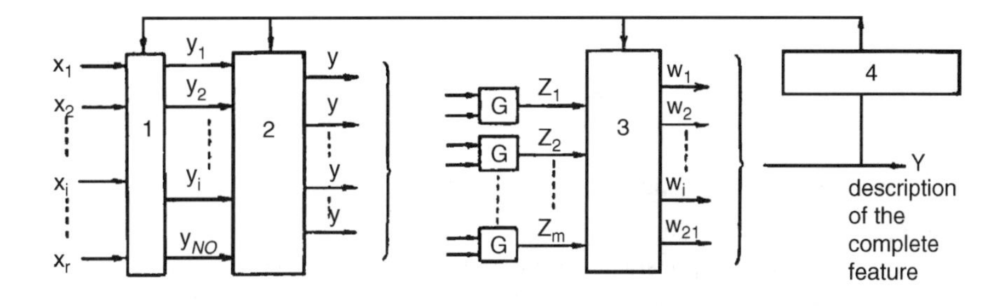

FIGURE 16.12 Structure of GMDH.

where the coefficients a_r ($r = 0, 1, \ldots, 5$) are determined by the training sequence, applying the least square method to minimize the error

$$e_k^2 = (y - Z_k)^2 \tag{16.19}$$

4. By using the coefficients obtained from procedure (3) and transforming the checking data by Eq. (16.18), the error criteria of Eq. (16.19) are evaluated in the same manner. Then, the intermediate variables are chosen in order of the small errors produced. Others are neglected.
5. Finally, a complete description of y is accomplished by repeating these procedures.

There is no specific rule for the selection of checking criteria. The great advantage of the GMDH is that it may show the best method for obtaining the polynomial description of a stochastic system from a small amount of experimental data.

16.4.3.3 Kalman Filter

The Kalman filter is one of the powerful techniques in the decision-making process.

Let a system be described by the difference equation of state

$$X(k+1) = P(k)\, X(k) + W(k) \tag{16.20}$$

and output equation

$$Y(k) = M(k)\, X(k) + U(k) \tag{16.21}$$

where $W(k)$ and $U(k)$ are zero-mean Gaussian noises, independent of each other, i.e., uncorrelated, and the expected values of the products are zero for all integers except for $k = j$:

$$E\,[W(k)\, W(j)] = 0, \quad E\,[U(k)\, U(j)] = 0, \quad \text{for all } k \neq j$$

and

$$E\,[W(k)^2] = \sigma_w^2,\ E\,[U(k)^2] = \sigma_u^2,$$

where σ_w^2 and σ_u^2 are rms values. The problem is to determine the best estimate $\hat{X}(k/k)$ of $X(k)$ in order to minimize the performance

$$J = [\{X(k) - \hat{X}(k/j)\}^{\mathrm{T}}\, \{X(k) - \hat{X}(k/j)\}] \tag{16.22}$$

where T is the transpose.

The best estimate is given by the following Kalman filter:

$$\hat{X}(k/k) = P(k-1)\,\hat{X}(k-1)/(k-1) + A(k)\,[Y(k) - M(k)\,P(k-1)\,\hat{X}(k-1)/(k-1)] \tag{16.23}$$

where $A(k) = S(k)\, M^{\mathrm{T}}(k)\, [M(k)\, S(k)\, M^{\mathrm{T}}(k) + R(k)]^{-1}$

$$C(k) = S(k) - A(k)\, M(k)\, S(k)$$

$$S(k) = P(k-1)\, C(k-1)\, P^{\mathrm{T}}(k-1) + Q(k-1)$$

Equation (16.23) implies a recursive relationship that yields a kth estimate $X(k)$ by using the $(k-1)$th parameters and estimate and kth observation $Y(k)$. Figure 16.13 represents a structure of the Kalman filter that is composed of a state transition system, an observation system, and an estimator.

16.4.3.4 Example of Sensor Fusion

16.4.3.4.1 Statement of the Problem

A multisensor framework may be preferable for autonomous robots to perceive and recognize the scene in a noisy environment. The redundancy of multiple observations stabilizes estimation of the object state. In sensor fusion the important issues are the formulation of constraints between observations and state

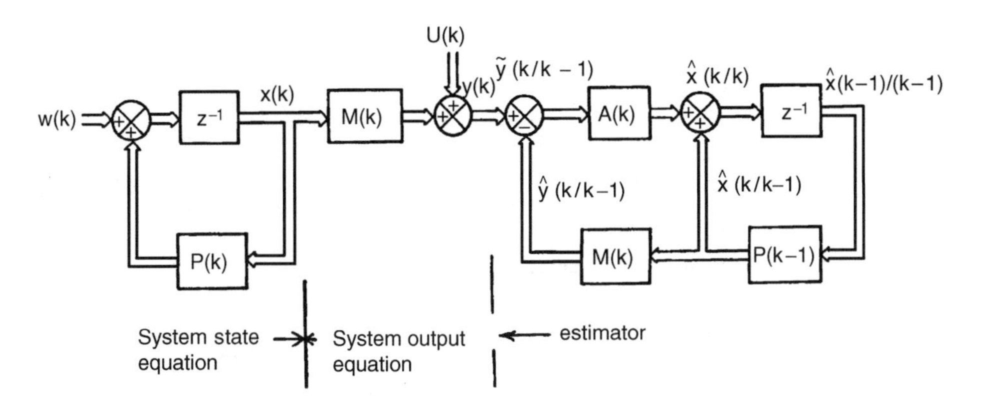

FIGURE 16.13 Structure of Kalman filter.

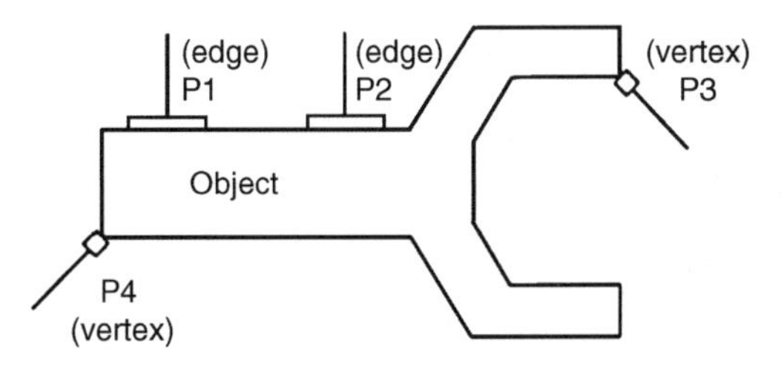

FIGURE 16.14 Observation of object with multiple sensor.

parameters, the consistency of integration, and the implementation of the algorithm to a real-time system. The problem considered is a model-based object tracking with sensors that contain measurement uncertainties [Kawashima et al., 1992]. In a model description a set of primitive names, such as edges, vertices, or planar surfaces, are recorded with their geometric feature parameters. An object model may be articulated, and this articulation is a sort of uncertainty. Uncertainties due to an object model are twofold: the unconstrained parameters of a shape, such as a joint, and the intrinsic size errors. Local features are observed by feature-tracking sensors with known error. The state of the target is then estimated by multiple sensors that observe the local features of an object. Each sensor autonomously tracks a local feature of a moving object.

Under these assumptions, the problem is to estimate the motion of the target from observations and object models.

16.4.3.4.2 Local Feature Sensor and Model Description

Sensor data contain the category of observed features of a primitive, the value of a feature parameter, and its variance-covariance matrix, which describes the uncertainty of the sensor data. Figure 16.14 shows optical sensors that observe the primitives of a two-dimensional shape; observation categories are edge or vertex. The parameters required for representing a local feature depend on the class of the primitive. A vertex-tracking sensor reports its position, and an edge tracker reports both the position of observation and the orientation.

A geometric model of an object is defined as a set of primitives, such as edges and vertexes, in the model database. If an object is rigid, any primitive must be strictly constrained by other primitives geometrically. In Figure 16.14, for example, the distance between edge P1–P2 and vertex P4 must be invariant for rotation or translation. Such geometric constraints can be formulated as equations of geometric parameters and used for position estimations. If the object has uncertainties, the constraint equations may be articulation and fluctuation. An articulated object composed of linked subparts (Figure 16.15) can be treated as partially constrained parameters. The degree of freedom corresponds to the articulation

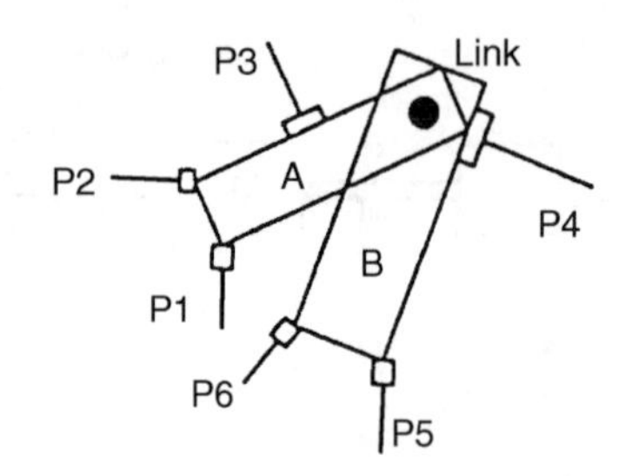

FIGURE 16.15 An object composed of linked subparts.

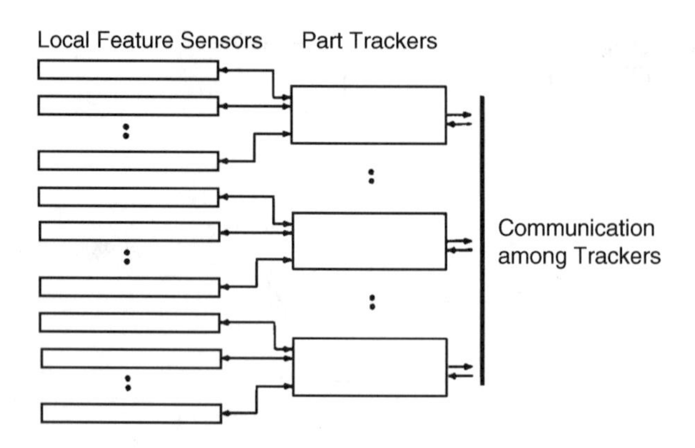

FIGURE 16.16 Architecture of the tracking system.

of the object. Fluctuation of parameters is defined with random variables. That is, the constraints are imposed on the mean values of observation.

16.4.3.4.4 Model-Based Tracking of an Articulated Object:

Part trackers collect data from local feature sensors and estimate each part state with a Kalman filter. A part tracker tracks the motion of a subject of the object. Each estimation is sent to other trackers to integrate part information through articulation constraints for maintaining consistency among subparts. The integrated results are then sent back to part trackers to update part information and used to estimate the state at the next observation. Observations are also carried out by local feature sensors with visual windows. In a Kalman filter, the state of a subpart is updated by the difference between expectation and observation.

The decentralized architecture system, which consists of processor nodes for multisensor data fusion, is used for integration in the tracking system. Each processor transmits only the difference between estimates and observations. Because the constraints between parameters are nonlinear, computation costs for transforming estimates are considerable, which necessitates implementing the decentralization at the part tracking level, as shown in Figure 16.16.

16.4.3.4.5 Implementation of the Multisensor Tracking

Local features of an object are observed by a multiwindow local-feature tracker, and each window reports the geometric parameters of an object (Figure 16.17(a)) to the part tracker. The windows track target features autonomously. Initially, each window is wandering in the scene to find an interesting local feature. Once a feature is found, however, the window is glued to it and tracks the motion. The window updates the position by itself so that the feature can be observed at the center of the window. The position can also be controlled by the part tracker that supervises the windows. Once the system enters the tracking mode, the window obeys the supervisor.

With several local feature reports the part tracker determines the state of the subpart (Figure 16.17(b)). A part tracker estimates the motion of the subpart by use of a Kalman filter. The static state of each subpart is described by its position, angle, and velocity. After the Kalman filter is implemented, the result

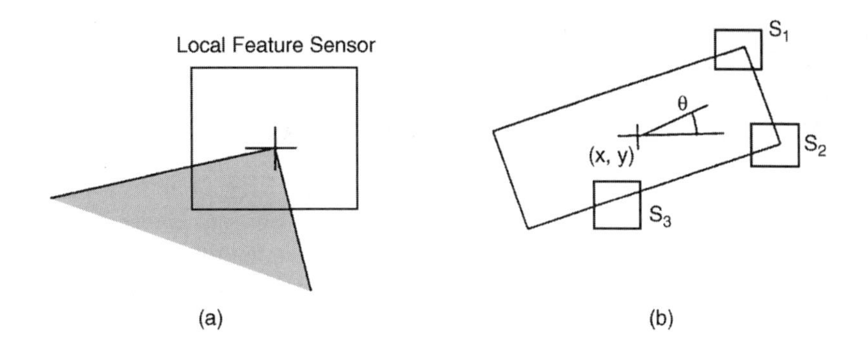

FIGURE 16.17 Local feature sensor (a) and part tracking by multiple local feature sensors (b).

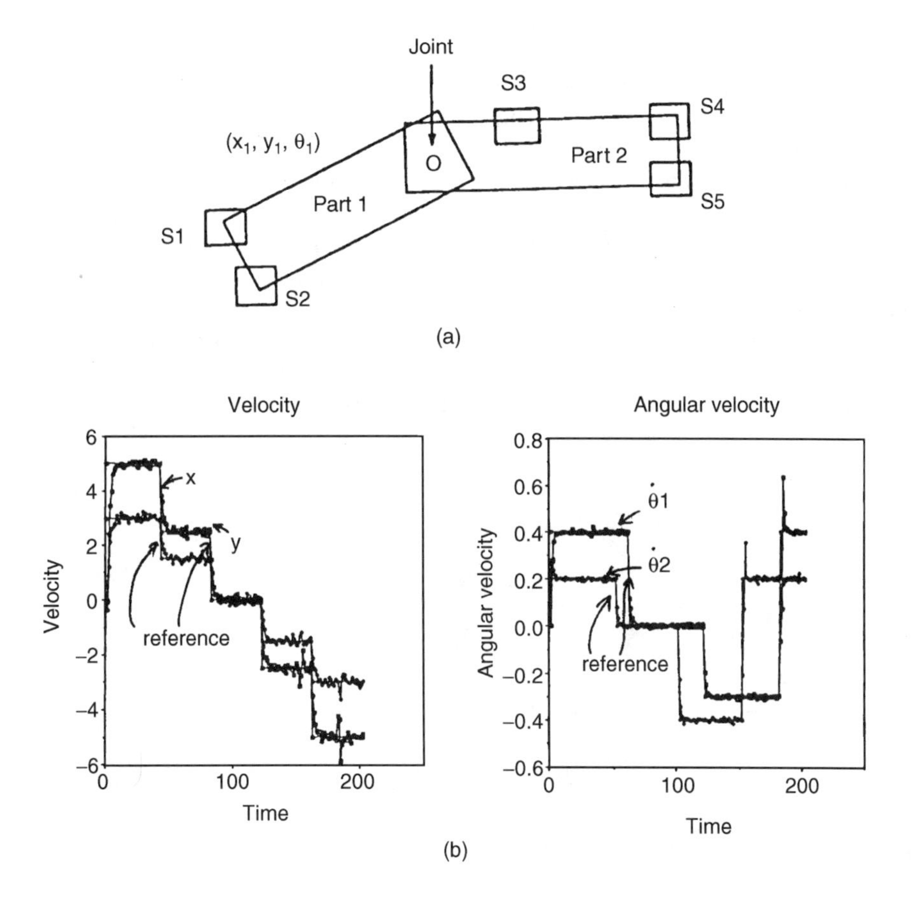

FIGURE 16.18 Tracking results: (a) target object; (b) results of tracking simulation.

is sent to other trackers to integrate results using the constraints of articulation. When the integrated data are received, the part estimate is updated to the new one. Part trackers communicate with each other to collect part estimates from other trackers and to refine the state.

16.4.3.4.6 Results

Figure 16.18(a) shows the test target object composed of subparts linked by each other. The sensor data are integrated for each rigid subpart, and then subparts are linked at the joint. Figure 16.18(b) demonstrates a simulated result of tracking the motion of an object that moves by changing its velocity and angular velocity. As shown in the figure, the estimation of positions and angles traces the accurate values exactly.

16.5 Outline of Optical System Construction

The basic construction of an opto-mechatronic system is illustrated in Figure 16.19. It generally consists of a light source, optics for light projection, receiving optics, a sensor, a signal processing system, a CPU (for data operation, for control, etc.), and a monitoring device or control system including actuator, mechanism, control devices, etc. These components are adequately combined, depending on the overall system operation. A representative example is a CD player [Morimoto et al., 1996].

16.5.1 CD Player

Automatic aperture control and camera shake stabilization are features that can often be found in autofocus and video cameras. Automatic aperture control is used to adjust the brightness of an image by operating a motor to mechanically change the aperture size. Camera shake stabilization is used to cancel out shakiness in a video image and is achieved either mechanically by moving a prism or electronically by shifting the video signal. In such cameras exposure control and shake stabilization are performed by continuously optimizing physical quantities, e.g., voltages, in the electronic circuits that modify settings such as the aperture, focus, and shutter speed. *Mechatronics* is the term used to describe mechanical systems that incorporate electronic elements for control purposes, and when such systems incorporate optical components, especially for the purpose of taking measurements, they can be referred to as opto-mechatronic systems. Control plays a crucial role in opto-mechatronics, and the development of new control devices that exploit the properties of light is already under way, as summarized in Figure 16.20. Such developments are resulting in control devices that surpass the capabilities of conventional electronic equipment in various ways. For example, they are capable of providing greater quantities of data for status acquisition, are more compact and lightweight, and are capable of providing a contact-free source of energy.

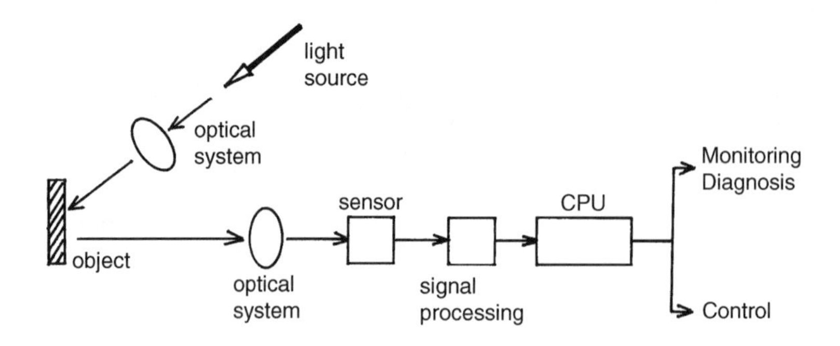

FIGURE 16.19 Basic construction of an opto-mechatronic system.

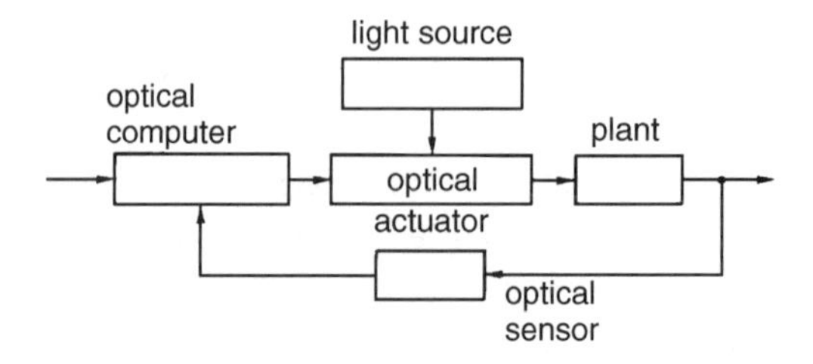

FIGURE 16.20 Construction of an optic-control system.

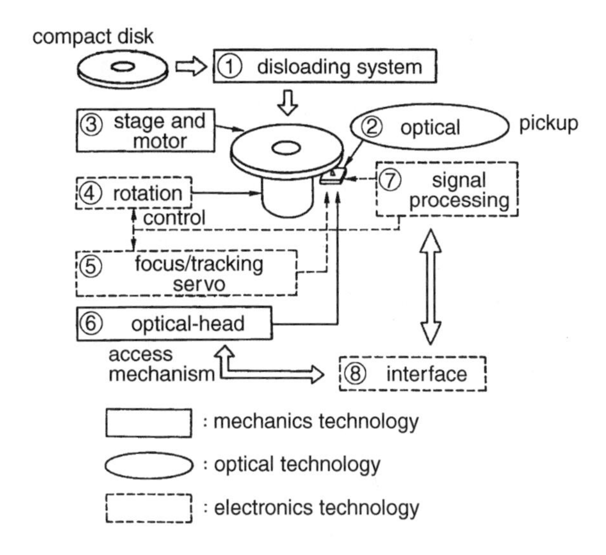

FIGURE 16.21 Compact disk player configuration.

A CD player is a typical example of a system in which opto-mechatronics plays a key role. An ordinary compact disk is capable of storing very large quantities of data, e.g., music or video data, in the form of digital signals that are recorded in a spiral-shaped line with a spacing of 1.6 μm. It is essential to follow this line correctly when reading data back from the disk. This is done by using an optical sensor to measure the position of the line and generating an electronic signal corresponding to the amount of offset at the current position. When the sensor is moved to correct this offset, the sensor can be maintained at the correct position, and the data stored on the disk can be accurately read back. In this way the system as a whole is made to operate optimally as a result of close cooperation between its constituent mechanical, electronic, and optical components. CD players are discussed in further detail below.

16.5.1.1 CD Player Configuration

Figure 16.21 shows the configuration of a CD player. It can be broadly divided into various modules: (1) a disk-loading mechanism, (2) an optical pickup, (3) a rotary stage, which includes a servo motor, (4) a rotation control system, (5) a focus/tracking servo system, (6) an optical head access mechanism, (7) signal processing circuitry, and (8) interfaces. Figure 16.21 illustrates how optical, mechanical, and electronic technologies are employed in these modules. For example, the optical pickup is an optical module, whereas the disk-loading mechanism, the rotary stage, and the optical head access mechanism are mechanical, and the focus/tracking servo system and signal processing circuitry are electronic. These modules are described in detail below.

16.5.1.1.1 Disk-Loading Mechanism

A compact disk is a thin disk (called an optical disk) with a diameter of 12 cm and a hole in the middle. When a disk is placed on the player's loading tray, the disk-loading mechanism slides the tray into the player and mounts the disk on the rotary stage.

16.5.1.1.2 Optical Pickup

The optical pickup lies at the heart of a CD player and corresponds to the needle in a record player. It uses a narrow beam of laser light to scan in the data written on the compact disk. The structure of an optical pickup is shown in Figure 16.22. It basically consists of a semiconductor laser, a beam splitter, a collimator lens, an objective lens, a condenser lens, and a photodetector. For improved performance it may also use components such as a diffraction grating, a quarter-wavelength plate, and a four-part photodetector. The principle of signal detection in a compact disk player involves the use of light scattering phenomena. As shown in Figure 16.22, two lenses are used to focus the light from a laser into a narrow

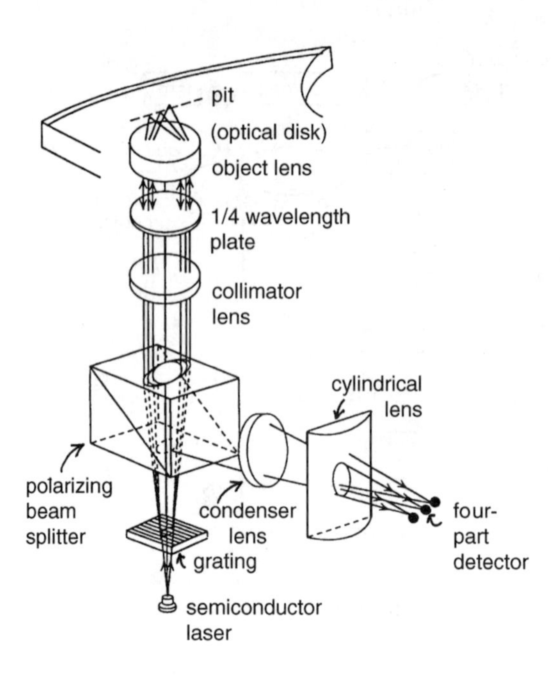

FIGURE 16.22 Optical pickup.

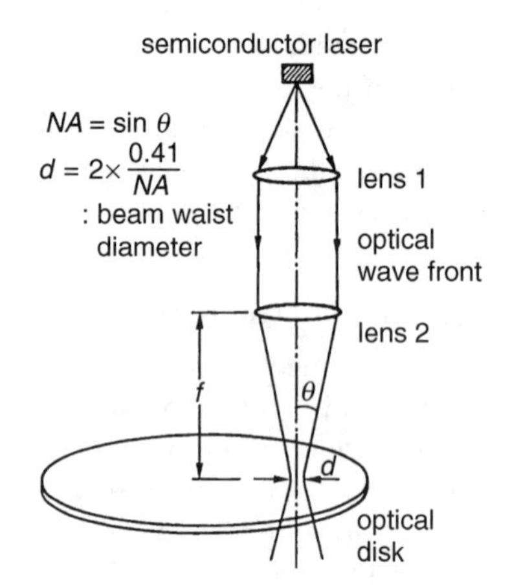

FIGURE 16.23 Collimate of semiconductor laser.

beam of laser light. Although the beam of light emitted from a semiconductor laser is only about 2 μm across, it spreads out to a width of some 300 μm after traveling just 1 mm. As Figure 16.23 shows, by placing a lens (1) at a certain fixed distance from the laser, it can be turned into a light beam of fixed width. When this fixed-width beam travels through the next lens (2), it can be focused to a sharp point about one wavelength across. This corresponds to a width of about 1 μm for the semiconductor lasers used in optical pickups. Figure 16.23 shows a formula for the beam waist diameter, which expresses how narrowly a beam of light can be collimated.

This sharply focused beam of light is aimed at the surface of the disk. If the disk surface has a mirror finish, then the light will be reflected straight back. However, data are recorded on the disk in the form of small indentations (called pits), which are 4 μm wide and range in length from 0.6 to 3 μm. When the light beam is directed at one of these pits, it is scattered, and the amount of light reflected straight back decreases (Figure 16.24). This is illustrated in Figure 16.25, where (a) shows the light being reflected straight back from a flat part of the disk, and (b) shows the light being scattered by a pit. These two states can be distinguished according to the amount of light that is reflected straight back, and it is thereby

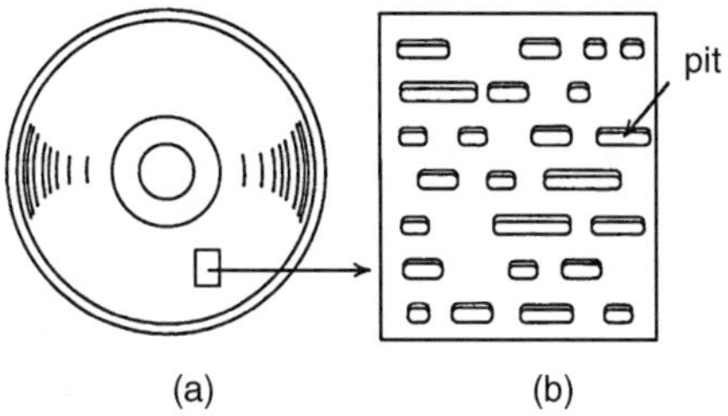

FIGURE 16.24 Compact disk and pits on its surface: (a) compact disk; (b) close-up view of disk surface.

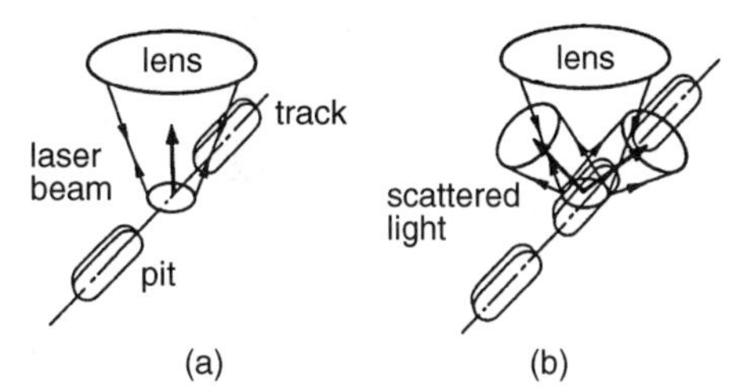

FIGURE 16.25 Reflection from the disk surface: (a) out of pit; (b) on pit.

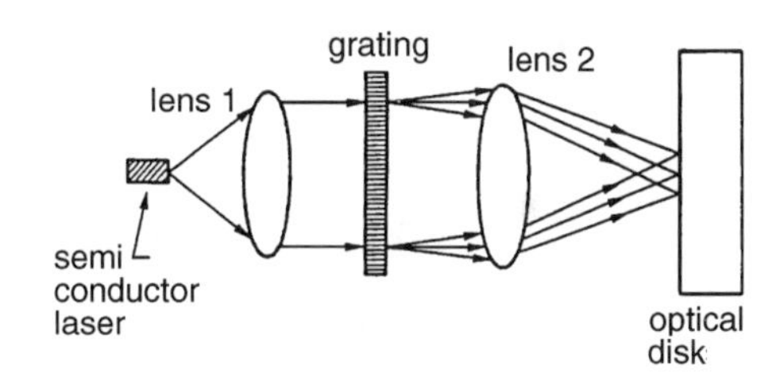

FIGURE 16.26 Three light beams through grating.

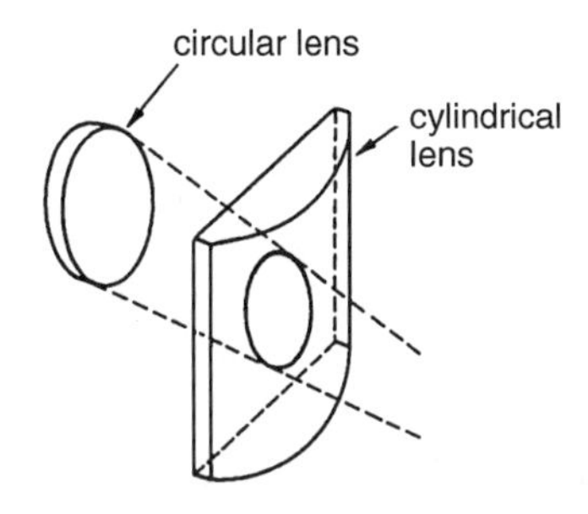

FIGURE 16.27 Circular and cylindrical lenses in a focus servo.

possible to read back the information stored on the disk. An optical pickup is based on this operating principle, but to facilitate accurate tracking of the sequence of pits written on the disk it also includes a diffraction grating and a four-part photodetector, as shown in Figure 16.22. The diffraction grating consists of a thin glass plate on which is inscribed a series of narrow parallel lines of about the same width as a wavelength of light. When a beam of light strikes this grating at right angles, three beams emerge from the other side, as shown in Figure 16.26. These three light beams have different intensities, with the central (zeroth order) beam being more intense than the two light beams on either side of it (the ±1st-order beams). The central beam is used to read out the information from the pits in the manner described above, while the other two beams are used for tracking.

16.5.1.1.3 Servo Mechanisms

Servo mechanisms are employed in CD players to facilitate precise operation. Typical examples are the focus servo that focuses the light beam onto the disk and the tracking servo that tracks the sequence of pits on the disk. The focus servo uses a lens system consisting of circular and cylindrical lenses (Figure 16.27) and a four-part photodetector. In the lens system a light beam that enters via the circular lens is synthesized into a beam that is circular at the focal point but elliptical elsewhere. This beam is directed toward the center of the four-part photodetector. At the focal point, the beam becomes circular, as shown

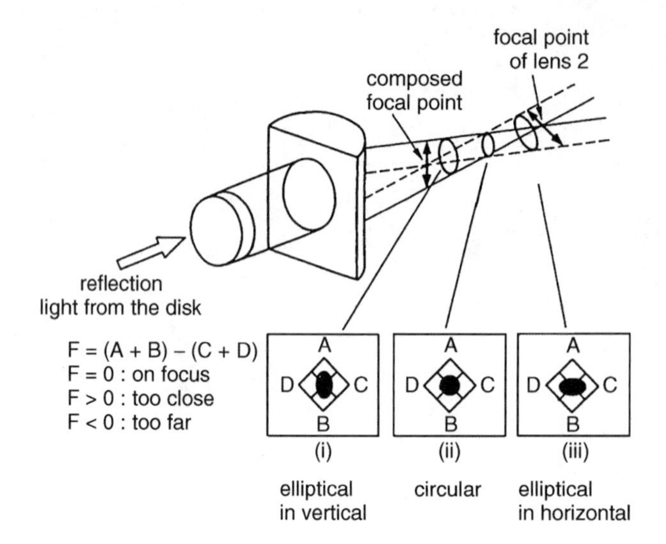

FIGURE 16.28 Signal detection for auto-focusing.

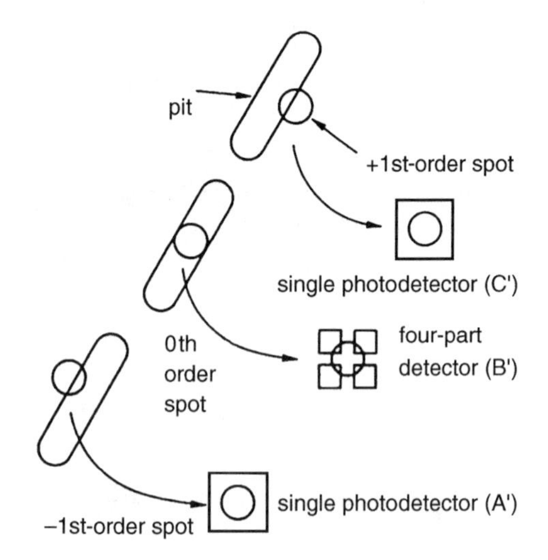

FIGURE 16.29 Three beam projections for tracking.

in Figure 16.28(ii), and an equal amount of light is incident on each part of the photodetector. However, when the light beam is elliptical, different amounts of light are incident on each photodetector. Suppose the four photodetector elements are called A, B, C, and D. When the detector is situated in front of the actual focal point position, the amount of light incident on elements A and B becomes greater than the amount of light incident on elements C and D (Figure 16.28(i)). The opposite situation arises when the detector is situated behind the actual focal point position (Figure 16.28(iii)). By comparing the respective outputs of the four photodetector elements A, B, C, and D, it is possible to detect the direction of any focusing discrepancy. Specifically, the sum of the outputs from elements C and D is subtracted from the sum of the outputs from elements A and B. When the beam is correctly focused, the result is zero. A positive result indicates that the pickup is too close to the disk, and a negative result indicates that it is too far away. This information can be used to automatically set the focal position of the optical pickup.

Tracking refers to the process whereby the spot of the optical beam is made to follow the series of pits in which the information is written. The pits are formed in a groove (or "track") formed circumferentially on the optical disk. The tracking process involves using the set of three light beams shown in Figure 16.26. As Figure 16.29 shows, the three beams are tilted slightly toward the track. The most intense zeroth-order beam is directed at the track on which the pits carrying the information to be read are situated,

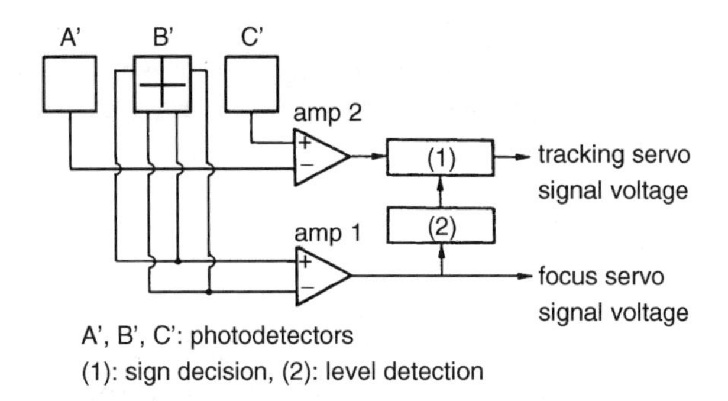

FIGURE 16.30 Electrical circuit for tracking.

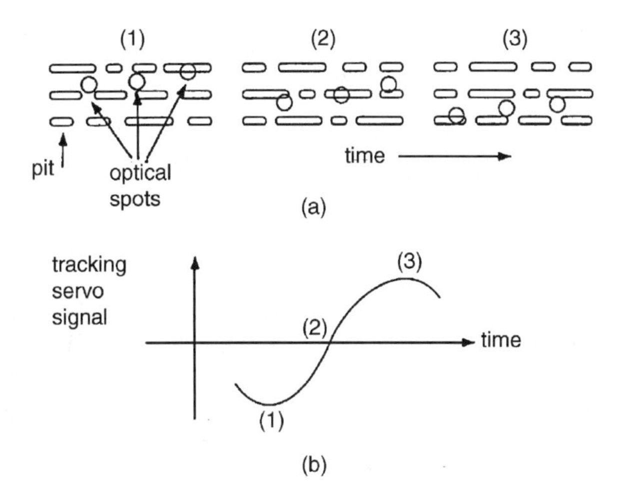

FIGURE 16.31 (a) Three beam positions and (b) tracking servo signal voltage vs. time.

while the two weaker first-order beams are offset to either side of the track. The reflected light originating from these three beams is received by three photodetectors (A′, B′, and C′). By comparing the outputs from these photodetectors, it is possible to detect whether or not the central beam is correctly positioned in the center of the track. The reflection of the zeroth-order light beam from the track on which the information is to be read is picked up by detector B′, and the reflections of the ±1st-order light beams on either side are picked up by detectors A′ and C′. Detector B′ is a four-part detector used for focusing, as shown in Figure 16.28. Figure 16.30 shows the electrical circuit used for detecting these signals. Figure 16.31(a) shows how the light beams appear when they are correctly positioned over the track and when they are offset. Figure 16.31(b) shows the tracking servo signal produced when the tracking position changes with time. In this figure, position (2) corresponds to the point where the difference between the outputs from detectors A′ and C′ is zero and the track is being read correctly.

16.5.1.1.4 Digitizing the Signals

This part explains how audio information is recorded on a compact disk. When an audio waveform is recorded, it can sometimes be affected by extraneous noise. This noise can be avoided by recording the audio signal as a sequence of digital quantities. The magnitude of a continuous signal is first sampled at regular time intervals and is then digitized by applying pulse amplitude modulation (PAM). In other words, it is subjected to pulse code modulation (PCM). To record music on an optical disk, the sound is pulse-code modulated and written as a sequence of pits. The sampling theorem is used to determine the size of the intervals at which the signal should be sampled. The range of frequencies audible to humans is roughly 20 Hz to 20 kHz, so the sampling frequency should be twice the necessary frequency range (sampling theorem). Accordingly, for audio data it is sufficient to sample at 40 kHz. In practice, a slightly higher frequency of 44.1 kHz is used.

16.5.1.1.5 Error-Correction Encoding

When recording on an optical disk the pulse-code-modulated signal is not written straight to the disk but is first subjected to a reordering process. This makes it easier for the optical pickup to track the data correctly and allows errors in the data to be corrected. When the sound level is small, performing pulse modulation would result in a signal with a continuous series of zeros. Any noise added to this would result in a completely different value. Furthermore, the sound will be affected if there are imperfections, such as scratches, on the disk. The signal is therefore subjected to error-correction encoding. This allows errors to be located and corrected, making the player resilient against scratches or dust on the surface of the disk. A laser disk is another form of optical disk on which pits are formed in a similar manner to record video signals. However, it differs from a compact disk in that the signals are recorded directly in analog form. This is because digitizing a video signal produces so much data that it would be impossible to record programs of appreciable length on a single disk. However, optical disks called DVDs (digital versatile disks) have been used that are the same size as a compact disk and on which video signals can be recorded by converting them into digital signals. These disks benefit greatly from signal-compression techniques that reduce the amount of data. Similar compression techniques are also being used for computer communication, especially in the field of multimedia.

16.5.1.1.6 Rotary Stage

A compact disk on which audio data have been recorded based on the above principles is mounted on the rotary stage by the disk loader. The rotary stage does not rotate at a fixed number of revolutions per minute like a record player, but instead rotates with a constant tangential velocity. As a result, the pits formed on the disk have the same constant pattern density over the entire surface. A motor servo is therefore used to make the disk rotate slower when reading data from pits at the outer edge, and faster when reading data from pits closer to the center of the disk.

16.5.1.1.7 Optical Head Access Mechanism

This is the mechanism that moves the optical pickup to the desired position. First, the optical pickup is moved at high speed to the position at which information is to be read out, and then the head is kept in a fixed position while a system of mirrors or lenses is adjusted to align the light beams with the desired position. A linear motor or pulse motor is used to move the head quickly to the target location. Fine adjustment is then performed, either by rotating a mirror attached to the head or by moving a lens slightly while reading an address written on the optical disk.

16.5.1.1.8 Interfaces

These are electronic circuits either for sending out the data read from an optical disk to external devices or for receiving information from the outside.

16.6 Conclusions

Optical-based in-process monitoring and control are key factors for real automation in industry. Achieving this goal implies developing high-quality monitoring and control techniques that ensure that the system outputs conform to target-value specifications in an online fashion. In this chapter the roles of in-process monitoring and control, sensor-selection, control-technique, sensor-fusion, and optical-system construction for in-process monitoring and control were comprehensively described through examples and theoretical approaches. Although these concepts have several problems for successful implementation, an optical-based approach is a promising technique for current and future technologies. To ensure the reliability of sensing and monitoring a sensor fusion seems to be one of the effective approaches for in-process technique.

Defining Terms

GMDH (group method of data handling): Identification and forecasting algorithm based on heuristic self-organization structures.

multi (multiple) sensor: Detection by using several sensors instead of a single sensor.

performance index: Criteria to evaluate condition or state.

sensor fusion: Comprehensive and intelligent detection system using multiple sensors and decision-making.

sensory feedback: Sensor-based feedback control.

servo mechanism: Motor-controlled system to follow a desired value.

smart sensor: Integrated hardware/software sensing system with intelligence.

References

Cho, H. S., Neural network applications to manufacturing processes: monitoring and control, *Computational Intelligence in Manufacturing Handbook*, in Wang, J. and Kusiak, A., Eds., CRC Press, Boca Raton, FL, 2001, pp. 12-1 to 12-33.

Kawashima, T., Nagasaki, T., and Aoki, Y., Sensor fusion system for model-based object tracking, in *Proc. 2nd Int. Sym. Measurement Control Robotics*, Tsukuba, Japan, 1992, pp. 265–269.

Morimoto, Y. et al., *Introduction to Opto-Mechatronics*, Kyoritsu Shuppan, Japan, 1996.

Shiraishi, M., Sensing and Control in Production and Manufacturing, private lecture notes at Ibaraki University, 1994–1998, pp. 1–230.

Yaskawa Electric Co., *Introduction to Servo Techniques for Mechatronics*, Nikkan Kogyo Shimbun, Japan, 1997.

V

Opto-Mechatronic Processes and Systems

17

Optical Methods for Monitoring, Modeling, and Controlling Semiconductor Manufacturing Processes

Gary S. May
Georgia Institute of Technology
Atlanta, Georgia

17.1 Introduction

New knowledge and tools are constantly expanding the range of applications for semiconductor devices and integrated circuits. The solid-state computing, telecommunications, aerospace, automotive, and consumer electronics industries all rely heavily on the quality of these methods and processes. In each

0-8493-1162-4/03/$0.00+$1.50

of these industries dramatic changes are underway. In computers, for example, the cost per unit of computing has fallen from $100,000/MIPS (million instructions per second) in mainframes to less than $10/MIPS for PCs and some workstations. In addition to increased performance, next-generation computing is increasingly being performed by portable, hand-held computers. A similar trend exists in telecommunications, where the user will soon be employing high-performance, multifunctional, portable units. In the consumer industry, multimedia products capable of voice, image, video, text, and other functions are also expected to be commonplace within the next decade.

The common thread that is pervasive in each of these trends is low-cost electronics, which represents a $300B market and is perhaps the single most important strategic technology in the world today. This multi-billion-dollar electronics industry is fundamentally dependent on the manufacture of semiconductor integrated circuits (ICs). However, the fabrication of ICs is extremely expensive. In fact, the last couple of decades have seen semiconductor manufacturing become so capital-intensive that only a few very large companies can participate. A typical state-of-the-art high-volume manufacturing facility today costs over $1 billion [Dax, 1996]. Because of rising costs, the challenge before semiconductor manufacturers is to offset large capital investment with a greater amount of automation and technological innovation in the fabrication process. In other words, the objective is to use the latest developments in computer hardware and software technology to enhance the manufacturing methods that have become so expensive. In effect, this effort in *computer-integrated manufacturing of integrated circuits* (IC-CIM) is aimed at optimizing the cost-effectiveness of integrated circuit manufacturing just as *computer-aided design* has dramatically affected the economics of circuit design.

Under the overall heading of reducing manufacturing costs, several important subtasks have been identified. These include increasing chip fabrication yield, reducing product cycle time, maintaining consistent levels of product quality and performance, and improving the reliability of processing equipment. Because of the large number of steps involved, implementing these procedures in an IC manufacturing facility requires strict control of thousands of process variables. The interdependent issues of high yield, high quality, and low cycle time have been addressed by the ongoing development of several critical capabilities in state-of-the-art IC-CIM systems: *in situ* process monitoring, process/equipment modeling, real-time closed-loop process control, and equipment malfunction diagnosis. Each of these activities increases throughput and reduces yield loss by preventing potential misprocessing, but each presents significant engineering challenges in effective implementation and deployment.

17.2 The Role of Optical Methods

In order for the unit processes used in fabricating an IC to repeatably yield reliable, high-quality products each unit process must be strictly controlled. Many diagnostic tools are used to maintain systematic control. Such control requires that the key output variables for each process step (i.e., those correlated with product functionality and performance) be carefully monitored.

Process monitoring enables operators and engineers to detect problems early on to minimize their impact. The economic benefit of effective monitoring systems increases with the complexity of the manufacturing process. Manufacturing line monitors consist of sophisticated metrology equipment that can be divided into tools that characterize the state of features on the semiconductor wafers themselves and those that describe the status of the fabrication equipment operating on those wafers. The issues involved in understanding and implementing both wafer-state and equipment-state measurements will be discussed in detail in this chapter.

Optical methods are extremely useful for process monitoring. Methods employing electron beams (such as electron diffraction techniques or electron microscopy) are equally useful. The primary advantage of these techniques is that they are almost always nondestructive and usually do not interfere with the process being monitored in any tangible way. Before discussing these techniques in greater detail, a brief description of the key unit processes to which optical measurement techniques are applied is warranted.

17.3 Key Processes

Monolithic ICs are fabricated through the repeated application of a number of basic processing steps in which layers of materials are deposited on semiconductor substrates (or "wafers"), doped with impurities, and patterned [Campbell, 2001]. Among these, the following processes are particularly amenable to optical measurement methods: oxidation, photolithography, reactive ion etching, ion implantation, molecular beam epitaxy, and rapid thermal processing. Because IC manufacturing requires a high degree of precision, this necessitates frequent and thorough process monitoring to assure high-quality final products. Depending on the process, optical measurements may be performed directly on product wafers after each step, or, alternatively, some measurements are performed *in situ* during a fabrication step.

17.3.1 Oxidation

Silicon dioxide (SiO_2) is a key component of many semiconductor technologies. SiO_2 is a high-quality electrical insulator formed by a chemical reaction of silicon and oxygen. Thermal oxidation of silicon is achieved by heating a silicon wafer to a high temperature (typically 900 to 1200°C) in an atmosphere containing pure oxygen or water vapor. This process is typically carried out in a high-temperature furnace tube, which is usually made of quartz, polysilicon, or silicon carbide, and fabricated to prevent contamination during oxidation.

Although mathematical models exist to predict oxide film thickness as a function of time, temperature, and reactant species [Campbell, 2001], these models are approximations and do not account for all observed phenomena or for nonuniformities in oxide thickness across a wafer surface. Therefore, oxide-thickness measurements are necessary for process characterization. Accurate and nondestructive thickness measurements can be achieved using a variety of optical methods, including interferometry and ellipsometry.

17.3.2 Photolithography

To produce integrated circuits thin films of various materials are used as barriers to the diffusion or implantation of impurity atoms, as insulators between conductive layers, or as conductive layers themselves. Patterns must be formed in these materials in order for the proper contact between films to be arranged. These patterns are formed and transferred through masks using photolithography. The basic steps in the photolithographic process include: wafer cleaning, deposition of a barrier layer, photoresist coating, soft baking, mask alignment, exposure, development, hard baking, etching, and photoresist stripping.

For a more thorough description of each of the steps involved in photolithography, the reader is encouraged to consult Campbell [2001]. In terms of the steps that are particularly relevant to the discussion of optical measurement methods, photoresist thickness is often measured optically after coating and prebaking. In addition, photoresist development can be characterized by optical dissolution-rate monitoring systems, such as the one described in Section 17.4.1.4 below. Finally, scanning electron microscopy (Section 17.4.1.8) is often used to characterize the critical dimension of a feature formed at the conclusion of the photolithography process.

17.3.3 Reactive Ion Etching

Reactive ion etching (RIE) is the removal of material by reactive gases in an electric field. RIE is a highly nonlinear, low-pressure form of plasma etching, which is a critical technology for current VLSI fabrication and also holds promise for future giga-scale integrated circuits [Meindl, 1993]. The quality of the etch may be characterized by its selectivity, anisotropy, uniformity, and etch rate. Selectivity defines relative material etching rates in the plasma. Anisotropy is the ability to etch vertically while minimizing horizontal etching. Uniformity implies a constant etch rate across the surface of the wafer and requires that the amount of material removed from the surface of the wafer be precisely monitored and controlled.

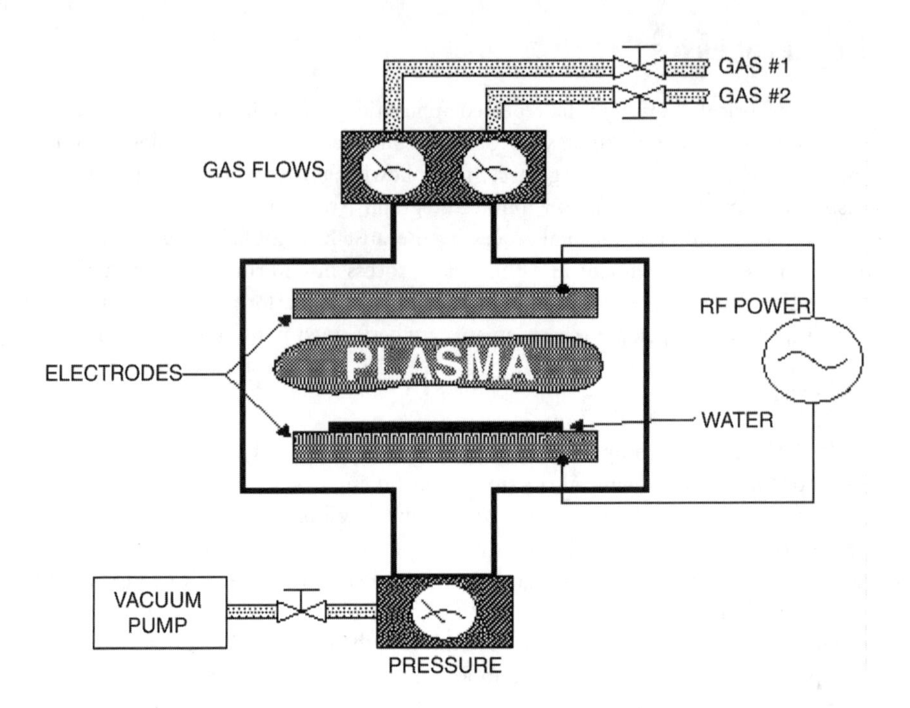

FIGURE 17.1 Typical reactive ion etching system.

The etching mechanism involved in RIE is both chemical and physical, and a synergy exists between those physical and chemical processes. As illustrated in Figure 17.1, applying RF power across two electrodes after a controlled flow of reactive gases is introduced into an evacuated chamber at low pressure generates a plasma. The chemical etching process has three phases. First, the molecular gases in the chamber, which may or may not be reactive, are dissociated in the plasma to create reactive species. This takes place through inelastic collisions between electrons excited by the RF power source and gas molecules. In the second phase, reactive species are absorbed and react with the exposed wafer to form volatile products. Finally, volatile products are desorbed and removed by the vacuum system. As an RF power source is applied, electrons, which have higher mobility than ions, are accelerated toward the electrode. This causes a DC self-bias voltage to form between the plasma and electrodes. Under this bias ions in the plasma are accelerated toward the wafer on the negative electrode. The resulting ion bombardment aids the etching process by breaking down stable molecules to facilitate surface reactions, enhancing absorption and desorption and removing material through the physical impact.

Real-time monitoring of RIE process conditions is essential for process control. Existing *in situ* monitoring techniques include optical emission spectroscopy (OES) and interferometry. OES is a bulk measure of the optical radiation of the plasma species (see Section 17.4.2.1). Because emissions can emanate from etch reactants as well as products, OES measurements are most often used to obtain the average optical intensity at a particular wavelength above the wafer. By setting an optical spectrometer to monitor the intensity at a wavelength associated with a particular reactant or by-product species, OES serves as an effective endpoint detector. There has also been progress in using OES to estimate *in situ* wafer conditions [White et al., 1997].

Laser interferometry is also frequently used for endpoint detection in RIE [Pope et al., 1993]. This technique gives direct etch rate measurements at the wafer surface by exploiting interference patterns of monochromatic light rays that occur when light is reflected from an upper and an underlying interface (see Section 17.4.1.1). The light source is usually a He–Ne laser, although white light or light emitted by the plasma can be used if analyzed with a filter or monochronomator. The reflected intensity is a periodic function of film thickness, and its amplitude is a function of the index of refraction of the ambient, film,

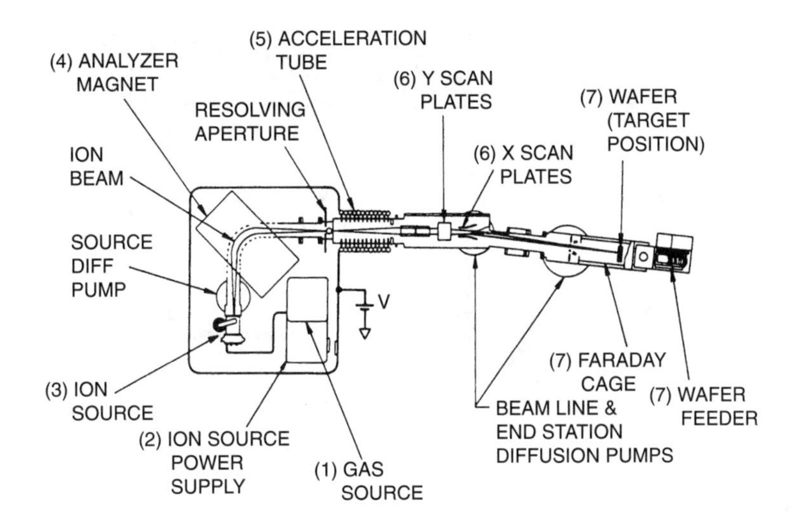

FIGURE 17.2 Schematic diagram of ion implantation system. (From Sze, S., *VLSI Technology*, McGraw-Hill, New York, 1983. With permission.)

and substrate. At normal incidence etch rate may be calculated by determining the time between minimum or maximum intensity peaks in an interferogram.

17.3.4 Ion Implantation

Ion implantation is a method for doping semiconductor substrates. In this process, ionized impurity atoms are accelerated through an electrostatic field to strike the surface of the substrate. The implantation dose can be tightly controlled by measuring the ion current. Doses can range from 10^{11} to 10^{18} cm^{-2}, and typical ion energies range from 5 to 200 keV. By controlling the electrostatic field, the penetration depth of the ions can also be controlled. Ion implantation thus provides the capability of precisely tailoring the dopant profile.

The schematic diagram of a tyipcal ion implantation system is shown in Figure 17.2. The dose of an implant may be measured using an optical density technique based on changes in the optical transparency of a dye-loaded polymer film (see Section 17.4.1.5). This change in transparency is proportional to the ion dose.

17.3.5 Molecular Beam Epitaxy

Molecular beam epitaxy (MBE) is a well-developed and versatile technology for growing various III-V, II-VI, and IV-IV semiconductor structures. MBE growth techniques are capable of achieving stringent composition and thickness control, theoretically limited only by the thickness of a single atomic layer of the material being deposited [Brown, 2000]. A typical MBE system is shown schematically in Figure 17.3. Each of the effusion cells (or "K-cells"), which serve to evaporate the constituent elements of the resultant solid film, consists of pyrolytic boron nitride crucibles and shutters to control molecular beam flow. The process chamber is shielded by a cryopanel that is cooled by liquid nitrogen to prevent contamination by residual gases. A few important features of MBE systems include [Herman and Sitter, 1989]:

1. An ultra-high-vacuum (UHV) environment ($<10^{-10}$ torr) to allow high-purity crystal growth without incorporation of unwanted impurities coming from residual ambient gases
2. The use of the reflection high-energy electron diffraction (RHEED) technique for *in situ* monitoring of crystal structure before, during, and after growth
3. Precise composition and dopant control (in principle to within an atomic layer)

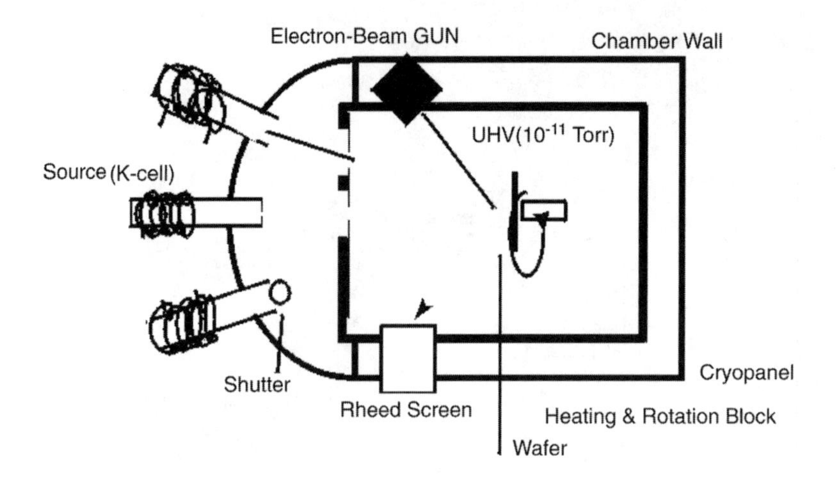

FIGURE 17.3 Schematic diagram of typical MBE system. (From Lee, K. et al., Using neural networks to construct models of the molecular beam epitaxy process, *IEEE Trans. Semiconductor Manufac.*, 13(1), 34–45, 2000. With permission.)

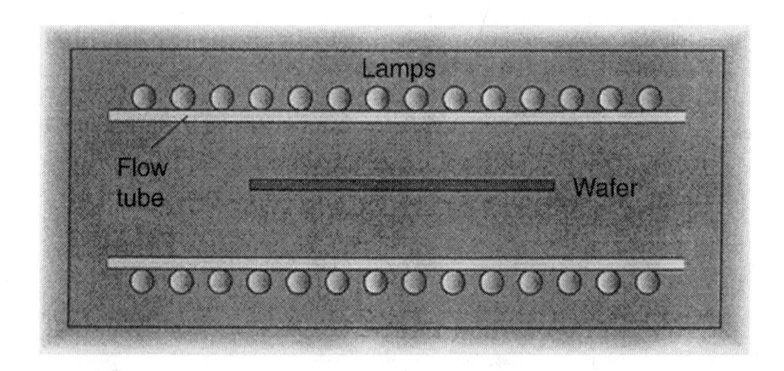

FIGURE 17.4 Typical RTP chamber design. (From Campbell, S., *The Science and Engineering of Microelectronic Fabrication*, Oxford Press, New York, 2001. With permission.)

4. A rotating substrate holder, providing extreme uniformity (as low as 1% thickness nonuniformity over a 3-in. diameter sample)

The MBE process benefits from a well-established *in situ* monitoring technique known as reflection high-energy electron diffraction (RHEED) [Parker, 1985]. RHEED intensity oscillations have been used extensively to extract *in situ* growth information during MBE, but they have yet to be fully exploited for use in process-control applications.

17.3.6 Rapid Thermal Processing

In semiconductor manufacturing there has been an increasing emphasis on single-wafer (as opposed to batch) processing. Single-wafer processes provide the best uniformity and reproducibility, particularly for large wafer sizes. Rapid thermal processing (RTP) collectively represents a suite of single-wafer, high-temperature processes that have been developed to minimize thermal budget by reducing time at temperature. The result of this reduction is the minimization of dopant redistribution. Although RTP was originally developed for implant annealing, in addition to rapid thermal annealing (RTA), RTP processes now also include rapid thermal oxidation (RTO), rapid thermal curing (RTC), and rapid thermal chemical vapor deposition (RTCVD), to name a few.

Most isothermal RTP systems use either tungsten-halogen lamps or noble-gas discharge lamps as heat sources. Various RTP chamber geometries have been proposed (see Figure 17.4). Central to all rapid thermal processes are a common set of problems: uniform wafer heating and cooling, maintenance of uniform temperature during processing, and measuring wafer temperature. Temperature measurement represents one of the most difficult tasks associated with RTP, but this task is critical as temperature is used in a feedback loop to control lamp power output and uniformity. Although temperature can be measured using a thermocouple, this method is intrusive, and the thermal impedance associated with the thermocouple contact point results in a substantial temperature difference between the wafer and thermocouple. For this and other reasons, indirect temperature measurements are the preferred method in RTP. Among the most popular technique is optical pyrometry, which is based on the measurement of radiant energy from the wafer. Optical pyrometry is described in greater detail in Section 17.4.1.7.

17.4 Optical Measurement Techniques

In Section 17.3, a few of the basic unit processes used in fabricating integrated circuits were discussed. In order for these processes to repeatably produce reliable, high-quality devices and circuits each unit process must be strictly monitored and controlled. Process monitoring enables operators and engineers to detect problems early on to minimize their impact. Manufacturing line monitors consist of extremely sophisticated metrology equipment that can be divided into tools that characterize the state of features on the semiconductor wafers themselves and those that describe the status of the fabrication equipment operating on those wafers. The issues involved in understanding and implementing wafer and equipment state measurements are discussed below.

17.4.1 Wafer State Measurements

There is no substitute for regular inspection of products during manufacturing to ensure high quality. Such investigations must not be limited to visual inspections, however, as not all processes have a visible effect on electrionic products. In addition, with ever-increasing levels of integration, features on wafers become smaller and more difficult to inspect. As a result, visual inspection must be supplemented by sophisticated optical measurements of various characteristics that describe the state of a wafer.

Wafer-state characterization includes the measurement of the physical parameters related to each manufacturing process step. The total collection of such measurements relates to the physical characteristics of product wafers, and these physical characteristics can be correlated with the electrical performance of devices and circuits. The following sections describe wafer-state measurement, and the corresponding measurement apparatus, in greater detail.

17.4.1.1 Interferometry

Optical metrology provides fast and precise measurements of film thickness and optical constants. In semiconductor manufacturing, *interferometry* (sometimes called *reflectometry*) is a widely used optical method for measuring such parameters. Single- or multiple-wavelength interferometers are commonly used for both *in situ* and post-process measurements of film thickness. In this method a light source, usually a laser, is focused on a semiconductor wafer while a detector measures the reflected light intensity. The wafer consists of a parallel stack of partially transparent thin films. The reflected-light intensity varies as a function of time depending on the thickness of the top layer due to constructive and destructive interference caused by multiple reflections.

To illustrate, consider Figure 17.5, which shows a film of uniform thickness d and index of refraction n, with the eye of the observer focused on point a. The film is illuminated by broad source of monochromatic light S. There is a point P on the source such that two rays (represented by the single and double arrows) can leave P and enter the eye after traveling through point a. These two rays follow different paths, one reflected from the upper surface of the film and the other from the lower surface. Whether point a appears bright or dark depends on the nature of the interference (i.e., constructive or destructive) between the two waves that diverge from a.

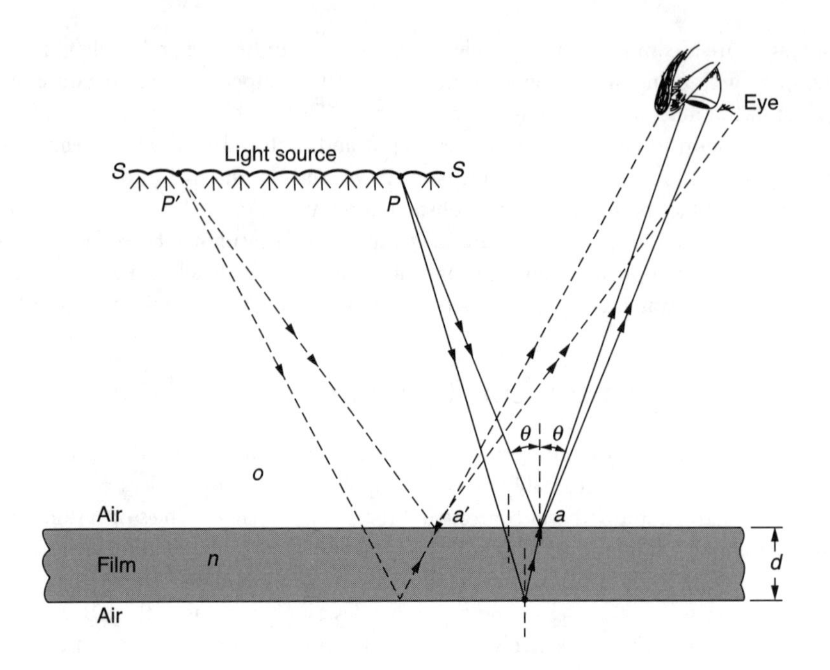

FIGURE 17.5 Interference by reflection from a thin film. (From Halliday, D. and Resnick, R., *Physics*, Wiley, New York, 1978. With permission.)

The two factors that impact the nature of the interference are differences in optical path length and phase changes upon reflection. For the two rays to combine to give maximum intensity, we must have

$$2dn\cos\theta = (m + 0.5)\lambda \tag{17.1}$$

where $m = 0, 1, 2, \ldots$, and θ is the angle of the refracted beam relative to the surface normal. The term 0.5λ accounts for the phase change that occurs upon reflection, as a phase change of 180° is equivalent to half a wavelength. The condition for minimum intensity is

$$2dn\cos\theta = m\lambda \tag{17.2}$$

Equations (17.1) and (17.2) hold when the index of refaction of the film is either greater or less than the indices of the media on *each* side of the film. Therefore, if the index of refraction is known, the thickness of the film may be computed by simply counting peaks or valleys in the reflected waveform. An example of such a waveform (or *interferogram*) appears in Figure 17.6.

Interferometry becomes more complex when applied to stacks of several thin films. The overall goal, however, is still to obtain film thickness information from the time-varying reflected intensity signal. The reflected light intensity is given by Yang et al. [2000]:

$$I_r(d,\lambda) = I_0(\lambda)r(d, \lambda, \phi_1, \phi_2, \ldots, \phi_N) \tag{17.3}$$

where I_0 is the incident-light intensity, r is the reflection coefficient, d is the thickness of the top layer, and ϕ_i are physical constants (i.e., thicknesses and refractive indices) associated with the lower films in the film stack.

The reflected intensity is monitored using a detector consisting of a light-sensitive transducer, such as a photodiode, in conjunction with an optical filter or diffraction grating to select the wavelength(s) of interest. The output of the detector corresponding to a particular wavelength is of the form

$$y_\lambda(kT) = \alpha(\lambda,kT)A(\lambda,kT)I_0(\lambda,kT)r(d(kT),\lambda) + e_\lambda(kT) \tag{17.4}$$

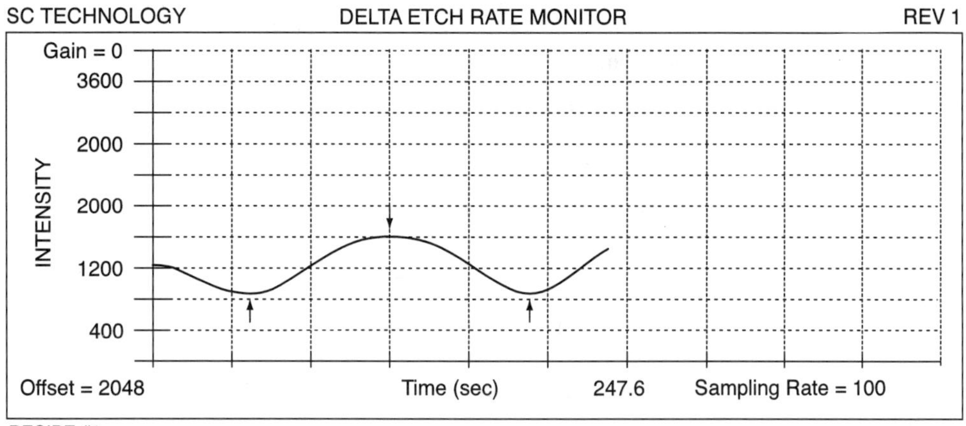

FIGURE 17.6 Sample interferogram used for RIE monitoring. (From Pope, J. et al., Manufacturing integration of real-time laser interferometry to etch silicon dioxide films for contacts and vias, *Proc. SPIE Conf. on Microelec. Processing*, 2091, 185–196, 1993. With permission.)

where T denotes the sampling period, k is an integer, α represents losses in the optical system, A is the gain of the detector, and e_λ is measurement noise. The physical parameters ϕ_i are considered to be known in this formulation and are not shown. For multiple-wavelength (or *spectroscopic*) measurements, this expression is repeated for each wavelength used. For p wavelengths, in matrix form, this is written as

$$\mathbf{y}(kT) = diag(\mathbf{h}(kT)\mathbf{r}(d(kT)) + e(kT) \tag{17.5}$$

where $diag(\mathbf{x})$ represents a matrix with the elements of the vector $\mathbf{x}$ along the diagonal and

$$\mathbf{y}(kT) = [y_{\lambda_1}(kT) \cdots y_{\lambda_p}(kT)] \tag{17.6}$$

$$\mathbf{h}(kT) = [\alpha(\lambda_1, kT)A(\lambda_1, kT)I_0(\lambda_1, kT) \cdots \alpha(\lambda_p, kT)A(\lambda_p, kT)I_0(\lambda_p, kT)]^T \tag{17.7}$$

$$\mathbf{r}(d(kT)) = [r(d(kT), \lambda_1) \cdots r(d(kT), \lambda_p)]^T \tag{17.8}$$

$$\mathbf{e}(kT) = [e_{\lambda_1}(kT) \cdots e_{\lambda_p}(kT)]' \tag{17.9}$$

where the superscript T represents the transpose operation.

To obtain film thickness or the rate of change of thickness (i.e., etch rate or deposition rate), the detector output is processed in one of two ways: (1) extrema counting or (2) least-squares fitting. Extrema counting takes advantage of the fact that the reflected light intensity varies approximately periodically with both the wavelength of the incident light and the thickness of the top film. The distance between peaks and valleys is a known function of the top film thickness. Thus, if many wavelengths are available, thickness can be determined by counting the peaks in a plot of reflectance vs. wavelength. If only a single wavelength is available, the movement of peaks and valleys over time during *in situ* measurements indicates that a specific amount of material has been etched or deposited. This provides the average etch or deposition rate between successive minima and maxima.

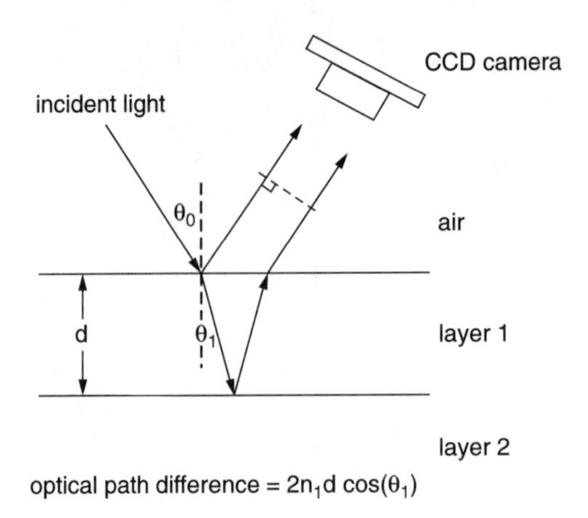

FIGURE 17.7 Schematic of full-wafer interferometry. (From Wong, K. et al., Endpoint prediction for polysilicon plasma etch via optical emission interferometry, *J. Vac. Sci. Tech. A*, 15(3), 1997. With permission.)

To use the least-squares approach, at each time point the following nonlinear optimization problem is posed:

$$\min_{d}[\mathbf{y}(kT) - diag(\mathbf{h})\mathbf{r}(d))^{T}(\mathbf{y}(kT) - diag(\mathbf{h})\mathbf{r}(d)] \tag{17.10}$$

The film thickness is then that for which the minimum is achieved. Etch rate or deposition rate is then calculated from the resulting thickness vs. time curve.

The final variation of interferometry we will discuss briefly is one that is particularly applicable to thickness monitoring during plasma etching. During etching, the emission from the plasma itself may be used as the light source. As this light is reflected from the etched film and underlying film surfaces while the thickness of the etched film decreases, the optical path difference between light rays varies, and the changing constructive and destructive interference results in periodic signals in the same manner as previously described. If a charge-coupled device (CCD) camera is placed in such a way that it can view these signals (see Figure 17.7), each pixel of the CCD camera then acts as an individual interferometer monitoring a different part of the wafer. This arrangement is called *full-wafer interferometry* [Wong, et al., 1997].

17.4.1.2 Ellipsometry

Ellipsometry is a widely used measurement technique that is based on the polarization changes that occur when light is reflected from or transmitted through a medium. Changes in polarization are a function of the optical properties of the material (i.e., its complex refractive indices), its thickness, and the wavelength and angle of incidence of the light beam relative to the surface normal. When multiple light beams of varying wavelength are used, the technique is referred to as *spectroscopic ellipsometry* (SE). SE, which can be used to make *in situ* or post-process measurements, is a fundamentally more accurate technique than interferometry for obtaining film thickness and optical dielectric function information. In general, SE measurements are performed at an off-normal angle with respect to the sample. In this configuration, the measurement is sensitive to the polarization state of both the incident and reflected waves.

Figure 17.8 shows an unpolarized beam of light falling on a dielectric surface. In this case, the dielectric is glass. The electric field vector for each wavetrain in the beam can be resolved into two components—one perpendicular to the plane of incidence (i.e., the plane of the figure) and the other parallel to this plane. The perpendicular component, represented by the dots, is called the σ-component (or "*s*-component"). The parallel component, represented by the arrows, is the π-component (or "*p*-component"). On average, for completely unpolaraized incident light, these two components are of equal amplitude. However, if the incident beam is polarized (as is the case in ellipsometry), this is no longer true.

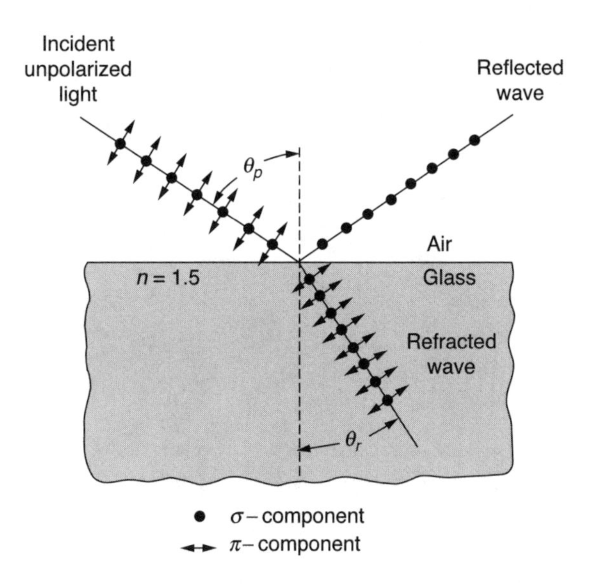

FIGURE 17.8 Illustration of components of polarization. (From Halliday, D. and Resnick, R., *Physics*, Wiley, New York, 1978. With permission.)

In the most common configuration, linearly polarized light is incident on the surface, and the elliptical polarization status of the reflected light is analyzed. Measured ellipsometry data are usually written in the form of the ratio (ρ) of the *total reflection coefficients* for *s* and *p* polarization (R^S and R^P, respectively). In other words

$$\rho = R^p/R^s = \tan(\psi)e^{i\Delta} \tag{17.11}$$

where $\tan(\psi)$ is the ratio of the magnitude of the *p*-polarized light to the *s*-polarized reflected light, and Δ is the difference in phase shifts upon reflection for the *p* and *s* polarizations, respectively.

Another set of expressions called the *Fresnel equations* relate Eq. (17.11) to the bulk complex dielectric function (ε). The dielectric function represents the degree to which the material may be polarized by an applied external electric field, and as a complex number it is often expressed as

$$\varepsilon = \varepsilon_1 + j\varepsilon_2 \tag{17.12}$$

where ε_1 and ε_2 are the real and imaginary parts, respectively. For heterogeneous samples consisting of multiple layers, the dielectric function determined by ellipsometry is an average over the region penetrated by the incident light called the *effective dielectric function*, $\langle\varepsilon\rangle$. If the sample structure is not too complicated, $\langle\varepsilon\rangle$ can be simulated by appropriate models (such as the "ambient-film-substrate" model). In this case, film and substrate properties can be separated, and film properties (i.e., thickness or dielectric function) can be determined as follows.

Because there are a maximum of two independent optical parameters (Ψ and Δ) measured at each wavelength, the maximum number of unknowns that can be determined from a single spectral measurement is 2*W*, where *W* is the number of wavelengths scanned. Thus far, we have discussed the index of refraction as if it were a single parameter. However, in general, the *complex index of refraction* (*N*) consists of a real part (*n*) and an imaginary part (*k*), or

$$N = n - jk \tag{17.13}$$

where k is called the extinction coefficient, which is a measure of how rapidly the intensity decreases as light passes through a material. The dielectric function is related to the complex index of refraction by the relationship

$$\varepsilon = N^2 \tag{17.14}$$

Therefore, we can obtain values for n and k in terms of ε_1 and ε_2 using

$$n = \sqrt{\frac{1}{2}\left[\left(\varepsilon_1^2 + \varepsilon_2^2\right)^{1/2} + \varepsilon_1\right]} \tag{17.15}$$

$$k = \sqrt{\frac{1}{2}\left[\left(\varepsilon_1^2 + \varepsilon_2^2\right)^{1/2} - \varepsilon_1\right]} \tag{17.16}$$

As mentioned above, the complex index of refraction is related to the total reflection coefficients by the Fresnel equations, which are given by Thompkins and McGahan [1999]

$$R^p = \frac{r_{12}^p + r_{23}^p \exp(-j2\beta)}{1 + r_{12}^p r_{23}^p \exp(-j2\beta)} \tag{17.17}$$

$$R^s = \frac{r_{12}^s + r_{23}^s \exp(-j2\beta)}{1 + r_{12}^s r_{23}^s \exp(-j2\beta)} \tag{17.18}$$

where the Fresnel reflection coefficients at the individual interfaces are of the form

$$r_{12}^p = \frac{N_2 \cos\phi_1 - N_1 \cos\phi_2}{N_2 \cos\phi_1 + N_1 \cos\phi_2} \tag{17.19}$$

$$r_{12}^s = \frac{N_1 \cos\phi_1 - N_2 \cos\phi_2}{N_1 \cos\phi_1 + N_2 \cos\phi_2} \tag{17.20}$$

and

$$\beta = 2\pi\left(\frac{d}{\lambda}\right) N_2 \cos\phi_2 \tag{17.21}$$

All subscripts and angles mentioned in Equations 17.17) through 17.21 are described in Figure 17.9.

Thus, materials with finite light absorption have two unknowns (ε_1 and ε_2, or equivalently, n and k) at each wavelength and one additional unknown in the film thickness. Thus, the total number of

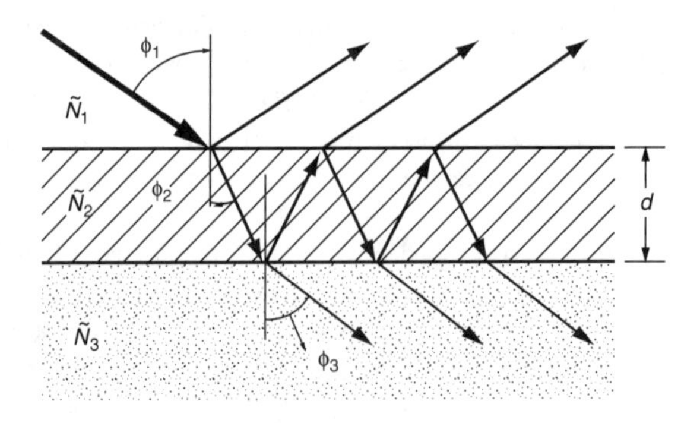

FIGURE 17.9 Reflections and transmissions in ambient (1), film (2), and substrate (3). (From Tompkins, H. and McGahan, W., *Spectroscopic Ellipsometry and Reflectometry*, Wiley, New York, 1999. With permission.)

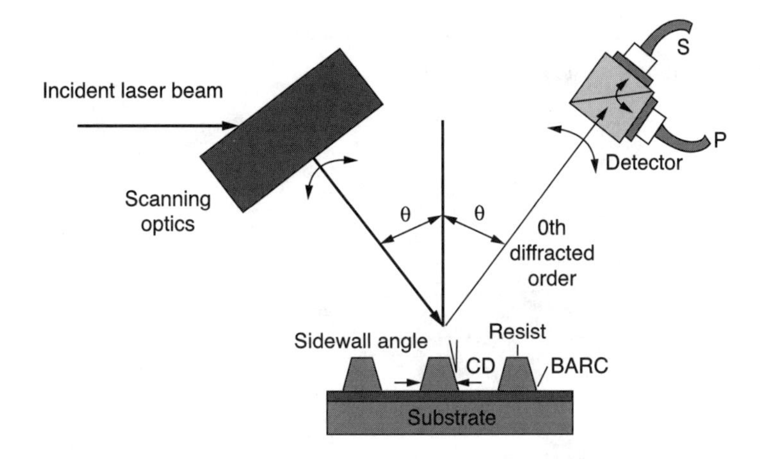

FIGURE 17.10 Schematic of a 2-θ angle-resolved scatterometer. (From Raymond, C., Angle-resolved scatterometry for semiconductor manufacturing, *Microlithography World*, 18–23, 2000. With permission.)

unknowns is $2W + 1$. Because this number of unknowns is one too many to be determined from spectroscopic ellipsometry data, it is necessary to employ a dispersion model. Such a model describes the functional dependence of n and k on λ based on P fitting parameters. Therefore, the total number of unknowns becomes $P + 1$. As long as $2W > P + 1$, film thickness and the optical constants may be determined simultaneously by numerically iterating the $P + 1$ fitting parameters to fit spectra [Yang et al., 2000].

For example, for a thin film on a substrate, the usual objective is to determine thickness d for a known substrate and film dielectric function. To do so, the value of d is found that minimizes the function

$$\sum_{\lambda} \left| \langle \varepsilon \rangle - \langle \varepsilon \rangle_{calc} \right|^2 \tag{17.22}$$

(or similar functions using ρ, or ψ and Δ) [McGilp et al., 1995]. Here, the first term represents measured values, and the second term represents theoretically calculated values. This expression can be minimized using well-known procedures such as Newton's method or the Levenberg-Marquardt algorithm [Press et al., 1988].

17.4.1.3 Scatterometry

Scatterometry is another optical measurement technique. It is used for patterned features based on an analysis of the light diffracted (or *scattered*) from a periodic structure such as a grating of photoresist lines. Figure 17.10 shows a schematic of an angle-resolved scatterometer, which measures the intensity of the light diffracted as a function of incident angle and polarization. Scatterometry is used to characterize surface roughness, defects, surface particle density, film thickness, or the critical dimension (CD) of the periodic structure.

The most common type of angle-resolved scatterometer is called a "2-θ" scatterometer due to the two angles (incident and measurement) associated with the method. An incident laser is focused on a sample and scanned through some range of incident angles (θ_i). The light is scattered by the periodic patterns into distinct diffraction orders at angular locations specified by the grating equation

$$\sin\theta_i + \sin\theta_n = n\lambda/d \tag{17.23}$$

where θ_i is taken to be negative, θ_n is the angular location of the nth diffraction order, λ is the wavelength of the incident light, and d is the spatial period (or pitch) of the periodic structure. Due to the complex interaction between the incident light and the periodic features, the fraction of power diffracted into each order is a function of the dimensions of the structure and thus may be used to characterize them.

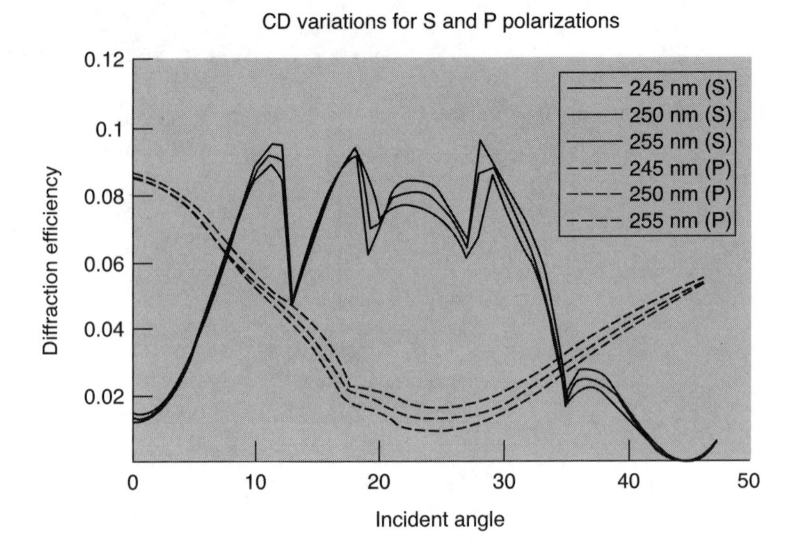

FIGURE 17.11 Scattterometry signatures for 5-nm photoresist CD variations. (From Raymond, C., Angle-resolved scatterometry for semiconductor manufacturing, *Microlithography World*, 18–23, 2000. With permission.)

Capturing diffracted light "signatures" (such as those depicted in Figure 17.11) is just the first phase of scatterometry. In the subsequent analytical phase, a diffraction model is used to interpret the experimental signatures in terms of key parameters such as CD or film thickness. Doing so requires a library of theoretical signatures to compare to the measured data. The generation of such a library is accomplished by first specifying nominal film stack dimensions and the expected variation of each parameter to be measured. A computerized diffraction model is then used to produce the library of scatter signatures that encompasses all combinations of these parameters for subsequent analysis.

17.4.1.4 Development Rate Monitoring

In photolithography exposure is one of the most important steps because the exposure dose and focus significantly impact film quality, resolution, and feature geometry. Underexposure results in insufficient cross-linking at the bottom of the photoresist, and overexposure causes light scattering off the underlying surfaces [Kim and May, 1999]. The development step is also critical. The development endpoint is very sensitive to resist thickness, developing solvent composition, delay between prebake and development, and the temperature of the solvent.

The *Lithacon process analyzer* is an online dissolution-rate monitoring and developing analysis tool for photolithography. It is designed for photochemical material and process performance characterization, allowing the user to collect, access, and analyze production wafer development rate data [Crisalle et al., 1998]. The *Lithacon* system consists of three main subsystems: the optical processing head (OPH), the signal-processing unit (SPU), and a computer with data analysis software (as shown in Figure 17.12).

The SPU monitors the track status through a track interface adapter and reports the results to the computer running the *Lithacon* software. When a trigger signal is received from the track, indicating the beginning of the development process, the computer initiates development endpoint detection by turning on a halogen lamp in the OPH. Light from the halogen lamp source passes through the OPH and illuminates the wafer being developed. As the light is passed through the OPH, it is filtered to remove wavelengths that would expose the resist, collimated for uniform illumination of the wafer, and circularly polarized to provide a means for rejection of scattering and ambient light. Rejected light from the wafer surface is then collected by the OPH and focused into a fiber optic cable. Only light that exhibits the proper polarization is preserved by the optical processing in the head. The light travels through the fiberoptic cable and is divided into eight channels at the SPU. The light in each of the channels must

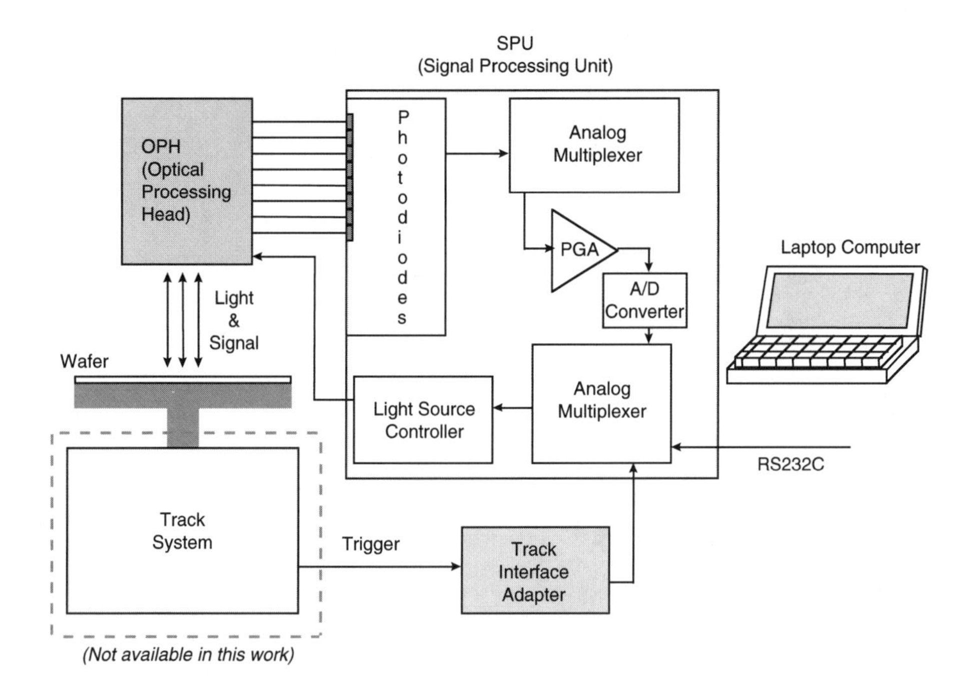

FIGURE 17.12 Block diagram of Lithacon system. (From Kim, T. and May, G., Time series modeling of photosensitive BCB development rate for via formation applications, in *Proc. 1999 Int. Electronics Manufacturing Technology Symp.*, 1999. With permission.)

pass through a narrow-band (10-nm) optical fiber before illuminating a photodiode. The center frequencies of the optical filters range from 700 to 1000 nm.

After monitoring dissolution rate during the developing step, the *Lithacon* uses the equations below to calculate a development rate [Crisalle et al., 1998]:

$$\textit{Development rate} = R_{\max}\frac{(a+1)(1-E)^n}{a+(1-E)^n} + R_{\min} \tag{17.24}$$

$$a = \frac{n+1}{n-1}(1-E_{TH})^n \tag{17.25}$$

where R_{max}, R_{min}, E_{TH}, and n represent maximum development rate, minimum development rate, threshold exposure dose energy, and selectivity, respectively.

17.4.1.5 Optical Density

Ion implantation dose may be measured by the optical density technique, which is based on changes in the optical transparency of a dye-loaded polymer film. The change in optical transparency is directly proportional to the ion dose. Doses in the range of 10^{11} to 10^{13} cm^{-2} can be measured using this method [Wolf and Tauber, 2000]. Monitor wafers are fabricated by spin-coating the dye-loaded polymer onto glass substrates. Prior to implantation, the film is scanned by an optical densitometer and its optical absorption pattern stored electronically. After implantation, the film is scanned again, and contour maps of the optical density changes are generated. These maps provide the spatial distribution of ions and the regions in which the dose uniformity is within a specified tolerance.

17.4.1.6 Reflection High-Energy Electron Diffraction

The MBE process benefits from a well-established *in situ* monitoring technique known as *reflection high-energy electron diffraction* (RHEED). In MBE monitoring using RHEED, high-energy (10- to 50-KeV)

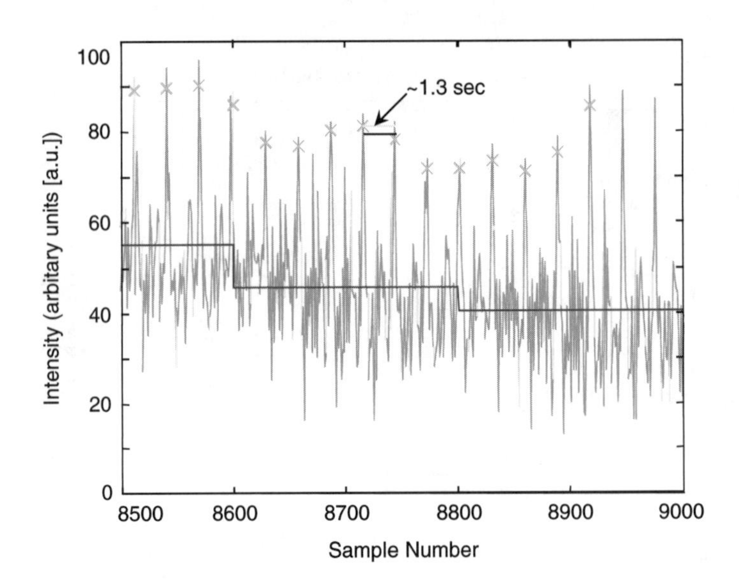

FIGURE 17.13 AlGaAs growth rate determination from RHEED oscillations. (From Lee, K. et al., Using neural networks to construct models of the molecular beam epitaxy process, *IEEE Trans. Semiconductor Manufac.*, 13(1), 34–45, 2000. With permission.)

electron beams strike the substrate with very shallow glancing angle (1 to 3 degrees). Those electrons undergo diffraction at the surface according to Bragg's law. The reflected beam is scattered and directed onto a screen lying opposite to the RHEED gun in the growth chamber. This screen then contains the RHEED oscillation pattern that corresponds to rows of atomic periodicity. In general, the resulting patterns displayed on the phosphor screen range from the smeared oval shapes or spots to streaks [Herman and Sitter, 1989].

As growth proceeds there is a maximum reflectivity in intensity for the starting smooth surface and the final smooth surface after an atomic layer is completely formed. A minimum intensity is reached for the intermediate stage, when the growing layer is approximately half complete. This establishes the RHEED intensity oscillations. By measuring a periodicity of these oscillations, the deposition rate of the growing film can be determined. Figure 17.13 shows a trace of intensity variation vs. time taken during AlGaAs film growth. The time span between two subsequent peaks corresponds to a growth rate of 1.3 atomic layers per second.

17.4.1.7 Optical Pyrometry

Thermal operations refer to any process step that occurs at an elevated temperature. Examples include epitaxial growth, chemical vapor deposition, evaporation, and annealing. *In situ* measurements of conditions such as temperature can be used to infer the quality of the wafers being produced in thermal processes. In many types of thermal processsing equipment, temperature is measured using a thermocouple embedded in the wafer holder. In rapid thermal processes the use of a thermocouple is not possible because there is no susceptor. Alternative temperature sensors used in such situations include *optical pyrometers*. Pyrometers operate by measuring the radiant energy received in a certain band of energies, assuming the source is a gray body of known emissivity. The input energy can then be converted to a source temperature using the Stefan–Boltzmann relationship [Campbell, 2001]. Most commercial systems monitor the mid-infrared band (3 to 6 μm). One major issue in using pyrometry is that the effective emissivity of the source must be accurately known. The effective emissivity includes both intrinsic and extrinsic contributions. Intrinsic emissivity is a function of the material, surface finish, temperature, and wavelength. Extrinsic emissivity is affected by the amount of radiant energy from other sources reflected back to the spot being measured (which can increase the apparent temperature). In addition, the presence of multiple layers of different thin film materials can alter the apparent emissivity due to interference effects.

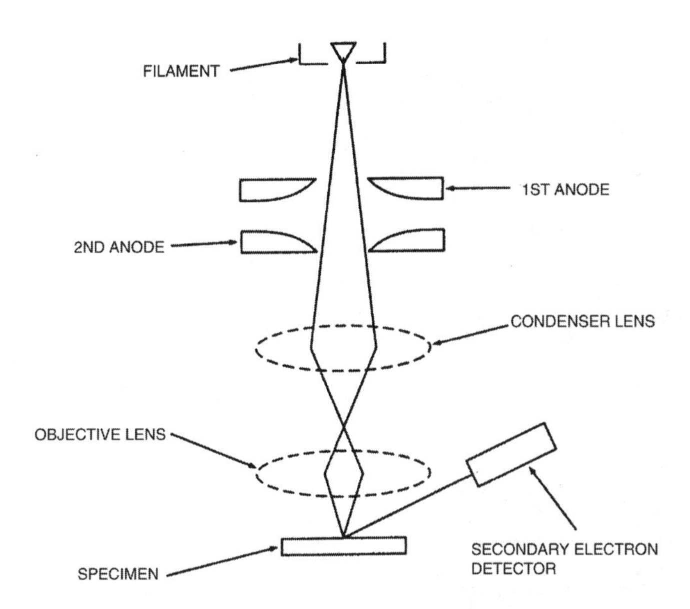

FIGURE 17.14 Schematic of field emission SEM optics. (From Landzberg, A., *Microelectronics Manufacturing Diagnostics Handbook*, Van Nostrand-Reinhold, New York, 1993. With permission.)

17.4.1.8 Scanning Electron Microscopy

Scanning electron microscopy (SEM) is a key technique for assessing minimum feature size in semiconductor manufacturing. The minimum feature size is often expressed in terms of the critical dimension (CD) or minimum *linewidth* that can be resolved by the photolithography system. The decrease in linewidths toward the scale of fractions of a micrometer has rendered conventional optical microscopes nearly obsolete. However, linewidth measurements based on SEM can overcome the limitations of optical techniques for submicron geometry features.

The fine imaging capability of the SEM is due to the fact that the wavelength of electrons is four orders of magnitude less than optical systems. At such small wavelengths, diffraction effects are usually negligible, and spatial resolution is excellent. Features as small as 100 nm can be readily resolved [Landzberg, 1993]. The electron beam may be based on thermionic or field emission sources. A schematic of a typical field emission SEM is shown in Figure 17.14.

As shown in this figure, the electron gun consists of a tip, first anode, and second anode. A voltage is established between the tip and first anode to facilitate field emission from the tip. An accelerating voltage is then applied between the tip and second anode to accelerate the electrons. The electron beam emitted from the tip passes through the aperture provided at the center of the first anode, is accelerated, and passes through the center aperture of the second anode to the condenser lens. Electron beams are collected by the condenser lens and formed into a small spot on the objective lens. The CD of the imaged feature is usually determined by an arbitrary edge criterion. While lateral resolution offers a tremendous benefit, it must be pointed out that SEM still suffers from several disadvantages, including high cost, low throughput (about 30 wafers per hour), and the destructive nature of the measurement (i.e., wafers must be cleaved to expose the feature to be imaged).

17.4.1.9 Defect-Inspection Systems

Contamination is a major concern in semiconductor manufacturing, and billions of dollars are spent annually by manufacturers in order to reduce it. Contamination often takes the form of particles that can appear on the surface of wafers and cause defects in devices or circuits. The fraction of the product that is sensitive to particles depends in part on the particle size. A general rule of thumb is that particles as small as one tenth the size of a structure can make the structure fail. With the industry currently immersed in manufacturing devices with submicron features, even nanometer-scale particles are of great concern.

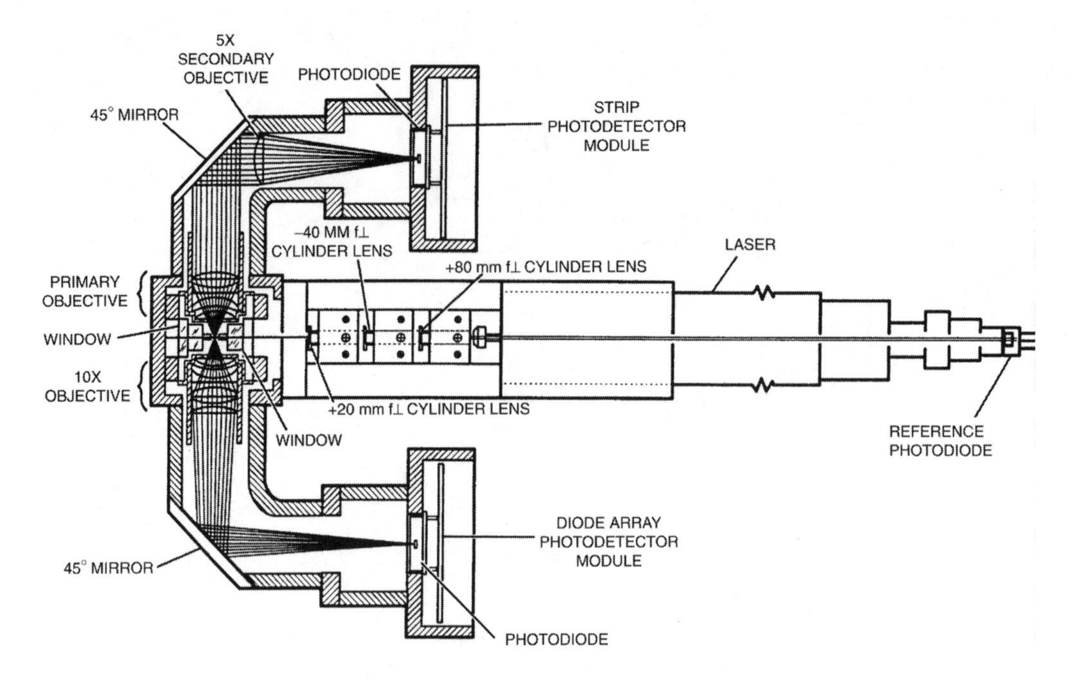

FIGURE 17.15 Optical particle counter. (From Landzberg, A., *Microelectronics Manufacturing Diagnostics Handbook*, Van Nostrand-Reinhold, New York, 1993. With permission.)

17.4.1.9.1 Cleanroom Air Monitoring

One method of controlling particulate contamination is performing manufacturing operations in a cleanroom environment. Despite the use of cleanrooms, however, semiconductor fabrication processes, as well as manufacturing personnel themselves, still generate materials that can contaminate products. Such contamination may originate from process gases and vapors, process liquids, processes that break up bulk material (such as sputtering), deposition processes, metallic impurities, wafer handling, or tool wear, just to name a few. One method for quantitatively determining cleanroom air quality involves sampling via optical particle counters.

For gases, liquids, and many types of surfaces, optical particle counters are used. Using these devices, particles are illuminated as they pass through a focused laser beam (see Figure 17.15). The light scattered from the particles is then measured and correlated with the number of particles present. The amount of light scattered into the sensing element will depend on the light (intensity, wavelength, polarization), the characteristics of the particles (size, shape, orientation, refractive index), and the measurement geometry (position and solid angle subtended by the optics with respect to the beam and the particle). In addition to cleanroom monitoring, this technique is also used for *in situ* particle monitoring inside processing equipment that produces particles such as ion implantation or sputtering equipment.

17.4.1.9.2 Product Monitoring

In addition to monitoring contamination in the ambient environment, it is perhaps more crucial to monitor particles that actually wind up on the wafer surface, as these are the particles that can cause circuit defects. To control the formation of such defects special in-line monitoring techniques are required. These techniques involve inspection of product wafers at various stages in the process. Two common approaches for local defects are "surfscan" and image evaluation. The surfscan technique uses scattered laser light and analyzes reflections to count the number of particles on the wafer surface (see Figure 17.16). Surfscan is usually applied to unpatterned wafers. Image-evaluation techniques, on the other hand, make use of automated inspection equipment to check the occurrence of local defects on patterned wafers at several critical points in the manufacturing process.

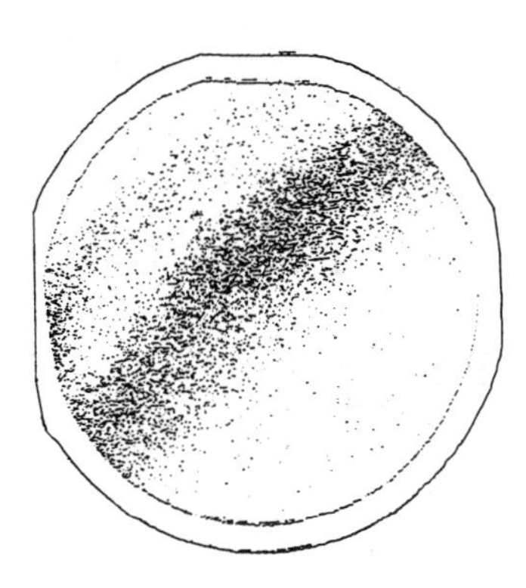

FIGURE 17.16 Sample surfscan. (From de Gyvez, J. and Pradhan, D., *Integrated Circuit Manufacturability*, IEEE Press, Piscataway, NJ, 1999. With permission.)

17.4.2 Equipment State Measurements

Rather than characterizing the state of the product wafers themselves, equipment monitors measure the status of tools while they are processing these wafers. Such monitors are the most immediate measure of process quality and, therefore, provide the shortest feedback loop for maintaining control. Certain phyiscal parameters are routinely measured as a part of equipment monitoring. The combined effects of these tool variables leads eventually to measurable impact on the characteristics of product wafers. The process engineer must therefore have available reliable methods for monitoring these variables in order to facilitate process control. The following sections describe several equipment state measurements used for monitoring such characteristics.

17.4.2.1 Optical Emission Spectroscopy

Optical emission spectroscopy (OES) is one of the oldest and most popular methods of plasma etch monitoring. Fundamentally, OES is a bulk measure of the optical radiation of the plasma species. Because emissions can emanate from etch reactants as well as products, OES measurements are most often used to obtain the average optical intensity at a particular wavelength above the wafer. By setting an optical spectrometer to monitor the intensity at a wavelength associated with a particular reactant or by-product species, OES serves as a noninvasive, real-time etch endpoint detector. Quantitative measurement of the species concentrations is not required for this purpose. Instead, the intensity of the emission from the key species, perhaps along with its time derivative, can be used empirically to determine the proper point to discontinue the etch process.

A series of such measurements for a particular etch process is referred to as an "endpoint trace," a curve representing the intensity of the optical emission of the key species over time. An example of such a trace is illustrated in Figure 17.17, which depicts fluorine- and CN-emission intensities during silicon nitride etching. At the beginning of the etch, the gas in the chamber consists of a mixture of process gas and that resulting from the etch. At the end of the etch the gas mixture again resembles its mixture prior to the start of the process. Therefore, the etch endpoint is characterized by a sharp change in the intensity of the endpoint trace.

OES measurements not only reflect the chemistry of the plasma, but also have embedded in them information concerning the status of the plasma equipment, pattern density on the substrate, and nonideal fluctuations in the processing conditions (i.e., gas flow, pressure, etc.). It is therefore possible to use OES signals to monitor and diagnosis etch equipment problems.

17.4.2.2 Fourier Transform Infrared Spectroscopy

Infrared (IR) spectroscopy is a widely used method for identifying organic compounds, such as those that may result from the etching of polymer films. This method is based on the absorption of infrared radiation by molecules at characteristic wavelengths. Radiation causes various components of such molecules to vibrate and rotate. Because the frequency of vibration/oscillation is dependent on the nature of the chemical bonds present, the presence or absence of absorption in certain well-defined regions of

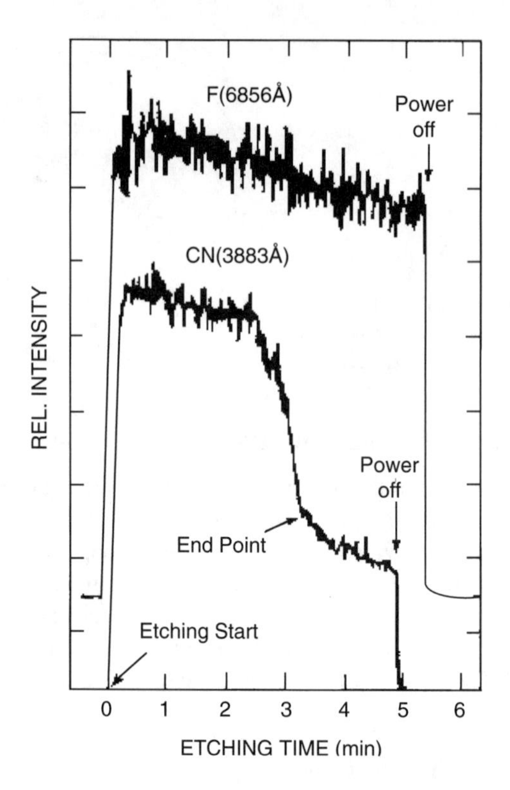

FIGURE 17.17 OES endpoint trace showing the intensity of the emission of key species in a silicon nitride etch process. (From Manos, D. and Flamm, D., *Plasma Etching: An Introduction*, Academic Press, San Diego, CA, 1989. With permission.)

the IR spectrum can be used to determine the presence or absence of chemical groups. The intensity of the absorption peaks is proportional to the amount of material present. Computer databases and search routines are usually used to identify compounds.

In Fourier transform infrared (FTIR) spectroscopy, an infrared source is sent through a beam splitter to the surface of the wafer being etched and to a movable mirror. The reflected radiation from both surfaces is added and sent to a detector. The distance of the mirror path is swept, and the intensity of the reflected beam as a function of the position of the mirror is monitored. The intensity of the IR peaks can then be used to determine the composition of the film on the wafer surface. An example of typical FTIR output is provided in Figure 17.18.

17.5 Optical Methods in Process Modeling

Once data have been obtained from an optical process monitor by measuring the response of interest under various combinations of processing conditions, the results may be summarized in the form of a process model. A process model is some form of fit to the measured optical data. This fit can be obtained using statistical techniques such as regression analysis [Box et al., 1978], response-surface methodology [Box and Draper, 1987], time-series modeling [Box and Jenkins, 1976], and principal-component analysis [Joliffe, 1986], or by using artificial-intelligence-based approaches such as neural networks [Dayhoff, 1990]. In each case the overall goal is to develop a quantitative model that predicts a relationship between process conditions and a given response. An accurate model should minimize the difference between the observed values of the response and its own predictions. In the following, several case studies describing the use of optical techniques for process modeling are presented.

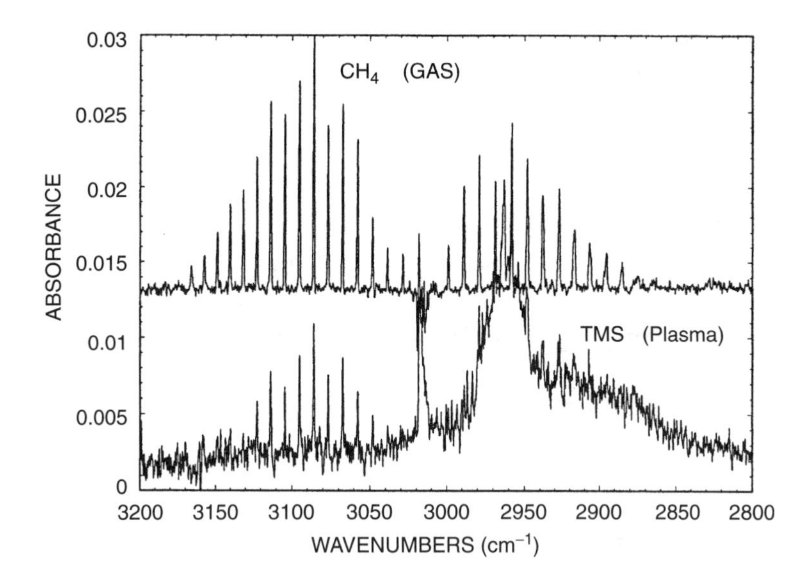

FIGURE 17.18 IR spectra of CH_4 and tetramethylsilane (TMS) in an electron cyclotron resonance plasma system. (From Raynaud, P. et al., Infrared absorption analysis of organosilicon/oxygen plasmas in a microwave multipolar plasma excited by distributed electron cyclotron resonance, *Appl. Surface Sci.*, 138–139, 285–291, 1999. With permission.)

17.5.1 Time Series Modeling of Polymer Development Rate in Photolithography

Kim and May of the Georgia Institute of Technology in Atlanta, GA, have developed a neural-network-based time series modeling scheme and applied it to determine the optimal endpoint for photosensitive benzocyclobutene (BCB) development [Kim and May, 1999]. Photosensitive BCB is a negative imaging material (i.e., unexposed areas are removed during development), and it is sensitive to 365-nm radiation (I-line and/or broadband exposure). The formation of vias using photosensitive BCB is very similar to a generic photolithography process. The basic unit process steps are polymer deposition, prebaking, pattern transfer (exposure and development), curing, and plasma descumming. The development step is critical, and the development endpoint is very sensitive to film thickness, developing solvent composition, delay between prebake and development, and the temperature of the solvent.

In Kim and May's study, for on-line dissolution-rate monitoring and development-step analysis, the *Lithacon 808 process analyzer* was used (see Section 14.4.1.4). To characterize development exposure dose energy and time series data consisting of previous film thickness measurements were used to model expected film thickness at future times. Data from a set of designed experiments were used to train feed-forward neural networks using the error back-propagation (BP) algorithm [Dayhoff, 1990]. Figure 17.19 shows the general structure of the BP neural networks used. To develop time series models, both auto- and cross-correlation in the data were considered. Autocorrelation among consecutive film thickness measurements was accounted for by using a series of such measurements as neural network inputs. Cross-correlation was accounted for by using both film thickness and exposure dose energy as network inputs. Model output consisted of the next value of film thickness or the thickness after n seconds. The forecasting capability of these models was evaluated by predictions of film thickness 0.02 to 6 s in the future. The development endpoint was determined by forecasting the exact time when the remaining film thickness was equal to zero.

Model prediction results were compared with experimental results, and it was shown that the neural time series model could effectively characterize the effects of changes in exposure dose energy on the development endpoint. For the best case, the prediction root-mean-square (RMS) error was 1.95%, and endpoint prediction result showed only a 1-s difference compared to measured results.

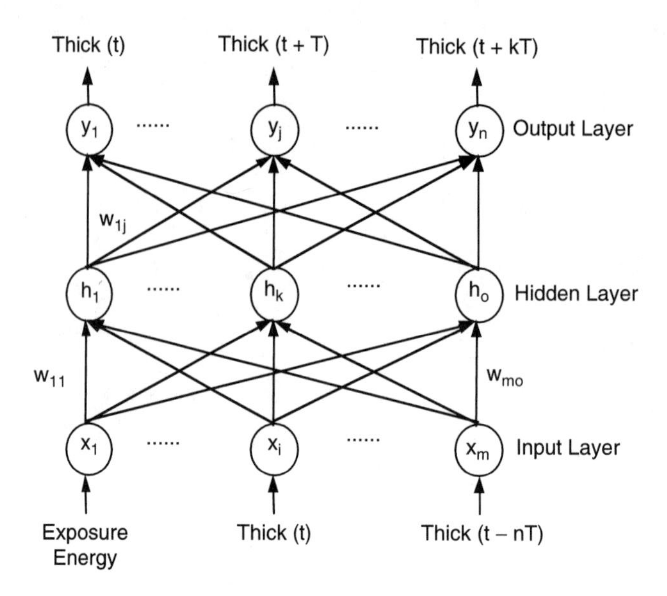

FIGURE 17.19 Neural network used for time series modeling. (From Kim, T. and May, G., Time series modeling of photosensitive BCB development rate for via formation applications, in *Proc. 1999 Int. Electronics Manufacturing Technology Symposium*, 1999. With permission.)

17.5.2 Modeling Reactive Ion Etching Using Optical Emission Spectroscopy Data

White et al. [1997] of the Massachusetts Institute of Technology in Cambridge, MA, have presented a method that uses multivariate OES data collected during reactive ion etching to estimate spatial asymmetries on the wafer surface. The key elements of this method are principal component analysis (PCA) for dimensionality reduction of the OES data and subsequent function approximation using principal component regression (PCR), partial least squares (PLS), or neural networks (NNs) for estimating spatial uniformity.

Principal component analysis is an established statistical technique for streamlining a multidimensional data set [Joliffe, 1986]. Consider a vector **x**, which consists of p random variables. Let $\mathbf{\Sigma}$ be the covariance matrix of **x**. Then, for $k = 1, 2, \ldots, r$, the kth principal component (PC) is given by

$$\mathbf{z}_k = \boldsymbol{\alpha}'_k \mathbf{x} \tag{17.26}$$

where $\boldsymbol{\alpha}_k$ is an eigenvector of $\mathbf{\Sigma}$ corresponding to its kth largest eigenvalue λ_k and $'$ represents the transpose operation. If $\boldsymbol{\alpha}_k$ is chosen to have unit length (i.e., $\boldsymbol{\alpha}'_k \boldsymbol{\alpha}_x = 1$), then the variance of $\mathbf{z}_k = \lambda_k$. Generally, if these eigenvalues are ordered from largest to smallest, then the first few PCs will account for most of the variation in the original vector **x**. Dimensionality reduction through PCA is achieved by transforming the original OES data to a new set of variables (i.e., the PCs), which are uncorrelated and ordered such that the first few retain most of the variation present in the original data set. Using principal-component regression, partial least squares, or neural network approaches, these PCs may be used as predictor variables to estimate the RIE process response.

The approach of the MIT team was verified experimentally for an aluminum etch in a Lam Research 9600 transformer-coupled power etching system. To characterize the process, a two-level full factorial experiment within a larger three-level full factorial was performed [Box et al., 1978]. Three factors were varied: top-coil power, RF power, and Cl_2/BCl_3 gas ratio. This design requried a total of 43 experimental trials. Optical emission spectra were acquired using a Chromex OES system. The aluminum underwent an etch for approximately 40 s, during which one spectrum, consisting of relative light intensity at each of 1166 wavelengths in the range of 240 to 400 nm, was recorded. Post-process electrical tests were then used to measure the line-width reduction, defined as the average difference between actual line width and the designed line wdith of 0.35 μm, for 27 die across each wafer.

For the PCA formulation, the raw 43 × 1166 matrix of OES data (**X**) was mean-centered with respect to each wavelength to produce the new data matrix **M** using the expression

$$M_{ij} = (X_{ij} - \overline{X}_j) \qquad \text{for} \quad 1 \le i \le 43 \text{ and } 1 \le j \le 1166 \tag{17.27}$$

The 1166 × 1166 sample covariance matrix **S** was then computed as

$$\mathbf{S} = \left(\frac{1}{r-1}\right)\mathbf{M'M} \tag{17.28}$$

where r is the number of experimental trials (43 in this case). **S** is an estimate of the true covariance matrix ($\mathbf{\Sigma}$). If **A** is the 1166 × 1166 matrix of eigenvectors of **S**, and $\mathbf{\Lambda}$ is the 1166 × 1166 diagonal matrix of eigenvalues, we can then project **M** onto **A** using

$$\mathbf{Z} = \mathbf{MA} \tag{17.29}$$

where the new matrix **Z** has orthogonal columns and represents all of the variance in **M**. PCA allows us, however, to reduce the dimensionality of **Z** by selecting only the first few largest eigenvectors in **A** (i.e., those corresponding to the first few largest eigenvalues in $\mathbf{\Lambda}$). In this case, it was found that the first four PCs accounted for 97% of the variance in the data, allowing a reduction of the 1166 × 1166 matrix **A** to a 1166 × 4 matrix $\hat{\mathbf{A}}$. The optical spectrum matrix **M** was then projected onto $\hat{\mathbf{A}}$ as in Eq. (17.29) to yield $\hat{\mathbf{Z}}$, thus reducing the the measured original experimental data (**X**) from a 43 × 1166 matrix to a 43 × 4 matrix ($\hat{\mathbf{Z}}$) with only a 3% loss in variation. The reduced matrix $\hat{\mathbf{Z}}$ was then used for subsequent process modeling.

Models for metal linewidth reduction were constructed, and the PCR, PLS, and NN function-approximation methods were compared. Both PCR and PLS utilize the decomposition of a matrix of input data (**U**) and the linear regression of an output data matrix (**Y**) upon **U**. In other words,

$$\mathbf{Y} = \mathbf{UB} + \boldsymbol{\varepsilon} \tag{17.30}$$

where **B** represents the regression coefficients and $\boldsymbol{\varepsilon}$ is the residual model error. In PCR, **U** is transformed by the orthogonal matrix of eigenvectors **U′U** to a matrix of scores **Z** as in Eq. (17.29):

$$\mathbf{Z} = \mathbf{UV} \tag{17.31}$$

The output matrix **Y** is then regressed upon the reduced matrix of scores $\hat{\mathbf{Z}}$. PLS, on the other hand, weighs the covariance matrix by the positive definite matrix **YY′**, resulting in a new covariance matrix **U′YY′U**. The new weighted covariance matrix is decomposed into a matrix of eigenvectors (**R**) and eigenvalues (**D**) using

$$\mathbf{U'YY'U} = \mathbf{RDR'} \tag{17.32}$$

The matrix **U** is transformed using the reduced eigenvector matrix $\hat{\mathbf{R}}$ to an orthogonal matrix of scores $\mathbf{U}\hat{\mathbf{R}} = \hat{\mathbf{Z}}$, and the output matrix **Y** is regressed upon $\hat{\mathbf{Z}}$. The NN approach uses the matrix $\hat{\mathbf{Z}}$ as the training data set for a back-propagation neural network [Dayhoff, 1990] that maps $\hat{\mathbf{Z}}$ into the output response space (where the output response is RIE line-width reduction).

All three function-approximation methods achieved similar results. The RMS prediction errors on test data withheld from training were 0.0134 μm for PCR, 0.014 μm for PLS, and 0.016 μm for NNs. The overall results demonstrate that OES in conjunction with PCA and function approximation can be effective in predicting wafer state characteristics in RIE.

17.5.3 MBE Process Modeling Using RHEED Signals

Lee et al. [2000] of Georgia Tech have used RHEED signals to systematically characterize the MBE process and quantitatively model the effects of process conditions on film qualities. To do so a five-layer, undoped AlGaAs-InGaAs single-quantum well (SQW) structure grown on a GaAs substrate was fabricated. The effect of six process conditions (time and temperature for oxide removal, substrate temperatures for AlGaAs and InGaAs layer growth, beam equivalent pressure of the As source, and quantum-well interrupt time) were then explored by means of a fractional factorial experiment. Defect density, x-ray diffraction, and photoluminescence (PL) were the responses modeled using RHEED data and neural networks.

Two novel approaches for characterizing RHEED signals used in the real-time monitoring of MBE were developed. In the first technique, principal-component analysis (see Section 17.5.2 above) was used to reduce the dimensionality of the RHEED data set, and the reduced RHEED data set was used to train neural nets to model the process responses. To accomplish this Lee and his colleagues first normalized RHEED intensity vs. time data to a common time scale and subsequently reduced the RHEED data into its principal components.

Normalization produced 100 intensity values over time for each of 16 RHEED data sets generated from the fractional factorial experiment. Dimensionality reduction through PCA was then achieved by transforming the normalized RHEED data to a new set of variables (i.e., the PCs). The reduced RHEED data were instead used to train BP neural networks to model the defect density, x-ray, and PL responses. The inputs to each of the neural networks consist of the first six PCs of the RHEED data, as they contained over 97% of the variance. By using only the first six PCs of the RHEED data set, the PCA-based neural network models achieved a 100:6 data reduction ratio while losing less than 3% of the variability in the available input information.

Of the 16 experimental trials, 12 were used for network training, and data from the remaining trials were used for testing. Figure 17.20 demonstrates the accuracy of the PCA-based neural network models. Models were constructed for defect density, GaAs and AlGaAs x-ray full-width-at-half-maximum (FWHM), x-ray peak separation, InGaAs and AlGaAs intensity FWHM, and InGaAs and AlGaAs PL peak position. The neural network models exhibited an average RMSE of 3% on training data and approximately 12% for test data.

A second technique developed by Lee and his colleagues uses neural nets to model RHEED-intensity signals as time series and matches specific RHEED patterns to ambient process conditions. The objective in this case is to facilitate the prediction of the RHEED-intensity variations as a function of time. Inputs to the neural network used for time series modeling included the six MBE growth conditions as well as five additional inputs representing the five immediate past values of RHEED intensity. Time series models were constructed using 75% of the data (100 intensity values for each of 12 trials) for network training and 25% (100 intensity values for each of the remaining 4 trials) for testing. Figure 17.21 shows the prediction capability of the trained neural time series models by comparing measured RHEED data to model output. These models yielded an average RMS error of 6.56 (arbitrary units) for training data and 6.03 for the test set. This error was less than 10% of total variation in RHEED intensity data.

17.5.4 RTP Process Modeling Using Optical Pyrometry Data

Gyurcsik et al. [1991] of North Carolina State University in Raleigh, NC, developed a first-principles model for a rapid thermal processor that used optical pyrometer readings to achieve temperature uniformity through lamp control. RTP systems usually have a bank of heating lamps, which can be individually controlled. However, temperature uniformity across an individual wafer can be difficult to attain, as a temperature gradient exists outward from the center of the wafer being processed. For a uniform heat flux at the surface of the wafer, its edge will be at a lower temperature than the center due to radial cooling. This nonuniformity can be counteracted by adjusting the relative power of the individual lamps, which is referred to *as lamp contouring*.

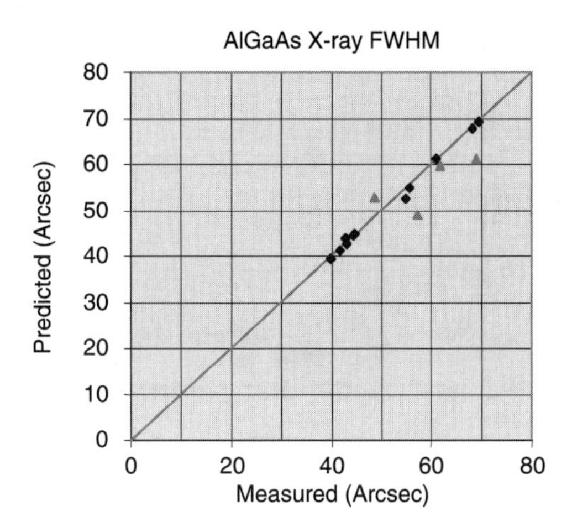

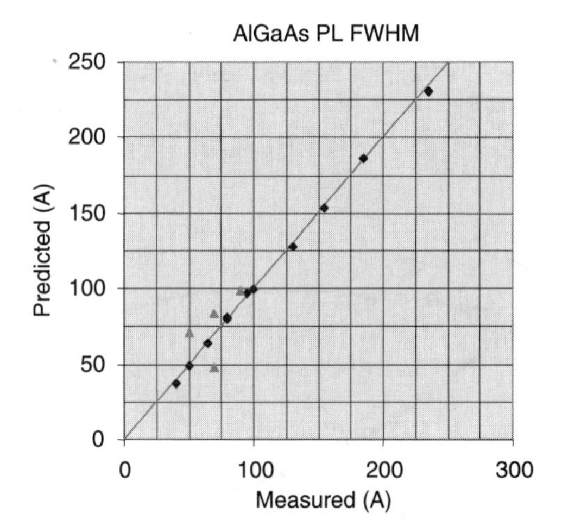

FIGURE 17.20 Neural process model predictions vs. experimental measurements for: (top) AlGaAs x-ray FWHM, and (bottom) AlGaAs PL FWHM (triangles represent test data not used during training). (From Lee, K. et al., Using neural networks to construct models of the molecular beam epitaxy process, *IEEE Trans. Semiconductor Manufac.*, 13(1), 34–45, 2000. With permission.)

Gyurcsik's model consists of two components. The first predicts the wafer temperature profile from individual lamp powers. This component models the heat balance of the wafer, which is used to determine the steady-state wafer temperature due to the heat flux density as a function of position on the wafer surface. The heat flux density at the surface is determined as a function of individual lamp power, lamp position, position on the wafer, and chamber geometry, including the reflectance of the chamber walls. The second component determines the relative lamp power necessary to achieve a uniform temperature profile. The end result is the following quadratic equation, which is used to relate the heat flux density due to the lamps at a radial position r on the wafer surface:

$$T_S(r) = C_0 I^2(r) + C_1(r) + C_2 \tag{17.33}$$

where T_S is the surface temperature, I is the heat flux, and C_0, C_1, and C_2 are coefficients determined by a least-squares fit of observed experimental data.

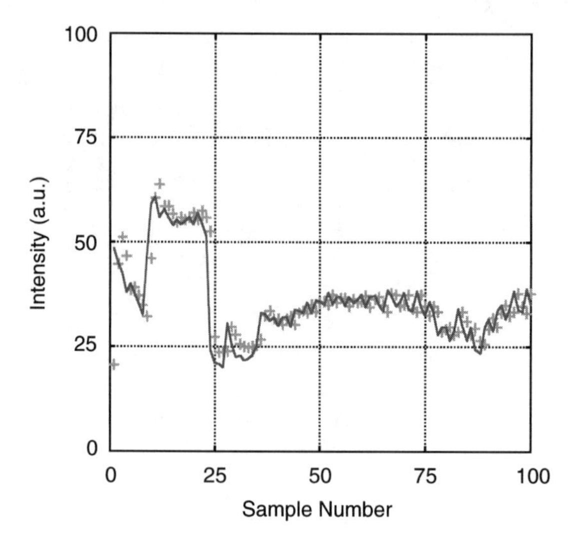

FIGURE 17.21 Typical prediction capability of the trained neural network time series models by comparing measured RHEED data to model output. (From Lee, K. et al., Using neural networks to construct models of the molecular beam epitaxy process, *IEEE Trans. Semiconductor Manufac.*, 13(1), 34–45, 2000. With permission.)

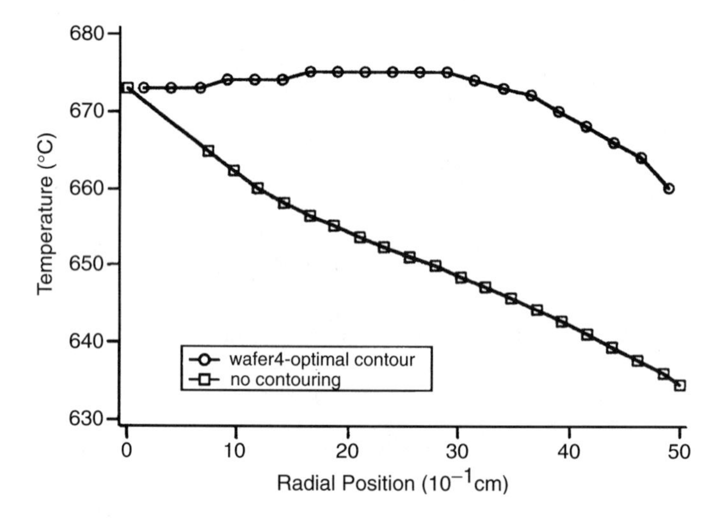

FIGURE 17.22 Comparison of experimental results for temperature uniformity at 675°C as a function of radial position with and without lamp contouring. (From Gyurcsik, R. et al., A model for rapid thermal processing: achieving uniformity through lamp control, *IEEE Trans. Semiconductor Manufac.*, 4(1), 9–13, 1991. With permission.)

These models were verified experimentally by RTCVD of polysilicon in a prototype RTP system similar to the schematic shown in Figure 17.4. A single-wavelength optical pyrometer was used to measure wafer temperature. The results presented the North Carolina State researchers showed a temperature uniformity of ±1%. A sample experimental temperature profile is shown in Figure 17.22. Overall, an average absolute temperature variation of 5.5°C and a worst-case absolute temperature variation of 6.5°C were achieved. These results compared quite favorably to the 30°C variation usually observed without control.

17.6 Optical Methods in Process Control

Semiconductor manufacturing processes must be stable, repeatable, and of high quality to yield products with acceptable performance. This implies that all individuals involved in manufacturing a product (including operators, engineers, and management) must continuously seek to improve manufacturing process output and reduce variability. Variability reduction is accomplished in a large part by strict process control. The application of process control in manufacturing continues to expand in the semiconductor manufacturing industry. In the following sections some case studies focusing on model-based statistical process control (SPC) [Montgomery, 1997] and run-by-run control techniques [Moyne et al., 2001] that use optical measurement methods as a means to achieve high-quality products are presented.

17.6.1 Control of Photoresist Properties in Photolithography Using Photospectrometry

Palmer et al. [1996] of the University of California at Berkeley have developed a means to reduce variations in photoresist parameters (such as film thickness and photoactive compound concentration) during manufacturing. A key issue in photolithography is CD control. In order to regulate fine line widths, it is necessary to correctly set the depth of focus and dose during exposure. These set points depend on the photoresist thickness and photoactive compound (PAC) concentration.

The approach of the Berkeley researchers, which consisted of a feedback scheme for accurately regulating these parameters, involved obtaining a static process model and recursively adjusting the model's coefficients based on previous wafer measurements. In so doing the adaptive model was used to determine the appropriate processing conditions for the next wafer. The equipment used to make the thickness and PAC measurements was a photospectrometer, a device that functions using principles similar to ellipsometry. The process variables actuated by the control scheme were the resist spin-coating speed and soft bake temperature.

To simultaneously regulate resist thickness (t) and PAC, these authors selected a static process model of the form

$$\log(t_k) = a_k\log(\omega_k) + b_k\log(T_k) + c_k + w_k \tag{17.34}$$

$$\log(\mathrm{PAC}_k) = d_k\log(\omega_k) + e_k\log(T_k) + f_k + w_k \tag{17.35}$$

where k is an index representing the wafer being processed, ω_k is the spin speed, T_k is the soft bake temperature, w_k is white measurement noise, and $a_k, \ldots, f_k$ are the model coefficients. The form of this model was validated by experiment. In order to regulate the model coefficients for the $(k + 1)$st wafer optimal estimates of the coefficients are obtained based on past wafer input–output measurements using Kalman filtering techniques [Kalman, 1960].

Experiments were conducted to verify this approach using unpatterned wafers consisting of silicon dioxide on a silicon substrate, which was subsequently coated with photoresist. Figure 17.23 shows a comparison of open-loop and closed-loop simultaneous control of photoresist thickness and PAC for a target resist thickness of 1.3 μm. Both the thickness and PAC concentration were within the limits determined by the measured value of sensor noise by the fifth wafer. These results support the conclusion that model-based feedback control using optical sensors is a promising approach for improving the consistency of photolithography. This can be accomplished with relative ease and minimal capital cost.

17.6.2 Statistical Feedback Control of Reactive Ion Etching Using OES

Mozumder and Barna [1994] of Texas Instruments in Dallas, TX, have developed a methodology for the automatic feedback control of a silicon nitride (SiN) reactive ion etching process. This methodology allows equipment operators to directly input the required process quality characteristic (such as etch rate or uniformity) rather than a recipe of process conditions (i.e., RF power, pressure, gas flow rates, etc.). The system relies on the following key components: (1) sensors for the measurement of process results,

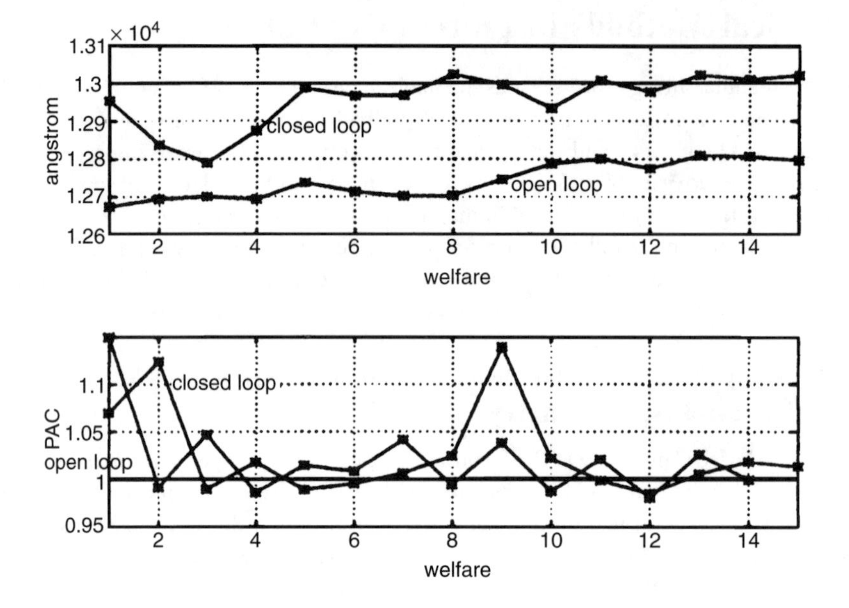

FIGURE 17.23 Open and closed-loop comparison of photoresist thickness and PAC concentration. (From Palmer, E. et al., Control of photoresist properties: a Kalman filter based approach, *IEEE Trans. Semiconductor Manuf.*, 9(2), 208–214, 1996. With permission.)

(2) in-line SPC techniques for ensuring these observables remain under statistical control, (3) a model to relate the observables to desired process outputs, and (4) an optimization tool to calculate new process settings for recentering an out-of-specification process.

The demonstration vehicle for this methodology was a SiN etch process using a $CHF_3/CF_4/O_2$ gas mixture. The requirements for this etch were a high nitride etch rate (ER) and low etch nonuniformity (NU). The control variables were chamber pressure, RF power, and the three gas-flow rates. The control algorithm was driven by *in situ* OES measurements. The slope of the OES endpoint curve was used as an indicator of etch rate and uniformity. The monochronometer of the OES system was tuned to the CN emission line at a 388.3-nm wavelength. Traces similar to that depicted in Figure 17.17 were typical.

The optimal settings of the process setpoints necessary to achieve the desired quality characteristic were determined using process models. In order to generate the necessary data for modeling ER and NU, a *D-optimal* experiment [Galil and Kiefer, 1980] requiring 31 trials was performed. Quadratic polynomial response surface models were subsequently fit to the experimental data using regression techniques.

The control strategy was based on the response surface models. The specific steps in the execution of this strategy were as follows:

1. Determine the target ER and NU.
2. Use an optimization package (such as *NPSOL* [Gill et al., 1986]) to solve the response surface models "in reverse" for the process settings necessary to achieve the target responses.
3. Measure the ER and NU using the OES sensor.
4. Perform standard SPC on the residuals (i.e., the difference between observations and model predictions).
5. If anomalous behavior is detected using SPC, then adjust the constant terms of the response surface models on a run-by-run basis to match the new state of the equipment.

Figure 17.24 shows a typical result obtained using this approach. This figure shows the moving average control chart of etch-rate measurements for a target etch rate of 5 nm/s. This chart, which plots a four-sample moving average of the etch-rate residuals vs. the sample number, illustrates the behavior of the output caused by RIE equipment state changes previously unknown to the control system, as well as the

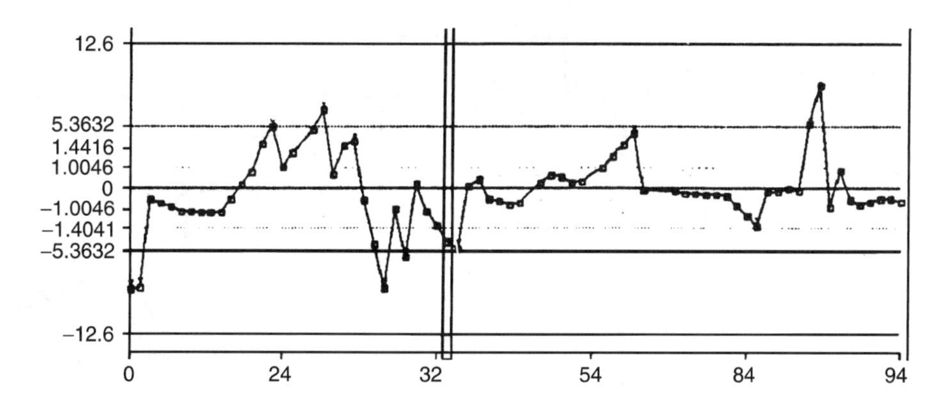

FIGURE 17.24 Moving average control chart for ER residuals. (From Mozumder, P. and Barna, G., Statistical feedback control of a plasma etch process, *IEEE Trans. Semiconductor Manuf.*, 7(1), 1–11, 1994. With permission.)

result of subsequent action by the system to recenter the process such that the actual output comes closer to the predicted value. The control algorithm was implemented, tested, and used to control the etching of thousands of product wafers under actual manufacturing conditions.

17.6.3 Run-by-Run Control of RIE Using Ellipsometry

Watts-Butler and Stefani [1994] of Texas Instruments in Dallas have used *in situ* ellipsometry to drive a run-by-run supervisory controller, which they have termed a *predictor corrector controller* (PCC), to alleviate the effect of machine and process drift in RIE. The process under investigation was the etching of polysilicon gates in a CMOS manufacturing line. This etching process determines the CD and thus the performance limits of the circuits produced. However, the process was known to drift due to aging of the reactor.

The response-surface modeling (RSM) technique was used to predict mean etch rate (MER) and uniformity from ellipsometry data. The process conditions that served as inputs to these models were RF power, chamber pressure, and total gas-flow (HCl + HBr) rate. Model coefficients were obtained from a central composite experimental design that required 21 trials. The etch uniformity was estimated by deriving relationships between the etch rate at the center of each wafer (as measured by ellipsometry) and at each of ten other specific sites on the wafer.

The predictive RSM models were employed by the PCC to generate optimal recipe settings to achieve etch rate and uniformity targets. The control system objectives were: (1) target tracking without lag, (2) disturbance compensation, and (3) noise rejection. The approach here represents an improvement over that of Mozumder and Barna described in Section 17.6.2 above. In Mozumder and Barna's system, the model is not tuned until *after* an SPC failure has been detected, which results in an unavoidable lag in the control action. The PCC system, however, exerts feedback control after each measurement, as opposed to after SPC failures.

The key component of the PCC is the *double exponential forecasting filter* (DEFF). This filter smooths current data (i.e., reduces noise) and provides a forecast. The DEFF consisted of one filter to estimate the output and another to estimate its trend. In other words,

$$\text{Current smoothed output} = (1-\alpha)(\text{current actual output}) + \alpha(\text{previous estimate}) \tag{17.36}$$

$$\text{Current smoothed trend} = (1-\beta)(\text{trend estimate}) + \beta(\text{previous trend}) \tag{17.37}$$

$$\text{Forecast} = \text{current smoothed output} + \text{current smoothed trend} \tag{17.38}$$

where α and β are tuning constants. The output data to be filtered were the RSM model residuals (i.e., measurements minus predictions). The equations for the DEFF that correspond to Eqs. (17.36)

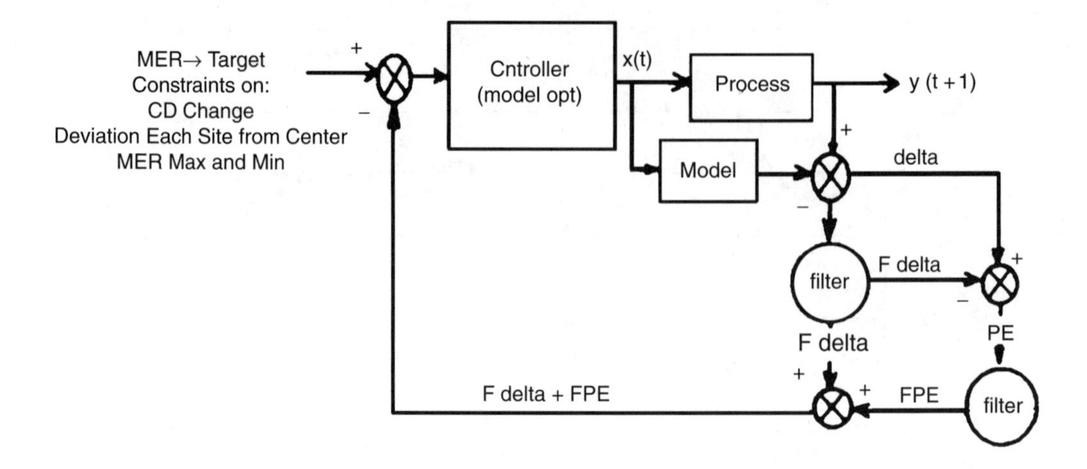

FIGURE 17.25 PCC system. (From Watts-Butler, S. and Stefani, J., Supervisory run-to-run control of polysilicon gate etch using *in situ* ellipsometry, *IEEE Trans. Semiconductor Manuf.*, 7(2), 193–201, 1994. With permission.)

to (17.38) are

$$\text{Fdelta}_t = (1 - \alpha)\text{delta}_t + \alpha^*\text{Fdelta}_{t-1} \quad (17.39)$$

$$\text{PE}_t = \text{delta}_t - \text{Fdelta}_{t-1} \quad (17.40)$$

$$\text{FPE}_t = (1 - \beta)\text{PE}_t + \beta^*\text{FPE}_{t-1} \quad (17.41)$$

$$\text{Prediction}_t = \hat{y}_t + \text{Fdelta}_t + \text{FPE}_{t-1} \quad (17.42)$$

where Fdelta_t is the filtered model error at time t, delta_t is the unfiltered model error at time t, PE_t is the unfiltered prediction error at time t (which serves as the trend estimate), FPE_t is the filtered prediction at time t, and $\hat{y}_t$ is the RSM model prediction at time t.

Figure 17.25 is a block diagram of PCC system. The "Controller" block in this figure is the commercial nonlinear optimization package *NPSOL* [Gill et al., 1986]. This package was used to solve the RSM equations "in reverse" to determine the optimal process recipe corresponding to targets. Because multiple solutions are possible, the controller chooses the solution closest to the current operating point. Set points for the last wafer are used if they produce predicted responses within one standard deviation of the target. If no solution is possible, the most recent recipe is repeated, or the system quits so that the problem may be diagnosed.

Implementation of the PCC initially occurred in a 200-wafer demonstration experiment in which half of the wafers used a standard recipe and the other half used PCC-generated optimal recipes. During this demonstration, two equipment faults were simulated: (1) a miscalibrated power supply and (2) neglecting the prior wafer-cleaning step. The controlled and uncontrolled measurement residuals for the process etch rate are shown in Figure 17.26. Overall, PCC resulted in a 36% decrease in the standard deviation from target for the mean etch rate, and similar results were achieved for uniformity. In addition, the natural variance of the process did not increase when PCC was used, indicating that continuous run-by-run control did not cause unnecessary control actions.

17.7 Optical Methods in Process Diagnosis

Optical measurement techniques are also useful for automated process or equipment diagnosis. When unreliable equipment performance causes operating conditions to vary beyond an acceptable level, overall product quality is severely jeopardized. Thus, timely and accurate malfunction diagnosis is a key to the success of the manufacturing process. Diagnosis involves determining the assignable causes for the equipment malfunctions and correcting them quickly to prevent the subsequent occurrence of expensive

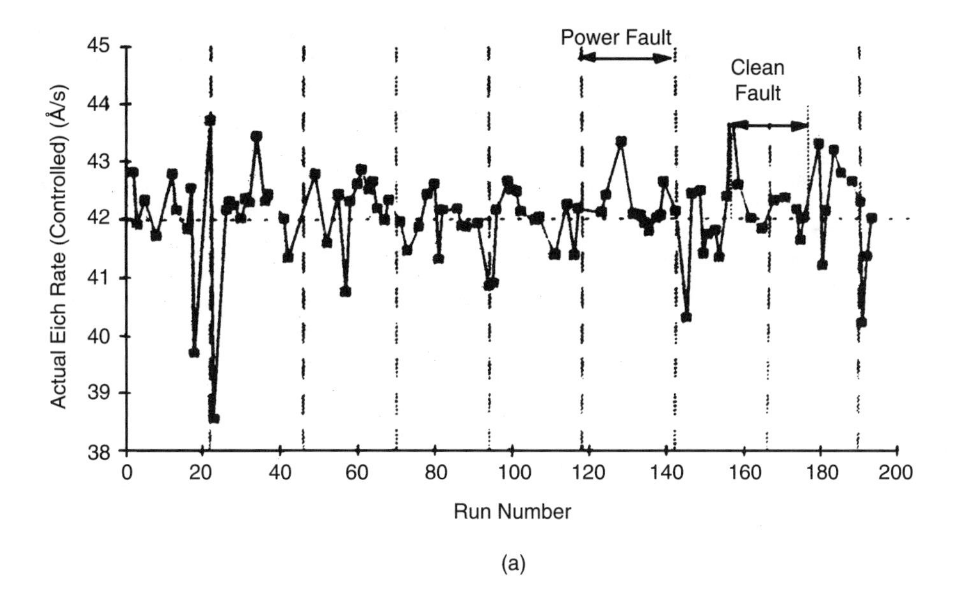

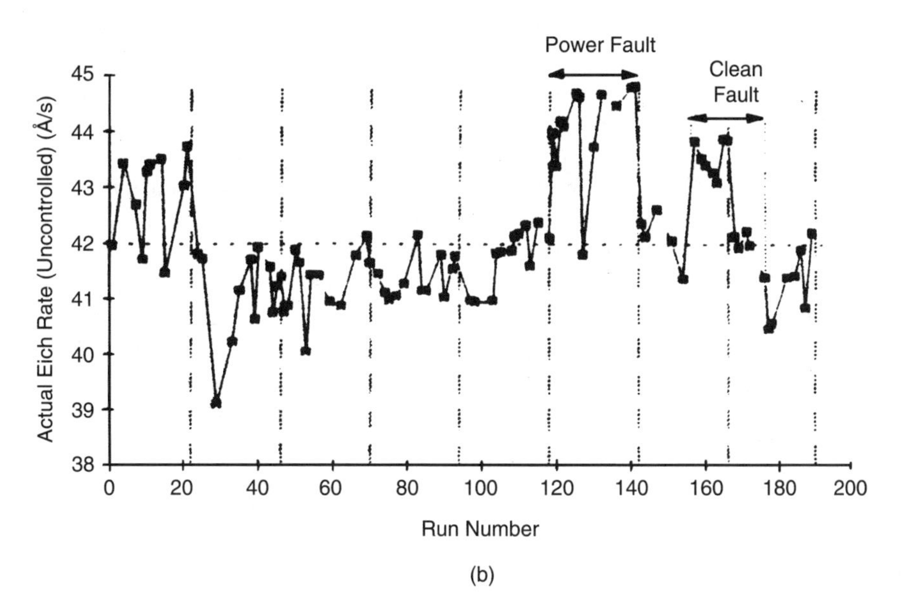

FIGURE 17.26 Actual etch rate (dotted line is the target): (a) controlled and (b) uncontrolled. (From Watts-Butler, S. and Stefani, J., Supervisory run-to-run control of polysilicon gate etch using *in situ* ellipsometry, *IEEE Trans. Semiconductor Manuf.*, 7(2), 193–201, 1994. With permission.)

misprocessing. Expert systems such as *PEDX* [Dolins et al., 1988] have been designed to use OES endpoint traces to develop qualitative models of RIE process behavior. In the following sections, two other case studies involving the use of optical methods for process diagnosis are presented.

17.7.1 An Equipment Diagnostic System for Photolithography Based on Photospectrometry

Leang and Spanos [1997] of the University of California at Berkeley have a general diagnostic system to assist photolithography equipment operators in finding causes of anomalous machine performance. Based on conventional probability theory, this diagnostic system incorporates both shallow- and

TABLE 17.1 Diagnostic Example For Photolithographic Exposure Fault

Fault	Probability (%)	Probability Range (%) (Confidence Level = 90%)
Wrong input thickness	3.19	0.65–7.23
Wrong input PAC	3.73	0.94–8.03
Wrong input dose	9.06	4.35–15.19
Bad lamp	34.73	26.14–44.21
Bad environmental temperature	1.29	0–4.13
Bad lamp strike	14.51	8.49–21.83
Damaged filter optics	0.22	0–1.71
Bad shutter timing circuit	7.78	3.44–13.55
Bad light integrating circuit	12.81	7.15–19.80
PAC measurement error	11.26	5.97–17.92
Miscellaneous fault	1.60	0–4.67
No fault	1.36	0–4.26

Source: Leang, S. and Spanos, C., A general equipment diagnostic system and its application on photolithographic sequences, *IEEE Trans. Semiconductor Manufac.*, 10(3), 329–343, 1997. With permission.

deep-level information. While sensor malfunctions and incorrect process settings are diagnosed from equipment models and measurements, environmental and maintenance problems are diagnosed from operator observations, sensor alarms, and maintenance logs.

From observed evidence (such as photospectroscopy measurements on photoresist) and from conditional probabilities of faults supplied by machine experts (and subsequently updated by the system), unconditional fault probabilities and their bounds are computed using rigorous *Bayesian inference* techniques [de Finetti, 1974]. The objective of the system is to provide a quantitative mapping between observable *evidence* (such as measurements, sensor readings, machine age, etc.) and potential equipment faults (such as incorrect film thickness, incorrect process settings, measurement errors, etc.).

As an example of the implementation of this system, consider the photoresist exposure step in the photolithography process sequence. In this example, a control alarm was triggered on an exposure tool because the measured photoactive compound (PAC) concentration in the resist drifted upward, triggering the alarm at a value of 0.43 when an equipment model predicted a value of 0.32. The input thickness and input PAC concentration were within specifications of 1.31 μm and 0.96, respectively. The input exposure dose was specified at a standard recipe value of 167 mJ/cm^2. The environmental temperature was within process tolerance. The ages of the exposure lamp and filter optics were 67 and 50 days, respectively, while their characteristic lifetimes were 45 and 120 days, respectively. The diagnostic system estimated the fault probabilities shown in Table 17.1. This diagnosis was corroborated by a maintenance technician who traced the actual cause of the problem to a weak lamp.

17.7.2 Fault Detection in RIE Systems Using OES Signals

Yue et al. [2000] of Tokyo Electron America, Inc., in Austin, TX, investigated the suitability of using OES for fault identification in plasma etching. A plasma etcher may operate under faulty conditions and produce off-specification wafers without detection until off-line metrology tests are performed. This delay results in misprocessed products.

These researchers used an OES system capable of collecting 512 different wavelengths. The large amount of data generated by this approach poses a difficulty in extracting the relevant information for fault detection and classification. This is because only a few wavelengths are sensitive enough to the relevant process and wafer states. If these key wavelengths can be identified, the amount of data required for process monitoring can be reduced.

These authors approached this problem using PCA to analyze the sensitivity of multiple wavelength scans within a wafer with respect to typical process faults. An example of such a fault is "etch stop," a fault that occurs in via etching when the polymer deposition rate in an open area is greater than the etch

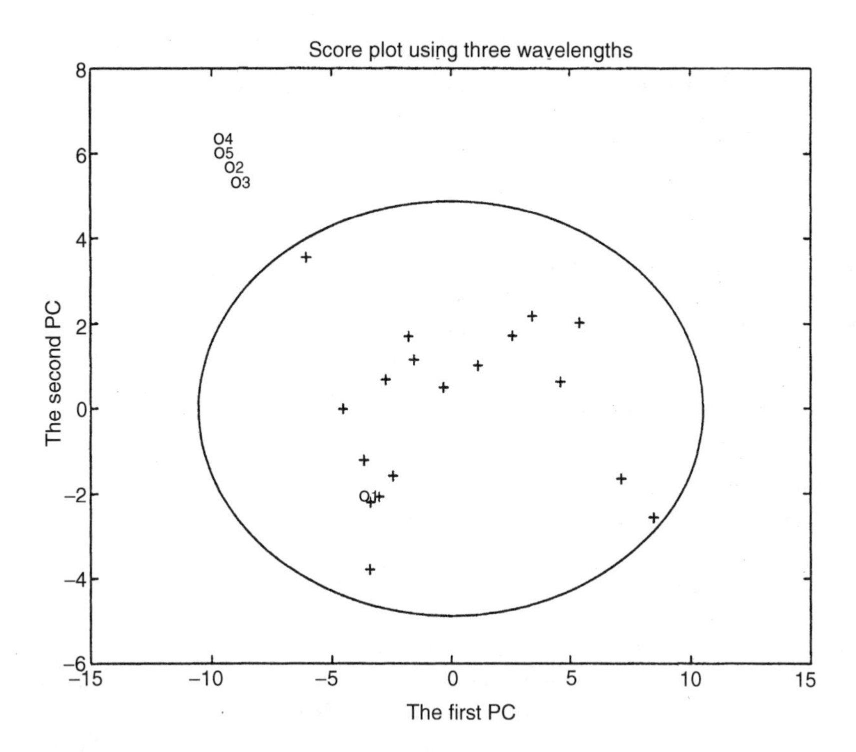

FIGURE 17.27 Scatter plot of first and second PCs indicating normal and faulty wafers. (From Yue, H. et al., Fault detection of plasma etchers using optical emission spectra, *IEEE Trans. Semiconductor Manuf.*, 13(3), 374–385, 2000. With permission.)

rate and ion bombardment is incapable of further removal of the polymeric material. The PCA implemented was *multiway* PCA (MPCA), a technique in which a two-dimensional data set is unfolded from a three-dimensional original data set. In this case, the three dimensions of the raw OES data set were the wafer number, time, and wavelength. In this approach the unfolded OES data matrix is arranged such that each row contains one time scan, and each column contains all the data at one wavelength. After the data are rearranged in this manner, standard PCA ensues.

The test vehicle for this research was a contact/via oxide etch with 1 to 2% open area on 2500 wafers in an *Applied Materials 5300* system. To obtain data from a known etch stop condition, an etch stop was manually induced for the final four wafers in the run. After PCA, four principal components, representing 99.5% of the variance in the raw OES data, were retained. Figure 17.27 shows the scores of the first PC plotted against those of the second PC. The points that fall outside of the ellipse are indicative of the faulty wafers. This plot shows that this approach is indeed an effective method of fault detection.

17.8 Summary

This chapter has provided an overview of issues relevant to optical methods for process monitoring, modeling, control, and diagnosis in semiconductor manufacturing. This included a description of some basic unit processes, a review of the measurement techniques necessary to characterize these manufacturing operations, and a presentation of several case studies.

In semiconductor manufacturing, process and equipment reliability directly influence cost, throughput, and yield. Over the next several years significant process modeling and control efforts will be required to reach projected targets for future generations of microelectronic devices and integrated circuits. Optical measurement methods will provide a strategic advantage in undertaking these tasks and have certainly proven to be a viable technique.

Defining Terms

CVD: Chemical vapor deposition; semiconductor fabrication process in which material is deposited on a substrate using reactive chemicals in the vapor phase.

ellipsometry: Optical measurement technique that is based on the polarization changes that occur when light is reflected from or transmitted through a medium.

expert systems: Experiential or algorithmic systems that attempt to encode and use human knowledge to perform inference procedures (such as fault diagnosis).

factorial designs: Experimental designs in which multiple input variables are varied simultaneously at two or more discrete levels in every possible combination.

fractional factorial designs: Factorial designs that reduce the number of experiments to be performed by exploring only a fraction (such as one half) of the input variable space in a systematic manner.

FTIR: Fourier transform infrared spectroscopy; method for identifying organic compounds, such as those that may result from the etching of polymer films.

IC-CIM: Computer-integrated manufacturing of integrated circuits.

interferometry: Optical method for measuring film thickness; consists of a light source focused on a semiconductor wafer while a detector measures the reflected light intensity.

ion implantation: Method for doping semiconductors in which controlled doses of ionized impurity atoms are accelerated through an electric field to strike the surface and penetrate the substrate.

least squares: Method of curve fitting in which a model is developed by finding coefficients that minimize the squared difference between model predictions and actual measured data.

MBE: Molecular beam epitaxy; well-developed and versatile technology for growing various III-V, II-VI, and IV-IV semiconductor structures capable of achieving stringent composition and thickness control.

neural networks: Artificial models that crudely mimic the functionality of biological neurological systems.

OES: Optical emission spectroscopy; a bulk measure of the optical radiation of plasma species most often used to obtain the average optical intensity at a particular wavelength above the wafer and infer etch endpoint.

oxidation: Growth of thin SiO_2 films in silicon substrates.

PCA: Principal component analysis; statistical technique for streamlining a multidimensional data set.

photolithography: Patterning of layers of materials used in integrated circuit fanrication.

pyrometry: Temperature measurement technique that monitors the radiant energy received in a certain band of energies, assuming the source is a gray body of known emissivity.

RHEED: Reflection high-energy electron diffraction; intensity of RHEED features shows oscillatory behavior directly related to the growth rate in MBE.

RIE: Reactive ion etching; method of removing material by reactive gases at low pressures in an electric field; also known as "plasma etching."

RSM: Response surface methodology; statistical method in which data from designed experiments are used to construct polynomial response models whose coefficients are determined by regression techniques.

RTP: Rapid thermal processing; a suite of single-wafer, high-temperature processes that have been developed to minimize thermal budget by reducing time at temperature.

scatterometry: Optical measurement technique used for patterned features based on an analysis of the light diffracted (or *scattered*) from a periodic structure such as a grating.

SEM: Scanning electron microscopy; imaging technique for assessing fine feature sizes.

SPC: Statistical process control; method of continuous hypothesis testing to ensure that a manufactured product meets its required specifications.

time series modeling: Statistical method for modeling chronologically sequenced data.

References

Box, G. and Draper, N., *Empirical Model-Building and Response Surfaces*, Wiley, New York, 1987.

Box, G. and Jenkins, G., *Time Series Analysis: Forecasting and Control*, Holden-Day, San Francisco, 1976.

Box, G., Hunter, W., and Hunter, J., *Statistics for Experimenters*, Wiley, New York, 1978.

Brown, A., Molecular beam epitaxy of III-V materials including migration-enhanced epitaxy, in *Encyclopedia of Advanced Materials*, Bloor, D. et al., Eds., Pergamon Press, Elmsford, NY, 2000.

Campbell, S., *The Science and Engineering of Microelectronic Fabrication*, Oxford University Press, New York, 2001.

Crisalle, O., Bickerstaff, C., Seborg, D., and Mellichamp, D., Improvements in photolithography performance by controlled baking, *Integrated Circuit Metrology, Inspection, and Process Control II*, SPIE 921, 1998, pp. 317–325.

Dax, M., Top Fabs of 1996, *Semiconductor Int.*, 19(5), 100–106, 1996.

Dayhoff, J., *Neural Network Architectures: An Introduction*, Van Nostrand-Reinhold, New York, 1990.

De Finetti, B., *Theory of Probability*, Wiley, New York, 1974.

De Gyvez, J. and Pradhan, D., *Integrated Circuit Manufacturability*, IEEE Press, Piscataway, NJ, 1999.

Dolins, S., Srivastava, A., and Flinchbaugh, B., Monitoring and Diagnosis of Plasma Etch Processes, *IEEE Trans. Semiconductor Manufac.*, 1(1), 23–27, 1988.

Galil, Z. and Kiefer, J., Time- and space-saving computer methods, related to Mitchell's DETMAX, for finding D-optimum designs, *Technometrics*, 22(3), 301–313, 1980.

Gill, P., Murray, W., Saunders, M., and Wright, M., *User's Guide for NPSOL (Smoothed Nonlinear Constrained Optimization) (Version 4.0): A Fortran Package for Nonlinear Programming*, Stanford University, Stanford, CA, 1986.

Gyurcsik, R., Riley, T., and Sorrell, Y., A model for rapid thermal processing: achieving uniformity through lamp control, *IEEE Trans. Semiconductor Manufac.*, 4(1), 9–13, 1991.

Halliday, D. and Resnick, R., *Physics*, Wiley, New York, 1978.

Herman, M. and H. Sitter, H., *Molecular Beam Epitaxy: Fundamentals and Current Status*, Springer-Verlag, New York, 1989.

Joliffe, I., *Principal Component Analysis*, Springer-Verlag, New York, 1986.

Kalman, R., A new approach to linear filtering and prediction Problems, *Trans. ASME J. Basic Eng.*, 82, 35–45, 1960.

Kim, T. and May, G., Time series modeling of photosensitive BCB development rate for via formation applications, in *Proc. 1999 Int. Electron. Manufact. Technol. Symp.*, 1999.

Landzberg, A., *Microelectronics Manufacturing Diagnostics Handbook*, Van Nostrand-Reinhold, New York, 1993.

Leang, S. and Spanos, C., A general equipment diagnostic system and its application on photolithographic sequences, *IEEE Trans. Semiconductor Manuf.*, 10(3), 329–343, 1997.

Lee, K., Brown, T., Dagnall, G., Bicknell-Tassius, R., Brown, A., and May, G., Using neural networks to construct models of the molecular beam epitaxy process, *IEEE Trans. Semiconductor Manuf.*, 13(1), 34–45, 2000.

Manos, D. and Flamm, D., *Plasma Etching: An Introduction*, Academic Press, San Diego, CA, 1989.

McGilp, J., Weaire, D., and Patterson, C., Eds, *Epioptics*, Springer-Verlag, New York, 1995.

Meindl, J., Evolution of solid-state circuits: 1958–1992–20??, *IEEE ISSCC Commemorative Supplement*, 1993.

Montgomery, D., *Introduction to Statistical Quality Control*, Wiley, New York, 1997.

Moyne, J., Castillo, E., and Hurwitz, A., *Run-to-Run Control in Semiconductor Manufacturing*, CRC Press, Boca Raton, FL, 2001.

Mozumder, P. and Barna, G., Statistical feedback control of a plasma etch process, *IEEE Trans. Semiconductor Manuf.*, 7(1), 1–11, 1994.

Palmer, E., Ren, W., Spanos, C., and Poolla, K., Control of photoresist properties: a Kalman filter based approach, *IEEE Trans. Semiconductor Manuf.*, 9(2), 208–214, 1996.

Parker, E., *The Technology and Physics of Molecular Beam Epitaxy*, Plenum Press, New York, 1985.

Pope, J., Woodburn, R., Watkins, J., Lachenbruch, R., and Viloria, G., Manufacturing integration of real-time laser interferometry to etch silicon dioxide films for contacts and vias, *Proc. SPIE Conf. Microelec. Processing*, 2091, 185–196, 1993.

Press, W., Flannery, B., Teukolsky, S., and Vetterling, W., *Numerical Recipes in C*, Cambridge University Press, Cambridge, MA, 1988.

Raymond, C., Angle-resolved scatterometry for semiconductor manufacturing, *Microlithography World*, 18–23, 2000.

Raynaud, P., Amilis, T., and Segui, Y., Infrared absorption analysis of organosilicon/oxygen plasmas in a microwave multipolar plasma excited by distributed electron cyclotron resonance, *Appl. Surface Sci.*, 138–139, 285–291, 1999.

Sze, S., *VLSI Technology*, McGraw-Hill, New York, 1983.

Tompkins, H. and McGahan, W., *Spectroscopic Ellipsometry and Reflectometry*, Wiley, New York, 1999.

Watts-Butler, S. and Stefani, J., Supervisory run-to-run control of polysilicon gate etch using *in situ* ellipsometry, *IEEE Trans. Semiconductor Manuf.*, 7(2), 193–201, 1994.

White, D., Boning, D., Butler, S., and Barna, G., Spatial characterization of wafer state using principal component analysis of optical emission spectra in plasma etch, *IEEE Trans. Semiconductor Manuf.*, 10(1), 52–61, 1997.

Wolf, S. and Tauber, R., *Silicon Processing for the VLSI Era*, Lattice Press, Sunset Beach, CA, 2000.

Wong, K., Boning, D., Sawin, H., Butler, S., and Sachs, E., Endpoint prediction for polysilicon plasma etch via optical emission interferometry, *J. Vac. Sci. Tech. A*, 15, 3, 1997.

Yang, F., McGahan, W., Mohler, C., and Booms, L., Using optical metrology to monitor low-K dielectric thin films, *Micro*, 5, 31–38, 2000.

Yue, H., Qin, S., Markle, R., Nauert, C., and Gatto, M., Fault detection of plasma etchers using optical emission spectra, *IEEE Trans. Semiconductor Manuf.*, 13(3), 374–385, 2000.

Further Information

The quarterly journal *IEEE Transactions on Semiconductor Manufacturing* is widely acknowledged as the definitive IEEE publication on semiconductor manufacturing. This publication often contains articles on optical measuremnt methods for semiconductor process monitoring, modeling, control, and diagnosis. For a subscription, contact the IEEE Service Center, 446 Hoes Lane, P.O. Box 1331, Piscataway, NJ 08855–1331; phone: (800) 678-IEEE.

18

Optical-Based Manufacturing Process: Monitoring and Control

Masatake Shiraishi
Ibaraki University
Ibaraki, Japan

18.1 Introduction

In an unmanned flexible manufacturing system (FMS) or computer-integrated manufacturing system (CIM) all functions must be automated and adaptively carried out without human intervention at the process–machine hardware level. Actually, a closed-loop adaptive control system needs sensors to feed back the real state of the working environment in the presence of disturbances. The objective is to realize a system in which process performance could be continuously reviewed and process variables varied for achieving optimal machining conditions with respect to maximum productivity, minimum cost, high accuracy, etc. The success of such systems relies heavily on the availability of online sensing devices. Therefore, the development of in-process measurement, including monitoring and control techniques, is the key factor for achieving the real goal of FMS and CIM.

In many processes, however, mechatronic components alone cannot achieve a desirable function or performance as specified for a system design because the measurements are difficult or not even feasible due to inherent characteristics of the systems. In some cases the measurement data obtained by conventional sensors are not accurate or reliable enough to be used for further processing. They may sometimes be noisy, necessitating some means of filtering or signal conditioning.

Due to this limitation in recent years optical sensors and measurement technologies have been increasingly incorporated at an accelerated rate into processes and systems.

This chapter focuses on current thinking on optical-based (noncontact) in-process measurement and control in such processes as machining and welding.

0-8493-1162-4/03/$0.00+$1.50

18.2 Machining Processes and Machines: Basic Concept [Shiraishi, 1994]

18.2.1 Basic Concept

The primary objective of in-process measurement in machining is to provide information on the basis of which corrections can be made during machining to ensure that the workpiece is manufactured to the desired size. Actually, however, it is also important to monitor tool failures such as cracking, chipping, and tool breakage in an online fashion, even though the control action may not always be taken in that case. In other words, monitoring the in-process measurement for size control ensures that the workpiece is manufactured to the desired size and monitoring machining conditions prevents tool failures such as cracking, chipping, and tool breakage. Accordingly, the role of in-process measurement must be widely interpreted as the online measurement for monitoring as well as control. FMS, for example, is usually comprised of cells of direct numerically controlled machine tools, conveyor, or robotic work handling and scheduling, all under the control of a central computer. Two or more machine tools are connected by robots for workpiece load/unload, interoperational inspection, tool changing, etc. With the rapid progress of such systems the removal of human participation from manufacturing processes naturally increases the demand for sensing and measurement to provide monitoring of both process and product.

18.2.1.1 Measurement Objects and Items

Machining accuracy depends on a number of factors involved in the machine-tool-workpiece structure and cutting process. In order to control and monitor manufacturing conditions it is necessary to sense the real states of corresponding factors in an online fashion. Figure 18.1 shows the measurement objects and items in machining operations. Among these, measurement objects such as the cutting tool, workpiece, and cutting process are directly or indirectly related to product quality, machining cost, or machining efficiency, while the measurement of the machine tool is mainly for ensuring safe operation. Most of the measurement items in the figure are measurable on-line, directly or indirectly, by contact or noncontact methods, depending on the type of machining processes involved. The selection of a sensor should be made considering the environment in which the instrument must operate. Also, the instrument must be capable of being accommodated in the space available without interfering with the basic operation of the machine tool.

In applying an optical technique countermeasures for disturbances such as a part vibration, swarf, cutting fluid, etc. should be taken to avoid their influence on the measurement. There is no general solution for this problem, and adequate countermeasures have been proposed case by case. An example of countermeasures in turning is provided in Table 18.1, where the proposed method seems to be basically applicable to other processes. In this approach a dimensional measurement of workpiece diameter was carried out by using a laser beam with a resolution of 0.2 μm. The basic principle of this measurement is shown in Figure 18.2. The laser unit is mounted on the tool post, and a photodetector that has a certain angle with respect to the tool is beside the tool post. To gauge the distance between the tool post and the workpiece surface a laser beam is illuminated just under the tool edge as close to the cutting position as possible. The beam passes by the side of the tool edge, which is set with a side cutting-edge angle of –5°. With the laser spot located 0.4 mm below the tool edge, the gap error caused by the delay between measurement and cutting point is suppressed within a very narrow region. The laser spot on the work surface is magnified optically, and the image of the spot is made on the photodetector. Thus, the information about the diameter change is conveyed by means of a displacement of this image. Serious problems in this measurement were those caused by the cutting fluid attached to the work surface and by the swarf around the laser beam.

First, in order to see how well the air blast method shown in Table 18.1 would remove the cutting fluid the measurement of reflection light was done by using two cutting oils with different viscosities. Figure 18.3 illustrates the relationship between air pressure (MPa) and relative intensity (I) of the reflection light. Plots were made with a photoconductive cell by measuring its resistance change. In the

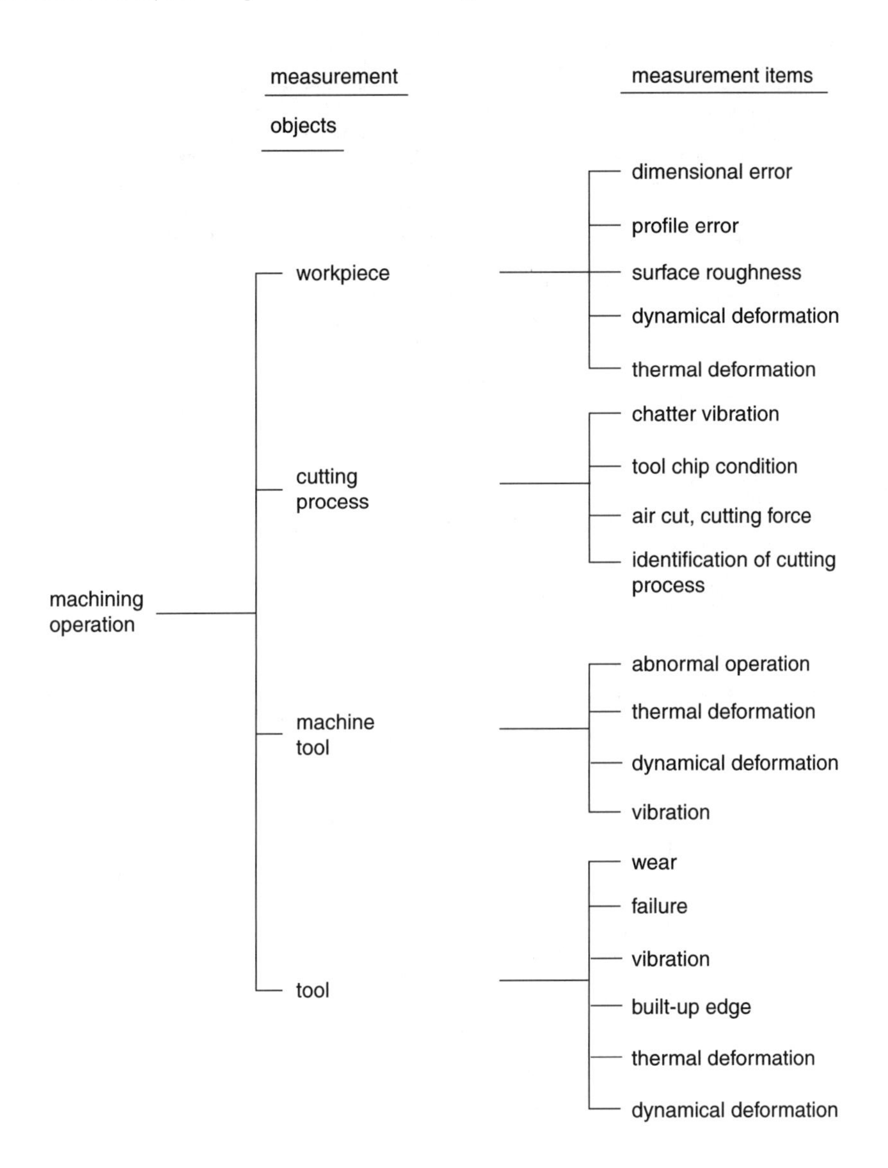

FIGURE 18.1 Measurement objects and items in machining.

figure, I = 1.0 corresponds to a resistance measured without oil. The data make clear that the intensity change for the spindle oil of low viscosity differs markedly from that of the machine oil of high viscosity, while at the same blasting air pressure. At a pressure of about 0.19 MPa (2.0 kg/cm^2), a low-viscosity oil provides a reflection light of nearly I = 1.0, while a high-viscosity oil provides at most I = 0.9 even at a pressure of 0.49 MPa (5.2 kg/cm^2). Therefore, the air blast is effective for cutting fluid with a viscosity of up to 1.8×10^{-5} m^2/s in this experiment.

Next, swarfs such as shear types and crack types could also be removed by the air blast. Figure 18.4 gives a comparison of the measured reflection lights with and without air blasts around the laser spot on the work surface. As demonstrated in Figure 18.4(a), the effects of swarfs are quite large, and it is almost impossible to continue the measurement. On the other hand, as shown in Figure 18.4(b), cutting a measurement off is an event that is short in duration, and serious effects are not evident from the record. The results of several experiments have suggested that this cut-off time t against the total machining time T should be within 10% in terms of (t/T) from the viewpoint of a real in-process measurement. This criterion was found to be effective at an air pressure of over 0.17 MPa (1.8 kg/cm^2).

TABLE 18.1 Countermeasure in Turning

Disturbances	Effects	Countermeasures
Forced vibration, chatter vibration	Instability of the measured signal	Add electrical low pass filter and bandpass filter to the measuring circuit
Cutting fluid	Irregular light reflection through oil film surface and oil splash to the measuring system	Add air blast close to the detecting point and the light pass
Swarf	Shielding the light pass	Add air blast around the light pass
Temperature	Drift of the measured signal	Provide constant low pressure air blast to the sensor and measuring circuit
Surface irregularities of workpiece	Instability of the measured signal	Increase the light spot size

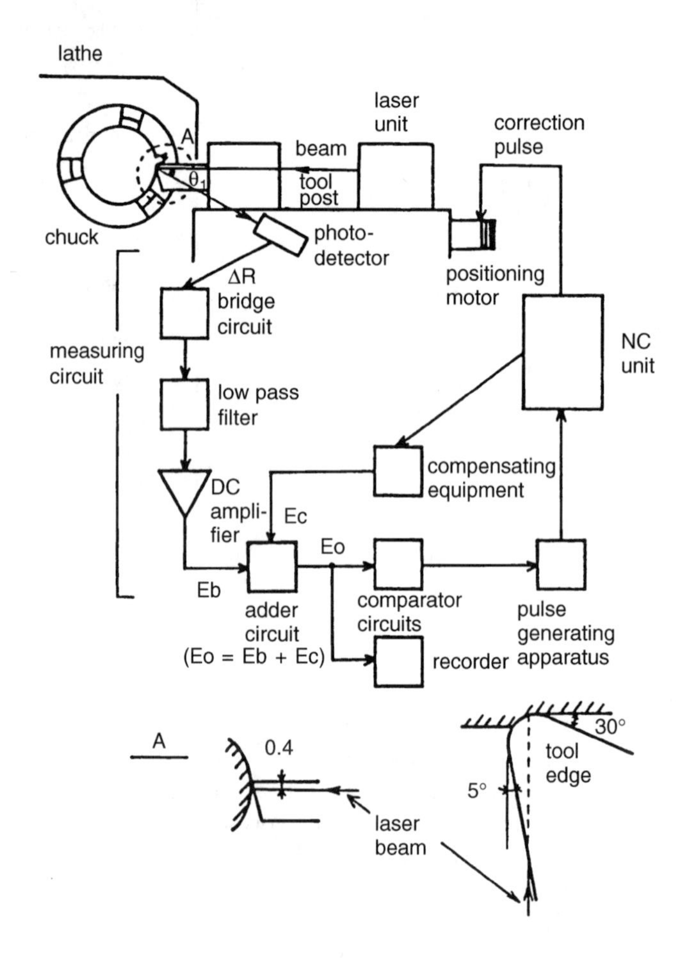

FIGURE 18.2 Countermeasure in dimension measurement.

Air blasting is very simple, and results obtained from it essentially indicate the capability of in-process measurement by the optical method in the production floor. At present, only a few sensors incorporate air-blast systems. A typical example of the commercially available proximity sensor is shown in Figure 18.5, where the air blast tube is incorporated to remove swarfs.

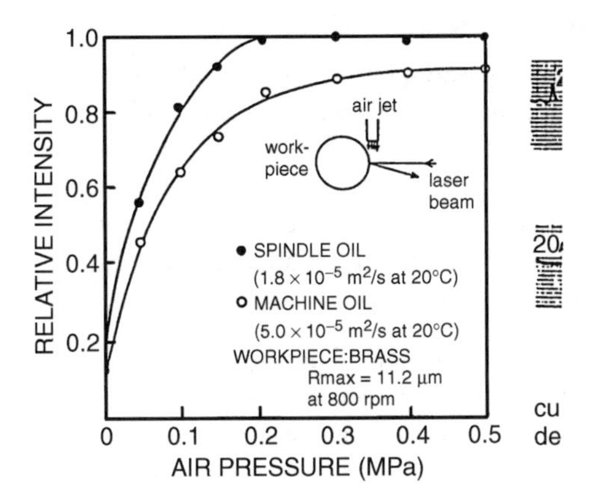

FIGURE 18.3 Intensity of reflection light vs. air pressure.

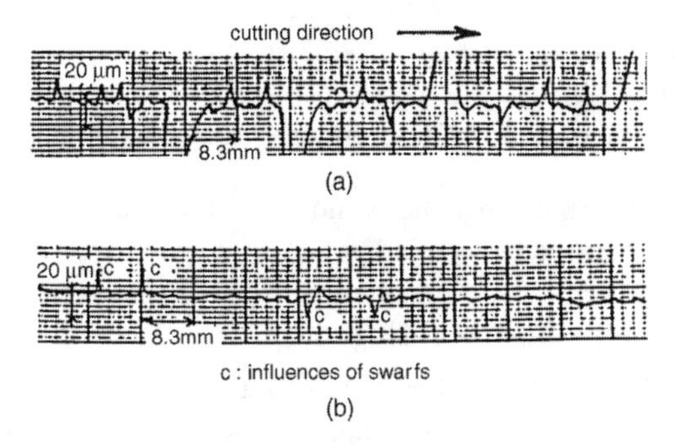

FIGURE 18.4 Reflection light during machining: (a) without air blast; (b) with air blast (0.2 MPa); cutting speed = 160 m/min, feed = 0.2 mm/rev., depth of cut = 0.4 mm, workpiece = S45C.

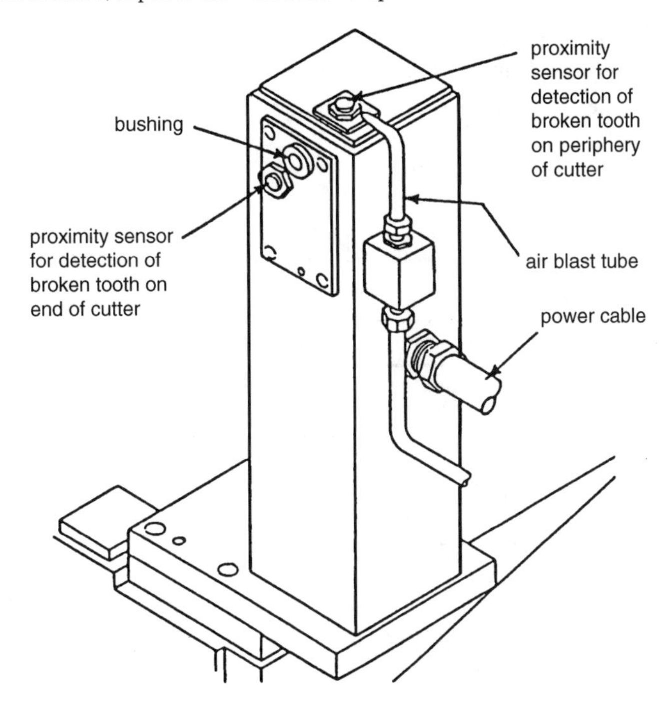

FIGURE 18.5 Proximity sensor with an air blast.

18.2.1.2 Other Types of Measurement

It is evident that correction and inspection by in-process methods are best for highly automated manufacturing systems. In case of the dimensional control of a workpiece, for example, not only systematic errors but also random errors can be corrected by means of an in-process measurement. In a real production floor, however, other types of measurements are also adopted in the FMS and CIM in the form of preprocess, interprocess, or postprocess because of some technical difficulties involved in the production environments. Figure 18.6 shows the relation between cutting time and measurement types in machining process, as mentioned below.

1. *Preprocess*: As a checking procedure, several measurements are conducted offline before machining. In a certain CIM, for instance, workpieces are monitored and checked to eliminate substandard products by using a two-dimensional image sensor before they are fed to the machining center.
2. *Interprocess*: The measurements are done offline between machining processes, as indicated in Figure 18.6. This measurement style may be taken in case of a rough cut because the information obtained during rough cut is effectively used for finish cut.
3. *Postprocess*: The measurements are done offline after machining, and then that information is used to make corrections for the subsequent workpieces. Systematic errors will often cause a relatively slow drift of size to one side of the tolerance band. Therefore, the postprocess measurement and control can provide a satisfactory accuracy for the next workpieces.

These measurement types have both advantages and disadvantages, consideration of which may sometimes be a preliminary step for in-process measurement.

18.2.1.3 Control Technology in Machining

There has been considerable research on control systems on the basis of an adaptive control (AC). However, these are not always successfully introduced into industry. The main difficulties lie in:

- Adequate sensors that can reliably measure the necessary process parameters
- Formulation of a PI (performance index) with an appropriate control policy
- Real-time evaluation of the proposed PI, such as a Taylor tool life equation
- Flexibility of the designed control systems over the entire range of operating conditions
- Reasonability of the result using advanced techniques (e.g., reduction of total production time)

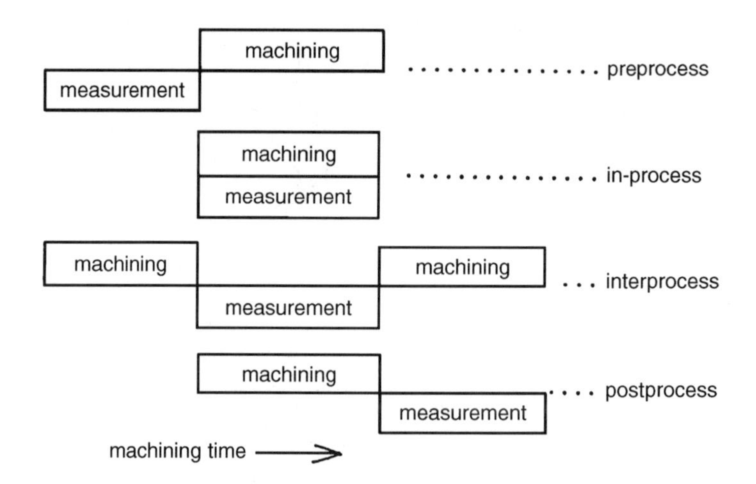

FIGURE 18.6 Measurement types in machining.

18.2.1.3.1 Definition of Adaptive Control

The term "adaptive control" has quite a different meaning in the field of manufacturing engineering than it does in the field of control engineering. First, the definition of adaptive control found in the literature must be clarified to avoid confusion.

In the manufacturing engineering field adaptive control is defined as a system in which the controller continuously adjusts the operating parameter by changing the setpoint on the basis of in-process measurement. Therefore, it is called a "feedback control system subject to the on-line measurement technique," as shown in Figure 18.7(a). In the control engineering field, it implies "self-redesign or self-organization of the system to compensate for unpredictable changes in the process, i.e., a change in the feedback controller itself," as illustrated in Figure 18.7(b).

Accordingly, most of the proposed adaptive control systems are categorized with the first definition. However, in recently developed systems the adaptive controls in machining processes have been applied by techniques based on the second definition, i.e., in the control engineering sense. Thus, the adaptive control may be summarized as shown in Figure 18.8.

18.2.1.3.2 Types of Adaptive Control System

Figure 18.9 shows an overall block diagram of adaptive control systems in the manufacturing engineering field. There are four primary feedback loops—A, B, C, and D—and each loop provides an adaptive control system itself, depending on the variable to be controlled.

Loop A: This control system is concerned with real states of workpiece such as dimension, geometry, and surface roughness. It is generally called a GAC (geometrical adaptive control), which refers to the control method for maintaining the stability of the dimensional accuracy and for improving shape accuracy and surface roughness. With environmental problems involved in the measuring method,

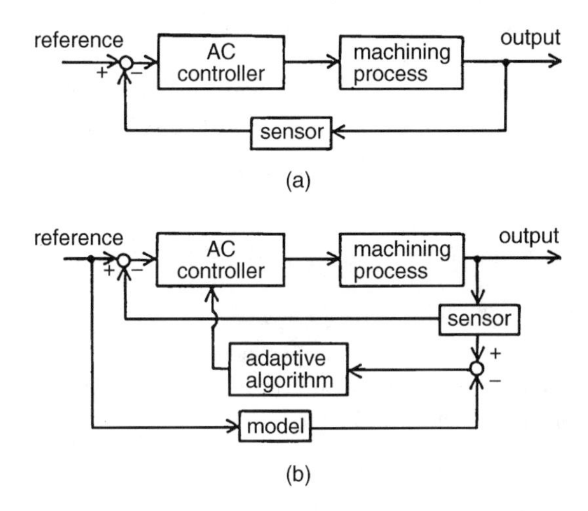

FIGURE 18.7 Structure of adaptive control: (a) in the manufacturing engineering field; (b) in the control engineering field.

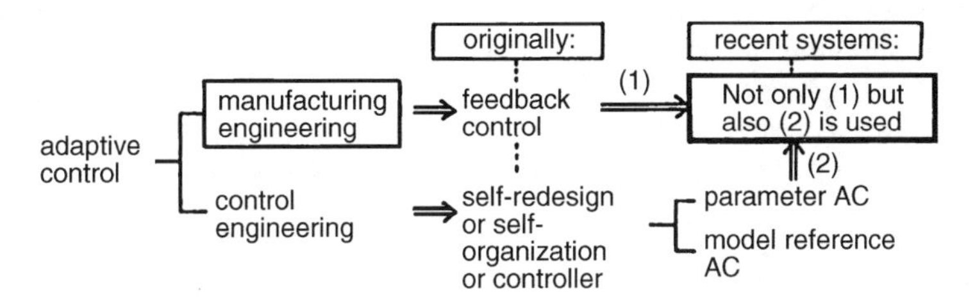

FIGURE 18.8 Transition of AC in the manufacturing field.

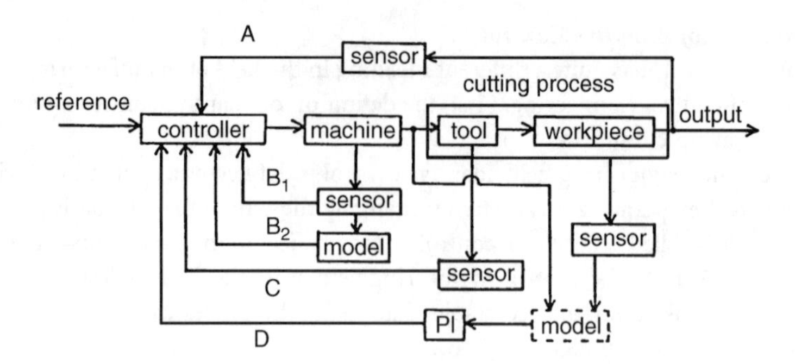

FIGURE 18.9 Types of AC in machining process.

applications of GAC systems are mostly seen in grinding. GAC is the way for controlling the machining accuracy because the real state of a workpiece can be directly obtained through a sensing device.

Loop B: Two loops of B_1 and B_2 are given in the figure. For safe operation of the machine tool, loop B_1 is used, in some cases for monitoring only and in other cases for controlling the states of the machine tool. Loop B_2 can indirectly contribute to the control of machining accuracy. For example, thermal and dynamic deformations of machine tools are related to machining accuracy. However, it is not easy to get information about machining error from sensor outputs in this case. Therefore, a model of thermal and dynamic deformations is prepared in advance, and then the machining error is indirectly estimated and corrected. The control results strictly depend on the accuracy of the model used.

Loop C: This loop has a function similar to that of loop B_1, i.e., for monitoring a tool damage or for maintaining normal tool operation.

Loop D: In this control system a certain PI is measured (sometimes through a model) using inputs, states, and outputs of the cutting process as indicated in the figure. Comparing the measured PI with a set of given ones, the controller modifies the parameters of the cutting process in order to maintain a PI close to the set of predetermined ones. This type of adaptive control is classified into two groups. ACC (adaptive control with constraints) focuses attention on maintaining safe operation under certain physical constraints such as a spindle torque or a cutting force. ACO (adaptive control for optimization) is implemented to realize minimum cost or maximum productivity by changing cutting conditions. This ACO type is typical in the use of the well-known Taylor equations for particular tool-workpiece combinations and economic criteria to optimize the cutting speed and feed rate. With the progress of CNC systems, ACC provides practical use in turning, milling, and drilling, while ACO application is limited mainly to grinding because of difficulties in the formation of PI.

18.2.2 In-Process Technique for Tools

With the arrival of adaptive control of machine tools, research in tool wear and tool failure has become quite active. One of the reasons is that the workpiece as a final output of the machining process is directly controlled by the tool itself. The ability to measure tool wear accurately and quickly during machining is quite important for optimizing the production rate. In addition, tool-failure detection is necessary to avoid damage to the workpiece or the manufacturing facilities. Considerable research has been conducted in the past for measuring tool wear and tool failure. However, the methods proposed are likely to be subject to at least one of the following drawbacks, which must be overcome:

- Measurements are greatly influenced by swarfs, cutting fluids, and vibrations.
- Response time is poor.
- Sensitivity is low and varies with the physical property of work materials.
- The technique is not easily applicable to various cutting conditions.

18.2.2.1 Measurement Method of Tool Wear

Tool wear measurements are divided into direct and indirect methods, as shown in Table 18.2. These methods are also applicable to tool-failure detection. The direct method involves measuring the wear directly and evaluating volumetric losses from the tool due to tool wear. In-process measurement is applicable to a machining process, such as milling, where the cutting edge does not contact the workpiece continuously. Among the direct measurements, the optical method is one of the promising approaches with a wide range of applications. Its reliability in practical use, however, is problematic, depending on circumstances.

Due to the difficulties involved in direct measurements, most of the proposed techniques are used in indirect methods in which a certain relationship is established between tool wear and other parameters. Based on criteria of reliability and ease of use, for example, among the indirect methods force detection and AE (acoustic emission) methods have industrial potential. However, most measuring methods of this category have the limitation that they are basically applicable only under constant cutting conditions.

18.2.2.1.1 Direct Method

Table 18.3 illustrates an optical-based three-measurement technique.

Light reflection: In the light reflection method, an adequate amount of light is projected onto the tool edge. The characteristics of the reflected light will depend on the surface wear of the tool. By in-process measurement, this simple technique is applicable to milling and grinding (Figure 18.10). Many studies in this category have been made in the laboratory, but commercially available sensors are not yet available. Considerable efforts will be required with respect to reliability in the actual environment before commercialization.

TABLE 18.2 Detection of Tool Wear

	Sensing Method	Applicable Machining Process	Main Features
Direct	Optical Contact resistance method Radioactive method (promising)	Grinding where the grinding wheel is continuously monitored optically Turning process by contact resistance or radioactive method Milling process, with limitations	Generally provides accurate results Can directly evaluate without being affected by changes in cutting conditions Often difficult to employ
Indirect	Cutting force or power detection (promising) Gap detection between tool and workpiece Temperature detection Electrical resistance method Vibration analysis method Sonic analysis method (AE: promising)	All machining processes	Ease of use Sensitivity and reliability lower than the direct method Quite useful under constant cutting conditions

TABLE 18.3 Optical-Based Measurement of Tool Wear

Sensing Method	Resolution (μm)	Accuracy (μm)	Ease of Use	Cost	State of Development
Light reflection	1	1–3	Easy	Inexpensive	Requires further effort for application
Reflection pattern	0.1	about 1	Easy but requires rather complicated signal processing	Rather expensive	Requires further effort for application
TV camera	2	3–5	Easy but requires rather complicated signal processing	Expensive	Requires further effort for application

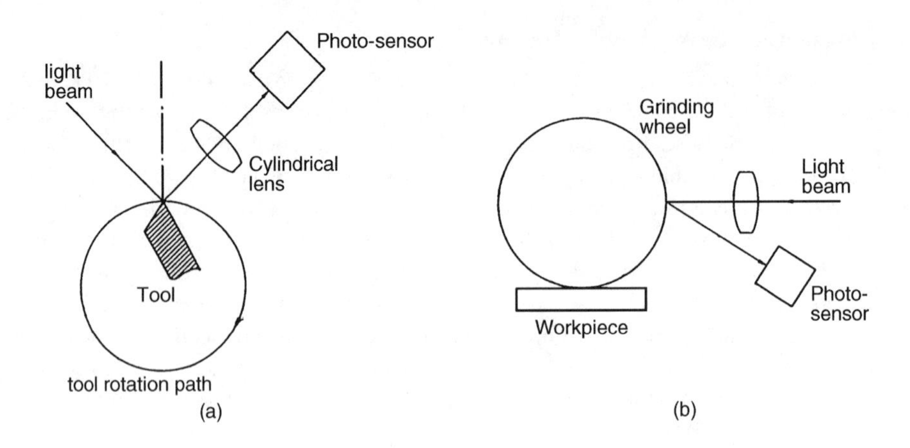

FIGURE 18.10 Light reflection method of tool-wear detection: (a) milling; (b) grinding.

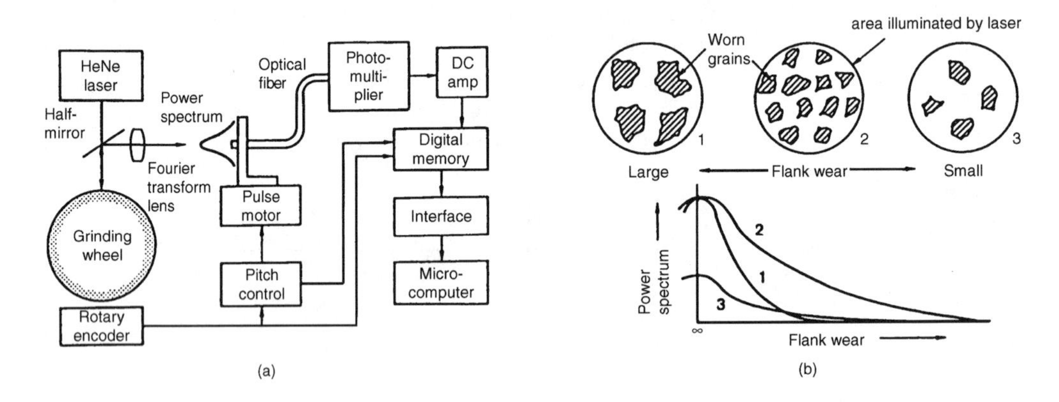

FIGURE 18.11 Reflection pattern method: (a) experimental setup; (b) relation between distribution grains and the power spectrum.

Reflection pattern: In the reflection pattern method, an interesting technique has been developed by using an optical Fourier transform to measure the properties of grinding wheel surfaces in terms of a power spectrum pattern. Figure 18.11(a) indicates the structural block diagram of the power-spectrum-measuring system in which the pattern is obtained at the focal plane of the Fourier transform lens. This optical pattern includes several kinds of information such as the width of grain wear flats, the area of worn grains, and the number of grains, as shown in Figure 18.11(b). The lifetime of the grinding wheel rotating at 3000 rev min^{-1} can be determined in real time from the average width of grain wear flats. The proposed method is one of the promising techniques among optical methods.

Another approach in milling shown in Figure 18.12 is based on laser diffraction pattern [Fan and Du, 1996]. A narrow slit is constructed between the cutting edge of a cutter and a sharp straight blade placed parallel to the cutting edge as a reference. A laser beam is projected onto the slit, which creates behind the slit a diffraction pattern consisting of a number of fringes. By comparing the fringe spacing variation, cutting edge wear can be obtained. This technique has high resolution of several microns and a wide range of measurement. The measurement can be correlated to the commonly used flank wear measure (VB) and is also applicable to drilling.

Television camera: Monitoring of tool wear by a television camera is basically an interprocess technique in turning. The image of the tool wear is displayed on a television screen and analyzed by a computer to provide information in terms of a wear pattern or quantity of wear. However, this system is rather expensive, does not have high accuracy, and cannot be applied by the in-process technique. Therefore, further progress is not expected in the future.

TABLE 18.4 Gap-Detection Method

Measuring Method	Representative Example	Resolution (μm)	Accuracy (μm)	Main Features	State of Development
Electric micrometer	primary detector detector for compensation	1	3–5	Reliable Wear of the probe head Rather expensive	Laboratory use
	measured distance tool stylus	1	5	Reliable Highly wear resistant Inexpensive	Research stage but promising
Air micrometer	workpiece tool nozzle	1–2	3–4	Reliable Ease of use Influenced by changes of temperature and supply pressure Inexpensive	Most promising
Ultrasonic micrometer	tool ultrasonic sensor	2	3–5	Ease of use Influenced by swarfs and cutting fluids Expensive	Laboratory use
Optical micrometer	laser beam photo-detector workpiece	0.5	2–4	High resolution Influenced by swarfs and cutting fluids Relatively inexpensive	Requires further investigation for reliability

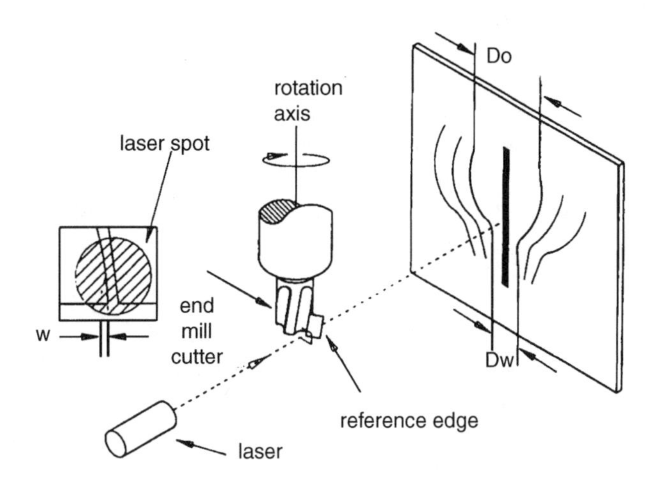

FIGURE 18.12 Setup for measuring rotating tool conditions.

18.2.2.1.2 Indirect Method

Gap-detection approaches are based on the change in the distance between the machined surface and some point on the tool flank or the tool holder. Table 18.4 compares the four methods, including electric, pneumatic, and ultrasonic detections developed so far.

In the electric micrometer method, a feeler from the primary detector is in contact with the work surface when turning immediately after being machined by the side cutting edge of the tool. With the progress of flank wear, the feeler drifts toward the tool axis, thus generating an electrical output proportional to the area of the tool. A compensating detector close to the tool edge is also provided to remove the error caused by the cutting temperature and cutting force. A more reliable and inexpensive sensor based on an electric approach has also been developed by using an eddy current transducer. A tungsten carbide stylus was used because it is highly wear-resistant. This method is applicable to straight turning with an accuracy of around 3 Ìm and will be promising among gap detection techniques.

A pneumatic proximity gauge can detect the gap between tool post and work surface. In the early days a nozzle was located within the cutting tool, but this produced poor machinability. In a more developed system the nozzle is either just below the cutting edge or within the tool holder. The pressurized air ensures that the work surface remains clear under fluid cooling conditions.

A prototype ultrasonic micrometer using the transit time of successive pulses has been developed that gives an accurate and continuous read-out of tool wear. The ultrasonic pulses strike the workpiece and are then reflected from its surface. Transit time gradually decreases with the progress of tool wear.

For the optical method, an optical micrometer mounted on a tool post consists of a He–Ne gas laser, an optical lens, and a photodetector. A laser spot on the work surface is located 0.4 mm below the tool edge. This spot is magnified optically, keeping its angle view at 10°. Its image is then created on the photodetector fixed to the tool post. Thus, the measurement is done by sensing the movement of the optical image on the detector in accordance with the progress of tool wear.

Of these methods the pneumatic proximity sensor has been commercially offered and will have promising potential because of its greater simplicity of operation. The problem is that the variation of temperature, pressure, and surface quality affect the magnitude of back pressure and system reliability.

18.2.2.2 Tool Failure Measurement

Figure 18.13(a) shows the in-process tool geometry sensing system in milling [Ryabov et al., 1996]. It consists of a support and location mechanism mounted on the main spindle, two laser sensors used to scan the tool shape and synchronize the signals to the tool rotation, a servo motor that moves the main sensor along the tool axis, laser and servo motor controllers, and a PC for the acquisition and processing of data. The position adjuster is used to immobilize the main laser head at a precise position in the x, y plane; the z position of the main laser head is determined by the servo motor, which controls the vertical motion of the support and location mechanism. The auxiliary laser synchronizes the main laser with tool rotation by detecting a marker on the tool that denotes the rotational origin of the tool. The main laser is equipped with a protective acrylic cover that has a trapezoidal cross-section and has openings at the bottom and top. The walls of the cover contain double-walled nozzles that are also open at the tops. The main laser sensor sits under the cover, and the laser beam passes through the top opening of the cover. Compressed air projected into the nozzles prevents obstacles from getting inside the cover and hitting the sensor.

The laser sensors with a deviation resolution of 0.5 μm are 670-nm, semiconductive-type, commercially available displacement meters that detect the angle and intensity of reflected light by means of photodetectors. An example of laser sensor displacement is illustrated in Figure 18.13(b). As the tool rotates the measurement system continuously measures the distance between the sensor and the tool. From the sensor output for a complete tool rotation, the system generates a cross-section of the tool shape at the Z level of the scanning laser beam, thus constructing a three-dimensional image of the tool geometry. In particular, the tool failure can be detected as a worn tool, as shown in the figure.

18.2.3 In-Process Technique for Workpieces

Given a particular part to be produced within a defined dimension or profile tolerance band, it is necessary to manufacture the part at the most economical rate while at the same time staying within specifications. The control of workpiece accuracy is, therefore, aimed at the use of a measuring device based on

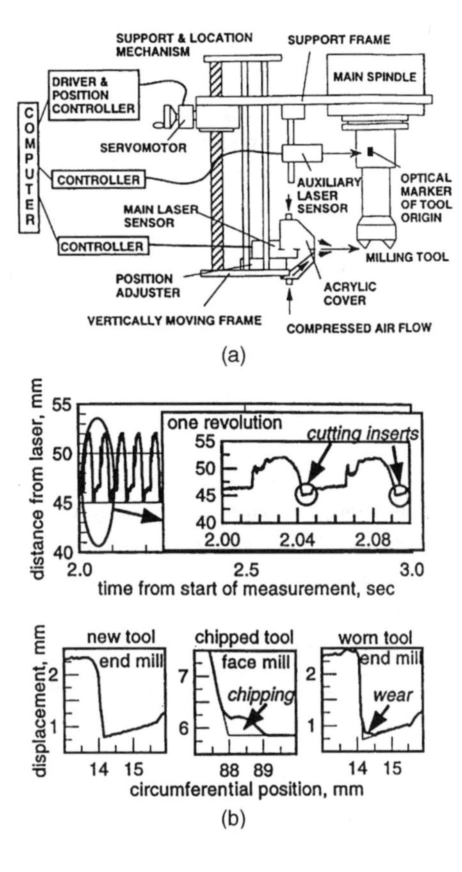

FIGURE 18.13 Tool-failure detection system: (a) system configuration; (b) displacement of output under air-cutting conditions.

in-process techniques. Although the in-process measurement of workpiece dimensions has a rather long history of development, promising methods have been rare in recent years. Surface finish of a workpiece also plays an important role in quality control. If in-process measurement of surface roughness gives a clear meaningful indication of the product quality, it is imperative that a continuous record of the roughness value be obtained. Several types of instruments have been developed in this field. However, most of the measuring instruments have been successful only in the laboratory, and only a few have emerged as working production systems.

18.2.3.1 Dimension and Profile Measurements

If high accuracy is required, it is necessary to measure the workpiece diameter directly by using an adequate instrument. On the other hand, the workpiece accuracy can be indirectly evaluated from radius measurements either by monitoring the motion of the carriage carrying the cutting tool or by noting the position of the tip of the cutting tool. In applying such an indirect method, the measuring device is generally set on the tool slide or on the upper side of the workpiece. Thus, the measuring technology is classified into direct and indirect methods, as shown in Table 18.5, which lists several advantages and disadvantages of both measurements. In high-precision machining, the deviation of geometrical profiles such as roundness and cylindricity substantially affects the shape accuracy of the workpiece. Although the profile measurement fundamentally obeys similar sensing techniques as the dimension measurement, it is not easy to improve the shape accuracy in the same order of magnitude as the dimensional one.

A number of optical methods have been developed and are listed in Table 18.6. However, due to the environmental problems involved in optical techniques, these methods are rather difficult to employ on production floors.

TABLE 18.5 Classification of Measuring Method in Workpiece Accuracy

	Direct	Indirect
	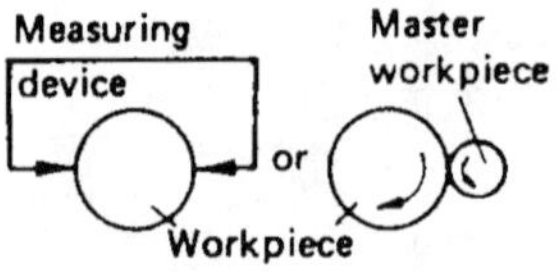	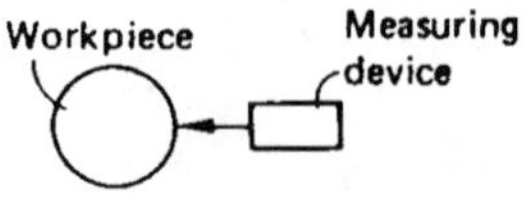
Main features	Method is applicable to diameters within the limited range of the measuring device.	Method is applicable to a wide range of diameters.
	Internal diameters and complex shapes are difficult to deal with.	Method is applicable to internal diameters and complex shapes with limitation.
	Effects such as tool wear, machine errors and distortion, and workpiece distortion are all taken into consideration.	These errors strictly affect the measurement.
	Measurements can deal with poor-quality machine tools (retrofitable).	Available techniques are very useful when applied to good quality machine tools.

TABLE 18.6 Direct and Indirect Method of Dimension and Profile Measurements

	Measuring Method		Representative Example: Resolution and Accuracy[a]	Representative Example: Main Features
Direct	Light project I		res: 10 μm acc: 30 μm	• Requires complicated signal processing • Low resolution and accuracy
Direct	Light project II	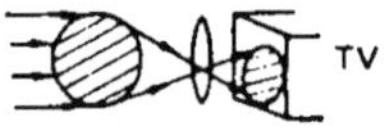	res: 2 μm acc: 10 μm	• Easy to operate • Rather expensive
Direct	Light gauging		res: 1 μm acc: 10 μm	• Requires careful system alignment • High resolution
Direct	Light focusing		res: 10 μm acc: 10–20 μm	• Very simple and easy to operate • Low resolution and accuracy
Indirect	Light spot detection	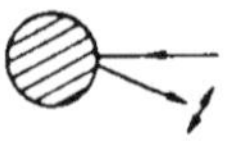	res: 0.5 μm acc: 5–10 μm	• High resolution • Easy to use • Requires further system reliability
Indirect	Light sectioning	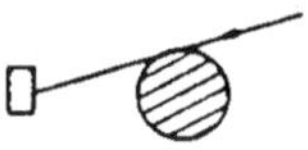	res: 1.2 μm acc: 5 μm	• Simple and easy to operate • Lack of feasibility and practical use

[a]res = resolution; acc = accuracy.

18.2.3.1.1 Direct Methods

Light projection: In the light-projection method, a parallel beam of light is projected from one side of the measuring area onto a detector on the other side. The beam strikes the detector, producing an output signal for the entire time it is being scanned across the measuring area. Alternatively, an image of the workpiece is focused onto the measuring grid on the face of an ITV camera. However, the systems work well only in good environments such as in a laboratory.

In the application of this approach dimensional and surface roughness controls for turning have been achieved by using a compensatory flat-bite tool. A simple optical measuring method was employed, as shown in Figure 18.14(a). A He–Ne laser light having a rectangular beam as a result of traversing lens L1 and concave lens C1 is projected onto the upper edge E of a workpiece. By adequately inserting another lens L2 and concave lens C2 into the measuring area, the laser beam passing through the edge makes a sharp profile pattern of the work surface onto the photo-array sensor Ds. A knife edge S between lens C1 and the workpiece is used to improve the sharpness of the pattern. As illustrated in Figure 18.14(b), this pattern consists of a shadow image corresponding to the surface profile and a bright part due to passing lights that change the sensor output. The photo array is composed of 32 photo diodes, and each photo diode can provide a current output signal. The total resolution in this measurement is around 2.0 μm, although it is dependent on the sharpness of the image and the magnification of the profile pattern by the optical system.

An experimental setup is shown in Figure 18.15 where a dimensional control is first accomplished by the regular cutting tool, then a roughness control is followed by the compensatory flat-bite tool immediately after the dimensional corrections. A positional relationship between two cutting tools is depicted in Figure 18.15. The laser and optical systems are mounted on the same tool post. The compensatory flat-bite tool is located on the opposite side of the tool post, 0.1 mm behind the regular tool. In general, the axis along which the tool post moves is not exactly parallel to the axis of the workpiece. Therefore, to improve the lack of axis parallelism the flat-bite tool is mounted on another axis driven by the precision slide unit with a dc motor and can travel at the same speed as the tool post.

Figure 18.16(a) illustrates an example of diameter changes with and without control for straight profiles. The dimensional control is carried out when the semidiameter change exceeds ±5.0 μm. Although the profiles with standard cut have positive and negative size deflections of several 10 m, these are nearly reduced by the control to the given tolerance limits of ±10.0 μm. To make clear the effectiveness of the roughness control, the traces of part of the surface profiles with and without the control are given in

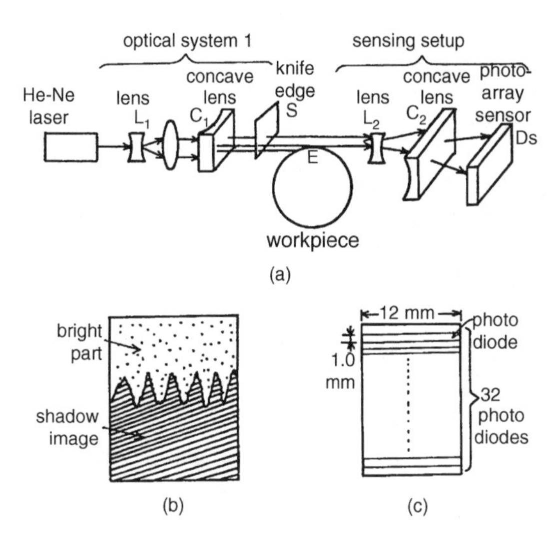

FIGURE 18.14 Principle of measurement for dimension and surface-roughness control: (a) optics; (b) profile pattern on the sensor; (c) construction of the photo-array sensor.

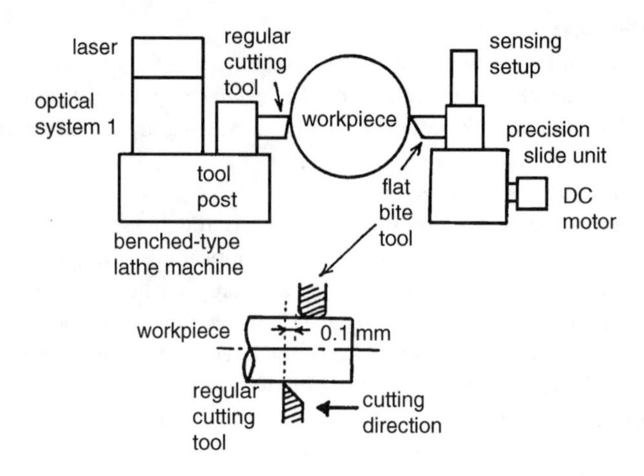

FIGURE 18.15 Experimental setup.

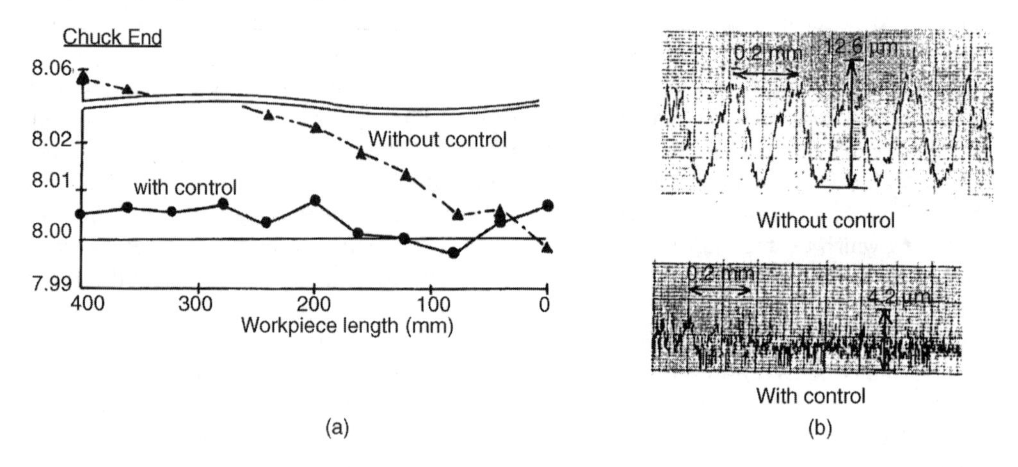

FIGURE 18.16 Control results: (a) diameter change with and without control; cutting speed = 100 m/min, feed rate = 0.1 mm/rev, depth of cut = 0.4 mm, P40 tool/545C workpiece; (b) surface profiles.

Figure 18.16(b), where the stylus method was used for comparison. This figure makes it clear that the machined surfaces are remarkably improved by the flat-bite control within several microns.

Light gauging: Figure 18.17 is an example of a commercially available technique for turning and grinding operations proposed by a light gauging method. A He–Ne laser beam is divided into two parallel measuring beams A and B. Each beam is screened by half of the incident beam at the workpiece edge, and the difference between the two measured light intensities corresponds to the change in workpiece diameter. General features of this method are illustrated in Table 18.6. The measuring system is compact and easy to operate, and in a modified form it can detect the internal diameter. Although the environmental problems have not yet been overcome, this system will become more useful in practical applications if it is incorporated with an air-blast unit.

18.2.3.1.2 Indirect Methods

Light focusing: Light focusing methods detect the focusing point of an incident beam on the work surface to produce the maximum intensity of the reflection light at the photodetector. A change of workpiece diameter yields a deflection of the focusing point, thereby generating less intensity of reflection light. The sensor output is then used to operate a servo mechanism that controls the tool position to maintain the diameter within a prescribed level. This method is very simple and has been applied to turning and grinding, but it has a low resolution, as shown in Table 18.6.

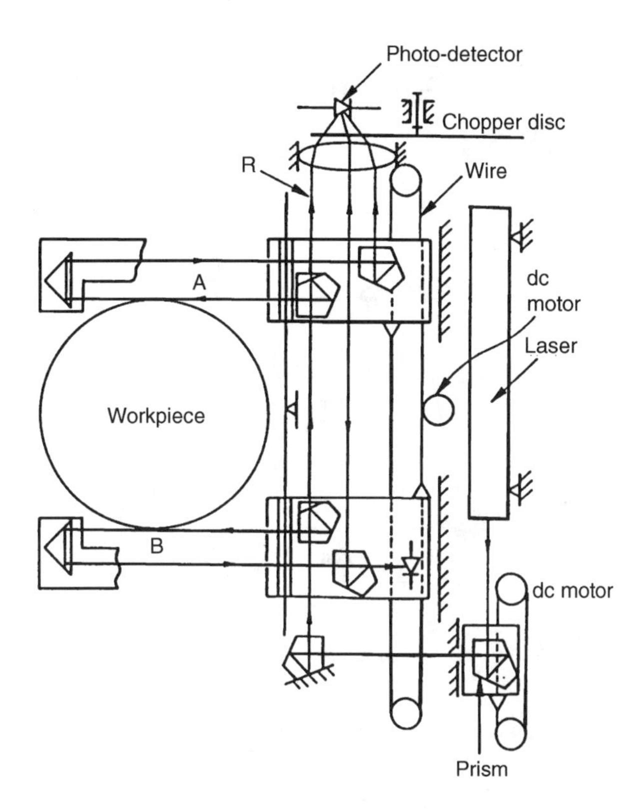

FIGURE 18.17 Light-gauging method.

The application of this category is shown in Figure 18.18 and consists of a light source, two collimators, exponential filter, Fourier transform lens, CCD area sensor, etc. [Miyoshi et al., 1996]. The conventional profile measurement of the small object is carried out by tracing a stylus or a probe precisely on the work surface. The new method, however, makes it possible to immediately reconstruct the three-dimensional form of the whole area illuminated by the laser beam spot without scanning. The object wave arriving from the work surface illuminated coherently has the phase information of the object. It is possible to reconstruct a complex field in the object plane by retrieving the phase information of the object. In Figure 18.18, a coherent beam of Ar laser is expanded by the first collimator, and then the expanded parallel light is returned to the spot size as small as the original laser beam by the second collimator. When an exponential filter is inserted between the two collimators, the incident intensity distribution is modulated when traveling through the exponential filter. Then the exponentially modulated Fraunhofer diffraction intensity is obtained on the CCD area sensor located at the focal distance of the Fourier transform lens. The intensity-distribution data are distributed with a personal computer and reconstruction of the work surface profile is performed. This method is applicable to in-process measurement of a fine form of a micromachined component with an accuracy of nanometer order.

Displacement detection: If a laser beam is projected just under the cutting position in turning, this spot will deflect according to the change in workpiece diameter. Therefore, its radius change can be evaluated by detecting the movement of the spot image on the photodetector. One of the approaches takes account of the changes in the position of the work axis and can improve the cylindricity as well. However, the problem is the measurement reliability. Another approach in turning is to use a fiber displacement sensor in conjunction with the neural network. A schematic diagram of this method is shown in Figure 18.19 [Choudhury et al., 2001]. A bifurcated optical fiber probe that detects the displacement between the sensor head and the work surface is held in front of the workpiece diametrically opposite to the cutting tool. The probe head is fixed in such a way that it lags the cutting tool tip by a

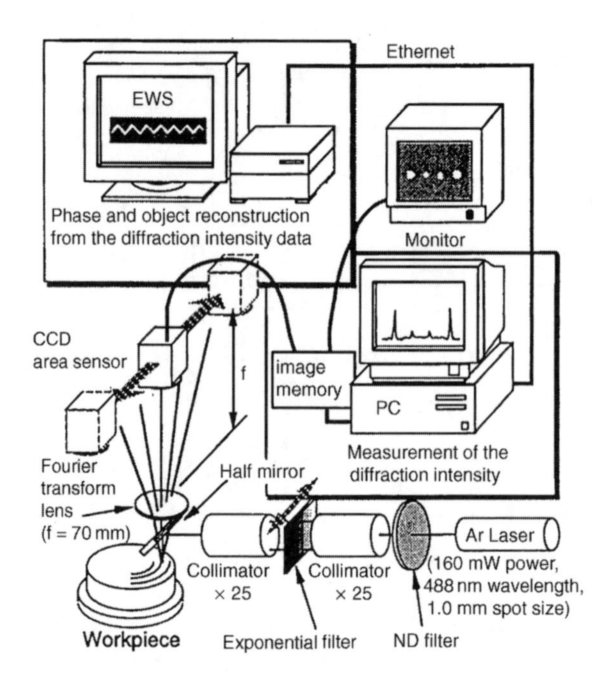

FIGURE 18.18 Inverse scattering phase reconstruction method.

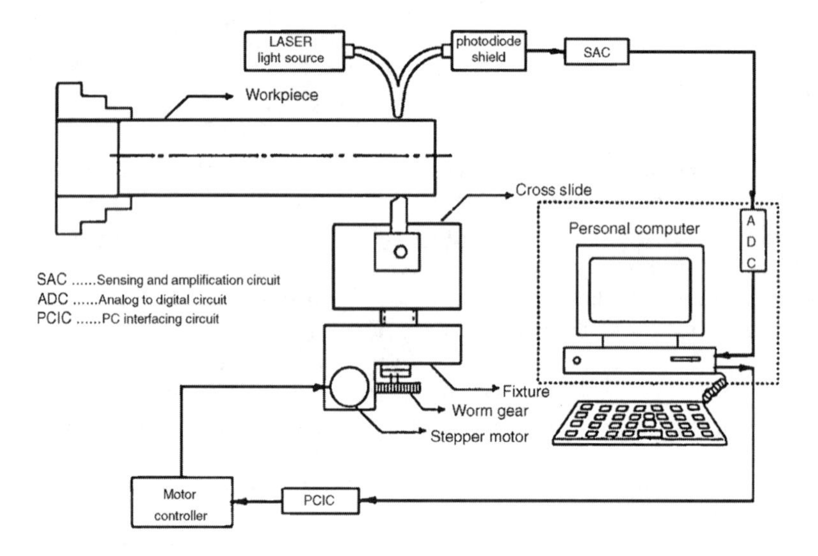

FIGURE 18.19 Monitoring and control of dimensional inaccuracy in turning.

certain distance and its axis intersects the axis of rotation of the workpiece. This is to ensure that the He–Ne laser beam, used as a light source, is incident on and reflected by the freshly produced work surface. The cutting conditions such as cutting speed and feed and depth of cut along with the sensor output are fed to the neural network, trained *a priori*. The neural network predicts the flank wear, in terms of its length, on the cutting tool flank face. Using this predicted value of flank wear, the feedback circuit is employed to compensate for this change in the workpiece dimensions. This is accomplished by an actuator consisting of a stepper motor and a combination of worm and worm gear coupled with the cross-feed lead screw of the lathe. According to the experiments, workpiece dimensional inaccuracy due to tool wear is controlled within 0.03 mm.

The profile accuracy of a workpiece is basically determined by the relative displacement between the workpiece and the cutting tool. A new proposed system called WORFAC (workpiece referred form accuracy control system) has three functions of waviness, profile, and surface roughness control systems in ultraprecision machining. Figure 18.20(a) demonstrates the WORFAC principle applied to the manufacture of a large mirror [Uda et al., 1996]. When the tool holder moves away from or toward the mirror surface, the microtool servo pushes the tool by the same amount as the gap detected, and vice versa. The machined surface error can generally be divided into three band frequencies: low (profile), middle

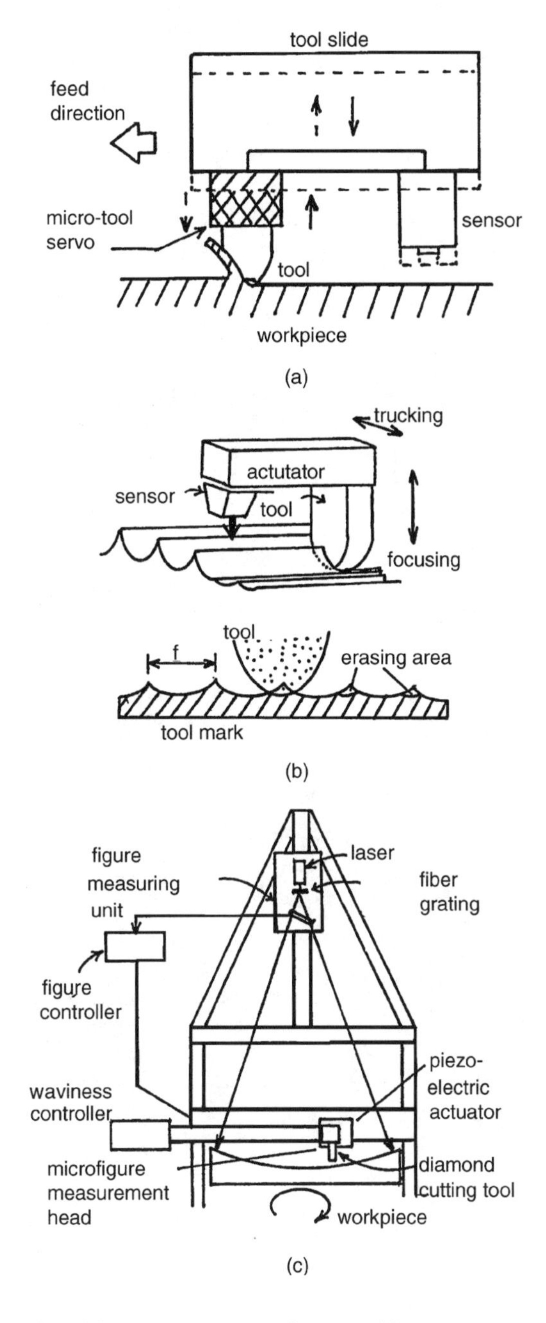

FIGURE 18.20 Workpiece referred from accuracy-control system: (a) WORFAC principle; (b) tool-mark erasing system; (c) precision turning system.

(waviness), and high (roughness). Figure 18.20(b) shows a schematic arrangement of these control systems. The procedure for three control modes is as follows. The mirror is turned with fast feed rate using the waviness control system with a piezoelectric actuator. The profile control system consists of a profile measuring unit and a controller and compensates for the low frequency surface error by moving the tool slide. The mirror is then turned again using the surface roughness system. A principle of a tool-mark erasing system is shown in Figure 18.20(c), where the roughness control is done by smoothing the local irregularities of a work surface.

Figure 18.21(a) is an optical measuring apparatus installed into the control system (Figure 18.20(b)). It detects a tool mark, and a tool controlled by the actuator removes the tool mark (erasing area). In this measuring apparatus the fiber grating is positioned in the center of the curvature of a concave mirror, and a cube-shaped half-mirror is employed to separate the illuminating and receiving optical systems. The profile error is determined by the ideal spot positions and those found through image processing from spots of the actual CCD image. On the other hand, an optical sensor for the tool-mark erasing system is shown in Figure 18.21(b), which is required to detect both vertical and horizontal displacements. A pick-up for compact disks has both of these functions, i.e., a function to detect a focusing signal and a function to detect a trucking one. A laser beam is divided into three beams by a diffraction grating, and three beams focus

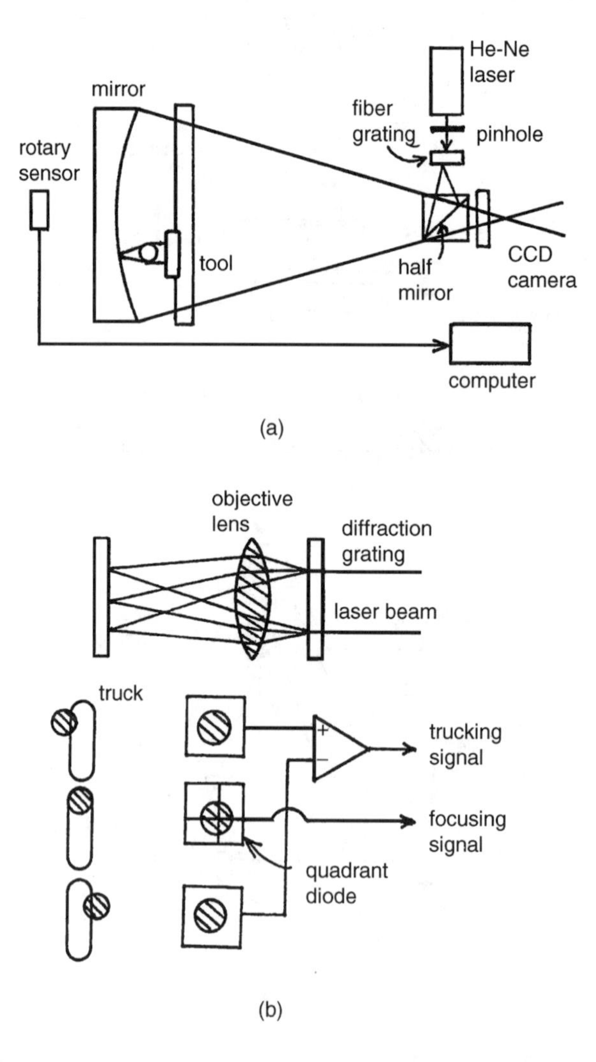

FIGURE 18.21 Optical measuring system of WORFAC: (a) optical measuring system; (b) optical sensor for tool-mark erasing system.

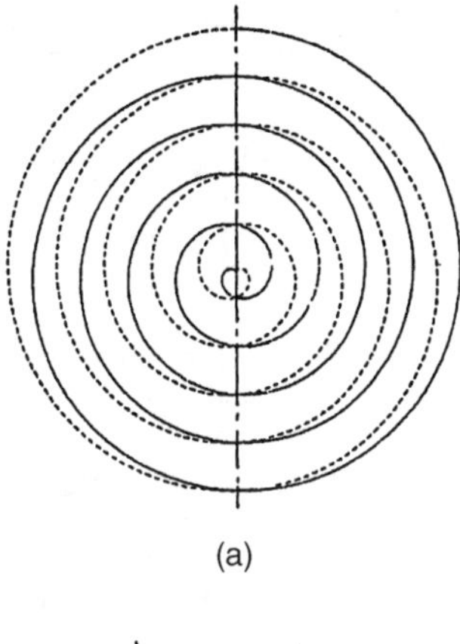

(b)

FIGURE 18.22 Control results: (a) traces of cutting tool on flat mirror; (b) results of roughness.

on a truck through an objective lens. The center spot is detected by a quadratic diode for a focusing signal. The other two spots are detected by two diodes individually for a trucking signal.

Figure 18.22(a) illustrates traces of a cutting tool on a flat mirror. The mirror is turned first radially from center to periphery (solid line) and then radially from periphery to center, retaining the same axial tool position (dotted line). Along two radial lines (dot and dashed lines) a tool mark erases. The peak-to-valley of the turned surface is improved to about 1/4 of the original one, as shown in Figure 18.22(b).

Light sectioning: An interesting technique in indirect method has been applied to milling by the light-sectioning method. Figure 18.23 outlines the sensor location and the measurement principle. This approach places a line of incident light across the workpiece and determines the coordinates of the work edge by analyzing the pixel array produced by a camera focused on the line. By mounting a photodetector on the spindle housing and focusing it behind the cutter, a relationship between work edge and spindle centerline is determined. The system operates effectively under the dry cut but requires further investigation to make the method feasible and practical.

18.2.3.2 Surface Roughness Measurement

It is clear that there is a need for techniques in surface assessment, preferably ones that are capable of high-speed assessment using noncontacting techniques. Optical techniques seem to be a natural solution to such demands. Many optical methods have been proposed that include specular reflectance, diffuseness, angular distribution, speckle, interferometry, etc., and many different devices have been offered commercially. However, the number is fairly limited in the application of in-process measurement because the instruments must operate in the manufacturing environment. This measuring method is classified into two groups: direct and indirect methods, as shown in Table 18.7. A direct method is derived from a point-by-point scan of surface height, y, as a function of distance, x, by using a very narrow light beam, while in the indirect method the surface roughness is evaluated by measuring the optical properties from the surface.

18.2.3.2.1 Direct Method

Measuring instruments based on the triangulation method have been developed that allow the profiles of relatively rough surfaces to be recorded. One of these has the advantage of being able to measure roughness not only along the circumferential path, but also along the feed direction during turning. In this method, double He–Ne laser beams, L_1 and L_2, having a diameter of several microns on the work

TABLE 18.7 Direct and Indirect Method of Surface Roughness Evaluation

	Direct	Indirect
	Point scan, y, x	Optical properties, y, x
General features	Topographic parameters such as R_a (roughness average) are obtainable.	It is difficult to connect the topographic parameters of the surface.
	It can measure the surface profile.	The surface profile cannot be obtained.
	System alignment is relatively difficult.	Technique is relatively easy to operate.
	Method is effective, especially for turning.	Technique is applicable to all machining processes.
	The resolution depends on the light spot size.	The sensor output depends on the work materials.

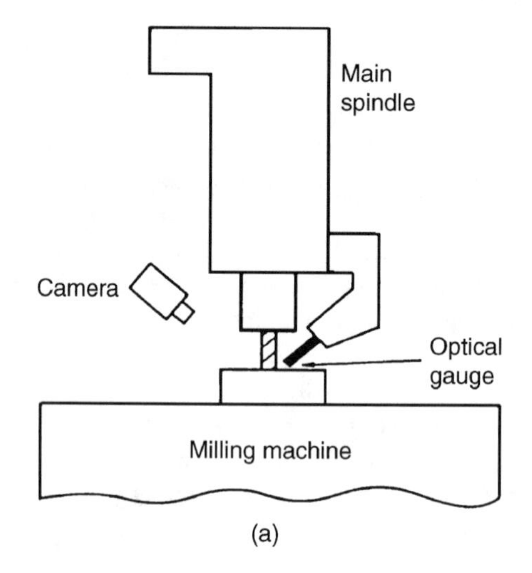

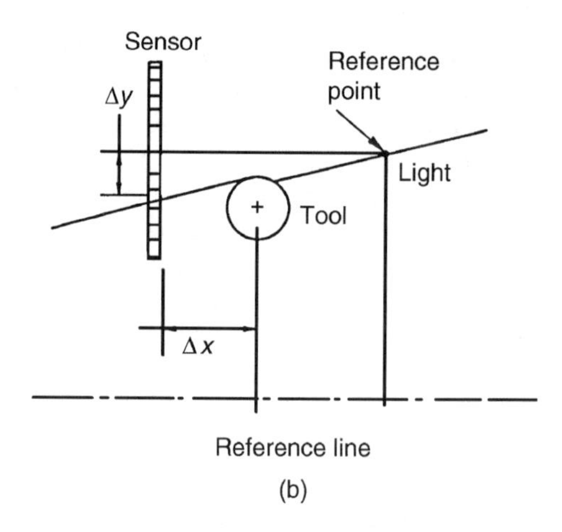

FIGURE 18.23 Light sectioning method in milling: (a) sensor location; (b) measuring principle.

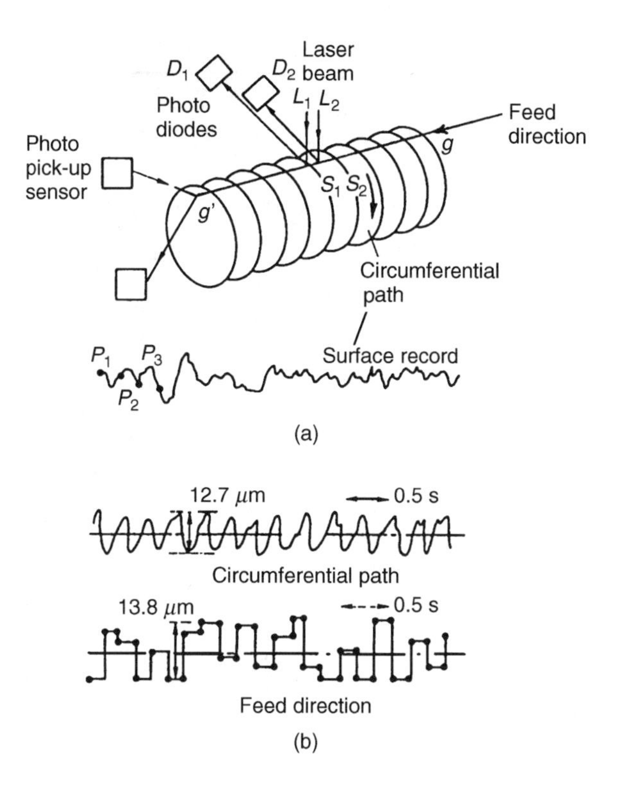

FIGURE 18.24 Double beam scanning of roughness measurement: (a) feed direction measurement; (b) vibration measurement.

surface, are incident on the peak S_1 and valley S_2 in the feed mark, respectively (Figure 18.24(a)). Each laser spot on the surface is then detected by two photodiodes, D_1 and D_2, using the triangulation method. Because the measuring device is set on the tool post, the roughness reading is carried out along the circumferential path. To obtain a feed direction measurement during machining, sampling pulses P_1 and P_2, corresponding to every one revolution of the workpiece, are generated along the feed direction gg′ by using another optical sensor. Thus, the roughness value at each sampling point on the circumferential path can provide the discrete peak-to-valley height in the feed direction gg′. Figure 18.24(b) represents an example of the surface trace produced under a certain chattering condition. The measuring system shows the vertical resolution of 0.5 μm and the measuring accuracy of about 3.0 μm. However, the system alignment is a little complicated and still at the research stage.

18.2.3.2.2 Indirect Method

Table 18.8 illustrates the measuring method of this category.

Diffuseness of scattered light and specular reflectance: Roughness can be assessed by characterizing the diffuseness of the scattered radiation pattern or by measuring specular reflectance from the surface. Several attempts have been made in this field, and some of them are promising. Table 18.9 shows a comparison of the advanced technologies. The advantage of the ratio measurements is that the characteristic curves between intensity and roughness do not produce large variations when different work materials are used. The characterization of surface microtopography by this method sometimes yields only a qualitative assessment or does not directly correspond to common roughness parameters. Therefore, the techniques listed in the table are primarily useful as a comparator to distinguish between the roughness of similar surfaces rather than as a metrological tool for measuring surface parameters.

TABLE 18.8 Indirect Method of Surface Roughness Measurement

	Best Example					
	Resolution (μm)	Accuracy (μm)	Range (R_a) (μm)	Speed	Main Feature	State of Development
Diffuseness	1	3–5	2–80	Fast	Effective for rougher surface Ease of system alignment	Laboratory use at present but promising
Specular reflectance	0.1	2	1–3	Fast	Effective for relatively smooth surface Accuracy depends on detector aperture	Laboratory use
Speckle	0.01	±0.2	0.1–0.2	Relatively fast	Effective for smooth surface Useful under Gaussian height distribution Narrow measuring range	Promising and commericially available
Ellipsometry	0.01	0.1	0.01–1.2	Relatively fast	Effective for smooth surface Need careful system alignment	Requires further investigation for measurement of engineering surface
Angular distribution	2	±5	Not specified	Slow	Not practical Need high-speed data processing	Not promising
Light sectioning	1	5	2–20	Slow	Effective for relatively rough surface Requires high-speed data processing	At the research stage but promising
[cf] direct method	0.5	3	2–80	Fast	Requires careful system alignment Accuracy depends on light spot size	At the research stage

Speckle: When a rough surface is illuminated with coherent light, the reflected beam consists partly of random granular patterns of bright and dark regions known as speckle. The roughness properties can be obtained from the speckle patterns by relating either the contrast or the degree of pattern correlation to the roughness of the surface. Figure 18.25(a) is an example of a commercially available in-process instrument. This measuring system uses the speckle contrast method in which the intensity variations are quantified in terms of an average contrast, defined as the normalized standard deviation of intensity variations at the observation plane. The light from a semiconductor laser is focused by an optical system onto the work surface. The scattered light produces a speckle pattern in the far field region of the illuminated surface whose intensity variations are detected by a camera. Finally, the output signal from the camera is processed by a microcomputer to calculate the contrast in terms of standard deviation/average contrast. As shown in Figure 18.25(b), the instrument works well for the roughness range of 0.1 R_a (μm) to 0.18 R_a (μm) (R_a = average roughness). Speckle techniques perhaps hold the most promise as an in-process tool for measuring surface roughness. They can easily be adapted to the measurement of moving parts. But expanding the measuring range presents a problem.

Another approach in this category is the so-called "speckle autocorrelation" method [Lehmann and Goch, 2000]. The schematic arrangement of polychromatic speckle autocorrelation is shown in Figure 18.26(a). The rough surface is illuminated by a trichromatic parallel beam using a fiber-optical coupling

TABLE 18.9 Advanced Technologies of Diffuseness and Specular Reflectance Methods

Measuring Method		Representative Example: Vertical Resolution (μm)	Representative Example: Accuracy (μm)	Representative Example: Range (R_a) (μm)	Notes
Specular intensity I_0	Intensity at one-off specular angle I_1	0.1	About 0.4	0.05–0.5	Incandescent light Grinding Promising
Specular intensity at the surface normal I_0	Back-scattered intensity off normal I_1	1.0	3	5–80	Halogen lamp Turning Most promising
Total scattered intensity I_1	Specular intensity I_0	1.0	3	2–20	He–Ne laser Grinding Under investigation
		0.2	0.2	0.2–2.5	Incandescent light Grinding Promising
Specular reflectance I_0		0.1	0.5	1–3	Incandescent light Turning Laboratory use

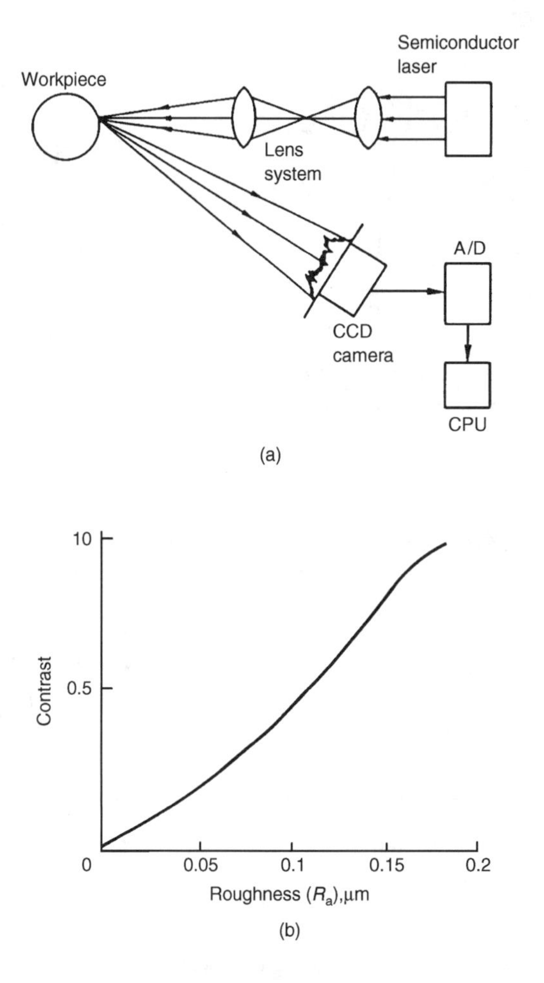

FIGURE 18.25 Speckle contrast method: (a) developed measuring system; (b) speckle contrast vs. roughness.

of the light emitted by three different laser diodes. The trichromatic light scattered from the workpiece is directed to a CCD array placed in the back focal plane of a Fourier transforming lens. In trichromatic speckle patterns the typical fibrous radial structure known as the speckle elongation phenomenon can be observed. This speckle elongation is a consequence of the angular dispersion in combination with a "similarity" of the monochromatic speckle patterns of the different wavelengths. The latter effect shows strong roughness dependence. A large roughness leads to three totally independent monochromatic speckle patterns. An efficient way to obtain an optical parameter for roughness characterization is based on the ratio of the mean speckle diameter in the horizontal direction of the speckle pattern to the mean speckle diameter in the vertical direction, calculated for different subsegments of each speckle pattern. The speckle diameters are estimated by use of local two-dimensional autocorrelation functions. The modulation depth expressed by the empirical standard deviation of the obtained data sets gives an appropriate optical roughness parameter for the roughness characterization as illustrated in Figure 18.26(b). The error bars in the figure represent the empirical standard deviation of independent measurements at different surface areas for ground and EDM (electrical discharge machined) surfaces.

Ellipsometry: This technique measures the change in the polarization state of a beam of light when it is reflected from a surface. A schematic diagram of the proposed ellipsometer is illustrated in Figure 18.27, where the optical unit is composed of a projection collimator that receives the specular reflection beam.

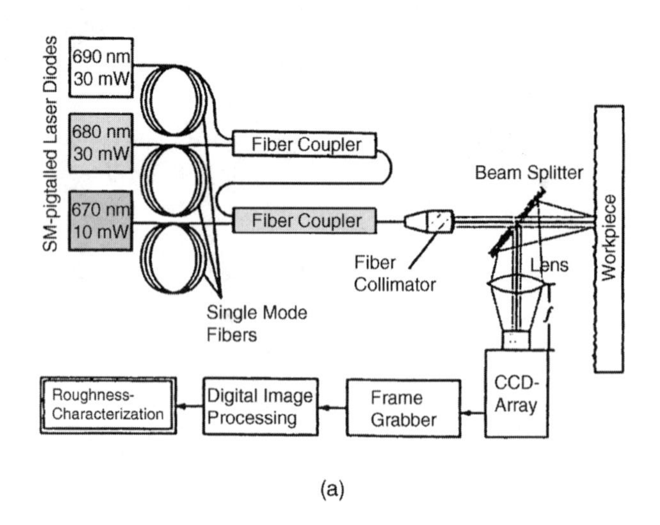

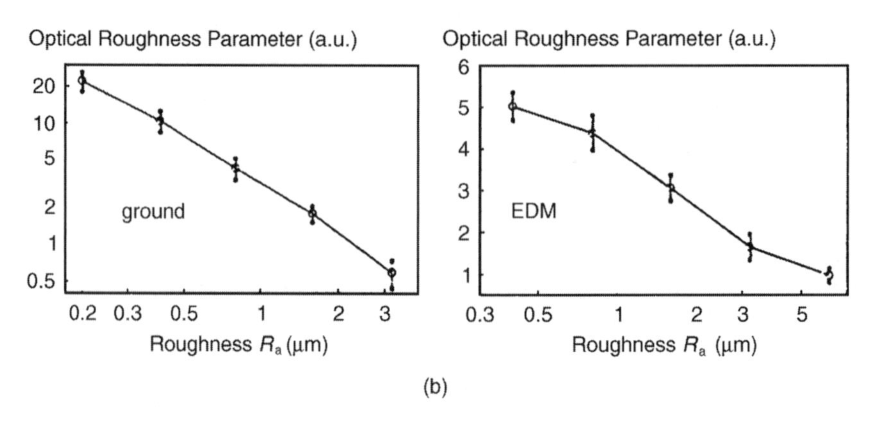

FIGURE 18.26 Polychromatic speckle autocorrelation method: (a) experimental setup; (b) ground and EDM surfaces.

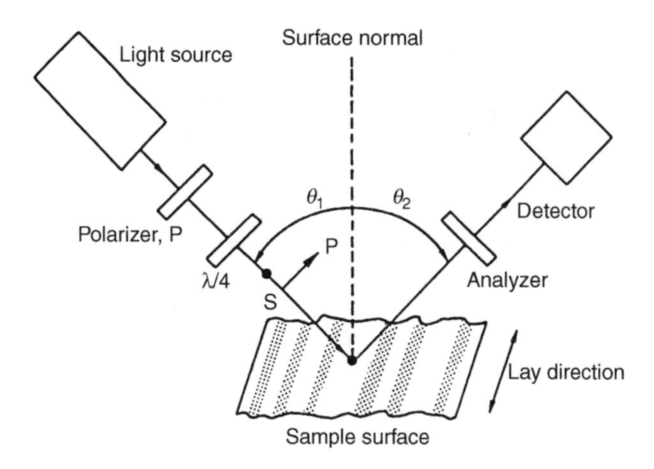

FIGURE 18.27 Ellipsometry method.

The important quantities in the measurement are the angle of the incident light beam and the rotational positions of the polarizing and analyzing elements in the light path. The surface roughness is evaluated by the ratio of the complex reflection coefficients for the P and S components of the electromagnetic field. The system shown in Figure 18.27 measures the ground and polished surfaces with R_a ranging from 0.01 to 1.1 μm. This method is a potential tool for in-process measurement of very fine surfaces, but it requires that the system is carefully aligned. The issue is whether ellipsometry can be used to measure the roughness of engineering surfaces directly.

Angular dependence of scattering: The entire angular distribution of the scattered radiation contains a great deal of information about surface topography. In the past the surface profile was obtained by detecting the positions of the reflected beams subject to the angular distributions. However, this method is not practical and is slow in detecting speed. The details will not be discussed here.

Light sectioning: Surface profiles are obtained by the light-sectioning method in which the height of the surface is determined from the lateral position of the spot cast on the surface by an obliquely incident light beam. Figure 18.28(a) illustrates an example of the system layout that consists of a He–Ne laser light, a lens system, an ITV camera, and a data-processing unit. The ITV camera catches the image of the cross-section through the microscope, then the image is transformed into the data-processing unit to provide its discrete configuration. Finally, it leads to a two-dimensional measurement of surface roughness along the feed direction of the workpiece, as given in Figure 18.28(b). The system shows the resolution of 1.0 μm and the measuring accuracy of several microns. However, high-speed data processing is required to satisfy a practical use.

Another method: Engineering surfaces can be evaluated in terms of fluctuation properties as a more global technique for measuring products. This is based on the 1/f property, in which the optical power spectrum is proportional to the inverse spatial frequency "f." The 1/f fluctuation is seen in such diverse members as river flow, nerve membranes, undersea currents, and fractals. For example, the spectral density of fluctuations in the loudness of many musical selections also varies approximately as 1/f down to 5×10^{-4} Hz, and compositions with such 1/f noise sources are said to sound pleasing. The 1/f behavior of quantities associated with surface textures is assumed to be interesting for future evaluation techniques. A quantity with a white power spectral density is uncorrelated with its past: a very random surface texture. A quantity with a $1/f^2$ spectral density strongly depends on its past: a periodic surface with randomness. A quantity with a 1/f spectral density has intermediate behavior with some correlation over all profiles yet is not too dependent on its past: an adequate random surface texture.

The in-process measurement in turning is done by using the optical system that is built around a Fourier transform lens, as shown in Figure 18.29(a). A 3-mm laser beam with a uniform intensity

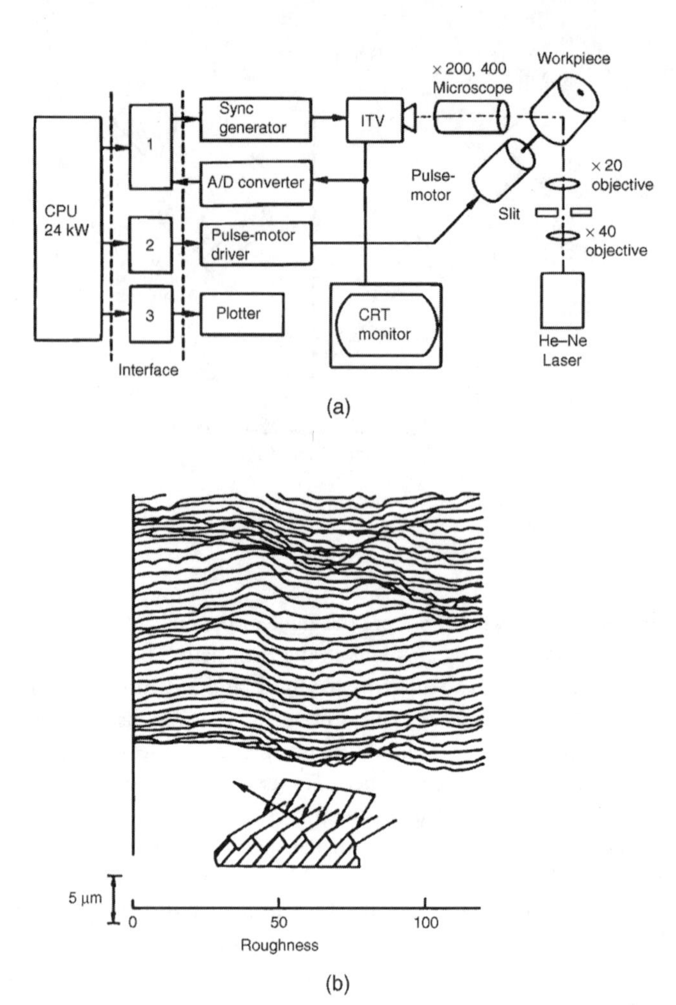

FIGURE 18.28 Light sectioning method for surface profile measurement: (a) system layout; (b) two-dimensional roughness.

distribution passing through the beam expander is projected onto the machined top surface to be examined. The reflected beam from the surface passes through the Fourier lens and then falls onto the screen as a circular power spectrum pattern with a diameter of 20 to 30 mm. The photosensor on the screen is a 320-cell array with each cell element 38 mm long and 0.1 mm wide. A current output from each cell is fed to the interface circuit and then processed in a personal computer to calculate the power spectrum. Because the feed motion of the measuring system is equal to that of the tool post the laser spot on the work surface scans along the helical path of a workpiece. The surface roughness in turning should be measured in the feed direction that will give the maximum roughness reading. To perform this measurement, therefore, a mechanical chopper synchronized with the spindle speed is used between the incident beam and the objective surface. A typical example showing the 1/f surface texture is given in Figure 18.29(b), where the roughness ratio between R_{max} and R_a (R_{max}: maximum peak-to-valley height, R_a: arithmetic average roughness) is around 1.89.

Table 18.10 lists the roughness parameters of 20 test pieces and the slope of power spectra obtained. Special attention should be paid to the roughness ratio of R_{max}/R_a. It may be concluded from these results that, as a whole, surface textures with adequate R_{max} and R_a roughness values seem to show 1/f-like behavior when the ratio is within 2.0.

TABLE 18.10 Roughness Parameters and Fluctuation Properties

No.	R_{max} (μm)	R_a (μm)	R_{max}/R_a	Slope of Power Spectra	Machining Conditions: Cutting Speed (m/min)	Feedrate (mm/rev)	Tool Nose Radius (mm)
1	1.92	1.2	1.60	1/f	60	0.02	0.8
2	5.0	2.8	1.81	≈1/f	85	0.05	0.8
3	7.7	3.8	2.02	$1/f^{1-5}$	100	0.08	0.8
4	8.7	5.5	1.58	≈1/f	110	0.1	0.5
5	8.8	4.1	2.14	$1/f^2$	110	0.1	0.8
6	20.7	6.4	3.23	White	165	0.25	0.5
7	14.1	5.2	2.71	White	120	0.16	0.5
8	7.1	3.6	1.98	$1/f^2$	120	0.1	0.5
9	4.4	2.4	1.86	≈1/f	120	0.08	1.0
10	4.6	2.4	1.92	≈1/f	120	0.1	1.0
11	11.3	4.9	2.31	$1/f^2$	120	0.25	0.8
12	7.3	3.6	2.02	$\approx 1/f^2$	110	0.16	0.8
13	12.8	5.0	2.56	White	140	0.2	0.5
14	5.6	3.1	1.89	≈1/f	110	0.16	0.8
15	8.9	3.7	2.40	White	140	0.16	0.5
16	5.7	2.9	1.96	$1/f^{1-5}$	100	0.1	0.8

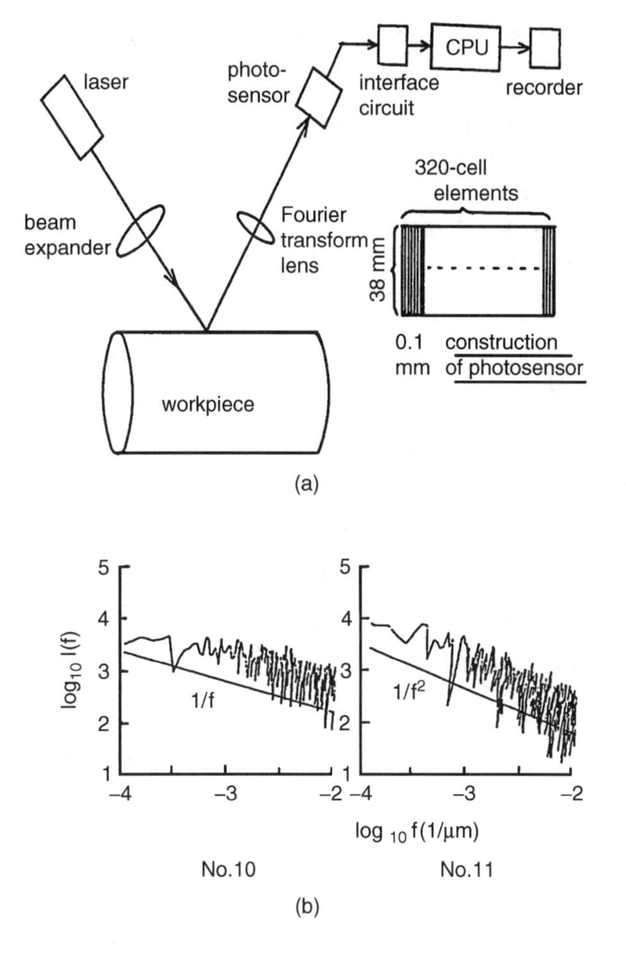

FIGURE 18.29 Evaluation by 1/f fluctuation: (a) optical system; (b) 1/f and other textures.

18.2.4 In-Process Technique for Cutting Process

The purpose of measuring the cutting process is to optimize the process with respect to cost, productivity, metal-removal rate, etc. There are many factors to be measured in the cutting process such as cutting force, machining chatter, tool chip condition, and tool/work collision. Among them, the cutting force detection is especially important for a variety of reasons: e.g., better surface finish and partial dimensional accuracy, prolonged tool life, better machine utilization, and, thus, high productivity. Aside from these measurements, a real-time identification of the cutting process is also important for better understanding of the cutting phenomena. Generally, an acoustic emission method will play an important role in this investigation. Unfortunately, however, an optical approach is not active and not easy to apply for the measurement of cutting process.

Chatter detection: Machined surfaces generally provide information such as chatter marks, cracks, and local irregularities. Chatter vibrations can, therefore, be sensed indirectly by monitoring the surface condition by means of optical techniques. In one case, for example, the reflection points subject to the chatter marks in grinding were detected by the laser scanning system. In a more accurate approach, both geometrical waviness and fluctuation of glossiness were monitored by sensing the reflection pattern from the surface. However, instruments of this type cannot detect the chatter vibration alone. Also included in their measurements are forced vibrations and surface irregularities due to other influences.

18.2.5 In-Process Technique for Machines

The developments in in-process techniques with respect to machine tools are not remarkable when compared with other measuring items such as tools and workpieces. Geometric accuracy of the machine tool strictly relates to the accuracy of the workpiece and depends on the geometric accuracy of its structural parts, guideways, and machine conditions. It is influenced by a number of functional effects such as weight deformations, deformations due to several kinds of forces, thermal effects, dynamic errors of moving systems, and so forth.

Measurement and control of driving systems: The leadscrew drive system serves as the basis of linear measuring accuracy in many machining tools. A higher quality machine tool, such as a precision finishing, might, therefore, require that the loop be truly closed around the leadscrew position or around the machine slide position. Improvement in the accuracy of this drive system is an important subject, and several techniques have been proposed so far. Table 18.11 shows some typical position feedback devices used in driving systems. The decision on which type of feedback device to use in a given application depends primarily on the required position resolution. With the advantage of high resolution, a laser interferometer has often been offered to compensate for positioning error.

Figure 18.30 illustrates an example of this method. Errors in thread-grinding operations are detected by a laser measuring system and automatically corrected by a DC servo motor on the modified thread-grinding machine. The measuring system consists of a dual-beam laser interferometer and an optical incremental encoder. The resolution of the total system is about 0.1 μm. The laser measures the linear distance travel (D) of the carriage, and the encoder measures the exact angle through which the leadscrew

TABLE 18.11 Position Feedback Devices

Sensor	Resolution	Main Features
Laser interferometer	Around 2 nm	Most accurate Requires careful system alignment Expensive
Rotary transducer	0.25–2.5 μm	Ease of use Reliable Inexpensive
Linear scale	0.1–0.5 μm	Somewhat complicated to operate Direct measurement Rather expensive

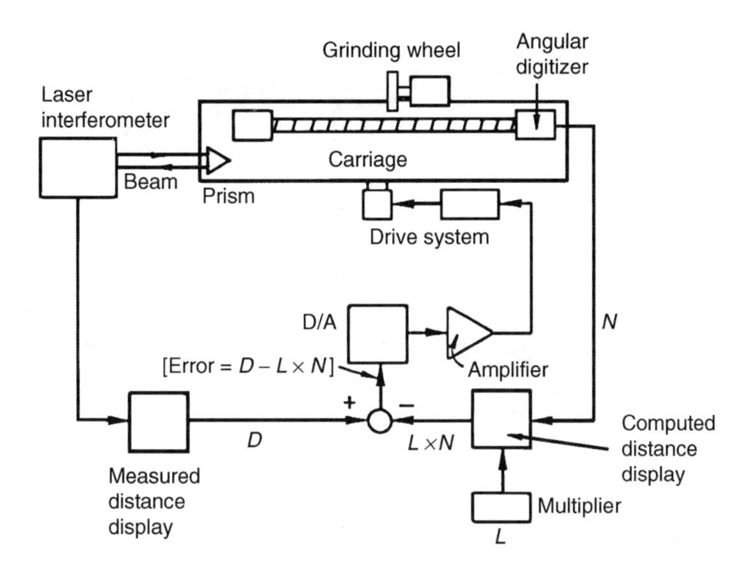

FIGURE 18.30 Positioning error compensation by laser interferometer.

is turned. If the thread grinding is completely accurate, then the linear movement of the leadscrew is equal to the number of resolutions (N) times the lead (L). A digital comparator circuit computes the error ($D - L \times N$) and the servo motor makes small correcting changes in the lead setting. Manufacturing errors are reduced by a factor of ten when compared with traditional machining. However, the interferometric measurement may be effective in carefully controlled environments.

18.3 Welding Processes

18.3.1 Basic Concepts

As in all fusion welding processes, the welding processes shown in Figure 18.31 involve melting and resolidification of base metal. There are two important problems associated with the welding task. One is automation of the process, and the other is control of weld quality, both of which are involved with in-process monitoring and control. The tasks associated with automation are welding processes, identification of the joint pattern to be welded (important because the joint shape determines the weld current and voltage needed to maintain satisfactory weld quality), and weld path tracking based on identification of the types of weld joints.

These problems actually represent a visual feedback control in which the weld torch must accurately follow the centerline of the weld joint. Technically, this problem is not easy to solve because automation suffers from the fixturing inaccuracies of the workpiece to be welded, workpiece-to-workpiece dimensional variations, process thermal distortions, variations in joint geometry, and offsetting of the welding machines' (robots') pretaught path.

The quality control problem is even more difficult because no direct in-process weld quality (specifically weld strength) sensing method is available. Therefore, only indirect methods are attempted to estimate instantaneous weld quality. There have been two approaches to this estimation problem: measurement of surface temperature at several points near the weld pool and measurement of the bead geometry produced at the rear part of the weld pool. One represents the visual measurement of the base shape (height and width) by a CCD camera [Vroman and Brandt, 1976]. The other represents the optical measurement of the multipoint surface temperatures by an infrared optical fiber sensor [Boo and Cho, 1994].

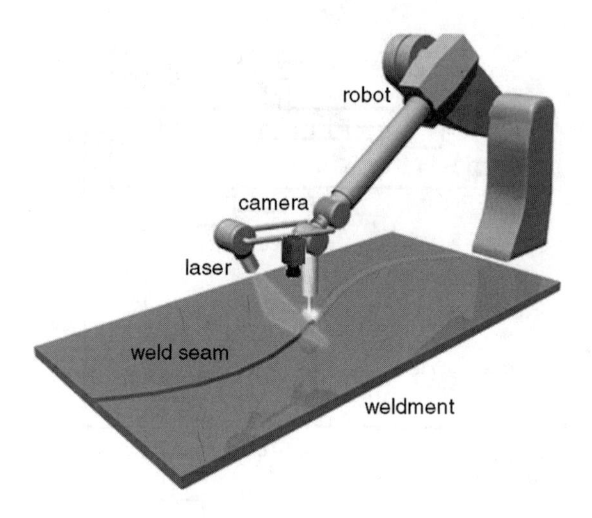

FIGURE 18.31 Tracking seam and welding of a welding robot.

18.3.2 Visual Feedback Control of Robotic Welding Processes

To automate welding processes using robots based on real-time monitoring and control requires use of the electrical noise-immure, noncontent properties of a sensing system, since the weld work environment is severely contaminated due to electric arc, arc spatter, arc noise, and high temperature of melted metal weld pool. In this hostile environment the most promising and powerful method of solving such problems is to use the visual information on the weld joint, weld path, and weld pool geometry [Kim et al., 1996, 2001]. In this section we will discuss a robotic welding based on a visual servoing technique.

In the robotic welding shown in Figure 18.31 an articulated robot with six degrees of freedom is carrying a welding torch and a vision sensor system. The vision sensor views the weld joint ahead of the welding torch as the robot travels along the seam. The images from the vision sensor are analyzed by a vision processing system, which computes the location of the weld seam with respect to the vision sensor. To compensate for variations in part or joint geometry this processing result is then used in the control module to correct the welding torch path.

18.3.2.1 Architecture of the Overall System

The overall system is composed of two modules: a vision-processing module and a robot control module, as shown in Figure 18.32. The vision processing module analyzes the image data obtained from the vision sensor to recognize the type of weld joint and compute the position of the weld joint center. This module communicates with the main processing unit of the robot control module through a PC-AT bus at every sensory feedback sampling period.

The robot control system consists of two units: main processing unit and joint servo unit. The main processing unit (INTEL 80486 CPU operating at 66 MHz) performs major processing operations such as coordinating transformation for trajectory control in cartesian coordinates, providing sensory feedback control for seam tracking, etc. The joint servo unit (TMS320C30 CPU operating at 33 MHz) receives the angular motion vector at each sampling period of 16 ms from the main processing unit and controls the AC servo motor of each axis by an angular-position servoing algorithm.

18.3.2.2 Vision Sensor

The vision sensor is composed of a CCD image sensor and a compact optical projection system that generates a plane of light beam, as shown in Figure 18.33. The camera is fitted with a narrow-band optical interference filter with a spectral bandpass of 10 nm centered at the 690-nm point. The 25-mW laser diode emitting at 690 nm is used as a source of light. The main advantage of illumination by light source with this band is its visibility, which considerably facilitates the calibration of the sensor in the vision

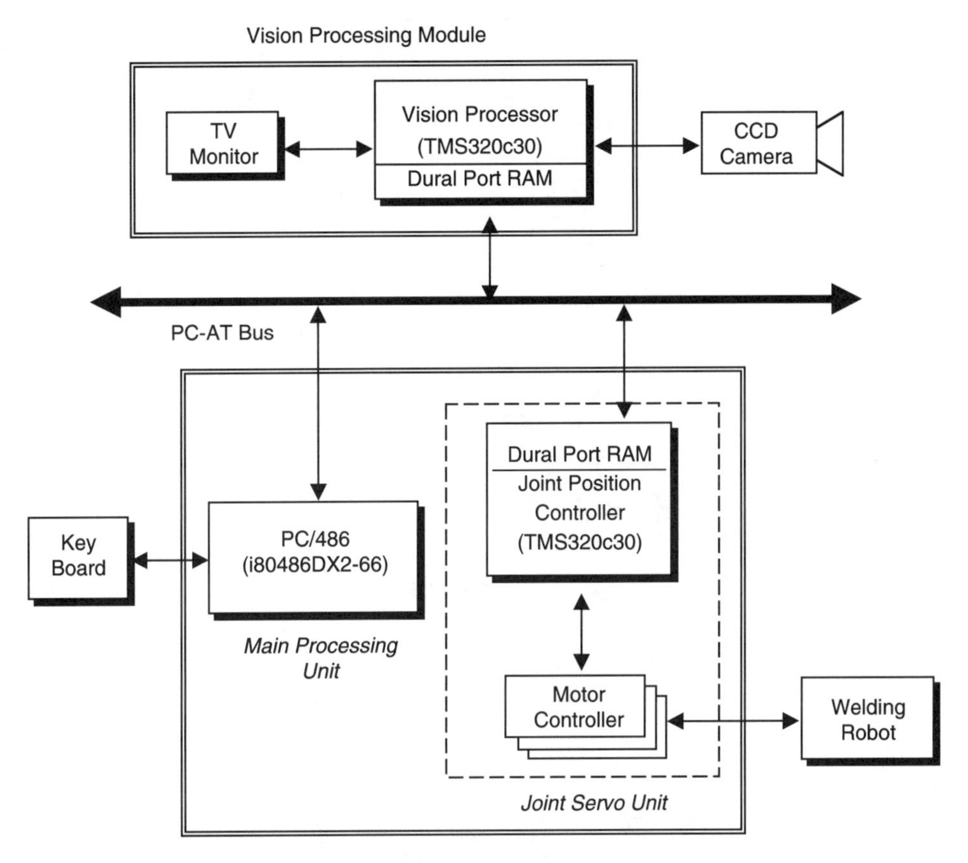

FIGURE 18.32 The overall system architecture.

sensor development stage. The sensor operates on a principle of active triangulation ranging that has been used to obtain information about the three-dimensional layout of surfaces.

18.3.2.3 Joint Feature Detection During Welding

The processing of the stripe images obtained during welding consists of profile segmentation and feature detection of the joints shown in Figure 18.34. To significantly decrease vision processing time, vision processing is conducted on only the windowed image surrounding the expected stripe location. To monitor the behavior of significant joint features in successive input images a windowed image of about 250 (H) × 200 (V) pixels is used. The expected location of the window center is recognized as the joint center position detected in the previous processing time.

The stripe centers are extracted by using the second derivative of a Gaussian filter built in the joint modeling stage. As in the joint modeling stage the extracted stripe data are fitted to straight lines, and then the incorrectly segmented, missing, or overrefined segments are refined through syntactic analysis.

To extract the joint features the segmented joint profile is matched with the template profile of the joint. The goal is to find the best match of the input profile data with the template;, the best match is when the template is attached to the breakpoints such that the total matching cost is minimized. Suppose that the ith branch of the template profile is matched to the breakpoints, b_q and b_r, of the input profile shown in Figure 18.34. In this case the matching cost $C_i(q, r)$ is defined by

$$C_i(q,r) = w_l L_i(q,r) + w_a A_i(q,r) + w_s S_i(q,r) \tag{18.1}$$

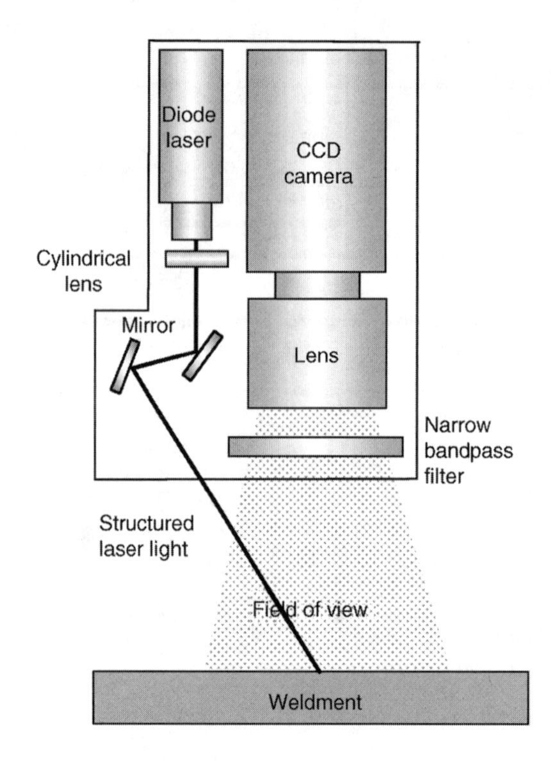

FIGURE 18.33 Configuration of the vision sensor.

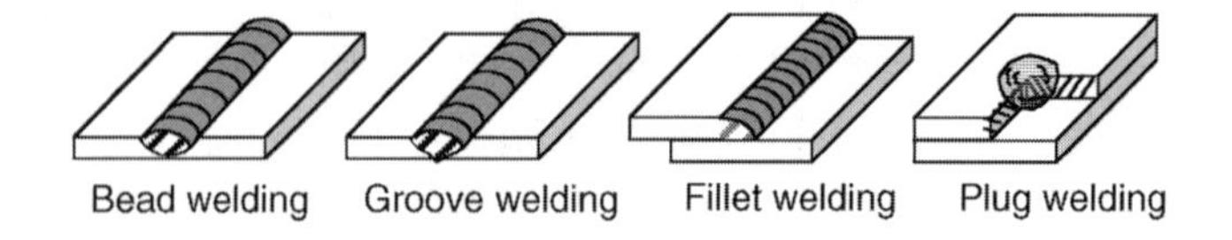

FIGURE 18.34 Four basic types of weld joints.

where w_l, w_a, and w_s are weighting factors, and $L_i(q,r)$, $A_i(q,r)$, and $S_i(q,r)$ denote the measures of similarity in, respectively, length, angle, and straightness between the pair to be considered for matching. The measures are given by

$$L_i(q,r) = \left| \frac{l_i - D(b_q, b_r)}{l_i} \right|$$

$$A_i(q,r) = \left| \frac{\theta_i - \angle(b_q, b_r)}{90} \right| \tag{18.2}$$

$$S_i(q,r) = \left| 1 - \frac{D(b_q, b_r)}{G(b_q, b_r)} \right|$$

where $D(b_q, b_r)$ is the distance between intersection points b_q and b_r, and $\angle(b_q, b_r)$ denotes the angle of an artificial line passing through these two points. Also, $G(b_q, b_r)$ is the sum of all the distances between two consecutive intersection points existing between the points b_q and b_r. The straightness similarity suggests that, as a proper matching candidate, a single line segment is preferred over an artificial line segment combining multiple line segments having different angles from each other. The total matching cost C is the sum of each matching cost over all the branches comprising the template profile. In the

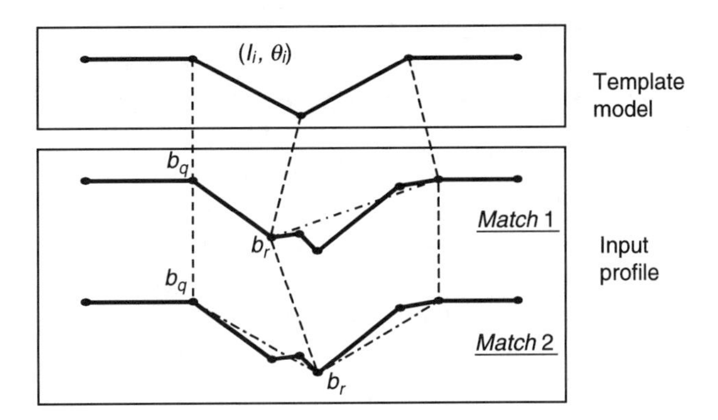

FIGURE 18.35 Typical examples for template matching.

example of Figure 18.35, "match 2" of a smaller total match cost is selected as a more probable case than "match 1." If the total matching cost at the best match is smaller than the threshold value, the selected intersection points serve to determine the joint features in the stripe data and replace the attributes of the template to cope with some possible variations of the joint features in the next incoming image.

18.3.2.4 Robotic Path Guidance

Feedback control of robot motion based on visual information is another important part of the vision-guided robotic welding. For the sake of simplicity the vision sensor coordinates are defined to coincide with the torch coordinates where the origin is attached to the torch tip. Assuming that a position U is the seam position relative to the torch coordinates and a matrix T is the transformation from the torch coordinates to the world coordinates, the absolute seam position P in the world coordinates can be easily calculated by $P = TU$. From the image coordinates of the joint features the seam position U is computed using the vision model previously calibrated.

Now, we define $P_w(\imath \Delta T)$ as the detected absolute seam position at the time $i \Delta T$ and $P_t(J \Delta t)$ as the absolute torch position at the time $j\Delta t$, where ΔT is the processing cycle time and Δt is the sampling period of the robot position control. Because the torch follows the actual seam at the uniform speed ν the robot motion required during one sampling time Δt at the time $j\Delta t$ can be obtained by the following vector ΔP:

$$\Delta P(j\Delta t) = \frac{E(j\Delta t)}{|E(j\Delta t)|} \cdot v\Delta t \qquad (18.3)$$

where E is the position vector from the current torch to the corresponding detected seam, and can be written as

$$E(j\Delta t) = P_w(i\Delta T) - P_t(j\Delta t). \qquad (18.4)$$

Once the motion vector ΔP is given the position for the torch to be reached at the next control cycle $(j + 1)\Delta t$ is evaluated by

$$P_t((j+1)\Delta t) = P_t(j\Delta t) + \Delta P(j\Delta t). \qquad (18.5)$$

The corrected torch position vector $P_t((j + 1)\Delta t)$ is transformed into the joint coordinates, and then the resulting joint angles are sent to the corresponding joint-servo driver to drive each joint synchronously.

18.3.3 Performance of the Visual Servoing System

To investigate the performance of this visual system two types of tests are carried out by using the actual robotic welding system. The first is to verify the robustness of the vision algorithms by detecting the significant joint features from the stripe images obtained during welding. The second is to examine the tracking performance of a robotic system guided by the vision processing results. The overall performance is affected by a degree of combined accuracy in vision processing, vision sensor calibration, and robot servoing. In each welding experiment, gas metal arc (GMA) welding has been processed on the workpieces of hot rolled steel 1025.

18.3.3.1 Joint Feature Detection

A series of tests was performed for a variety of butt, fillet, lap, and vee joints. Figure 18.36 shows vision processing results for the four welding joint types. The stripe images captured from the vision sensor are exhibited in the first column of the figure; these images are obtained with the arc turned on and possess some optical noises caused by arc glares, spatters, etc. To obtain the images with various levels of optical

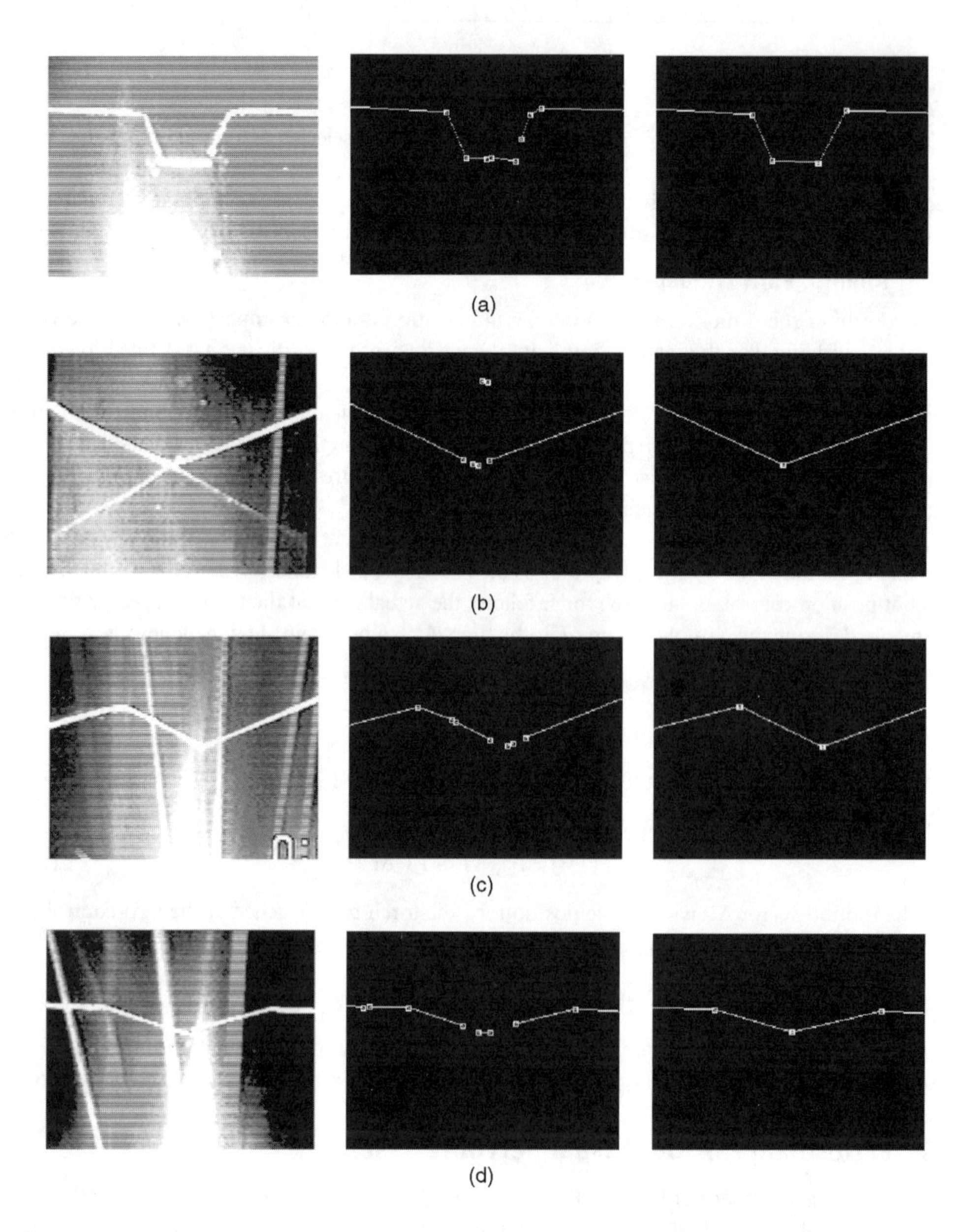

FIGURE 18.36 Detection results of various joint features: (a) butt type; (b) fillet joint; (c) lap joint; (d) vee joint.

noises they were taken from the vision sensor ahead of the welding torch by different distances ranging from 10 mm to 50 mm.

The right images in each row show the joint features that are finally detected. The result of the first column was obtained using the vision processing algorithm for joint identification and modeling, while the results of two other columns were obtained according to the vision processing steps for joint feature detection during welding. Here, the joint profile yielded in the first row was used as a template for the joint feature detection in two other columns. Squares indicate the locations of the resulting feature points. All of the results show that the significant joint features were successfully extracted regardless of intensive arc glares or spatters.

Although a number of overrefined, missing, or isolated line segments are found in the images of the second column showing the results after line feature extraction, the successfully extracted joint features are shown in the right images after template matching.

18.3.3.2 The Robotic Seam Tracking

Once the joint features are located in the image the coordinates of the three-dimensional location of the joint center are determined to guide the welding torch. As described in the previous section, they can be obtained with the vision sensor model. For this, the vision sensor model was obtained offline by observing laser stripes on an accurately machined block of known dimensions and their images. From a set of observed data the 4 × 4 homogeneous transformation matrix characterizing the sensor model was estimated by using a standard linear least-squares fit.

To test the seam-tracking ability of the system a number of robotic weldings guided by the vision processing results have been conducted on various types of weld joints. One of the results is shown in Figure 18.37, in which the vee-grooved joint of the curved shape on the flat plates (300 mm × 220 mm × 8 mm) slanted upward along the welding direction was used. The vision sensor was located ahead of the torch by a distance of 35 mm. Figure 18.37(a) shows the weld bead after robotic welding in which the torch path was guided by the vision processing results. In the welding experiment the welding speed and input power were fixed at 8 mm/s and 3000 W, respectively. This inappropriate welding environment yielding insufficient penetration was intentionally considered for enhancing the view of the torch positions and the joint configurations in the cross-sections of the weld bead. Figure 18.37(b) shows the

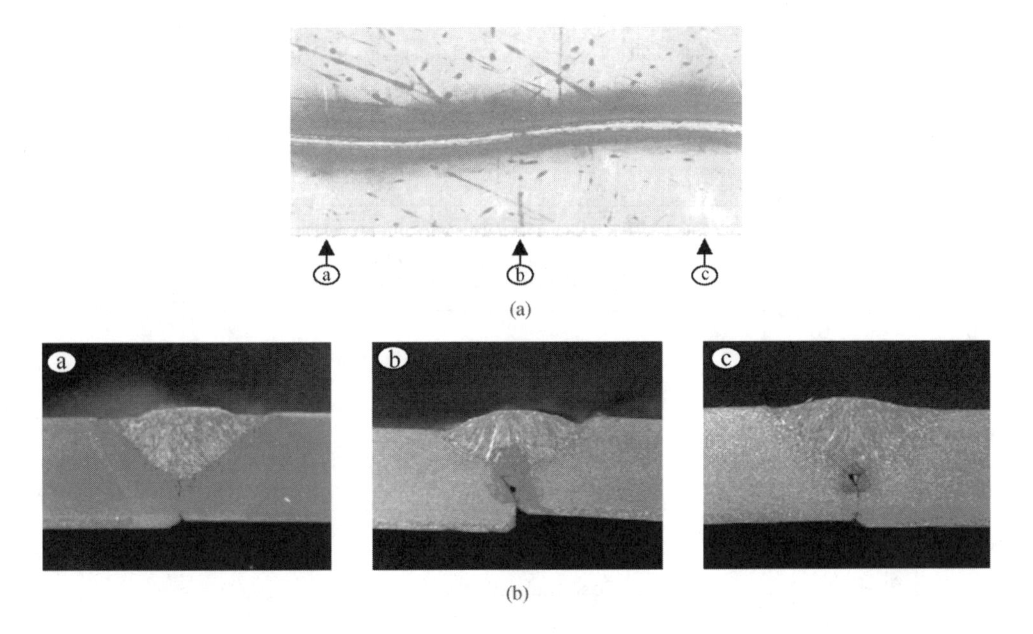

FIGURE 18.37 Weld bead after vision-guided robotic welding: (a) appearance of a weld bead; (b) cross-sections of weld bead after vision-guided robotic welding.

macro-photographs of the cross-sections of the weld bead at the positions noted in Figure 18.37(b). The results demonstrate that the tracking has been performed with reasonable accuracy regardless of variations in joint arrangement and groove preparation.

In the experiment the vision sensor was also located ahead of the torch by a distance of 35 mm, and the welding input power was increased to 4300 W. In Figure 18.38(a) and (b), the *x* axis is aligned with

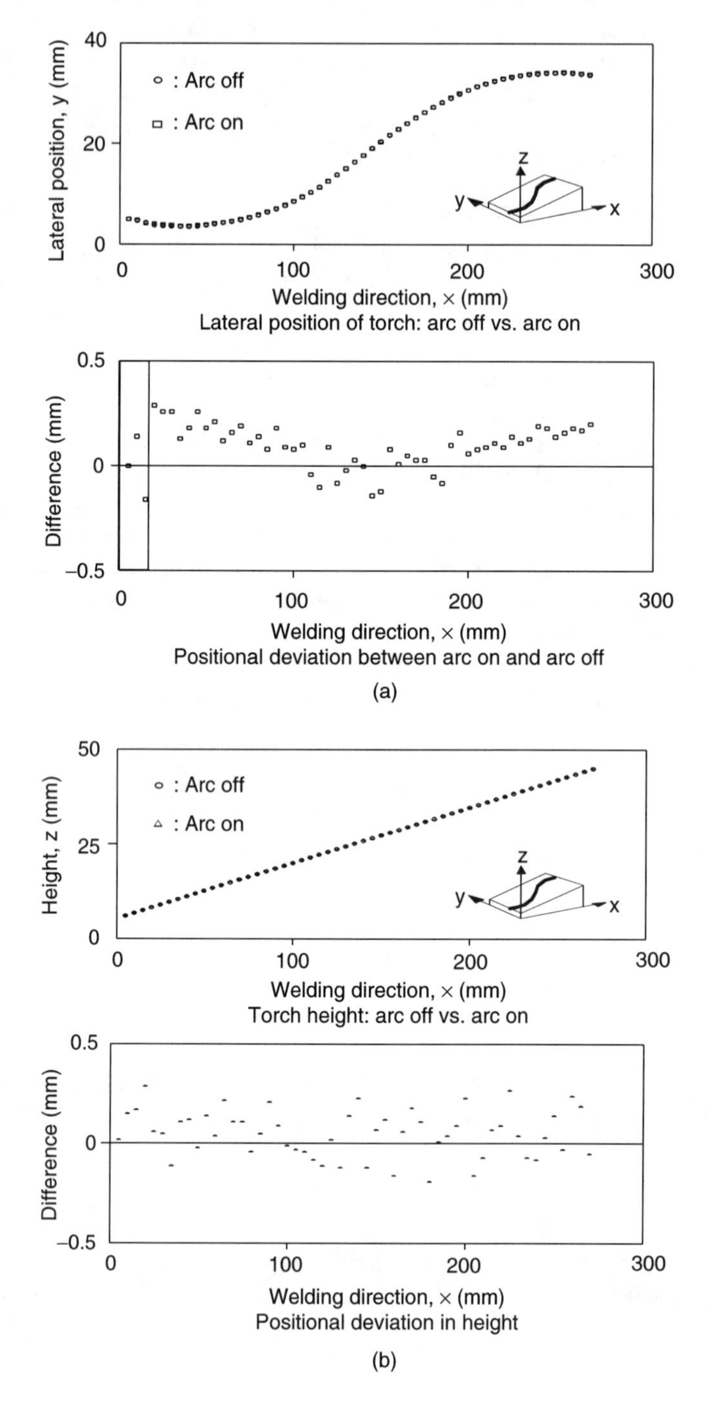

FIGURE 18.38 Seam tracking results of the vee joint: (a) *x*, *y* plane; (b) *x*, *z* plane.

the welding direction, while the z axis represents the height of the torch with respect to the workpiece coordinates. Figure 18.38(a) shows the torch paths in the x, y plane, while Figure 18.38(b) shows the x, z plane. In both figures the results show that additional distortion of the torch position by the optical noises added during welding is always below ±0.3 mm.

18.4 Conclusions

The study of optical-based process monitoring and control has a long history and has been seldom used in practical applications. The reason is that the optical approach can hardly withstand disturbances and the surrounding atmosphere. This problem is, however, gradually diminishing with advances in both hardware and software approaches. A representative example is the application of the optical approach to the manufacturing process for monitoring manufacturing conditions. In particular, in-process approaches for workpiece quality are quite important because the final output in manufacturing represents workpiece quality itself. This chapter introduced several optical techniques using a laser beam with examples, and these techniques successfully demonstrated an ability to evaluate workpiece surface roughness. A welding process is also a field that typically requires optical-based monitoring because of the nature of the process itself. A successful implementation of a welding process using a laser beam was introduced in cooperation with a CCD camera as a robotic welding machine.

Defining Terms

adaptive control: Self-redesign or self-organization control of a system.
adaptive control for optimization: Control for minimum cost or maximum productivity.
adaptive control with constraints: Control system to maintain safe operation under constraints.
arithmetic average roughness: Average value of surface roughness (R_a).
characteristic equation: Equation to characterize the behavior of a control system.
disturbance observer: Observer to estimate disturbance applied to a control system.
1/f fluctuation: Fluctuation in which the power spectrum is proportional to the inverse frequency.
geometrical adaptive control: Control method to maintain a workpiece quality.
R_{max}**:** Maximum peak-to-valley height roughness.
VB: Criteria to evaluate flank wear of tool.
visual feedback: Vision-based feedback control such as a CCD camera.

References

Boo, K. S. and Cho, H. S., A self organizing fuzzy control of weld pool size in gma welding processes, *Control Eng. Practice*, 2(6), 1007–1018, 1994.

Broman, A. R. and Brandt, H., Feedback control of gia welding using puddle width measurement, *Welding J.*, 8, 742–749, 1994.

Choudhury, S. K., Jain.V. K., and Krishna, S. R., On-line monitoring of tool wear and control of dimensional inaccuracy in turning, *Trans ASME, J. Manuf. Sci. Eng.*, 123, 10–15, 2001.

Fan,Y. and Du, R., Monitoring rotating tools using laser diffraction, *Trans ASME J. Manuf. Sci. Eng.*, 118, 664–667, 1996.

Kim, J. S., Son, Y. T., Cho, H. S., and Koh, K. I., A robust visual seam tracking system for robotic arc welding, *Mechatronics*, 6(2), 141–163, 1996.

Kim, J. S., Koh, K. C., and Cho, H. S., An adaptive tracking of weld joints using active contour model in arc welding processes, *Proc. SPIE Opto-Mechatronic Systems*, Boston, 2001.

Lehmann, P. and Goch. G., Comparison of conventional light scattering and speckle techniques concerning an in-process characterisation of engineered surfaces, *Ann. CIRP*, 49(1), 419–423, 2000.

Miyoshi, T., Takaya, Y., and Saito, K., Micromachined profile measurement by means of optical inverse scattering phase method, *Ann. CIRP*, 45(1), 497–500, 1996.

Ryabov, O., Mori, K., and Kasashima, N., An in-process direct monitoring method for milling tool failures using a laser sensor, *Ann. CIRP*, 45(1), 97–100, 1996.

Shiraishi, M., Sensing and control in production and manufacturing, Lecture Notes, 1–230, 1994–1998.

Uda, Y., Kohno, T., Yazawa, T., Suzuki, T., and Soyama, A., Concept and basic study of improvement system of surface roughness, waviness and figure accuracy by WORFAC, *J. Mat. Proc. Tech.*, 62, 423–426, 1996.

19

Inspection and Control of Surface Mount Processes for Electronic Part Assembly

Hyungsuck Cho
Korea Advanced Institute of Science and Technology
Taejeon, South Korea

19.1 Introduction

A printed circuit board (PCB) is a board on which an electrical circuit is assembled. It has two functions: to hold the electronic components in place and to transmit electrical signals between devices on the board. The board itself is usually made of fiberglass and is called a substrate. The electrical signals are transmitted by a copper pattern etched on the surface of the substrate from component to component.

The rapidly growing demand for microelectronic equipment such as camcorders, cellular phones, portable stereos, etc. has accelerated the rate of miniaturization of components, the dense packing of boards, and high performance, resulting in the development of highly automated PCB manufacturing lines having high productivity and high reliability. Surface mount technology (SMT), by which electronic parts can be placed on a printed circuit board (PCB) automatically and not by hand soldering, has played a key role in this development. The trend toward increasing density of electronic components on PCBs is accelerating: the lead pitch of SMTs, such as quad flat packs (QFPs), will become much narrower than 20 parts/cm [Hata, 1990]. However, ever more stringent requirements on density have demanded new packing formats—ball grid array (BGA) and chip scale package (CSP)—that have allowed significant reductions in component size compared to conventional surface mount devices (SMDs) [Furuno et al., 1999]. Figure 19.1 shows the density of electronic components used in PCBs [Hata, 1990; Kim, 1996].

To accommodate the rapid change of electronic components and requirements for increased density on PCBs, PCB manufacturing equipment has been forced to become very precise, highly reliable, and very productive. In particular, to maintain a desired level of PCB quality, stringent requirements on

0-8493-1162-4/03/$0.00+$1.50

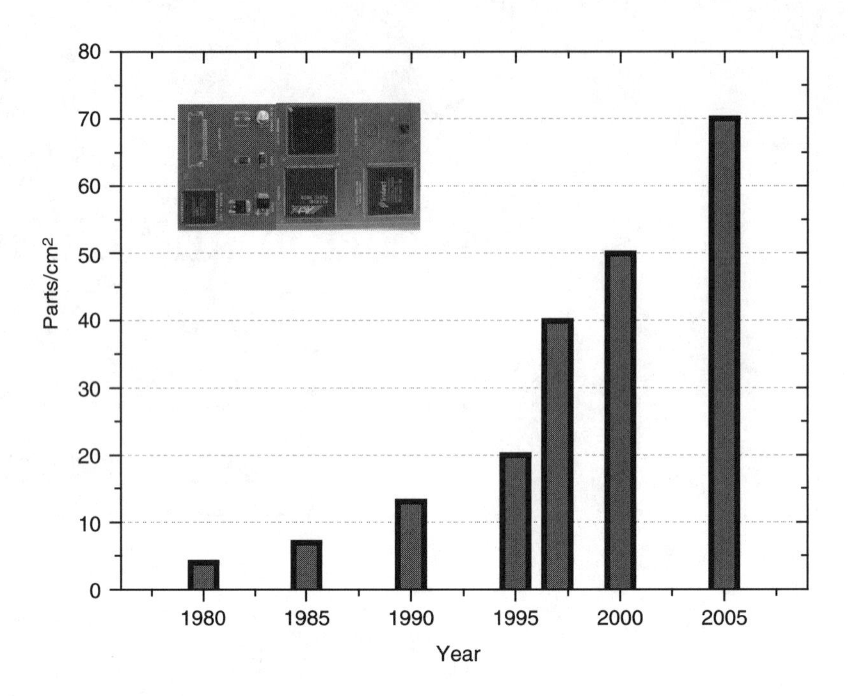

FIGURE 19.1 Density of electronic components used in PCB.

fabrication processes have been imposed. Accordingly, fabrication processes such as mounting and inspection have been evolving continuously, as shown in Figure 19.2. However, due to limitations of equipment performance each process often produces many undesirable defects that should be detected and corrected before the board proceeds to the next processing operation.

As shown in Figure 19.3, basically four processes are involved in the assembly of PCBs. First, a bare board is placed in a screening press, and the solder paste is stenciled on the board in a process much like silk screening. Second, the components are placed onto the device footprints, which should be covered with solder paste. This process is called "on placement." Third, the board is passed through an infrared oven to slowly heat the board and components on it to the reflow temperature. This is done to avoid thermal shock at the next stage. Finally, the board is heated above the melting point of the solder. This may take place either in an infrared oven or a vapor phase reflow bath, although an infrared oven may not heat the bright solder pads and pins very effectively. Alternatively, a vapor phase reflow may heat the device too quickly and cause shock, which would crack the chips. Because these four processes are complex and interconnected sequentially, affecting one after another, inspection of each process becomes critical.

The complexity involved in the assembly of PCBs and interaction between the processes necessitated the development of an automated inspection and diagnosis system to automate inspection processes and to utilize the inspection results generated from the systems to automatically control the fabrication processes. This system is usually capable of sensing components and defects, detecting and identifying them, and performing automatic feedback control based on the information gained from the recognized results. Most defects occur during the fabrication processes because they are microscopic in size and exhibit characteristics that are difficult to measure in a noncontact and on-line manner. Although several other inspection methods are available, to date the majority of those in use employ the optical visual method. This is also true with electronic component inspection and measurement of the component pose to be used for placement onto PCBs. The reason is that this sensing method provides advantages in accuracy, simplicity, speed, and low cost over existing conventional noncontacting methods. In fact, the method seems to be a unique approach in solving the problem of PCB quality, which is difficult to assess.

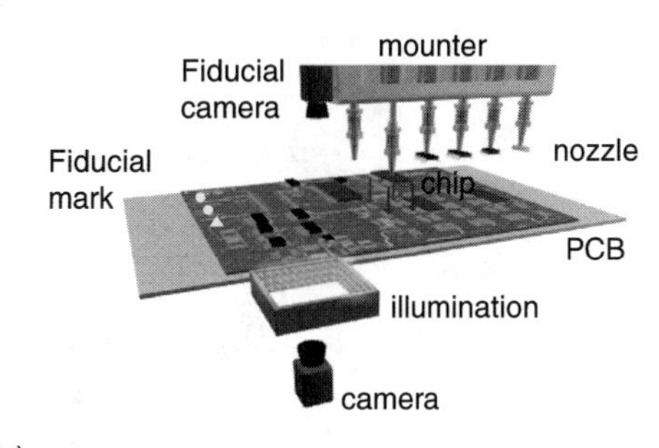

(a)

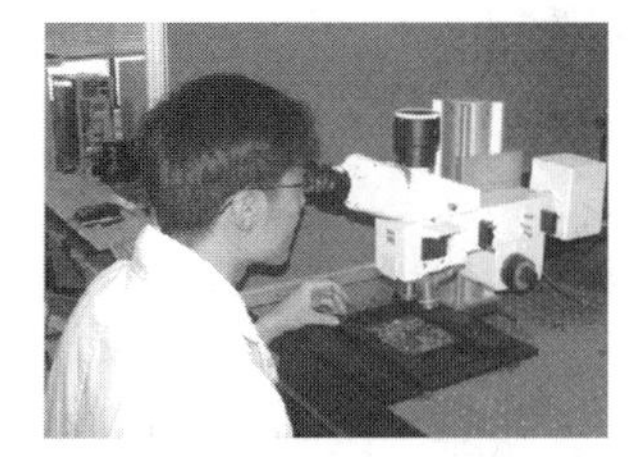

(b)

FIGURE 19.2 Evolutionary change in surface mount technology: (a) chip/SMD mounting; (b) PCB inspection.

Therefore, optical/visual inspection and diagnosis technology becomes indispensable for situations associated with PCB production systems. The aim of this chapter is to show how to embed optical/visual elements and devices in the above processes by incorporating other relevant functional components to achieve an intelligent, optically embedded inspection system. We begin with an introduction to the state-of-the-art technology of PCB process inspection available today.

19.2 Optically Embedded Inspection and Control System

To facilitate understanding of opto-mechatronic technology associated with an PCB inspection system we will introduce the PCB production process in more detail and explain the inspection tasks of each individual process.

19.2.1 Manufacturing Process Inspection Tasks

Previously, when optical/visual systems had not been developed and were not available, most inspection, part identification, and decision-making were done by human workers. Chip part mounting was performed by a chip mounter, a "mechanical positioning mounting unit." Most fabrication lines today have optical/visual devices and components embedded in chip mounting and inspection equipment. As a result of these developments the number of defects is significantly reduced, thereby resulting in a much higher production yield and PCB quality. Figure 19.3 indicates the manufacturing processes that utilize optical/visual sensing and signal processing. These include (1) bare PCB production, (2) solder paste printing, (3) SMD mounting component pose inspection, and (4) reflow soldering.

Copper conductor patterns shown in the figure are inspected before solder masking because the solder mask obscures the conductor. Defects appearing on these patterns are primarily caused at the stages of etch, exposure and development, and artwork. Zhang and Weston [1994] and Moganti [1996] provide a brief overview of bare PCBs of this automatic optical inspection (AOI) system.

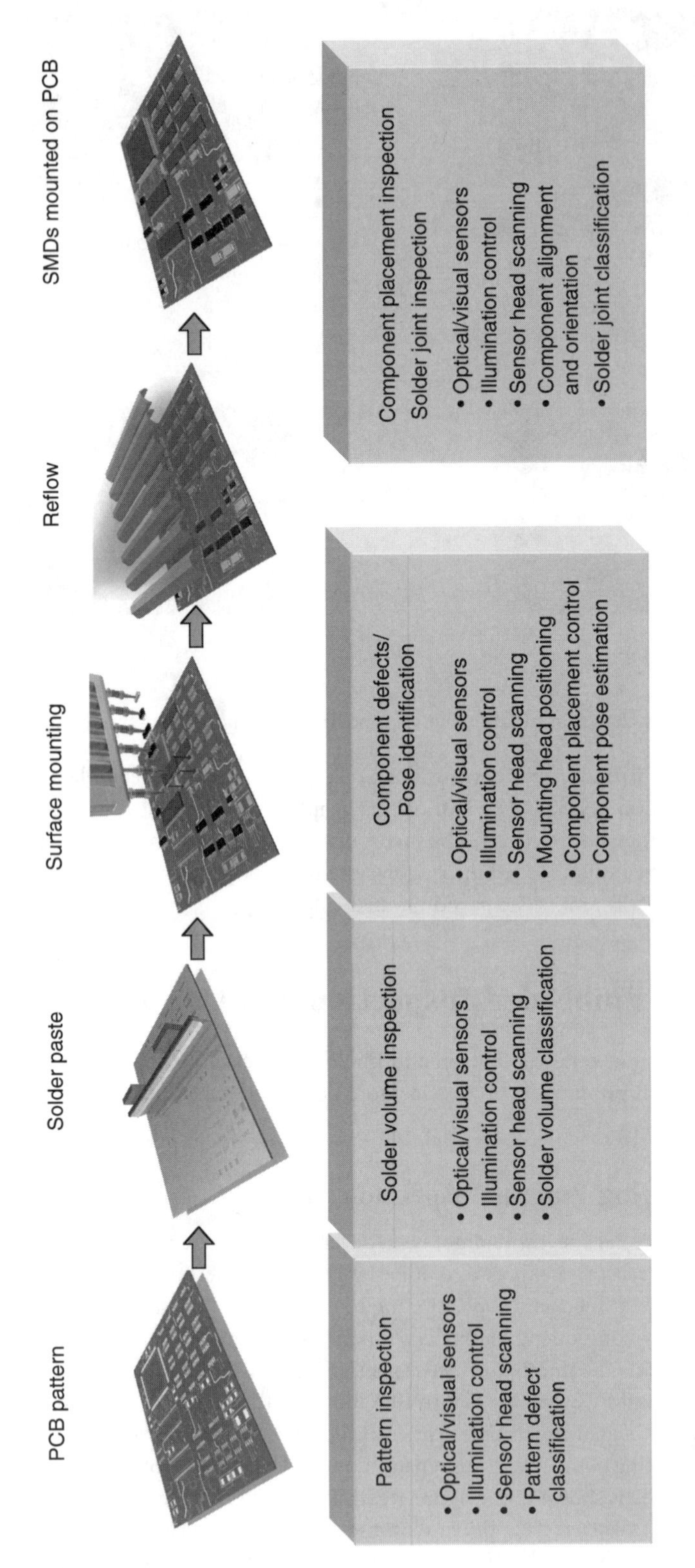

FIGURE 19.3 Various tasks needed for each stage of the SMT process inspection.

The second figure from the left shows the state of solder cream pasted onto the surface of the copper pattern. Several problems can occur when depositing the solder paste. The silk screen may be deformed so that the solder paste is deposited in the wrong place. The silk screen may clog or tear, resulting in incorrect amounts of solder being deposited on the pads. Additionally, the layer of solder may be too thick or too thin. If too much solder is deposited, it may form bridges, and if too little solder is deposited, the joints may not be sound. The third figure from the left illustrates the state in which all electronic parts are properly mounted by an automatic chip mounter machine. Major defects in this state are missing components, reversed components, and wrong components.

The figure on the right shows the soldered state of a loaded PCB after the solder reflow process. Many defects arise in the soldering process, any of which can cause the entire PCB to misfunction [Oyeleye and Lehtihet, 1998], including: (1) open circuit/no solder-no conduct made at a pad; (2) solder bridging unwanted solder between pads/component leads; (3) solder balling/splashing solder beads around the pad; (4) pits, holes, and voids decreasing joint strength, current conducting capacity; (5) solder deficiency causing failure of the joint structure; (6) excess solder decreasing flexibility of the component lead; (7) prolonged heating causing a thick, brittle intermetallic layer in the solder joint, leading to shearing; (8) webbing, in which solder sticks to surrounding insulation; (9) icicling, in which sharp peaks appear on joints; (10) nonwetting, in which pad or lead is not wetted; (11) dewetting, in which solder retracts from pad or lead; (12) grainy solder caused by contaminants in solder; (13) disturbed solder, disturbed joint during solidification; (14) cold solder or insufficient solder temperature, causing unstable joint; and (15) cracked solder joint, caused by PCB flexing and metal fatigue. Additionally, there are contamination problems, placement problems, etc.

Solder joint inspection can be expected to become increasingly important as chip sizes are reduced and as it becomes increasingly difficult to inspect the joint manually. These processes can be categorized into two types. One process can be monitored online. Solder paste printing, SMD mounting, and component pose estimation belong to this category. The other process can only be monitored offline. At present, PCB pattern generation and reflow soldering are examples of this process.

If we examine in more detail the characteristics of the processes and the defects generated during the processes, we realize that in all processes optical/visual elements, part positioning and identification, and measurement of process output are essential for inspection. As listed below the drawings in Figure 19.3, each process requires an optical/visual sensing that requires an illumination control, sensor head scanning, data or signal processing, decision-making software, and an I/O interface unit. These elements are combined and fused together to constitute an optically embedded system.

19.2.2 Optical Devices Embedded in a System

A generic structure of this system is schematically illustrated along with its functional block in Figure 19.4. There are four modules that carry the optical/visual sensing, signal processing, decision-making, and control mechanisms. The sensor module is composed of optical/visual sensors, light sources including illumination and structured light, and the sensor-positioning control unit. The illumination and light sources are controlled to adjust their intensity and position by an independent control mechanism. Typical applications are: an illumination control provides variation of incident light angle and intensity, while a galvanometer unit provides scanning action of the light source and optical units such as mirror, beam splitter, lens, etc. The optical signal (data)-processing module performs on-and offline acquisition of the signal (data) obtained from the processes. Based on these acquired data the module operates on the acquired raw data and performs filtering thresholding and transforming, either in a deterministic or stochastic domain, to obtain the edge or features of objects to be measured.

The decision-making module receives from the signal-processing module the information necessary to determine the state of the processes being monitored. Based on that information the module then carries out fault identification and classification associated with each process, pose estimation of electronic

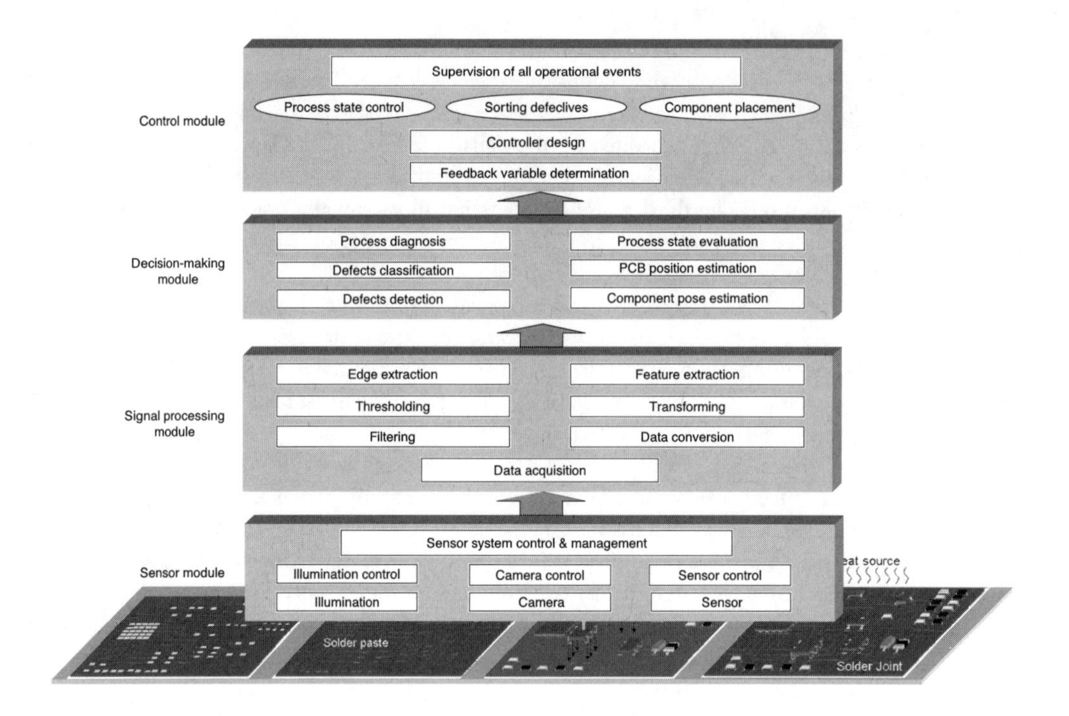

FIGURE 19.4 Hierarchical structure of basic task modules for PCB assembly.

components to be mounted, and control signal generation for the part mounting head and the solder paste printing head.

The main functions of the control module are to control the position of the PCB loaded in a conveyor, to accurately place electronic components to be aligned with printed solder pastes, to control the solder paste printing speed, and, finally, to supervise all states of the process. In addition to these functions, the module executes sorting out the defective components and PCB assemblies that need to be transported for automatic removal and replacement to a rework station.

Each of the detailed functional descriptions of PCB assembly discussed above reveals that (1) a PCB assembly station could not perform assembly work without the assistance or utilization of optical/visual elements that provide the sensory information; (2) this information is utilized for multiple purposes including motor control, optical sensory element control, electronic component quality inspection, in-process and offline PCB process monitoring, electronic component pose identification and control, etc.; and (3) each process that constitutes SMD assembly work affects the quality of subsequent processes. Also, subtasks performed within a process influence the quality of subsequent subtasks. This dependency deteriorates overall PCB quality and can be significantly reduced via an iteration of optical sensing, signal processing, and control technologies. Further developments of technology fusion associated with opto-mechatronics will enable next-generation surface mount technology to be truly intelligent, autonomous, reliable, and adaptable to changes in SMD components, mounting density, and batch size.

19.2.3 Monitoring and Control of PCB Manufacturing Processes

More than 50 process steps are required to fabricate a printed circuit board (PCB) [Moganti et al., 1996]. Human operators monitor the results after each process and simply inspect the work visually against prescribed standards. Then, they make a decision on the quality of each process output and take necessary actions by adjusting the process parameters relevant to the cause of the resulting output. This procedure is a repetitive open-loop control system. But if the processes are monitored and controlled online, then the control system can be regarded as a feedback control system. In any case, optical/visual sensors provide

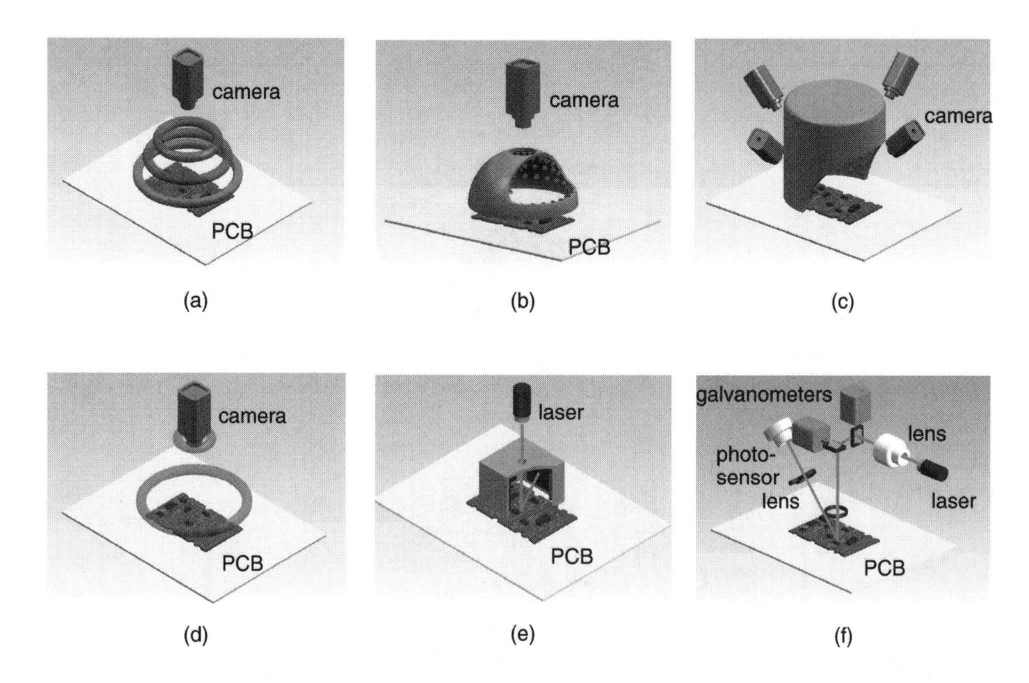

FIGURE 19.5 Typical visual inspection configurations: (a) RGB color lighting; (b) multi-LED lighting; (c) multi-light and multi-camera; (d) upper/lower lighting; (e) laser spot reflection; (f) laser triangulation.

the controller with information relevant to the process output to make a decision regarding parameter adjustment. Therefore, these optical elements are essential to process monitoring and control.

19.2.3.1 The Optical/Visual Inspection System

To carry out visual inspection tasks various configurations of inspection system have been developed. Some of the typical ones are shown in Figure 19.5 [Hata, 1990]. The vision-based inspection system consists of a CCD camera, an illumination system, an image acquisition system, and a processing system. In the case of the laser triangulation method, the system consists of a laser source, optical components (beam splitters and adjustable mirrors), and detectors. A typical inspection process involves the acquisition of optical/visual data over the entire area of interest to be inspected.

In the vision-based system digitization of the acquired image data processes a raw image that needs to yield clear and accurate enough imagery to do either an analysis of the processed image requiring feature extraction or operation of the image. Finally, pattern classification and recognition based on the analysis images are performed to detect relevant defects.

19.2.3.2 Illumination System

Illumination is one of the critical factors that must be considered in acquiring an image of good quality. The choice of this lighting method largely depends on surface characteristics of the PCBs and the constraints imposed by the camera. For example, the commonly adopted lighting techniques for finding edge image of electronic parts include direct lighting, vertical lighting, and bidirectional lighting, as shown in Figure 19.6 [Moganti et al., 1996]. The method of determining an appropriate lighting condition requires experience or experimental analysis, as reliable analytical methods are not available to date.

19.2.3.3 Image Acquisition

Images are acquired by use of a CCD (charge-coupled device) camera of the area or linear type. Depending on applications, laser scanner cameras are also frequently used. The choice of a camera should consider

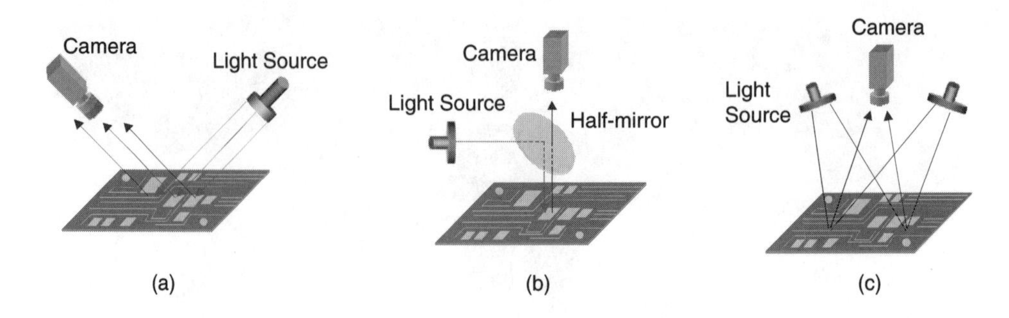

FIGURE 19.6 Typical illumination methods: (a) direct lighting; (b) vertical lighting; (c) bidirectional lighting.

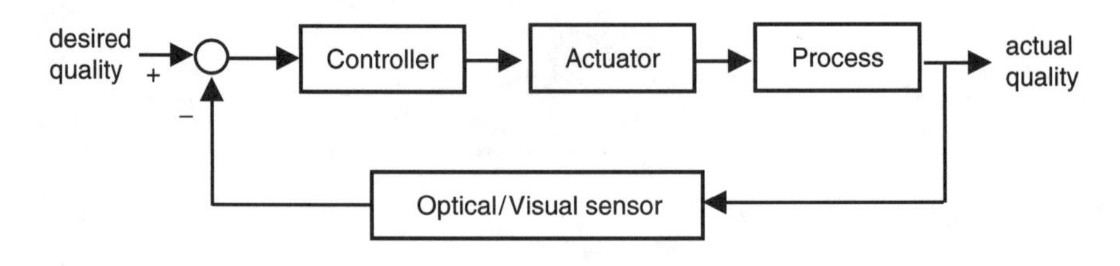

FIGURE 19.7 General block diagram for quality control.

the maximum resolution and depth of field (DOF), so that the sensor details the defects under inspection as accurately as possible.

19.2.3.4 Image Processing

Image processing involves enhancement of the acquired image in which any noise is removed, image contrast is enhanced, and edges of the image are contrasted against the image background. Enhancement techniques include thresholding, convolution, and picture processing.

19.2.3.5 Position Control System

As seen in Figure 19.6, acquisition of optical/visual data is usually required over a large area of objects to be inspected. In the case of a vision system, the inspection area is sequentially scanned by using a positioning control system. In the optical-based inspection system, the light path is usually controlled to scan over the entire inspection area by a positioning mechanism (scanning device) such as a galvanometer.

19.2.3.6 Quality Control System

Several control systems are involved with the control of output quality obtained at each stage of the PCB manufacturing processes. In general, these systems can be expressed as shown in Figure 19.7. The characteristics of these systems are distinct from other control systems normally encountered in practice in that the quality measurement with these systems can be done only by optical/visual sensing methods. This means that no other sensing techniques enable one to monitor the process state effectively.

In the PCB pattern generation process, the quality of the patterns is inspected in a run-to-run manner, and the quality information is properly fed back to fabrication process controllers. In the process of solder paste printing, solder volume, geometry, and location are measured optically either on-line or in a run-to-run manner and fed back to a stencil nozzle controller. In the mounting process output quality means the positioning/orienting accuracy of electronic parts. This objective can be fulfilled by sensing their pose, correcting any errors by a control mechanism, and comparing these with their printed pose. We will discuss this in more detail in the next section.

19.3 Inspection and Control of Various Processes

19.3.1 Solder Paste Printing Process

Solder paste printing is one of the most critical processes in surface mount part assembly, as the most common defects are caused by improper printing of solder paste. This necessitates fully automated in-line inspection methods for the quantification of printability. Typical types of solder paste defects are listed in Table 19.1. The solder paste printability is found to depend on the choice of stencil opening geometry, the matching of solder paste, fixing of the waiting time, and the parameter setting of the printing machine and printing surface (baseboard topology) [Owen, 2000].

Some measurement methods, such as microscopic visual, laser triangulation, and machine vision, have been developed over the past several years. These measurement systems identify specific types of visible defects by measuring solder paste area, height, placement, and volume. Two major methods are machine vision and laser triangulation. Two laser triangulation methods are illustrated in Figure 19.8. As shown

TABLE 19.1 Types of Solder-Paste Defects

Defect Type	Inspection Type	Possible Cause
Improper paste-to-pad offset	Two-dimensional	• Misaligned stencil • Bad stencil or bad PCBs
Solder bridge	Two-dimensional	• Excess paste • Damaged printer apertures
Solder smear	Two-dimensional	• Poor board handling • Solder paste on the back of the stencil
Excess solder area	Two-dimensional	• Poor aperture gasket due to excessive squeegee pressure • Debris on the PCBs • Damaged aperture
Undersized solder area	Two-dimensional	• Dried solder paste in stencil apertures • Printer paste volume too low • Squeegee speed too fast
Excess solder volume and height	Three-dimensional	• Contamination at the PCB-printer-interface • Warped stencil
Solder slump	Three-dimensional	• Squeegee speed too fast • Paste temperature too high • Moisture absorbed by solder paste
Large height variations	Three-dimensional	• Warped stencil • Separation control speed too fast • Squeegee speed too fast
Solder volume too low	Three-dimensional	• Polymer blades scooping paste out of apertures • Squeegee speed too fast

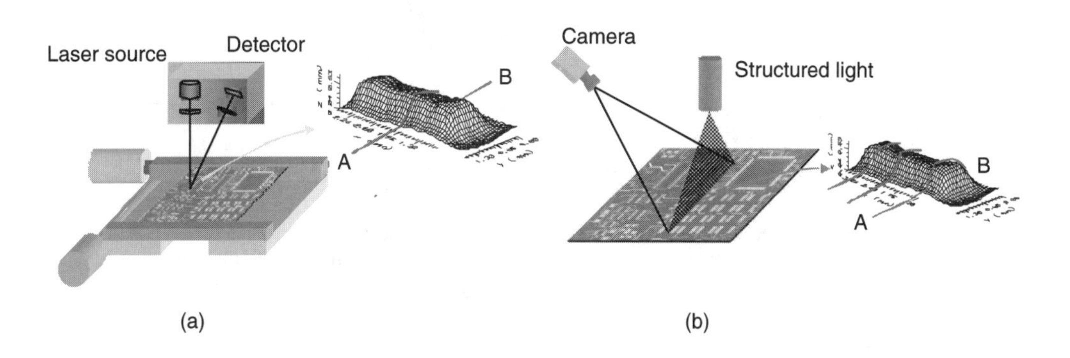

FIGURE 19.8 Solder paste inspection: (a) point scanning method; (b) laser strip scanning method.

in Figure 19.8(a), a point of light is projected from a laser diode onto the surface of the printed solder joints being measured. The light scattered from the surface is imaged onto a light-sensitive digital detector array. Because the height of the paste along line A–B varies, the imaging point varies accordingly, and this variation is detected by the sensor. Figure 19.8(b) shows three-dimensional (3D) laser triangulation using a slit laser beam and a scanning device.

Here, we introduce the method of employing a scanning point laser, as shown in Figure 19.8(a) [Lathrop, 1997]. The sensor is a high-speed point range specular sensor that nominally generates 200 readings per second. The beam diameter for this sensor is 10.2 μm with a triangulation angle of 45°. The measurement range of this sensor is 0.3 mm with a resolution of 0.76 μm. The printed solder paste is scanned by the sensor, in such a way that the *X-Y* table moves the PCB to a designated location. Printing process variables are limited to the following for the printer: steel squeegee at 45° angle, squeegee speed of 7.63 mm/sec, stencil wiped after every 20th print. The measurement is limited to the 0.5-mm pitch deposits. The measurement is made every 12.5 μm along the slice, and ten slides are obtained per printed pad (see Figure 19.8). Collecting all these are 64 pads per print on this test board. The total number of height measurements is 140,800. In this measurement, the resolution of height is 0.07 μm, whereas width resolution is 12.5 μm. The massive amount of height data can be plotted for various cases of different mesh paste, as illustrated in Figure 19.8. To evaluate printability under various conditions, quantitative values such as volume pad height and pad width are needed. To this end, the measuring system computes the slice area, average pad width, and average pad height for each slide of pad data. Figure 19.9 shows average height and width and their standard deviation for the case of 325 mesh. Pad height is calculated as the average height of the "plateau" of the deposit profile. Volume is approximated by slice area multiplied by the pitch of the slices. Although the measurement results are not shown here for other conditions, the volume, height, and width of pads obtained under various conditions of paste material and mesh powder show no significant variation in their value if pitch and squeegee speed are fixed.

19.3.2 Surface Mounting Process

The optical-based system plays a key role in surface mounting part placements by locating PCB fiducial marks, assuring part alignment, conducting part defect inspection, and performing tolerance checking. Although these functions have been greatly enhanced to accommodate changes in new component types and higher densities in PCBs, even further changes are expected in the near future [Woolstenhulme and Lubofsky, 2000]. Surface mounting of electronic parts is accomplished through five main processes:

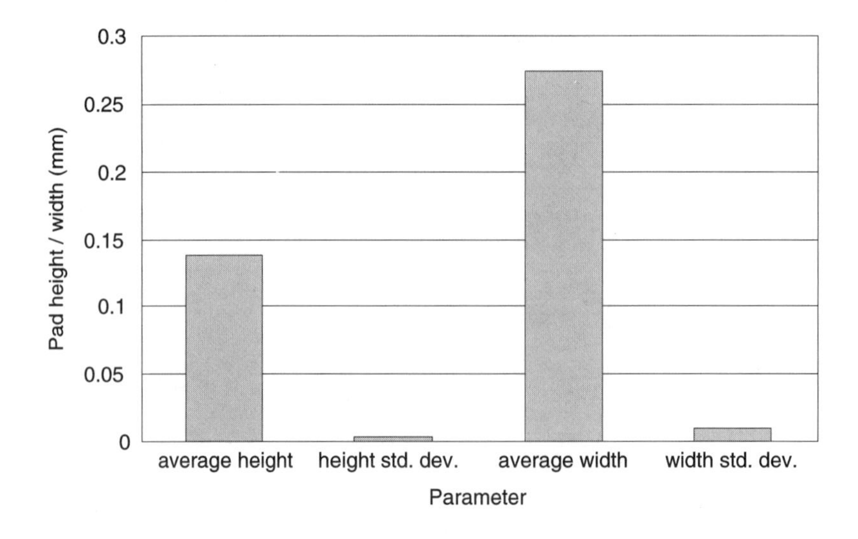

FIGURE 19.9 Pad height and width measuring results. (From Lathrop, R. R., in *IEEE Transaction on Components, Packaging, and Manufacturing Technology—Part C*, 1997, p. 20. With permission.)

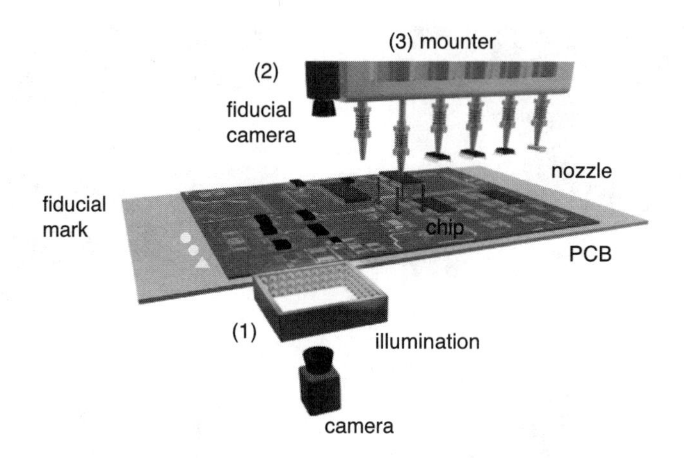

FIGURE 19.10 Part mounting processes: (1) part pose estimation; (2) PCB fiducial mark inspection; (3) part placement.

(1) feeding and pick-up of parts, (2) quality inspection of parts, (3) estimation of part pose, (4) pose measurement of PCB to be loaded, and (5) placement of electronic parts in designated locations. The processes are all involved with opto-mechatronic technology because they utilize microprocessors, controllers, servo motors, and machine vision system and information processing. Process (1) involves control; processes (2), (3), and (4), image and information processing; and process (5), control technology. Therefore, we will confine ourselves to describe processes (3) through (5). These three processes are schematically shown in Figure 19.10.

19.3.2.1 Pose Measurement of Parts and PCB

In the part pose estimation stage, multiple mounting heads (pickers) pick up parts from part feeders or matrix trays. Because the location and orientation of the picked-up parts relative to the center of the head are not known exactly, this information needs to be determined by a sensory system.

Figure 19.10 shows a visual sensing system adopted by Mirae [Cho et al., 2001]. The system acquires an image of the part, processes the acquired image, and determines the orientation and center location of the part based on edge information. The system is composed of a CCD camera, a lighting system, an image frame grabber, and a microprocessor. The accuracy of the acquired image depends largely on how the lighting is conditioned. This is especially true with electronic parts, as their shape and geometry are quite complex and vary within a wide range. To accommodate this situation three lighting methods are adopted and combined together here. They are the direct, indirect, and back lighting systems, as shown Figure 19.11. Because the three lighting systems perform differently on various dimensions and materials of parts, the choice or combination of the three is very critical to obtaining a visual depiction of parts to be inspected. This characteristic of the lighting system is clearly indicated for the case of the PLCC-IC chip part in Figure 19.11. Figure 19.12 indicates a summary of the information on shape, dimension, position, and orientation of SMD parts that can be obtained by the light systems.

The pose recognition task must determine the center location and orientation of a part relative to those of the gripping tool (picker), as illustrated in Figure 19.13 [Cho et al., 1989]. The recognition system has tight accuracy and processing time requirements—a measuring accuracy of less than 0.01 mm in the center and 0.10 μm in orientation, and a processing time, including that of defect inspection, of less than 0.1 s. Figure 19.14 shows four different methods of obtaining the center and orientation of a part. Figure 19.14(a) uses four corner points, (b) all edge information, (c) boundary information of the part body with all leads eliminated, and (d) all coordinate points intersected by leads and a box-type mask. Here, for more accurate measurements, we introduce a method that combines methods (c) and (d) as stated below:

1. Obtain the image of a part body with all lead images eliminated by thresholding gray level at G_{t1}, as shown in Figure 19.15(a).

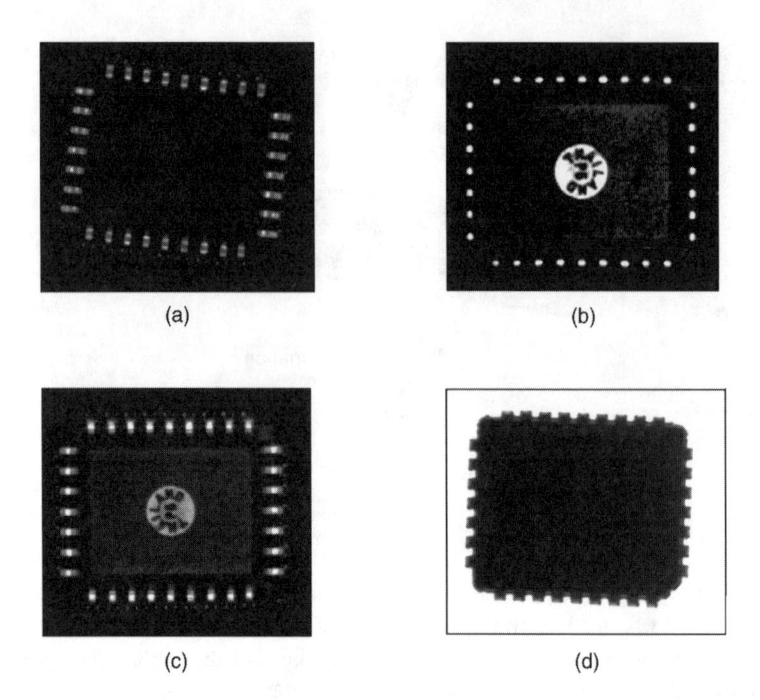

FIGURE 19.11 Image characteristics obtained by various lighting methods: (a) indirect lighting; (b) direct lighting; (c) direct + indirect lighting; (d) back lighting.

2. Determine the center of the part body by placing it where a cross template onto the obtained body image. Then determine a more accurate center value and part orientation by utilizing the extended cross template.
3. Determine four corner points from the obtained information.
4. Obtain an image containing the part body and the leads by thresholding at G_{t2}.
5. Mask a box into the obtained image by using the corner points determined in step 3.
6. Calculate all intersection points between the box mask and the lead edge.
7. Calculate the average center point and orientation based on the data determined in step 6.

The procedure recognizing the pose of a flat IC is summarized in Figure 19.15(b).

The pose measurement of PCB is obtained by detecting fiducial marks printed on several locations of its surface. A machine vision system views the marks and takes their image. Based on the image, the system determines their edge and finally obtains their centers. In this way the orientation and position of the PCB to be loaded can be determined. Figure 19.16 illustrates some of the typical fiducial marks used in finding the PCB pose. In each mark characteristic parameters to be detected are indicated. These parameters are usually used as the information to be measured; this facilitates fast detection of the marks.

19.3.2.2 Control of Part Mounting Head

Denser packing and component miniaturization require not only delicate handling of components but also delicate placement of them into boards. Therefore, a mount head must be designed in such a way that it places them with the desired mounting force in order to avoid any breakage or warp of the components and to generate the required adhesive force for it all to be properly soldered together. From this point of view identification of the dynamics characteristics and control of the mount head are critical to the performance of the mounter [Lee et al., 2001]. To characterize its dynamics identification needs to be made to estimate the modeled parameters by using a genetic algorithm (GA). This identification model is then used to compensate for the friction force existing in an actual system for controller design, and the compensated model control part is feed forwarded to cancel out the friction force. The mounter shown in Figures 19.17 makes a linear motion within a linear guide, which is provided by a rotary motor.

SMD Part	Indirect Light	Direct Light	Back Light
BGA 	Number of ball, ball dimension, grid dimension		Body size, body location orientation
SOP	Lead type, lead size, lead location, number of lead	Lead foot length, lead size, lead location, number of lead	Body size, body location orientation
QFP 	Lead type, lead size, lead location, number of lead	Lead foot length, lead size, lead location, number of lead	Body size, body location orientation
SON		Lead type, lead size, lead location, number of lead	Body size, body location orientation
RC		Lead type, lead size, lead location, number of lead	Body size, body location orientation
Capacitor 		Lead type, lead size, lead location, number of lead	Body size, body location orientation
SOIC 		Lead type, lead size, lead location, number of lead	Body size, body location orientation
TR 		Lead type, lead size, lead location, number of lead	Body size, body location orientation
SOIJ 	Lead type, lead size, lead location, number of lead		Body size, body location orientation
PLCC	Lead type, lead size, lead location, number of lead		Body size, body location orientation

FIGURE 19.12 The information acquired by three lighting methods.

A suction-type nozzle pick-up device is attached to the bottom of the head. This part-holding device releases a part when it hits the surface of the solder paste with the desired contact force. The system has another rotational motor to align the rotational direction of the mounting chip. The figure illustrates the rotational motor and the translation mechanism. x is the displacement of the mount head, and $U = K_a u$ represents the supplied voltage, where K_a stands for the amplifier gain and u is the controller output entering into the amplifier. Using the parameter shown in the figure, the system dynamics governing the

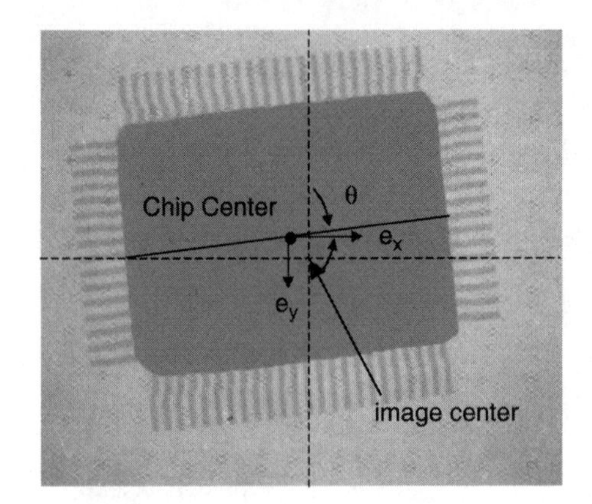

FIGURE 19.13 Information to be determined.

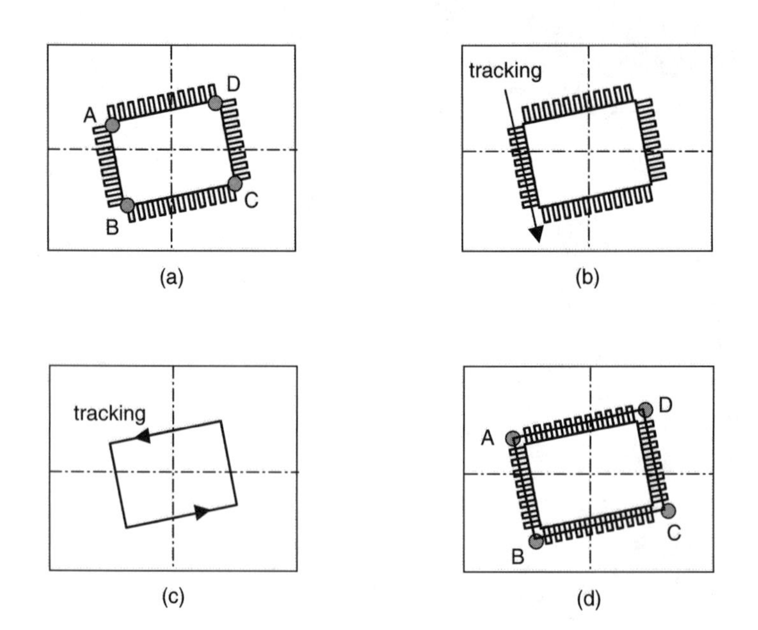

FIGURE 19.14 Various methods to obtain the part pose information: (a) corner detection method; (b) lead tracking method; (c) IC body detection method; (d) extended rectangle detection.

mounting head in free motion can be simplified by the following equation:

$$M\ddot{x} + F_f + mg = \alpha u \tag{19.1}$$

where M is the effective mass, F_f is the total friction force, and mg is the gravity force, while α is a constant. The total friction force F_f in a ball screw and linear guide is given by

$$F_f = \left\{(F_s - F_c)e^{-\eta|\dot{x}|} + F_c\right\}\mathrm{sgn}(\dot{x}) + F_v\dot{x} \tag{19.2}$$

where F_s and F_c are the coefficients of static friction and Coulomb friction, respectively; η is the coefficient to denote the effect of Stribeck; and F_v is the coefficient of viscous friction. In this chapter a genetic

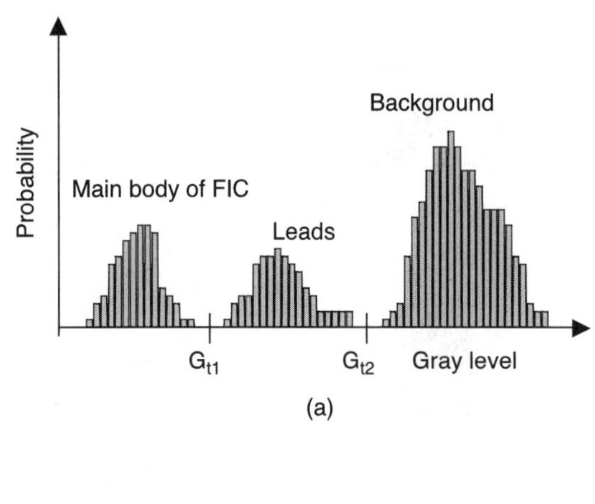

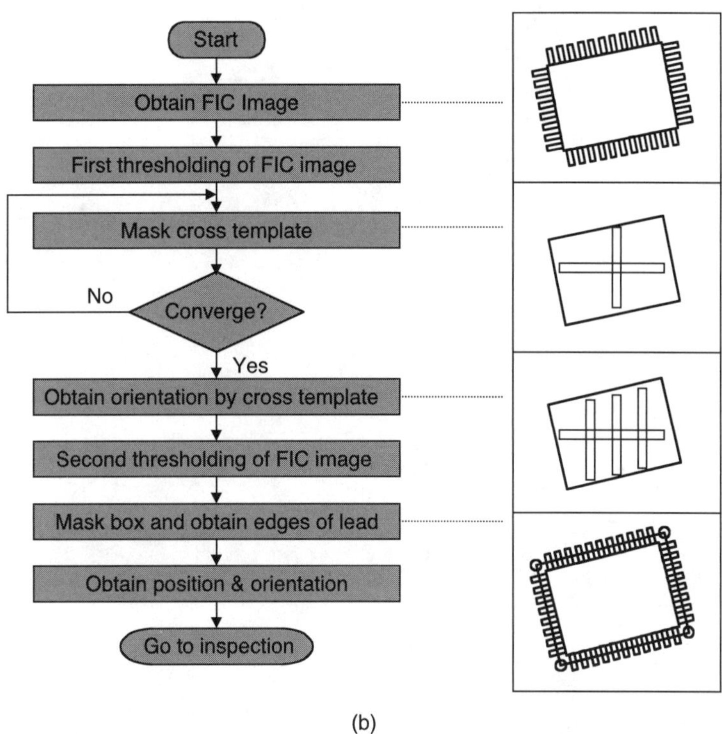

FIGURE 19.15 Procedure of the obtaining position and orientation of FIC chip: (a) typical histogram of a flat IC; (b) flow chart for part pose estimation.

algorithm through a series of experiments identifies the unknown parameters, $\bar{s}$.

$$\bar{s} = (F_s, F_c, F_v, \eta). \tag{19.3}$$

19.3.2.3 System Identification via GA

This method of identification is discussed here briefly using a block diagram shown in Figure 19.18. The proposed algorithm uses feedback P control according to various reference positions and minimizes the error between the model response and the actual response. The error between the actual system and

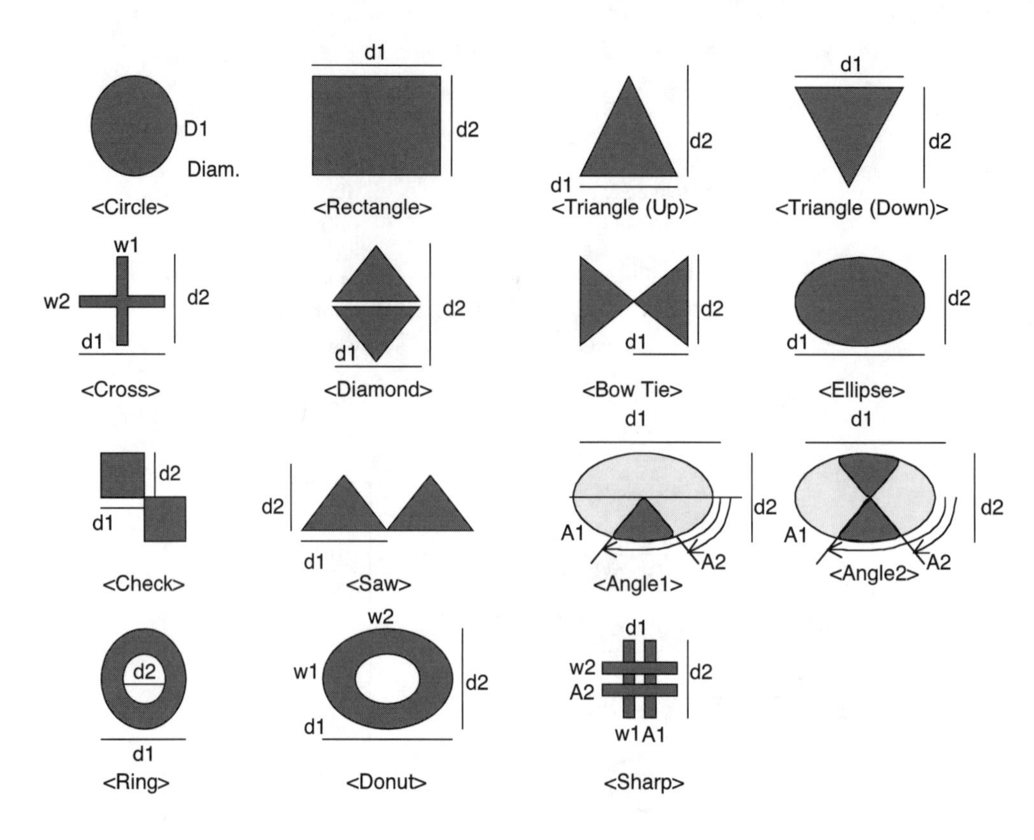

FIGURE 19.16 Various types of fiducial marks.

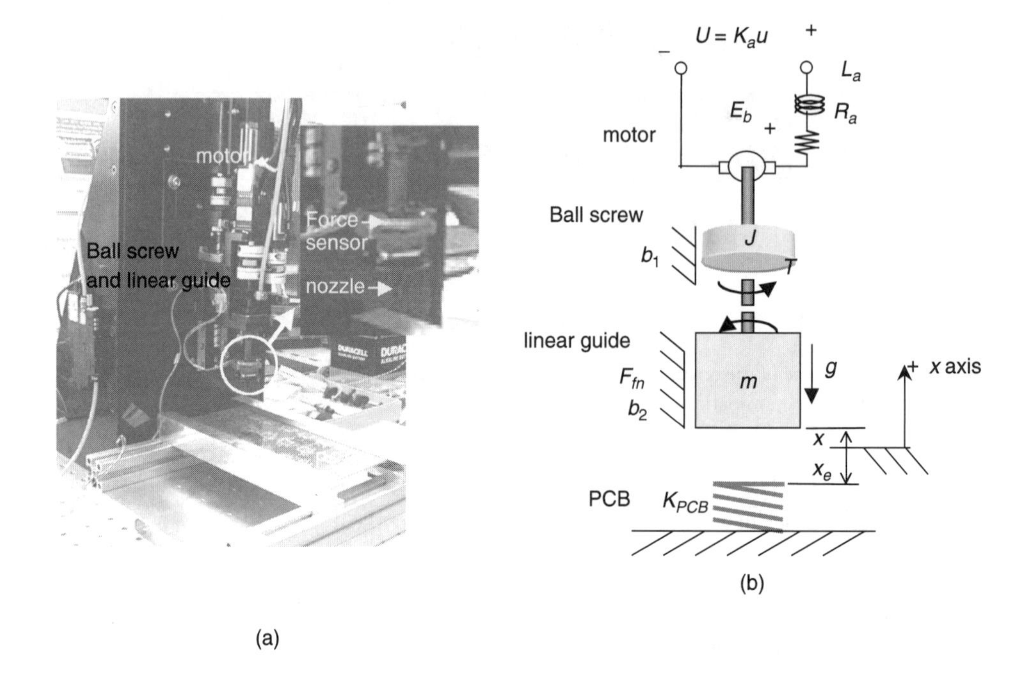

FIGURE 19.17 The surface mounting system: (a) photograph of surface mounting system; (b) schematic of the system model.

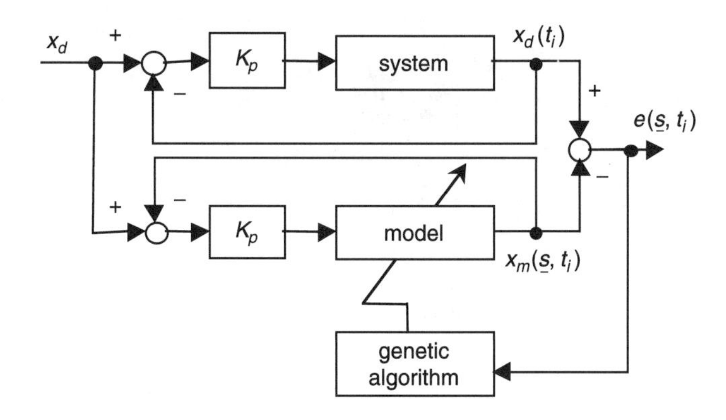

FIGURE 19.18 System parameter identification using GA.

the mathematical model is defined by

$$e(\bar{s},t_i) = x_a(t_i) - x_m(\bar{s},t_i), \tag{9.4}$$

where $\bar{s}$ denotes the estimated system parameters, $x_a(t_i)$ is the actual system output at the ith instant $t = t_i$, and $x_m(\bar{s}, t_i)$ is the model output with the identified parameters, $\bar{s}$ at $t = t_i$. A genetic algorithm is initialized to estimate the parameters contained in the dynamics.

To represent parameters in the GA binary vectors are used, and gray coding is used as a coding method. Each parameter has a finite bit string. All parameters are coded as one long string, called population string, for the late calculation. Such strings can be lengthened to provide more resolution or shortened to provide less resolution for the representation of the parameters.

To evaluate the performance of the identified system parameters, its fitness function is defined by

$$J_e = \sum_{i=1}^{N} \{e(\bar{s},t_i)\}^2$$

$$\textit{fitness function} = \frac{1}{J_e} \tag{19.5}$$

where N is the number of training data. J_e is the summation of square error between the actual system output and the modeled system output. To maximize the fitness function in a GA the fitness function is defined by the inverse of J_e. To code the parameters as used in GA each parameter is normalized within the meaningful regions and assigned to binary 10 bits, and the total chromosome length is 40 bits. Each generation consists of 30 populations, and the maximum generation number is set to 30. Crossover and mutation rates are 0.75 and 0.03, respectively. The fitness value for each generation is calculated by the total error defined in Eqs. (19.4) and (19.5), the experiment on step and ramp reference inputs. Based on the identification using GA, the estimated friction force $\hat{F}_f$ is given by

$$\hat{F}_f = (1.88 - 0.65)e^{-7740\dot{x}_d^2}\mathrm{sgn}(\dot{x}_d) + 0.65\mathrm{sgn}(\dot{x}_d) + 8.89\dot{x}_d. \tag{19.6}$$

Having identified the system dynamics, we now consider the control problem for the mounting head in order to accurately and smoothly place electronic parts into the specified region of the PCB surface. Here, a feed forward control like that shown in Figure 19.19 is considered. In the feed forward control

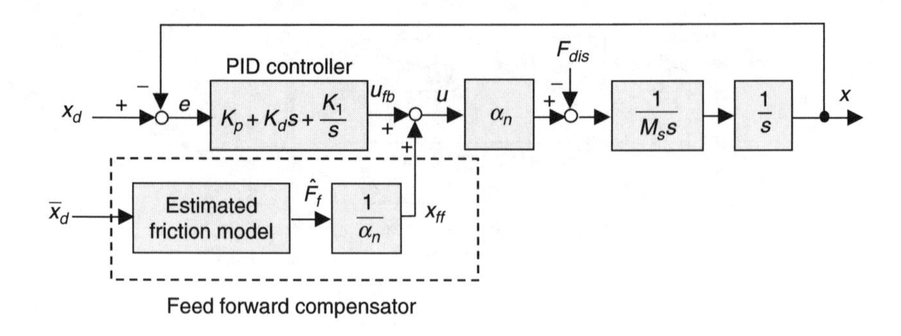

FIGURE 19.19 The mounting control systems based on feed forward compensator.

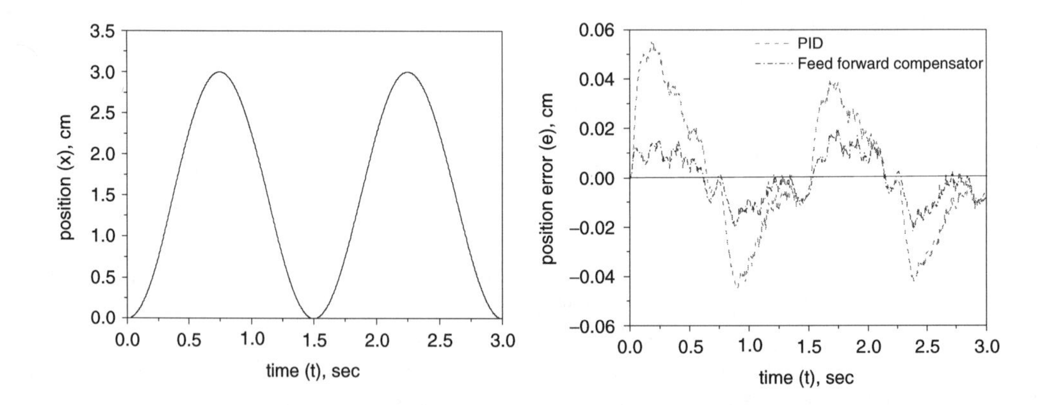

FIGURE 19.20 Controlled responses of the mounting head: (a) desired motion; (b) position error.

the friction force is compensated in the control loop, and the friction force is feed forwarded to cancel out the friction force in the actual system.

19.3.2.4 Performance Test

Figure 19.20 illustrates the test results of the PID control and the feed forward compensator using the estimated friction force. The reference position was given by a sine wave, as shown in Figure 19.20(a), and PID control gains were designated by $K_p = 2.5$, $K_i = 0.9375$, and $K_d = 1.28$. In the PID control result, when the mount head moves backward, the friction force becomes highly nonlinear at the region with zero velocity due to static and Coulomb friction effect, so that the position error becomes large. Moreover, a large position error in the start phase results from a gravity force because the system has stayed at $t = 0$ without control. In the feed forward control using the friction model the position error is reduced to within a ±0.2 mm error bound. In the case of the PID control the mount head cannot overcome the static friction at the starting position. This results from the small control input from the slight position error and the large static friction force. However, the proposed feed forward control method adds predicted static friction to the PID control input. Actually, because friction is a very complex phenomenon, Eq. (19.2) is limited to describing it using a simple and fixed model. Therefore, this method may not be a very effective one. An adaptive and robust method may be required.

19.3.3 Optical Solder Joint Inspection

A number of technologies have been developed in the area of solder joint inspection. These can be categorized into: (1) visual inspection, (2) optical sensing method, (3) x-ray imaging method, and (4) ultrasonic imaging method. In this section we will consider the optical sensing and the x-ray imaging

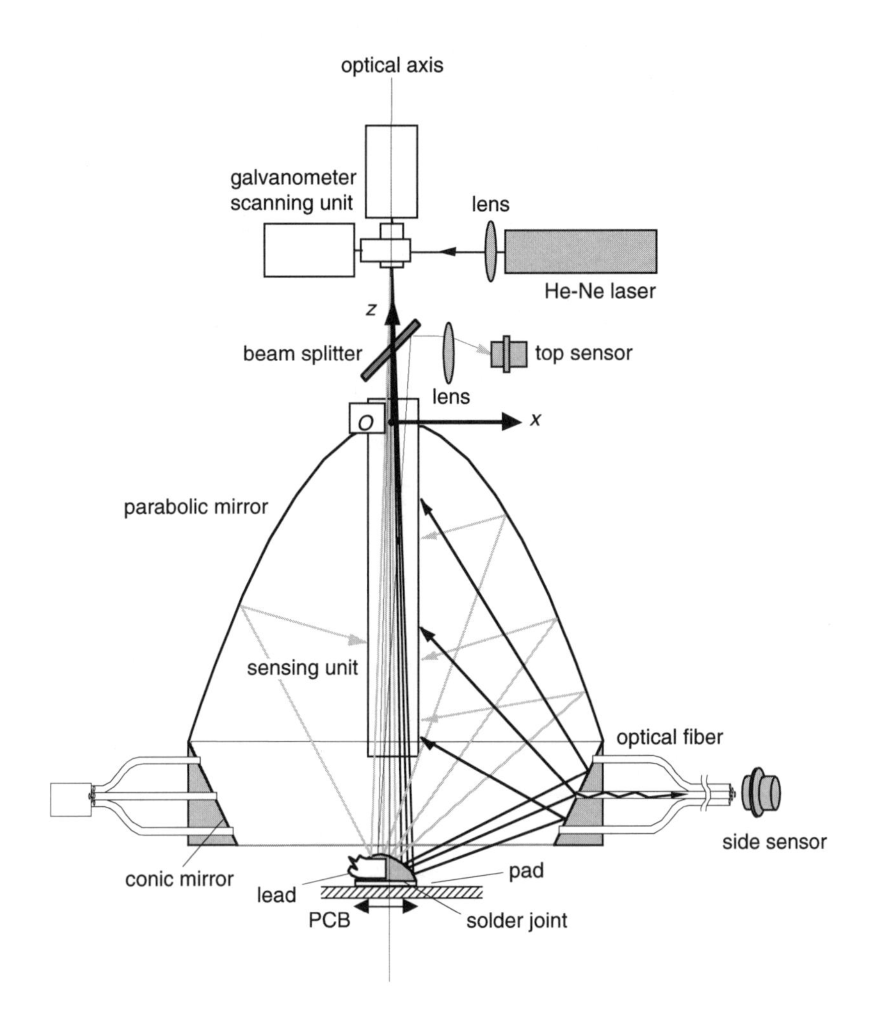

FIGURE 19.21 Sensing principle of the optical solder joint inspection system.

methods. We will see how optical sensing units are incorporated into mechatronic components to determine the 3D shapes of solder joints.

An optical sensing system is utilized to acquire the 3D shape of solder joints, as shown in Figure 19.21 [Ryu and Cho, 1997]. The sensing system is composed of a GSU (galvanometer scanning unit), which steers the laser beam's incident position, a parabolic mirror, which gathers the reflected beam to its center, and a sensing unit whose outer surface is covered by photodiode arrays. In the figure the ray depicted by the solid line represents a ray trace of the reflected light whose intensity is maximum. The sensing unit is positioned at the center of the parabolic mirror and detects the intensity of reflected light. If the laser is projected onto the curved surface of an object by a GSU, the reflected rays from each different surface's normal vectors are incident to the different position of the sensing unit. In other words, the direction of the reflected ray can be detected by the sensing unit, and accordingly the vector normal to the surface of an object can be estimated. To distinguish the beam leaving the light source from the beam reflected back to the source a beam splitter and a top sensor are placed on top of the parabolic mirror. When the rays are reflected by the conic mirror, small components of reflected light are detected by the side sensor of an optical fiber positioned on the conic mirror. Therefore, by monitoring the optical signal detected on the side sensor the reflected ray on the conic mirror is distinguished from that on the parabolic mirror.

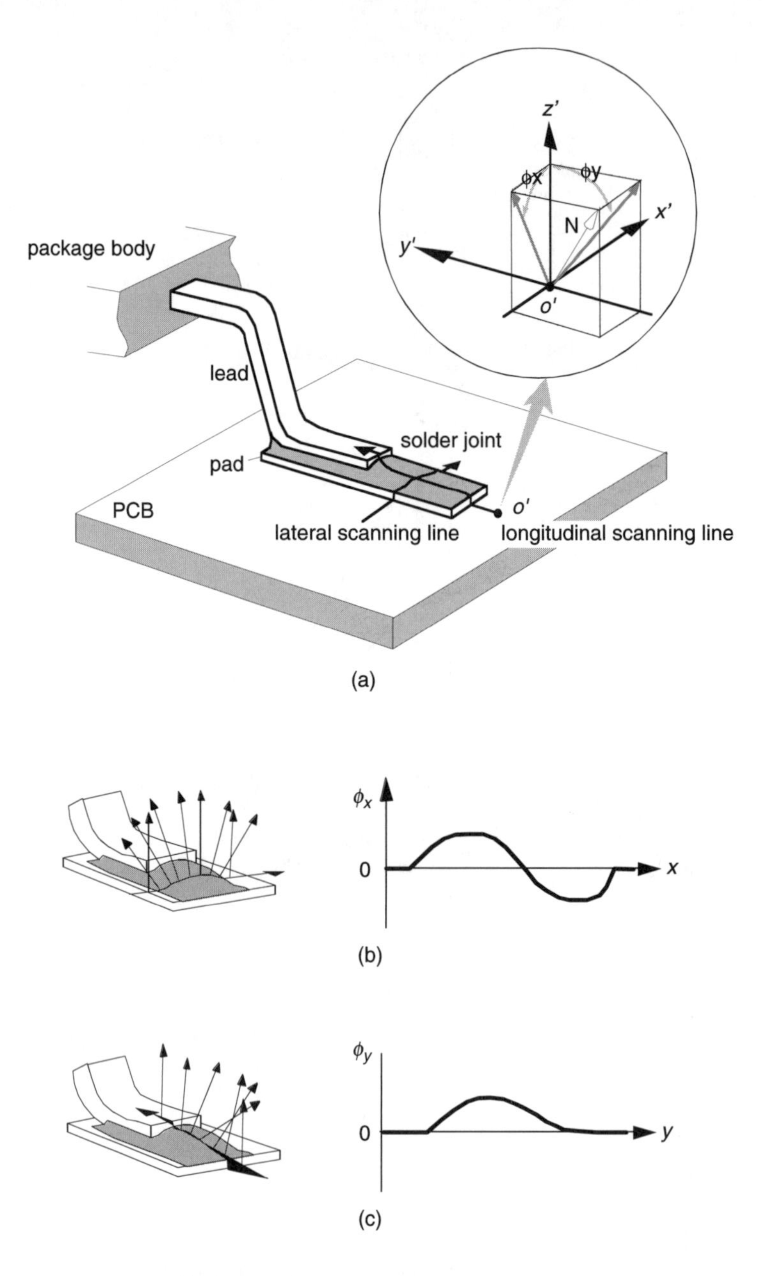

FIGURE 19.22 Surface-mount solder joint and surface orientation curve: (a) configuration of surface-mount solder joint; (b) ϕ_x surface orientation along the lateral scanning line; (c) ϕ_y surface orientation along the longitudinal scanning line.

19.3.3.1 Solder Joint Shape Recognition

Figure 19.22(a) illustrates the shape parameter of a typical mount solder joint. The flow of solder around the perimeter of the lead relates to the quality of the joint. Here, we will deal with the inspection strategies such as feature extraction method and classification algorithm. To describe the orientation of a surface normal vector we employ local coordinates x', y', z' whose origin O' is positioned at the end of a pad. Let the x' axis be perpendicular to the direction of the lead. When the laser beam is traversed along the lateral scanning line, the zenith angle of a surface normal vector projected in the x', z' plane is denoted by ϕ_x. In the case of the longitudinal scanning line the zenith angle of a surface normal vector projected in y', z' plane is denoted by ϕ_y. As an example, the trends of the surface normal vector orientations ϕ_x and ϕ_y are shown in Figure 19.22(b) and (c).

Solder joints are classified by the features extracted from the surface orientation curves measured along the longitudinal and lateral scanning lines. Selection of features is critical to the success of any classification algorithm. Here, nine features were chosen from the orientation curves of solder joint surface for each class. The classification scheme identifies two types of acceptable solder joints (QFP, SOP) and three different classes of joints (normal, insufficient, excess). Figure 19.22(b) and (c) show typical examples of the surface orientation curves of solder joint fillets. Figure 19.22(b) represents the detected zenith angle during lateral scanning, while Figure 19.22(c) represents the detected zenith angle during longitudinal scanning. To extract the trend of the surface orientations the curves of Figure 19.22(b) and (c) are fitted as lines. Defining,

$\bar{F}$: feature vectors $(=[f_1, f_2, \ldots, f_9])$\
$\phi_{x,i}$: zenith angle of solder joint surface at ith position along the lateral scanning line
$\phi_{y,j}$: zenith angle of solder joint surface at jth position along the longitudinal scanning line
$\phi_y^{\max}$: maximum value among the ϕ_y
N_x: number of ϕ_x during the lateral scanning
N_x, N_y: number of ϕ_y during the longitudinal scanning
y^{int}: y axis-intercept point of the fitted line
y^{int}, θ^{slp}: slope of the fitted line

and using these variables the following nine features can be obtained. In the lateral scanning line, one feature value was estimated as an absolute area of ϕ_x:

$$f_1 = \sum_{i=1}^{N_x} \left| \phi_{x,i} \times x \right|. \tag{19.7}$$

In the case of the longitudinal scanning line, eight features are chosen as the area of ϕ_y:

$$f_2 = \sum_{j=1}^{N_y} (\phi_{y,j} \times y) \tag{19.8}$$

the absolute area of ϕ_y:

$$f_3 = \sum_{j=1}^{N_y} \left| \phi_{y,j} \times y \right| \tag{19.9}$$

the absolute area divided by $\phi_y^{\max}$:

$$f_4 = \frac{f_3}{\phi_y^{\max}} \tag{19.10}$$

where the number of crossing the 0° of ϕ_y is denoted by f_5.

The centroid of ϕ_y is:

$$f_6 = \frac{f_2}{\sum_{j=0}^{N_y} \phi_{y,j}} \tag{19.11}$$

the moment of ϕ_y:

$$f_7 = \frac{\sum_{j=0}^{N_y} \phi_{y,j} \times y^2}{\sum_{j=0}^{N_y} \phi_{y,j}} \tag{19.12}$$

the y axis-intercept point of the fitted line:

$$f_8 = y^{int} \tag{19.13}$$

the slope of the fitted line:

$$f_9 = \theta^{slp}. \tag{19.14}$$

These nine features allowed defects to be identified based on the characteristics of the orientation curves. A decision-making problem of defect identification may be solved using simple classifiers. These classification algorithms require no assumptions as to the probability density function of the data and use simple distance metrics for classification. Included among these are the K-nearest neighbor (K-NN) algorithm, the minimum distance classifier, and others. Here, a minimum-distance classification algorithm [Parker, 1994] is adopted to classify the solder joint.

19.3.3.2 Inspection System Performance

System performance was demonstrated with results obtained from the inspection of two commercially manufactured PCBs. There were 360 inspection points per PCB. To obtain the 3D shapes of the solder joint, a series of tests was performed on the three classes such as normal, excess, and insufficient soldering conditions. The normal joints were wave-soldered with flaws produced by manual modification of normal joints. Figures 19.23 and 19.24 show, respectively, the 3D shape of SOP and QFP recovered from the detected orientation data and the photograph of the corresponding actual fillet shape. In this case, the window size was 1.6×1.0 mm^2 and scanning interval($\Delta x'$, $\Delta y'$) was 50 μm. Also, the recovered shape resembles closely the actual shape. In parts (b) and (c), the arrows indicate the region of a steep slope. For the same reason as the above-mentioned limitation, the recovered shapes of these solder areas do not accord very well with the actual shapes.

To inspect solder joint quality a series of experiments was performed for SOPs and QFPs in insufficient, normal, and excess soldering condition. To inspect the solder joints automatically an X-Y table was used in the sensing system. Approximately 100 training samples of the solder joints for each type were used for the test. Figure 19.25 shows examples of the estimated orientation curves ϕ_x and ϕ_y SOP along the lateral and longitudinal scanning lines. The figure shows that the defective solder joints—the insufficient and excess joints—have a different form of orientation curve compared with the normal case. The features extracted from the orientation curve of the SOPs were used to create the 3D map shown in Figure 19.26. The figure shows that it is not easy to classify the classes of the solder joints with only three features because the distances between the mean value of the three classes are not sufficient if separated. For this reason the nine features introduced in the previous section were used. By using the minimum-distance classification algorithm, we have performed the solder joint inspection. To measure the performance of the statistical techniques using the nine features and the three classes the following three quantities are calculated:

CC—the number of joints correctly classified
GB—the number of acceptable joints (normal case) classified as unacceptable (insufficient and excess case)
BG—the number of unacceptable joints classified as acceptable

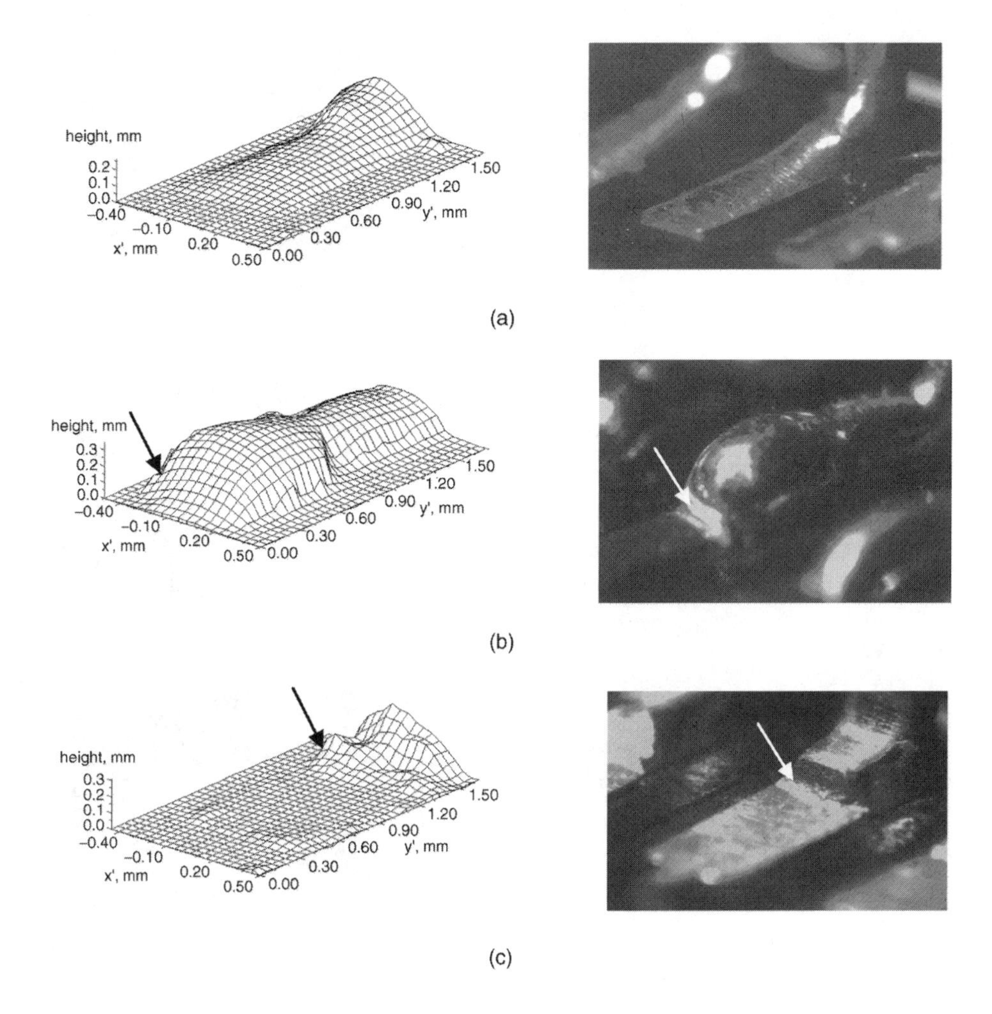

FIGURE 19.23 Recovered three dimensional shape of the SOP solder joints: (a) normal solder joint; (b) excess solder joint; (c) insufficient solder joint.

The inspection results for two commercially manufactured PCBs are listed in Table 19.2. The table shows that the percentage of correct classification is about 95%, which is rather high. Therefore, this proposed sensing system is considered to achieve reasonable accuracy of classification performance for solder joint inspection.

19.3.4 X-Ray Imaging for BGA Solder Joints Inspection

The inspection and measurement methods employing x-ray imaging are extensively applied in measuring the density and shape characteristics of assembled electronic parts. In particular, when the ball-grid array (BGA) chips for high-density printed circuit boards are solder jointed and inspected, the use of x-ray imaging is inevitatable because the joints are located between the packages and the board. With the J-lead and the gull-wing chip, part of the solder joint is occluded from the view of the vision system. In these circumstances an x-ray cross-sectional imaging method, such as laminography and digital tomo-synthesis (DT), which can form a cross-sectional image of 3D objects, is needed to image and inspect their solder joint parts [Roh et al., 1999].

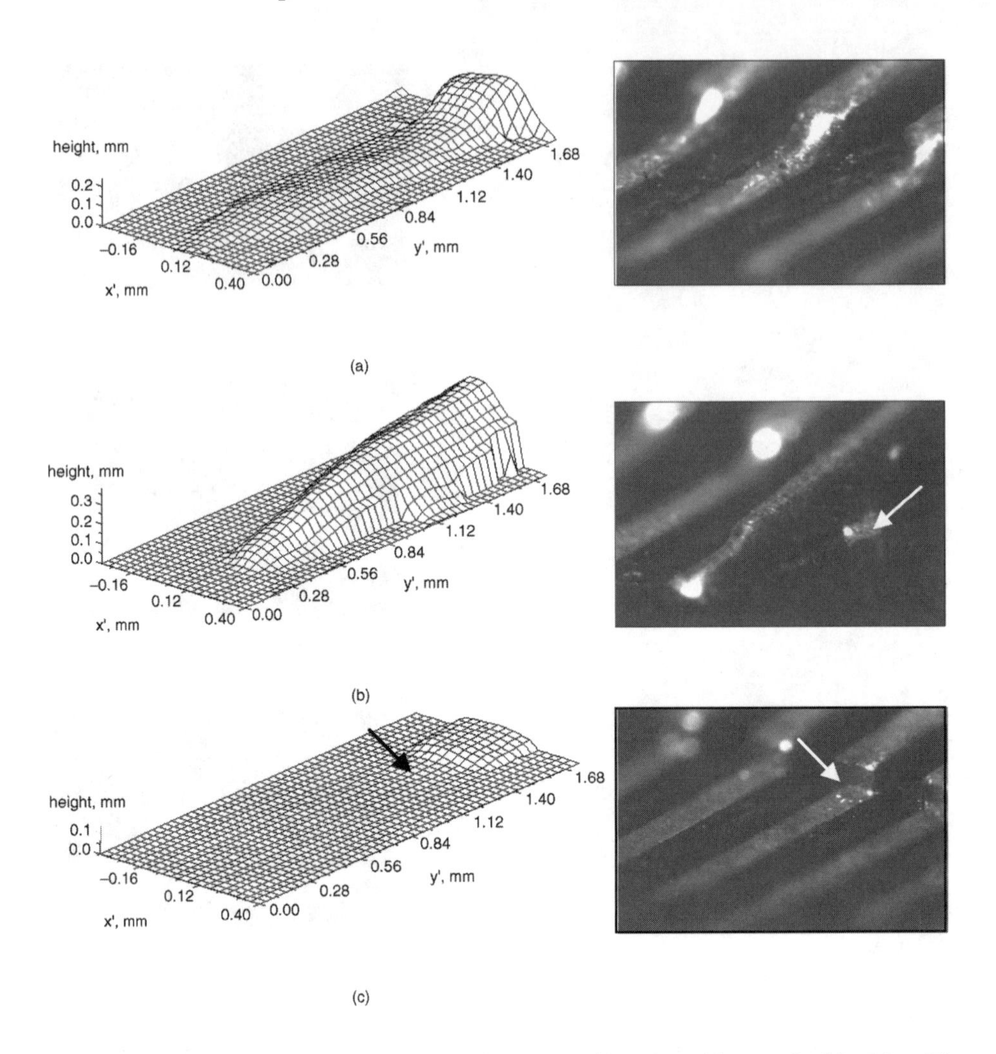

FIGURE 19.24 Recovered three-dimensional shape of the QFP solder joints: (a) normal solder joint; (b) excess solder joint; (c) insufficient solder joint.

19.3.4.1 The Principle of X-Ray Cross-Sectional Imaging

X-ray laminography is one of the methods used for acquiring cross-sectional images and has been used for years in both medicine and technology for looking through opaque materials to find the underlying material's structure.

Figure 19.27 shows a typical DT system for obtaining the cross-sectional images of the jointed electronic parts. As illustrated in Figure 19.27(a), the DT system consists of a scanning x-ray tube, an image intensifier, a rotating prism, and a camera with zoom lens. The scanning x-ray tube is designed to control the position of an x-ray spot electrically and to project an x-ray beam into an object (PCB) from different directions. The x-ray spot is steered onto the PCB with a predefined circular trajectory. The x-ray passing through the PCB is collected by an image intensifier. It plays an important role as an x-ray detector. Photons of light are emitted on its screen in proportion to the x-ray's intensity. At eight or more predefined positions a zoom camera sequentially acquires images. Two types of image-capturing methods are utilized: (1) rotating prism and (2) galvanometer. A prism rotates to synchronize with the x-ray's position to catch the projected image on the screen of an image intensifier, as shown in Figure 19.27(b) and (c). The control of the prism rotation is achieved by an AC servo motors, shown in the figure, whose angular velocity is regulated by a velocity control loop. With the galvanometer, a two-axis galvanometer controls the angles of the x and y mirrors to track the x-ray position. The two independent servo motors are

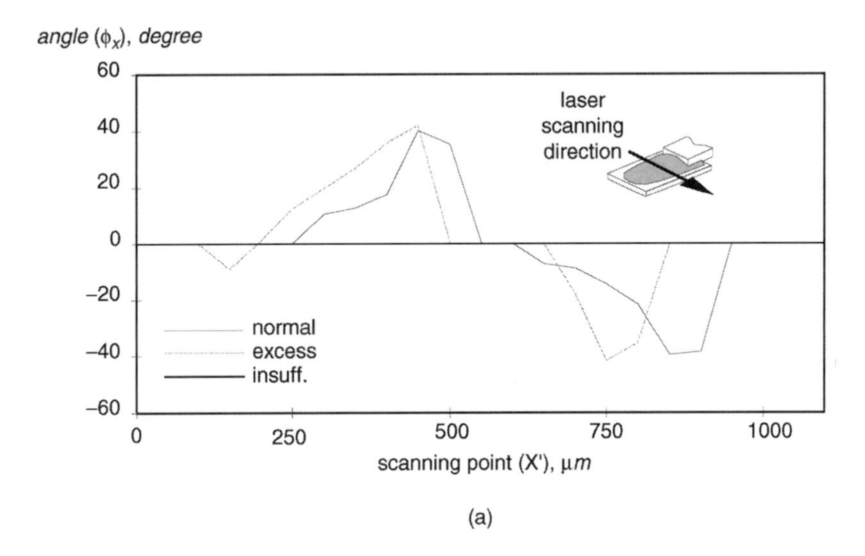

(a)

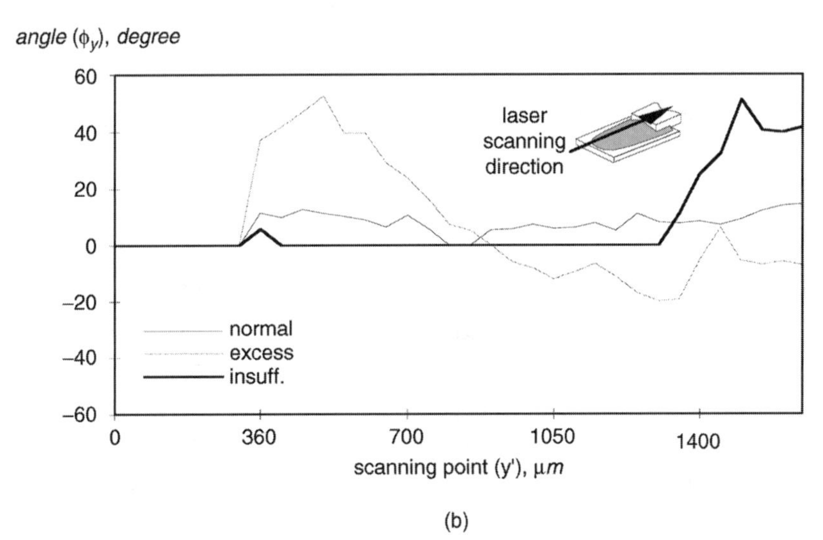

(b)

FIGURE 19.25 Variation of surface orientations of SOP: (a) surface orientations along the lateral scanning line; (b) surface orientations along the longitudinal scanning line.

controlled similarly according to the feedback control loop, as shown in the rotating prism. Captured images at eight different positions are saved in a PC's digital memory and averaged to generate a cross-sectional image at the focal plane.

The schematic of ball-grid array joint and cross-sectional images at three differently located focal planes are shown in Figure 19.28. The focal plane is located at the carrier, ball center, and pad. For the BGA inspection the focal plane at the carrier is selected. An x-ray cross-sectional image of the BGA solder joint obtained by using laminography or a DT method has inherent blurring and artifact features, as shown in the figure. This problem has been a major obstacle in the extraction of suitable features for quality classification. Several quality classification methods have been reported: the neural network approach adopted herein does not require effort in order to extract suitable geometric features from solder joint images. Accordingly, the neural network method can be expected to be superior over the conventional geometric-feature-based method.

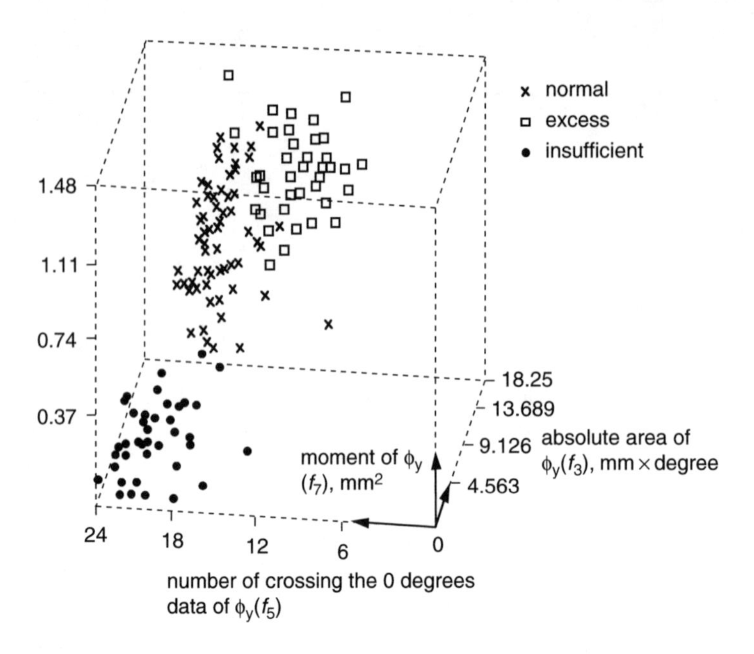

FIGURE 19.26 The 3D map of the extracted feature data of SOP.

TABLE 19.2 Inspection Results of the Developed System[a]

Board Number	Part Type	Solder Joint Class	Number of Joints	Classification Result	
				CC (CC%)	GB (GB%)/ BG (BG%)
B1	QFP	Normal	83	69 (83.1%)	14 (16.9%)/
		Excess	57	54 (94.7%)	3 (2.5%)
		Insufficient	60	60 (100.%)	
	SOP	Normal	77	75 (97.4%)	2 (2.6%)/
		Excess	38	36 (94.7%)	3 (3.6%)
		Insufficient	45	44 (97.8%)	
B2	QFP	Normal	80	75 (93.8%)	5 (6.3%)/
		Excess	60	50 (83.3%)	9 (7.5%)
		Insufficient	60	60 (100.%)	
	QFP	Normal	75	70 (93.3%)	5 (6.7%)/
		Excess	40	39 (97.5%)	1 (0.0%)
		Insufficient	45	45 (100.%)	

[a]CC: the number of joints correctly classified; GB: the number of acceptable joints classified as unacceptable; BG: the number of unacceptable joints classified as acceptable; B1, B2: serial number of PC main board.

19.3.4.2 Neural Network-Based Pattern Classification

A neural network structure suitable for inspecting x-ray cross-sectional images of BGA solder joints is introduced here. As shown in Figure 19.29(a), the neural network structure consists of four LVQ (learning vector quantization) neural networks [Zurada, 1992] and a multilayered neural network hierarchically. The four LVQ neural networks shown in Figure 19.29(b) are used as a preprocessor to reduce the dimension of input images. They can cluster gray-level profiles acquired from four different directions (0, 45, 90, and 135°), as shown in Figure 19.29(b), and generate prototypes. They can convert gray-level profiles of input images into Euclidean distances. The multilayered neural network then learns only each Euclidean similarity distance between four directional gray-level profiles and their prototypes.

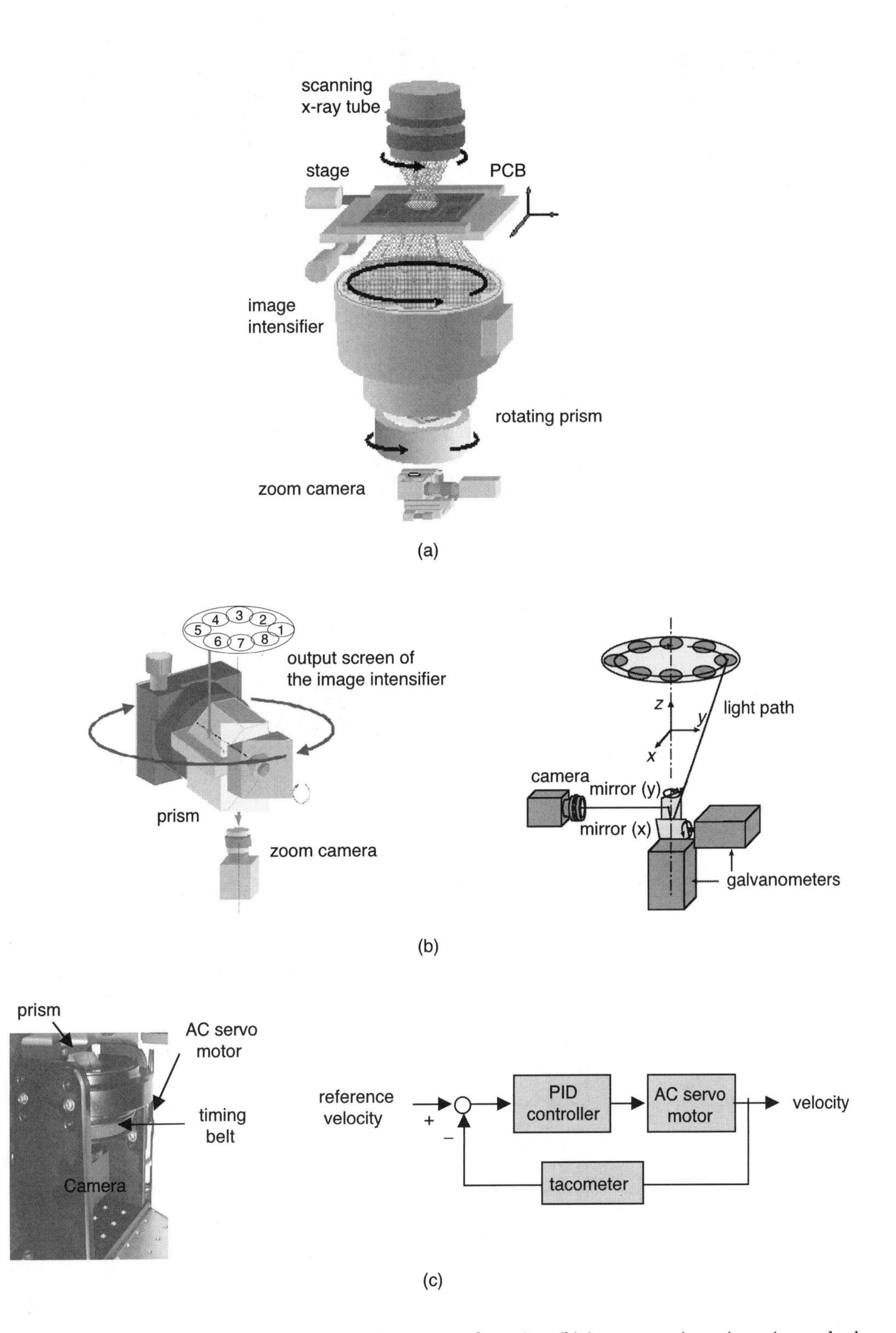

FIGURE 19.27 Digital tomosynthesis system: (a) system configuration; (b) image scanning using prism and galvanometers; (c) AC-servo for rotating prism.

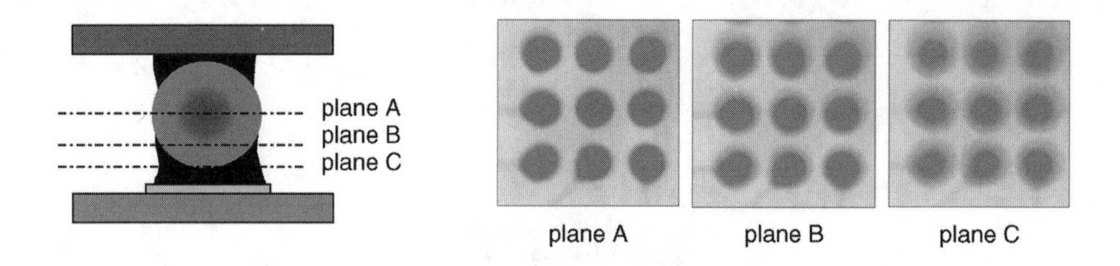

FIGURE 19.28 A schematic of BGA solder joints and their cross-sectional images.

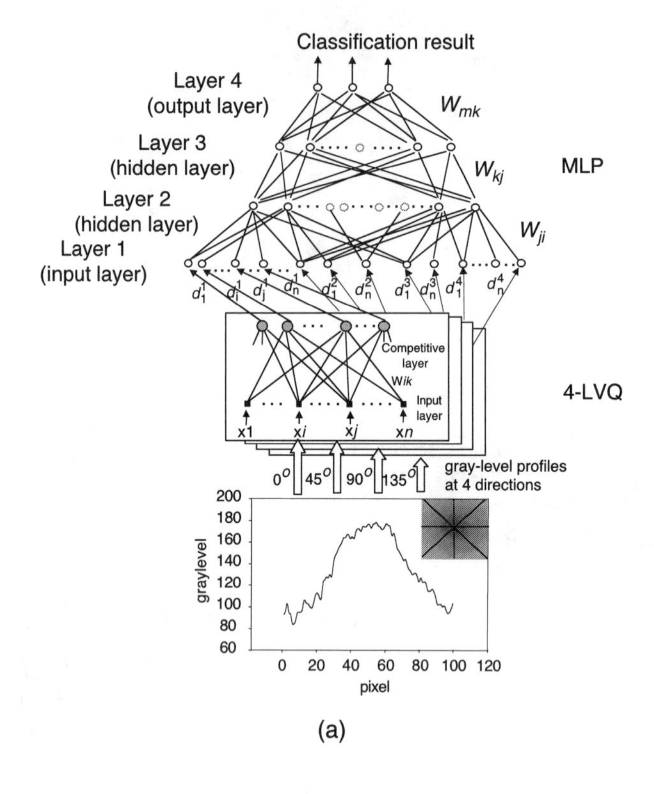

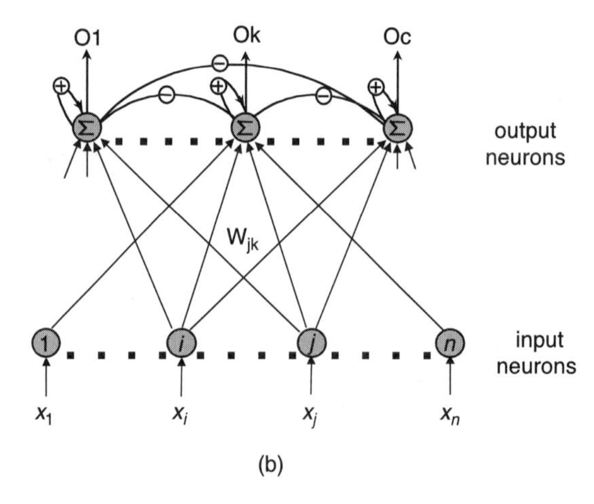

FIGURE 19.29 The neural network architecture: (a) overall architecture of the neural network; (b) LVQ (learning vector quantization) neural network structure.

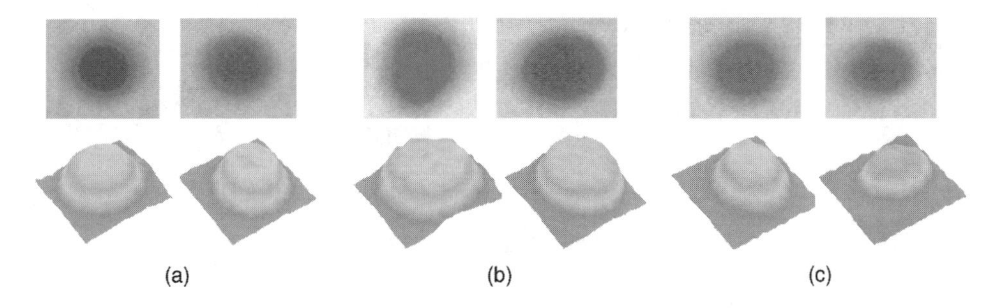

FIGURE 19.30 Examples of BGA solder joints quality: (a) normal soldering; (b) excessive soldering; (c) insufficient soldering.

The input of an LVQ neural network is the gray-level profile of the cross-sectional image BGA. The number of neurons in the input layer corresponds to the pixel number in the line mask within a predefined inspection window. The number of neurons in the competitive layer is the same as the cluster number. It is difficult to find the optimal number of output nodes. To find the number for various applications various experiments with a changing number of output neurons have been conducted. During recall, a nearest-neighbor classification technique is used.

A multilayer perceptron network is shown schematically in Figure 19.29(a). It is implemented to learn the mapping characteristics of a human inspector and then to classify the solder joint quality from the cross-sectional image of BGA. The network consists of a large number of neurons arrayed in layers. The inputs of an MLP network are the set of Euclidean distances in output nodes of 4-LVQ. The output data sets are the corresponding classification results of input data sets performed by a human inspector. The neural network can learn the human classification rule during training procedures.

19.3.4.3 The Performance of Quality Inspection

The sample images of the 220 BGA solder joints for experiments are collected from PCBs. These include various solder joint shapes according to the amount of solder. Of the 220 samples, 160 samples selected randomly are used for network training, and the remaining 60 are used for testing. All samples are grouped by an expert inspector into three classes according to their qualities: insufficient soldering ($\boldsymbol{I}$), acceptable soldering ($\boldsymbol{A}$), and excess soldering ($\boldsymbol{E}$). Each sample image is 120×120 pixels in size, as shown in Figure 19.30.

19.3.4.4 Training the Networks

The input of an LVQ clustering network is a gray-level profile from a BGA image in a predefined window. The gray-level profile is collected from each of the four directions. The centroid of the ball is calculated so that each profile center is identified. The number of input nodes is set to 120, and the initial dynamic learning rate is set to 0.3 and decreases to 0.05 as the learning epoch increases. The maximum learning epoch is set at 5000. Several experiments put the number of output nodes at seven. Each final weight vector is labeled corresponding to its soldering quality by a human inspector. These weight vectors are the representatives of the class and are called prototypes. The LVQ clustering neural network can distinguish solder joint profiles according to the variations between their shapes and generated prototypes. The main objective of training the multilayer neural network is to learn the relationship between inputs and outputs. The LVQ neural network is used for clustering of input gray-level profiles as a preprocessor. The number of neurons in the MLP input layer is set to four times the output neurons in one LVQ neural network because the neural network can use the clustering results of four-directional gray-level profiles as an input.

The input data set of MLP neural network consists of the Euclidean distance of LVQ expressed by

$$\left[d_1^1, d_2^1, \ldots, d_7^1 \vdots d_1^2, d_2^2, \ldots, d_7^2 \vdots d_1^3, d_2^3, \ldots, d_7^3 \vdots d_1^4, d_2^4, \ldots, d_7^4\right] \tag{19.15}$$

TABLE 19.3 Classification Result for Training and Test Samples[a]

	Q	NS	I	A	E	NC
Result for training samples	I	64	65	0	0	0
	A	50	0	50	0	0
	E	56	0	0	46	0
	Total success rate: 1 – 0/60 = 1.00 (100.0%)					
	Q	NS	I	A	E	NC
Result for test samples	I	17	17	0	0	0
	A	21	2	19	0	2
	E	22	0	0	22	0
	Total success rate: 1 – 2/60 = 0.967 (96.7%)					

[a]Q: quality, NS: number of samples, NC: number of confusion, I: insufficient, A: acceptable, E: excessive.

where d_a^b is the Euclidean distance between an input gray-level profile of the *a*th direction and *b*th weight vector, expressed as follows:

$$d_a^b = \left\| \overline{X} - \overline{W}_a{}^b \right\|. \tag{19.16}$$

The number of neurons in a hidden layer is selected experimentally. According to this result, a 28–20–10–3 structure is finally selected to classify the solder joint images.

After the training the classification performance was tested for the remaining 60 ball images. These test data images were not used for the training. Table 19.3 shows the results of the test data. Only two samples in the acceptable (A) class are misclassified in the insufficient (I) class. The total success rate is therefore found to be 96.7%, which is good for use in real industrial applications. In the solder joint inspection the insufficient solder is a crucial defect because insufficient solder joints can be cracked easily under vibrations or shocks. For this reason the success rate of the classification of insufficient defects should be near 100% for purposes of designing the classifier. From these experimental results the success rate of classifying correctly the insufficient solder joints appears to be very high, which implies that the adopted classifier can be effectively used for actual application.

19.3.5 In-Circuit Testing Process

The in-circuit test is one of the inspection tasks for assembling a circuit board. This test probes the conductivity integrity of the soldering of the electronic components mounted on a PCB at high speed. Such probing enables us to detect faults in the soldering and check the positioning correctness of the assembled electronic parts. The most advanced type of commercially available probing devices for this test is the flying probing system, which uses a test probe installed at the end-effector of a robot. The probe contacts actively with the surface of a solder joint while absorbing the impact energy only by passive springs, as shown in Figure 19.31. However, the probing devices currently used in industry are incapable of controlling the contact force generated when the rigid probe contacts with the less rigid solder joint at high speed (0.2 to 0.4 m/sec).

19.3.5.1 Robotic Probing System

The probing system for the in-circuit test requires the ability to successfully complete the given contact task with minimum impact force, minimum oscillation during contact, no-slip motion of the probe tip, and the ability to approach from an arbitrary direction. A probing system achieving some of these requirements is schematically shown in Figure 19.32 [Shim and Cho, 1997, 1999]. The system consists of a macro-motion device, a micro-motion device, an optical force sensor, and an optical sensor measuring the relative distance between the probe tip and the solder joint surface.

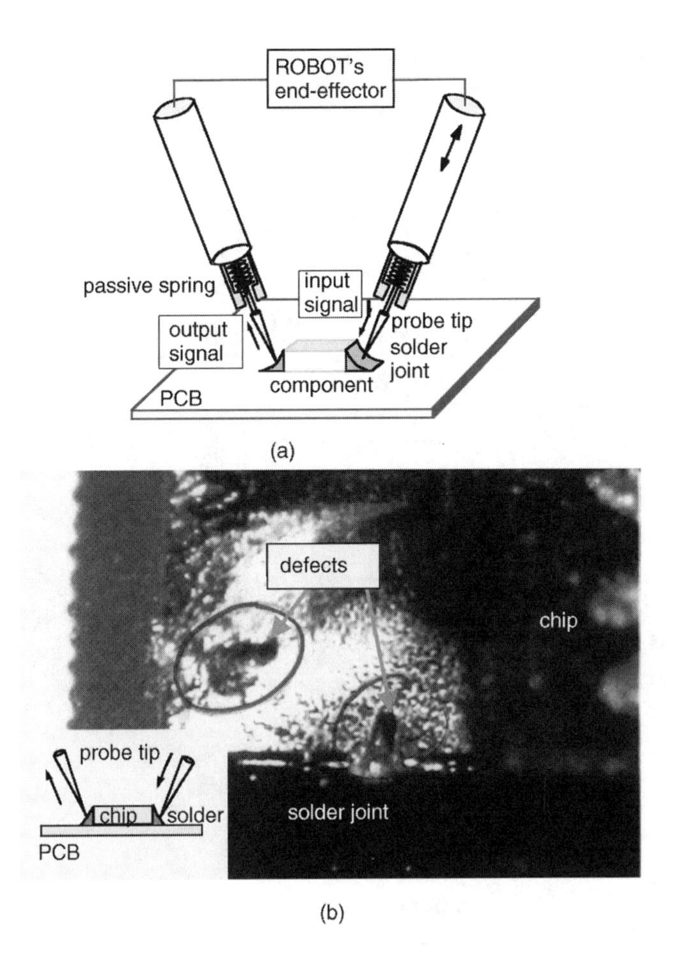

FIGURE 19.31 Conventional flying probing method for in-circuit test of a circuit board: (a) flying probing mechanism; (b) deformed surface of solder joint of a chip after contact by probe tip.

19.3.5.2 The Micro-Motion Device

The micro-motion device utilizes the Lorentz force—the force generated by a current-carrying conductor in a static magnetic field. The actuator is designed to act on only one degree of freedom of translational motion by adopting linear guide bearings. As shown in Figure 19.33(a), the conductor, a moving coil, is positioned among four rectangular neodymium iron boron (NdFeB) magnets, which provide a high gap field. The position of the micro-motion device is measured by an optical sensor, shown in the figure [Shim et al., 1997]. The sensor is comprised of a diode laser, two mirrors, and a one-dimensional PSD (position sensitive device). Light rays from the fixed diode laser are projected on a fixed mirror and reflected 90° into the direction of movement. The rays are projected onto the other mirror attached to the moving coil and reflected into the PSD sensor. The centroids of the reflected light spots are easily obtained by measuring the output current of the PSD. The sensing resolution of this device is detected by a commercially available laser displacement sensor (Keyence LC-2220) and results in approximately ± 6 μm.

19.3.5.3 Acquiring *a Priori* Information about the Solder Joint

The schematic diagram of the optical sensor is shown in Figure 19.33(b), which is utilized in order to acquire in advance information about the solder joint in real time. Suppose that the Lambertian of an object to be measured is more dominant than the specular properties. If the laser is projected onto the surface of an object by a mirror mounted on the probe, diffusely reflected rays from the surface are incident to a two-dimensional PSD through a focusing lens.

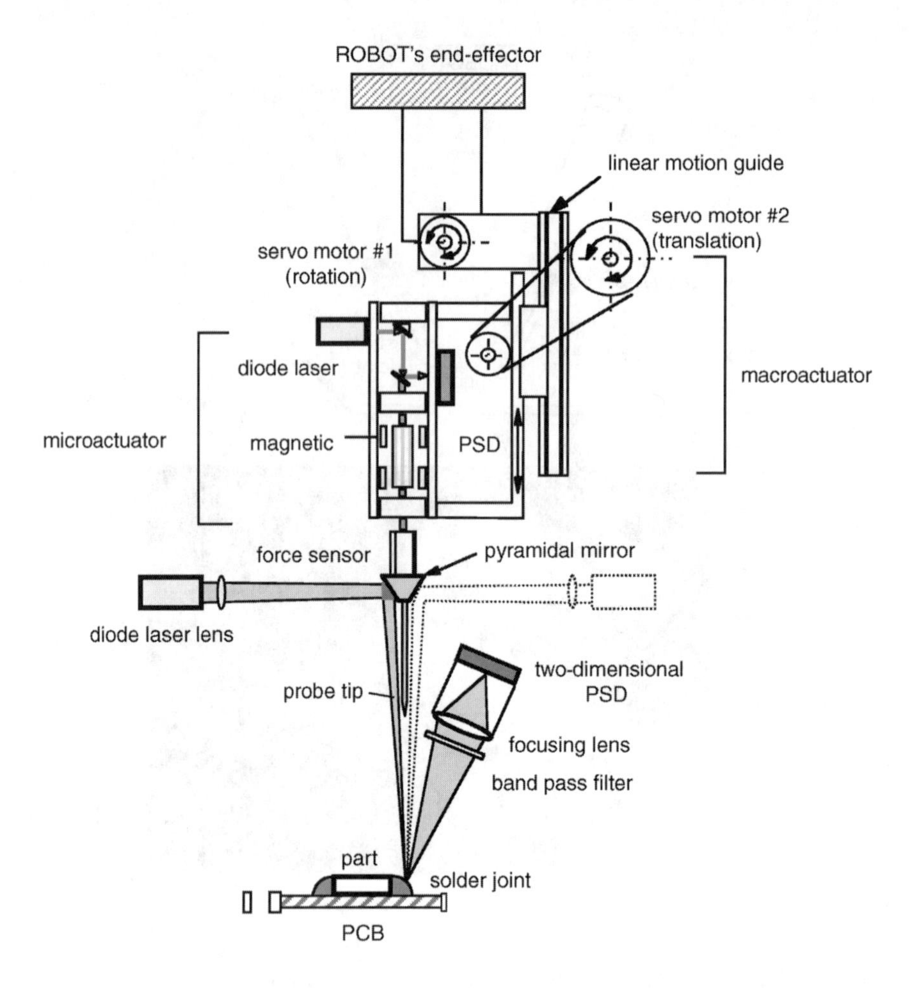

FIGURE 19.32 A schematic of the proposed sensor-guided probing mechanism.

The ray from the laser i, which is deflected at a point $\mathbf{p}_{0i} = [x_{0i}, y_{0i}, z_{0i}]^T$ by a pyramidal mirror, is incident to a point $\mathbf{p}_{li} = [x_{li}, y_{li}, z_{li}]^T$ of the surface of the solder joint in the direction of $\hat{\mathbf{u}}_i$. At this time, $\hat{\mathbf{u}}_i$ crosses the point $\mathbf{p}_{2i} = [x_{2i}, y_{2i}, z_{2i}]^T$. $\mathbf{p}_{2i}$ is the known point obtained from the design specification of the optics. The subscript i represents the order of the laser source. Then, the reflected ray is incident on the sensing unit PSD $\mathbf{p}_{si} = [x_{si}, y_{si}, z_{si}]^T$ through the center of the focusing lens $\mathbf{p}_l = [x_l, y_l, z_l]^T$. The point of the object $\mathbf{p}_{li}$ is determined from the intersection of two straight lines, $\overline{\mathbf{P}_{0i}\mathbf{P}_{2i}}$ and $\overline{\mathbf{P}_{si}\mathbf{P}_l}$

$$\mathbf{p}_{li} = \left[x_{li}, y_{li}, z_{li}\right]^T = \overline{\mathbf{P}_{0i}\mathbf{P}_{2i}} \cap \overline{\mathbf{P}_{si}\mathbf{P}_l} \tag{19.17}$$

where $\overline{\mathbf{P}_{0i}\mathbf{P}_{2i}}$ is a line crossing points $\mathbf{p}_{0i}$ and $\mathbf{p}_{2i}$, and $\overline{\mathbf{P}_{si}\mathbf{P}_l}$ is a line crossing two points $\mathbf{p}_{si}$ and $\mathbf{p}_l$. Here, the average gap distance z_a is obtained by averaging the gap distance z_{li} from

$$z_a = \frac{\sum_{i=1}^{N_j} z_{li}}{N_d} \tag{19.18}$$

where N_d is the number of the projecting laser sources. In a similar manner the next corner points $\mathbf{p}_{12}$, $\mathbf{p}_{13}$, and $\mathbf{p}_{14}$ of the plane patch can be obtained. The normal vector $\hat{\mathbf{n}}_{pl}$ of the plane patch is obtained by

$$\hat{\mathbf{n}}_{pi} = \hat{S}_{ij} \times \hat{S}_{ji} \quad (1 \le (i, j) \le 4) \tag{19.19}$$

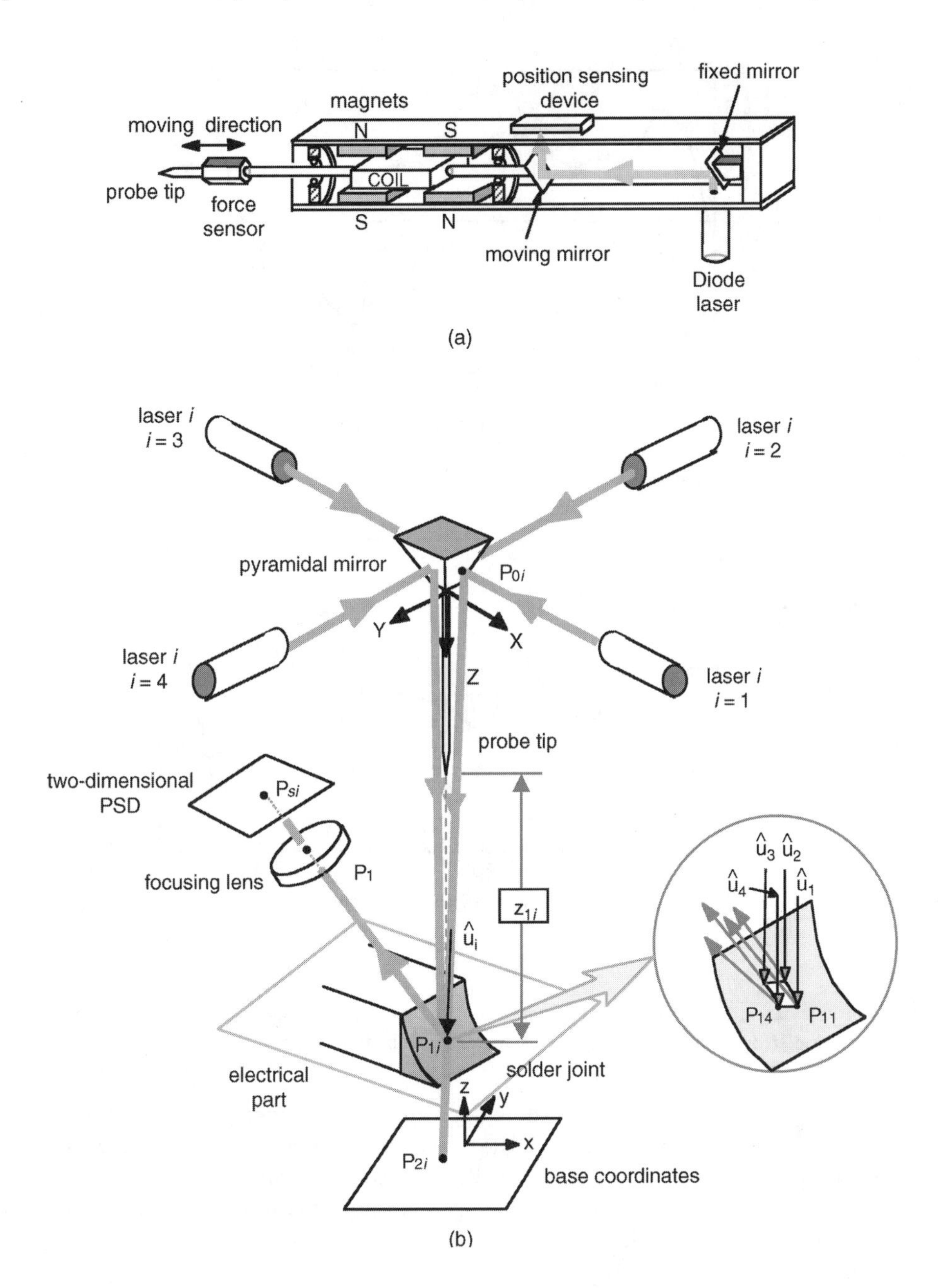

FIGURE 19.33 The configurations of the microactuator and optical sensor: (a) micro-motion device; (b) optical gap sensor.

where $\hat{S}_{ij}$ is a line vector from a point $\mathbf{p}_{li}$ to a point $\mathbf{p}_{lj}$. The average normal vector $\mathbf{n}_p$ given by

$$\hat{\mathbf{n}}_p = \frac{\sum_{i=1}^{N_d} \hat{\mathbf{n}}_{pi}}{N_d}. \tag{19.20}$$

Figure 19.34(a) shows the measured average deviation error between the gap distance provided by the reference displacement sensor and the gap distance detected by the sensor. The reference sensor is a commercially available laser displacement sensor (Keyence LC-2220). From the test result we see that the

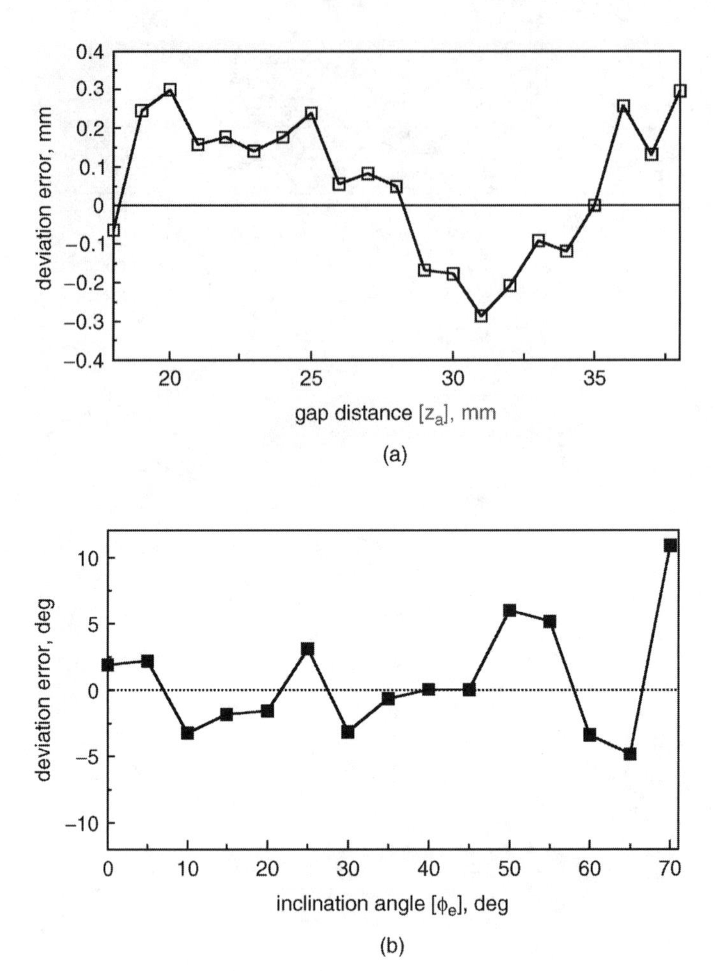

FIGURE 19.34 Deviation errors of the gap distance and the inclination angle measured by the optical sensor: (a) average deviation errors of the measured gap distance; (b) inclination angle.

gap distance with errors can be obtained within approximately ±0.3 mm. Figure 19.34(b) shows the results of the measurement of an inclination angle ϕ_e at a gap distance $z = 5$ mm. The measurement of the inclination angle was set from 0 to 70° at intervals of 5°. In the experiment the inclination angle with errors can be obtained up to 65° within ±5°. Although the sensing performance of the sensor is not as good as expected due to specular reflectance of the solder joint, it seems to satisfy the sensing requirements for the probing task.

19.3.5.4 Contact Control Algorithm

The contact force control regulates the speed on impact by carefully controlling the probe tip position. The required working time for contact transition permits just 0.1 sec; thus, a fast and, therefore, stable contact control must be achieved. After an adjustment is made in the probing direction of the probing manipulator, its trajectory is modified by considering the gap distance measured by the optical sensor. Figure 19.35 shows the controller structure employed to control the force occurring at contact. The control law is given by

$$u_i = \mathrm{sgn}(e_f)\left[\left(k_p + k_v \frac{d}{d_t} + k_I \int dt\right)(\dot{x}_{md} - \dot{x}_m) + M\dot{x}_{md}\right] + (1 - \mathrm{sgn}(e_f))\left[\left(k_{pf} + k_{If} \int d_t\right)(f_d - f')\right] \quad (19.21)$$

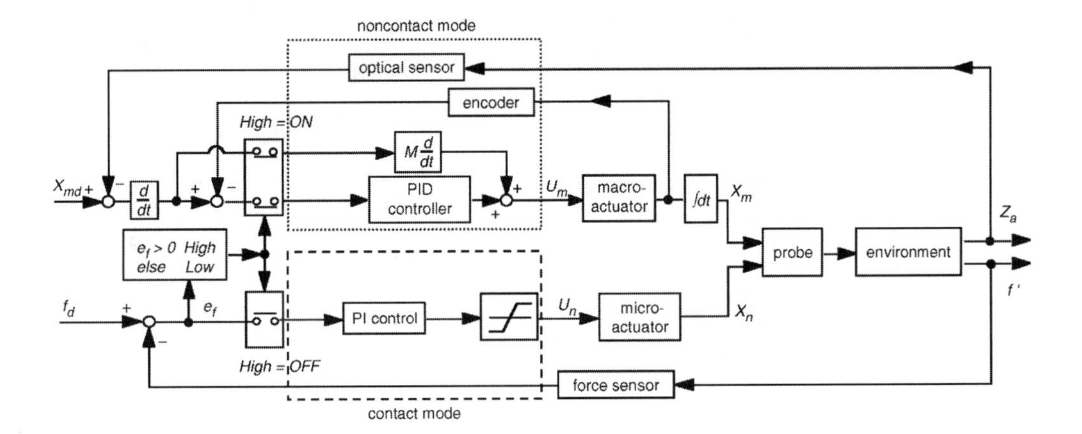

FIGURE 19.35 A probing force control scheme.

where

$$\mathrm{sgn}(e_f)\begin{cases} =0, & i=m, \quad if_{ef} \le 0 \\ =1, & i=n, \quad if_{ef} \le 0 \end{cases} \tag{19.21}$$

$$e_f = f_d - f'. \tag{19.22}$$

In the above equations $\dot{x}_m$ and $\dot{x}_{md}$ are the actual and the desired velocity of the macroactuator, respectively, and k_p, k_v, and k_I denote, respectively, the proportional gain, the derivative gain, and the integral gain of the velocity controller in the macroactuator. The M, k_{pf}, and k_{If} represent, respectively, the mass of the probe, the proportional gain, and the integral gain of the force controller in the microactuator. Also, u_i and f_d are the control input and the desired contact force, while f' is the sensed contact force. Subscripts m and n denote the macroactuator and the microactuator, respectively. In the noncontact mode when $e_f < 0$, only the velocity control of the macroactuator is needed; this can be easily implemented by a PID controller. Usually, the position of the robot manipulator is determined by its velocity profile. Most velocity trajectories use the trapezoidal type with time intervals consisting of acceleration, constant velocity, and deceleration periods.

19.3.5.5 The Probing System Performance

To verify the effectiveness of the sensing method and the control method the system was tested under conditions of an actual environment. The force sensor (PCB Piezotronics, 209A12) provided a reading up to 10 N with a resolution of 2.22×10^{-4} N. The system was capable of 1 KHz servo rates, where the sampling rate used for the control of the micro-motion device was 10 KHz. The solder material used was a type of Sn–Pb (composition ratio 63:37). We set $\alpha_{\text{limit}} = 0.2$ mm and $f_d = 0.3$ N. These values were selected from the experimental data of actual production.

For the purpose of comparing the performance of the designed controller with that of the passive compliance method of conventional probing manipulators and the PID force feedback control method the three controllers were tested under the same conditions. Figure 19.36(a) shows the contact force responses in a normal direction to the surface of the solder joint for three different controllers at the approach velocity $v_0 = 0.1$ m/s. The PID force feedback control method decreased the peak impact force by approximately 25% and attenuated the contact oscillations much more than the passive compliance method. Compared with the conventional passive compliance method, the designed method can reduce the magnitude of peak impact force by approximately 50%. In addition, it shows stable contact after the initial contact. Because the force controller has low gains it is less susceptible than former methods to noise.

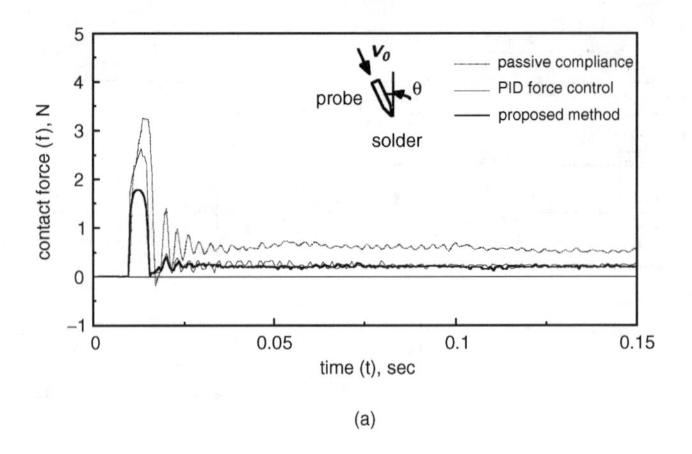

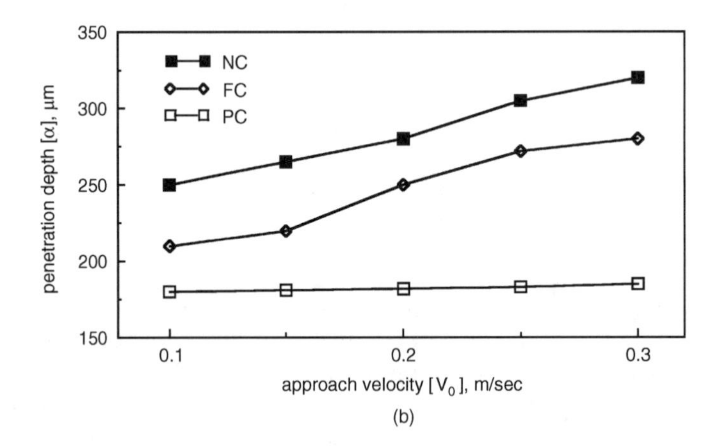

FIGURE 19.36 (a) The test results of the control performance, $v_0 = 0.1$ m/sec; (b) test results for the penetration depth according to variation of the approach velocity and the control method (NC: no control or passive compliance method, FC: force feedback control method, PC: proposed control method).

Figure 19.36(b) shows the measured results of the depth of penetration for various approach velocities and control methods. The depth of penetration by the proposed method is constantly maintained below the allowable magnitude with respect to the change of the approach velocities, while that of two other methods increases according to the increase of the approach velocity. For $v_0 = 0.3$ m/s, compared with the passive compliance method, the proposed method is found to decrease the depth of penetration by approximately 50%.

Summary

Surface mount technology has been playing a key role in accommodating the requirements of miniaturization, high performance, and high productivity. The problem associated with PCB assembly using this technology is that the fabrication processes often contain many undesirable defects that need to be strictly corrected for quality control. With a view toward providing guidelines on keeping electronic part assembly under good quality control this chapter addresses the quality inspection problems connected with defects in various assembly processes including the solder paste process, the surface mounting process, the solder joint inspection process, and the circuit testing process. The role of opto-mechatronic technology for monitoring and control of the processes is reviewed. Developments to date make it clear that mechatronic technology alone cannot solve complex inspection problems. This necessitates the integration of mechatronic elements with optical elements to enhance the performance of inspection systems. Due to the complexity associated with the problems and the trend of miniaturization, however, further developments of precision sensing and control are ultimately needed for the devices and methods in order to upgrade the inspection performance; such developments will include resolution, speed, and reliability. The developments should also elaborate on autonomy and intelligence of SMT systems for future applications.

Defining Terms

contact control: A control method that regulates the input force or speed upon contact between two objects.

feed forward control: A control method that forwards *a priori* or sensory knowledge to the control signal computation.

fiducial mark: A mark engraved on the surface of PCBs for locating electronic parts.

galvanometer: A device that steers light beams in a sequential motion.

image convolution: Linear computation of numerical filters in image application, where the operation becomes the weighted sum of the pixels in the neighborhood around an image point with a convolution mask.

neural network classifier: A nonparametric, neural-network-based analytical tool that discriminates dissimilar data, clustering similar ones into groups.

SMT: Surface mount technology.

solder paste printing: An SMT process that pastes solder creams onto a printed circuit board.

References

Cho, H. S., Cho, Y. C., and Kim, J. H., Development of Automated Flat Integrated Circuit (FIC) Chip Mounting System with Visual Recognition, research report to Samsung Electronics Co., Korea Advanced Institute of Science and Technology, Taejeon, Korea, 1989.

Cho, H. S., Ko, K. W., Roh, Y. J., and Kim, J. W., Development of Vision System and Image Processing Algorithms for Surface Mount Technology, research report to Mirae Co., Korea Advanced Institute of Science and Technology, Taejeon, Korea, 2001.

Furuno, M., Masuda, T., Doi, K., and Nomura, H., Flux free flip chip attach technology for BGA/CSP packages, *IEEE Conf. on Electron. Components and Technol.*, San Diego, 1999, pp. 404–414.

Hata, S., Vision systems for PCB manufacturing in Japan, IECON '90, *16th Annual Conference of IEEE*, 1, 792–797, 1990.

Kim, S. K., Factory automation technology for electronic product manufacturing, *J. Korean Soc. Mech. Engineers*, 36, 443–453, 1996.

Lathrop, R. R., Solder paste print qualification using laser triangulation, *IEEE Transaction on Components, Packaging, and Manufacturing Technology—Part C*, 1997, p. 20.

Lee, D. Y., Kim, B. M., and Cho, H. S., System identification and control of a mount head for surface mounting systems, SICE/ICASE Joint Workshop, *Control Theory and Applications*, Nagoya, Japan, 2001, pp. 43–48.

Moganti, M., Ercal, F., Dagli, C. H., and Tsunekwa, S., Automatic PCB inspection algorithms: a survey, *Computer Vision and Image Understanding*, 63(2), 287–313, 1996.

Owen, M., Two-dimensional and three-dimensional inspections catch solder-paste problems, *Test Measurement J.*, February 2000.

Oyeleye, O. and Lehtihet, E. A., A classification algorithm and optimal feature selection methodology for automated solder joint defect inspection, *J. Manuf. Sys.*, 17(4), 251–262, 1998.

Parker, J. R., *Practical Computer Vision Using C*, John Wiley & Sons, New York, 1994.

Roh, Y. J., Ko, K. W., Cho, H. S., Kim, H. C., Joo, H. N., and Kim, S. K., Inspection of BGA (ball grid array) solder joints using x-ray cross-sectional images, *SPIE Intelligent Sys. Advanced Manuf.*, 3836, 168–178, 1999.

Ryu, Y. K. and Cho, H. S., New optical measuring system for solder joint inspection, *Opt. Lasers Eng.*, 26, 487–514, 1997.

Shim, J. H. and Cho, H. S., An actively compliable probing system, *IEEE Control Sys. Mag.*, 17(1), 14–21, 1997.

Shim, J. H. and Cho, H. S., A new macro/micro robotic probing system for the in-circuit test of PCBs, *Mechatronics*, 9, 589–613, 1999.

Shim, J. H., Cho, H. S., and Kim, S., An optical sensor of a probing system for inspection of PCBs, *SPIE Int. Symposium Intelligent Sys. Advanced Manuf.*, Pittsburgh, 1997, pp. 14–17.

Woolstenhulme, J. and Lubofsky, E., Machine vision placement considerations, *Surface Mount Technol. Mag.*, 14, 62–63, 65–66, 2000.

Zhang, J. B. and Weston, R. H., Reference architecuture for open and integrated automatic optical inspection systems, *Int. J. Prod. Res.*, 32, 1521–1543, 1994.

Zurada, J. M., *Introduction to Artificial Neural Systems*, West Publishing, Minneapolis, MN, 1992.

20

Opto Skill Capturing and Visual Guidance for Service Robots

Shiu Kit Tso
City University of Hong Kong
Kowloon, Hong Kong, China

King Pui Liu
City University of Hong Kong
Kowloon, Hong Kong, China

20.1 Introduction

An important motivation for developing a skill-capturing and guidance system based on optical means is to program and operate service robots, which serve users in a variety of ways such as feeding paralyzed or handicapped people at home and in hospitals or cleaning toxic air filters in laboratories. Because service robot users are usually ordinary people, an easy-to-use robot programming approach is vital if these kinds of robots are to become popularized. The traditional textual and key-in programming approach requires considerable learning and preparation time that is not suitable for service robots. Over the last decade an easy-to-use alternative approach, which has attracted much research attention, is **robot programming by human demonstration** [Delson and West, 1994; Ikeuchi and Suehiro, 1994; Yang et al., 1997]. This approach involves two phases of operation—programming and execution— that may usefully apply skill-capturing and visual-guidance technologies, respectively.

Generally speaking, skill capturing by optical means is a convenient way of transferring human skills to a robot manipulator for performing a certain task in the form of a program, while visual guidance is a technique to endow the robot manipulator with the ability to reach a target effectively using a given program. With this kind of service robot programming technique the preparation of a list of instruction codes for the robot to follow is no longer a tedious chore. It is achieved by allowing the human to demonstrate the intended task using opto-mechatronic motion-capturing devices; the recorded multi-dimensional trajectory information is preprocessed, recognized, and converted directly into robot instruction codes. Relying on visual guidance for robot execution, the target position with respect to the robot can be determined. The robot is then guided to reach the target and manipulate objects in the target position according to the prepared instruction codes.

0-8493-1162-4/03/$0.00+$1.50

Consider a scenario in which a service robot must turn a panel thumb-wheel in a radioactive control station. The human operator demonstrates the task in a safe physical environment with an optical skill visual capturing device. When the robot takes up the actual task, there is spatial uncertainty between the robot and the wheel. Vision-based techniques can be applied to determine the transformation matrix between the robot and task coordinate systems. The power of machine vision enables visual guidance to reduce this spatial uncertainty to the point where the robot can effectively execute the program already prepared to carry out the intended task.

This chapter addresses opto-mechatronic approaches to dealing with two major issues related to the practical application of service robots.

20.1.1 Opto Skill Capturing

Opto skill capturing for programming robots has become an important research topic in recent years because of its potential to facilitate user-friendly robot operations. The basic idea may be traced back to the "teaching by lead-through" method introduced some decades ago. It has been applied to specific tasks such as robot spraying. The main limitation of this lead-through approach is that only the demonstrated spatial information is recorded, and it cannot be reused for any other similar tasks with different geometrical environments. Alternatives aimed at improving this programming approach have been reported. To improve flexibility and convenience during demonstration some researchers propose using an alternative opto-sensor-based programming tool instead of the robot itself to introduce the robot instruction [Tso and Liu, 1993; Delson and West, 1994].

In 1994 Ikeuchi and Suehiro proposed the **assembly plan from observation** (APO) method to realize natural programming by allowing the human teacher to do the demonstration manually wearing a cyber-glove. However, the main problems with APO are its requirement for extremely intensive computations and the fact that each image frame takes more than one minute of processing time with present-day computer capabilities. Besides, the inherent trajectory-based nature of the method limits reusability of information. A more advanced and recent approach of human behavior modeling is the transfer of human skills to robots through skill learning [Yang et al., 1997; Tso and Liu, 1998]. This approach emphasizes human control strategy and action learning rather than merely capturing specific spatial information. However, either method of robot programming by human demonstration faces the problem of segmentation, particularly when either the demonstration involves two-hand cooperation or the process itself is long. When the master hand is identified during a demonstrated two-hand cooperation segment, data processing for further segmentation and segment recognition will be done on the master hand only. This greatly reduces the complexity of handling two-hand spatial information, as the combination of two rigid bodies gives rise to 12 degrees of freedom altogether. With proper segmentation on each master-hand segment the size of storage space for prototype forms and computational effort for data processing can be further reduced. Some researchers simply propose using the velocity profile as the basis for segmentation [Kang and Ikeuchi, 1994]. Others propose manually adding a break signal during the demonstration. As the consideration for segmentation is task specific, each human demonstration processing system may have to develop its own segmentation rules. Therefore, developing a sophisticated segmentation process is an important component for an easy-to-use robot programming approach by demonstration. Section 20.2 provides more details on an approach to acquiring human skills by an opto-capturing system as well as on the modeling of the demonstrated trajectory and segmentation process.

20.1.2 Visual Guidance

After the robot-programming (by human demonstration) phase, visual guidance is employed in the robot execution phase, mainly to reduce spatial uncertainty, e.g., to ensure reliable steering of the robot so that it reaches its specified target. In fact, research on visual guidance has attracted considerable attention for more than two decades [Longuet-Higgins, 1981; Tsai, 1987; Espiau et al., 1991; Enrico et al., 1996].

Consider the simplest case in which the precise position of a target object relative to the robot coordinate system is predetermined. Then, the motion-control algorithm that allows the robot to reach the target position can be constructed and fixed. However, it is more often the case that the target's spatial information cannot be accurately determined in advance, or it may change from case to case. Accordingly, some kind of sensor technology is relied upon to assist determination of the position information for steering the robot end-effector so that it approaches the target correctly. Particularly as the robot gets close to the target machine vision will provide a powerful sensor technique to give the required information.

Visual guidance can be categorized into two major paradigms: the **3D metric paradigm** and the **visual servoing paradigm** [Enrico et al., 1996]. In the former, robot motion is controlled in the robot's joint or Cartesian space. Therefore, the error-based control function driving the robot is also defined in this space. However, visual information about the target is obtained in a fixed-image frame. Hence, the transformation matrix between the robot frame and the image frame must be accurately determined. This is done by a calibration process. In the visual servoing approach, on the other hand, active vision technique is normally employed where the camera coordinate system is not fixed, as the camera is itself mounted on the moving robot's end-effector. Therefore, it is more convenient to express the error-based control function in the image frame [Papanikolopoulos and Khosla, 1993]. Because the transformation matrix between the robot and the image keeps changing when the robot end-effector moves, accurate calibration to find the varying transformation matrix is not meaningful; the calibration process would be too time consuming, and the reference information so obtained is valid only momentarily. Table 20.1 summarizes the advantages and disadvantages of the two visual guidance approaches applied to steering the robot to the target object.

The selection of either approach depends on the required accuracy and the reaching strategy. If the robot's motion is preprogrammed with reference instruction codes and only the target object's position relative to the robot is unknown, then the accurate three-dimensional metric paradigm is more suitable because of its ability to obtain precise information about relative position. On the other hand, if the robot is not supplied with a preprogrammed code, and it is only necessary for the robot to reach and grasp a target object (and the way to grasp it is unimportant), then the visual servo approach is more suitable because of its robustness with respect to uncertainty. The aim of this chapter, however, is to apply visual guidance to a robot provided with predetermined instruction codes generated by human demonstration; thus, the chapter assumes the 3D metric paradigm, and the discussion that follows will focus on this paradigm.

TABLE 20.1 Comparison between Three-Dimensional Metric Paradigm and Visual Servoing

	Advantages	Disadvantages
Three-dimensional metric paradigm	Reaches target efficiently and quickly when the target is stationary Target reached directly in three-dimensional sense	High-precision camera-calibration process required Unmoved target assumed once its three-dimensional spatial information is determined
Visual servoing	Precise camera-calibration process not needed Moving object allowed because image space information is directly fed back to the robot motion control algorithm	Slow convergence for robot steering to reach target Very slow sampling rate for robot control because image processing and motion control are executed alternately, and the image processing time is very long, unless extra (redundant) sensory information is allowed

20.2 Basic Opto Skill-Capturing Approach

This section describes the hardware design and system configuration as well as analytical and software development associated with the proposed skill capturing by optical means.

A marker-based opto-measurement device will be introduced to capture and record rigid-body motion. The human operator demonstrates a task with a pair of rigid-object tools, which are similar in shape and structure to the service robot end-effectors. The tools are conveniently called the robot end-effector representatives (REERs). The demonstration by the operator is recorded and stored in the form of a multidimensional trajectory. Details of this opto-measurement device and the consequent trajectory modeling technique will be discussed as follows.

20.2.1 Design of an Opto-Measurement Device

The measurement device is composed of three major components: marker-based position sensor unit, control unit, and host computer, which are connected as shown in Figure 20.1.

A typical marker-based position sensing system, e.g., the OPTOTRAK system [Krist et al., 1990], may be used for this application. It consists of three linear (one-dimensional) CCD cameras mounted evenly and firmly along a rigid bar. The middle camera at the center of the bar is aligned with its pixel array orthogonal to the length of the bar. The other two cameras are mounted near the two ends of the bar. Their pixel arrays are aligned with the length of the bar and angled slightly inwards (say at 9°). Each CCD camera is equipped with a programmable-gain amplifier, a high-speed analog-to-digital converter, and an on-board dedicated RISC processor. The programmable gain compensates for the wide variation in signal strength received from the infrared markers to be detected at different distances away from the camera assembly. The on-board RISC processor is responsible for controlling the sensors and performing the centroid position calculation. The control unit is a command-generating, interfacing, and processing unit. There is a high-speed parallel processor to handle the sequencing of the markers and act as a link between the host computer and the three-camera position sensor unit. The host computer is an ordinary personal computer (PC) that acts as a user input/output interface. When a frame showing the observable marker emitters is captured by all three CCD cameras, the frame information is first processed locally inside each camera unit to determine the image position of each marker projected onto

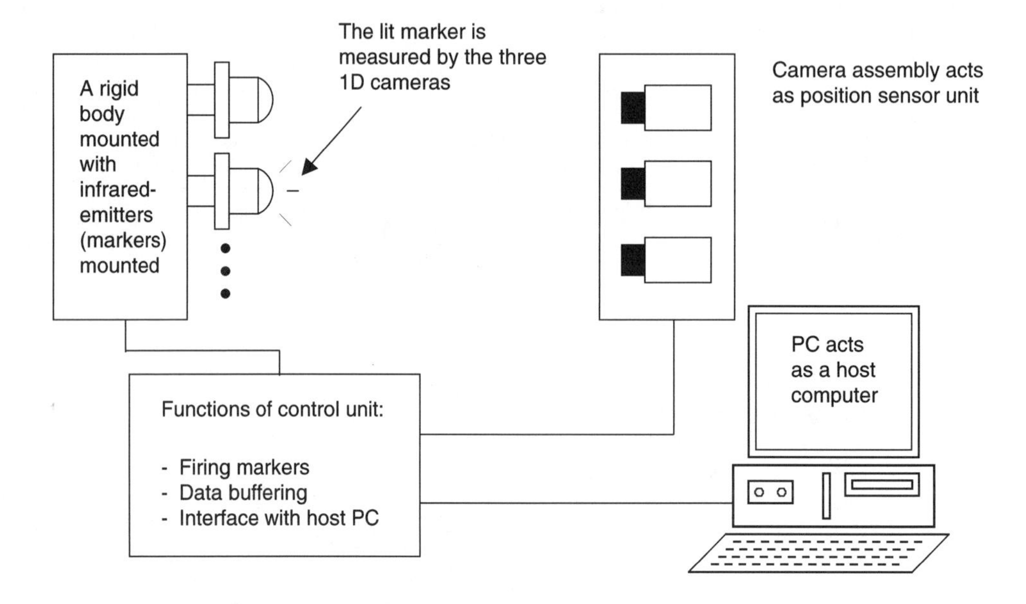

FIGURE 20.1 Marker-based visual measurement device.

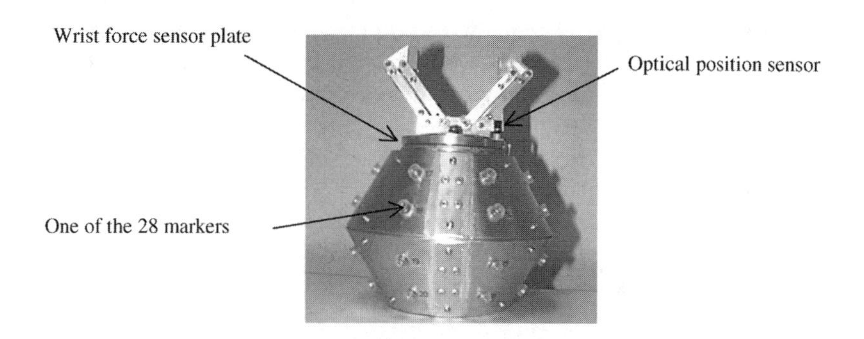

FIGURE 20.2 Appearance of one robot end-effector representative (REER).

the linear CCD sensor. Then the control unit gathers the pixel information obtained from all three cameras and performs further processing, the purpose of which is to obtain the 3D information of all the observable markers and hence the 6D location data of the rigid body containing the markers, given the relative mounting arrangement of the markers. A total of 28 infrared emitters (markers) mounted evenly on the lateral surface of the double conical rigid frames as part of each REER (Figure 20.2) are used to record the spatial information of each REER. Treating each REER as a rigid body, we can determine the rigid 6D values provided that we know the 3D position values of at least three markers. This condition can be generally satisfied with only 28 markers in the given geometrical arrangement. In addition, an optical position sensor is included to indicate the gripper open/close states during operation.

20.2.2 Fundamental Theory of Trajectory Modeling Technique

As mentioned in Section 20.1.1, a reasonably generalized segmentation process capable of handling demonstrated two-hand (or two-REER) cooperation trajectory is important for an easy-to-use robot programming system. During two-hand cooperation, it is reasonable to assume that the cooperation task is basically a master–slave process. It is assumed that the master REER executes the task and the slave REER just follows and coordinates with the master action. Therefore, at the start of the segmentation process we extract all the portions/segments considered to be associated with master–slave cooperation and label them with the right/left REER interpreted as the master. The interpretation is based on the principle that the master action dominates the slave action in terms of spatial variation, as reflected in the velocity profile.

By comparing the velocities of the two REER motions, if one is marked with a higher profile and the other is relatively low, they signify a master and a slave segment, respectively. If both have low (or high) profiles, the right REER is treated by default as the master because people are usually right-handed. If the operator is left-handed, then the position of the right REER is replaced by the left one, while the rest of the procedure remains the same. After the master segment extraction, only the master REER spatial information is used for further segmentation.

As mentioned, the way to determine the master–slave segments is by examining the velocities. However, as velocity information is usually noisy, a moving-average technique, called wavelet filtering [Gilbert and Truong, 1996], is first applied to filter out the spikes so that it becomes more reliable to distinguish between the low- and high-velocity profiles. After the segmentation based on the master–slave principle, the whole process is divided into a number of segments, each of which is labeled with the left or right REER treated as the master.

In the second part of the segmentation process we obtain further segmentation for each master REER segment only. We make use of information available from more than one source so that a broader consideration caters to a wider choice of criteria. Again, the wavelet filtering technique is applied to the raw master spatial trajectory so that only the macro trend is retained and the less important details are filtered out. The reason for applying this filtering treatment is that the segmentation principle is based on the trajectory macro trend but not the local intended or unintended quivering movements. Mathematical treatment is next applied to the filtered trajectory so that the differential-geometric features [Carmo, 1976], including local curvature and velocity, are extracted. Then, the additional information about the gripper-status signal is also evaluated. The two sources of information are input to a set of AND/OR logical operations to generate a simple segmentation index for efficient partitioning purpose.

The subsections are organized as follows. Subsection 20.2.2.1 provides more details on the basic theory and procedure of the segmentation process. Subsection 20.2.2.2 describes the method of grouped-segment comparison and recognition by using a hidden Markov model. Subsection 20.2.2.3 deals with the composition of robot instruction code based on the nature of the recognized segments.

20.2.2.1 Trajectory Segmentation on Master REER Information

The trajectory-segmentation process is based on only the master REER demonstrated information. As mentioned in the last section, the segmentation process makes use of the information of differential geometric properties and the gripper-status signal. Basically, two types of segments are generated by the above information. They are the "transition segment" and the "manipulation segment." The former represents a purely REER transition from one location to another, with the intervening path details considered to be unimportant. The latter segment type represents a movement with all the path details preserved, to be followed in full by the robot end-effector.

20.2.2.1.1 Segmentation by Using Gripper-Status Signals

The segmentation process basically partitions a complete demonstration trajectory into a number of portions, each with recognized characteristics expressible in terms of the geometric properties and the gripper-status signal combined. The segmentation procedure starts from the gripper-status signal. When the gripper-status signal of an REER is 0, it means that the gripper is not grasping any object; it is then considered to be undergoing a purely tool movement, and the segment is assigned to be a transition type. When the gripper-status signal is 1, the differential geometric features of the 3D trajectory are extracted for evaluation.

20.2.2.1.2 Segmentation by Using Differential Geometric Properties

The theory of segmentation by using differential geometric properties is based on the relationship between "human mental intention" and "demonstration speed and trajectory curvature." For any master demonstration portion when the speed is high (higher than a threshold setting), it implies that the operator is not concerned with the trajectory path itself, so that this portion is considered to be a "transition segment." However, when the speed is low (but is still higher than the slave REER), it is interpreted as a "manipulation segment." Empirical investigation shows that a REER-carrying operator who considers the path details important will carefully trace the required path. In this way, demonstration speed will generally be low and the trajectory curvature usually high. Because the REER is not too light the human carrying the weight while tracing a path will give rise to a slow and nonsmooth (containing low-amplitude ripples) trajectory. The mathematical background of the trajectory velocity and curvature is described below.

Consider a general 3D curve α, parametrized by arc length s and given by a mapping such that $\alpha(s): R^1 \rightarrow R^3$. Because s is the arc length, the first derivative $\alpha'(s)$ gives a tangent vector with unit length. The magnitude of the second derivative $|\alpha''(s)|$ measures the rate of the neighboring tangents pulling away from the tangent line at s. So $\alpha''(s)$ is also called the centripetal vector. This suggests the definition of the curvature of the parametric curve α to be given by Eq. (20.1):

$$k(s) = |\alpha''(s)| \tag{20.1}$$

The local velocity of the trajectory is defined as $\frac{\Delta s}{\Delta t}$, where Δt is the constant sampling time interval and $\Delta s = \sqrt{\Delta x^2 + \Delta y^2 + \Delta z^2}$. Without loss of generality, we set Δt equal to unity, so that the normalized local velocity v is equal to Δs, as shown in Eq. (20.2):

$$v = \sqrt{\Delta x^2 + \Delta y^2 + \Delta z^2} \tag{20.2}$$

TABLE 20.2 Summary of Segmentation Rules on Master REER Information

	High Demonstration Speed	Low Demonstration Speed
$k(s) + h(s)$ is high	Transition segment (TS) interpreted	Manipulation segment (MS) interpreted
$k(s) + h(s)$ is low	Transition segment (TS) interpreted	Transition segment (TS) interpreted

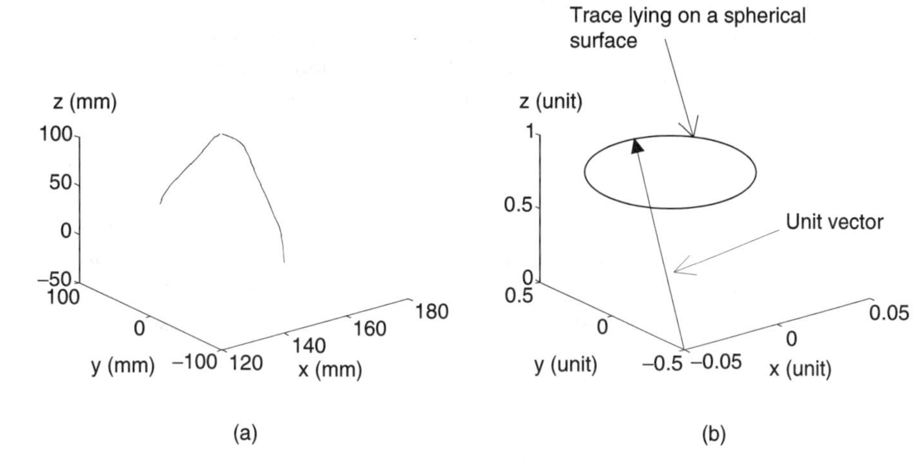

FIGURE 20.3 A typical pair of 3D translation and orientation trajectory traces: (a) translation trace; (b) orientation trace.

Although the trajectory itself is a list of 6D spatial information, human demonstration provides two sequences of 3D trajectories representing translation and orientation, respectively. Figure 20.3(a) and (b) show a typical pair of 3D translation and orientation trajectory traces from an experimental demonstration. Each 3D trajectory trace can be expressed in the arc-length domain. For the parametric curve describing "translation," Eqs. (20.1) and (20.2) are directly applied to extract the curvature and velocity information of the trajectory.

Consider the extraction of differential geometric features for orientation segments. We need to first convert the orientation data from, say, the roll–pitch–yaw data type to Euler parameters, represented by an angular rotation ϕ around a unit vector $\mathbf{r}$. The movement of the endpoint of the unit vector $\mathbf{r}$ describes a trace lying on a spherical surface with unit radius $|\mathbf{r}|$, as shown in Figure 20.3(b). Thus, the curvature function $h(s)$ is also calculated by Eq. (20.1). However, the local velocity of the trace described by $\mathbf{r}$ is not considered in the segmentation process because it is not related to the demonstration speed but only to the fluctuation of $\mathbf{r}$ due to hand quivering. Also, the component ϕ (angular rotation) itself is used not in this segmentation process but in the subsequent grouped-segment-recognition processes.

The rules for recognizing segmentation types applied to the master REER are summarized in Table 20.2. It is worth mentioning that only the master hand motion is needed to do further processing in order to separate the transition and manipulation subsegments. However, the spatial information of the slave hand is always considered important because its position is needed to coordinate with the master hand. Therefore, the slave arm is always under position control that corresponds to the movement of the master arm.

After the segmentation process we obtain a number of transition and manipulation segments for the master hands. However, in some cases a simplified combination of the two types of segments can be described by a single high-level task, such as SCREW, INSERT, etc. The recognition of such grouped segments (a combination of the transition and manipulation segments) will be discussed in the next section.

20.2.2.2 Grouped-Segment Recognition

The grouped-segment trajectory is described by combining adjacent segments to form a more understandable subtask such as "turn and screw nut onto bolt." As mentioned before, only the master REER spatial information will be considered for both segmentation and recognition. Therefore, the information to be processed is based only on 6D information of the master REER, even though two REERs may be cooperating as the slave hand will follow the master hand with similar characteristics. In this section we will discuss how to use the hidden-Markov-model (HMM) technique to perform trajectory-segment recognition. This is done with reference to Yang [1994]; also, a brief introduction is given in Appendix A.

20.2.2.2.1 Preprocessing of Spatial Trajectory

The raw data of the trajectory segment comprise a list of 6D spatial information in time sequence. They provide the fundamental observations for an HMM. To facilitate efficient computer operation we utilize the so-called discrete HMM technique, meaning that the observation is recorded as a list of discrete symbols. Therefore, we need to preprocess position trajectories so that they are converted to a finite number of discrete symbols.

The first part of preprocessing is to transform the data from the time domain to the frequency domain. The idea is based on the fact that shifting a waveform in the time domain will affect only the phase and have no effect on the amplitude of the spectrum in the frequency domain. Therefore, two demonstrated trajectories of similar shape but shifted in the time scale will be interpreted as being the same in the frequency domain. This is what we want to do with the similarity measurement. The technique applied is the so-called short-time Fourier transform (*STFT*) [Hlawatsch and Boundeaux-Bartels, 1992]. The trajectory segment is processed increment by increment, as shown in Eq. (20.3):

$$STFT_x^{\gamma}(t,f) = \int_{t'} [x(t')\gamma^*(t'-t)]e^{-j2\pi f t'}dt' \tag{20.3}$$

where the signal $x(t')$ is multiplied by a moving "analysis window" $\gamma^*(t'-t)$ centered around the time t. It is in fact a local spectrum of the signal $x(t')$ around time t. The frequency spectrum reflects the shape and amplitude of a short-time portion of the trajectory segment.

In the second part of preprocessing the frequency spectra are quantized to a limited number of spectrum-vector units. This part is processed differently for modeling and for evaluating the trajectories.

In the case of modeling prototype trajectory segments the lists of spectrum vectors, obtained from the *STFT* applied to all possible prototype trajectory segments, are quantized to a finite number of spectrum-vector units. As the quantization is multidimensional, it is called vector quantization (VQ). The algorithm chosen is the LBG algorithm [Linde et al., 1980]. The steps are summarized below:

1. *Initialization*: Set the number of partitions to $K = 1$ and find the centroid of all spectrum vectors in the partition.
2. *Splitting*: split K into $2K$ partitions.
3. *Classification*: accept the kth partition C_k of each spectrum vector $v^{(l)}$ of dimension l, according to the specified condition, i.e.,

$$v^{(l)} \in C_k \text{ iff } d(v^{(l)}, \bar{v}_k^{(l)}) \leq d(v^{(l)}, \bar{v}_{k'}^{(l)}) \quad \text{for all } k \neq k' \tag{20.4}$$

   where $\bar{v}_k^{(l)}$ is the centroid vector of C_k and d is a distortion measure to be defined.
4. *Centroid updating*: recalculate the centroid of each accepted partition.
5. *Termination*: steps 2 to 4 are repeated until the decrease in the overall distortion at each iteration relative to the value at the previous iteration is below a selected threshold. The number of partitions is increased to a value that meets the required level.

After termination, we will have a number of centroids, $\{\bar{v}_k^{(l)}\}$, of all the partitions. These centroids are in fact the spectrum vector units. In a discrete HMM the multidimensional spectrum-vectors are represented by observation symbols $\{O_k^l\}$. These symbols are applied to train the HMM, as explained in the next subsection.

For evaluation purposes the frequency spectra for each individual trajectory segment are mapped to prototype spectrum vectors $\{\bar{v}_k^{(l)}\}$. The mapping is based on the minimum-distortion principle, and the

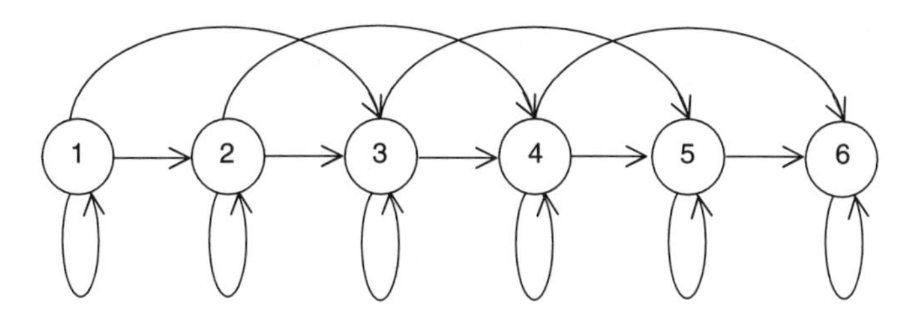

FIGURE 20.4 A six-state left-to-right HMM.

distortion measure is given by Eq. (20.5):

$$d(v^{(l)}, \bar{v}_k^{(l)}) = \left\| v^{(l)}, \bar{v}_k^{(l)} \right\| = \sum_{r=1}^{R} (v_r^{(l)} - \bar{v}_{k,r}^{(l)})^2 \tag{20.5}$$

where R is the total spectrum vector dimension. After mapping is completed, the trajectory segment is converted to a list of observation symbols, which will be sent through the tuned HMM for evaluating the likelihood index.

20.2.2.2.2 Training of HMM Parameters

The type of HMM [Rabiner, 1989; Tso and Liu, 1997] applied in our application is assumed to be a six-state left-right model, as shown in Figure 20.4. The HMM consists of three elements:

- {S} is a collection of states.
- **A** is the transition probability matrix.
- **B** is the multidimensional output probability matrix. $\mathbf{B} = \{b_j(O_k^l)\}$ where $b_j(O_k^l)$ is the probability of the output being O_k^l at state j.

Suppose the initial state has a probability distribution $\pi = \{\pi_i\}$. An HMM can be written in a more compact form as $\lambda = (\mathbf{A}, \mathbf{B}, \pi)$ with the state sequence {S} hidden in the notation.

According to Figure 20.4, the transition matrix of the six-state left-to-right HMM is represented by:

$$\mathbf{A} = \begin{bmatrix} a_{11} & a_{12} & a_{13} & 0 & 0 & 0 \\ 0 & a_{22} & a_{23} & a_{24} & 0 & 0 \\ 0 & 0 & a_{33} & a_{34} & a_{35} & 0 \\ 0 & 0 & 0 & a_{44} & a_{45} & a_{46} \\ 0 & 0 & 0 & 0 & a_{55} & a_{56} \\ 0 & 0 & 0 & 0 & 0 & 1 \end{bmatrix}.$$

The six Markov states are time-sequencing states. The model must start from state 1, i.e., $\pi_1 = 1$ and $\pi_i = 0$ for all $i \neq 1$. As time progresses a state may change from state i to state i, $i+1$ or $i+2$, but it is not allowed to jump three states to state i + 3; therefore, $a_{i(i+u)} = 0$ for $u \geq 3$.

There are two basic processes involved in training an HMM to model the human-generated trajectory segment: the evaluation and training processes. The former is concerned with obtaining the probability factor of each trajectory satisfying the HMM, given the model parameters. The latter is concerned with training the model parameters of an HMM so that the HMM well represents the human-generated trajectories using the probability factors obtained in the evaluation process. The evaluation and training processes are repeated recursively until the improvement meets a certain predefined saturation level.

20.2.2.2.3 Evaluation Process

The evaluation process serves to calculate the conditional probability, $P(O \mid \lambda)$, i.e., the probability of outputting an observation sequence $O = O_1O_2 \cdots O_T$, provided that the HMM model parameter, λ, is known. There are two possible approaches, called the forward and backward procedures, to evaluate the probability.

Forward Procedure

Consider the forward variable $\alpha_t(i)$, which is defined by Eq. (20.6):

$$\alpha_t(i) = P(O_1O_2 \cdots O_t, S_t = i \mid \lambda) \tag{20.6}$$

It is the probability of having the observation sequence $O_1O_2 \ldots O_t$ until time t, and of being in state i, given that λ is known. The steps of calculating the probability inductively are summarized below [Rabiner, 1989]:

1. Preparation of multidimensional probability:

$$b_j(O_t) = \prod_l b_j^l(O_t^l), \quad 1 \le j \le N \tag{20.7}$$

2. Initialization:

$$\alpha_1(i) = \pi_i b_i(O_1), \quad 1 \le i \le N \tag{20.8}$$

3. Induction:

$$\alpha_{t+1}(j) = \left[\sum_{i=1}^{N} \alpha_t(i) a_{ij}\right] b_j(O_{t+1}), \quad \begin{matrix} 1 \le t \le T-1 \text{ and} \\ 1 \le j \le N \end{matrix} \tag{20.9}$$

4. Termination:

$$P(O \mid \lambda) = \sum_{i=1}^{N} \alpha_T(i) \tag{20.10}$$

Backward Procedure

Consider the backward variable $\beta_t(i)$, which is defined as:

$$\beta_t(i) = P(O_{t+1}O_{t+2} \cdots O_T \mid S_t = i, \lambda) \tag{20.11}$$

that is, the probability of having the partial observation sequence from $t + 1$ to the end. Given state $s_t = i$ and model parameter λ, similar to the forward procedure, the backward variable can be calculated by induction in three steps:

1. Initialization:

$$\beta_T(i) = 1, \quad 1 \le i \le N \tag{20.12}$$

2. Induction:

$$\beta_t(i) = \sum_{j=1}^{N} [a_{ij} b_j(O_{t+1}) \beta_{t+1}(j)], \tag{20.13}$$

$$\text{where} \quad T-1 \ge t \ge 1 \quad \text{and} \quad 1 \le i \le N$$

3. Termination:

$$P(O \mid \lambda) = \sum_{i=1}^{N} \pi_i b_i(O_1) \beta_1(i) \tag{20.14}$$

Basically, only either the forward procedure or the backward procedure alone can find the required probability. However, the forward and backward variables are both useful for expediting the training process.

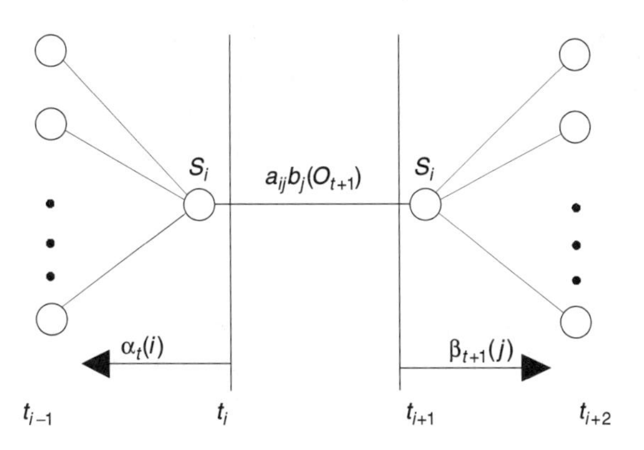

FIGURE 20.5 Sequence of states around time instant *t*.

20.2.2.2.4 Training Process

Given the model construction of the HMM, the aim of training is to adjust the model parameters $\lambda = (\mathbf{A}, \mathbf{B}, \pi)$ to maximize the probability for a given observation sequence. There is no known method to solve this kind of problem analytically. However, based on an iterative procedure, similar to the Baum–Welch reestimation method [Huang et al., 1990], we can choose the parameter λ, such that $P(O|\lambda)$ is locally maximized. The basis of the training procedure is summarized below:

Let $\xi_t(i,j)$ be the probability of being in state i at time t, and in state j at time $t+1$, that is,

$$\xi_t(i,j) = P(S_t = i, S_{t+1} = j | O, \lambda) \tag{20.15}$$

The sequence of states around time t leading to the derivation of the formula to find $\xi_t(i,j)$ in Eq. (20.15) is illustrated in Figure 20.5.

By the definition of the forward and backward variables Eqs. (20.6) and (20.11), and with the aid of Figure 20.5, it is not difficult to see by counting occurrences that

$$\xi_t(i,j) = \frac{\alpha_t(i) a_{ij} b_j(O_{t+1}) \beta_{t+1}(j)}{P(O|\lambda)}, \quad 1 \le t \le T-1 \tag{20.16}$$

Second, let $\gamma_t(i)$ be the probability of being in state i at time t, i.e.,

$$\gamma_t(i) = P(S_t = i | O, \lambda). \tag{20.17}$$

By the definition of $\xi_t(i,j)$ and $\gamma_t(i)$, we have the following relations:

$$\left.\begin{aligned} \gamma_t(i) &= \sum_{j=1}^{N} \xi_t(i,j), \quad 1 \le t \le T-1, \text{ and} \\ \gamma_T(i) &= \frac{\alpha_T(i)\beta_T(i)}{\sum_{i=1}^{N} \alpha_T(i)} \end{aligned}\right\} \tag{20.18}$$

$$\sum_{t=1}^{T-1} \gamma_t(i) = \text{expected number of transitions from } S_i \tag{20.19}$$

$$\sum_{t=1}^{T} \gamma_t(j) = \text{expected number of times that } S_j \text{ is visited} \tag{20.20}$$

$$\sum_{t=1}^{T-1} \xi_t(i,j) = \text{expected number of transitions from } S_i \text{ to } S_j \tag{20.21}$$

By the concept of counting event occurrences, and by using Eqs. (20.19) and (20.21), we can establish the following formula to reestimate the HMM parameters:

$$\tilde{a}_{ij} = \frac{\sum_{t=1}^{T-1} \xi_t(i,j)}{\sum_{t=1}^{T-1} \gamma_t(i)} \tag{20.22}$$

An effective estimation of the elements of the output probability matrix $\mathbf{B} = \{b_j(O_k^l)\}$ is given by the division of "the expected number of times in state j with observation symbol O_k^l" by "the expected number of times in state j." From Eq. (20.20), it follows that:

$$\tilde{b}_j^l(O_k^l) = \frac{\sum_{t \in O_t = O_k^l} \gamma_t(j)}{\sum_{t=1}^{T} \gamma_t(j)} \tag{20.23}$$

where the numerator is the total number of occurrences of O_k^l in state j and l is the dimension. Then the training processes can be summarized as follows: if we initialize the model by $\lambda = (\mathbf{A}, \mathbf{B}, \pi)$, then we can calculate the forward and backward variables, $\alpha_t(i)$ and $\beta_t(i)$ by Eqs. (20.8), (20.9), (20.12), and (20.13). Then $\xi_t(i,j)$ and $\gamma_t(i)$ can be found by Eqs. (20.16) and (20.18). Hence, the reestimated model, $\tilde{\lambda} = (\tilde{\mathbf{A}}, \tilde{\mathbf{B}}, \tilde{\pi})$, can also be calculated by Eqs. (20.22) and (20.23). It has been proved by Baum et al. [1970] that the above reestimation will lead to either (1) the initial model λ corresponding to a critical point so that the new estimated $\tilde{\lambda}$ equals the old one, or (2) the new estimated model $\tilde{\lambda}$ being better than model λ in the sense that $P(O|\tilde{\lambda}) > P(O|\lambda)$. Usually, the use of evenly distributed initial parameters is good enough to avoid settling at a null point; otherwise, we just repeat with another guess of the initial parameters. The successful reestimation process can be continued until a preset limit is met.

20.2.2.2.5 Measurement of Similarity for Segment Recognition

The process of training the HMM to obtain the appropriate parameters λ is first applied to a finite number of prototype trajectory segments. Then, a set of $\{\lambda\}$ representing the prototype trajectory segments is stored. The basic principle of the similarity measure is to evaluate the probability $P(O|\lambda)$ where $O = O_1 O_2 \cdots O_T$ is the observation sequence of an unknown trajectory segment to be recognized. The values $P(O|\lambda)$ of all prototype segments are evaluated so that the highest value representing the unknown trajectory is matched to the corresponding prototype trajectory segment. As shown in Section 20.2.2.2.3, probabilities of $P(O|\lambda)$ can be found by the forward or backward procedure.

20.2.2.3 Instruction-Code Composition

Instruction codes are divided into two categories: a low-level and a high-level command set. Typical commands are summarized in Tables 20.3 and 20.4.

The low-level command set contains mainly the two commands MOVE (new position) and PATH (via point list) for the transition segment and manipulation segment, respectively. The argument "New-position" is the new position for the arm end-effector to move to. It is directly determined by the segment endpoint. The argument "Via point list" gives the spatial details of the whole manipulation segment to be tracked by the PATH command.

Each element of the high-level command set is in fact a combined sequence of low-level commands. Once a grouped segment is recognized to be a prototype task, the corresponding high-level command will be extracted. The command arguments are determined by the segment via points, which are in fact the endpoints of the transition and manipulation segments inside the grouped-segment.

The expressions of the high-level commands in terms of the low-level commands are summarized in Table 20.5. The arguments are extracted in different ways for the various high-level commands. However, they are all derived from the segment endpoints.

TABLE 20.3 Low-Level Command Set

Commands	Arguments	Description
LH-MOVE	New position	Left hand moves from current position to the new position.
RH-MOVE	New position	Right hand moves from current position to the new position.
LH-PATH	Via point list	Left hand tracks the path details.
RH-PATH	Via point list	Right hand tracks the path details.
LH-GRASP	—	Close the left-hand gripper.
RH-GRASP	—	Close the right-hand gripper.
LH-UNGRASP	—	Open the left-hand gripper.
RH-UNGRASP	—	Open the right-hand gripper.

TABLE 20.4 High-Level Command Set

Commands	Arguments	Description
LH-SCREW	Position1, No. of strokes, Angle per stroke, Pitch per stroke	Left hand moves to Position1 and performs the screw operation.
RH-SCREW	Position1, No. of strokes, Angle per stroke, Pitch per stroke	Screw operation for right hand
LH-INSERT	Position1, Distance1	Left hand moves to Position1 and along the gripper principle axis moves Distance1.
RH-INSERT	Position1, Distance1	Insert operation for right hand.
LH-WOBBLE-INSERT	Amp1, Amp2, Amp3, Position1, Distance1	Left hand enables the superposition of 2HZ sinusoidal wobble to the insert operation. The sinusoidal amplitudes specified in the three Euler parameters for orientation are Amp1, Amp2, and Amp3.
RH-WOBBLE-INSERT	Amp1, Amp2, Amp3, Position1, Distance1	Right hand does the wobble insertion operation.

TABLE 20.5 High-Level Commands in Terms of Low-Level Commands

High-Level Commands	Low-Level Commands
LH-SCREW	LH-MOVE(Position1) + (No. of strokes)* (LH-GRASP + LH-MOVE(Angle) + LH-MOVE(Pitch) + LH-UNGRASP + LH-MOVE(-Angle))
RH-SCREW	RH-MOVE(Position1) + (No. of strokes)* (RH-GRASP + RH-MOVE(Angle) + RH-MOVE(Pitch) + RH-UNGRASP + RH-MOVE(-Angle))
LH-INSERT	LH-MOVE(Position1) + LH-MOVE(Position1 + Distance1)
RH-INSERT	RH-MOVE(Position1) + RH-MOVE(Position1 + Distance1)
LH-WOBBLE-INSERT	LH-MOVE(Position1) + For Distance from 0 to Distance1: LH-MOVE(Distance * (Amp1*Sin(4π)*Euler_parameter1+ Amp2*Sin(4π)*Euler_parameter2 + Amp3*Sin(4π)*Euler_parameter3))
RH-WOBBLE-INSERT	RH-MOVE(Position1) + For Distance from 0 to Distance1: RH-MOVE(Distance * (Amp1*Sin(4π)*Euler_parameter1+ Amp2*Sin(4π)*Euler_parameter2 + Amp3*Sin(4π)*Euler_parameter3))

The low-level commands are the primitives, while the high-level commands can be updated if new prototype tasks are added.

20.3 Basic Visual-Guidance Approach

As mentioned above, visual guidance is considered important in the robot execution phase. After human demonstration of the required task the output of the programming phase is a list of robot instruction codes for the robot to follow. However, as mentioned in Section 20.1, there is spatial uncertainty between the robot and the target object. Consider a two-arm service robot composed of a pair of robot arms (say,

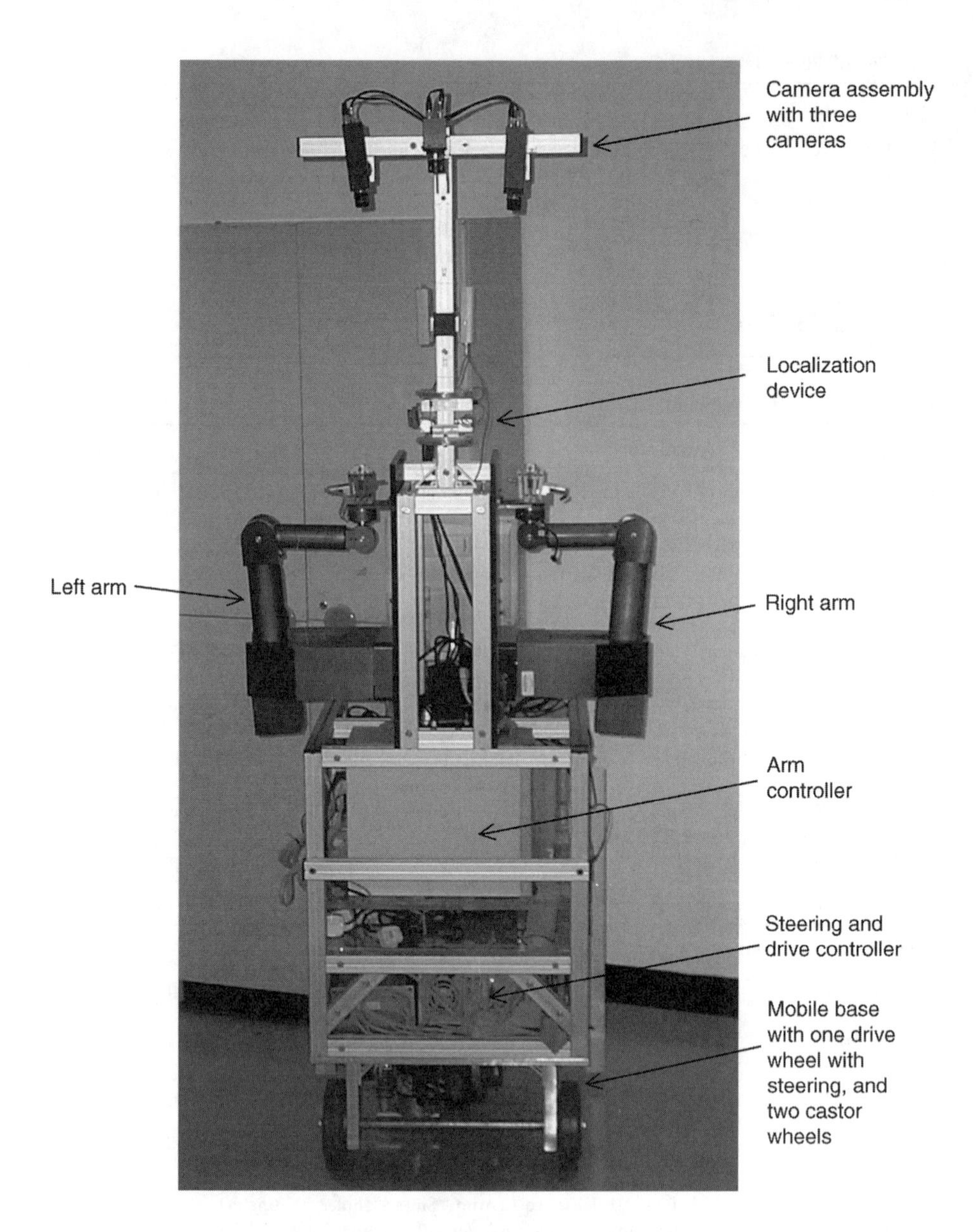

FIGURE 20.6 A two-arm service robot.

Zebra-ZERO) and a mobile base with three wheels, as shown in Figure 20.6. The visual guidance approach is designed first to steer the robot base wheel to within reach of the target and then to drive the robot arm-effector to manipulate on the object reliably.

In this section a laboratory-built localization device will be introduced. It is used to steer the robot to reach the target object reasonably closely. The robot then stops, and a 3D metric paradigm will be applied to accurately determine the target location with respect to the robot body. As stereo vision is involved, at least two cameras are required. As mentioned later in this section, three cameras will be employed to increase the measurement accuracy as well as to solve the corresponding point problem [Xu, 1992]. To perform visual spatial measurements two calibration processes are carried out: robot arm calibration and camera calibration. The former calibration is needed to determine the fixed transformation matrix relating the robot arm coordinate system and the robot base reference. The latter calibration is used to determine the fixed transformation matrix relating each camera frame and the robot base reference.

The robot arm is usually mounted horizontally or vertically to the robot body, as shown in Figure 20.6. As there is no rotational transformation taking place, the calibration will be rather simple and involve only tape and caliper measurements. However, the three cameras are mounted at certain inclination angles to the vertical axis of the robot body. There are nine unknown rotational parameters in the transformation matrix between each camera and the robot body. Together with the uncertainty of the lens center position and other kinds of parameters, this camera calibration process is rather complicated and will be discussed in greater detail in this section.

After these two calibration processes are performed, spatial relationships between the camera system and the two robot arms are determined by multiplying the corresponding transformation matrices generated by the two calibration processes. The spatial information of an object seen by the camera system can be determined in terms of Cartesian coordinates with respect to the robot arm coordinate frame. Because the robot arms are controlled in joint space, the transformation from the Cartesian coordinates to the joint-angle values is in fact the inverse kinematics problem and will be discussed in Appendix B. The practical issues of object recognition and location determination will be discussed in this section.

20.3.1 Localization Device for Robot Positioning

A service robot is able to move autonomously close to the target before it stops to manipulate on the object. In order to steer the robot to approach the target robustly, it is necessary to determine the global position of the robot at a rate as high as the video rate. Therefore, a localization device should be introduced for this position determination. It is similar to a global positioning system (GPS) applied to a small and local area. Like the REER mentioned in Section 20.2.1, the localization device is a rigid body with two separate circular disks. There are 16 markers mounted evenly on each disk, as shown in Figure 20.7.

This localization device is defined as a rigid body and is viewed by the marker-based position sensor unit. The 6D spatial information of the rigid body can be determined by the position sensor in accordance with the principle described in Section 20.2.1. Because the localization device is mounted on the vertical axis of the service robot, the global position of the service robot can also be determined.

FIGURE 20.7 Localization device mounted on vertical axis of service robot.

20.3.2 Camera Calibration

The camera calibration process provides the transformation matrix relating each camera coordinate system and the robot base reference. Although the transformation matrix is fixed, it is difficult to directly calibrate the camera with respect to the robot base reference because all cameras are tilted to view the working envelope of the robot end-effectors in front of the robot base. As there is no visual connection between the cameras and the robot base, calibration between them is difficult. Therefore, the camera calibration process is carried out with respect to a fixed origin, which is located at the central portion of the camera views. In this section the coordinate system based on this origin is regarded as the real-world coordinate system. The transformation from this real-word coordinate system to the robot reference is rather simple. It is given by a pure translation between the origins of the two systems, which can be measured accurately. Therefore, the camera calibration to be discussed is mainly based on the process of determining the transformation matrix from each camera's local coordinate system to the real-world coordinate system. The process outlined below basically follows the method described by Tsai [1987] and Xu [1992]. First of all, we define parameters that will be calibrated as follows:

1. Intrinsic parameters
   - Focal length of the camera, f_c
   - Horizontal image scale factor, s_x
   - Lens distortion factor, k_1
2. External parameters
   - Translation vector **T** relating world coordinates to camera coordinates
   - Rotation matrix **R** relating world coordinates to camera coordinates

As there are three and nine unknown components in **T** and **R**, respectively, together with f_c, s_x, and k_1, a total of 15 parameters must be calibrated.

Figure 20.8 shows the camera model with perspective projection and assuming lens distortion. The point O_c is the origin of one camera's local coordinate frame, O is the origin of the camera image frame, and O_w is the origin of the real-world coordinate frame. Figure 20.8 explains graphically how a 3D real-world coordinate point $P(x_w, y_w, z_w)$ is projected to a camera's two-dimensional image frame (O-X-Y), with possible camera distortion.

Apart from the geometrical transformation, we also need to consider the feature of computer sampling of the image data. In fact, the camera image frame will be stored as a two-dimensional memory frame in the computer. Apparently, we can consider the camera image frame as a computer memory frame with a horizontal image scale factor s_x. The transformation mechanism can be described as follows.

Consider an arbitrary point P in the Cartesian space and its coordinates referred to the two 3D coordinate systems. We define:

- $\mathbf{x_w} = [x_w\ y_w\ z_w]^T$ as the real-world coordinate vector of P.
- $\mathbf{x_c} = [x_c\ y_c\ z_c]^T$ as the camera local coordinate vector of P.
- $[X_u\ Y_u]^T$ as the undistorted image point P_u coordinate vector on the image frame.
- $[X_d\ Y_d]^T$ as the distorted image point P_d coordinate vector on the image frame.
- $[X_f\ X_f]^T$ (not shown in the figure) as the computer-sampled value corresponding to $[X_d\ Y_d]^T$.

The details of each transformation follow.

20.3.2.1 Transformation from Real-World Coordinates to Camera Coordinates

As every coordinate transformation can be considered as a combination of a translation **T** and a rotation **R**, we can relate the two coordinate systems by Eq. (20.24):

$$\mathbf{x} = \mathbf{R}\mathbf{x_w} + \mathbf{T} \tag{20.24}$$

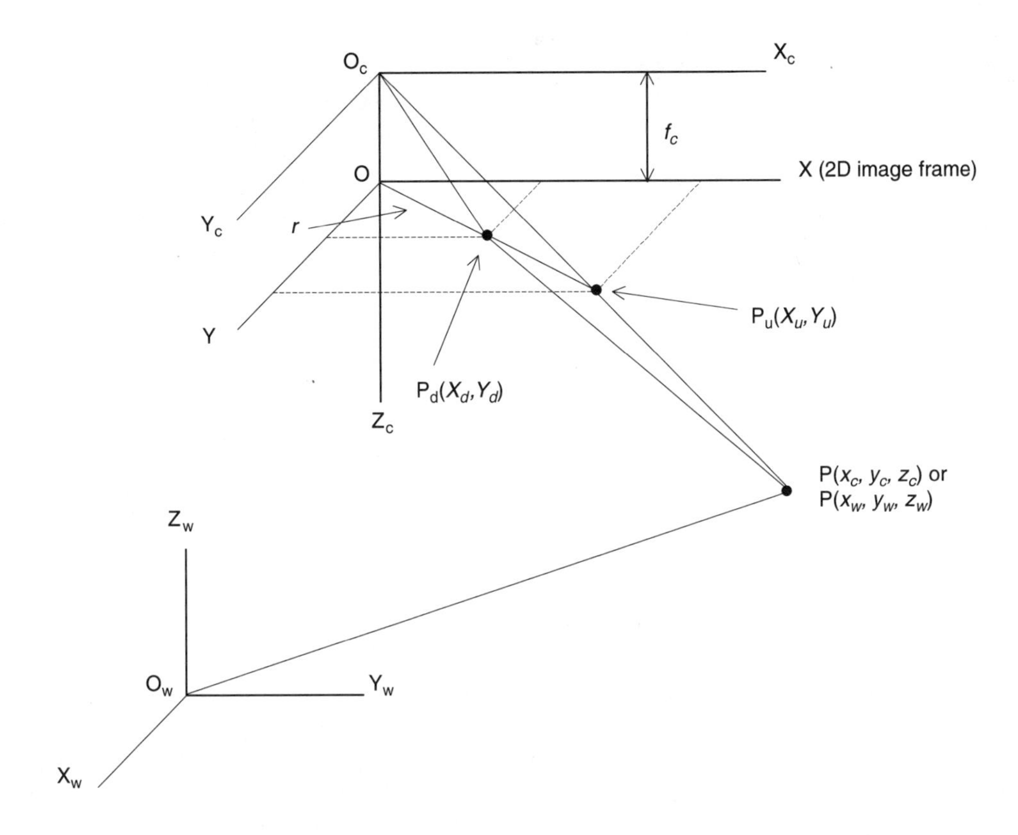

FIGURE 20.8 Camera model with perspective projection and radial lens distortion.

where **R** and **T** are defined by:

$$\mathbf{R} \equiv \begin{bmatrix} r_1 & r_2 & r_3 \\ r_4 & r_5 & r_6 \\ r_7 & r_8 & r_9 \end{bmatrix} \quad \text{and} \quad \mathbf{T} \equiv \begin{bmatrix} T_x \\ T_y \\ T_z \end{bmatrix}$$

20.3.2.2 Perspective Transformation from Camera Coordinates to Two-Dimensional Undistorted Image Coordinates

Applying the pin-hole camera principle, the perspective transformation of a point *P*, in a camera's 3D coordinate system, as shown in Figure 20.8, to the 2D image frame can be found by the similar-triangle method:

$$\begin{bmatrix} X_u \\ Y_u \end{bmatrix} = \frac{f_c}{z_c} \begin{bmatrix} x_c \\ y_c \end{bmatrix} \tag{20.25}$$

where $[x_c\ y_c]^T$ is the *x*-*y* coordinate vector of x_c, z_c is the *z* coordinate value of x_c, and f_c is the focal length of the camera, which is an intrinsic parameter to be estimated.

20.3.2.3 Transformation from Undistorted Image Coordinates to Distorted Image Coordinates

The lens is normally produced while rotating around its center. Therefore, the nonlinearity of the *x* and *y* axes of the lens is ignored. Only the radial nonlinearity is considered. To express the model nonlinearity mathematically, referring to Figure 20.8, consider the case in which there is no distortion, $[X_d\ Y_d]^T = 1 \times [X_u\ Y_u]^T$. However, if there is distortion, the difference between $[X_u\ Y_u]^T$ and $[X_d\ Y_d]^T$ will become larger with an increase in r ($\equiv \sqrt{X_d^2 + Y_d^2}$) along the radial axis. We define the radial lens distortion by Eq. (20.26):

$$\begin{cases} X_d + D_x = X_u \\ Y_d + D_y = Y_u \end{cases} \tag{20.26}$$

where D_x and D_y are the differences (or distortions) between $[X_u \; Y_u]^T$ and $[X_d \; Y_d]^T$, modeled by assuming the distortions are linearly proportional to the square of the of radius r, as expressed below:

$$\begin{cases} D_x = X_d(k_1 r^2) \\ D_y = Y_d(k_1 r^2) \end{cases} \tag{20.27}$$

where k_1 is the camera distortion factor, assumed uniform in all radial directions.

From Eqs. (20.26) and (20.27) we can rearrange to get the transformation between the undistorted and distorted image coordinates:

$$\left(1 + k_1 r^2\right)\begin{bmatrix} X_d \\ Y_d \end{bmatrix} = \begin{bmatrix} X_u \\ Y_u \end{bmatrix} \tag{20.28}$$

20.3.2.4 Transformation from Image Coordinates to Computer Image Coordinates

Based on the geometrical considerations described above, we can get the distorted camera image coordinates of P. However, when the computer samples this image point, there are a number of factors that we must consider.

First, the image coordinates on the CCD image frame are discretized and recorded by small finite sensor elements as pixels. The ratios between the image coordinates and the pixel number are, respectively, d_x and d_y, where d_x is the center-to-center distance between adjacent sensor elements in the X (scan line) direction, and d_y is the center-to-center distance between adjacent sensor elements in the Y direction. However, there is in general a difference between the number, N_{cx}, of CCD sensor elements along the X axis of the image frame and the number, N_{fx}, of computer-sampled pixels along the same axis. Therefore, X_f and X_d can be related according to Eq. (20.29). Along the Y direction, there is no difference between the numbers of CCD sensor elements and sampled data. The vertical scan line can exactly match the computer-sampling interval. Therefore, Y_f and Y_d are related by a simpler Eq. (20.30):

$$X_f = \frac{X_d N_{fx}}{d_x N_{cx}} \tag{20.29}$$

$$Y_f = \frac{Y_d}{d_y} \tag{20.30}$$

However, an additional uncertainty parameter must be introduced. This is due to a variety of factors, such as slight timing mismatch between the image acquisition hardware and camera scanning hardware, or the imprecision of the timing of the scanning itself. Even a 1% difference can cause a three- to five-pixel error for a full-resolution frame. Therefore, an unknown parameter s_x is added, as shown in Eq. (20.31). Because no computer image coordinates are stored as negative values, there is a center offset for the computer image coordinates. The center-pixel offset term $[C_x \; C_y]^T$ is added in Eq. (20.31) to keep $[X_f \; Y_f]^T$ always positive.

$$\begin{bmatrix} X_f \\ Y_f \end{bmatrix} = \begin{bmatrix} s_x / d'_x & 0 \\ 0 & d_y^{-1} \end{bmatrix} \begin{bmatrix} X_d \\ Y_d \end{bmatrix} + \begin{bmatrix} C_x \\ C_y \end{bmatrix} \tag{20.31}$$

where $d'_x \equiv d_x \dfrac{N_{cx}}{N_{fx}}$.

20.3.2.5 Summary of Transformations

The transformations are summarized by the following relation:

$$\begin{bmatrix} x_w \\ y_w \\ z_w \end{bmatrix} \xrightarrow[\text{transformation}]{\text{rigid body}} \begin{bmatrix} x \\ y \\ z \end{bmatrix} \xrightarrow[\text{transformation}]{\text{perspective}} \begin{bmatrix} X_u \\ Y_u \end{bmatrix} \xrightarrow[\text{distortion}]{\text{lens}} \begin{bmatrix} X_d \\ Y_d \end{bmatrix} \xrightarrow{\text{sampling}} \begin{bmatrix} X_f \\ Y_f \end{bmatrix}$$

By cascading Eqs. (20.24), (20.25), (20.28), and (20.31), we obtain a single 3×4 transformation matrix **M** to relate $[x_w \; y_w \; z_w]^T$ and $[X_f \; Y_f]^T$:

$$\begin{bmatrix} X_f \\ Y_f \\ 1 \end{bmatrix} = \mathbf{M} \begin{bmatrix} x_w \\ y_w \\ z_w \\ 1 \end{bmatrix} \tag{20.32}$$

where the components of **M** are obtained by simple matrix multiplication, and the result is shown by Eq. (20.33):

$$\mathbf{M} = \begin{bmatrix} \dfrac{f_c}{z_c(1+k_1r^2)}\left(\dfrac{s_x}{d'_x}\right) & 0 & C_x \\ 0 & \dfrac{f_c}{z_c(1+k_1r^2)d_y} & C_y \\ 0 & 0 & 1 \end{bmatrix} \begin{bmatrix} \mathbf{R} & \mathbf{T} \end{bmatrix} \tag{20.33}$$

As mentioned at the beginning of this section, there are 3 intrinsic parameters f, s_x, and k_1, and at most 12 parameters included in the **R** and **T** matrices to be calibrated. Therefore, it is necessary to use a minimum of 15 measurement points with accurately known coordinates (x_w, y_w, z_w). In practice, 100 points, obtained from a laboratory-built high-precision calibration block (see Figure 20.9), are used, and the least-squares method [Press et al., 1992] is applied to reduce measurement error.

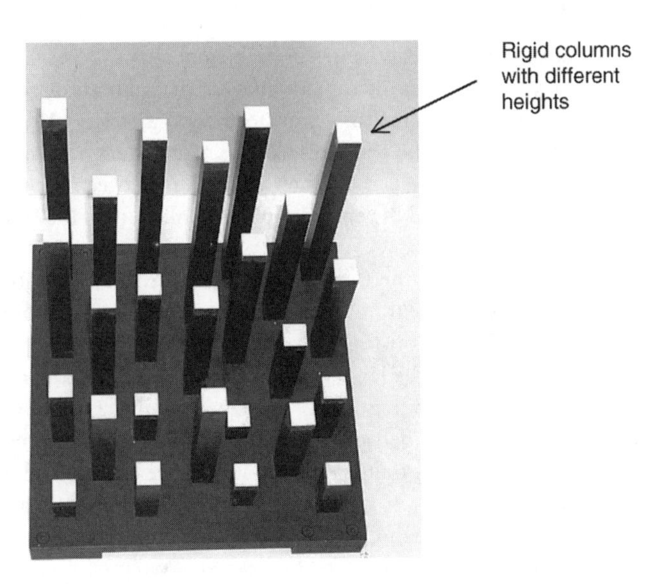

FIGURE 20.9 Calibration block.

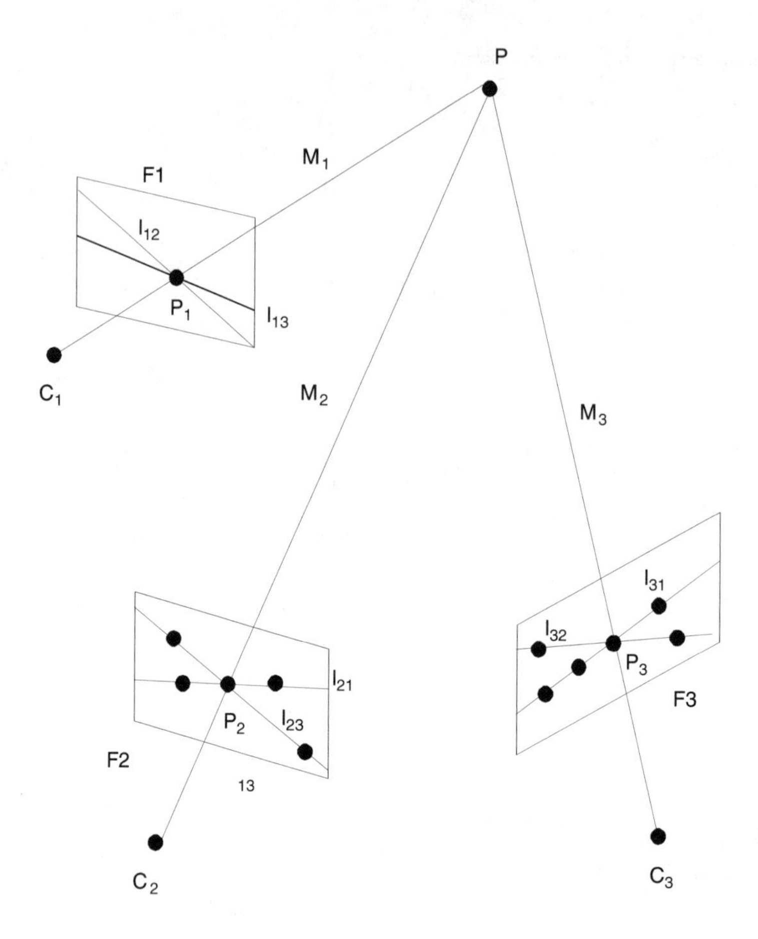

FIGURE 20.10 Geometry of trinocular stereo system.

20.3.3 Corresponding-Point Matching

Consider a trinocular stereo system. The lens center of each camera is C_i. A real point P is mapped to three corresponding points P_1, P_2, and P_3 in the camera images. The calibration processes are applied to the three cameras independently to obtain the three transformation matrices, $\mathbf{M}_1$, $\mathbf{M}_2$, and $\mathbf{M}_3$ relating the three camera coordinate systems to the real-world coordinate system according to Eq. (20.32), as shown graphically in Figure 20.10. The line l_{ij} is the intersecting line between the respective CCD image frame F_i (including the image point P_i) and the **epipolar plane** [Bolles et al., 1987] defined by the points P, P_i, and C_j.

The corresponding-point matching for the three-camera system can be described as follows. When we arbitrarily select a point P_1 on the image frame F_1 and want to find the corresponding point on frame F_2, we can construct an **epipolar line** l_{21} on frame F_2 by finding the intersection between F_2 and the epipolar plane defined by P_1, C_1, and C_2 (the lens center of the second camera). However, all the points lying on l_{21} may possibly be the corresponding point. Therefore, we make use of another epipolar line l_{31} on frame F_3 by finding the intersection between F_3 and the epipolar plane defined by P_1, C_1, and C_3. Although there are still several points lying on the epipolar line l_{31}, we can use each possible point on l_{31} to construct another epipolar line l_{23} on F_2. If l_{23} and the previous l_{21} intersect at a certain image point, say P_2, then P_2 will be the corresponding point.

In contrast to corresponding-point resolution for two-camera systems [Ayache, 1991], much faster matching results have been obtained from the trinocular approach.

20.3.4 Object Recognition and Location Determination

To facilitate object recognition and location determination edges and corners in the images with polyhedral objects are normally extracted. The corner position of such objects can be more accurately found by the intersection of two edges. Therefore, a good edge-detection algorithm is important for this purpose.

To find the edges the first step is to threshold the grabbed image to turn it into a binary image. The second step is to apply a chain-code technique [Fu et al., 1987] to locate roughly where the edge is. The third step is to transform the raw image to an intensity-gradient diagram so that there will be a clear intensity distribution across the edge. As the edge is roughly located in the second step, much more accurate edge location can be obtained from the mean of the intensity distribution curve. With the intensity-gradient information of a number of pixels along the line segment perpendicular to the edges, it is possible in fact to obtain the edge and hence the corner estimation in a least-squares error sense because more pixels are used to determine the corner.

The method can yield results more accurate than those based on a simple Sobel operator [Fu et al., 1987]. The cycle time of running the algorithm is much faster than for algorithms that process all the pixel points in an image. With chain-code searching the cycle time is proportional to the perimeter of the polyhedral object image. In this scheme the three-camera system efficiently provides the corresponding points for extracting the important required features.

The object-recognition technique described in this section is much simpler than general 3D object recognition on a 2D image frame. Because we will have the real-world coordinates of the vertices of a 3D object, recognition can be done simply by comparing the object's edge lengths with the values stored in the database. It is fast and reliable because we are using 3D machine vision to recognize 3D objects. Once the object is recognized as the target, the defined local frame information can be extracted, and the object translation and orientation can be determined by using the information of the coordinates of the object vertices.

After the target is recognized and its location determined, the spatial information of the target and the corresponding robot end-effector posture to manipulate on it will be converted from the Cartesian coordinates to the robot joint-angle values. As mentioned in the early part of this section, the transformation is described in Appendix B.

20.4 An Illustrative Example

The scenario used for illustration is that of a two-arm service robot mixing two volatile and toxic solutions to form a stable and safe solution. The two solutions are originally stored in two separate test tubes with caps. First, the human operator demonstrates the mixing task (with water) in two test tubes, which are originally kept on a fixture stand. The operator picks up one of the test tubes with the right REER and passes it to the left REER, which is also held by the operator. Then the operator starts turning the cap of the test tube with the right REER until the cap can be separated from the test tube. The operator pours the water of the uncapped test tube into a beaker lying on the table. Next, the operator replaces the cap on the test tube and then places the capped tube back to the fixture stand. A similar operation is next applied with a second test tube. The liquid contained originally in the two test tubes is eventually mixed in the beaker.

The whole process described above is recorded by the marker-based opto-measurement device. The demonstrated procedural information is processed and recognized by the system. The generated robot instruction codes can then be applied to the two-arm service robot so that it can finish the same task with volatile and toxic solutions instead inside the test tubes. First, the robot, steered and guided by the localization device, approaches the table. When the robot reaches the region for manipulation, it momentarily stops and starts executing the task with the two end-effectors. However, as there may be spatial uncertainty between the end-effectors and the test tube, the visual guidance technique is applied to finely adjust the end-effectors' position to ensure robust object picking and to transfer with sufficiently high spatial precision. The experimental results are recorded, and these are discussed in the following section.

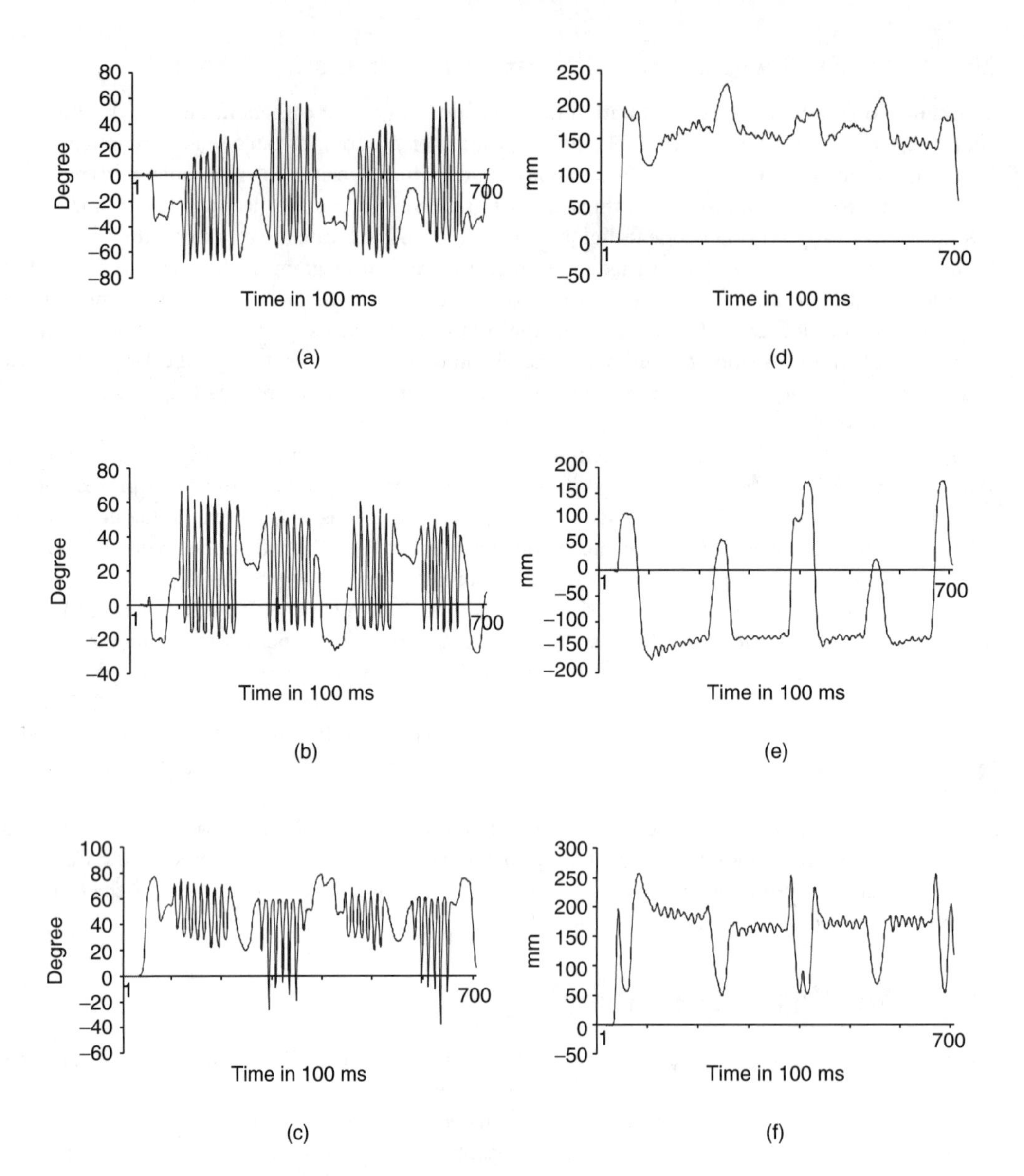

FIGURE 20.11 6D information of demonstrated trajectory of the right REER: (a) Euler parameter 1 of orientation; (b) Euler parameter 2 of orientation; (c) Euler parameter 3 of orientation; (d) x of translation; (e) y of translation; (f) z of translation.

20.4.1 Experimental Results and Discussions

The task described above is first demonstrated by a human operator and recorded by the opto skill-capturing device. The recorded information is in the form of two 6D spatial data listings. They are presented graphically for the complete task, as shown in Figures 20.11 and 20.12. The first segmentation process is applied to the two 6D spatial listings so that different master segments are extracted. The separation of different master segments is represented by a segmentation index, where index 0 implies the master segment belonging to the right hand (REER), while index 1 implies the master segment belonging to the left hand.

The segmentation index profile is shown in Figure 20.13, where the whole process is divided into right REER master segments and left REER master segments. A second segmentation process is applied to each of these master segments, and each segment displays six degrees of freedom. The velocity and

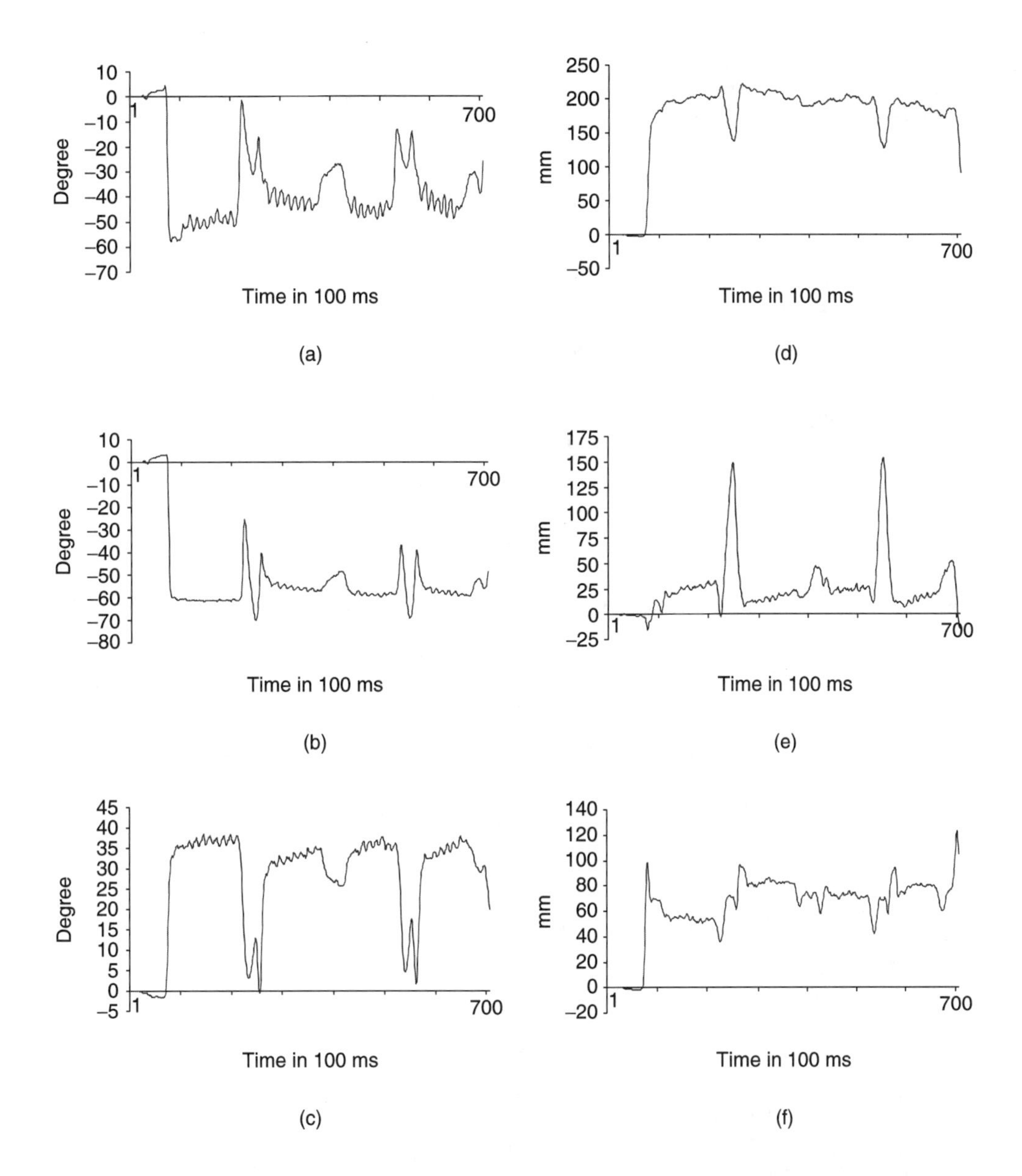

FIGURE 20.12 6D information of demonstrated trajectory of the left REER: (a) Euler parameter 1 of orientation; (b) Euler parameter 2 of orientation; (c) Euler parameter 3 of orientation; (d) x of translation; (e) y of translation; (f) z of translation.

curvature information as well as the gripper open/close information is applied to assist this second segmentation process, as mentioned in Section 20.2.2.1. For simplicity, only the segmentation result of the first master segment of the right REER master is shown in Figure 20.14. The segmentation results of the other master segments give results similar to the first one. The extraction of the corresponding translation and manipulation segments is indicated by the segmentation index shown in Figure 20.14.

The combined information of translation and manipulation segments is sent to a trajectory preprocess algorithm so that the temporal spatial information is converted to frequency spectrum variations, as shown in Figure 20.15. These frequency spectra are quantized to give a finite number of symbols for HMM use. Each grouped segment is input to the prestored HMMs in turn so that the model giving the highest-likelihood index is recognized as the prototype task for the segment. Then the corresponding command codes are extracted, and the code arguments are extracted from the corresponding

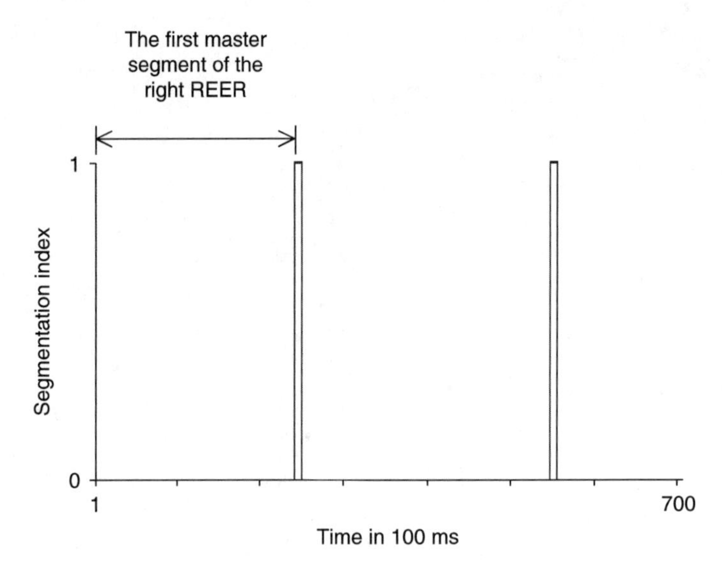

FIGURE 20.13 First-segmentation result: Index 0 implies assigning the right REER as master; index 1 implies assigning the left REER as master.

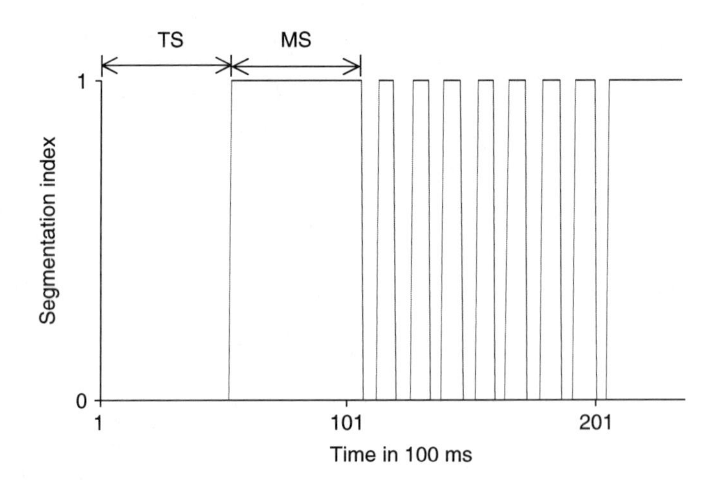

FIGURE 20.14 Second-segmentation result on the first master segment of the right REER: Index 0 implies a translation segment (TS); index 1 implies a manipulation segment (MS).

segment endpoints, which are the results obtained from the second segmentation process. The case shown in Figures 20.14 and 20.15 is recognized as a task consisting of MOVE, PATH, and SCREW command codes, and the completed program codes corresponding to just this segment are shown in Table 20.6.

Following the programming phase the prepared instruction codes are downloaded to the robot controller for execution. During the execution phase, the robot is first driven by the localization device and steered to approach the target, which in this case is the test tube. The error function of the position difference between the service robot and the target is shown in Figure 20.16. Once the robot reaches

TABLE 20.6 Robot Program Codes Generated by the Programming Phase of the System on the First Master Segment of the Right REER and the First Master Segment of the Left REER

Steps	Commands
1	RH-MOVE([−32.45, 64.79, −20.25, 192.68, 106.88, 90.56])
2	RH-GRASP
3	RH-INSERT([−32.45, 64.79, −20.25, 192.68, 106.88, 90.56], −100)
4	RH-MOVE([−3.28, 55.93, 17.82, 113.19, −169.04, 213.96])
5	LH-MOVE([−56.62, −60.72, 35.44, 183.21, 9.29, 68.20])
6	RH-PATH([−1.67, 38.01, −3.95, 135.15, −152.27, 195.11], [2.64, 36.33, −9.19, 139.16, −149.96, 193.84], [2.63, 37.46, −10.13, 141.45, −149.484, 193.238525], [0.42, 38.92, −9.11, 142.37, −149.97, 192.55])
7	LH-PATH([−52.31, −61.12, 35.44, 191.02, 13.80, 60.39], [−52.34, −61.17, 35.54, 192.04, 17.02, 59.244438], [−52.38, −61.19, 35.56, 193.36, 20.15, 57.92], [−51.92, −61.17, 35.48, 194.67, 22.13, 56.90])
8	RH-SCREW([0.42, 38.92, −9.11, 142.37, −149.97, 192.55], 8, 40, 2)
9	RH-MOVE([−26.61, 33.30, 23.73, 202.15, 8.03, 116.01])
10	LH-PATH([−26.61, 33.30, 23.73, 202.15, 8.03, 116.01], [−16.62, −45.66, 3.05, 181.53, 52.04, 52.63], [−18.49, −48.53, 3.15, 176.73, 59.92, 56.31], [−19.90, −50.99, 3.24, 172.28, 68.18, 60.16], [−21.05, −53.21, 3.39, 168.34, 76.86, 63.92], [−22.24, −55.27, 3.99, 164.51, 85.43, 66.99], [−23.45, −57.23, 4.90, 160.66, 93.74, 69.13], [−24.53, −59.23, 5.99, 157.19, 101.75, 70.46], [−25.52, −61.20, 6.85, 154.24, 109.74, 71.03], [−27.43, −64.66, 8.54, 147.71, 123.79, 71.85])
11	LH-MOVE ([−26.52, −61.82, 4.46, 150.87, 128.31, 71.10])
12	RH-MOVE([0.59, 22.13, 21.81, 225.20, 52.30, 59.60])

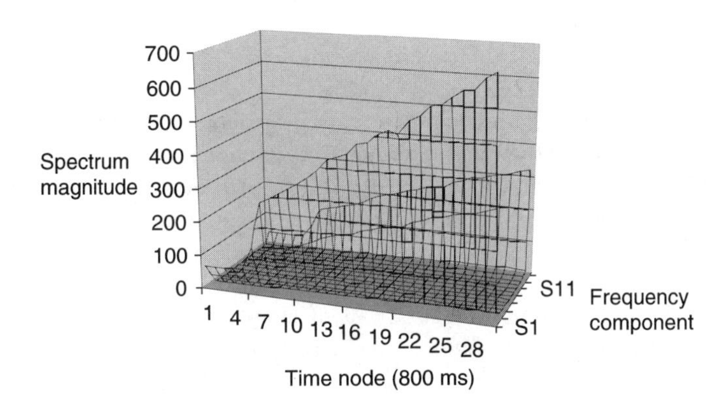

FIGURE 20.15 Frequency spectrum information of the first master segment of the right REER based on short-time Fourier transform.

the target within a reachable distance of the grippers, the service robot base stops, and the visual guidance system starts to function. As the camera calibration process is carried out beforehand, the coordinate transformation matrices between the robot and image frames of the three cameras are known. Once the target appears within the field of vision of the three cameras, a typical instance of which is shown in Figure 20.17, the center positions of the images of the objects to be fetched are calculated, and the 3D position of the target with respect to service robot can be determined by simple triangulation method. The data measured by the vision system are considered to be an offset to the

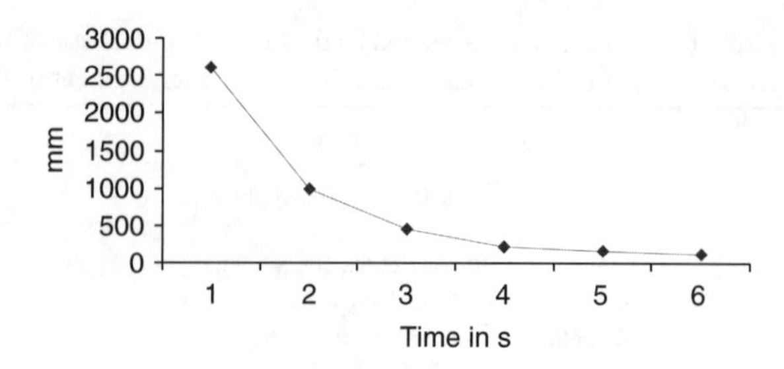

FIGURE 20.16 Error function of the position difference between the service robot and the target with steering aided by the localization device.

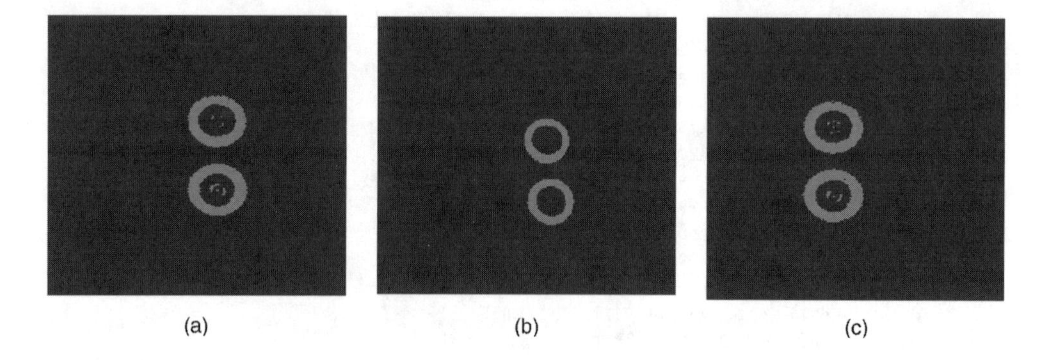

FIGURE 20.17 Test-tube images of the three different cameras: (a) camera 1; (b) camera 2; (c) camera 3.

robot instruction code to be executed. Then, the robot can perform the task precisely with the aid of visual guidance.

The camera calibration process has to be completed before robot execution. Each camera will have its own calibration. For simplicity, we show only the camera 1 calibration process. The image of the calibration block on one camera frame is shown in Figure 20.18(a). Images after edge detection and corner extraction are shown in Figure 20.18(b) and (c), respectively. The corner coordinates are recorded as $[X_f\ Y_f\ 1]^T$. On the other hand, the physical corner *x-y-z* coordinates of the calibration block can be precisely measured by a microaccuracy measurement device. As there are altogether 100 corner coordinates, 100 equations are generated by using Eq. (20.32). As there are only 15 unknown parameters in **M** in Eq. (20.33), the least-squares approach is applied to obtain the best-fit transformation matrix **M** to relate the image coordinates and the real-world coordinates.

20.5 Conclusion

An opto skill-capturing and visual guidance scheme for service robots is introduced. It is an example of an opto-mechatronic application to robot systems. By using the opto skill-capturing device, the human demonstration can be recorded at a frequency as high as 20 Hz. The segmentation procedures can divide the complicated human demonstration into a number of simpler subsegments. There are two functions of the subsegment extraction. The first function is to obtain the subsegment endpoints needed for the program-code argument information. The second function is to break down the complicated human demonstration into a number of simpler subtasks so that they can be compared to the finite number of known prototype subtasks. Once the subtask is recognized as a certain prototype kind, the corresponding

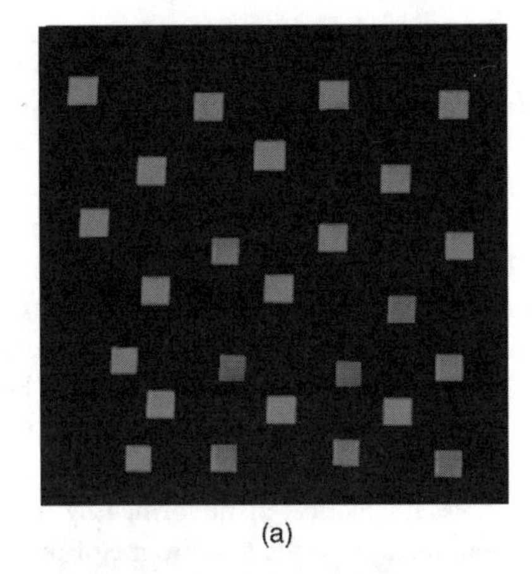

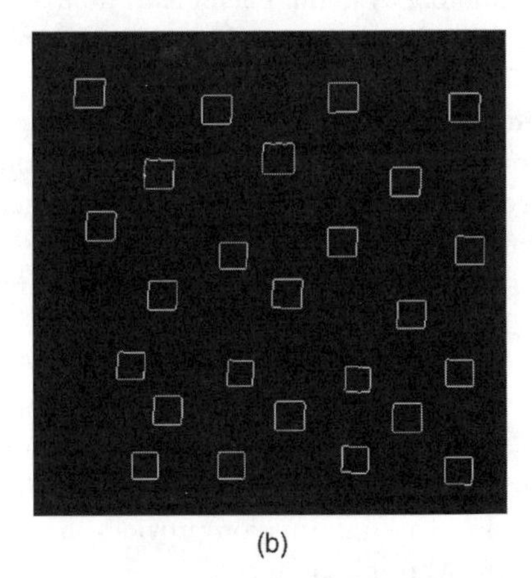

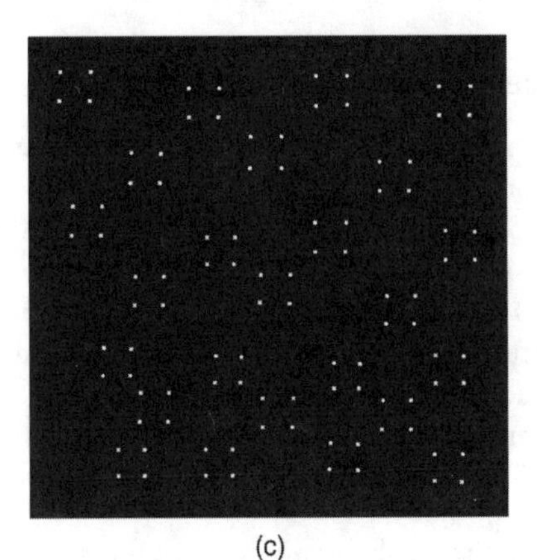

FIGURE 20.18 Image processing for camera calibration: (a) original image of calibration block; (b) edge image of calibration block; (c) corners determined by intersection of extended-edge straight lines.

program instruction codes can be extracted as a component of the complete program. Visual guidance is a means to enhance the system robustness for executing a service task. With its introduction the system can effectively work despite the inevitable uncertainty of the target position in most practical cases. However, a limitation of this visual-guidance approach is that it cannot cater to very fast moving targets because computer vision algorithms cannot update the target position at a rate higher than allowed by physical constraints imposed by the processor.

Defining Terms

assembly plan from observation (APO): A type of robot-programming and task-planning method based on human demonstration. In the first step the human provides the intelligence in creating the desired hand trajectory physically (corresponding to the robot end-effector motion, the grasping strategy, and hand-target interaction). In the second step the robot repeats the task based on the previously captured information. The sensor required for this system is usually machine vision in the main.

epipolar line: Mathematical line formed by the intersection of an epipolar plane and the plane of the CCD frame of one camera (e.g., A). The epipolar plane is formed by three points—the two lens centers of two separate cameras (A and B) and an image point on the CCD frame of camera B.

epipolar plane: Mathematical plane formed by the two fixed camera-lens centers and an image point under examination on the CCD frame of one camera.

robot programming by human demonstration: A modern approach to robot programming. The robot program is generated by information directly demonstrated by the human operator or user. The research issues of this approach involve the utilization and integration of concepts and ideas from the fields of artificial intelligence, computer vision, and signal processing.

3D metric paradigm: A traditional computer-vision approach to guiding the robot end-effector to reach a target object by accurate visual measurement. Camera calibration is required. The raw image information in the camera frame is processed and mapped to 3D Cartesian space. This is the common space where the error information of robot control and the location information of the target are referenced, hence its name.

visual servoing paradigm: Another approach to visually guiding the robot end effector to reach a target object by analyzing directly image data received from the camera. It is the method used in the active-vision approach where the camera is mounted on the moving end effector. As the camera is not fixed, the image of the target object will change with respect to the camera movement. The change in image information provides the relative target location information for the robot to determine its own movement.

References

Ayache, N., *Artificial Vision for Mobile Robots: Stereo Vision and Multisensory Perception*, MIT Press, Cambridge, MA, 1991.

Baum, L. E., Petrie, T., Soules, G., and Weiss, N., A maximization technique occurring in the statistical analysis of probabilistic functions of markov chains, *Ann. Math. Stat.*, 41(1), 164–171, 1970.

Bolles, R., Baker, H., and Marimont, D., Epipolar-plane image analysis: an approach to determining structure from motion, *Int. J. Computer Vision*, 1(1), 7–55, 1987.

Carmo, M. P., *Differential Geometry of Curves and Surfaces*, Prentice-Hall, Englewood Cliffs, NJ, 1976.

Delson, N. and West, H., Robot programming by human demonstration: the use of human variation in identifying obstacle free trajectories, *Proc. IEEE Int. Conf. on Robotics and Automation*, San Diego, 1994, pp. 564–571.

Denavit, J. and Hartenberg, R. S., A kinematic notation for lower-pair mechanisms based on matrices, *J. Appl. Mech.*, 77, 215–221, 1955.

Enrico, G., Giorgio, M., Andrea, O., and Giulio, S., Robust visual servoing in three-dimensional reaching tasks, *IEEE Trans. Robotics Automation*, 12(5), 732–742, 1996.

Espiau, B., Chaumette, F., Rives, P., and Espiau, B., Positioning of a robot with respect to an object, tracking it and estimating its velocity by visual servoing, *Proc. IEEE Int. Conf. Robotics Automation*, Sacramento, 1991, pp. 2248–2253.

Fu, K. S., Gonzalez, R. C., and Lee, C. S. G., *Robotics: Control, Sensing, Vision, and Intelligence*, McGraw-Hill, New York, 1987.

Gilbert, S. and Truong, N., *Wavelets and Filter Banks*, Wellesley-Cambridge Press, Boston, 1996.

Hlawatsch, F. and Boundeaux-Bartels, G. F., Linear and quadratic time-frequency signal representations, *IEEE SP Magazine*, 9(2), 21–67, 1992.

Huang, X. D., Ariki, Y., and Jack, M. A., *Hidden Markov Models for Speech Recognition*, Edinburgh University Press, Edinburgh, 1990.

Ikeuchi, K. and Suehiro, T., Towards an assembly plan from observation, Part I: task recognition with polyhedral objects, *IEEE Trans. Robotics Automation*, 10(3), 368–385, 1994.

Kang, S. B. and Ikeuchi, K., Determination of motion breakpoints in a task sequence from human motion," *Proc. IEEE Int. Conf. Robotics Automation*, San Diego, 1994, pp. 551–556.

Krist, J., Melluish, M., Kehl, L., and Crouch, D., Technical description of the Optotrak three-dimensional motion measurement system, *Proc. World Congress Biomech.*, La Jolla, CA, August 1990, pp. 23–39.

Lee, C. S. G. and Ziegler, M., A geometric approach in solving the inverse kinematics of puma robots, *IEEE Trans. Aerospace Electronic Sys.*, 20(6), 695–706, 1984.

Linde, Y., Buzo, A., and Gray, R. M., An algorithm for vector quantizer design, *IEEE Trans. Comm.*, COM-28, 84–95, 1980.

Longuet-Higgins, H. C., A computer algorithm for reconstructing a scene from two projections, *Nature*, 293, 133–135, 1981.

Papanikolopoulos, N. P. and Khosla, P. K., Adaptive robotic visual tracking: theory and experiments, *IEEE Trans. Automatic Control*, 38(3), 429–445, 1993.

Press, W. H., Teukolsky, S. A., Vetterling, W. T., and Flannery, B. P., *Numerical Recipes in C*, 2nd ed., Cambridge University Press, Boston, 1992.

Rabiner, L. R., A tutorial on hidden Markov models and selected applications in speech recognition, *Proc. IEEE*, 77(2), 257–286, 1989.

Tsai, R., A versatile camera calibration technique for high-accuracy three-dimensional machine vision metrology using off-the-shelf TV cameras and lenses, *IEEE Trans. Robotics Automation*, 3(4), 323–344, 1987.

Tso, S. K. and Liu, K. P., Visual programming for capturing of human manipulation skill, *Proc. IEEE/RSJ Int. Conf. Intelligent Robots Sys.*, Yokohama, Japan, 1993, pp. 42–48.

Tso, S. K. and Liu, K. P., Demonstrated trajectory selection by hidden Markov model, *Proc. IEEE Int. Conf. Robotics Automation*, Albuquerque, NM, 1997, pp. 2713–2718.

Tso, S. K. and Liu, K. P., General representation of human demonstration using differential-geometry properties, *Proc. IEEE/RSJ Int. Conf. Intelligent Robots Sys.*, Victoria, Canada, 1998, pp. 1950–1955.

Xu, Z., Polyhedron recognition based on the models in the robot trinocular vision system," *Proc. 2nd Int. Conf. Automation, Robotics, Computer Vision*, Singapore, Vol. 1, 1992, pp. 1–5.

Yang, J., Hidden Markov Model for Human Performance Modeling, Ph.D. dissertation, University of Akron, Akron, OH, 1994.

Yang, J., Xu, Y., and Chen, C. S., Human action learning via hidden Markov model, *IEEE Trans. Systems, Man, Cybernetics*, 27(1), 34–44, 1997.

Zebra-ZERO, *Zebra-ZERO User's Manual*, Integrated Motion, Berkeley, CA, 1992.

For Further Information

Proceedings of the IEEE International Conference on Robotics and Automation are published annually. This conference is among the most important ones in this field. Another important conference is *IEEE/RSJ International Conference on Intelligent Robots and Systems*. Another relevant conference is *International Conference on Field and Service Robots*. The journal *IEEE Transactions on Robotics and Automation* reports advances in robot programming and practical computer vision. Contact information for subscription and reprint: IEEE Operations Center, 445 Hoes Lane, P.O. Box 1331, Piscataway, NJ 08855–1331; telephone (NJ): +1 (732) 981-0060.

Appendix A: Fundamental Concept of Discrete Hidden Markov Model

A hidden Markov model consists of three components. They are a set of states $\{S\}$, a state transition probability matrix $A=\{a_{ij}\}$, and an output probability matrix $B=\{b_j(O_k)\}$, where O_k stands for an observation symbol. Understanding how a hidden Markov model represents a physical system is made easier by looking at a bag-and-ball experiment. Consider a simple three-bag four-ball model in the following diagram:

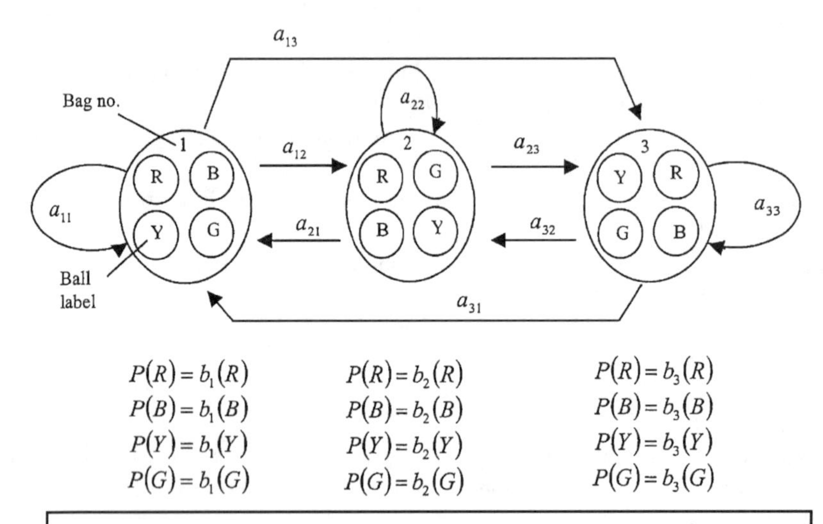

The elements of the model above are unknown to the observer. During the experiment, the bag being chosen is also hidden to the observer. That is why it is called a 'hidden' Markov model. The only observation is the actual selected ball.

Observation output sequence: (R) (B) (G) (Y) (R) (G) • • •

Applying a hidden Markov model to the above system, the bag number is the state number, and the ball label is the observation symbol. When the experiment starts, an initial bag is chosen, e.g., a bag with the number j = 1, 2, or 3, and a ball is selected from this bag based on the probability $b_j(O_k)$. After the ball label is recorded, the ball is put back into the bag. Then another bag is selected based on the last bag's number, according to the probability a_{ij}. Then another ball is picked. The process continues until the number of observation points is enough to investigate the unknown model. The way to estimate the model is described in Section 20.2.2.2.

Appendix B: Inverse Kinematics of Robot Arm

According to the Denavit–Hartenberg representation [Denavit and Hartenberg, 1955], the parameters of the robot arm used [Zebra-ZERO, 1992] are summarized in the following table:

Joint i	θ_i (degree)	a_i (degree)	a_i (mm)	d_i (mm)	Joint Range (degree)
1	90	−90	0	0	−180 to +180
2	0	0	279.4	0	−135 to +45
3	90	90	0	0	+90 to +225
4	0	−90	0	228.6	0 to +360
5	0	90	0	0	−100 to +100
6	0	0	0	156	−180 to +180

Note: θ_i is the joint angle from the x_{i-1} axis to the x_i axis about the z_{i-1} axis (using the right-hand rule); d_i is the distance from the origin of the $(i-1)$th coordinate frame to the intersection of the z_{i-1} axis with the x_i axis along the z_{i-1} axis; a_i is the offset distance from the intersection of the z_{i-1} axis with the x_i axis to the origin of the ith frame along the x_i axis (or the shortest distance between the z_{i-1} and z_i axes); and α_i is the offset angle from the z_{i-1} axis to the z_i axis around the x_i axis (using the right-hand rule).

Let the translation of the wrist point (hand-held tool) with respect to the $x_0 - y_0 - z_0$ frame be given by $[T_x\ T_y\ T_z]$ and the Euler angles of rotation be yaw ψ, pitch θ, and roll ϕ. The joint angles are Joint1, Joint2, Joint3, Joint4, Joint5, and Joint6. Using the geometric approach for solution [Fu et al., 1987]:

A. Joint1 solution

$$\text{Joint1} = \tan^{-1}\left(\frac{T_y}{T_x}\right)$$

B. Joint2 and Joint3 solutions

$$\text{Joint2} = \text{Ang1} - \cos^{-1}\left(\frac{a_2^2 + (T_u^2 + T_z^2) - d_4^2}{2a_2\sqrt{T_u^2 + T_z^2}}\right)$$

where

$$T_u = T_x \cos(\text{Joint1}) + T_y \cos(\text{Joint1})$$

and

$$\text{Ang1} = \tan^{-1}\left(\frac{T_z}{T_u}\right)$$

$$\text{Joint3} = 270° - \cos^{-1}\left(\frac{a_2^2 + d_4^2 - (T_u^2 + T_z^2)}{2a_2 d_4}\right)$$

C. Joint4 solution—As the Euler angles yaw ψ, pitch θ, and roll ϕ are determined by the visual system, the elements of the rotational matrix of the robot hand can be found. They are represented by the three vectors **a**, **s**, and **n**. They are the base vectors of the frame attached to the robot hand

[Lee and Ziegler, 1984], expressed as follows in terms of the Euler angles:

$$\mathbf{a} = \begin{bmatrix} \cos\phi \cdot \sin\theta \cdot \cos\psi + \sin\phi \cdot \sin\psi \\ \sin\phi \cdot \sin\theta \cdot \cos\psi - \cos\phi \sin\psi \\ \cos\theta \cdot \cos\psi \end{bmatrix}$$

$$\mathbf{s} = \begin{bmatrix} \cos\phi \cdot \sin\theta \cdot \sin\psi - \sin\phi \cdot \cos\psi \\ \sin\phi \cdot \sin\theta \cdot \sin\psi + \cos\phi \cdot \cos\psi \\ \cos\theta \cdot \sin\psi \end{bmatrix}$$

$$\mathbf{n} = \begin{bmatrix} \cos\phi \cdot \cos\theta \\ \sin\phi \cdot \cos\theta \\ -\sin\theta \end{bmatrix}$$

The vector useful for finding Joint4 is

$$\mathbf{z}_3 = \begin{bmatrix} \mathrm{C1 \cdot S23} & \mathrm{S1 \cdot S23} & \mathrm{C23} \end{bmatrix}^T$$

where

$$\mathrm{C1} = \cos(\mathrm{Joint1})$$

and

$$\mathrm{S23} = \sin(\mathrm{Joint2} + \mathrm{Joint3})$$

$$\mathrm{Joint4} = \tan^{-1}\left(\frac{M(\mathrm{C1}\mathbf{a}_y - \mathrm{S1}\mathbf{a}_x)}{M(\mathrm{C1C23}\mathbf{a}_x + \mathrm{S1C23}\mathbf{a}_y - \mathrm{S23}\mathbf{a}_z)} \right)$$

When the wrist orientation is restricted to point downward:

M = the sign of $\mathbf{s} \bullet \mathbf{y}_5$ or M = the sign of $\mathbf{n} \bullet \mathbf{y}_5$ if $\mathbf{s} \bullet \mathbf{y}_5 = 0$, and

$$\mathbf{y}_5 = \mathbf{z}_3 \times \mathbf{a}$$

When the wrist orientation is restricted to point upward, the sign of M is reversed. Therefore, there are two possible solutions to Joint4.

D. Joint5 solution

$$\mathrm{Joint5} = \tan^{-1}\left(\frac{\sin(\mathrm{Joint5})}{\cos(\mathrm{Joint5})} \right)$$

where

$$\sin(\mathrm{Joint5}) = \mathbf{a} \bullet \mathbf{x}_4$$

$$\cos(\mathrm{Joint5}) = -\mathbf{a} \bullet \mathbf{y}_4$$

$$\mathbf{x}_4 = \begin{bmatrix} \mathrm{C1 \cdot C23 \cdot C4 - S1 \cdot S4} \\ \mathrm{S1 \cdot C23 \cdot C4 + C1 \cdot S4} \\ \mathrm{-S23 \cdot C4} \end{bmatrix}$$

E. Joint6 solution

$$\mathrm{Joint6} = \tan^{-1}\left(\frac{\sin(\mathrm{Joint6})}{\cos(\mathrm{Joint6})} \right)$$

where

$$\sin(\text{Joint6}) = \mathbf{a} \bullet \mathbf{y}_5$$

$$\cos(\text{Joint6}) = \mathbf{n} \bullet \mathbf{x}_5$$

$$\mathbf{y}_4 = \begin{bmatrix} -C1 \cdot C23 \cdot S4 - S1 \cdot C4 \\ -S1 \cdot C23 \cdot S4 + C1 \cdot C4 \\ S4 \cdot S23 \end{bmatrix}$$

21

Optical Pick-Up Devices for Disk Storage Systems

Osamu Matsuda
Sony Corporation
Tokyo, Japan

Noriaki Nishi
Sony Corporation
Kanagawa, Japan

Takeshi Mizuno
Sony Corporation
Tokyo, Japan

21.1 Introduction

The first mass-produced optical disk storage system, the **CD** or compact disk, was introduced in 1982 as a new music-reproduction system. Since then, optical storage systems have been widely and vigorously applied not only as music-reproduction systems such as the **CD** and **MD**, but also as computer-data storage systems such as the **CD-ROM**, **MO**, and **DVD-ROM**.

Recent research and development on optical pick-up systems can be classified into two main categories. One is investigation into high-density recording [Kishima et al., 2001] using near-field optics and short-wavelength radiation sources. The other investigation is into optical integration technology [Suhara, 1996; Ukita et al., 1991] to provide the disk system with high performance and allow it to be mass-produced. In this chapter we focus on the latter technology, focusing particularly on optical pick-up devices and optical head devices for recordable disk systems. We will show that the integrated optical pick-up and head devices are good examples of opto-mechatronic devices.

In Section 21.2 we consider the history of the development of optical pick-up devices, their function and fabrication process, and the variety of hybrid optical integrated devices now available. In Section 21.3 we focus on integrated optical pick-up devices for the portable and recordable audio disk system called the *MD* or *MiniDisc* system. In Section 21.4 we report our new monolithic optical pick-up that takes advantage of a new device technology.

0-8493-1162-4/03/$0.00+$1.50

21.2 Optical Integrated Pick-Up Devices: An Overview

21.2.1 Developmental History and Future Prospects

Figure 21.1 shows the developmental history of optical pick-up devices from 1976 and their prospects to 2008. The early optical pick-up consisted of individual optical and electronic components, i.e., two sets of glass lenses, a bulk glass prism, a packaged photodiode, a packaged laser diode, and other mechanical parts such as a two-axis magnetic actuator device. This arrangement is called *discrete optics*, as shown in Figure 21.1. This device was one of the first mass-produced opto-mechatronic devices combining precise optical parts with fast mechanical actuator devices. The precisely aligned packaged photodiode detects servo error signals and digital signal information. The assembly of the optical pick-up using discrete optics is continuously being refined, as shown in Figure 21.2, in which the pick-up has an oblique glass plate instead of a bulk glass prism.

The digital signal of the CD is obtained by detecting the beam reflected from the compact disc's pits as the sum of segmented photodiode signals. The two types of servo error signals are obtained as the difference between the segmented photodiode signals. One is the focus error signal for obtaining the pit signal by the diffraction-limited small light spot, and the other is the tracking error signal for precise tracking of the spot onto the pit array of the disk. Controlling the position of the objective lens operates the servo as a pick-up. This theory of operation is essentially the same as that used by the four types of optical pick-up devices shown in Figure 21.1 and will be explained in detail as applied to the CD in Section 21.2.2.

Two different device-integration techniques were introduced around 1990, both developed in Japan. We call the devices produced using these techniques hybrid optical integrated devices, or simply *hybrid*

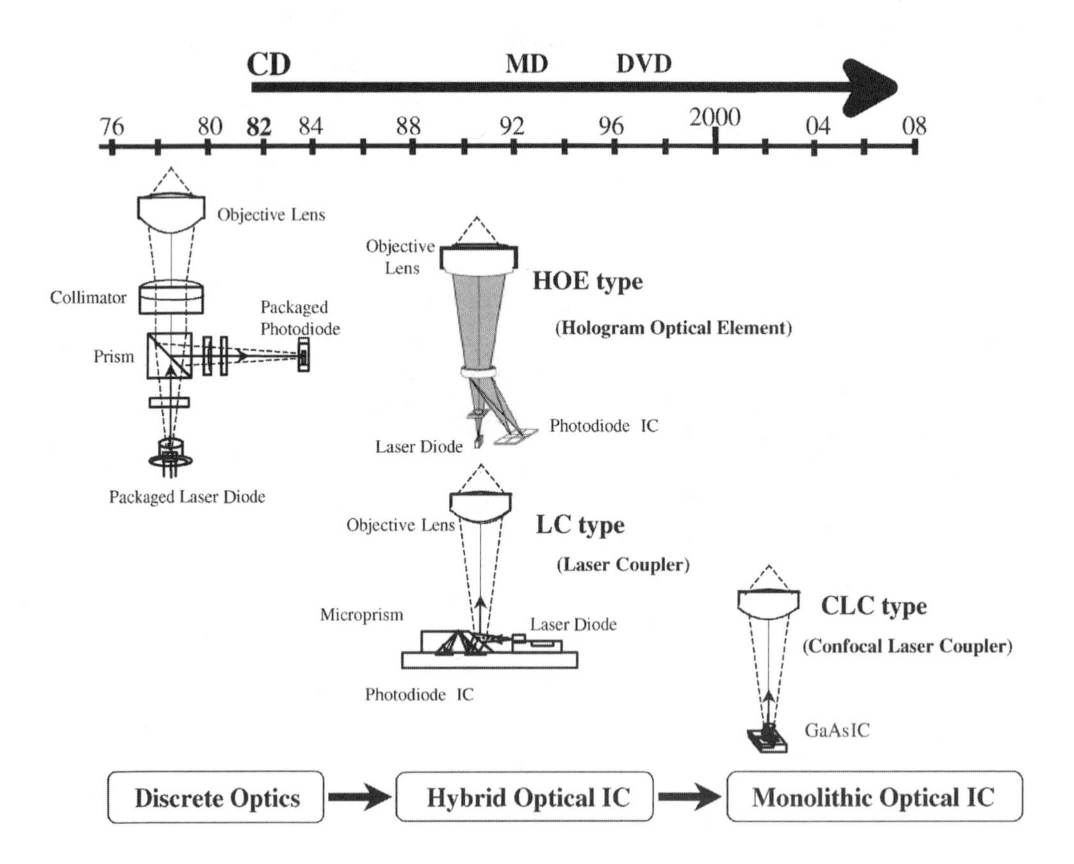

FIGURE 21.1 Innovation trend of optical pick-up devices.

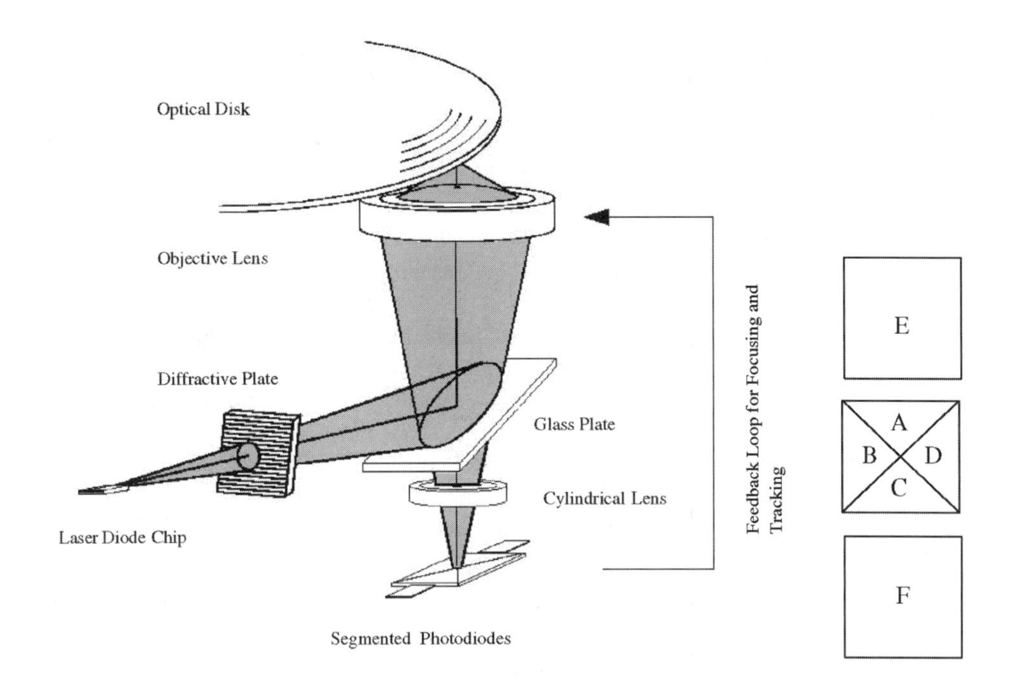

FIGURE 21.2 Schematic representation of discrete optics and PD pattern.

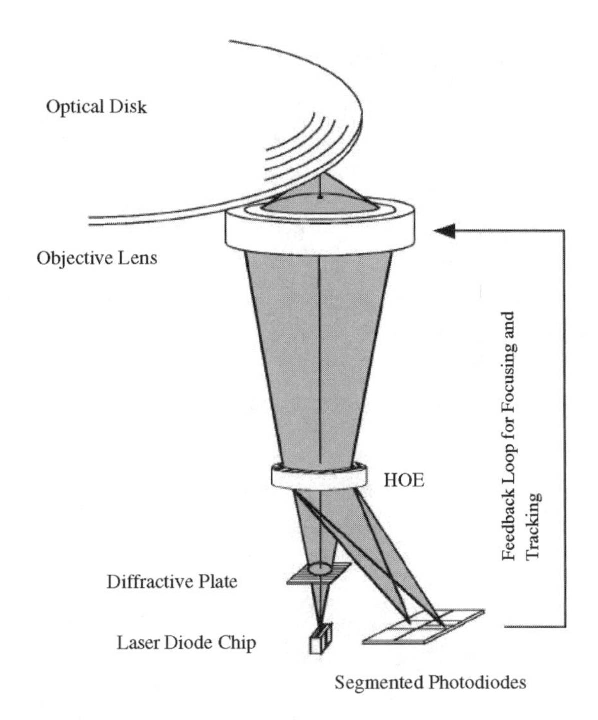

FIGURE 21.3 Schematic representation of HOE-type optics.

optical ICs, as shown in Figure 21.1. One hybrid optical IC was realized [Yoshida et al., 2000; Yoshikawa, 1995] using a *holographic optical element* (**HOE**) as a beam splitter. The HOE-type device is schematically illustrated in Figure 21.3. The other device, called a *laser coupler* (**LC**), was realized [Matsumoto et al., 1996] by application of a microprism as a beam splitter. Both the HOE and LC devices are now being mass-produced. A new integrated monolithic optical pick-up device called the *confocal laser coupler* (**CLC**) has been developed [Doi, 1997], as shown in Figure 21.1. We will describe the LC and CLC devices in detail in the following sections.

The HOE-type device was developed for CD players in 1990, and for MO [Yoshida et al., 1998] and DVD players subsequently. Figure 21.1 shows the evolution of the optical disk system and the history of the development of integrated optical pick-up devices of the microprism type. The optical head for the MD system was developed [Nishi, 1998] and is described in Section 21.3. The device for DVDs was refined and is now being used in a video-game player [Nemoto and Miura, 2001] that can play both DVD-ROM and CD-ROM.

For monolithic devices the CD pick-up takes two different configurations—the **3 CLC** type [Narui et al., 1998] and the **CKE** types [Sabert et al., 1998], both of which will be described in detail in Section 21.4. The work to obtain magneto-optical detection [Tamada et al., 1997] was carried out simultaneously with the work on monolithic optical ICs.

To integrate optical devices much research [Hong, 1992] has been done to combine an optical discrete device with electronic circuits in the communication area. The goal is to provide a fast response and low noise characteristics of the integrated device. In the case of an optical pick-up, our goals were:

Small, thin, and lightweight pick-up mainly for portable use
Ultra-low cost in fabrication
Reliability over long operation times
Stability in servo operation (so-called *playability*)

The situation is somewhat different from that of electrical ICs and communication optical ICs.

21.2.2 Theory of Optical Pick-Up Operation

For CD players that use discrete optics, the optical paths are as shown in Figure 21.2. Light from the laser diode is split by the diffractive plate into three beams to obtain the tracking error signal. The three beams are one main 0th-order beam and two ±1st-order diffracted beams. After passing through the beam splitter of the oblique glass plate, three beams are focused on the disk by the objective lens. The reflected light from the disk is returned back to the beam splitter. Passing through the oblique glass plate and objective lens, these beams increase in astigmatic aberrations and are detected by the segmented photodiodes as the well-known astigmatic servo error signal. The pattern of the photodiodes is also shown in Figure 21.2. The 0th-order diffracted main beam is detected by the four photodiodes of $PD_{A\sim D}$. The focusing error signal and the RF signal (digital signal) are derived from the photocurrent of these photodiodes $I_{A\sim D}$ and described as follows.

$$\text{Focusing error signal} = (I_A + I_C) - (I_B + I_D)$$

$$\text{RF signal} = I_A + I_B + I_C + I_D$$

The ±1st-order diffracted subbeams are detected by the photodiodes PD_E and PD_F and the tracking error signal is described by the two photocurrents as follows.

$$\text{Tracking error signal} = I_E - I_F$$

In conventional HOE-type optics the HOE works as a beam splitter, and the optical geometry is similar to that of the conventional discrete optics shown in Figure 21.2.

With laser couplers [Matsumoto et al., 1996] the optical paths are shown in Figure 21.4. The light from the laser diode is reflected at the slope of the microprism and focused on the disk. The reflected beam returns to the microprism and passes through the slope into the microprism. The slope of the microprism works as a nonpolarized beam splitter or half-mirror. This single spot beam is detected at the first set of photodiodes located on the bottom of the microprism. The beam is nearly focused at the top surface of the prism and reflected. Then the beam is launched at the second set of photodiodes. The focusing servo signal is obtained by comparing the two sets of photodiode signals, a method known as the spot size detection method. A new "push–pull" method has been introduced [Matsumoto et al., 1996] as a single spot detection method.

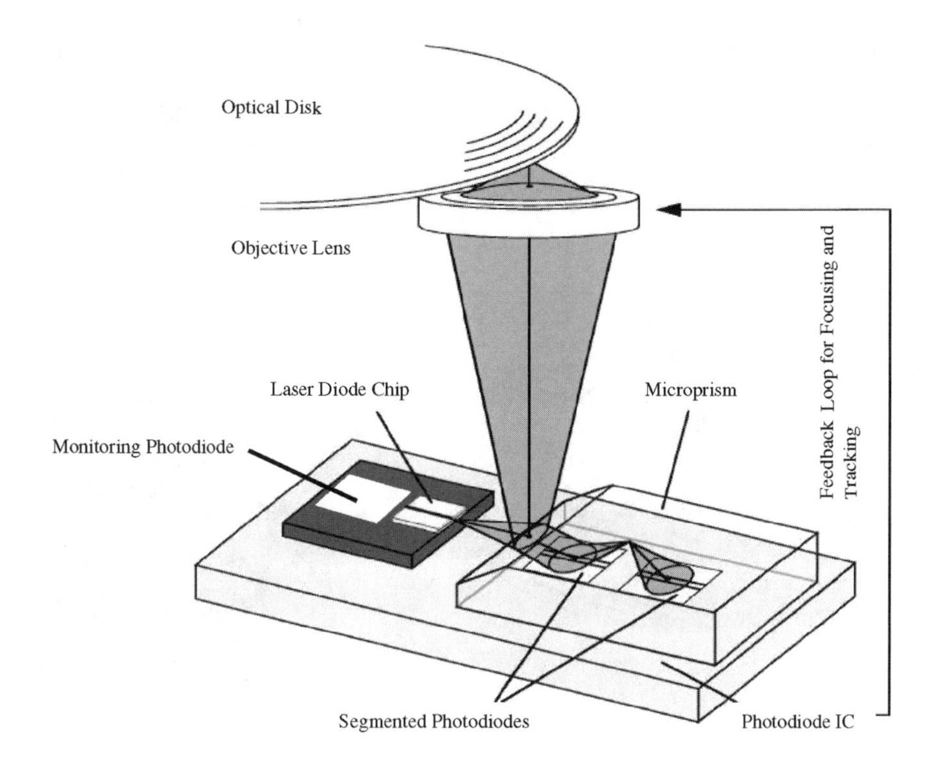

FIGURE 21.4 Schematic representation of laser coupler optics.

Using two servo error signals we can obtain a good opto-mechatronic servo system by controlling the position of an objective lens with a two-axis-type galvanomagnetic actuator device. Elastic metal plates or plastic hinges are suspended and guided to drive the objective lens in two directions. One direction is perpendicular to the disk plane (corresponding to the focusing servo). The other is parallel to the radial direction of the disk (corresponding to the tracking servo). When it comes to CD systems the focusing servo must suppress the vertical disk deviation of ±0.2 mm into less than ±0.2 μm, which is 20% of the 3.9-μm focal depth. Tracking servo suppresses the disk eccentricity of ±0.1 mm into less than ±50 nm, which is ±3% of the 1.6-μm tracking pitch. A servo frequency range of several kHz is typically required for commercial optical disk systems.

21.2.3 Fabrication and Alignment

Figure 21.5 is a schematic representation of the fabrication process and the structure of a kind of hybrid optical device called a *packaged laser diode* with a monitoring photodiode. First, a large number of laser diode chips are precisely positioned on the patterned solder of a silicon photodiode wafer, then the laser chips are μ heated and soldered to the wafer in a batch process. After the soldering, the silicon wafer is diced mechanically in a conventional IC flow, and each laser diode chip with its monitoring photodiode—called an **LOP** (*laser on photodiode*)—is packaged on a heat sink in the same manner as a conventional power IC chip. This fabrication process is essentially our approach to mass production.

A similar fabrication technique is applied to the microprism-type of integrated optical pick-up device called the *laser coupler* (**LC**). The size of the LC chip in Figure 21.4 is as small as $0.5 \times 2 \times 3$ mm^3.

The hybrid integrated laser diode with a photodiode, the LOP chip, is also mounted precisely in reference to the optical axis. In this case the photodiode IC acts as an optical bench. This alignment process of laser coupler thus described is essentially different from that employed in the discrete optics because there is no feedback loop for alignment using *in situ* detection of disk signal in the device fabrication process. To fabricate a discrete optical pick-up precise positioning of optical components is

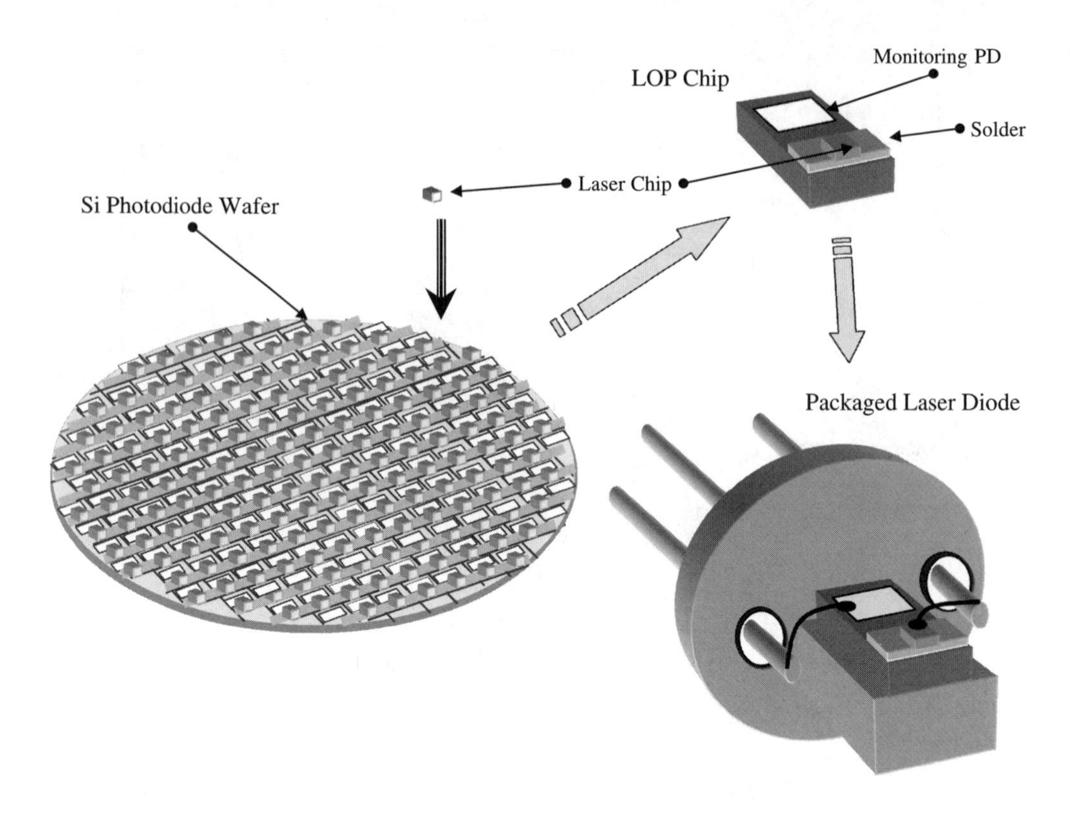

FIGURE 21.5 Schematic representation of laser on photodiode (LOP).

essential for the diffraction-limited reading of the CD pits. After assembling discrete optics devices, we tune the position of the grating, objective lens, and photodetector using the disk pit signals. We are also aware that this precision should endure in the face of environmental change.

With discrete optics, where each component is fixed to its bulk optical base by a UV cohesive agent, there are considerable limits to attaining the four goals listed in the previous section, particularly reliability. Precise mounting of the laser diode and the prism are both carried out in a mass-productive laser coupler process. This allows the optical alignment much easier laser alignment in the optical pick-up assembly process.

For monolithic optical pick-up devices the assembly process resembles that of the semiconductors. Photolithography is used for the two-dimensional optical alignment of each active element. Batch processes are used for most fabrication processes such as crystal growth and dry etching. The built-in mirror formed by crystal growth works as a micron-size optical element, and the GaAs substrate works as an optical bench in this case. The "crystallographic facet" is utilized as an optically flat surface and a precisely angled reflection mirror. The size of CLC device is typically as small as $0.2 \times 1 \times 2$ mm^3.

21.3 Hybrid Optical Integrated Head Devices

21.3.1 Introduction

As already noted, a head device integrated into an optical disk system must have portability, low cost, reliability, and stability. In integrating the device there are several approaches following each format of the systems—CD, MD, DVD-ROM, etc. In particular, integration of an optical head in the MD system [Sony Corp. and N.V. Philips, 1992] is more complicated than that of the other formats. In addition, there is more demand for integrating a head device for the MD system because it is more suitable for portable use.

In this section the MiniDisc laser coupler (**MDLC**) [Nishi et al., 1998] is presented as an example of integration of the optical head for the MD system, following the concept of *hybrid optical IC*.

21.3.2 Optical Head for the MD System

The MD system, which is one of the most popular rewritable optical disk formats, is capable of recording 140 MB of user data—i.e., 74 minutes of compressed audio data—on a 64-mm-diameter optical disk in a cartridge using the magneto-optical (MO) recording method. In addition, the read-only optical disk, which is similar to the CD medium, is also supported.

Figure 21.6 shows a schematic representation of the conventional optical head for an MD system. The optical head consists of a laser diode (**LD**), a grating, a polarizing beam splitter (**PBS**), a Wollaston prism (three-beam type), a multilens, an integrated eight-photodiode-unit photodetector (1), and a photodiode photodetector (2).

The light beam emitted from the AlGaAs-based stripe LD is focused through the grating, the PBS, and the objective lens on the optical disk surface. On the forward path the beam is divided into two beams by the PBS; one beam propagates to the photodetector (2) for the front automatic power control (FAPC) to stabilize the LD power. After obtaining the RF signal on the disk using the other beam, the returned light beam is reflected by the PBS and then propagates to the photodetector (1) through the Wollaston prism and the multilens to detect the servo, addressing, and RF signals. The grating and the multilens are used to realize tracking-servo using the three-beam method and focusing-servo using the astigmatic method [Bouwhuis et al., 1985]. When the grating is inserted, the beam emitted from the LD is divided into three beams –0th and ±1st-order diffraction beams and separated in the tangential direction. For the multilens, on one side there is a cylindrical plane and on the other a concave plane. The three-beam Wollaston prism functions as a polarizing beam splitter to separate the returned beam into three components—two orthogonally polarized components and a nonpolarized component in the radial direction. The two polarized beams, whose polarizations are orthogonal to each other, irradiate on PD_i and PD_j in Figure 21.6(b) to detect the MO signal. The nonpolarized beam irradiates on the quadrant photodiode PD_a to PD_d in Figure 21.6(b) to detect focusing-error signals. As a result of diffraction due to the grating and separation by the Wollaston prism, nine beams appear on the photodetector (1) as shown in Figure 21.6(b).

In the MD system two kinds of **RF** signals—the MO signal for a rewritable medium and the pit signal for a read-only medium, which is similar to the CD medium—are supported. In addition, an address in pregroove (**ADIP**) signal generated by the embedded wobbled groove is detected for the addressing and clock signals.

When using this conventional optical configuration for the MD system, tracking-error (**TE**) signal S_{TE1}, focusing-error (**FE**) signal S_{FE1}, ADIP signal S_{ADIP1}, RF signal for the MO medium S_{RF11}, and RF signal for the read-only medium S_{RF12} are given by

$$\begin{aligned} S_{\text{TE1}} &= I_e - I_f \\ S_{\text{FE1}} &= (I_a + I_c) - (I_b + I_d) \\ S_{\text{ADIP1}} &= (I_a + I_d) - (I_b + I_c) \\ S_{\text{RF11}} &= I_i - I_j \\ S_{\text{RF12}} &= I_\text{I} + I_\text{j}, \end{aligned} \qquad (21.1)$$

where I_a, I_b, I_c, I_d, I_e, I_f, I_i, and I_j are the signals detected on PD_a, PD_b, PD_c, PD_d, PD_e, PD_f, PD_i, and PD_j, respectively.

As we see, a complicated optical configuration is required for the MD system. With such a complicated system multifunctions should be assigned to each element of the optical pick-up to minimize the number of elements and to realize simple assembly and adjustment, instead of "one function for one element," as shown in Figure 21.6(a). To realize the integration based on the hybrid optical IC, several configurations

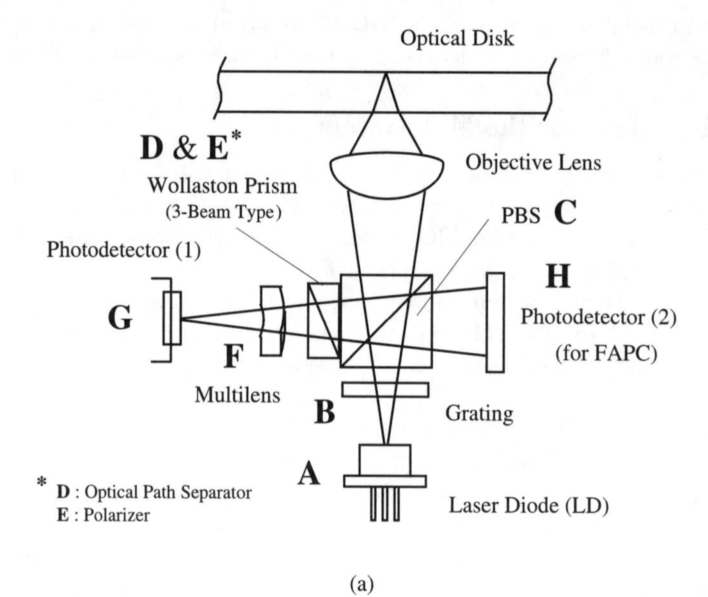

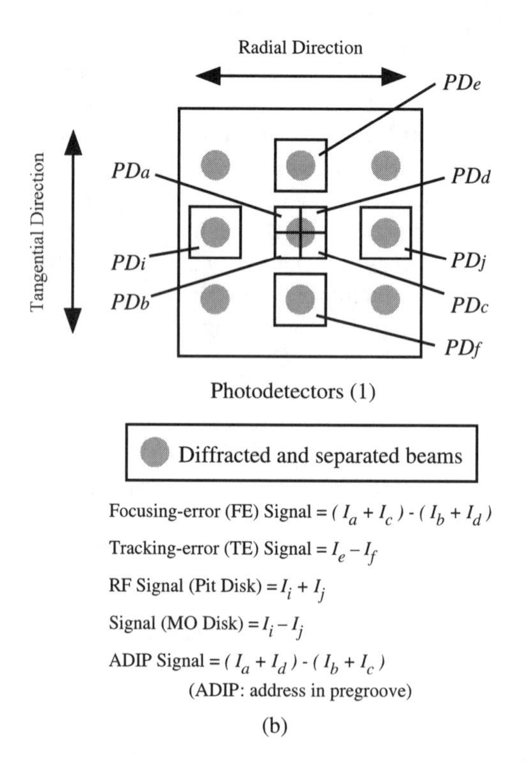

FIGURE 21.6 Schematic representation of conventional "discrete-type" optical head for MD system: (a) optical setup; (b) arrangement of photodiodes on photodetector (1), top view.

have been proposed and developed for the practical MD system, as shown in Figure 21.7. This figure shows three approaches—"laminated-prism" type [Horinouchi et al., 1995], "HOE/prism" type [Arai et al., 1997], and "biaxial microprism" type.

In the laminated-prism type shown in Figure 21.7(a), each optical element is fabricated on the surface or both surfaces of the wafer using thin film or HOE techniques and laminated for each wafer. Then, the

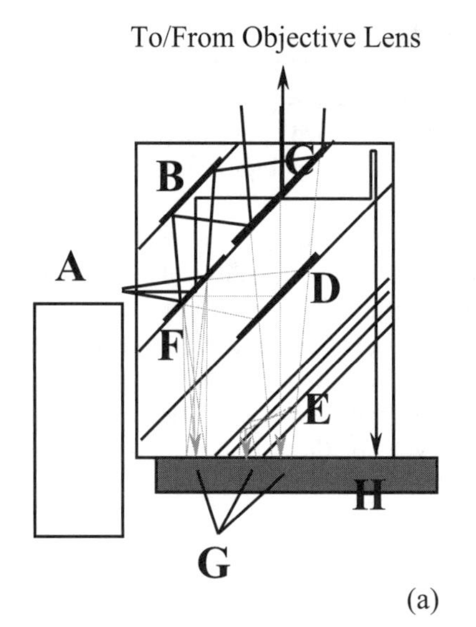

(a)

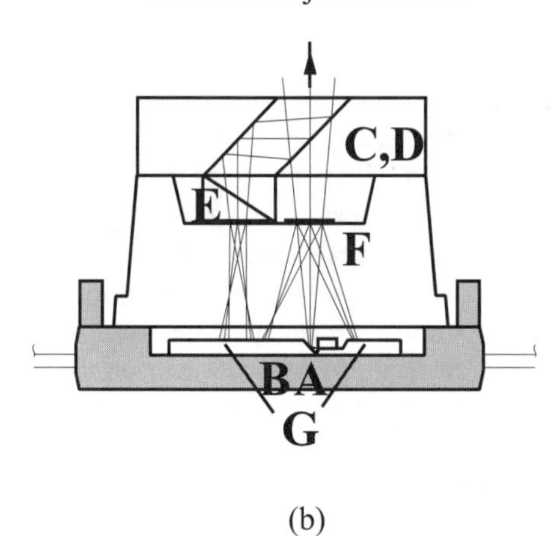

(b)

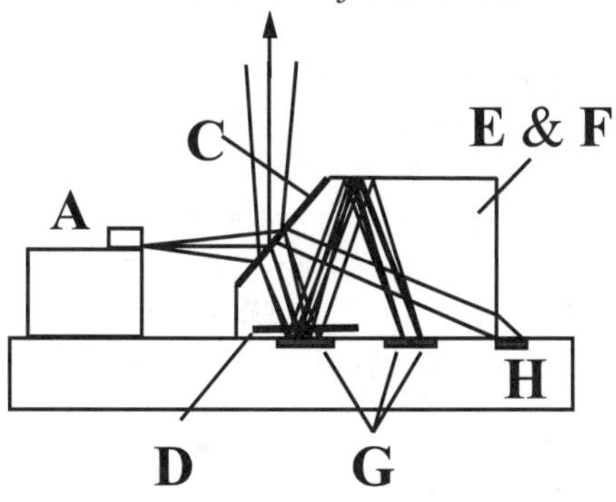

(c)

FIGURE 21.7 Schematic representation of approaches to "hybrid-type" integration for MD system. (a) "Laminated-prism" type; (b) "HOE/prism" type; (c) "biaxial microprism" type of MiniDisc laser coupler (MDLC).

sealed wafers are diced for each head device. As a result, the optical elements B to F in Figure 21.6(a) are integrated. However, the number of optical elements is not decreased in this type. In the HOE/prism type, as shown in Figure 21.7(b), only tracking- and focusing-servo functions are integrated using an HOE. In this type the miniaturization and rearrangement of each optical element enable the integration of the head device, rather than the simplification of the configuration obtained from decreasing the number of elements.

On the other hand, in the biaxial microprism type, as shown in Figure 21.7(c), a microprism made of a biaxial crystal merely covers all functions of the MD system. Additional techniques such as wafer-batch processing are needed to produce the laminated-prism type. However, the biaxial microprism-type device structure is essential for obtaining not only good integration of optical elements B to F in Figure 21.6, but also fair simplification of the optics. We call this optical head device for the MD system a MiniDisc laser coupler (**MDLC**). In the following paragraph details of the MDLC are presented as a practical example of integrating an optical head device for the MD system, based on the concept of hybrid optical IC.

21.3.3 MiniDisc Laser Coupler

21.3.3.1 Optical Configuration

Figures 21.8(a) and (b) show a schematic representation of an MDLC and the arrangement of the photodiodes on the Si substrate. Figure 21.11 is a photograph of the MDLC fabricated and packaged. The MDLC consists of an LD, a biaxial microprism that is made of $KTiOPO_4$ (**KTP**), and two kinds of photodetector—an integrated 11-photodiode unit (photodetector (1)) for detecting FE, TE, ADIP, and RF signals, and photodetector (2) for FAPC, located on the same Si substrate. Using the thin-film technique, three functions are enabled on the surface of the microprism—a polarizing beam splitter (PBS) on the inclined surface, a half-mirror on the back surface positioned immediately above the photodiodes designated by P_1 in Figure 21.8, and a reflective mirror on the top surface.

After being reflected on the inclined surface of the biaxial microprism, the light beam emitted from the AlGaAs-based stripe LD is focused through the objective lens on the optical disk surface. Because there is a PBS film on the inclined surface of the microprism the forward beam is divided into two beams by the PBS film, and one propagates to photodetector (2) for FAPC signaling through the microprism. Here, the transmittances of p-polarized beam T_p and s-polarized beam T_s are 70% and 40%, respectively. These values were optimized analyzing the dependence of jitter on T_p and T_s, as shown in Figure 21.9. After reading out the RF signal on the disk using the other forward beam, the light beam returns to the PBS film through the objective lens, then propagates into the biaxial microprism. When propagating inside the biaxial microprism, the beam is divided into two extraordinary beams. These two beams, whose polarizations are orthogonal to each other, are detected on the photodiodes designated by P_1, P_2, and P_3 in Figure 21.8. The portion of the intensity of the beams is divided by the half–mirror and detected on P_1. After being reflected on the half-mirror above P_1, the rest of the beams irradiate on P_2 and P_3 via the reflective mirror on top of the biaxial microprism. Although these two orthogonal beams impinge on almost the same position on P_1, they are completely divided and separately detected on P_2 and P_3. Here, the crystal axis of the biaxial microprism is chosen, as shown in Figure 21.10, in order to achieve the appropriate separation of the propagating beams according to each polarization.

As noted above, the biaxial microprism with three kinds of functional thin film achieves all functions obtained using the optical elements C to F in Figure 21.6(a), without any additional techniques such as wafer batch processing. Here, the grating B in Figure 21.6(a) for the tracking servo is eliminated by devising the pattern of the photodiodes on photodetector (1) together with additional signal processing.

The spot size detection method [Oka et al., 1987] and push–pull method [Bouwhuis et al., 1985] are employed for focusing- and tracking-servos, respectively.

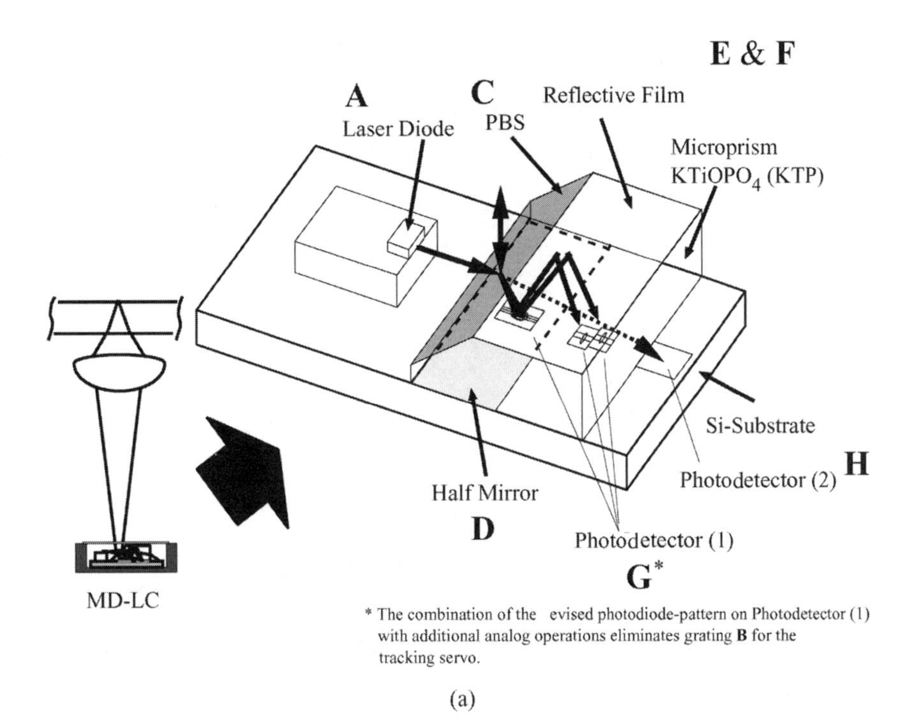

* The combination of the evised photodiode-pattern on Photodetector (1) with additional analog operations eliminates grating **B** for the tracking servo.

(a)

Tangential Direction

Radial Direction

Photodetector (1)

Photodetector (2)

Si-Substrate

PDa2, PDb2, PDc2, PDd2, PDix21, PDiy2, PDix22, PDjx21, PDjy2, PDjx22, PDf2

P_1 P_2 P_3 P_4

FE Signal = $K\{(I_{a2} + I_{d2}) - (I_{b2} + I_{c2})\} - \{(I_{ix21} + I_{ix22} - I_{iy2}) + (I_{jx21} + I_{jx22} - I_{jy2})\}$

TE Signals: derived from $(I_{a2} + I_{b2}) - (I_{c2} + I_{d2})$ K: Constant

RF Signal (Pit Disk) = $(I_{ix21} + I_{ix22} + I_{iy2}) + (I_{jx21} + I_{jx22} + I_{jy2})$

RF Signal (MO Disk) = $(I_{ix21} + I_{ix22} + I_{iy2}) - (I_{jx21} + I_{jx22} + I_{jy2})$

ADIP Signal = $(I_{a2} + I_{b2}) - (I_{c2} + I_{d2})$

(ADIP: address in pregroove)

(b)

FIGURE 21.8 MiniDisc laser coupler (MDLC). (a) Schematic representation of MDLC; (b) arrangement of photodiodes on Si substrate (top view).

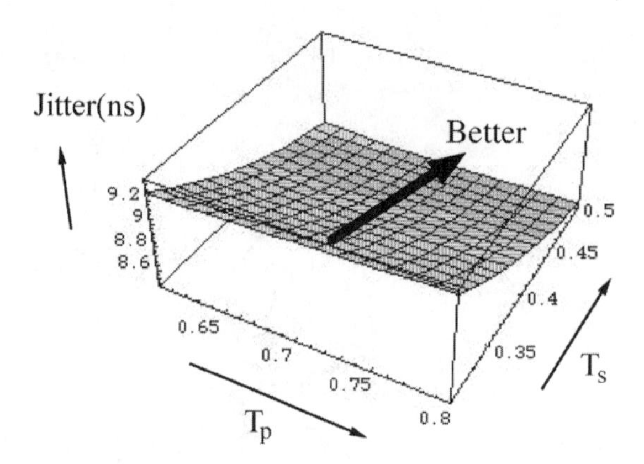

FIGURE 21.9 Dependence of jitter on T_p and T_s.

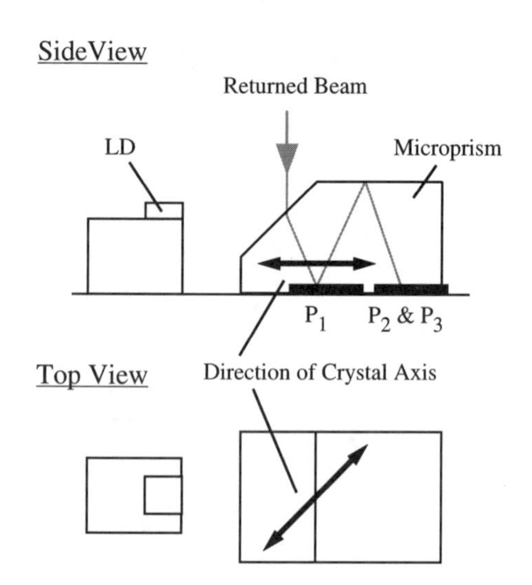

FIGURE 21.10 Direction of crystal axis (n_C-axis) of KTP.

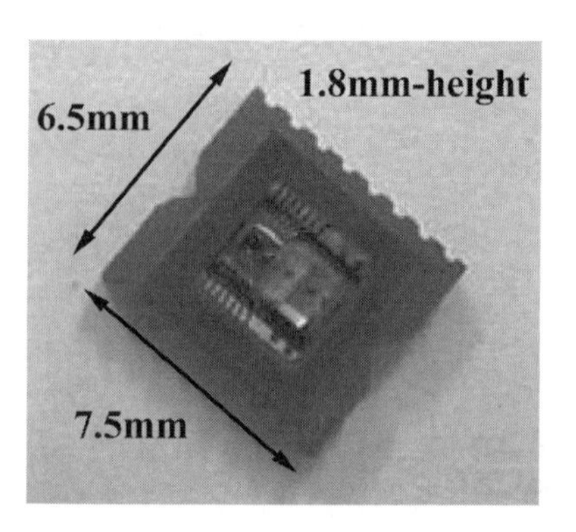

FIGURE 21.11 Photograph of packaged MDLC.

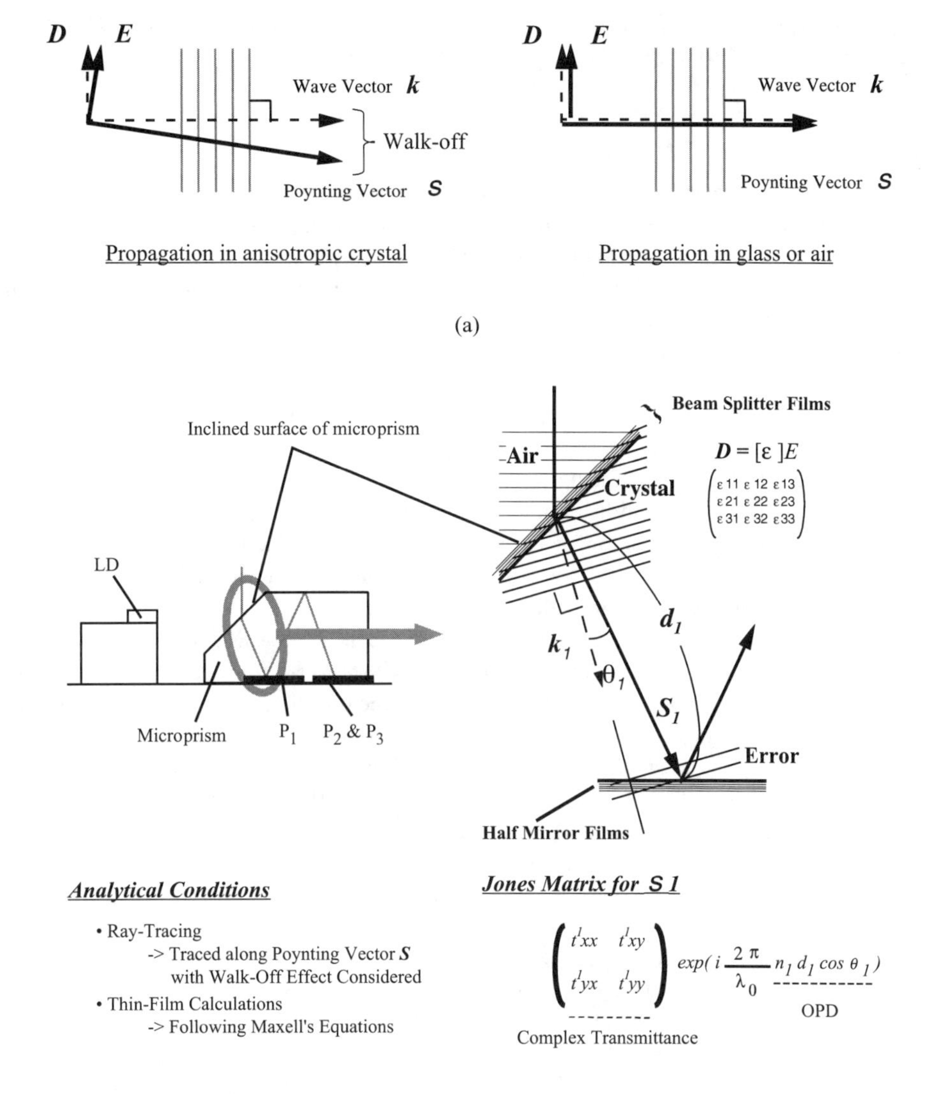

FIGURE 21.12 Schematic representation of ray-tracing for MDLC: (a) walk-off effect in propagation in anisotropic crystal; (b) ray-tracing considering walk-off effect and thin film conditions for biaxial microprism (made of anisotropic crystal).

When using the MDLC, TE signal S_{TE2}, FE signal S_{FE2}, ADIP signal S_{ADIP2}, RF signal for the MO medium S_{RF21}, and RF signal for the read-only medium S_{RF22} are given by

$$
\begin{aligned}
S_{TE2} &= (I_{a2} + I_{b2}) - (I_{c2} + I_{d2}) \text{ with additional signal processing} \\
S_{FE2} &= \mathrm{K}\ \{(I_{a2} + I_{d2}) - (I_{b2} + I_{c2})\} \\
&\quad - \{(I_{ix21} + I_{ix22} - I_{iy2}) + (I_{jx21} + I_{jx22} - I_{jy2})\},\ \text{K: constant} \\
S_{ADIP2} &= (I_{a2} + I_{b2}) - (I_{c2} + I_{d2}) \\
S_{RE21} &= (I_{ix21} + I_{ix22} + I_{iy2}) - (I_{jx21} + I_{jx22} + I_{jy2}) \\
S_{RE22} &= (I_{ix21} + I_{ix22} + I_{iy2}) + (I_{jx21} + I_{jx22} + I_{jy2}),
\end{aligned}
\tag{21.2}
$$

where I_{a2}, I_{b2}, I_{c2}, I_{d2}, I_{e2}, I_{f2}, I_{ix21}, I_{ix22}, I_{iy2}, I_{jx21}, I_{jx22}, and I_{jy2} are the signals detected on PD_{a2}, PD_{b2}, PD_{c2}, PD_{d2}, PD_{e2}, PD_{f2}, PD_{ix21}, PD_{ix22}, PD_{iy}, PD_{jx21}, PD_{jx22}, and PD_{jy}, respectively.

21.3.3.2 Optimization of MDLC

In fabricating the MDLC there is an advantage for alignment and arrangement of each element against the other optical configurations, as shown in Figure 21.7(a) and (b), in addition to the simple structure. When we employ the optical configurations in Figure 7(a) and (b), each element must be precisely aligned and arranged together, observing the various signals. In fabricating the MDLC, however, an LD, a biaxial microprism, and photodiodes are simply aligned and arranged on the same Si substrate without observing any signals. To achieve such a simple alignment fabrication process, it is essential to optimize (1) the relative position between each element on the Si substrate, (2) the shape and its accuracy in fabrication of the microprism itself, and (3) the size and the arrangement pattern of each photodiode, through precise beam-propagation analysis [Nishi et al., 1999]. The combination of diffraction analysis for the optical path in the MDLC and ray-tracing consideration of the walk-off effect [Wiechman et al., 1995; Saito et al., 2000] is carried out, and it becomes evident that the three kinds of thin films deposited on the microprism determine the position of the propagating beam inside the microprism and the spot diagram and distribution of the light intensity on each photodiode. Using the analyzed results, the MDLC is optimized. It results in a simple alignment fabrication process.

Figure 21.12 shows a schematic representation of ray tracing for the MDLC. Using the ray-tracing technique with the walk-off effect considered, the positions of the propagating beam inside the microprism and the spot diagram on each photodiode are derived along the direction of Poynting vector **S**, instead of wave vector ***k***. In addition, the optical path distances (OPDs) between each optical plane are derived along **S**. The result of the OPDs is used when deriving the distribution of light intensity on each photodiode using the diffraction analysis. Further, the thin-film conditions on the biaxial microprism are considered in ray tracing to obtain an accurate ratio of the redundant orthogonal two eigen-polarizations occurring by the transmittance of the inclined surface and reflectivities on the bottom (around P_1 in Figure 21.8) and top surfaces of the microprism. Each condition is estimated according to the electromagnetic boundary conditions given by Maxwell's equations.

Figure 21.13 shows the procedure for analyzing beam propagation and the distribution of light intensity on each photodiode in the MDLC. The analysis includes calculating the diffraction pattern on the photodiodes and the results of ray tracing — OPDs and the transmittance and reflectivities for p- and s-polarizations. The distribution of light intensity on each photodiode is derived using Jones vector analysis and considering each condition: (1) the polarized condition and the Gaussian distribution of the light beam emitted from an LD, (2) disk conditions including the deviation of the polarized condition due to the Kerr effect, (3) thin-film conditions on the microprism including the results of the above ray-tracing, and (4) beam propagation in the microprism.

Following these analyses, the appropriate relative position between the spot diagram (or the distribution of the light beam) and each photodiode is optimized, as shown in Figure 21.14. Based on this relationship, the relative position between each element on the Si substrate and the shape and its accuracy in fabrication of the microprism itself are also optimized. As a result, the simple alignment fabrication process is realized in the MDLC.

21.3.3.3 Performance of MDLC

Using the fabricated device shown in Figure 21.11, FE, TE, and RF signals were received. Figure 21.15 shows the experimental results of FE, TE, and nonequalized MO-RF signals. In the experiment the wavelength of the LD and the numerical aperture (NA) of the objective lens were 0.785 and 0.45, respectively. The peak-to-valley range of the FE signal was about 15 μm. Figure 21.16 shows the experimental results of the tolerance of jitter, BLER, and ATER against the focus offset, where BLER is

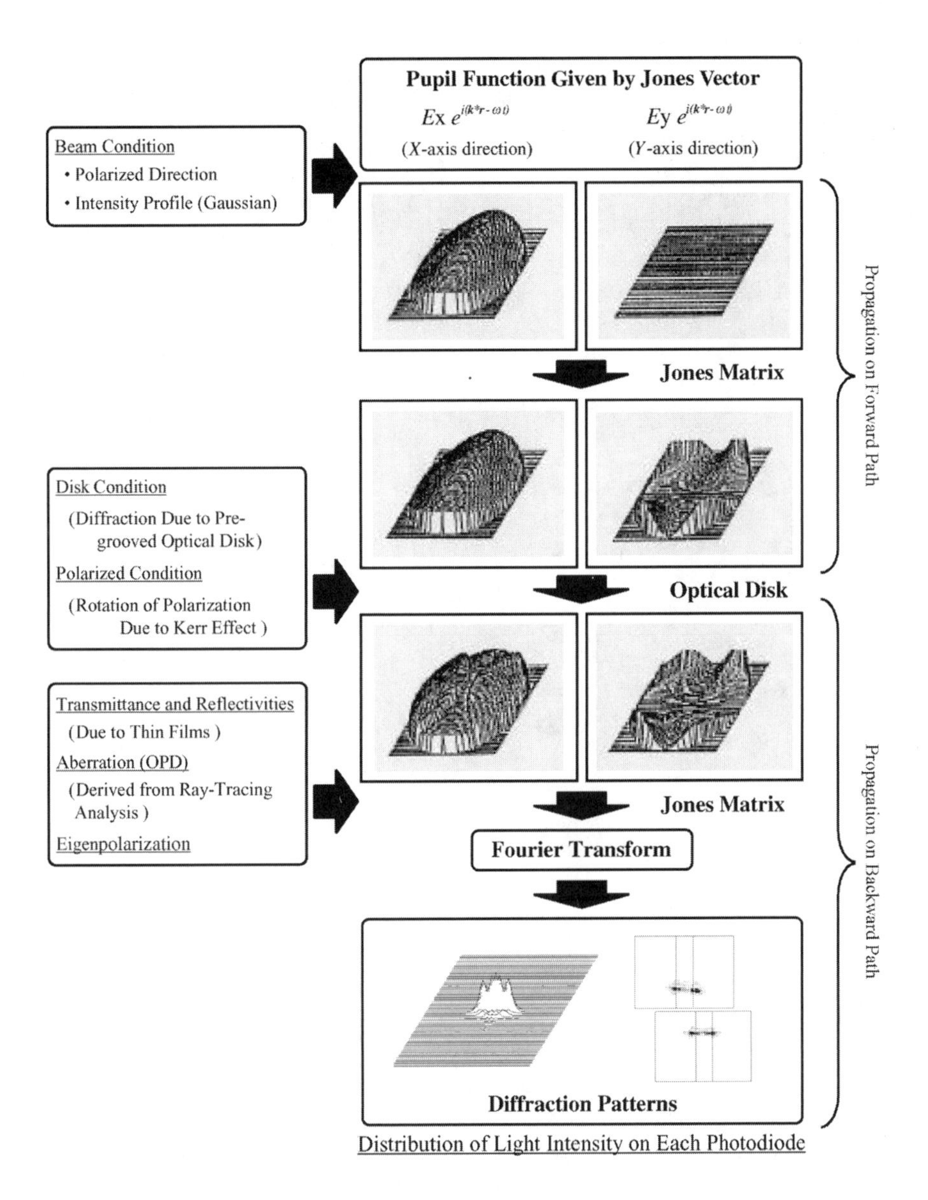

FIGURE 21.13 Procedure of analyzing beam propagation and distribution of light intensity on each photodiode in MDLC.

the error rate of the RF signal (the block error rate) and ATER the error rate of the ADIP signal. The minimum values of jitter of the nonequalized MO signal and of the equalized MO signal, BLER, and ATER were 9.3, 8.5, 0.01, and 0.00%, respectively (linear velocity, 1.4 m/s; modulation, EFM, detecting window width = 231.5 ns). Considering that the acceptable values of BLER and ATER against the focus offset are 3% in either case, the tolerances of the focus offset for BLER and ATER are more than ±3 μm, respectively. The tolerances for BLER and ATER using the MDLC are almost the same as those using the conventional optical configuration. This means that the MDLC provides practical stability for the MD system.

To investigate long-term reliability a thermal shock test was employed. The detail of the thermal shock pattern is shown in Figure 21.17(a). In the test the heat-cycle from +80 to −40° (and vice versa) was repeated 500 times for 500 hours. Figure 21.17(b) shows the experimental results of the thermal shock

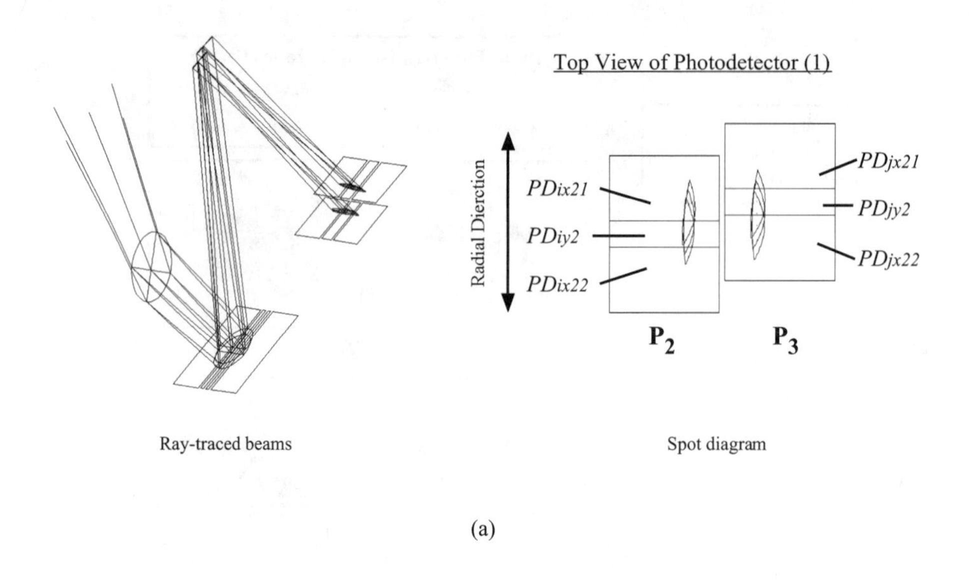

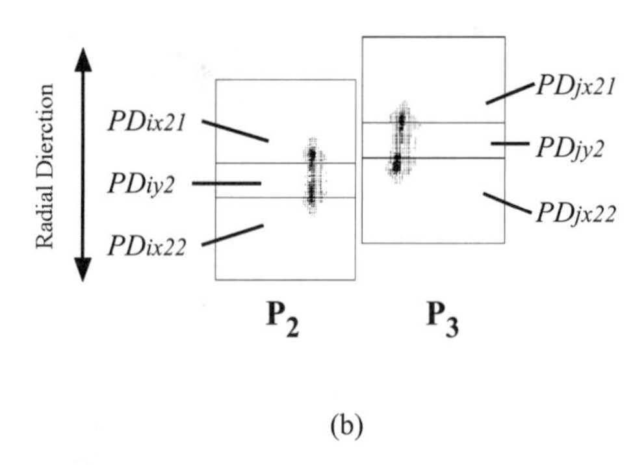

FIGURE 21.14 Optimized results of spot diagram and distribution of light intensity on photodetector (1): (a) ray-traced beams and spot diagrams; (b) diffractive patterns on photodetector.

test. As shown in this figure, there is almost no degradation of the tolerance of jitter against the focus offset, even after the test. Considering that the test is for confirming whether the sample is capable of shipping to market, we can easily understand that the MDLC offers long-term reliability for practical use.

These results indicate that integration of an optical head device using the MDLC is a workable approach to achieve portability, stability, and reliability for a practical MD system.

21.3.4 Summary

An optical head device, which is based on the concept of hybrid optical IC and which we call MiniDisc laser coupler (MDLC), has been presented for a practical MD system. The entire optical elements for the optical head except a laser diode and an objective lens are replaced with a one-piece biaxial microprism. This results in not only integration of the optical head but also simplification. A method to design the MDLC based on ray tracing with diffraction in the biaxial crystal has also been reported. In addition, the demonstration of the MDLC indicates that the MDLC achieves reliability and stability in practical use.

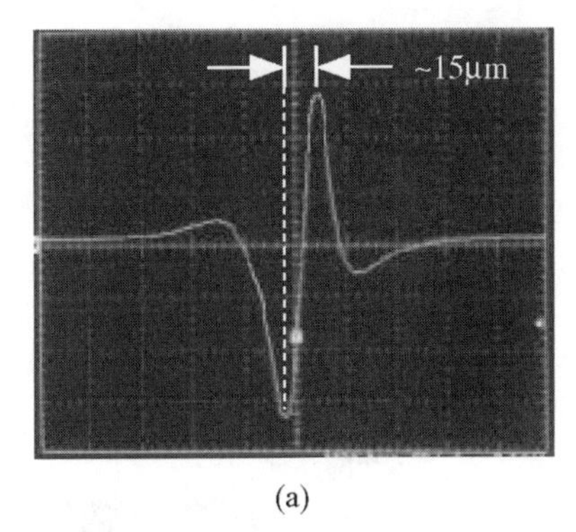

(a)

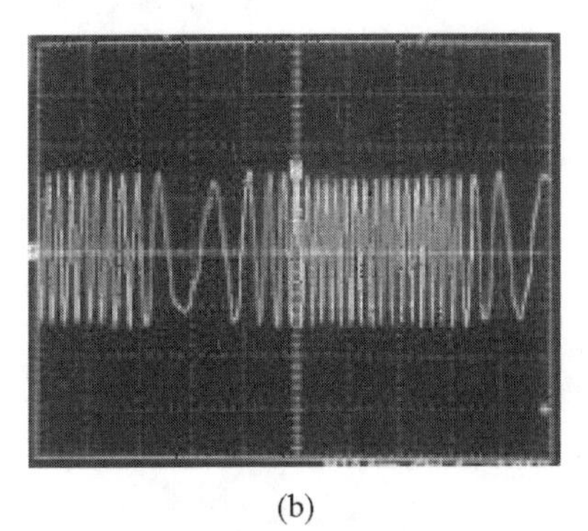

(b)

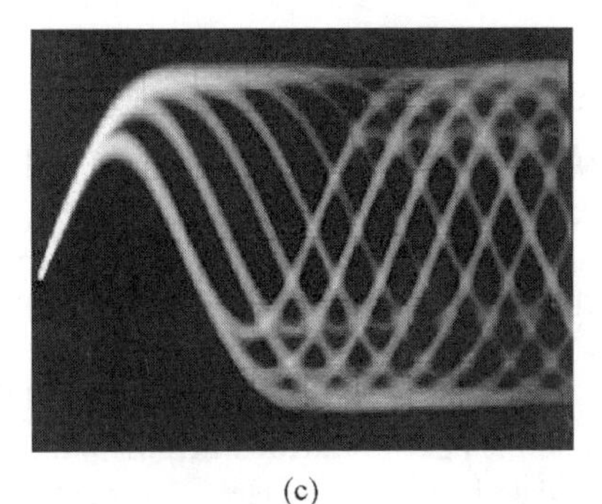

(c)

FIGURE 21.15 Experimental results of servo and RF signals: (a) focusing-error signal; (b) tracking-error signal; (c) eye-pattern of nonequalized MO-RF signal.

21.4 Monolithic Optical Pick-Up Device

21.4.1 Introduction

As presented above, various optical disk systems have been developed, and small, lightweight, low-cost pick-up devices using a microprism [Matsumoto et al., 1996] or a hologram [Yoshida et al., 1994] have been developed since the compact disk (CD) system [Sony Corp. and N.V. Philips, 1980; Carasso et al., 1982] appeared on the consumer market. These pick-up devices, however, have a limited application to modern systems with higher accessing rates and larger capacity.

With this technology serving as the background, the next-generation optical pick-up, which features a confocal optical configuration and can be manufactured using the monolithic fabrication process based on semiconductor batch processing, is presented in this section. This configuration not only enables a small, lightweight, and low-cost pick-up, but also generates more stable servo signals.

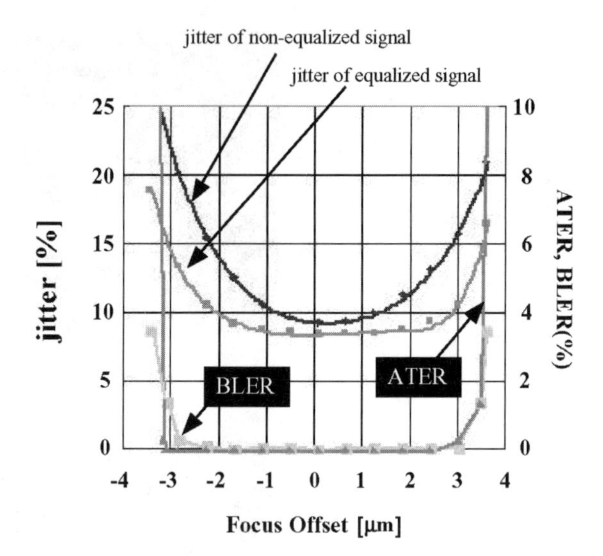

FIGURE 21.16 Focus offset tolerances.

21.4.2 Confocal Laser Coupler (CLC)

Figure 21.18 shows a schematic representation of the optical pick-up using a confocal laser coupler (CLC) [Narui et al., 1998]. The CLC consists of a monolithic optical element, which includes a laser diode (LD), two photodiodes (PD), and a built-in mirror positioned near the confocal plane. The light beam emitted from the AlGaAs-based stripe LD is reflected by the built-in mirror and focused through the objective lens on the optical disk surface. After reading out the pit signals on the disk, the light beam returns on the same path. The returned light beam is spread by diffraction to a spot in the order of 1.22 λ/NA ~ 10 μm around the built-in mirror when the numerical aperture (NA) is 0.1 and the laser wavelength is 0.78 μm; the light beam is then detected by the two PDs, whose size is almost the same as the diffraction spot size on the confocal plane, to obtain the tracking-error (TE) signal and the RF signal. Because the CLC device can be fabricated using the semiconductor batch process, it is expected to enable a low-cost, highly reliable optical pick-up that does not require optical alignment.

To apply this device to an actual optical pick-up, a three-CLC configuration is used to obtain not only TE and RF signals but also the focusing-error (FE) signal, as shown in Figure 21.19(a). Two LDs, i.e., LD1 and LD3, as well as two PDs, i.e., PD1 and PD3, located near the confocal plane are used for FE signal detection using the spot-size method. The built-in mirror produced by crystal growth makes an angle of 54.7° instead of 45°, so this method is achieved by creating both positive and negative differences in path length between symmetrical optical paths (for LD1 and LD3) against the main beam emitted from LD2, as shown in Figure 21.19(b).

TE signal detection is performed by placing PD2a and PD2b near the confocal plane so as to be symmetric with respect to the laser stripe. We call this TE signal-detection method the confocal push–pull (CPP) method. Because a confocal optical configuration is adopted in the CPP method, no DC offset results even if there is a displacement in the lens position. Therefore, there is stable detection

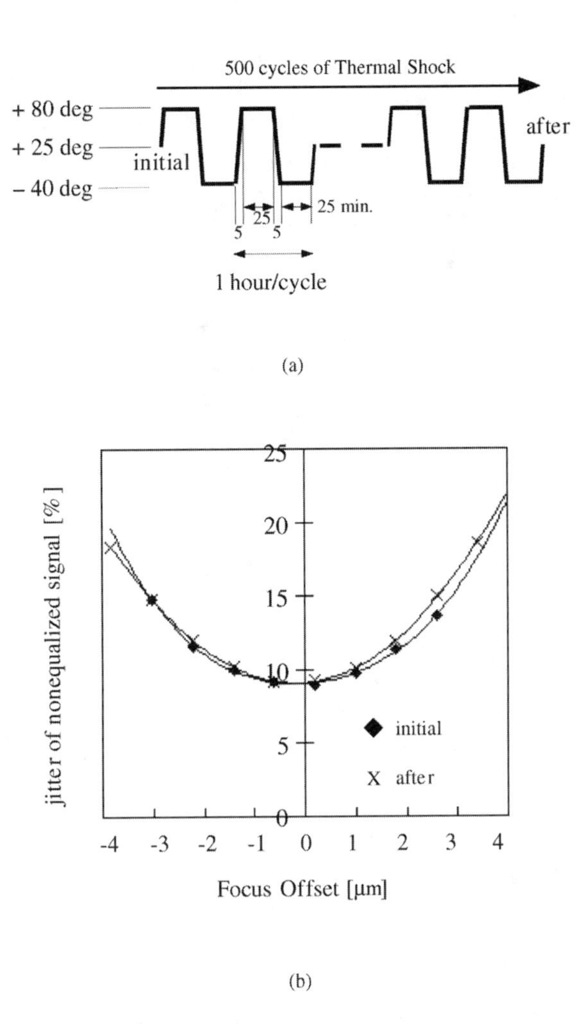

FIGURE 21.17 Experimental results of thermal shock test: (a) thermal shock pattern; (b) experimental results.

of the TE signal. Figure 21.20 shows the calculated results for the dependence of conventional push–pull signals and CPP signals on lens displacement in the radial direction. In the calculations the well-known scalar theory for the optical disk [Hopkins, 1979] was employed, and a simple groove disk and the typical values of each parameter for the CD player shown in Figure 21.23 were assumed. DC offsets of conventional push–pull signals are about 40% for the amplitude, and those of CPP signals are less than 8% for the amplitude for a radial lens displacement of ±400 μm. This result shows that CPP signal detection is superior to conventional push–pull signal detection, even if there is lens displacement.

When using the three-CLC configuration, TE signal S_{TE}, FE signal S_{FE}, and RF signal S_{RF} are given by

$$\begin{aligned} S_{TE} &= I_{PD2a} - I_{PD2b} \\ S_{FE} &= I_{PD1} - I_{PD3} \qquad (21.3) \\ S_{RF} &= I_{PD2a} + I_{PD2b}, \end{aligned}$$

where I_{PD1}, I_{PD2a}, I_{PD2b}, and I_{PD3} are the signals detected on PD1, PD2a, PD2b, and PD3, respectively.

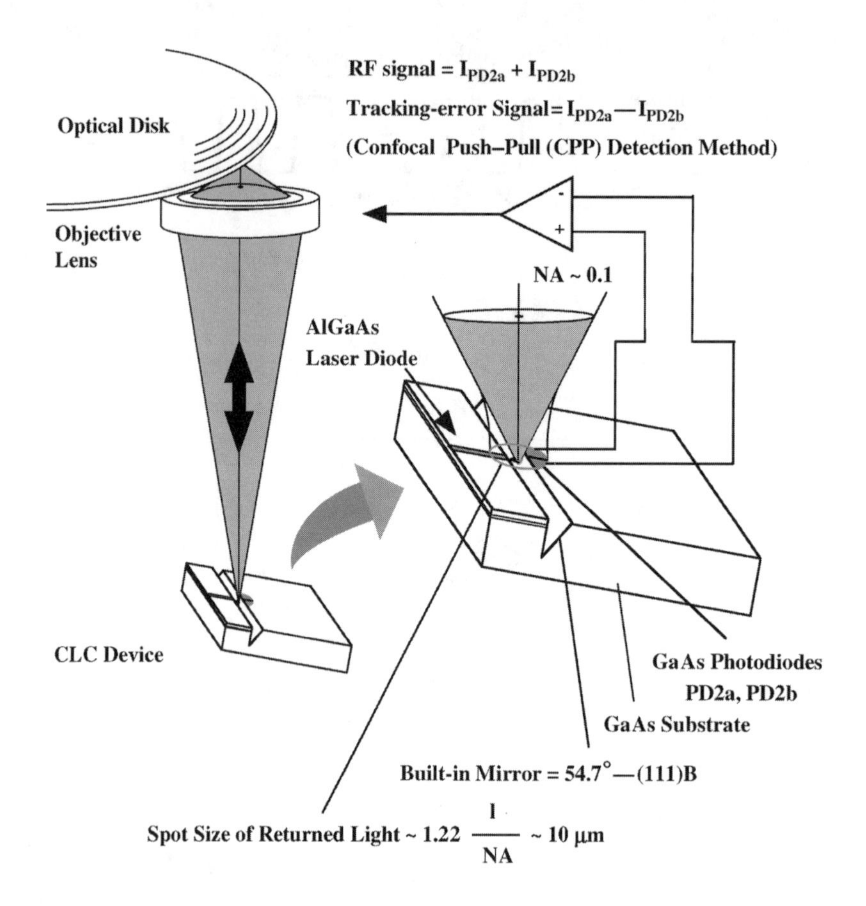

FIGURE 21.18 Schematic representation of confocal laser coupler (CLC).

21.4.3 Technical Issues Concerning Tracking-Error-Signal Detection Using CPP Method

As noted above, the CLC device can be manufactured using a monolithic process and is expected to be able to generate stable servo signals. Because of its simple structure, however, the three-CLC configuration is required to detect not only RF and TE signals but also the FE signal. In addition, the calculated results of the dependence of the CPP signal on the groove/land ratio (duty ratio) of the optical disk and the focus condition, which are shown in Figure 21.21 and calculated using the same parameters as the results in Figure 21.20, suggest the following two problems specific to CPP detection [Mizuno, 1999b].

In the CPP method, the +1st-order diffraction beam from the optical disk lies exactly on the 1st-order beam on the photodiode positioned near the confocal plane, as shown in Figure 21.22, although the ±1st-order diffraction beams are split in the farfield region in the conventional push–pull method. As a result, the CPP signal includes the half-cycle component of the original TE signal generated by the interference between the ±1st-order diffractions. It should be noted that only this half-cycle component appears if the duty ratio is 50%.

The CPP signal strongly depends on the focus condition, even if the focus deviation is within the focal depth. The CPP signal changes from plus to minus or vice versa depending on whether this small deviation is far from or near the on-focus position. This is due to the influence of the deviation from the complete Fourier transform by the objective lens.

However, such a small deviation in the focus condition may occur even when the focusing servo is activated, so this may cause a serious problem in TE signal detection.

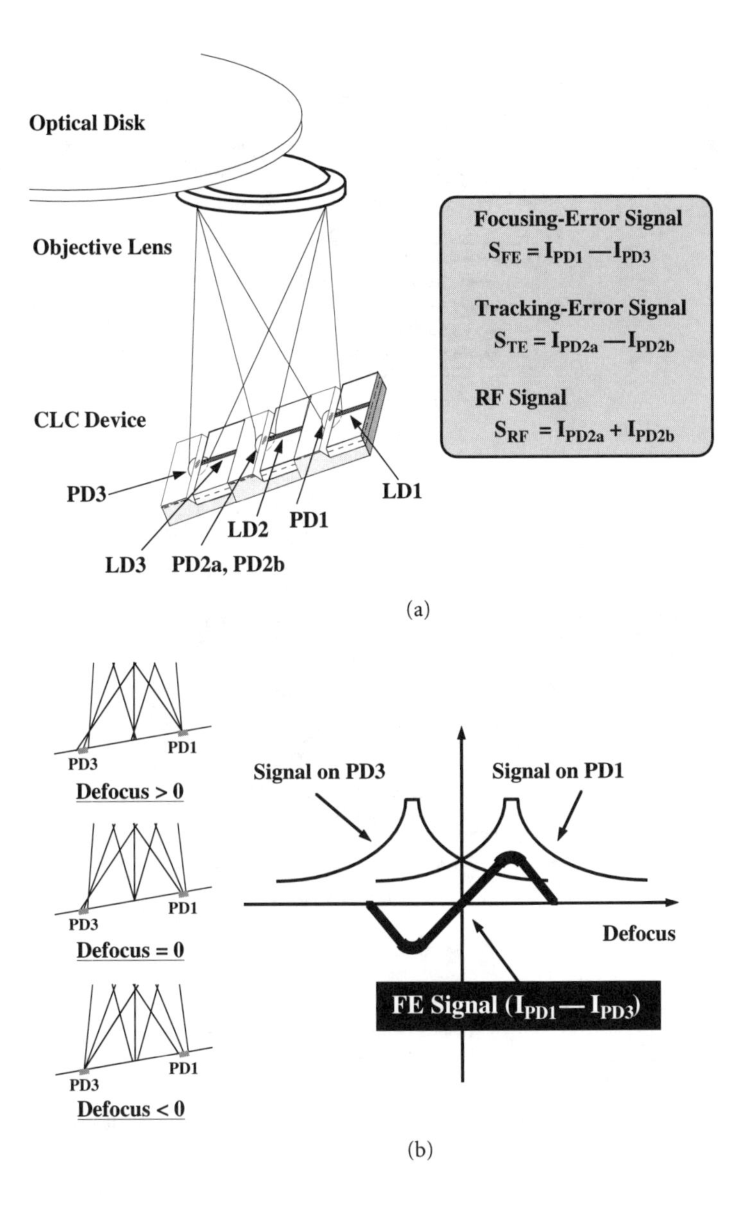

FIGURE 21.19 Optical pick-up using 3-CLC: (a) 3-CLC configuration; (b) spot size detection method for focusing servo.

We can easily understand the above substantial problems by referring to the following expression for CPP signal S_{cpp}, which is derived analytically using some approximations [Mizuno et al., 1999a].

$$
\begin{aligned}
S_{cpp} &\approx M_0 + M_1 \times df \\
M_0 &= C_{S1} \times \sin\left(2\pi \cdot dtr / g_{pt}\right) + C_{S2} \times \sin\left(2 \times 2\pi \cdot dtr / g_{pt}\right) \\
M_1 &= C_{S3} \times \sin\left(2k \cdot g_{pt}\right) \cdot \sin\left(2 \times 2\pi \cdot dtr / g_{pt}\right)
\end{aligned}
\tag{21.4}
$$

where C_{S1}, C_{S2}, and C_{S3} are constants independent of defocus, detrack, and media parameters; df is the deviation within focal depth; dtr is detrack; and g_{pt} is groove pitch.

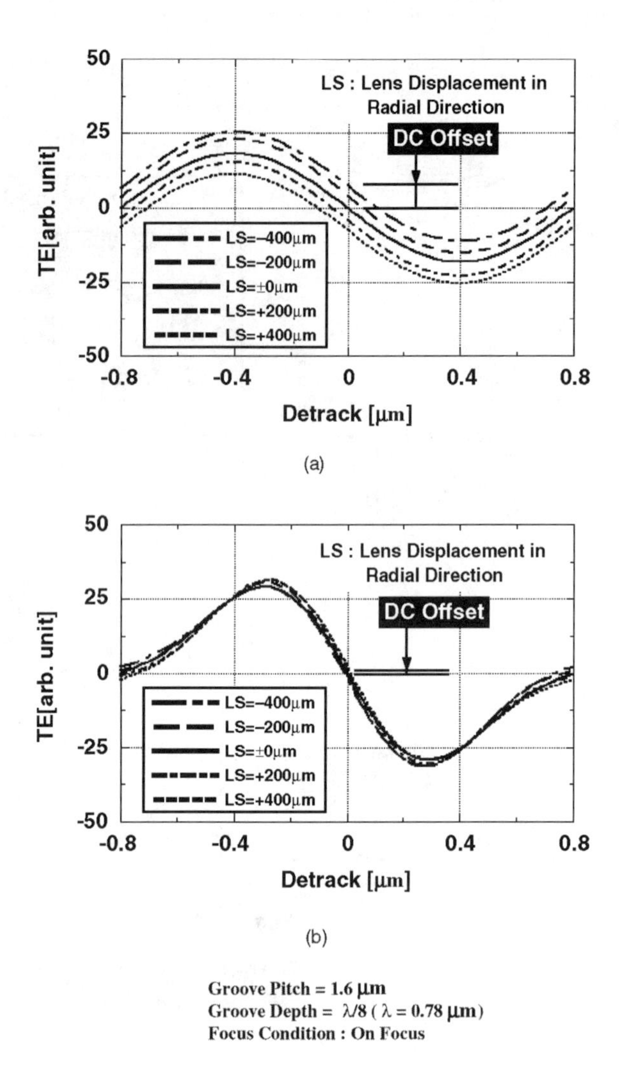

FIGURE 21.20 Characteristics of CPP tracking-error signals; dependence of lens displacement in the radial direction: (a) conventional push–pull method; (b) confocal push–pull (CPP) method.

As a result of this investigation, it is apparent that CPP signals must be improved if they are to be used as TE signals.

21.4.4 Improved CPP Signals

Based on the analytical solution for the CPP signal given by Eq. (21.4), an improved CPP detection method is introduced here [Mizuno et al., 1999c].

As easily understood from Eq. (21.4), the coefficient M_1 of Eq. (21.4) is equivalent to the conventional push–pull signal. Therefore, another TE signal, $S_{cpp'}$, can be obtained by calculating M_1 using Eq. (21.5):

$$S_{cpp'} = dS_{cpp}/ddefocus \tag{21.5}$$

This TE signal will be stable even in the face of variations in objective lens displacement. This is because the positional relationship between the 0th-order diffraction beam and the ±1st-order diffraction beams

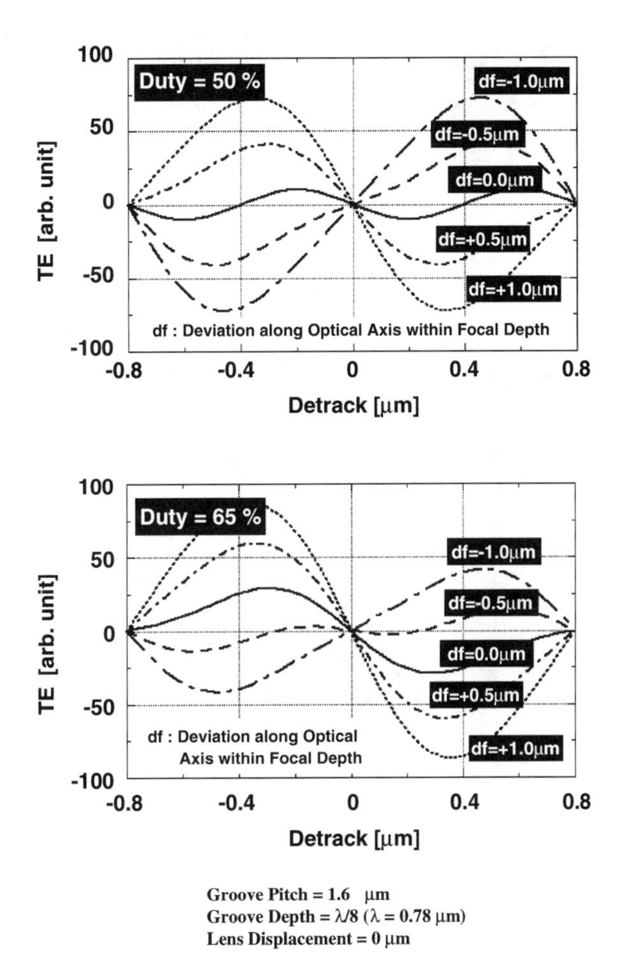

FIGURE 21.21 Characteristics of CPP tracking-error signals; dependence of focus condition and duty ratio.

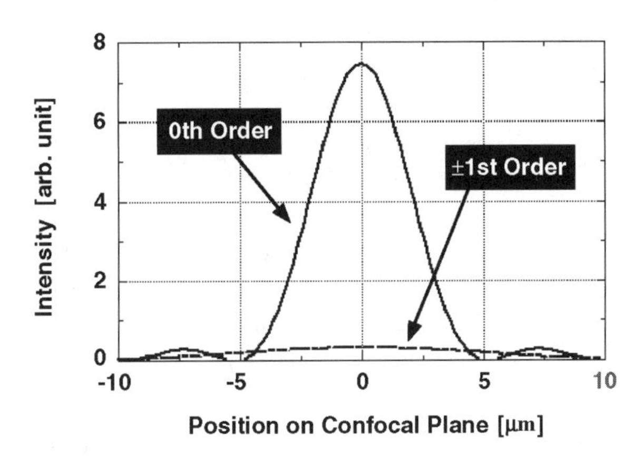

FIGURE 21.22 Cross-section of intensity of each diffraction beam on confocal plane.

on the photodiodes located on the confocal plane remains the same even if there is a displacement of the objective lens. This means that coefficients M_0, M_1, and M_2 have a constant value regardless of the lens displacement. Therefore, the signal given by Eq. (21.5) can be expected to be an improved CPP signal.

Optical configuration	
Wavelength (λ)	780 nm
NA (Objective Lens)	0.45
Optical Magnification	5
Working Distance	5.0 mm
EPD of Objective Lens	5.0 mm
Media parameters	
Track Pitch	1.6 μm
Groove/Land Ratio (Duty Ratio)	50, 65%
Groove(Pit) Depth	λ/6 ~ λ/8

FIGURE 21.23 Optical specification.

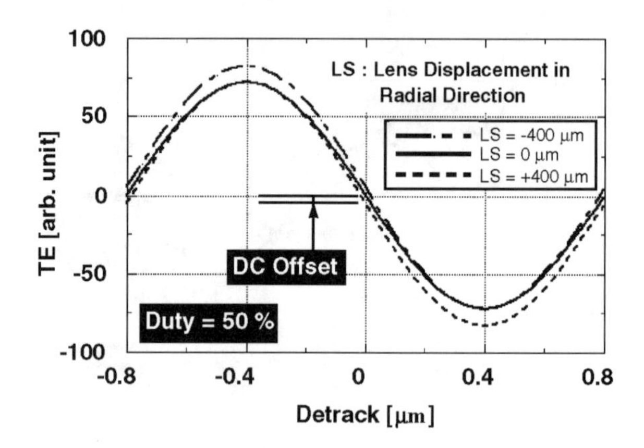

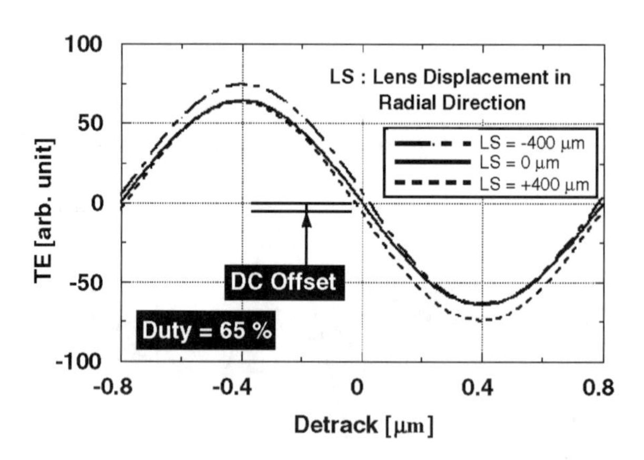

Groove Pitch = 1.6 μm
Groove Depth = λ/8 (λ = 0.78 μm)
Focus Condition : On Focus

FIGURE 21.24 Calculation result of dependence of lens displacement in radial direction: tracking-error signals by coefficient M_1.

To confirm this hypothesis, the results shown in Figures 21.20 and 21.21 were used again to estimate stability through numerical analysis using the least-squares method. Figure 21.24 shows the calculated results of the TE signals derived using this new method. The DC offsets for a lens displacement of ±400 μm in the radial direction derived using this method are about 10%, while those derived using the conventional push–pull method in Figure 21.20(a) are more than 40%. These results suggest that this new method is more stable than the CPP method for the disk parameters, such as duty ratio, while maintaining the advantage as regards lens displacement. Here, the reason Figure 21.24 does not show the result for the deviation along the optical axis within the focal depth is that the $S_{cpp'}$ is inherently independent of the focus condition.

This kind of differential operation given by Eq. (21.5) is easily performed using the well-known signal-processing method. It is efficient to measure variations in the CPP signal S_{cpp} while perturbing the lens in the direction of the optical axis in a constant cycle. In the tracking servo with this method actually employed, for example, in the case of a CD system, a 1-MHz or higher frequency is appropriate as the constant cycle, considering that the bandwidth of the FE and TE signals is 50 kHz or less.

Even if there is a small deviation of focus due to detecting $S_{cpp'}$, any conventional FE detection method, such as the astigmatic method or the spot size detection method, can be used because the focusing condition in this situation is still within the focal depth.

21.4.5 CLC with Confocal Knife Edge (CKE) Structure

As noted above, the CPP signal can be improved by additional signal processing. However, it is desirable to overcome this technical problem involving the tracking servo when using a CLC device without any additional servo or processing. In this section, a new type of CLC device is introduced. Using this new CLC device, the issue is optically solved while the advantages of the original CLC device are retained [Mizuno et al., 1999a].

Figure 21.25 shows a schematic representation of an optical pick-up using the new type of CLC, and Figure 21.26 shows the laser scanning microscope photographs and scanning electron microscope (SEM) photographs of the optical pick-up device actually fabricated. This pick-up device consists of a monolithic optical element (which includes an AlGaAs LD, eight GaAs PDs, and a pyramid-shaped prism mirror positioned near the confocal plane) and a glass window.

As in the first CLC device, the light beam emitted from the LD is reflected by the built-in mirror, which is one of the prism mirror surfaces—{111}B plane fabricated by the crystal growth—and is focused through the objective lens onto the disk surface. Here, the beam emits at 90° using the 9.7° of [001] off-substrate, although the angle between the {111}B and the (100) plane is 54.7°. After reading out the pit signals on the disk, the light beam returns on the same path. The returned light beam, spread by diffraction, is split into two beams by the other two prism mirror surfaces—the {110} planes. The split light beams then irradiate on the two quadrant photodiodes on the GaAs substrate to obtain the FE, TE, and RF signals.

The glass window positioned immediately above the prism mirror functions as a thin slit to block the portion of the incident light beam that is impinging directly on the photodiodes when the focusing condition is out of the peak-to-valley range of the FE signal.

This pick-up device is called a CLC with a CKE structure—simply a CKE device, where CKE means confocal knife edge, because the pyramid-shaped prism mirror positioned near the confocal plane acts as a knife edge to generate the FE and TE signals.

The FE signal $S_{\mathrm{FE\_CKE}}$, TE signal $S_{\mathrm{TE\_CKE}}$, and RF signal $SR_{\mathrm{RF\_CKE}}$ are given by the following equations.

$$\begin{aligned} S_{\mathrm{FE\_CKE}} &= (I_{\mathrm{PD3L}} - I_{\mathrm{PD2L}}) + (I_{\mathrm{PD3R}} - I_{\mathrm{PD2R}}) \\ S_{\mathrm{TE\_CKE}} &= [(I_{\mathrm{PD3L}} + I_{\mathrm{PD1L}}) - (I_{\mathrm{PD4L}} + I_{\mathrm{PD2L}})] - [(I_{\mathrm{PD3R}} + I_{\mathrm{PD1R}}) - (I_{\mathrm{PD4R}} + I_{\mathrm{PD2R}})] \\ S_{\mathrm{RF\_CKE}} &= [I_{\mathrm{PD1L}} + I_{\mathrm{PD2L}} + I_{\mathrm{PD3L}} + I_{\mathrm{PD4L}}] + [I_{\mathrm{PD1R}} + I_{\mathrm{PD2R}} + I_{\mathrm{PD3R}} + I_{\mathrm{PD4R}}], \end{aligned} \tag{21.6}$$

where I_{PD1L}, I_{PD2L}, I_{PD3L}, I_{PD4L}, I_{PD1R}, I_{PD2R}, I_{PD3R}, and I_{PD4R} are the signals detected on each photodiode PD1L, PD2L, PD3L, PD4L, PD1R, PD2R, PD3R, and PD4R, respectively (see Figure 21.25).

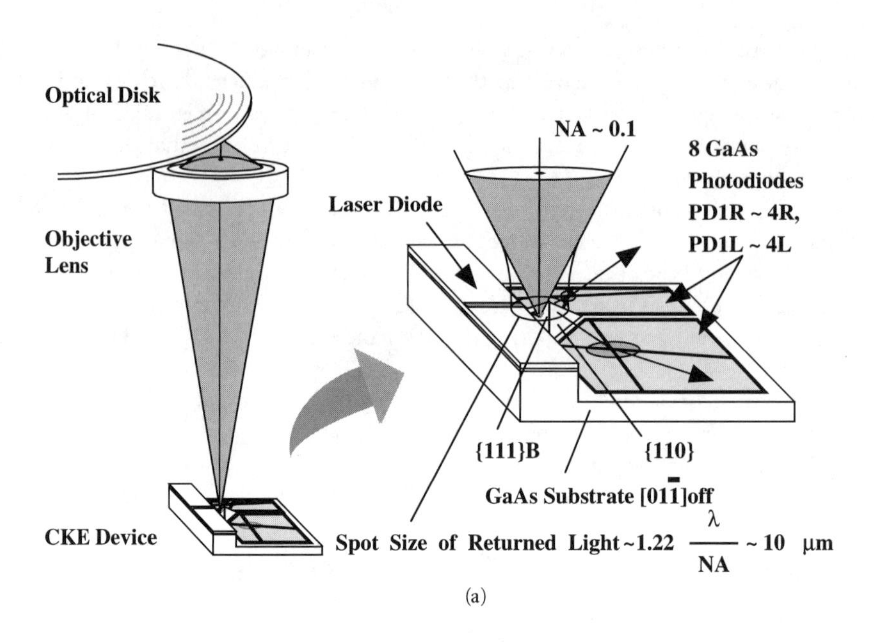

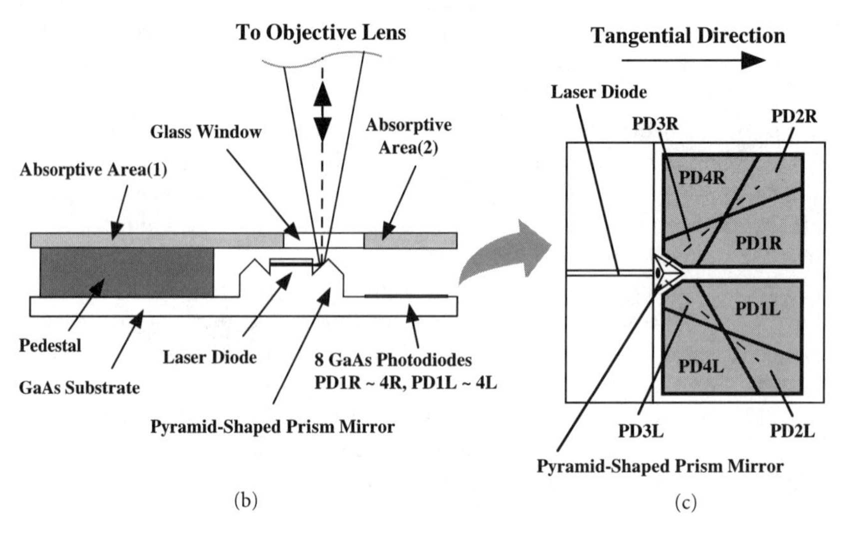

FIGURE 21.25 Schematic representation of CLC with confocal knife edge (CKE) structure: (a) overview; (b) cross-section; (c) top view.

Focusing error detection is based on Foucault's method. Considering the moving direction of the impinging beams on each quadrant PDs for radial lens displacement, we can understand that this FE signal-detection method is not influenced by radial lens displacement.

The TE detection method based on the differential push–pull detection, which we call the CKE push–pull method, solves the technical problems of the CPP method while retaining the advantages of the CPP method. The problems with the CPP signal, caused by the interference between the ±1st-order diffractions from the pits on the optical disk, can be solved by making the split light beams impinging on the pyramid-shaped prism mirror propagate to the farfield region. As a result of this propagation of several tens of microns or more, the ±1st-order diffraction light beams are separated onto each quadrant's PDs. Therefore, the interference term is eliminated. In addition, the influence of the deviation on the focusing condition is suppressed, just as it is with conventional push–pull detection in the farfield region.

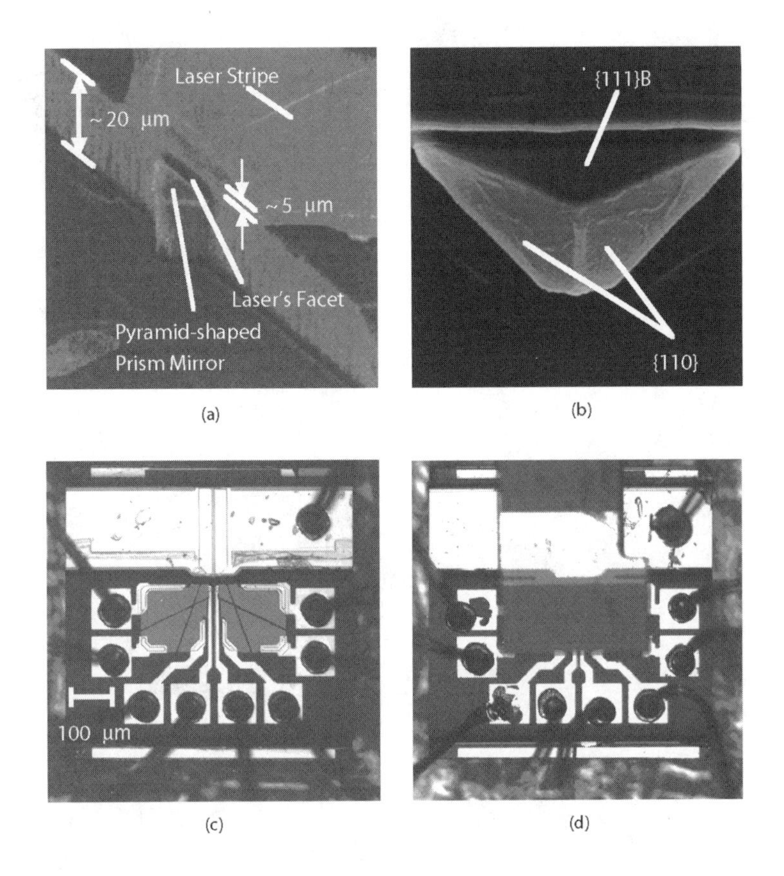

FIGURE 21.26 Fabricated CKE device: (a) pyramid-shaped prism mirror beside the laser's facet; (b) pyramid-shaped prism mirror; (c) original CKE device; (d) CKE device with a glass window.

Figure 21.27 shows the calculated results of the TE signals for various parameters. The calculations are based on the same scalar theory used to derive the results in Figures 21.20 through 21.24 using the parameters in Figure 21.32. The results presented in this figure suggest that this TE signal-detection method can be used as a stable tracking servo method.

As noted above, a glass window is fitted on the CKE device to yield an appropriate FE signal. Figures 21.28 and 21.29 show the details of the fabricated glass window and a schematic representation of the role of this window. As can be seen in Figure 21.29, this glass window functions as a shutter to eliminate the direct incident beam onto the two quadrant PDs not via the pyramid-shaped prism mirror, when the focusing condition is out of the peak-to-valley range of the FE signal. Therefore, appropriate FE and pull-in signals can be obtained. In future mass-production processes, these glass windows will be incorporated into the wafer batch process using existing technologies such as rear electrode, via hole, etc., although the glass windows are provided in pairs for each CKE device, as in Figure 21.26. This allows production of this device without degrading its monolithic aspects. Using the fabricated device shown in Figure 21.26, the FE and pull-in signals, TE signal, and RF signal were demonstrated.

Figure 21.30 shows the experimental results of the FE and pull-in signals. Comparing the signals using a CKE device with and without a glass window, it is apparent that the fitting of a glass window to the CKE device helps to obtain appropriate FE and pull-in signals. The peak-to-valley range of the improved FE signal was about 3 μm. On the other hand, Figure 21.31 shows the experimental results of the TE signals. Using the CKE push–pull method, the DC offsets were less than 20% of the amplitude for a radial lens displacement of ±400 μm. This indicates that the DC offsets were reduced to less than 1/3 of those of the conventional push–pull signal $S_{PP,}$ which is obtained by the CKE device using the

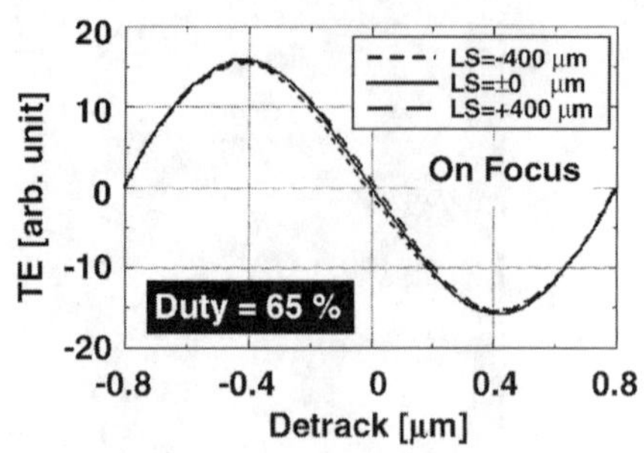

Groove Pitch = 1.6 μm, Groove Depth = λ/8 (λ = 0.78 μm)
Focus Condition: On Focus
LS: Lens Displacement in Radial Direction

(a)

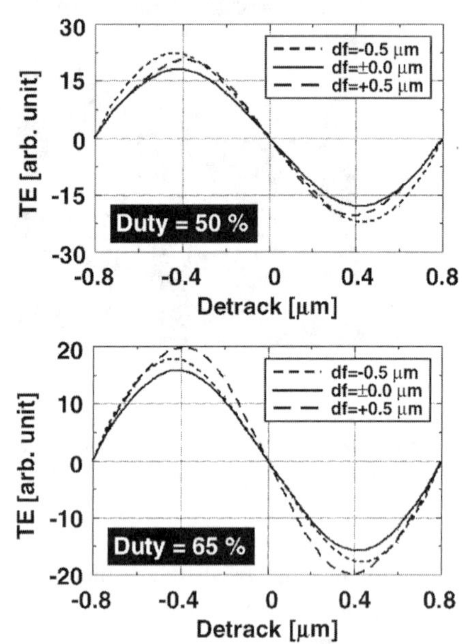

Groove Pitch = 1.6 μm, Groove Depth = λ/8 (λ = 0.78 μm)
Lens Displacement = 0 m
df: Deviation along Optical Axis within Focal Depth

(b)

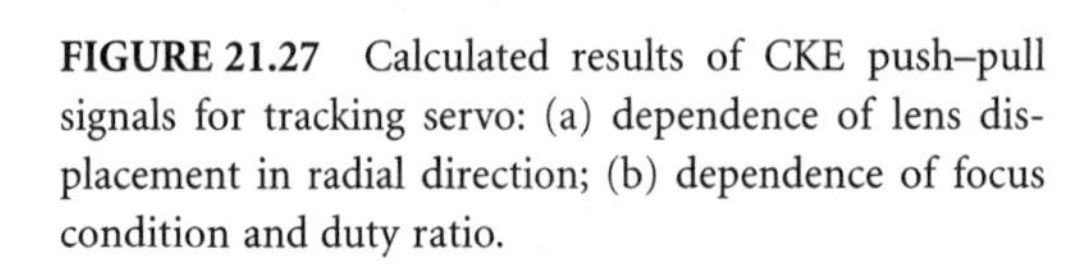

FIGURE 21.27 Calculated results of CKE push–pull signals for tracking servo: (a) dependence of lens displacement in radial direction; (b) dependence of focus condition and duty ratio.

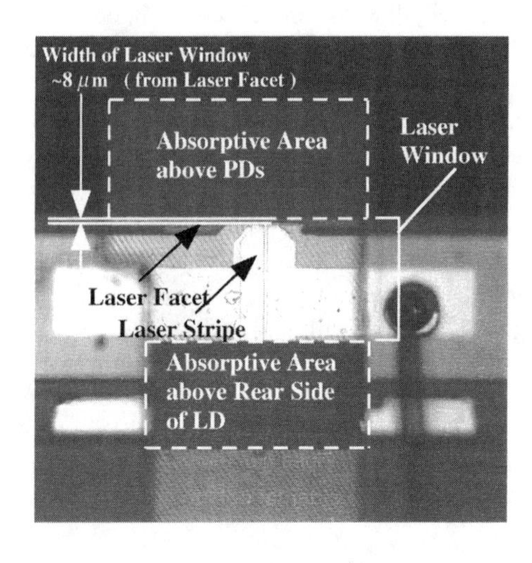

FIGURE 21.28 Relative position between CKE device and glass window.

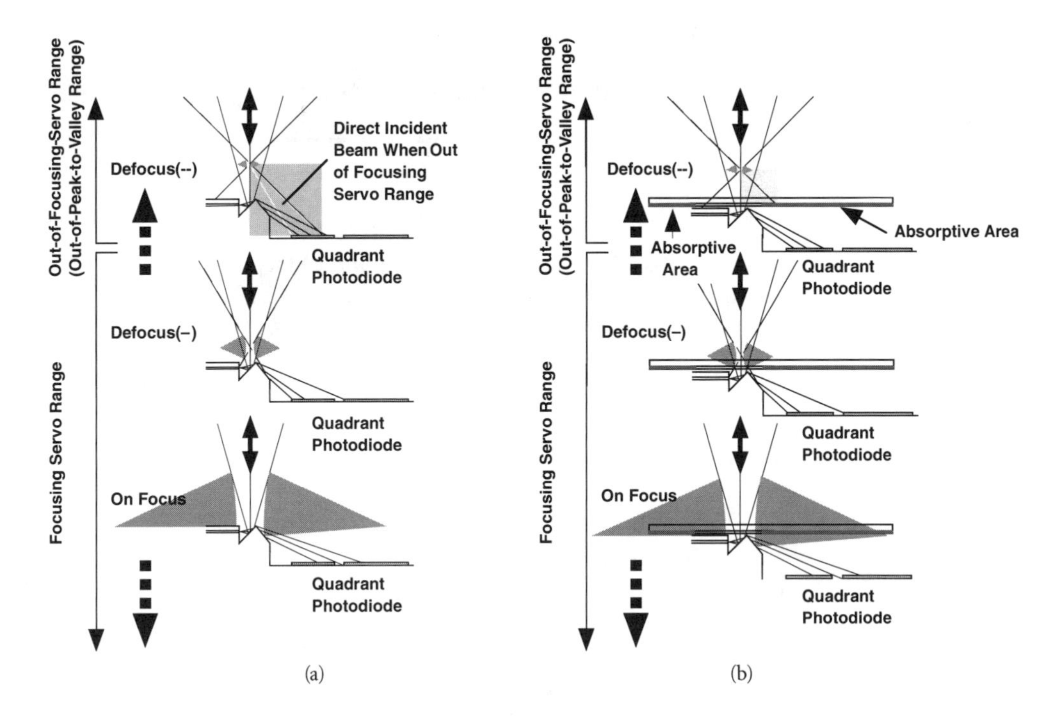

FIGURE 21.29 Schematic representation of the effect of a glass window: (a) original CKE device; (b) CKE device with glass window.

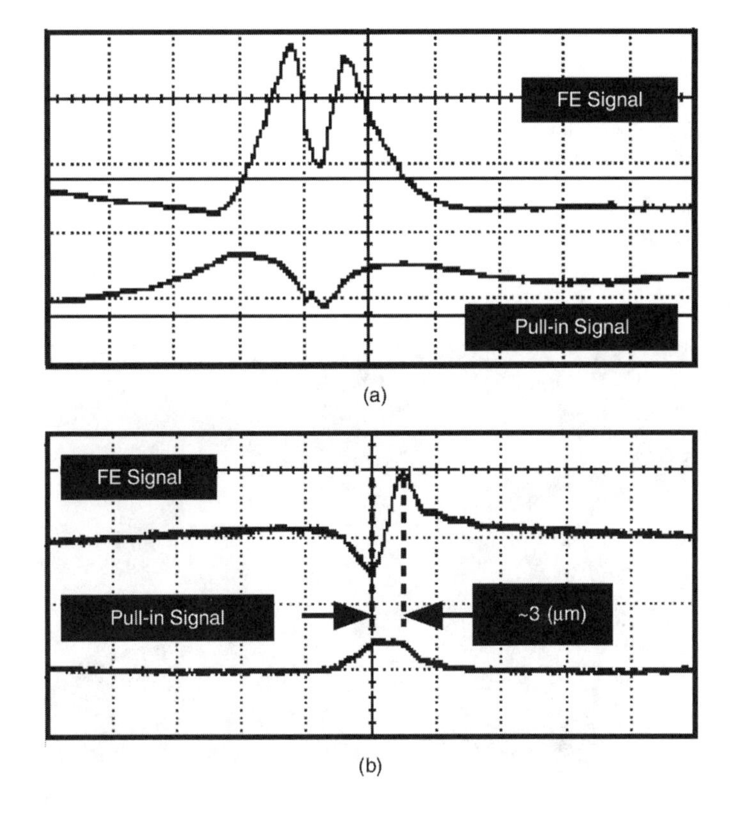

FIGURE 21.30 Experimental result of focusing error and pull-in signals using CKE device: (a) original CKE device; (b) CKE device with glass window.

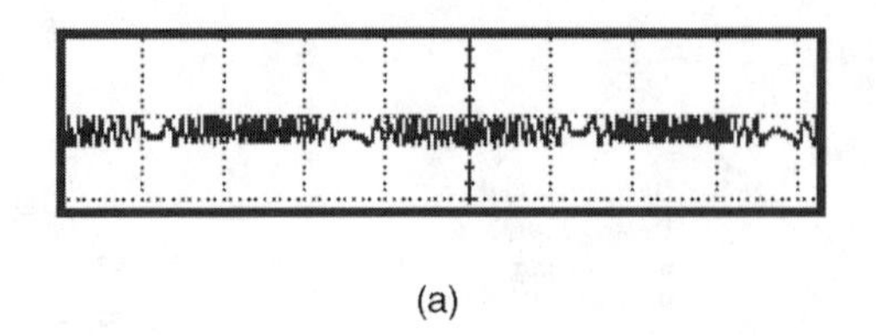

(a)

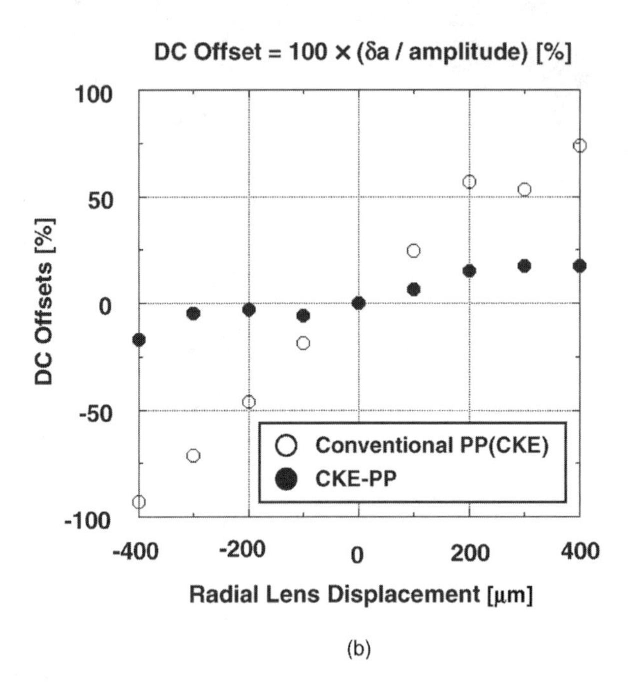

(b)

FIGURE 21.31 CKE push–pull tracking-error signal detection method: (a) CKE push–pull signal; (b) CKE push–pull tracking-error signal detection method.

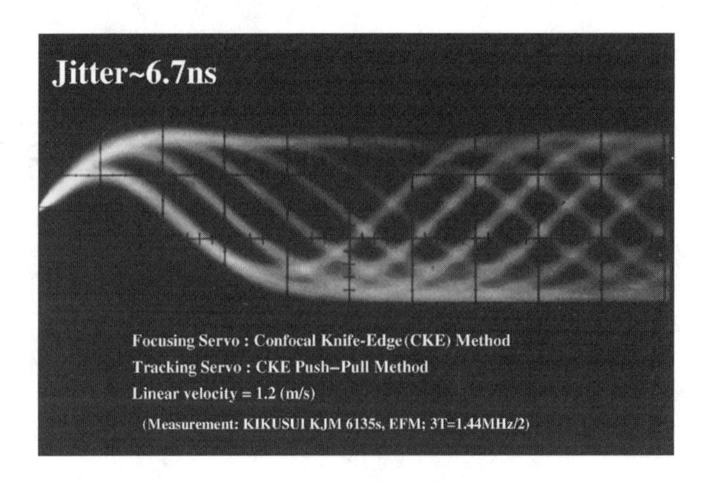

FIGURE 21.32 Eye patterns of compact disk.

following Eq. (21.7):

$$S_{PP} = [(I_{\mathrm{PD3L}} + I_{\mathrm{PD4L}}) - (I_{\mathrm{PD1L}} + I_{\mathrm{PD2L}})] - [(I_{\mathrm{PD3R}} + I_{\mathrm{PD4R}}) - (I_{\mathrm{PD1R}} + I_{\mathrm{PD2R}})] \tag{21.7}$$

These results suggest that a stable TE signal in the face of radial lens displacement can be obtained using the CKE push–pull method.

With these focusing and tracking servos in a CKE device containing a glass window, CD playback was demonstrated as shown in Figure 21.31. The jitter was 6.7 ns with a line velocity of 1.2 m/s (measurement: KIKUSUI KJM6135s, EFM; 3T = 1.44MHz/2 = 0.72MHz). These results indicate that the device can be applied to a practical optical pick-up.

21.4.6 Summary

An optical pick-up device, which is based on a confocal optical configuration and is produced using a monolithic fabrication process, is presented as a candidate for the next-generation optical pick-up device. The inherent characteristics of the servo and RF signals using the confocal optical configuration are described based on the results of numerical analysis. The demonstration on a compact disk system using a confocal laser coupler (CLC) with a confocal knife-edge (CKE) structure indicates that this type of device can be applied to a practical optical pick-up.

21.5 Conclusion

We have discussed integrated optical pick-up devices, focusing on two particular devices. One is the hybrid optical pick-up shown in Figure 21.1. In addition to its application to CDs, the microprism type has been applied to the magneto-optical recordable disk system using an MD, or MiniDisc. The device has a simple structure using a KTP crystal to analyze magneto-optical modulation. A hybrid integrated MD head showed excellent performances and enabled us to reach our goals, as discussed in Section 21.2.3.

The other device is the monolithically integrated device. A new concept of optical device, the confocal laser coupler, was fabricated by using a submicron controlled crystal growth technique and photolithography. We fabricated a monolithic device and confirmed its operation as an optical pick-up for the first time. Fabrication technology is still under investigation with an eye to mass production. The operation of the confocal optical device was verified experimentally, and solid agreement with theory was demonstrated. The device's potential high performance and low cost suggest it may replace the current optical pick-up devices. We propose various potential applications [Matsuda et al., 1998] such as a new class of sensor and a microscope optical device.

The recent video game player PS2 (Play Station 2) is another example of the successful application of an integrated optical pick-up device and is a good combination of hybrid integration and monolithic integration. The laser diode chip consists of two monolithically integrated laser diodes whose emission wavelengths are 780 and 650 nm. This laser array is mounted on a single photodiode IC with a microprism and can read CD and DVD signals using a single optics. This two-wavelength laser coupler [Nemoto and Miura, 2001] is already in mass production and will be applied to a recordable system.

The task at hand is to continue recent work on hybrid optical ICs, monolithic optical ICs, and their combination with work on high-density recording technology.

Defining Terms

ADIP: **Ad**dress **i**n **P**regroove.

APC: **A**utomatic **P**ower **C**ontrol.

CD: **C**ompact **D**isk, for music reproduction. Recently, the **CD-R** has been used in the single-write mode of recording.

CKE: **C**onfocal **K**nife **E**dge device; another type of monolithic device that uses a crystal prism as a microprism and a knife edge for focusing.

CLC: Confocal Laser Coupler; a basic optical device unit of monolithic devices. In combination with a micromirror and a photodiode, CLC works as a radiation source with a confocal detector for unique optical pick-up.

CPP: Confocal Push–Pull.

DVD: Digital Versatile Disk; a video disk read-only memory system with a capacity of 4.7 GB.

DVD-ROM: DVD Read-Only-Memory for computer use.

FAPC: Front Automatic Power Control.

FE: Focusing Error servo signal.

HOE: Holographic Optical Element; produces a three-dimensional image of an object and is applied to an optical pick-up system as a beam splitter.

KTP: Biaxial-refractive crystal of $KTiOPO_4$.

LC: Laser Coupler; a hybrid integrated optical pick-up device with a microprism on a silicon photodiode IC.

LD: Laser Diode.

LOP: Laser on Photodiode; the basic hybrid integrated optical chip developed for manufacturing packaged laser diodes.

MD: MiniDisc, a recordable small disk that uses a magneto-optical recording system. MD is a small-size disk 64 mm in diameter and has a memory capacity of about 140 MB.

MDLC: MiniDisc Laser Coupler.

MO: Magneto-Optical disk for computers; 3.5″ ISO system has maximum capacity of 1.3 GB.

PBS: Polarizing Beam Splitter.

PD: Photodiode.

RF: radio-frequency signal.

TE: Tracking Error servo signal.

References

Arai, A., Hayashi, T., Nakamura, T., Nagata, T., Takashima, M., Aikoh, H., and Tomita, H., Integrated optical head with new one-beam tracking detection for magneto-optical disk, *Tech. Dig. Joint Magneto-Optical Recording Int. Sym./Int. Symp. Optical Memory*, 1997, pp. 202–203.

Bouwhuis, G., Braat, J., Huijser, A., Pasman, J., van Rosmalen, G., and Schouhamer-Immink, K., *Principles of Optical Disc Systems, Philips Research Laboratories*, Adam Hilger, Ltd., Boston, pp. 70–80, 1985.

Carasso, M. G., Peak, J. B. H., and Sinjou, J. P., The Compact Disc Digital Audio System, *Philips Tech. Rev.* 40, 1982.

Doi, M., Sahara, K., Narui, H., and Matsuda, O., Optical Device Having a Light Emitter and a Photosensor on the same Optical Axis, U.S. Patent No. US5,679,947,1997.

Hong, C. S., Integrated optoelectronics for communication and processing, *Proc. SPIE*, 1582, 1992.

Hopkins, H. H., Diffraction theory of laser read-out systems for optical video discs, *J. Opt. Soc. Am.*, 69, 4–24, 1979.

Horinouchi, S., Yoshinaka, H., Takeuchi, S., Higo, K., Koga, T., and Kobayashi, F., Integrated magneto-optical disk head using laminated glass plates, *Tech. Dig. Int. Symp. Optical Memory*, 1995, pp. 53–54.

Kishima, K., Ichimura, I., Yamamoto, K., Osato, K., Kuroda, Y., Iida, A., Masuhara, S., and Saito, K., Demonstration of 45 Gbit/in^2 in near-field phase-change recording, Optical Data Storage (ODS) 2001, *Tech. Dig.*, WB5, 280–282, 2001.

Matsuda, O., Doi, M., Narui, H., Sahara, K., and Nakao, T., A monolithically integrated confocal laser and detector, *Tech. Dig. Laser & Electro Optics '98 Conf.*, San Francisco, 1998, pp. 57–58.

Matsumoto, Y., Yamamoto, E., Taniguchi, T., Maeda, K., Kume, H., Matsuda, O., and Hata, I., Optical features of an integrated "laser coupler" optical pickup, *Proc. 6th Sony Research Forum*, 1996, pp. 541–544.

Mizuno, T., Doi, M., Higuchi, Y., Taniguchi, T., Okano, N., Nakao, T., Narui, H., and Matsuda, O., A monolithic confocal laser coupler for an optical pick-up, *J. Appl. Phys.*, 38(4A), 2001–2006, 1999a.

Mizuno, T., Doi, M., Narui, H., and Matsuda, O., Analysis of confocal push detection of tracking-error signals of optical disk player, *Jpn. J. App. Phys.*, 38(7A), 4073–4078, 1999b.

Mizuno, T., Narui, H., and Matsuda, O., Deviation of analytical solution for confocal push–pull (CPP) signals and improved CPP detection method using signal processing, *Jpn. J. App. Phys.*, 38(7A), 4079–4083, 1999c.

Narui, H., Doi, M., Nakao, T., Sahara, K., and Matsuda, O., A monolithic confocal pickup devices, *Topical Meet. Optical Data Storage, Aspen, Tech. Dig.*, 8, 50–51, 1998.

Nemoto, K. and Miura, K., A laser coupler for DVD/CD playback, *JSAP Int.*, 3, 9–14, 2001.

Nishi, N., Toyota, K., Okamatsu, K., Saito, K., Horie, K., Tanaka, K., and Nemoto, K., Integrated Optical Device—MiniDisc Laser Coupler, *Topical Meet. Optical Data Storage 1998, Aspen, Tech. Dig.*, 8, 52–54, 1998.

Nishi, N., Toyota, K., and Saito, K., A method to design an integrated MO head using a diffraction calculation in biaxial crystals, *J. Magnetic Soc. Jpn.*, 23(Suppl. s1), 245–246, 1999.

Oka, M., Fukumoto, A., Osato, K., and Kubota, S., A new focus servo method for magneto-optical disk systems, Proc. Int. Symp. Optical Memory (1987), *Jpn. J. Appl. Phys.*, 26 (Suppl. 26-4), 187, 1987.

Sabert, H., Mizuno, T., Doi, M., and Narui, H., Confocal knife edge focus servo for optical disk systems, *Proc. 8th Sony Research Forum*, 1998, p. 310.

Saito, K., Sato, S., Shino, K., and Taniguchi, T., Device for magneto-optic signal detection with a small crystal prism, *Appl. Opt.*, 39(8), 1315–1322, 2000.

Sony Corp., MiniDisc System Description, *Rainbow Book*, 1992.

Sony Corp. and N.V. Philips, Compact Disc System Description, *Red Book*, 1980.

Suhara, T., Integrated-optic disk pickup devices: hybrid to monolithic integration, *Tech. Digest Int. Symp. on Optical Memory and Optical Data Storage*, Maui, 1996, pp. 284–286.

Tamada, H., Yamaguchi, T., Doumuki, T., Matsumoto, S., Nemoto, K., Narui, H., Nakao, T., and Matsuda, O., Aluminum-wire grid polarizer for a compact magneto-optic pickup device, *1997 Optical Data Storage Topical Meet., Tucson, Tech. Dig.*, 1997, pp. 22–23.

Ukita, H., Sugiyama, Y., Nakada, H., and Katagiri, Y., Read/write performance and reliability of a flying optical head using a monolithically integrated LD-PD, *Appl. Opt.*, 30, 3770–3776, 1991.

Wiechmann, W., Eguchi, N., and Kubota, S., A walk-off plate for magneto-optical disk system, *Tech. Dig. Int. Symp. Optical Memory*, 1995, pp. 55–56.

Yoshida, Y., Miyake, T., Sakai, K., and Kurata, Y., Optical pickup using blazed holographic optical element for video disk players, *Jpn. J. Appl. Phys.*, 33(7A), 3947–3951, 1994.

Yoshida, S., Minami, K., Okada, K., Yamamoto, H., Ueyama, T., Sakai, K., and Kurata, Y., Integrated hologram pickup with optical waveguide device for magnetooptical disk players, *Jpn. J. Appl. Phys.*, 37, 4401–4404, 1998.

Yoshida, S., Minami, K., Okada, K., Yamamoto, H., Ueyama, T., Sakai, K., and Kurata, Y., Optical pickup employing a hologram-laser-photodiode unit, *Jpn. J. Appl. Phys.*, 38(2B), 877–882, 2000.

Yoshikawa, A., Laser-detector-hologram unit for thin optical pick-up head for CD player, *IEEE Tran. Components, Pack. Manuf. Technol.*, B18, 245, 1995.

22

Optical MEMS: Light Source Array

Sukhan Lee
Samsung Advanced Institute of Technology
Yongin, Kyungghi-Do, South Korea

Jideog Kim
Samsung Advanced Institute of Technology
Yongin, Kyungghi-Do, South Korea

Yongkwon Kim
Seoul National University
Seoul, South Korea

22.1 Introduction

MEMS (Microelectromechanical systems) technologies make it possible to construct sensors and actuators on the micron scale, using micromachining technologies that are derived from semiconductor process technologies [Motamedi, 1994]. In the 1970s integrated sensors that incorporated driving circuitry were first developed from the semiconductor processes. In the early 1980s micromechanical components such as the spring, cantilever, and bridge were developed. In the late 1980s micromechanical components separated from the substrate were developed; examples include micro tweezers, micro motors, and micro gears. The 1990s witnessed the integration of logic circuits with actuators and sensors.

Micromachining technology has many advantages such as miniaturization, low cost, automatic alignment, and reliability through batch processes. Application areas include magnetic heads for hard disk drives, inkjet printer heads, micro displays (information technology), cell manipulators, medical equipment for diagnosis and surgery, artificial organisms (biotechnology), sensors, automatic controlling systems (sensing and controlling), optical switches, variable foci mirrors, micro lenses (optical components), filters, nozzles, valves, and motors (ultra-small devices).

A MOEMS (micro-optoelectromechanical system) has many natural advantages over conventional methods in micro-optical systems that are derived from micromachining technology. It can produce a variety of optical components such as micro mirrors, micro lenses, micro stages, micro gratings, etc. Among these, micro mirrors play an important role in manipulating the direction of a laser beam in free space. For example, micro mirrors are used to reflect beams in a plane or to steer light in any other direction [Wu et al., 1997].

A micro mirror or its array can also be used to change the direction of light upward to the substrate at a right angle so that it can be used as a multiple-position light source [Jang et al., 2001]. The principle of a multiple-position light source is illustrated in Figure 22.1. A divergent beam from an optical fiber is first collimated by, for example, a ball lens. (Ideally, the light source itself would be integrated into the device. Here it is assumed that the light source itself is externally supplied by an optical fiber.) Then, the collimated

0-8493-1162-4/03/$0.00+$1.50

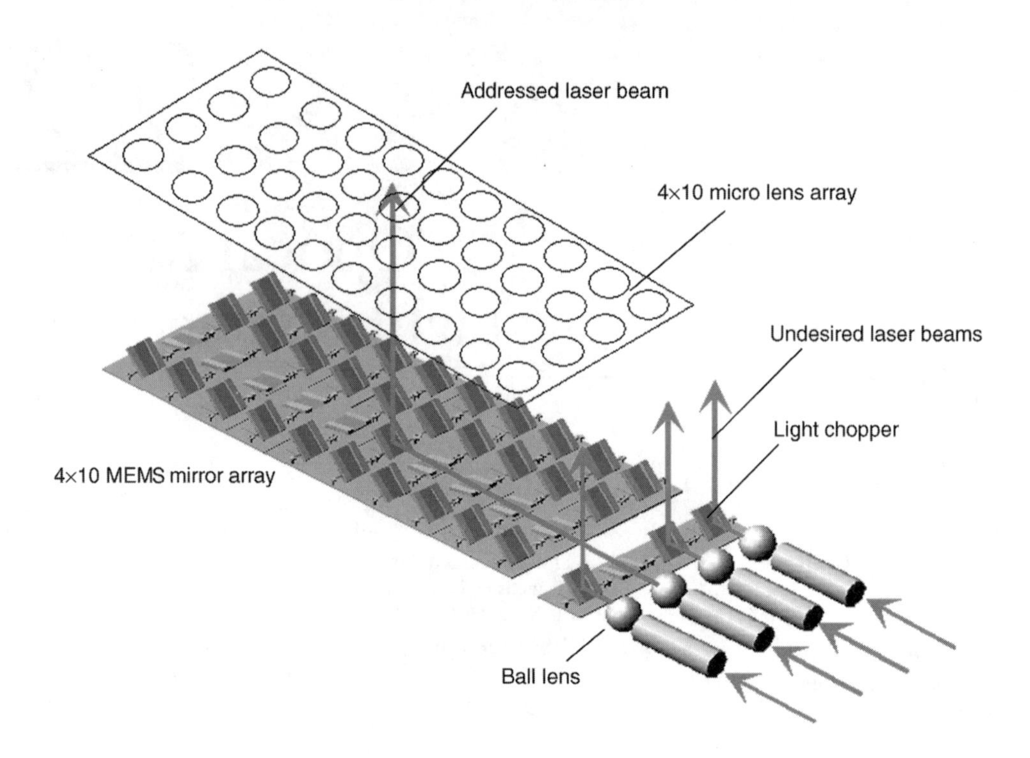

FIGURE 22.1 Configuration of a 4 × 10 light source array.

light impinges on an addressed micro mirror that is rotated by 45 degrees and is deflected upward to the substrate. Finally, the reflected light is appropriately shaped, for example, by a micro lens, as shown in Figure 22.1, to be used in a particular application. This kind of device was originally introduced as a multiplexer for the reference beam in a volume holographic data storage system [Kim et al., 1999]. When developing a small-sized storage system for mobile devices, components made by MEMS or MOEMS technology would be especially advantageous.

In general, a MEMS mirror can be actuated by electrostatic force or magnetic force. In the application considered here a combination of electrostatic and magnetic forces is used to maintain 0 or 45°. Magnetic force is used to actuate a mirror by 45°, and electrostatic force is used to hold mirrors that are not to be rotated. Ball lenses are assembled with optical fibers in order to collimate the divergent beam from the optical fibers. A micro lens array is located on top of the MEMS mirror arrays to shape the reflected light beam from the MEMS mirror array. These lenses are fabricated by reflow process using heat treatment of polymer patterned on glass. The MEMS mirror array and micro lens array can be integrated by a bench fabricated using a silicon-etching process. As a result, an addressable light source array for an angle multiplexed holographic memory could be developed [Jang et al., 2001].

22.2 Design in Optical MEMS

22.2.1 Multiplexing Device

A schematic view of a multiplexing device is shown in Figure 22.2. It is necessary to align the center of the ball lens with the center of the horizontally reflected beam. The etching process is adopted for aligning the ball lens, and the fiber bench is designed to be fabricated simultaneously with mirror fabrication. Ideally, a light source (for example, a semiconductor laser diode) should be integrated with the multiplexing device.

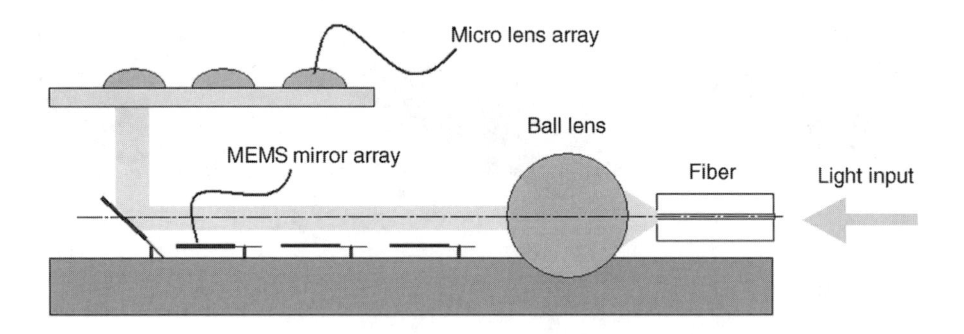

FIGURE 22.2 Schematic view of the multiplexing device.

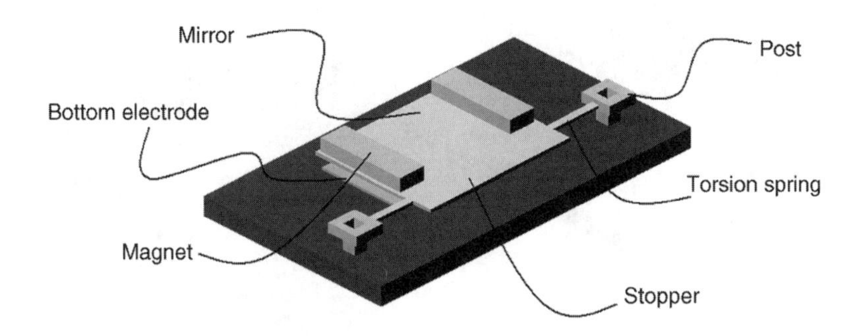

FIGURE 22.3 Schematic view of a MEMS mirror.

However, it is connected with a pigtailed fiber here for the sake of simplicity. The beam of light from an optical fiber is modeled as a Gaussian beam. The optical design of the ball lens is optimized by requiring that the sizes of the collimated Gaussian beam does not much change over the distances from the first mirror to the last mirror along the direction of beam propagation.

22.2.2 MEMS Mirror

A MEMS mirror is designed to reflect the collimated beam by 45 degrees. Al is used for the spring and mirror and Ni is used for the soft magnet. As shown in Figure 22.3, the shape of the magnet is determined such that it occupies a reasonable area in the mirror and increases the anisotropy in the magnet in order to improve the actuating characteristics with the external magnetic field. Each mirror has a bottom electrode for electrostatic clamping, a mirror plate for light reflection, a magnetic material for magnetic actuation, a stopper for angular deflection control, and springs for a restoring mechanism.

The right-angle reflection perpendicular to the substrate requires 45° of angular deflection of a micro mirror. Therefore, sufficiently long springs have been designed so as not to be plastically deformed within full angular operation. Another consideration is an actuation method for large angular deflection. Electromagnetic actuation is known to be more advantageous than other methods involving large angle actuation, and Ni is used as a soft magnetic material for electromagnetic actuation [Judy and Muller, 1997].

The last consideration regarding the device is a mechanical stopper, which aims to control angular deflection precisely using the dimension ratio of the mirror. The function of the stopper is to touch the substrate and prevent the mirror from further rotation, when the full 45° rotation is achieved.

A simplified view of an operation sequence is shown in Figure 22.4. As shown in Figure 22.4(a), all of the mirrors are parallel to the substrate in the initial state. To address a specific mirror, all the other mirrors are clamped by applying a voltage, Vc, as in Figure 22.4(b). Then, if a magnetic field B is applied, this specific mirror, which is not clamped by electrostatic force, is deflected, as in Figure 22.4(c).

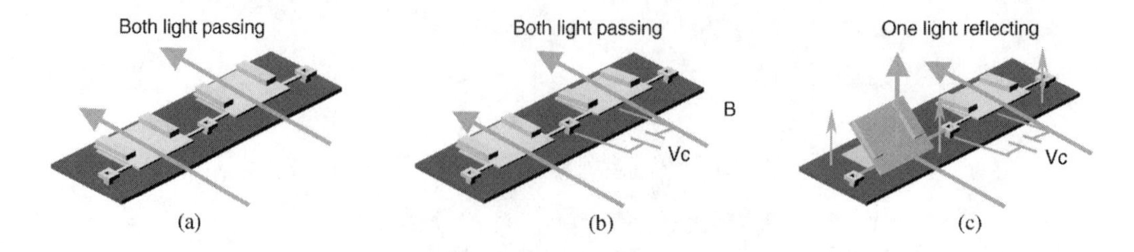

FIGURE 22.4 Operation sequence of a 2 × 1 micro mirror: (a) initial state; (b) addressing state; (c) actuating state.

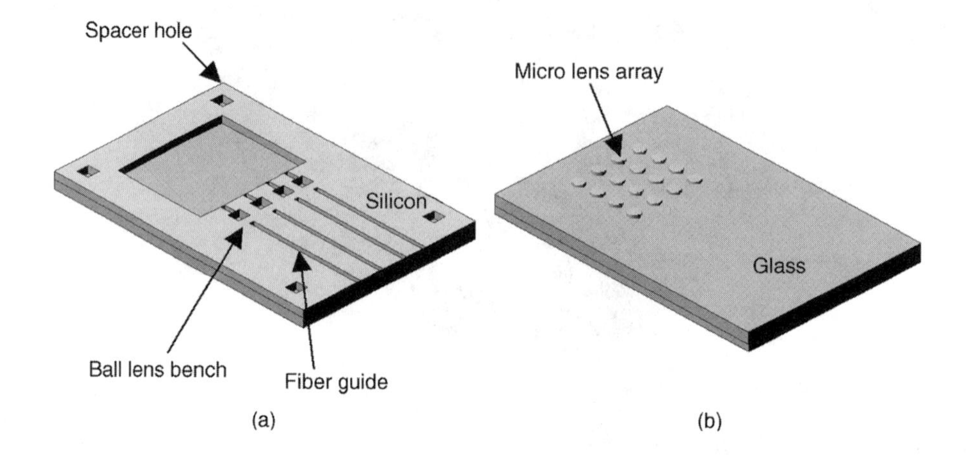

FIGURE 22.5 Simplified view of top plate for micro lens array, ball lens bench, and fiber guide: (a) bottom side view; (b) topside view.

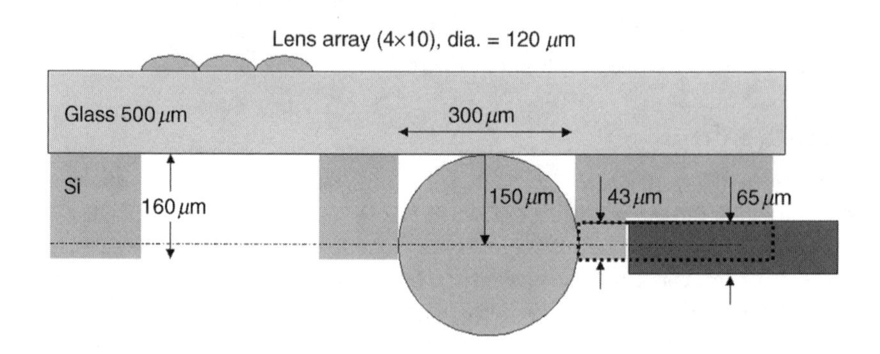

FIGURE 22.6 Schematic design of micro lens array.

The micro lens array is fabricated on the top plate whose simplified view is shown in Figure 22.5. On the bottom side are the ball lens bench and fiber guide using the silicon deep-etch process. On the glass side the micro lens array is formed using a polymer reflow process for collimating the output beam. Figure 22.6 shows a cross-sectional view of the top plate after integration with the micro lens array, ball lens, and optical fiber. The top plate has three functions: (1) the micro lens array shapes the output light, (2) the silicon bench holds the ball lenses, and (3) the silicon bench acts as fiber guide. The top plate is required for maintaining an accurate space for the movable mirrors beneath it. In this system ball lenses are also used as spacers to maintain the gap between the mirror array and the micro lens array. A useful process for making benches for fibers and ball lenses is the silicon and glass bonding process. This process offers relatively accurate dimensions patterned by a deep silicon etching system, strongly built structures made from silicon, and a transparent substrate of glass.

22.3 Theoretical Analysis of Mirror Actuation

22.3.1 Magnetic and Mechanical Analysis

To understand the actuating mechanism, relevant torques and the relationship between them must be known. Figure 22.7 shows a simplified view of various torques generated in a MEMS mirror under an applied external magnetic field *H*. When a magnetic material on the mirror with a net magnetization vector *M* is placed in an external magnetic field, *H*, torque T_H exerted on *M* by *H* can be expressed as

$$T_H = VMH\sin\alpha, \tag{22.1}$$

where *V* is the magnet volume and α is the angle between *H* and *M* [Cullity, 1972]. Torque T_H rotates M by an angle θ away from its equilibrium direction, called the easy axis.

However, when *M* rotates away from the easy axis, a magnetic anisotropy torque, T_a, is generated that attempts to realign *M* and the easy axis. T_a can be expressed as

$$T_a = -K_a \sin 2\theta, \tag{22.2}$$

where K_a is called the magnetic-anisotropy constant [Cullity, 1972]. In the MEMS mirrors the dominant magnetic anisotropy is due to the shape of the magnetic material, allowing K_a to be expressed as [Cullity, 1972]

$$K_a = K_{shape} = \frac{1}{2\mu_0}(N_c - N_a)M \tag{22.3}$$

where N_a and N_c are the length and thickness shape anisotropy constants given by expressions derived by Osborn [1945], μ_0 is the permittivity of free space, and *M* is the magnetization. Because the anisotropy torque T_a attempts to bring *M* and the easy axis back together, an equal but opposite torque $-T_a$ is exerted on the easy axis and, hence, on the magnetic material itself.

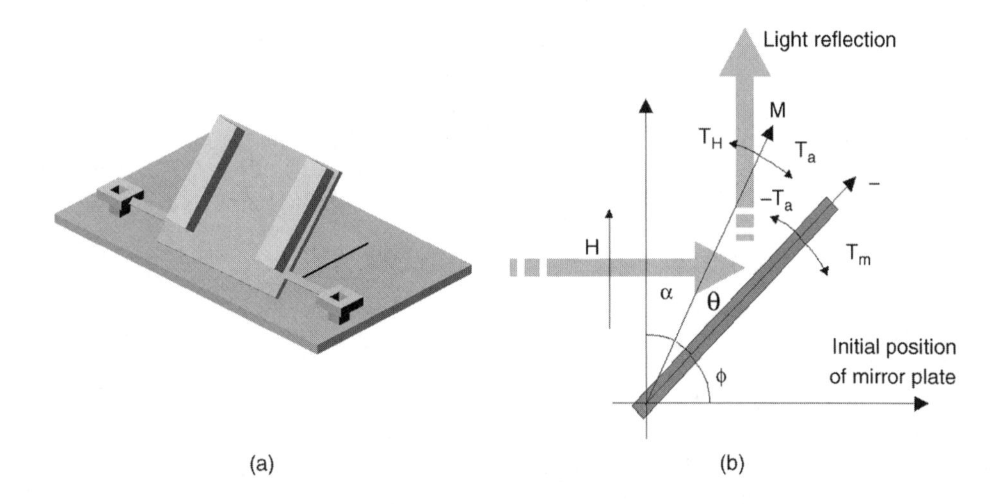

FIGURE 22.7 Relevant torques for magnetic actuation: (a) rotated mirror with magnet; (b) torques exerted on a rotated MEMS mirror and its magnetization *M* under magnetic field *H*.

If the magnetic sample is attached to a structure with torsional stiffness k_ϕ, $-T_a$ will cause the magnetic sample to rotate by an angle ϕ from its original orientation. As ϕ increases, the mechanical restoring torque, T_μ, given by

$$T_m = -k_\phi \phi \tag{22.4}$$

also increases. The torsional stiffness can be expressed under the assumption that each torsional beam of length l_{beam} is straight, has a uniform cross section, and is made of a homogeneous isotropic material that obeys Hooke's law:

$$k_\phi = 2\left(\frac{K_{beam}G + \sigma J_{beam}}{l}\right) \tag{22.5}$$

with cross-section-dependent factor K_{beam}, shear modulus $G = E/2(1 + \nu)$, elastic modulus E, Poisson's ratio ν, residual stress σ, and polar moment of inertia J_{beam} [Roark, 1989].

For a beam with a rectangular cross section of width w_{beam} and thickness t_{beam}, the cross-section shape-dependent factor is expressed as

$$K_{beam} = \frac{a^3 b}{3}\left(1 - \frac{192a}{\pi^5 b}\sum_{n=1,3,5\ldots}^{\infty}\frac{1}{n^5}\tanh\left(\frac{n\pi b}{2a}\right)\right) \tag{22.6}$$

with $2a = \min(w_{beam}, t_{beam})$ and $2b = \max(w_{beam}, t_{beam})$ [Timoshenko et al., 1970]; the polar moment of inertia J_{beam} is expressed as [Roark, 1989]

$$J_{beam} = \frac{1}{12}\left(wt^3 + w^3 t\right). \tag{22.7}$$

The residual stress-dependent factor of k_ϕ can be ignored because even for a stress as high as 1 GPa, the σJ_{beam} term is a very small part of the $K_{beam}G$ term.

If the direction of the external magnetic field remains at a constant angle γ to the original direction of the easy axis, then although $\alpha = \gamma$ initially, it is reduced by θ and ϕ, allowing it to be rewritten as

$$T_H = VMH\sin(\gamma - \theta - \phi) \tag{22.8}$$

In equilibrium the field torque T_H is balanced with the anisotropy torque T_a. The T_H rotates magnetization vector M away from the easy axis, and the T_a acts to align M with the easy axis. In turn, the torque on the magnetic material $-T_a$ is balanced with the mechanical restoring torque T_m. The resulting equilibrium condition is

$$|T_H| = |T_a| = |T_m| = T. \tag{22.9}$$

If the magnitude of magnetization vector M is known, this equilibrium condition can be used to solve for ϕ, θ, and T by Eqs. (22.3), (22.4), and (22.8).

In a simple permanent-magnet analysis it is assumed that the magnitude of magnetization vector M has a fixed value and that $\theta = 0$. However, in a soft-magnetic material analysis M varies as a function of the applied field. As a sample is magnetized by an increasing field H_a, applied along M and given by $H_a = H_{ext}\cos(\gamma - \theta - \phi)$, the magnitude of magnetization vector M increases, and magnetic poles form on the ends of the sample. These magnetic poles generate an H_a applied along M. The magnitude of the demagnetizing field is $H_d = -\frac{N_M}{\mu_0}M$, where N_M is the shape-anisotropy coefficient of the sample in the direction of M [Cullity, 1972]. In out-of-plane devices like those being considered here N_M can be expressed as

$$N_M^2 = N_l^2 \cos^2\theta + N_t^2 \sin^2\theta. \tag{22.10}$$

The net field inside a magnetic sample H_i is the sum of the applied field and the demagnetization field:

$$H_i = H_a + H_d. \tag{22.11}$$

Domain walls in a soft magnetic material move to reduce H_i to the coercive field of the magnetic material H_c, resulting in

$$\pm H_c + H_a = -H_d = \frac{N_M M}{\mu_0}, \tag{22.12}$$

from which the following relation is obtained:

$$M = \frac{\mu_0(\pm H_c + H_a)}{N_M}. \tag{22.13}$$

Substituting H_a and N_M into Eq. (22.13), choosing H_c to be positive (implying that the sample had been magnetized in the positive direction by a previously applied field), and recognizing that M cannot exceed M_s, we can obtain M as in Eq. (22.14) [Judy, 1997]:

$$M \approx \min\left(\frac{\mu_0(H_c + H\cos(\gamma-\theta-\phi))}{\sqrt{N_l^2\cos^2 + N_t^{\,2}\sin^2\theta}}, M_s\right). \tag{22.14}$$

By solving Eq. (22.9) with M described in Eq. (22.14), the deflecting angle of mirrors with the applied external magnetic field can be evaluated.

22.3.2 Analysis of Electrical Clamping for Individual Operation

The addressable actuation of a micro mirror array is made by the electrostatic force between the mirror and the bottom electrode. In the mirror clamped by electrostatic force four kinds of torques—electrostatic torque, eletromagnetic torque, mechanical torque, and gravitational torque—are exerted. These torques are schematically shown in Figure 22.8.

When the mirror is deflected downward, the electrostatic torque and the gravitational torque are exerted in a counterclockwise direction, and the electromagnetic torque and the mechanical torque are exerted in a clockwise direction. If the angular deflection is small, the electromagnetic torque can be expressed as

$$T_H = VMH\cos\theta \tag{22.15}$$

and the gravitational torque can be expressed as

$$T_g = mgr\cos\theta, \tag{22.16}$$

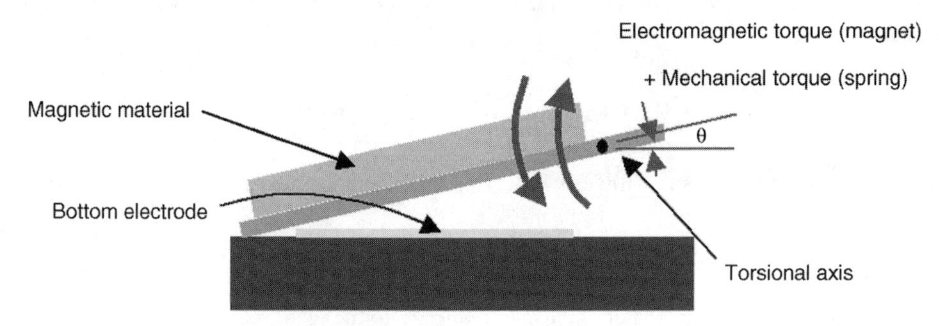

FIGURE 22.8 Torques exerted on a clamped mirror.

where m is mass of the mirror and magnet, g is the acceleration of gravity, and r is the distance from rotation axis to the center of gravity. Gravitational torque is so small compared with electrostatic torque that the effect of gravity can be dismissed. If the mirror is to be clamped under the applied external magnetic field, which is required to deflect the mirror by 45°, the sum of the electrostatic torque and the gravitational torque should be greater than the sum of the electromagnetic torque and the mechanical torque. That is, in order to clamp the mirror Eq. (22.17) must be satisfied:

$$T_e + T_g > T_H + T_m. \qquad (22.17)$$

22.4 Fabrication in Optical MEMS

22.4.1 MEMS Mirror Array

The fabrication process for a MEMS mirror array is shown in Figure 22.9. The fabrication process begins with a thermally oxidized Si wafer. On this substrate the bottom electrode (Al, 2000 Å) is formed to apply the voltage for clamping the mirror (Figure 22.9(a)). Then, the sacrificial layer (AZ4620) is coated on the wafer by a spin coater and baked to prevent any other deformation in postprocessing. The thickness of a sacrificial layer was controlled carefully because the gap-length ratio affects the full angular deflection. Because the sacrificial layer lost its photosensitivity after baking, a hard mask and dry etching process was adapted. As a hard mask plasma-enhanced CVD (PECVD) oxide (3000 Å) was deposited on the sacrificial layer. After the oxide was etched for a contact hole, the sacrificial layer was vertically etched by O_2 plasma, and then the remaining oxide was also removed by blanket etching (Figure 22.9(b)). After these processes Al (5000 Å) for the spring was deposited, and this Al deposited in the hole plays a role in forming electrical contact with the bottom electrode and post. Then, PECVD oxide (5000 Å) is deposited on Al again and patterned as a spring (Figure 22.9(c)). The role of oxide on Al is for etching the stop layer in the post-etching process for the mirror. After this process Al for the mirror (10,000 Å) was deposited on the oxide and Al (Figure 22.9(d), (e)). By using reactive ion etching (RIE), Al for the mirror and spring was patterned, and the oxide on Al for the spring was removed by dry etching. As a result, an Al mirror (15,000 Å) and Al spring (5000 Å) could be made through etching of the stop layer and dry etching.

A soft magnet must be deposited to actuate the mirrors with an external magnetic field. For this purpose, electroplated Ni is adapted. To improve the adhesion and to electroplate the Ni, seed layer Cr/Au (150 Å/1800 Å) was deposited. Thick Ni can reduce the intensity of actuation field because electromagnetic force is proportional to the volume of magnetic material. But thick Ni increases the total mass of a micro mirror, so Ni with a height of 10 μm was fabricated using a negative PMER mold (Figure 22.9(f)). The schematic view of the electroplating bath used is illustrated in Figure 22.10.

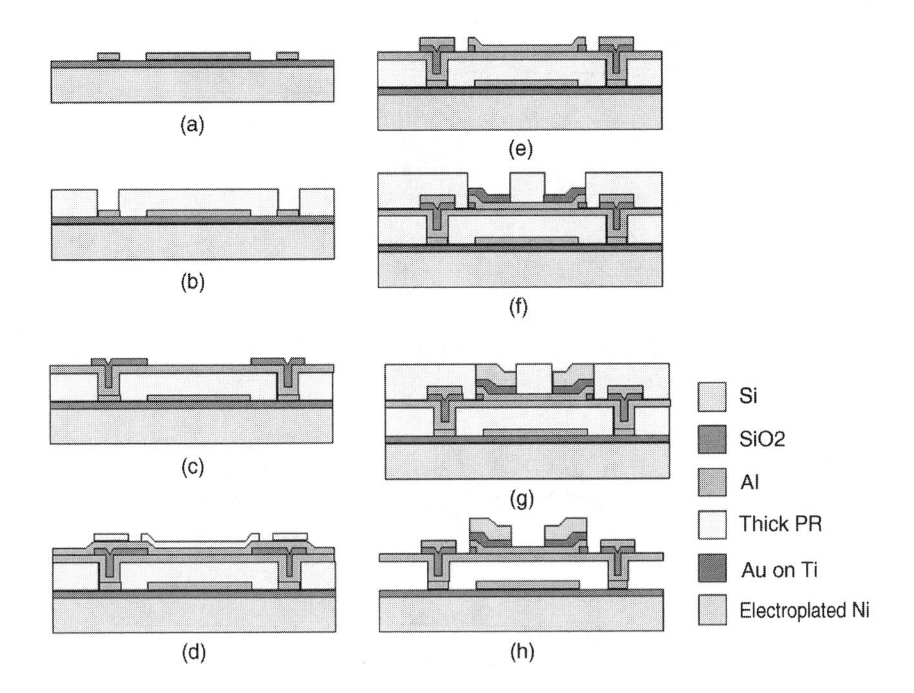

FIGURE 22.9 Detailed process for MEMS mirror array: (a) bottom electrode forming; (b) sacrificial layer coating and semicure; (c) spring Al deposition and etch stop patterning; (d) mirror Al deposition and mask patterning; (e) spring and mirror etching; (f) Cr and Au deposition and electroplating mold formation; (g) electroplating; (h) sacrificial layer etching.

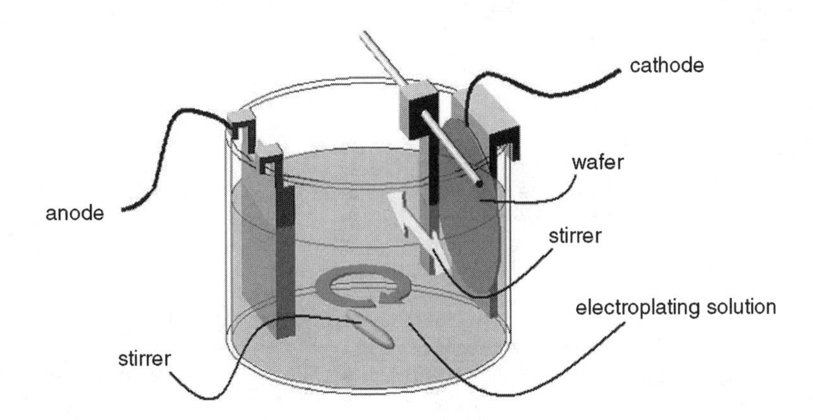

FIGURE 22.10 Schematic view of electroplating bath.

A magnetic stick was used to stir the solution in the bottom of the electroplating bath. In front of the wafer is a stick moving right and left to improve uniformity. After the mold and seed layer were removed, the sacrificial layer was removed by isotropic O_2 plasma, then the MEMS mirror array could be fabricated (Figure 22.9(g)).

The entire fabrication process for coupling the ball lens and the fiber with a MEMS mirror array is shown in Figure 22.11. Before fabricating the mirror array, a groove was fabricated with wet etching to align the beam axis between the mirrors and the ball lens. In fabricating the MEMS mirror array the bench for aligning the fibers was also made as shown in Figure 22.11. Vacuum holes were made after electroplating Ni to affix the ball lens to the substrate.

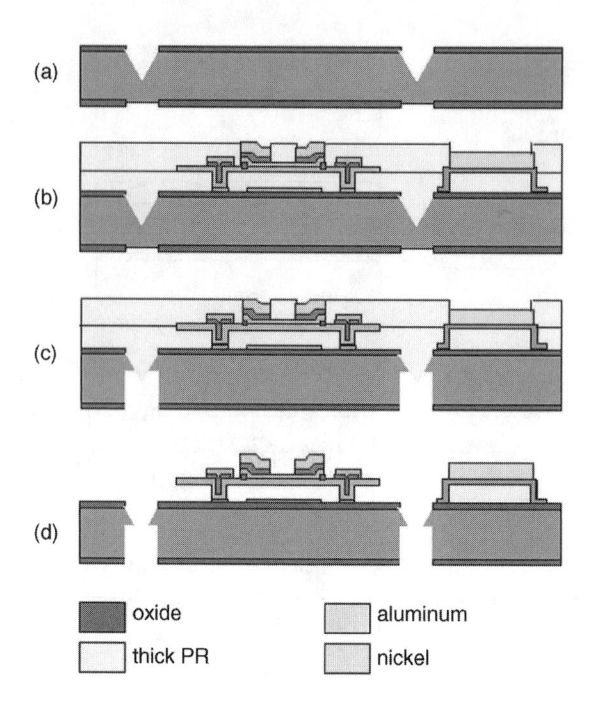

FIGURE 22.11 Simplified process for MEMS mirror array with bench: (a) groove formation for ball lens; (b) MEMS mirror array formation; (c) vacuum hole; (d) removal of sacrificial layer.

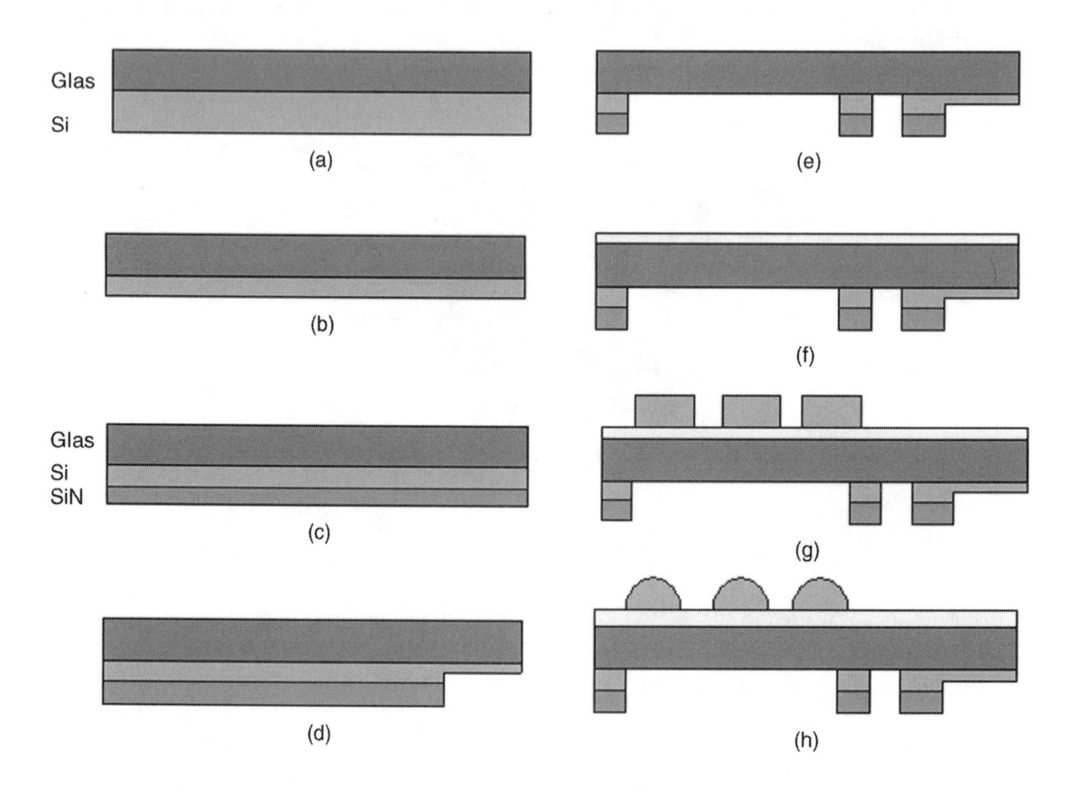

FIGURE 22.12 Detailed process for micro lens array: (a) surface modification and bonding; (b) Si thinning; (c) SiN deposit on Si; (d) deep RIE for fiber trench; (e) deep RIE for ball lens bench and window; (f) glass surface modification; (g) photolithography with thick PR; (h) thermal reflow.

22.4.2 Micro Lens Array

The fabrication process for the micro lens array begins with anode bonding between the glass and Si substrate. Because the fiber guide and hole for fixing the ball lens are to be fabricated on the Si substrate the thickness of the Si substrate must be reduced. After the anodic bonding process, the thickness of the Si substrate was reduced by chemical-mechanical polishing (CMP). Because making the fiber guide and holes for the ball lens requires a deep RIE process the SiN layer was deposited as a masking material for the deep RIE process. Then, to make the micro lens array, one of the thick photoresists—PMER (P-LA900PM)—was spin-coated and patterned on glass. In the process of heat treatment thermal reflow occurs in the thick photoresist. As a result, its shape is changed from that of a circular cylinder to a spherical lens. In general, the surface of a glass wafer is hydrophilic, so its property must be changed to be hydrophobic through surface treatment.

22.4.3 Results

22.4.3.1 Fabrication of MEMS Mirror Array and Micro Lens Array

Figure 22.13 shows a SEM picture of a MEMS mirror array fabricated through the process described above. As shown in Figure 22.13, the spring and the mirror with the magnet were fabricated well, and the sacrificial layer was removed without any other problems with the O_2 plasma. After the releasing process, sticking between mirror and substrate occurred in some mirrors, but this problem could be solved by coating the antisticking layer.

Figure 22.14 shows the fabricated hole for the fiber, ball lens, and micro lens array. The size of the holes for fixing the fiber and ball lens was designed at 60 and 300 μm, respectively. The sizes of the fabricated objects were measured to be 60 and 299 μm, respectively. The micro lens was designed to have a diameter of 120 μm and a thickness of 17 μm (depth of focus, 192 μm), but the size of the fabricated micro lens was measured to have an average diameter of 104 μm (standard deviation, 1.8 μm) and a thickness of 23 μm (standard deviation, 0.4 μm). As a result, the depth of focus was changed from 192 to 117 μm.

22.4.3.2 Assembling Results

The multiplexing device can finally be made by assembling the MEMS mirror array and the micro lens array after the ball lens and fiber are affixed to the micro lens array. To assemble these devices an *x-y* stage, microscope, and CCD camera can be used. UV epoxy is used for affixing the fiber to the micro lens array.

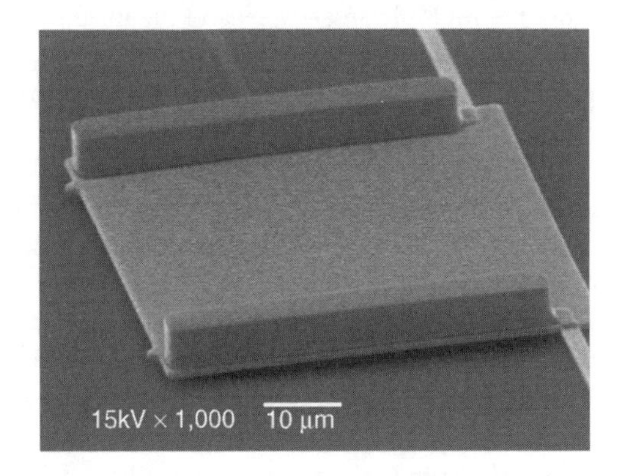

FIGURE 22.13 SEM view of a fabricated mirror array.

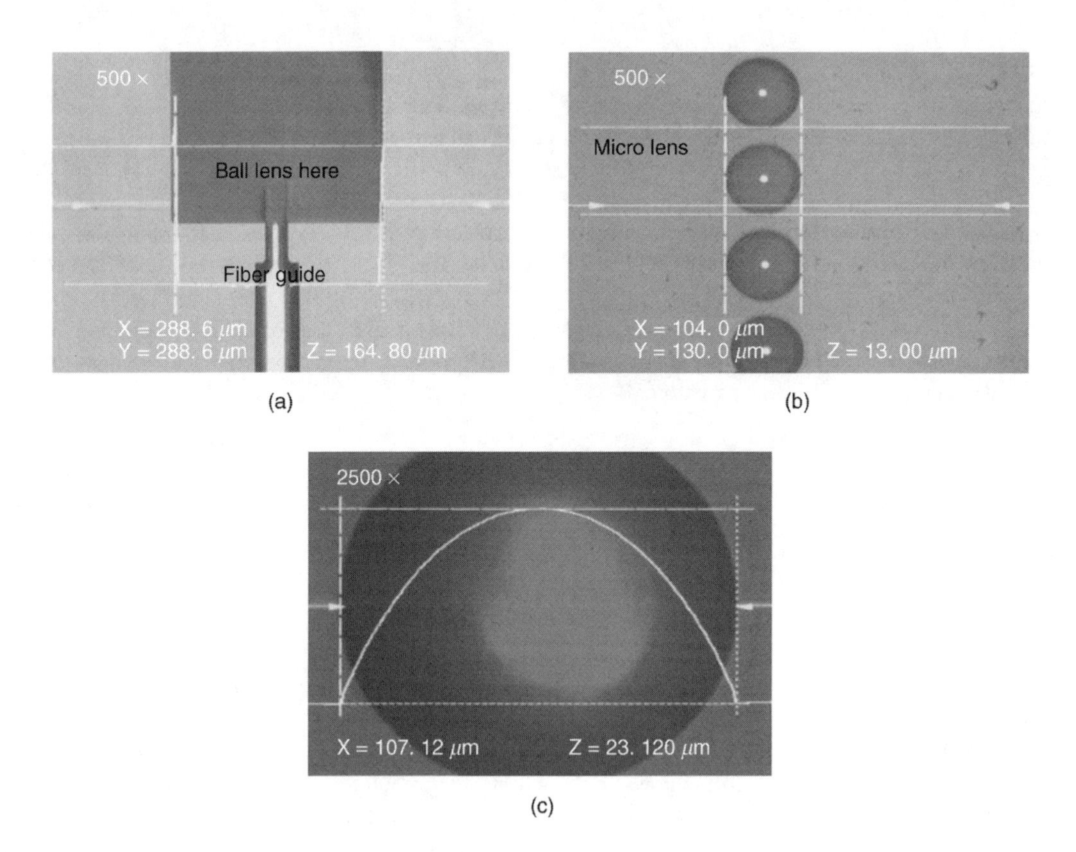

FIGURE 22.14 SEM views of a fabricated micro lens array: (a) hole for fiber and ball lens; (b) micro lens array; (c) micro lens.

Figure 22.15 shows one of the assembled devices. After the micromachining processes the device is fixed on the PCB, and wires are connected for electrostatic clamping. A coil generating the magnetic field is then attached under the PCB.

22.4.3.3 Results of Actuation

The deflection angle of a fabricated mirror was measured under the applied external magnetic field. A concentric coil was used to apply the magnetic field under the multiplexing device, and the magnitude of the magnetic field was measured using a Gauss meter (F. W. Bell series 9900). To estimate the angle of the deflected mirror a laser profiler was used. As shown in Figure 22.16, the measured data properly coincided with the theoretical value. In addition, the coercive force turned out to have a great effect in the range of the small deflection angle.

The calculated value of the clamping voltage is about 14 V at the magnetic field 13,000 Å/m for deflecting a mirror by 45°, but the measured value was about 50 V. This great discrepancy may be caused by the reduction of mirror area and by the unwanted deformation of the mirror. This deformation of the mirror resulted in the increase of the gap between the mirror and the bottom electrode; this increased gap between the mirror and the electrode can greatly affect the clamping voltage. Presumably, this problem can be solved by tuning the process of Al deposition. But, in the system considered here, an evaporator for Al deposition was used. In general, evaporator systems have limitations when it comes to controlling process parameters. If the sputter for Al deposition is used, Al deformation can be reduced by controlling process parameters.

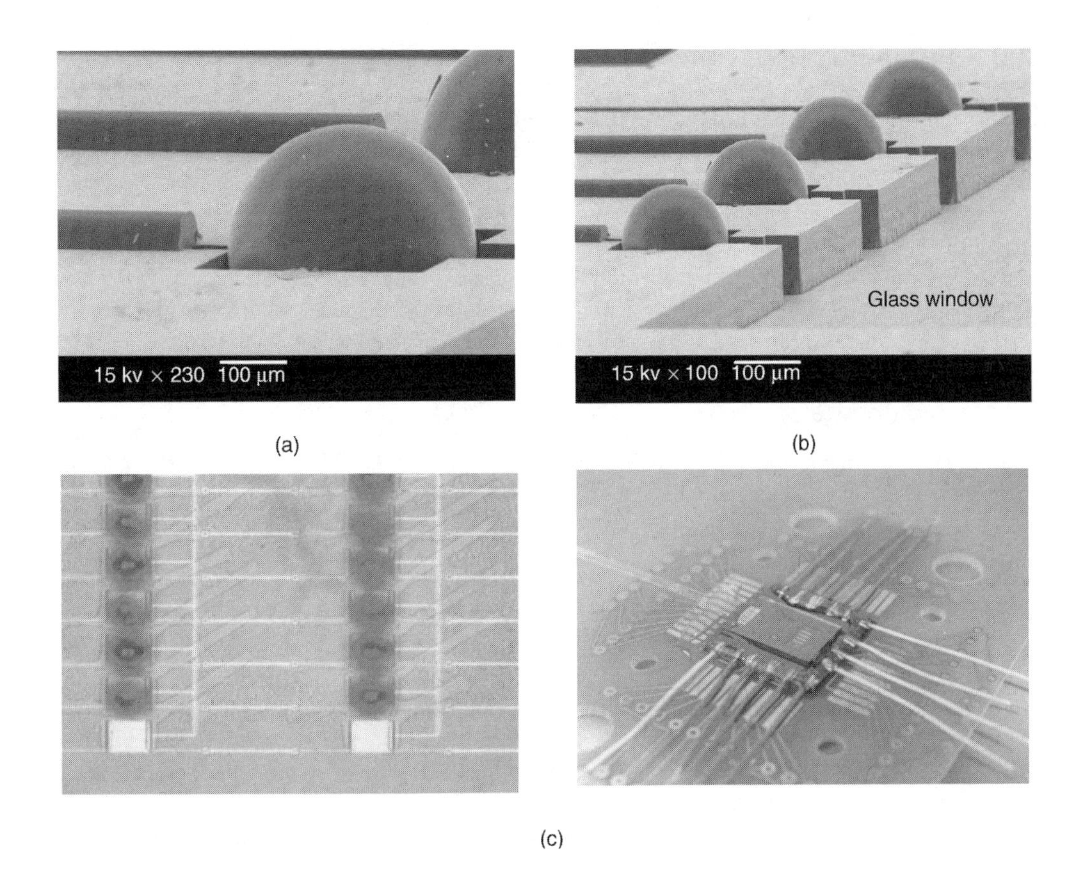

FIGURE 22.15 The fabricated multiplexing device: (a) SEM view of ball lens; (b) SEM view of coupled fiber and ball lens; (c) multiplexing device assembled with MEMS mirror array and micro lens array.

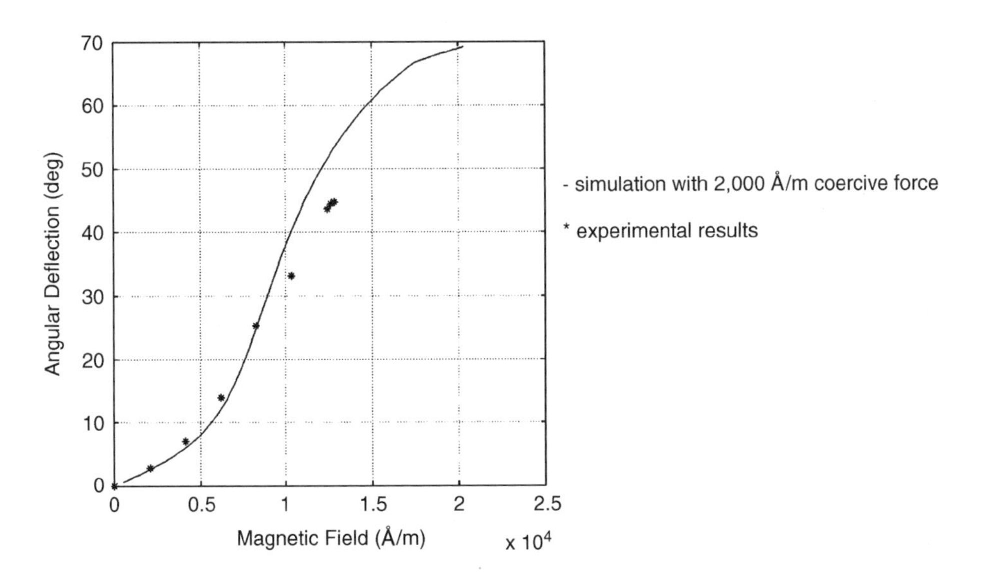

FIGURE 22.16 Deflection angle of mirror with external magnetic field.

22.5 Perspectives

The micro mirror has been applied in optical data storage [Judy and Muller, 1997; Miller and Tai, 1997], where it was used as a scanning mirror for changing the direction of light propagation. In the application considered here mirrors in a two-dimensional array are used as a two-dimensional light source. Of course, the light source itself is not incorporated into the device yet but is fed from outside through, for example, optical fibers. As a possible direction of future research, the integration of light sources with MEMS mirror arrays would be interesting. Smaller mirror dimension is one of the compelling requirements for making a storage system with a reasonable capacity for mobile devices; other requirements include lower power consumption, lower driving voltage, precision of the deflected angle, reliability, etc.

Acknowledgments

The authors would like to thank Mr. Yun Ho Jang, Mr. Haesuk Park, and Dr. Hong-Seok Lee for their help in the preparation of this manuscript and for their technical contributions.

References

Cullity, B. D., *Introduction to Magnetic Materials,* Addison-Wesley, Reading, MA, 1972, p. 527.

Jang, Y. H. and Kim, Y. K., Fabrication of electromagnetic micromirror array, *Proc. SPIE*, 4557, 395–402, 2001.

Judy, J. W. and Muller, R. S., Magnetically actuated addressable microstructures, *J. MEMS*, 6(3), 249–256, 1997.

Kim, J. and Lee, S., MEMS applications in three-dimensional data storage systems, *Proc. of the 2nd KIEE MEMS Symposium*, 1(2), 27–32, 1999.

Miller, R. A. and Tai, Y. C., Micromachined electromagnetic scanning mirrors, *Opt. Eng.*, 36(5), 1399–1407, 1997.

Motamedi, M. E., Micro-opto-electro-mechanical systems, *Opt. Eng.*, 33(11), 3505–3517, 1994.

Osborn, J. A., Demagnetizing factors of the general ellipsoid, *Phys. Rev.*, 67(11, 12), 351–357, 1945.

Roark, R. J., in *Roark's Formulas for Stress and Strain*, 6th ed., Young, W. C., Ed., McGraw-Hill, New York, 1989.

Timonshenko, S. P. and Goodier, J. N., *Theory of Elasticity*, 3rd ed., McGraw-Hill, New York, 1970.

Wu, M. C., Lin, L. Y., Lee, S. S., and King, C. R., Free-space integrated optics realized by surface-micromachining, *Int. J. High Speed Electron. Sys.*, 8(2), 283–297, 1997.

23

Optical/Vision-Based Microassembly

Bradley J. Nelson
University of Minnesota
Minneapolis, Minnesota

Barmeshwar Vikramaditya
Seagate Technology
Bloomington, Minnesota

23.1 Introduction

If a microdevice, for example, a **MOEMS** (microoptoelectromechanical system), must be made of different materials, has a complicated geometry, or is manufactured using incompatible processes, assembly of the device is required. The scaling of microparts leads to complications in their assembly. Key factors contributing to the increased complexity include the need for high-precision endpoint positioning, vastly different mechanics of manipulation, tolerance stackup due to thermal effects, errors and approximations in the modeling of sensors and manipulators, and the poor relative tolerance of micromachined parts. The physics of part interaction is only approximately understood for simplified assembly scenarios [Nelson et al., 1998], and *a priori* modeling of the complicated interactions is not possible. The combined uncertainty involved in a typical assembly task, therefore, requires the use of sensor feedback.

Vision is a useful robotic sensor because it provides dense information about the task space while being a noncontact sensing modality. Since the early work of Shirai and Inoue [1973], considerable research effort has been devoted to the visual control of robotic manipulators. This research can ideally be leveraged toward automation of microassembly tasks. However, there are significant differences in the characteristics of the sensors used in the macro and micro domains.

The use of vision feedback has disadvantages as well. The dense information content requires high data throughput, on the order of 8 to 24 MB/s. This information must be interpreted at a high speed if a feedback loop is to be implemented. The observed motion is nonlinear and highly coupled, resulting in a nonlinear multi-input, multi-output (**MIMO**) control architecture. This greatly complicates the design of feedback compensators.

The use of visual feedback for microassembly faces a fundamental trade-off in the requirement of high resolution and large depth of field. The need for high resolution demands the use of high numerical

0-8493-1162-4/03/$0.00+$1.50

aperture lens systems, which consequently have a very small depth of field and a limited field of view. This poses severe challenges to a typical three-dimensional microassembly task.

The need for high numerical aperture microscopes for microrobotics introduces some challenges unique to the micro domain. The depth of field of the optics is small, ranging from 0.2 μm and below to 120 μm. As a result, perception of depth is quite difficult, and some means of automatic focusing is essential. This makes manipulation along the optical axis challenging.

The high-magnification requirement results in a very small field of view. Although the parts being assembled are small, they generally need to be transported relatively large distances prior to assembly. This requires either the use of another large field-of-view sensor or some means of encoding global-position information in a CAD model.

Visual feedback is used extensively by human operators in assembling MOEMS devices and in remote teleoperation. The control decisions are made by a trained human operator based on visual feedback from the task space. There are quite a few disadvantages to this scheme of assembly, the primary one being that the operators need to be extensively trained. This results in increased production costs and more than offsets the gains of batch fabrication of the components. Manual assembly directly leads to large variations in assembled devices, and as a result there is a need for large-scale testing and strict quality control. Production volumes are quite small, which again negates the advantages of batch fabrication. Design practices such as design-for-assembly have not been fully developed for MOEMS devices, so microassembly tasks are quite complicated in some cases. A large number of parameters must be varied in a typical assembly scenario due to the constraints imposed by limited field of view and depth of field. Most assembly tasks are highly repetitive and, as a result, quite monotonous. All these factors demonstrate the need for automation and the need for intelligent visual feedback for microassembly tasks.

This chapter describes the use of visual feedback for conducting automated microassembly. First, a variety of proposed architectures are described and contrasted. Key components of the architectures are described, including visual tracking and object recognition. Next, a coarse-to-fine visual servoing scheme is developed, which requires calibration of a zoom lens. Depth estimation and autofocusing are then described. These components combine to form an intelligent sensor-guided microassembly workcell.

23.2 Visual Servoing

The use of visual information in a closed-loop feedback control scheme is known as **visual servoing**. A comprehensive classification of the various control schemes based on visual feedback has been developed by Sanderson and Weiss [1982, 1983] and Weiss [1984]. The various control schemes are classified based on three key parameters:

- Whether the vision control system generates set-points for the robot joint-level controllers, or the controller directly computes joint-level torques
- Whether the error signal is defined in three-dimensional (task space) coordinates or in terms of image features and image space
- Whether the camera remains static or moves along with the arm

Based on these parameters some fundamental control structures have been identified. If the vision control system eliminates the robot controller entirely and directly generates joint torque commands, then the architecture is called *visual servo* or *direct visual servo* [Hutchinson et al., 1996]. If the control structure is hierarchical and uses the vision system to provide set-point updates for the joint-level controller, then the scheme is referred to as *look-and-move*. The joint level controller uses joint feedback to internally stabilize the robot and responds to updates of the set point. This results in a multirate control architecture. Furthermore, based on the rate of set-point updates the scheme may be divided into *static look-and-move* and *dynamic look-and-move*.

There are several reasons why the dynamic look-and-move scheme (Figure 23.1) is preferred over the direct visual servoing scheme (Figure 23.2). The sampling rate of area-scan vision sensors is generally limited to

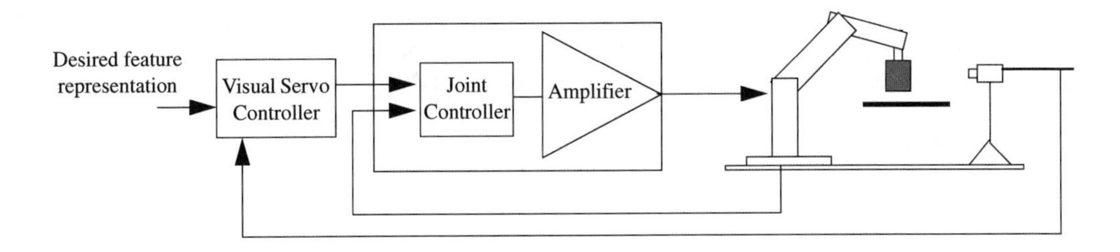

FIGURE 23.1 Dynamic image-based look-and-move structure.

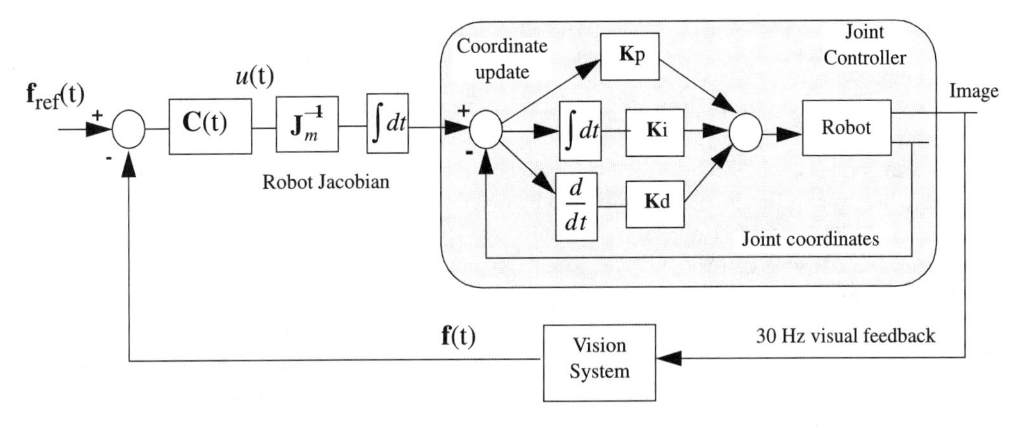

FIGURE 23.2 Dynamic image-based visual servoing.

30 to 60 Hz. This limited bandwidth makes it impossible to compensate for the complex nonlinear dynamics inherent in a joint, including joint friction, gearing backlash, and joint compliance. A joint controller with a high sampling rate is able to compensate for these dynamics and disturbances and presents the vision controller with highly simplified axis dynamics. Second, from an implementation standpoint most industrial controllers allow for dynamic set-point update and Cartesian increments at the joint level. This makes the implementation of a dynamic look-and-move scheme much easier. Third, look-and-move separates the kinematic singularities of the mechanism from the visual controller, allowing the robot to be considered as an ideal Cartesian motion device. This reduces complexity in the design of the visual servo controller.

The second major classification is based on space in which error signals are computed. In *position-based control* [Allen, 1989; Koivo and Houshangi, 1991; Wilson, 1993] features are extracted from the image and used in conjunction with a geometric model of the object and the known camera model to estimate the pose of the object with respect to the camera. The estimated pose is used to generate the control signal in Cartesian space. In *image-based control* [Corke and Paul, 1989; Feddema and Lee, 1990; Papanikolopolous et al., 1991; Hashimoto and Kimura, 1993], control signals are generated directly in terms of features in the image space. This eliminates the requirement for a precise sensor model and calibration and reduces computation delays and errors. This in turn simplifies the observer dynamics and makes for a simpler implementation.

Features in an image must be tracked in order to implement an image-based scheme. The feature vector $\mathbf{f}(t)$ may be comprised of coordinates of vertices, areas of faces, or any quantifiable parameter of the object being viewed. This feature vector obtained from the sensor is to be controlled to achieve a $\mathbf{f}_{ref}(t)$ position in the feature space. For an end-effector-mounted camera, the feature in general would be some function of the relative pose, $\mathbf{X}_c(t)$, of the camera to the target. In general this function is nonlinear and coupled, such that motion of the camera along one degree of freedom results in a complicated motion of all features. This relationship can be linearized about some operating point as

$$\delta\,\mathbf{f}(t) = J_c(\mathbf{X}_c(t))\delta\,\mathbf{X}_c(t) \tag{23.1}$$

where $J_c(X_c(t)) = \frac{\partial f(t)}{\partial X_c(t)}$ is referred to variously as the *feature Jacobian, image Jacobian, feature sensitivity matrix,* or *interaction matrix* [Hutchinson et al., 1996]. In order for the Jacobian to be square we must have the number of features equal to the number of permitted degrees of freedom. The Jacobian is a function of the extrinsic and intrinsic parameters of the sensor as well as the type and number of features being tracked. Assuming that the Jacobian is invertible, we then have

$$\dot{\mathbf{x}}_c(t) = J_c(\mathbf{X}_c(t))^{-1}\dot{\mathbf{f}}(t) \tag{23.2}$$

and based on this we can design a simple proportional velocity feedback controller as

$$u(t) = \mathbf{K}J_c(\mathbf{X}_c(t))^{-1}(\mathbf{f}_{ref}(t) - \mathbf{f}(t)) \tag{23.3}$$

where $\mathbf{K}$ is a diagonal gain matrix and is time varying, and $\mathbf{f}(t)$ is the current feature vector. This simple control law is represented in the time domain and can be easily discretized for digital implementation. This is known as a *resolved-rate control scheme.*

The system is typically MIMO, and the controller formulation, especially adaptive time-varying control, is difficult. Usually, the system is controlled by decomposing the system into multiple single-input, single-output (**SISO**) systems. Implemented results and simulations have been reported for the **MRAC** (model reference adaptive controller) [Weiss, 1984; Koivo and Houshangi, 1991], controllers incorporating Kalman filters [Wilson, 1993], and LQ optimal controllers [Hashimoto and Kimura, 1993].

Finally, a classification is made based on the location of the camera. If the camera is mounted on the manipulator arm and moves with it, then the system is called *eye-in-hand.* Most initial work began with this scheme [Allen, 1989; Corke and Paul, 1989; Feddema and Lee, 1990; Papanikolopolous et al., 1991]. However, now most researchers use a static camera configuration [Koivo and Houshangi, 1991; Nelson et al., 1993; Hager, 1994].

The use of dynamic sensor placement [Tarabanis et al., 1990] for visual servoing was developed by Nelson [1995]. The sensor is placed in order to optimize some control objective, which is determined by the task to be executed. Parameters such as the level of focus, achieved magnification, resolvability of the sensor, and required field of view are candidates for the augmented control objective being optimized. Nelson [1995] proposes the use of an augmented control objective, which is minimized in order to determine control input to the manipulators for both task execution and sensor placement. A supervisory logic-based controller with high-level task information is used to dynamically reconfigure and locate the sensor. Employing a zoom lens within this framework allows one to resolve the conflicting requirements of field of view and high magnification. The supervisory-logic-based controller selects the optimal sensor and manipulator configuration for the task to be executed.

23.2.1 Visual Servoing for Micromanipulation

As pointed out earlier, extending visual servoing concepts to the micro domain incurs some additional challenges due to the significantly different sensor characteristics. The key differences are the small depth of field and limited field of view. The small depth of field introduces the need for autofocusing, especially with the multiple task planes typical of an assembly task. In addition, focus must be dynamically maintained throughout most microassembly tasks. A number of research groups have developed visual control schemes for micromanipulation. Some key efforts are summarized and contrasted in this section.

Visual control of a microrobot under a stereo optical microscope and SEM has been investigated by the Nanorobotics research group at ETHZ, Swiss Federal Institute of Technology, Zurich [Codourey et al., 1994; Pappas and Codourey, 1996; Rodriguez et al., 1996; Rodriguez and Codourey, 1997]. The objective of the project was to develop a visually guided microassembly system with a 1-cm^3 workspace and a positioning resolution of 10 nm. A dynamic position-based look-and-move architecture was developed for the vision controller which allowed for teleoperation and semiautomatic position control of five-degree-of-freedom microrobots. A pattern-matching algorithm locates features in the image space with subpixel accuracy; these features are then used to extract the pose of the microrobot under the microscope.

The position update commands are generated in Cartesian space and sent to the microrobot, which then executes the moves in an open-loop fashion. A Kalman-filter-based estimator is used to extract position parameters from the pattern-matching results. Though it is stated that semiautomatic visual control is feasible, the system is used primarily in a teleoperative fashion guided by reference commands generated by a human operator through an integrated graphical user interface (GUI). To enable full three-dimensional assembly tasks, the operator is presented with views of a three-dimensional model and visual feedback of the task space.

Automatic assembly using visual feedback has also been investigated at EPFL, the Swiss Federal Institute of Technology, Lausanne [Sulzman et al., 1995; Allegro, 1998]. The assembly is carried out both teleoperatively and automatically. A virtual-reality-based microteleoperation interface has been developed that allows the user to guide microrobots in a teleoperative fashion. The automatic assembly is controlled by a dynamic position-based look-and-move visual controller, guided by visual feedback from an optical microscope. Two optical sensors are used to guide the assembly. A global sensor is used to locate the manipulators and for long-range moves, while an optical microscope guides the fine manipulation task. The system is precisely calibrated, and a significant amount of offline processing is performed in order to carry out automatic assembly. The system does not achieve real-time image processing and, as a result, operates in a strictly look-and-move fashion. Position information is updated by performing connected component analysis on captured images. This information is used along with "prior knowledge" to generate command updates for the micromanipulators. Image-based autofocusing is used to determine depth and to establish the location of the task planes along the optical axis. However, focus is not maintained dynamically throughout the assembly task. Average execution speeds for a pick-and-place task are between 12 and 17 sec.

Research at the University of Karlsruhe, Germany, has led to the development of an automatic desktop micromanipulation station [Fatikow, 1996; Fatikow et al., 1999]. A global vision sensor is used to guide large-scale manipulation. Micromanipulation is carried out under an optical microscope guided by a static image-based vision controller. A fuzzy-logic-based neural network is used for determining image-based commands. A novel laser range finder is used to estimate the depth of objects in the task space. Typical cycle times for manipulation tasks are not available.

Research at the Lawrence Livermore Berkeley National Laboratory [Parvin et al., 1996, 1997] has led to the development of a visually servoed micromanipulation system for biological systems. The system framework has been implemented as a parallel distributed computing environment and is primarily designed for networked micromanipulation tasks. The capabilities of the system include tasks such as microdissection of immobilized DNA molecules and dynamic studies of crystal structure formation. The visual servoing system is comprised of multiple threads that execute functions including object recognition, tracking, servo-control, and autofocusing. The tracking subsystem is able to track features at 4 Hz, resulting in a lower closed-loop system bandwidth servo controller.

Microassembly research at Sandia National Labs [Feddema and Simon, 1998] has led to the development of a CAD-based microassembly workcell. The optical system is modeled using Fourier optics to determine its defocusing characteristics. Offline synthetic images are generated using the developed optical model to test various image-processing schemes and for determining path-planning strategies. Depth estimates can be made from the radius of the blur circle, and a gradient-based approach can be used to maximize the focus in the image. Dynamic image-based visual servoing based on a four-degree-of-freedom image Jacobian with a resolved-rate servo controller is used to guide the final assembly task. The assembly of a LIGA gear is demonstrated using this approach, achieving a positioning repeatability of 1 μm.

A dynamic image-based look-and-move control architecture with a dynamically reconfigurable sensor is described in this chapter. An image-based Jacobian is developed for the optical microscope. The inverse-image Jacobian is used to transform differential motion in the sensor space to differential motion in the task space. This differential motion is then transformed through individual manipulator Jacobians to generate command updates for the manipulators and stages.

A supervisory logic-based controller is used in this framework. The supervisory controller is guided by high-level task information and optimally reconfigures the dynamic parameters in the assembly

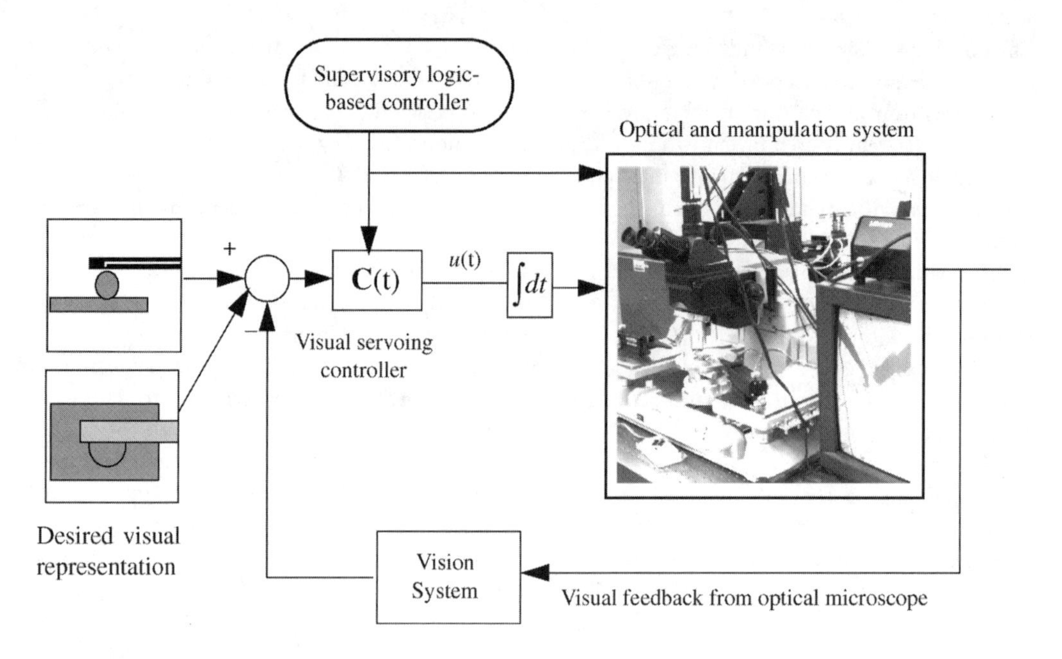

FIGURE 23.3 System architecture.

workcell to carry out the assembly task. The controller also serves as the task planner and updates system configuration for the task to be executed. The system architecture is shown in Figure 23.3.

23.3 Controller Formulation

A visual servoing controller is developed in this section. To develop the controller an image Jacobian must first be derived for the optical microscope. The Jacobian is then used to formulate an optimal controller [Nelson et al., 1993; Nelson, 1995].

A discrete state equation for a visual servoing system can be written as

$$\frac{x(k+1)-x(k)}{T} = J_c(k)u(k) \tag{23.4}$$

$$\begin{bmatrix} \dot{x}_s \\ \dot{y}_s \end{bmatrix} = TJ_c(k)\left[\dot{X}_c\ \dot{Y}_c\ \dot{Z}_c\ \omega_{Xc}\ \omega_{Yc}\ \omega_{Zc}\right]^T \tag{23.5}$$

where kT is represented as k for simplicity, $x(k) \in R^{2N}$ represents the state vector of N features being tracked, $u(k) = \dot{X}_c(k)$ is the control input to the joint controller, T is the sampling time, and $J_c(k)$ is the image Jacobian. Assuming optical parameters of the system remain constant throughout the task, the Jacobian is given as

$$J_c(k) = \begin{bmatrix} -\frac{m}{s_x} & 0 & 0 & 0 & -\frac{Z_c m}{s_x} & \frac{y_s s_y}{s_x} \\ 0 & -\frac{m}{s_y} & 0 & -\frac{Z_c m}{s_y} & 0 & -\frac{x_s s_x}{s_y} \end{bmatrix} \tag{23.6}$$

A state-space representation for the system in Eq. (23.4) is given as

$$\begin{aligned} x(k+1) &= Ax + Bu(k) \\ y(k) &= Cx(k) \end{aligned} \tag{23.7}$$

where $A = C = I_2$. B represents a partitioned Jacobian for the system corresponding to the active degrees of freedom of the manipulator and varying intrinsic parameters.

The depth of field the microscope prevents the observation of motion along the optical axis of microscope. This can be seen from the form of the Jacobian in Eq. (23.6). For the case where rotation of the sensor with respect to the current frame is neglected and the zoom and the objective are not changed during the course of servoing, B is given as

$$B = \begin{bmatrix} \frac{-m}{s_x} & 0 \\ 0 & -\frac{m}{s_y} \end{bmatrix} \tag{23.8}$$

where m is the magnification for the optical system, and s_x and s_y are the horizontal and vertical dimensions, respectively, of the pixels of the CCD array. If the depth of the object being viewed is constant with respect to the sensor, then we can formulate a controller for optimal control and trajectory generation based on one observed feature. A cost function $F(k + 1)$ is formulated and is given by

$$F(k+1) = [\boldsymbol{x}(k+1) - \boldsymbol{x}_D(k+1)]^T Q[\boldsymbol{x}(k+1) - \boldsymbol{x}_D(k+1)] + \boldsymbol{u}^T(k) L \boldsymbol{u}(k) \tag{23.9}$$

where $\boldsymbol{x}_D(k + 1)$ represents the desired values for the measurands, and Q and L are diagonal gain matrices with constant terms. The Q matrix must be positive semidefinite, while the L matrix must be positive definite for a bounded response. These weighing matrices allow the user to weigh the significance of control action and feature error. Their selection affects the stability and response of the system. This cost function is optimized by minimizing it with respect to $\boldsymbol{u}(k)$ which yields

$$\boldsymbol{u}(k) = -(B^T Q B + L)^{-1} B^T Q[\boldsymbol{x}(k) - \boldsymbol{x}_D(k+1)] \tag{23.10}$$

A key advantage of this controller is that it eliminates the need for a trajectory generator as the control gain matrix is optimized to reduce the error given by $[\boldsymbol{x}(k) - \boldsymbol{x}_D(k + 1)]$ in every cycle while optimally reducing the controller output signal. However, this controller does not allow for path following and can be used for point-to-point visual servoing only.

This control law does not account for delays in the system due to processing and communication latencies. Delays in a feedback loop impact system performance and stability. System delays may be modeled and incorporated in the state Eq. (23.7) as

$$\begin{aligned} x(k+1) &= Ax(k) + Bu(k-d) \\ y(k) &= Cx(k) \end{aligned} \tag{23.11}$$

leading to a new control law of the form [Nelson, 95]

$$u(k) = -(B^T Q B + L)^{-1} B^T Q\left[x(k) - x_D(k+d) + \sum_{m=1}^{a-1} B(k-m)u(k-m)\right] \tag{23.12}$$

By selecting a different cost function it is possible to derive a controller that introduces integral action as well.

23.4 Feature Tracking

As an object being viewed under a camera undergoes motion, or if the camera undergoes motion with respect to the object, there is a change in the intensity profile for the object on the image sensor. The change is usually nonlinear and coupled to the motion along various degrees of freedom. Sensor feedback for image-based visual servoing is generated based on data gathered from the features being viewed in the scene by the camera. The features are tracked continuously within the image, and their location, brightness profile, etc. provide the requisite information for generation of control signal for closed- loop servoing. As a result, it is necessary to develop and implement a fast, robust, and accurate tracker.

Two feature-extraction and tracking schemes have been investigated. Modified SSD, or *sum-of-squared-differences*, and blob tracking methods are both employed for different tasks.

23.4.1 SSD Tracking

The implemented tracker uses a scheme of tracking similar to the proposed scheme by [Nelson et al., 1993] called the *modified sum-of-squared-differences* method. In general, brightness patterns can be represented by three variables, two space variables x and y and a time variable t, as $I(x, y, t)$. If the patterns are assumed to be time invariant, and there exists a feature in a window of size $N \times N$ in the initial image A, $\mathbf{P}_A$ (x_A, y_A), and it moves in a subsequent image B and is located at some point, $\mathbf{P}_B$ (x_A+dx, y_A+dy), then the objective is to determine this movement of the image pattern. Assuming that intensity values are time invariant and constant in the neighborhood of $\mathbf{P}_A$ (x_A, y_A), and that the sampling time is fast enough so that dx and dy are within some limit L, then the SSD estimator selects the displacement $\Delta\mathbf{X}$ $(dx, dy)^T$ in order to minimize the SSD measure given by

$$e(\mathbf{P}_A, \Delta\mathrm{X}) = \sum_{m,n \in N} \left| I_A(x_A + m, y_A + n) - I_B(x_A + m + dx, y_A + n + dy) \right| \tag{23.13}$$

such that $\Delta\mathbf{X} \in$ L and I_A and I_B are the intensity functions in images A and B, respectively. The pattern I_A is selected either by the user or by a task-planning module using offline knowledge of the objects being tracked. The key parameters are the values set for N and L. They determine the speed of the tracker and robustness to noise. In this work, window size $N \times N$ is selected as 16×16, while the search region given by $L \times L$ is set as 48×48. The SSD tracker is accurate if the feature window shows image gradients and the image pattern has sufficient disparity because the error function would show a distinctive peak, even in the presence of image noise.

Sum-of-squared-differences tracking allows for tracking multiple features at frame rate due to the short processing time. Currently, five features can be tracked at frame rates using a PC-based framegrabber. Further optimizations have been made in order to speed up tracking using pyramidal schemes [Nelson et al., 1993], loop short-circuiting [Smith, 1996], and velocity estimation. SSD tracking is highly susceptible to brightness variations and noise in the video input. Feature adaptation has been suggested as an alternative to handle these situations [Smith, 1996]. However, in practice this scheme does not show promising results.

The step response of the visual servoing scheme is illustrated in Figure 23.4. The input corresponds to a step move of 50 pixels in the X and Y directions. Objectives with different magnifications are used to achieve high precision (0.17 μm) and high speeds (2.2 mm/sec) of manipulation. However, this requires switching the discrete objectives on a microscope turret that can be obviated by the use of a zoom lens system using a coarse to fine visual servoing scheme.

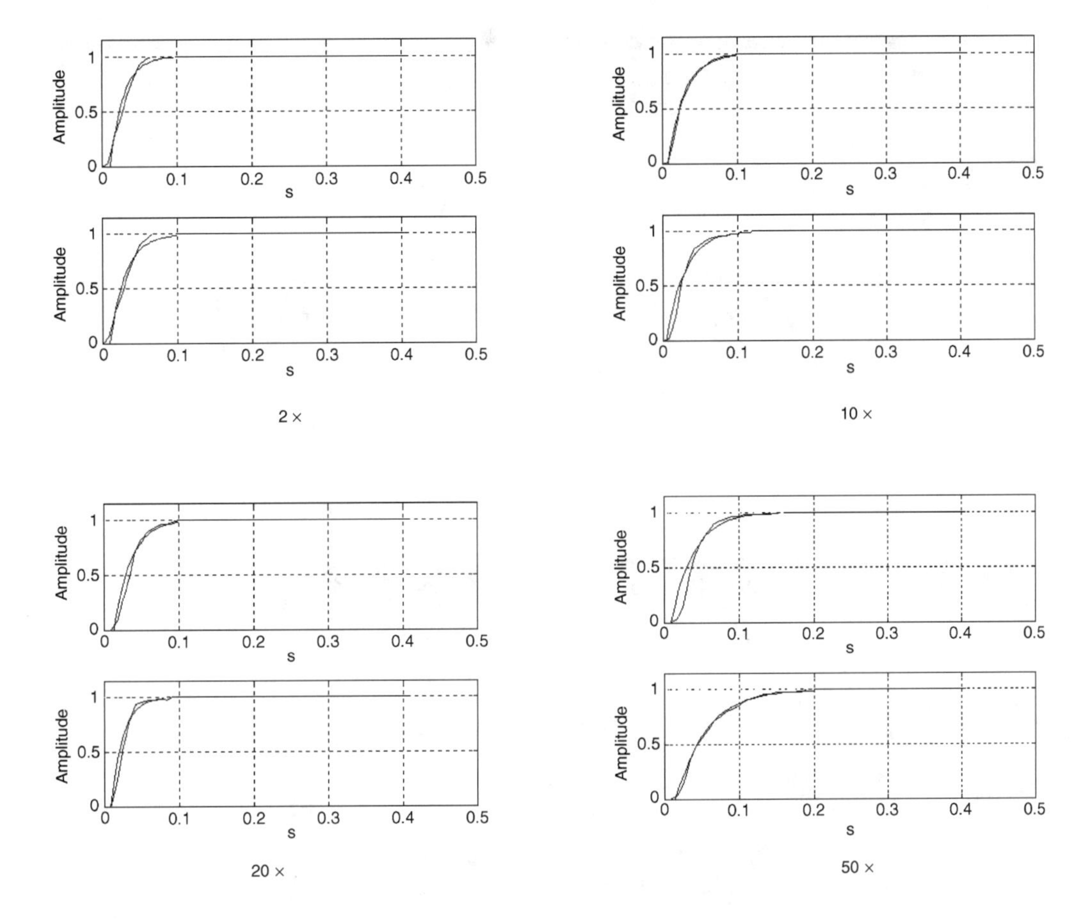

FIGURE 23.4 Step response of visual servo controller with various objectives.

23.4.2 Connected Component Analysis

Connected component, or blob, analysis is a commonly used statistical pattern-recognition technique in visual inspection (Figure 23.5). The analysis provides a list of connected components and their properties such as centroid, moment of inertia, perimeter, size of bounding rectangle, number of holes in objects, and major and minor axis orientation, among others. This information can be used for task-level planning and object recognition. Connected component analysis is much more computationally intensive than the SSD feature-tracking scheme. However, by optimizing the implementation connected component analysis can be used to track features in order to implement an image-based visual servo controller.

The algorithm requires the following sequential processing steps:

- Thresholding the image to generate a binary image
- Labeling of the generated image
- Information extraction

23.4.2.1 Thresholding

Thresholding is the processing step that converts a gray-scale image in this case to a binary image. A dynamic threshold is selected based on the image content. This provides some robustness to the image-processing steps in the presence of illumination variations that are necessary when an image is zoomed. However, this adds computational intensity to the overall process.

There are two commonly used schemes for dynamically selecting an optimum threshold value—Otsu's method and the Kittler–Illingworth measure. According to Haralick and Shapiro [1992], for gray-scale

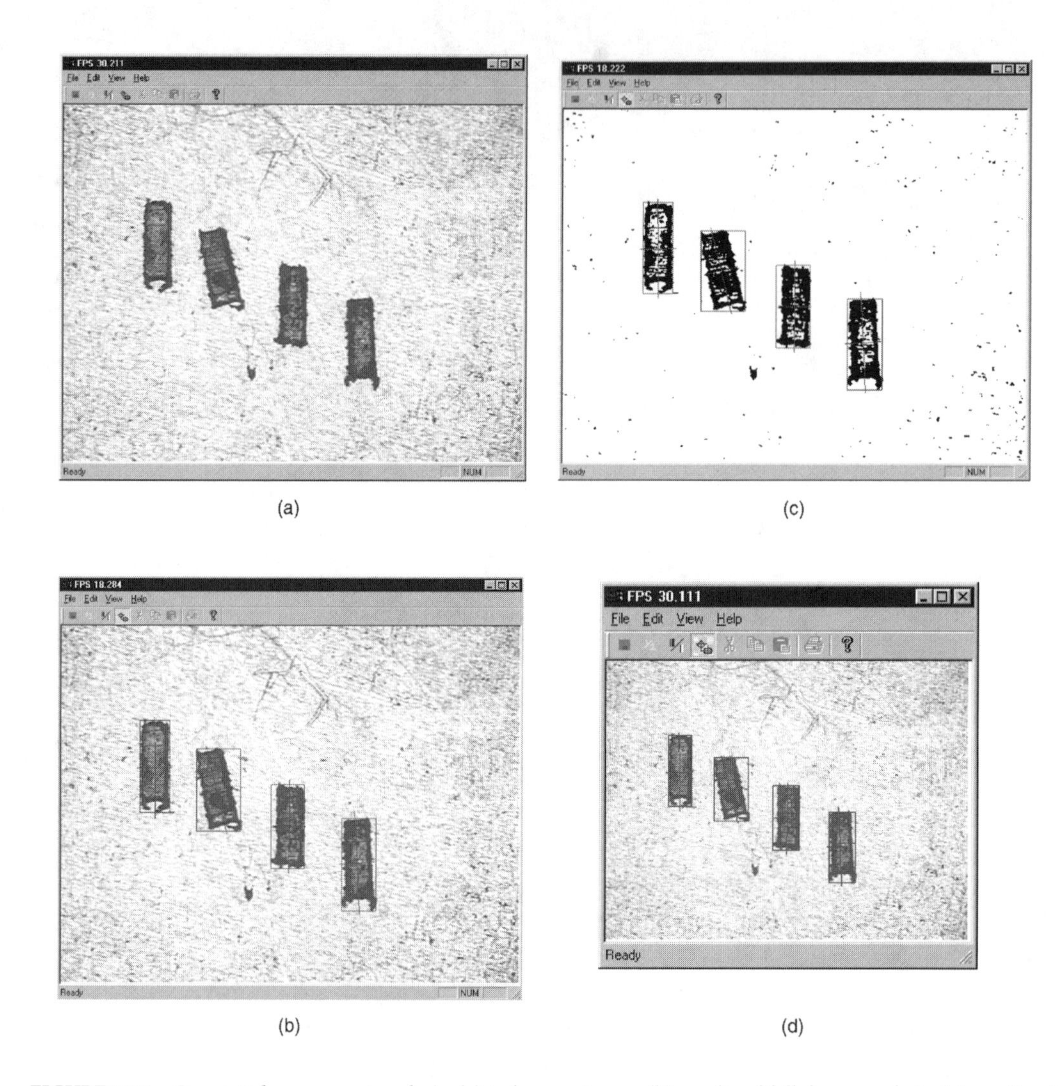

FIGURE 23.5 Connected component analysis: (a) task space image; (b) results of full-frame analysis; (c) results of dynamic thresholding and connected component analysis; (d) results of half-frame analysis.

images with a bimodal histograms the Kittler–Illingworth measure provides superior input images for binary-image-processing methods. Most microassembly tasks present a bimodal image histogram, the Kittler–Illingworth threshold is used in this implementation.

The iterative implementation minimizes a quantity termed the *Kullback distance measure* [Haralick and Shapiro, 1992] and selects an optimum threshold value for the input image. Some assumptions are made in order to decrease the execution time for this iterative scheme. The search for a minimum is conducted only for gray-scale values between 90 and 160 instead of the entire range from 0 to 255.

23.4.2.2 Connected Component Labeling

The binary image generated by the thresholding operation is labeled using a connectivity rule. An eight-connectivity scheme is used in this implementation, whereby all eight adjacent neighbors of a pixel are considered as part of its neighborhood.

Several labeling algorithms have been proposed in the literature. In this implementation a space-efficient, two-pass algorithm is used [Haralick and Shapiro, 1992, p. 37]. This requires two passes over the binary image in order to label the image, but the algorithm is efficient both in terms of computational and storage requirements. Certain parameters of the connected components are computed simultaneously to reduce computational load for subsequent steps.

23.4.2.3 Information Extraction

The labeled image is passed to the information-extraction subroutine. Only the minimum bounding rectangle is used to search for the connected components, thereby significantly increasing computational speed. Currently the information collected includes:

- Minimum bounding rectangle
- Major and minor axis
- Area and centroid of the regions
- Second-order moments for the regions
- Gray-level mean and variance

The latter two parameters are used primarily for object recognition, while the first three parameters are used primarily in task planning and look-and-move control schemes. The centroid serves as a robust measure in feature tracking and calibration of the optical system, as described in the zoom-calibration section.

Currently, the entire image processing requires 60 to 75 ms for a 640 × 480 input image using a PC-based framegrabber (Sensoray 611). However, by processing a reduced-scale image frame-rate tracking is easily achieved. The higher-level information generated by connected component analysis is invaluable in runtime path planning for manipulation; as a result, this scheme is used only in the look-and-move control scheme and for calibration purposes.

23.5 Coarse-to-Fine Visual Servoing

A trade-off resulting from the requirement of a high numerical aperture and magnification optical system is resolved by using a coarse-to-fine visual servoing scheme [Ralis et al., 1998]. Based on the task requirements, a supervisory task-level controller dynamically varies the parameters for the optical system to optimally provide the requisite field of view, magnification, and feature resolvability.

A number of dynamic sensor-placement schemes have been proposed in the literature [Cowan and Bergmann, 1989; Tarabanis et al., 1990, 1991, 1994; Nelson, 1995]. Most schemes developed for robotics use a potential function to develop a control scheme for the placement of the sensor to satisfy requirements including focus, field of view, depth of field, occlusions, or spatial resolution. However, for most micromanipulation tasks sensor placement is rather constrained, generally limited to Cartesian axes overlooking the task space. This is due to the small depth of field of the lens systems. However, the intrinsic sensor parameters can be varied to increase the capability of the sensor system for the task instead of dynamically relocating it [Tarabanis et al., 1994; Hosoda et al., 1995].

A number of factors must be considered simultaneously in a typical visually servoed micromanipulation task, including

- Visual resolution must be sufficient for the task
- Features must lie within the field of view
- Features must stay in focus or can be brought into focus when needed
- Illumination must remain stable across all optical parameter variations to enable tracking

23.5.2 Visual Resolution Constraint

For a through-lens illumination system the objective lens generally functions as the primary aperture for the lens system and sets the upper limit on achievable resolution. Choice of objective lens for a given sensor system is, therefore, critical for a microassembly task. The wave nature of light determines the maximum optical resolution of the microscope lens system. The first step in this process is to determine the resolving power of the microscope. The ultimate limit on the spatial resolution of any optical system is set by light diffraction; an optical system that performs to this level is termed *diffraction limited.* In this case, the spatial resolution is given by

$$d = \frac{0.61\lambda}{NA} \tag{23.14}$$

where d is the smallest resolvable distance in the object space, λ is the wavelength of light being imaged, and NA is the numerical aperture of the microscope objective. This is derived by assuming that two point sources can be resolved as being separate when the center of the airy disc from one overlaps the first dark ring in the diffraction pattern of the second (the Raleigh criterion).

It should be further noted that, for microscope systems, the NA to be used in this formula is the average of the objective's numerical aperture and the condenser's numerical aperture. Thus, if the condenser is significantly underfilling the objective with light, as is sometimes done to improve image contrast, then spatial resolution is sacrificed. Any aberrations in the optical system, or other factors that adversely affect performance, can only degrade the spatial resolution past this point. However, most microscope systems do perform at, or very near, the diffraction limit. For the employed microscope the objective also functions as the condenser lens.

The formula above represents the spatial resolution in object space. At the detector the resolution is the smallest resolvable distance multiplied by the magnification of the microscope optical system. It is this value that must be matched with the CCD. The most obvious approach to matching resolution might seem to be simply setting this diffraction-limited resolution to the size of a single pixel. In practice, what is really required of the imaging system is that it be able to distinguish adjacent features. If optical resolution is set equal to single-pixel size, then it is possible that two adjacent features of like intensity could each be imaged onto adjacent pixels on the CCD. In this case, there would be no way of discerning them as two separate features.

Separating adjacent features requires the presence of at least one intervening pixel of disparate intensity value. For this reason, the best spatial resolution that can be achieved occurs by matching the diffraction-limited resolution of the optical system to two pixels on the CCD in each linear dimension. This is called the *Nyquist limit*, and the magnification necessary is

$$m = \frac{2s_{x,y}NA}{0.61\lambda} \tag{23.15}$$

where s_x and s_y are pixel spacings on the CCD sensor.

Specifications for the lens system, objective lens, and CCD sensor are given in Table 23.1. Based on the system parameter and Eq. (23.2), the desired magnification and numerical aperture for a suitable lens can be determined from Figure 23.6, where $\lambda = 0.5$ μm. It can be seen that a 5× objective satisfies the condition while offering the largest field of view. Hence, a 5× objective is used for all micromanipulation tasks.

Task-space spatial resolution and directional dependence of resolution is determined by an analysis of the image Jacobian. The supervisory controller varies this resolution optimally for the task to be executed. The *resolvability ellipsoids* [Nelson, 1995] for the image Jacobian can be used for this process.

A dynamic sensor-placement scheme using a visual servoing controller has been developed [Nelson, 1995]. The placement of the sensor is guided by objective functions maximizing resolvability, among

TABLE 23.1 Objective Lens Specifications

	M Plan Infinity-Corrected Apochromat ULWD Objectives				
M	N.A.	W.D. (mm)	*f* (mm)	*R* (μm)	DoF (μm)
2×	0.055	34	100	5.0	184.4
5×	0.14	34	40	2.0	28.5
10×	0.28	33.5	20	1.0	8.6
20×	0.42	20	10	0.7	3.3
SL50×	0.42	20.5	4	0.7	2.2

M = magnification; N.A. = numerical aperture; W.D. = working distance; *f* = focal length; *R* = resolution; DoF = depth of field.

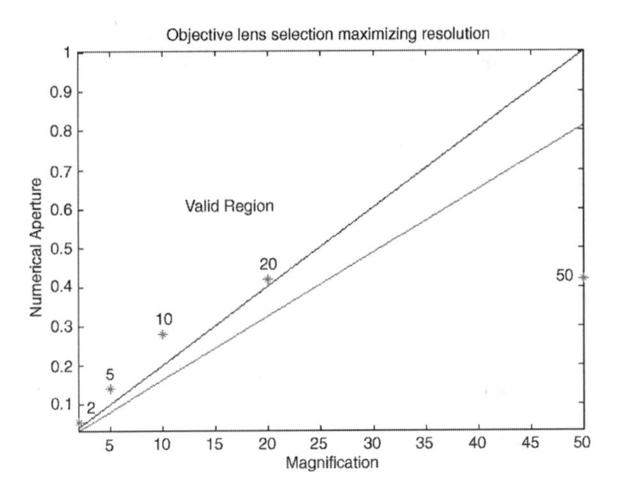

FIGURE 23.6 Objective selection parameter.

other parameters. Resolvability ellipsoids can be determined from a singular value decomposition of the image Jacobian indicating the directional dependence of the resolution capability of the imaging system [Nelson, 1995]. Considering a partitioned Jacobian that accounts for translation and rotation of the feature with respect to the sensor frame

$$\begin{bmatrix} \dot{x}_s \\ \dot{y}_s \end{bmatrix} = \begin{bmatrix} -\frac{m}{s_x} & 0 & 0 & 0 & -\frac{Z_c m}{s_x} & \frac{y_s s_y}{s_x} \\ 0 & -\frac{m}{s_y} & 0 & \frac{Z_c m}{s_y} & 0 & -\frac{x_s s_x}{s_y} \end{bmatrix} \begin{bmatrix} \dot{X}_c \\ \dot{Y}_c \\ \dot{Z}_c \\ \omega_{Xc} \\ \omega_{Yc} \\ \omega_{Zc} \end{bmatrix} \tag{23.16}$$

and further partitioning the Jacobian into translational and rotational components yields

$$\begin{bmatrix} \dot{x}_s \\ \dot{y}_s \end{bmatrix} = \begin{bmatrix} -\frac{m}{s_x} & 0 & 0 \\ 0 & -\frac{m}{s_y} & 0 \end{bmatrix} \begin{bmatrix} \dot{X}_c \\ \dot{Y}_c \\ \dot{Z}_c \end{bmatrix} \quad \begin{bmatrix} \dot{x}_s \\ \dot{y}_s \end{bmatrix} = \begin{bmatrix} 0 & -\frac{Z_c m}{s_x} & \frac{y_s s_y}{s_x} \\ -\frac{Z_c m}{s_y} & 0 & -\frac{x_s s_x}{s_y} \end{bmatrix} \begin{bmatrix} \omega_{Xc} \\ \omega_{Yc} \\ \omega_{Zc} \end{bmatrix} \tag{23.17}$$

A singular value decomposition, where $USV^T = J$, of the Jacobian partitions determines the resolvability ellipsoids. The eigenvectors corresponding to $J^T J$, i.e., the columns of V, are the set of basis vectors for the range of the Jacobian, and these vectors indicate the directionality of resolvability. The corresponding singular values are a measure of the ability of the sensor to resolve position and orientation in the task space along these directions. The directional gradients of the eigenvectors determine the direction of motion for optimal sensor placement [Nelson, 1995].

An examination of the singular value decomposition clearly shows that resolvability can be increased by increasing the magnification m for the lens system. This can be accomplished by the use of a zoom-lens subsystem. However, the zoom-lens system affects the field of view, magnification, and illumination simultaneously, and the supervisory controller thus needs to optimally determine the configuration for a task based on task requirements.

Incorporating a zoom element into the controller requires the determination of the Jacobian. However, as the zoom is varied in a quasi-static manner a gain-scheduling control scheme is used by the supervisory controller. The zoom lens needs to be calibrated in order to implement the scheme. The calibration is discussed in a later section. The modified Jacobian, including the zoom element, is given as

$$\begin{bmatrix} \dot{x}_s \\ \dot{y}_s \end{bmatrix} = \begin{bmatrix} -\frac{z_x m}{s_x} & 0 & 0 & 0 & -\frac{Z_c z_x m}{s_x} & \frac{y_s z_y s_y}{s_x} \\ 0 & -\frac{z_y m}{s_y} & 0 & \frac{Z_c z_y m}{s_y} & 0 & -\frac{x_s z_x s_x}{s_y} \end{bmatrix} \begin{bmatrix} \dot{X}_c \\ \dot{Y}_c \\ \dot{Z}_c \\ \omega_{Xc} \\ \omega_{Yc} \\ \omega_{Zc} \end{bmatrix} \tag{23.18}$$

where $z_{x,y}$ is the observed image zoom. Optical zoom is uniform for the lens system; however, the effective zoom shows directional dependence, and this is incorporated in the zoom factors z_x and z_y.

23.5.3 Field-of-View Constraint

Features must constantly lie within the field of view (FoV) in order to be able to implement visual servoing. With the incorporation of a zoom-lens system, the effective FoV varies with zoom, and the supervisory controller must either relocate the manipulators or sensor system or vary the zoom in such a manner as to keep the features within the FoV.

The system takes as its input the desired task-space spatial resolution and outputs zoom and sensor relocation control outputs. The desired spatial resolution determines the final zoom level for the lens system. This state is achieved in a quasi-static manner by a combination of manipulator and sensor relocations without violating the FoV constraint. An inherent assumption in the implementation is that features are not diverging during the course of a task.

The relocation and zoom components are handled separately by the controller. The relocation controller ensures that features remain in the FoV for manipulation tasks by locating the image center at the mean of the feature vectors. Subsequently, the microscope stage is moved, if necessary, to center the features within the FoV. The controller precomputes the location of features in the new coordinate system and performs sensor relocation, if the features do not violate the FoV constraint. The employed algorithm is illustrated in Figure 23.7. A low-priority execution thread is responsible for ensuring that the FoV constraint is not violated during the course of a manipulation task.

When the features converge, zoom is increased till the spatial resolution requirements are met. This requires a calibration of the zoom lens system.

23.6 Zoom Calibration

To implement the visual servo controller the intrinsic and extrinsic parameters in the Jacobian given by Eq. (23.18) need to be determined for the zoom lens. The effective magnification and FoV vary during the course of a manipulation task with changes in zoom settings.

In this section the parameters required for the Jacobian are estimated for varying optical system parameters. While formulating the Jacobian it was assumed that the optical axis coincides with the center of the image and remains fixed. However, for most zoom lenses the optical axis shifts during the process of zooming, and this assumption is no longer valid. Additionally, the zoom center does not coincide with the sensor coordinate frame, and this needs to be characterized to implement a look-and-move-based visual servoing control scheme.

Considering a feature located at (x, y) in the sensor coordinate frame prior to a zoom change, the transformed feature coordinates $(\tilde{x}, \tilde{y})$ are given as

$$\begin{bmatrix} \tilde{x} \\ \tilde{y} \\ 1 \end{bmatrix} = T(c_x, c_y) T_s(z_x, z_y) T(-c_x, -c_y) \begin{bmatrix} x \\ y \\ 1 \end{bmatrix} = \begin{bmatrix} z_x & 0 & c_x(1-z_x) \\ 0 & z_y & c(1-z_y) \\ 0 & 0 & 1 \end{bmatrix} \begin{bmatrix} x \\ y \\ 1 \end{bmatrix} \tag{23.19}$$

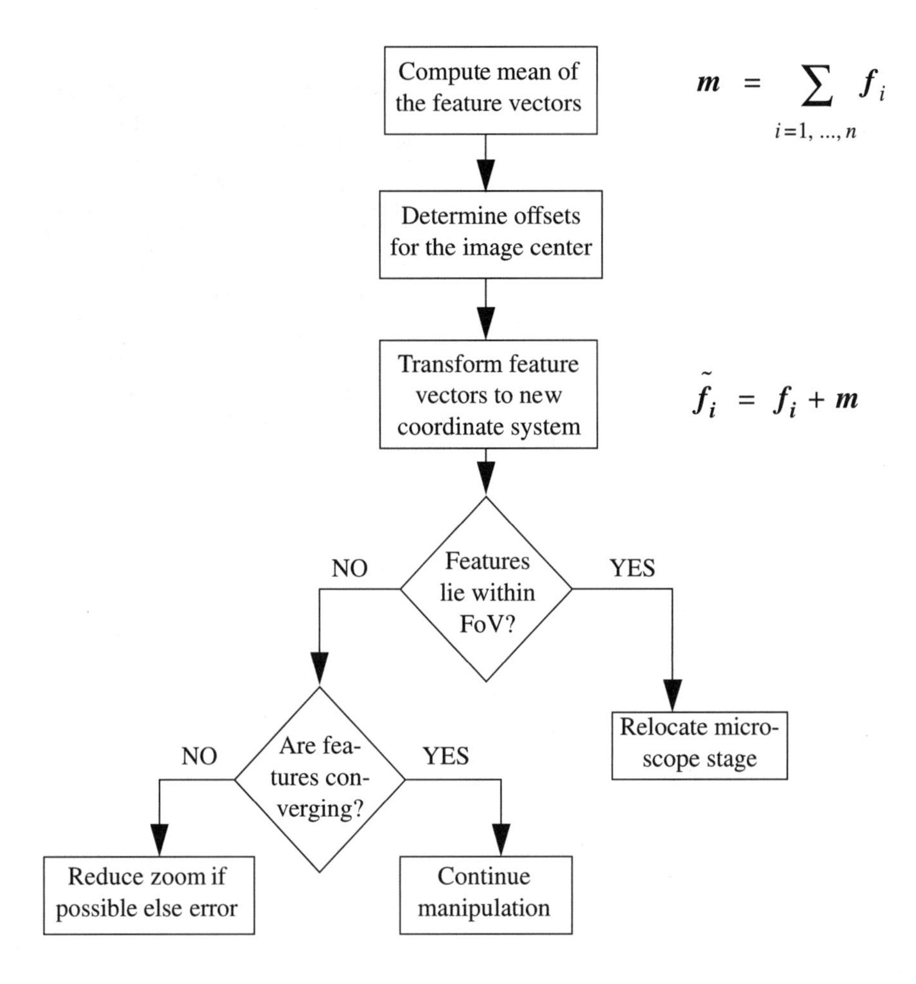

FIGURE 23.7 Sensor relocation algorithm.

where (z_x, z_y) are the zoom factors about the X and Y axes by zooming, and (c_x, c_y) are the coordinates of the zoom center, expressed in the sensor coordinate frame, about which scaling occurs. For implementing a look-and-move-based scheme both these sets of parameters need to be determined, whereas for the dynamic visual servoing controller only the zoom factors need to be characterized.

Nonlinear distortion due to zooming is not incorporated in this formulation. This assumption is justified by experimental results where the measured distortion is insignificant across the entire zoom range. While optical zoom is uniform along both axes, the observed factors show directional dependence. According to Tsai [1987], digitization of sensor input by the framegrabber results in an uncertainty scale factor along the rows (X) of the digitized image. This scale factor is incorporated into the zoom factor z_x resulting in the variation.

A calibration grid (Figure 23.8) fabricated using LIGA is used for the experiment. The grid consists of 20 × 20-μm blocks evenly spaced at intervals of 20 μm. The zoom is varied in steps of five units, and images are captured. The locations of the centroids of the connected components are recorded. A least-squares approach is used to solve the overdetermined system of Eq. (23.20) to extract the calibration parameters. The procedure is repeated with decreasing zoom to characterize the bidirectional repeatability of the zoom-lens system. Subsequently, a third-order polynomial is curve-fitted to the scaling factors in order to express them in terms of the zoom number that is sent to the A-Zoom 40 zoom controller.

FIGURE 23.8 LIGA microfabricated calibration grid.

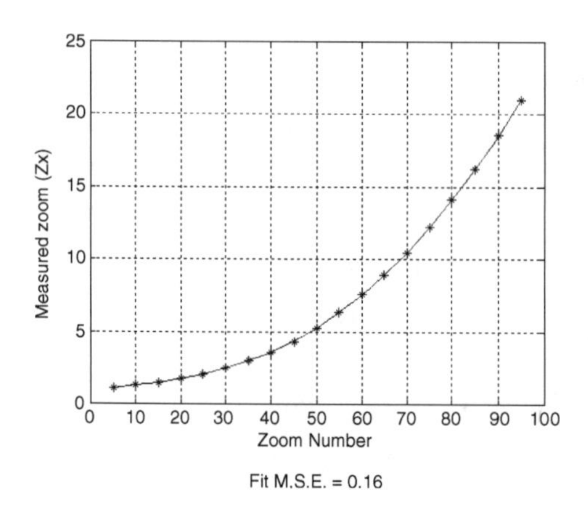

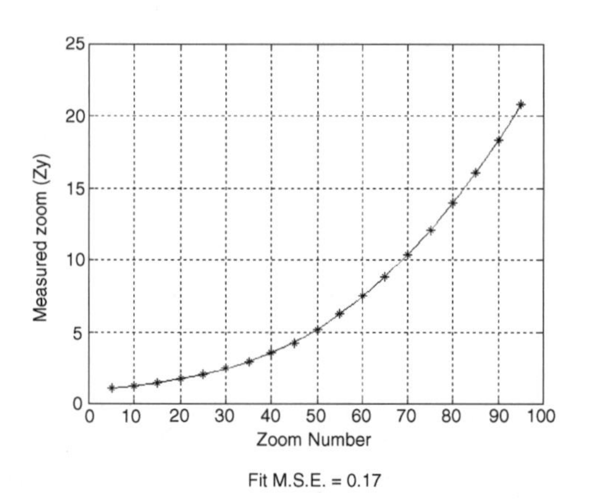

FIGURE 23.9 Observed zoom factors vs. zoom number.

$$\begin{bmatrix} \tilde{x}_i \\ \tilde{y}_i \\ 1 \end{bmatrix}_{i=1,n} = \begin{bmatrix} z_x & 0 & c_x(1-z_x) \\ 0 & z_y & c(1-z_y) \\ 0 & 0 & 1 \end{bmatrix} \begin{bmatrix} x_i \\ y_i \\ 1 \end{bmatrix}_{i=1,n} \Leftrightarrow \tilde{X} = TX \qquad (23.20)$$

$$\therefore\ T = \tilde{X}X^T(XX^T)^{-1}$$

The fitted polynomials are as follows:

$$\begin{aligned} z_x &= 1.2917\times10^{-5}z^3 + 9.3635\times10^{-4}z^2 + 4.3013\times10^{-3}z + 1.1359 \\ z_y &= 1.2688\times10^{-5}z^3 + 9.4341\times10^{-4}z^2 + 3.4955\times10^{-3}z + 1.1298 \end{aligned} \qquad (23.21)$$

where z is the zoom number. The estimated zoom factors are plotted in Figure 23.9 along with the curve-fit polynomial. The behavior is highly repeatable based on the results of multiple experiments. The movement of the image center as a function of zoom number is plotted in Figure 23.10. This is due to the mechanical movement of the lens systems, and the erratic movement observed in the zoom range from 0 to 20 can be attributed to the deadband in the gearing system. While the movement is erratic in this range, it is bounded and thus can be compensated for by relocating the sensor system after zooming,

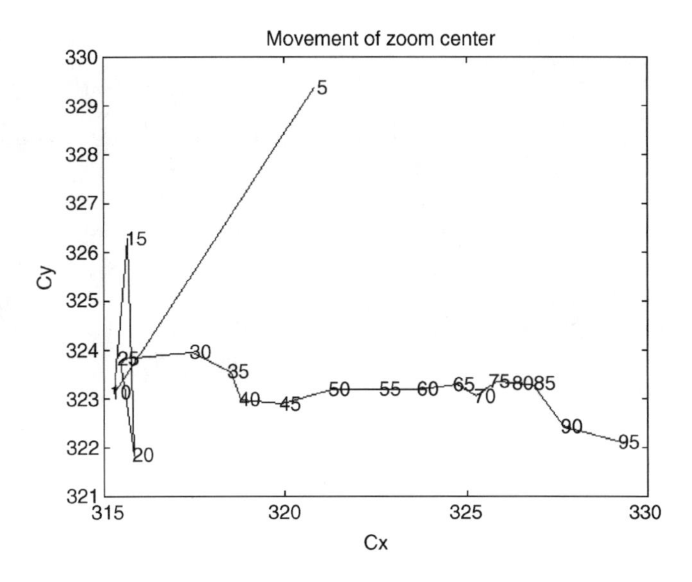

FIGURE 23.10 Observed movement of zoom center (pixels) with zoom.

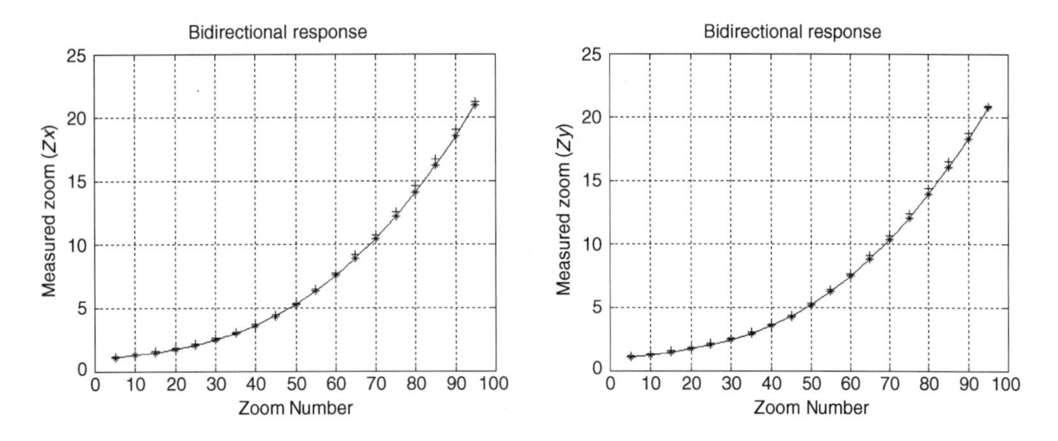

FIGURE 23.11 Observed bidirectional repeatability of zoom lens system. *, forward; +, reverse.

if necessary. The movement of the image centers is stored in a look-up table, and the sensor relocation controller compensates for the movement when operating in the look-and-move scheme. The bidirectional repeatability can be seen in Figure 23.11.

The results of a zoom transformation between two images at zoom levels of 5 and 85 are shown in Figure 23.12. The estimated and computed centroids are accurate to within subpixel values, verifying the validity of the computed zoom parameters.

23.7 Focusing and Depth Estimation

With the limited depth of field (DoF) of high numerical aperture microscope lens systems, establishing and maintaining focus become paramount in order to conduct any visually guided micromanipulation. However, this limited DoF can also be used for depth estimation using the depth-from-focus technique. The task of autofocusing, maintaining focus, and estimating depth using depth from focus are detailed in this section.

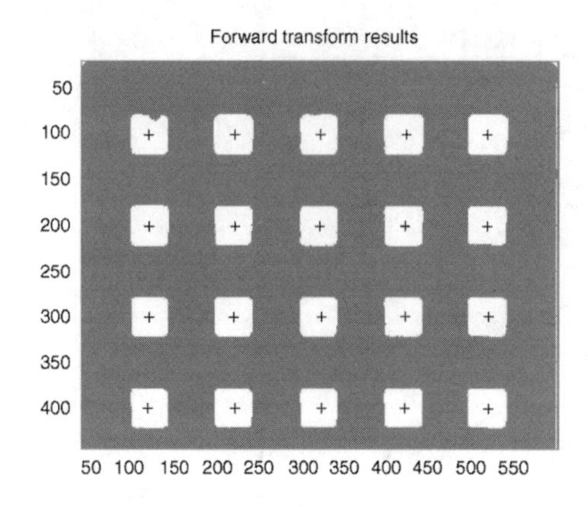

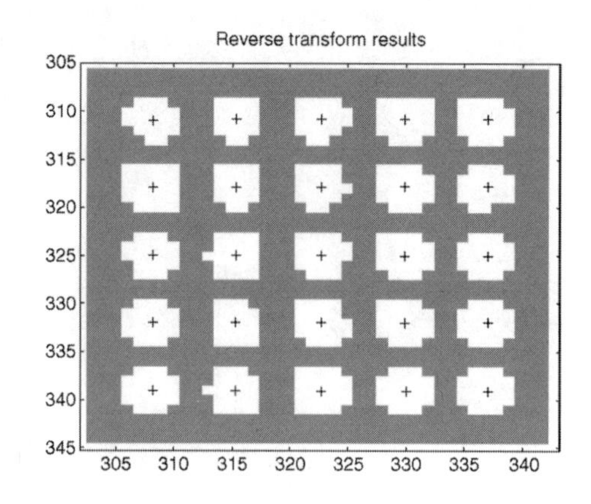

FIGURE 23.12 Results of transformation using calibrated zoom parameters.

23.7.1 Autofocusing

General microassembly tasks have multiple task planes that span depths beyond the DoF of the lens system. To conduct a visually guided microassembly task in this situation two possible alternatives may be employed. A strongly calibrated and thermally stabilized workcell allows for the use of *a priori* determined task plane information that is used directly to move along the optical axis. This strategy is highly susceptible to calibration and measurement errors and requires thermal stability. Thermal drift alone can exceed the DoF of the lens system. Additionally, tolerance of microfabricated parts often exceeds the DoF as well. Thus, there is a need to dynamically establish focus in order to conduct a general microassembly task. This is the approach adopted in this work.

Defocus acts as a low-pass filter that attenuates high-frequency content in an image [Horn, 1986; Born and Wolf, 1965]. The level of focus in an image can thus be estimated by computing the frequency content in an image. Various focus measures have been proposed and developed based on this [Tenenbaum, 1970; Schlag et al., 1983; Jarvis, 1983; Krotkov, 1987, 1989; Grossman, 1987; Pentland, 1987; Nayar and Nakagawa, 1990]. These measures include Fourier transforms, Tenengrad, high-pass filtering, histogram entropy, gray-level variance, and sum-modulus-difference.

The quality of focus can be estimated by the magnitude of intensity gradients. Edges in images generate high-frequency content; thus, edge detectors directly reflect the level of focus in an image. A focus measure, termed the *Tenengrad*, based on gradient magnitude has been proposed by Schlag et al. [1983]. Directional gradients are computed by convolving the image with Sobel kernels, and the magnitude of the resulting gradient within the window of interest is computed as an average of the square of the gradients.

Histogram-based techniques rely on some heuristically established threshold values. As a result, these techniques are image specific unless the threshold is adaptively determined. In addition, these techniques are data intensive, so computation times are high. It has been reported [Schlag et al., 1983] that histogram entropy techniques are unable to handle scenes with complicated texture. The applicability of high-pass filtering and modified Laplacians has been shown by Nayar and Nakagawa [1990] and Noguchi and Nayar [1996]. Laplacians involve second-order intensity gradients, and this makes them more susceptible to image noise. In addition, the computation cost is double that of first-order derivatives. The use of Fourier-transform-based focus measures has been demonstrated by Bove [1989]. However, discrete Fourier transforms are computationally intensive, and the phase information is neglected by the focus measure, limiting the applicability for a real- time implementation.

Invariably images contain random noise. Differential operators amplify this noise if the image signal is directly passed through such operators. However, for most microscopy this noise is greatly reduced due to the quality of the sensors employed; therefore, first-order derivatives can be employed without preprocessing the images, making them computationally attractive.

Autofocusing is thus accomplished by moving the lens system along the optical axis to maximize a measure of focus. The range of motion for the lens system is about five orders of magnitude larger than the DoF, which makes the search space quite large. In addition, the measure of focus should necessarily be unimodal in order to be able to locate a maxima. The focus measure needs to be able to handle diverse image content and should have low computational costs.

23.7.2 Focus Metric

The focus metric adopted in this work is based on the Tenengrad proposed by Schlag et al. [1983]. The computational cost of this metric is minimal, and the metric is predominantly unimodal and able to handle diverse image content, provided there is some texture in the image. This is seldom a problem for microscopy due to the high spatial resolution, which delineates edges and surface roughness, making the resulting images textured.

The computation of the Tenengrad is modified to reduce computational costs, and the employed algorithm is shown in Figure 23.13. Temporal averaging is employed to reduce the effect of noise in the input image. A window of interest is averaged over five frames. Image gradients (G_x, G_y) are computed at each pixel (i) within this window (W) of interest by convolving the image with a Sobel kernel (K). For most focus measurements a 40×40 pixel window is employed with a 3×3 Sobel kernel. The absolute values of the gradients are summed for the pixels within the window of interest to generate G. This sum of gradients is normalized with respect to the size of the window and compared to a predetermined threshold (T). The threshold is determined at runtime to compensate for baseline noise in a featureless window. If the computed gradient exceeds the noise threshold, it is assigned to the focus measure F.

The response of the focus metric under varying image zoom levels as a function of focus motor position is shown in Figure 23.14. The response is primarily unimodal with a sharp peak and can be approximated by a Gaussian curve in a region near the peak. The peak response is centered within a small range of the total search space, complicating a search for the peak; however, this allows for accuracy in establishing focus.

$$G_x = I(x, y) \otimes K$$

$$G_y = I(x, y) \otimes (-K^T)$$

$$K = \frac{1}{4}\begin{bmatrix} -1 & 0 & 1 \\ -2 & 0 & 2 \\ -1 & 0 & 1 \end{bmatrix}$$

$$G = \frac{1}{W}\sum_{i \in W} |G_x|_i + |G_y|_i$$

$$F = \begin{cases} G \text{ if } (G \geq T) \\ 0 \end{cases}$$

FIGURE 23.13 Computation of focus measure.

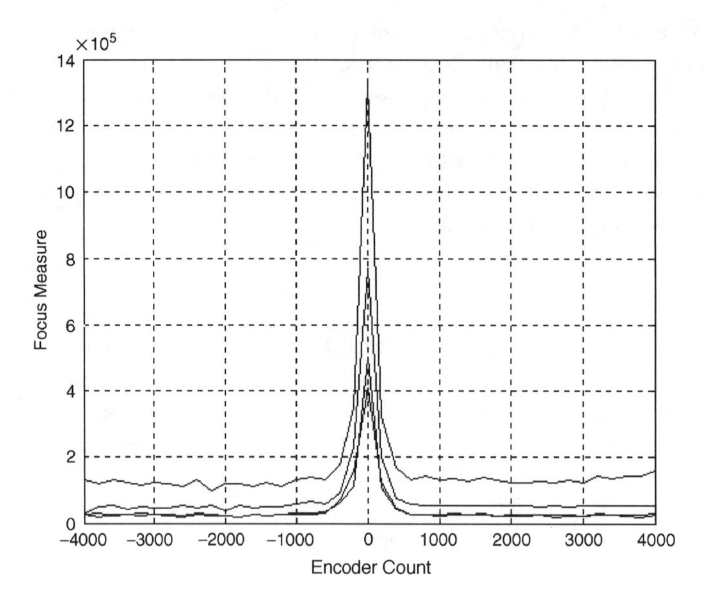

$$F \approx F_p e^{-\frac{1}{2}\left(\frac{z - z_0}{\sigma}\right)^2}$$

where z_0 is the peak focus motor count

FIGURE 23.14 Response of focus measure with increasing magnification.

23.7.3 Search for Focus Maxima

The search space for the focus maxima is about five to six orders of magnitude larger than the DoF of the lens system. In addition, the focus peak is located in a small band in this large search space, and large-focus motor steps can easily miss the peak completely. Thus, the task of locating the focus peak in a robust and time-optimal fashion is quite challenging.

A number of search algorithms have been developed in the past, including hill-climbing (gradient ascent) [Schlag et al., 1983; Jarvis, 1983], adaptive control [Weiss, 1984], Fibonacci search [Krotkov, 1987, 1989], and coarse-to-fine search schemes [Boddeke et al., 1994; Allegro, 1998].

Gradient-ascent search schemes work well for uniformly increasing or decreasing functions. However, these schemes are susceptible to local extrema in the search region. The focus measure has regions that are relatively flat with local extrema, and this leads to a degeneration of the search in practice.

The coarse-to-fine search schemes employ a heuristic-based search, wherein the search begins by moving in large steps till an extremum is located. Thereafter, fine steps are employed in the vicinity of the extremum, and finally some refinement is carried out to estimate the extremum using curve-fitting techniques. The choice of employed step sizes in each phase of the search is critical to its success. However, this scheme is not as susceptible to local extrema in the search space. Computationally, the coarse-to-fine search scheme is not guaranteed to be time optimal.

The Fibonacci search scheme is an optimal search scheme for locating an extremum of a unimodal function of a single variable [Krotkov, 1987]. The Fibonacci search scheme successively narrows the search interval until this interval is a given fraction of the previous interval. Thus, the number of steps required in a search can be determined *a priori* for a desired accuracy in the search space. In addition, the search scheme requires the computation of the focus measure at only one point in the search interval subsequent to the first step, making it computationally optimal. The pseudocode for the search algorithm is given in Krotkov [1987]. A minor correction is needed in the algorithm for implementation purposes.

The Fibonacci search technique serves the purpose of a coarse search; thereafter, Gaussian curve fitting is employed to locate the maxima, and the lens is directly moved to this position. This makes the search robust to local extrema and time optimal while eliminating any heuristics in determining the step sizes for implementing a coarse-to-fine search scheme. The search space is limited to 20,000 motor counts. This translates to the 22nd Fibonacci number. The commands sent to the focus motor can range within ±9999 and, as a result, the lens is moved by 5000 counts to ensure that the true focus lies below the current lens

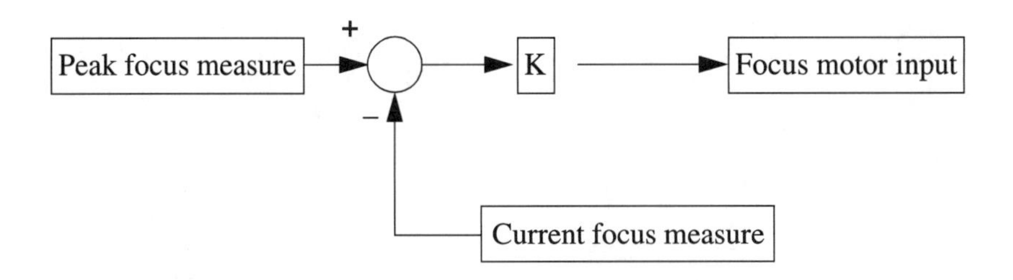

FIGURE 23.15 Dynamic focusing scheme.

position. Hereafter, the Fibonacci search scheme is employed with the placement accuracy set to 300 counts. When the search interval reduces to this limit, i.e., after 8 to 9 iterations of the Fibonacci search scheme, the search is refined. The refinement is based on the fact that in this interval the focus function can be approximated by a Gaussian distribution. Based on focus measures at three discrete points in this search region, the parameters (F_p, z_0, σ) of the Gaussian fit $F = F_p e^{-\frac{1}{2}\left(\frac{z-z_0}{\sigma}\right)^2}$ are estimated. The peak of this fit, z_0 is then used to directly move the lens to the true focal plane, thereby establishing focus.

In practice this scheme has proven to be highly robust for general microassembly tasks. The low computational requirement and high accuracy allow for the use of this scheme for determining relative depth of objects in the task space as well.

Once focus has been established, a proportional control scheme (Figure 23.15) is employed to dynamically maintain focus. This is similar to the approach suggested by Schlag et al. [1983]. The input to the controller is the difference between the focus measure at a given instance from the focus measure at the true focal plane. This error is used to move the focus motor to maintain focus in the image. If the error exceeds 25% of the peak focus measure, then the control scheme is stopped and autofocusing is repeated with a smaller search region.

23.7.3.1 Depth Estimation

The limited DoF of an optical lens system can be exploited to estimate depth accurately using techniques such as depth from focus [Jarvis, 1983; Grossman, 1987; Krotkov, 1987, 1989; Nayar and Nakagawa, 1990; Ens and Lawrence, 1991; Noguchi and Nayar, 1996] and depth from defocus [Pentland, 1987; Subbarao, 1989; Subbarao and Surya, 1992; Xiong and Shafer, 1993]. These techniques are based on the inherent inability of optical systems to focus at all distances. The image from an optical system exhibits regions of sharp focus and regions of blur or image degradation. A focused image is characterized by high spatial-frequency content, while a blurred image has attenuated high-frequency content. This fact provides a robust method to characterize level of focus in an image, and this can be used to develop dense depth maps. An important advantage of this system is that there is no problem of correspondence, as with binocular stereo, making real-time implementation feasible.

A defocused lens system is characterized as a low-pass filter (Figure 23.16); therefore, in order to successfully implement the technique of depth from focus/defocus, suitable high-spatial-frequency content is needed in the object space. This requires that the object be textured, which is a key requirement for the implementation of this system. This is applicable for microassembly because surfaces that appear smooth at normal magnification appear to be highly textured under a high-magnification and high-resolution lens system.

The accuracy of measurements using these techniques is limited to the DoF of the optical system. A number of expressions for DoF have been proposed in literature. According to Born and Wolf [1965], the DoF, Δz, is given as

$$\Delta z = \frac{\lambda}{2(NA)^2} \tag{23.22}$$

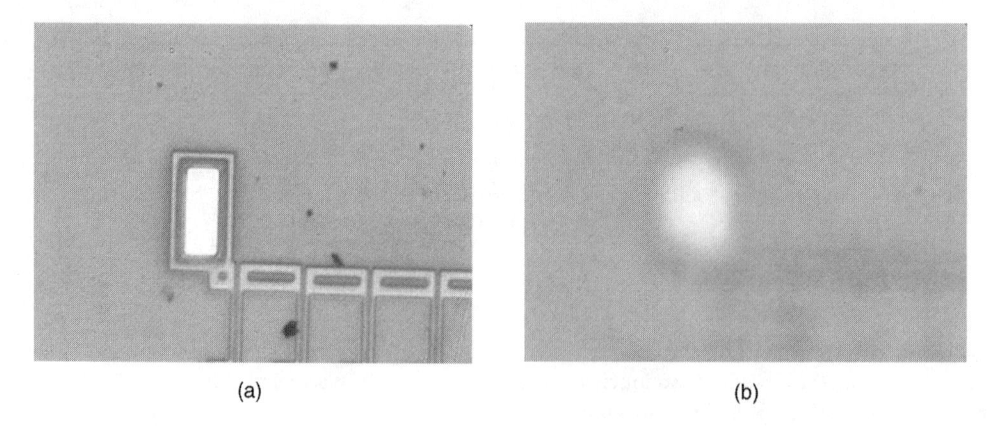

FIGURE 23.16 Behavior of a defocused lens system: (a) object in focus under a 5× objective; (b) same object when the lens is moved by 20 μm from focus plane.

where λ is the wavelength of the illumination and *NA* is the numerical aperture for the lens system. According to Martin [1966],

$$\Delta z = \frac{\lambda}{2(NA)^2} + \frac{n}{7m(NA)} \tag{23.23}$$

where n is the refractive index of the lens and m is the magnification of the lens, while according to Young et al. [1993],

$$\Delta z = \frac{\lambda}{4n\left(1 - \sqrt{1 - \left(\frac{NA}{n}\right)^2}\right)}. \tag{23.24}$$

In practice it has been observed that the DoF shows dependence on the magnification of the lens employed, and the expression given by Martin is largely valid. The accuracy of depth estimation does not improve with increase in zoom and depends on the objective lens magnification alone.

For most microassembly tasks establishing the focus planes is critical. This is achieved by using the autofocusing technique developed earlier. The distance between the focal planes is determined based on their relative separation in the focus motor count space, which has been calibrated to obtain depth information directly. Multiple planes can also be isolated, and a coarse depth map can thus be created using the autofocusing technique.

For developing fine-resolution depth maps, a technique similar to Nayar and Nakagawa [1990] is employed. The autofocusing routine is executed to establish focus for a single point on the object of interest and for a point belonging to the background. This information is then used to determine the step size for the movement of the lens system. Images are collected at ten discrete steps, and for each point within the window of interest a 10 × 10 neighborhood of pixels is used to determine the response of the focus measure. Based on three focus measures, corresponding to the peak and one each to either side, a Gaussian curve $F \approx F_p e^{-\frac{1}{2}\left(\frac{z-z_0}{\sigma}\right)^2}$ is fitted, and the depth of the point, z_o, is extracted. Runtime optimization using look-up tables as suggested by Nayar and Nakagawa has not been implemented, as depth extraction is generally limited to separating multiple assembly planes.

The results of an experiment on a fine cut file (inset) are shown in Figure 23.17. A 400 × 6 pixel section of the file was used for depth estimation, and the computed depth map is shown. The dimensional accuracy of this technique is limited only by the depth of field of the employed objective lens.

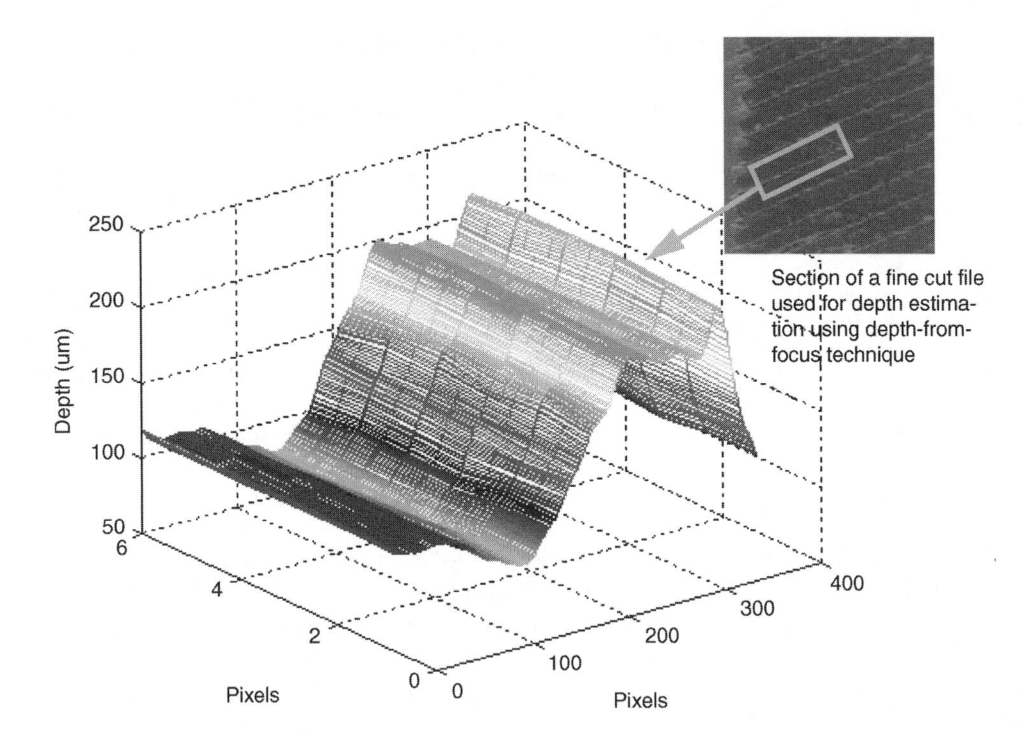

FIGURE 23.17 Computed depth map using depth-from-focus technique.

Summary

The microassembly of MOEMS devices generally requires the use of computer vision feedback in order to overcome tolerance stackup issues associated with automated assembly equipment. This chapter presented an overview of the field of visual servoing and its application in the domain of microassembly. Research continues in this field, with the eventual goal being the development of intelligent microassembly workcells capable of assembling complex microsystems that exhibit true three-dimensional geometries and microsystems assembled from subcomponents that are the result of incompatible manufacturing processes.

Defining Terms

MIMO: Multi-input, multi-output system.
MOEMS: Microopticalelectromechanical systems.
MRAC: Model reference adaptive control.
SISO: Single-input, single-output system.
visual servoing: The use of vision sensor feedback in a closed-loop feedback control scheme.

References

Allegro, S., Automatic Microassembly by Means of Visually Guided Micromanipulation, Ph.D. thesis, EPFL, Lausanne, 1998.

Allen, P. K., Real-time motion tracking using spatio-temporal filters, *Proc. DARPA Image Understanding Workshop*, 1989, pp. 695–701.

Boddeke, F. R., van Vliet, L. J., Netten H., and Young, I. T., Autofocusing in microscopy based on the OTF and sampling, *Bioimaging*, 2(4), 193–203, 1994.

Born, M. and Wolf, E., *Principles of Optics: Electromagnetic Theory of Propagation, Interference and Diffraction of Light*, Pergamon Press, London, 1965.

Bove, Jr., M. V., Discrete Fourier transform-based depth-from-defocus, *Image Understanding and Machine Vision, Tech. Digest Series*, 14, 118, 1989.

Codourey, A., Zesch, W., Buchi, R., and Siewgart, R., High precision robots for automated handling of micro objects, *Seminar on Handling and Assembly of Microparts*, Vienna, November 14, 1994.

Corke, P. I. and Paul, R. P., Video-rate visual servoing for robots, in *Lecture Notes in Control and Information Science*, in Hayward, V. and Khatib, O., Eds., Springer-Verlag, New York, 1989, pp. 429–451.

Cowan, C. K. and Bergman, A., Determining the camera and light source location for a visual task, *Proc. 1989 IEEE Int. Conf. on Robotics and Automation*, 1989, pp. 509–514.

Ens, J. and Lawrence, P., A matrix based method for determining depth from focus, *Proc. CVPR*, 1991, pp. 600–606.

Fatikow, S., An automated micromanipulation desktop-station based on mobile piezoelectric microrobots, *Proc. SPIE*, 2906, 66–77, 1996.

Fatikow, S., Buerkle, A., and Seyfried, F., Automatic control system of a microrobot-based microassembly station using computer vision, *Proc. SPIE*, 3834, 11–22, 1999.

Feddema, J. T. and Lee, C. S. G., Adaptive image feature prediction and control for visual tracking with a hand-eye coordinated camera, *IEEE Trans. on Systems, Man, and Cybernetics*, 20(5), 1172–1183, 1990.

Feddema, J. T. and Simon, R. W., CAD-driven microassembly and visual servoing, *Proc. 1998 IEEE Int. Conf. on Robotics and Automation*, May 16–21, 1998, pp. 1212–1219, Leuven, Belgium.

Grossman, P., Depth from focus, *Pattern Recogn. Lett.*, 5, 63–69, 1987.

Hager, G. D., Robot feedback control based on stereo vision: towards calibration free hand-eye coordination, *Proc. 1994 IEEE Int. Conf. on Robotics and Automation*, 1994, pp. 2850–2856.

Haralick, R. M. and Shapiro, L.G., *Computer and Robot Vision*, Addison-Wesley, Reading, MA, 1992.

Hashimoto, K. and Kimura, H., LQ Optimal and nonlinear approaches to visual servoing, in *Visual Servoing*, Hashimoto, K., Ed., World Scientific Series on Robotics and Automation, Vol. 7, World Scientific, Singapore, 1993, pp. 165–198.

Horn, B. K. P., *Robot Vision*, MIT Press, Cambridge, MA, 1986.

Hosoda, K., Moriyama, H., and Asada, M., Visual servoing utilizing zoom mechanism, *Proc. IEEE Int. Conf. on Robotics and Automation*, 1995, pp. 178–183.

Hutchinson, S., Hager, G. D., and Corke, P. I., A tutorial on visual servo control, *IEEE Trans. on Robotics and Automation*, 12(5), 1996.

Jarvis, R. A., A perspective on range finding techniques for computer vision, *IEEE Trans. PAMI*, 5(2), 122–139, March 1983.

Koivo, A. J. and Houshangi, N., Real-time vision feedback for servoing of a robotic manipulator with self-tuning controller, *IEEE Trans. on Systems, Man, and Cybernetics*, 21(1), 134–142, 1991.

Krotkov, E. P., Focusing, *Intl. J. Computer Vision*, 1, 223–237, 1987.

Krotkov, E. P., *Active Computer Vision by Cooperative Focus and Stereo*, Springer-Verlag, New York, 1989.

Martin, L. C., *The Theory of the Microscope*, Blackie, London, 1966.

Nayar, H. K. and Nakagawa, Y., Shape from focus: an effective approach for rough surfaces, *Proc. IEEE Int. Conf. on Robotics and Automation*, 1990, pp. 218–225.

Nelson, B. J., Object Schemas and Port-Based Agents for Assimilating Disparate Sensory Feedback, Ph.D. thesis, Robotics Institute, Carnegie Mellon University, Pittsburgh, PA, 1995.

Nelson, B., Papanikolopoulos, P., and Khosla, P. K., Visual servoing for robotic assembly, in *Visual Servoing*, Hashimoto, K., Ed., World Scientific Series in Robotics and Automation, Vol. 7, World Scientific, Singapore, 1993, pp. 139–164.

Nelson, B. J., Zhou, Y., and Vikramaditya, B., Sensor-based microassembly of hybrid MEMS devices, *Control Sys.*, 18(6), 35–45, 1998.

Noguchi, M. and Nayar, H. K., Microscopic shape from focus using a projected illumination pattern, *J. Math. Computer Modeling*, 24(5/6), 31–48, 1996.

Papanikolopoulos, N., Khosla, P. K., and Kanade, T., Vision and control techniques for robotic visual tracking, *Proc. IEEE Int. Conf. on Robotics and Automation*, April 1991, pp. 857–864.

Pappas, I. and Codourey, A., Visual control of a microrobot operating under a microscope, *Proc. IEEE/RSJ Int. Conf. on Intelligent Robots and Sys., IROS '96*, November 4–8, 1996, pp. 993–1000.

Parvin, B., Callahan, D. E., Johnston, W., and Maestre, M., Visual servoing for micro manipulation, *Proc. Int. Conf. on Pattern Recognition*, August 1996.

Parvin, B., Taylor, J., Callahan, D. E., Johnston, W., and Dahmen, U., Visual servoing for online facilities, *IEEE Computer Magazine*, July, pp. 56–62, 1997.

Pentland, A. P., A new sense for depth of field, *IEEE Trans. PAMI*, 9(4), 523–531, 1987.

Ralis, S. J., Vikramaditya, B., and Nelson, B. J., Visual servoing frameworks for microassembly of hybrid MEMS, *Proc. SPIE Int. Symp. on Intelligent Sys. and Advanced Manuf.*, 3519, 1998.

Rodriguez, M. and Codourey, A., Graphical User Interface to manipulate objects in the micro world with a high precision robot, *Proc. IEEE Int. Conf. on Robotics and Automation*, Albuquerque, NM, 1997.

Rodriguez, M., Codourey, A., and Pappas, I., Field experiences on the implementation of a graphical user interface in microrobotics, *SPIE Conf. on Microrobotics: Components and Applications*, Boston, November 21–22, 1996.

Sanderson, A. C. and Weiss, L. E., Image based visual servo control or robots, *26th Annual SPIE Technical Symp.*, San Diego, 1982.

Sanderson, A. C. and Weiss, L. E., Image-based servo control of robots, in *Robot Vision*, Pugh, A., Ed., IFS International Publications, Springer-Verlag, New York, 1983.

Schlag, J. F., Sanderson, A. C., Neumann, C. P., and Wimberly, F. C., Implementation of automatic focusing algorithms for a computer vision system with camera control, Tech. Report CMU-RI-TR-83–14, August 1983.

Shirai, Y. and Inoue, H., Guiding a robot by visual feedback in assembling tasks, *Pattern Recognition*, 5, 99–108, 1973.

Smith, C., Visually Guided Manipulation of Static and Moving Objects, Ph.D. thesis, University of Minnesota, Minneapolis, 1996.

Subbarao, M., Efficient depth recovery through inverse optics, in *Machine Vision for Inspection and Measurement*, Freeman, H., Ed., Academic Press, San Diego, CA, 1989.

Subbarao, M. and Surya, G., Application of spatial-domain convolution/deconvolution transform for determining distance from image defocus, *SPIE Conf. OE/Technol. 92*, 1822, 159–167, 1992.

Sulzman, A., Breguet, H. M., and Jacot, J., Microvision system (MVS): a three-dimensional computer graphic-based microrobot telemanipulation and position feedback by vision, *Proc. SPIE*, 2593, 38–49, 1995.

Tarabanis, K., Tsai, R. Y., and Allen, P. K., Satisfying the resolution constraint in the MVP machine vision planning system, *Proc. 1990 DARPA Image Understanding Workshop*, 1990, pp. 850–860.

Tarabanis, K., Tsai, R. Y., and Allen, P. K., Automated sensor planning for robotic vision tasks, *Proc. 1991 IEEE Int. Conf. on Robotics and Automation*, 1991a, pp. 76–82.

Tarabanis, K., Tsai, R. Y., and Allen, P. K., Analytical characterization of feature detectability constraints of resolution, focus and field-of-view for vision sensor planning, *CVGIP-Image Understanding*, 59(3), 340–358, 1991b.

Tenenbaum, J. M., Accommodation in Computer Vision, Ph.D. thesis, Stanford University, Stanford, CA, November 1970.

Tsai, R. Y., A versatile camera calibration technique for high-accuracy three-dimensional machine vision metrology using off-the-shelf TV cameras and lenses, *IEEE J. Robotics Automation*, RA3(4), 323–344, 1987.

Weiss, L. E., *Dynamic Visual Servo Control of Robots: An Adaptive Image-Based Approach*, Ph.D. thesis, CMU-RI-TR-84–16, Carnegie Mellon University, Pittsburgh, PA, 1984.

Wilson, W. J., Visual servo control of robots using Kalman filter estimates of robot pose relative to work-pieces, in *Visual Servoing*, Hashimoto, K., Ed., World Scientific Series on Robotics and Automation, Vol. 7, World Scientific, Singapore, 1993, pp. 71–104.

Young, I. T., van Vliet L. J., Mulliken, J., Boddeke, F., and Netten, H., Depth-of-focus in microscopy, *Proc. SCIA'93*, 1993, pp. 493–498.

Xiong, Y. and Shafer, S. A., Depth from focusing and defocusing, Tech. Report CMU-RI-TR-93–07, Carnegie Mellon University, Pittsburgh, PA, 1993.

For Further Information

Several journals publish research related to microassembly, including the *Journal of Micromechatronics, IEEE Transactions on Electronic Packaging Manufacturing, IEEE Transactions on Robotics and Automation*, and *IEEE/ASME Journal on Microelectromechanical Systems.*

Conferences that maintain a large focus on microassembly include the *International Workshop on Microfactories* (*IWMF*), held in even years in the fall, and the annual *SPIE Microrobotics and Microassembly* conference series, held annually in conjunction with the *SPIE Photonics East* in the fall.

Sessions on microassembly often appear at the annual *IEEE International Conference on Robotics and Automation* (*ICRA*) and the *IEEE/RSJ International Conference on Intelligent Robots and Systems (IROS).*

Index

A

B

C

D

E

F

G

H

I

J

K

L

M

N

O

P

Q

R

S

T

U

V

W

X

Y

Z